Sources and Studies
in the History of Mathematics and Physical Sciences

Sources and Studies in the
History of Mathematics and Physical Sciences

Continued after Index

Gert Schubring

Conflicts between Generalization, Rigor, and Intuition

Number Concepts Underlying the Development of Analysis in 17–19th Century France and Germany

With 21 Illustrations

Springer

Gert Schubring
Institut für Didaktik der Mathematik
Universität Bielefeld
Universitätstraße 25
D-33615 Bielefeld
Germany
gert.schubring@uni-bielefeld.de

Sources and Studies Editor:
Jed Buchwald
Division of the Humanities
and Social Sciences
228-77
California Institute of Technology
Pasadena, CA 91125
USA

Library of Congress Cataloging-in-Publication Data
Schubring, Gert.
   Conflicts between generalization, rigor, and intuition / Gert Schubring.
     p. cm. — (Sources and studies in the history of mathematics and physical sciences)
   Includes bibliographical references and index.
   ISBN 0-387-22836-5 (acid-free paper)
     1. Mathematical analysis—History—18th century.   2. Mathematical
analysis—History—19th century.   3. Numbers, Negative—History.   4. Calculus—History.   I.
Title.   II. Series.
   QA300.S377 2005
   515′.09—dc22   2004058918

ISBN-10: 0-387-22836-5   ISBN-13: 978-0387-22836-5   Printed on acid-free paper.

Printed in the United States of America.   (EB)

9 8 7 6 5 4 3 2 1   SPIN 10970683

springeronline.com

Preface

This volume is, as may be readily apparent, the fruit of many years' labor in archives and libraries, unearthing rare books, researching *Nachlässe*, and above all, systematic comparative analysis of fecund sources. The work not only demanded much time in preparation, but was also interrupted by other duties, such as time spent as a guest professor at universities abroad, which of course provided welcome opportunities to present and discuss the work, and in particular, the organizing of the 1994 International Graßmann Conference and the subsequent editing of its proceedings.

If it is not possible to be precise about the amount of time spent on this work, it is possible to be precise about the date of its inception. In 1984, during research in the archive of the *École polytechnique*, my attention was drawn to the way in which the massive rupture that took place in 1811—precipitating the change back to the synthetic method and replacing the limit method by the method of the *quantités infiniment petites*—significantly altered the teaching of analysis at this first modern institution of higher education, an institution originally founded as a citadel of the analytic method. And it was in a French context, so favorably disposed to establishing links between history and epistemology, that I first presented my view that concept development is culturally shaped; it was at the third *Ecole d'Eté de Didactique des Mathématiques* of July 1984 in Orléans that I presented my paper "Le retour du réfoulé— Les débats sur 'La Méthode' à la fin du 18ème et au debut du 19ème siècle: Condillac, Lacroix et les successeurs."

When the work was eventually completed in 2002, it was accepted as a *Habilitationsschrift* by the Mathematics Department of the University of Bielefeld. I am grateful to Jesper Lützen, editor of the series *Sources and Studies in the History of Mathematics and Physical Sciences*, and to the publishing house of Springer, for publishing the script in their series. The relatively independent former Chapter C, investigating the context of the 1811 switch in the basic conceptions at the *École polytechnique* is published separately as "Le Retour du Réfoulé—Der Wiederaufstieg der synthetischen Methode an der École polytechnique" (Augsburg: Rauner, 2004). Its principal results are summarized here in Chapter IV.3.

In an effort to accelerate the publication of this volume, the sheer size of the manuscript led me to organize its translation from German into English, practically as a collective endeavor. I am indebted to the commitment shown by Jonathan Harrow and Günter Seib, Chris Weeks, and Dorit Funke as translators.

I wish to thank the following archives for their kind permission to reproduce documents: Archiv der Berlin–Brandenburgischen Akademie der Wissenschaften (Berlin), Archives de l'École polytechnique (Paris/Palaiseau), Archives de l'Académie des Sciences (Paris).

In conclusion, I should like to make two points. Firstly, all the translations of non-English quotations—both from the original sources and from publications—are ours, except in cases where English translations already existed, which are so indicated. In a few cases, the original quotation is preserved when the context makes its meaning sufficiently clear.

The second point concerns terminology. The French reflections on foundations constitute the major focus of this study; consequently, the "triad" of basic concepts used in French mathematics to "span" the contemporary concept fields—namely *quantité*, *grandeur*, and *nombre*—is also used here, as the configuration of basic terms. Since *quantité* was understood to be the key foundational concept (cf. the *Encyclopédie* defining mathematics as the science of *quantité*), and since my English terms are intended to recall the French understanding of the time, *quantité* is rendered here throughout as "quantity," and *grandeur* as "magnitude."

University of Bielefeld   Gert Schubring
September 2004

Contents

Illustrations

Chapter I

Question and Method

1. Methodological Approaches to the History of Science

This study presents a historical analysis of how specific mathematical concepts developed—not confined to small regions and brief periods, but as a *long-term* history. The historiography of mathematics cannot be said to abound with methodological reflections. Thus it is appropriate to begin by examining which tools and methods are available for studies of the above design.

Concept development may at first seem unproblematic, since tracing concepts has traditionally been a major pattern of historiography that is usually termed "history of ideas." Although this method considers itself "unbiased," since it appears to merely study "facts," it nevertheless continually passes judgment: first by how it selects its objects, and then by highlighting those among its results that it declares to be novel and meaningful.

What is problematical in this accepted method is that its standards of judgment are never explicitly stated, let alone reflected upon. Any historical achievement will be measured by its end, by the outcome. The status of today is depicted as the historical development's necessary result. Earlier historical periods are judged according to which elements belonging to modern times can be discerned in them already. While the history of mathematics is understood to have been cumulative, it is also seen to have evolved in an essentially linear way. According to this view, everything rational and significant in contributions of earlier epochs has been "dissolved and raised in higher synthesis" (*aufgehoben*) within the concepts of today. Each case will thus present but a single result and just one "rightful" path to it. The problem with such for the most part implicit assumptions is that concepts tend to change meaning over time even if their terms remain identical. Who, or what, will supply the "standard," the tape to measure that change and to evaluate it? To judge an isolated instant of development from its present-day result may be of heuristic value as a tentative approximation, but it will become ever more questionable the longer the period under investigation.

The contrast between internalist and externalist approaches is familiar in the historiography of mathematics as well as representing a methodological controversy. If the "internalist" approach, which conceives of development as based on elements intrinsic to the concepts, was viewed critically above, perhaps the externalist approach, which views development as based on "extrinsic" (e.g., social) factors, will yield more power of explanation. What has been criticized above was less the internalist approach per se, but rather how it is traditionally realized. Besides, it is obvious that exclusively externalist approaches do not offer greater power of explanation than the internalist approaches, at least not in the case of long-term analyses of concept formation.

At the same time, it becomes clear from the present debate that a historical analysis of concept developments requires notions about the nature of mathematical development—that it needs to be supported by an epistemology of mathematics. Since there have been a number of methodological debates about the qualitative nature of the development of science, we may now inquire whether relevant clues are to be obtained from the best known discussion of this subject, from the debate on "scientific revolutions." Indeed, Thomas Kuhn's book of 1962 bearing this phrase in its title has initiated heated debates in quite a number of disciplines about what is characteristic for their development. In mathematics, this debate started relatively late and has been by far not as intense as in other disciplines. An essential difference between mathematics and Kuhn's own model disciplines of physics and chemistry can be derived from Donald Gillies's anthology revisiting significant original contributions to the discussion of how the category of "scientific revolution" can be applied to mathematics (Gillies 1992). While revolutions in physics and in chemistry have taken place in the sense of rendering the respective previous paradigm totally invalid, or of transforming it into a restricted special case of a basically differently structured paradigm, similarly fundamental "overthrows" cannot be ascertained in mathematics. Mathematics, therefore, seems to evolve more cumulatively than, say, physics or chemistry. In mathematics, a change of paradigm occurs rather in the form of a change in epistemology. A typical example of such an epistemologically determined change of paradigm is the transition from a substantialist view of mathematical objects to a relational view, as Ernst Cassirer established it for nineteenth century mathematics, a situation Michael Otte has investigated in numerous publications.

This is why I will take as a guiding idea for analyses concerning the history of mathematics that generalization is an essential criterion for the evolution of mathematical concepts: as a way to extend and deepen the concepts' ranges of application, as a way to determine the already known by what is theoretically new: "the known by the unknown" (cf. Mormann 1981).

On the basis of these assumptions about concept development in mathematics, I have elaborated a new methodological approach to the analysis of the history of mathematics. It starts from the fact that the "ideas" or concepts of mathematics are objectified in written form, in *texts*. Since no single text speaks directly for itself, since it will not immediately reveal its meaning upon inspection, a well-reflected method of interpreting texts is necessary for analyses into the history of mathematics.

In a paper for the 1983 annual conference of the *Deutsche Mathematiker Vereinigung*, I proposed to establish *hermeneutics* as a methodological guideline that might make it possible to resolve some of the problems discussed above:

> In fact, an established methodology like that of *hermeneutics* is indispensable for interpreting a text: even a historical mathematical text will not speak for itself; it cannot be deciphered as it were on a *micro*scopic level alone, but must be simultaneously analyzed *macro*scopically, with explicit reference to the knowledge available to contemporaries, to how they viewed problems, and to the scientific methodologies at their disposal (Schubring 1986a, 83).

I had distinguished this concept of hermeneutics from the model of "geisteswissenschaftliche" (humanistic) hermeneutics established at the end of the nineteenth century after Wilhelm Dilthey, which advocates comprehension as "empathy," in contrast to scientific explanation. Against that, I referred to the elaboration of hermeneutics in critical philology at the turn to the nineteenth century, a current of which Friedrich August Wolf (1759–1824) is characteristic. He defined hermeneutics, or the art of explanation, as the method that "teaches us to understand and to explain the thoughts of an Other from their signs." This method, he says, requires an acute gift of judgment that "penetrates the analogy of the Other's mode of thought upon which one establishes principles to explain his ideas," and it requires scholarly knowledge of the author's language. Wolf continued:

> Knowledge of language alone will not do, however. We must have knowledge of the customs of the period about which we read; we must have knowledge of its history and literature and must be familiar with the spirit of the times (Wolf 1839, 272 ff.).

This is to say that philologists like Wolf considered social history an indispensable basis for interpreting texts. My conclusion in 1983 was that the hermeneutical method taken in the wide sense necessary for the comprehensive interpretation of mathematical texts is not too far distant from historical–empirical methods as presently applied in the so-called social history of mathematics (Schubring 1986a, 83).

The *leitbild* of hermeneutics was later taken up at different points of the historiography of mathematics, in a particularly marked way by Laugwitz and Spalt's paper of 1988, in which they presented the *Method of Conceiving* as their own approach to historical sources—in that instance to Cauchy's works, describing this approach to understanding rather cursorily. It required, they said, "to stay inside the conceptual universe of the source under inspection" (Laugwitz/Spalt 1988, 7). In particular, they refrained from explaining what they meant by "conceptual universe." In their use of the term, it is transmuted to "universe of discourse" and eventually to "Cauchy's universe of discourse" (ibid., 10), which they opposed to a "universe of discourse of today" (ibid., 16). While the latter discourse means the communicative system of today's mathematics, they reduced the former to a hermetic monologue by Cauchy. In particular, they did not use Cauchy's communication with contemporary mathematicians as a means to uncover what the respective concepts meant in their own period. As the texts concerned were textbooks intended for an engineering college, their evident purpose being to convey the respective meanings of the concepts to students, and not to hermetically close some *private* mathematics, the *geisteswissenschaftliche* approach to "understanding" is particularly badly suited for mathematics.

Dilthey's "geisteswissenschaftliche" approaches have been so much intensified and popularized by Heidegger and Gadamer that hermeneutics is almost identified with them today. Hence, it is essential to stress earlier conceptions, all the more so because these were more open for mathematics and the sciences. While hermeneutics came into being originally as a theological discipline applied to interpreting Biblical texts, it saw its first radical extension into a scientific–critical method by the Jewish philosopher Spinoza (1632–1677): as historical critique of the Bible. Daniel Friedrich Schleiermacher (1768–1834), generalizing previous approaches from theology, philology, and philosophy, developed a *general* hermeneutics. He aimed at reproductively reiterating what the mind had originally produced. The element of "elevating the interpreter," already implicitly contained in this, was enhanced in the twentieth century by Heidegger and Gadamer, for whom "understanding" did not at all mean to refer *objectively* to anything originally produced, but rather to constitute meaning from the position of the reader. This subjectivist claim was raised for the so-called historical *Geisteswissenschaften*, in contrast to and to the exclusion of the exact sciences.

Since the 1960s, the philologists Peter Szondi and Jean Bollack, sharply criticizing the philosophical "humanistic" version of hermeneutics and taking up Spinoza's radical ideas, have developed a "material" version of hermeneutics, which recognizes the "alterity" of historical texts and which tries to make them objectively readable by situating them in the precise space and time of their production while at the same time striving to reduce the reader's subjectivity as far as possible (cf. Bollack 1989).

There have been recent approaches in the historiography of mathematics pleading in favor of a hermeneutics that also emphasizes this objective relation and draws on Schleiermacher's conception. In a discussion on methodology, Epple recalls Schleiermacher's distinction between a laxer practice and a more rigorous one concerning the "craft of understanding the linguistic testimonies." While

> the laxer practice in the art [...] assumes that understanding arises by itself, and expresses the goal negatively: misunderstanding is to be avoided, [...] the more rigorous practice assumes that misunderstanding arises by itself and that understanding must be striven for and sought everywhere.

Epple noted

how simple it is to be blinded by the modern mathematical level of knowledge and to perceive in such texts as valid knowledge only what reveals itself as such from today's view. One overlooks the strange, incomprehensible passages with which these texts are larded without noting that much gets lost by doing so (Epple 2000, 140).

In my paper of 1983, I had concluded that interpreting texts in the sense of material hermeneutics requires a broad interdisciplinary approach using contributions in particular of science research, sociology, and the science of history (Schubring 1986, 85). It was realized that such an approach corresponds to an approach in the general history of science that is known as the *social history of ideas*. The first to intensely elaborate it was probably the historian of science Peter Gay, in his research into the Enlightenment and into Voltaire since the 1950s. Looking back, Gay characterized his own approach in 1985:

> I wanted to discover [...] how mental products—ideas, ideals, religions and political and aesthetic postures—had originated and would define their shape under the impress of social realities (Gay 1985, xiii).

His focus was

> to place the principles of the Enlightenment into their natural environment: the scientific revolution, medical innovation, state making, and the impassioned political debates of the eighteenth century.

Gay's intention was

> to grasp ideas in all their contexts. A moral imperative, an aesthetic taste, a scientific discovery, a political stratagem, a military decision and all the countless other guises that ideas take are, as I have said, soaked in their particular, immediate, as well as in their general cultural surroundings. But they are also responses to inward pressures, being, at least in part, translations of instinctual needs, defensive maneuvers, anxious anticipations. Mental products in this comprehensive sense emerge as compromises (ibid., xiii f.).

Gay's approach has been taken up in different areas of the history of science. A noteworthy concretization was presented by Brockliss in 1987, in the latter's study of how the transmission of philosophical ideas in the seventheenth and eighteenth centuries was shaped by the institutional structures in France, and what kind of "compromises" were thus materialized in the ideas (Brockliss 1987). Anthony Grafton made this approach eminently concrete for the classical humanities ("*Altphilologie*") in a series of studies exploring in depth how this discipline was professionalized and institutionalized (see, e.g., Grafton 1983). Book titles like *Text in Context* (1992) show how diversely this program was implemented and how attractive it is.

It must be admitted, however, that these approaches—in particular that of Gay—are less concrete in showing how social history led to "compromises" in the ideas under formation. What is meant by "social" remains too vague, and does not become operational. Collins's 1998 volume on the history of philosophical ideas is interesting in that it assumes the concept of the *social production of ideas* using sociological communication theories to define the social *setting* as the relation between group and individual. Then, the traditional "cult of the genius or the intellectual hero" proves to be unwarranted: "We arrive at individuals only by abstracting from the surrounding context" (Collins

1998, 3). Intellectual creativity occurs mostly not in groups, he says, but in individual work. Nevertheless, production of ideas never occurs in an isolated manner:

> The group is present in consciousness even when the individual is alone: for individuals who are the creators of historically significant ideas, it is this *intellectual* community which is paramount precisely when he or she is alone (ibid., 7).

Collins thus represented the social production of philosophical ideas as the history of different contexts of communication.

Indeed, the concept of communication provides a fertile approach to mold a hermeneutically focused conceptual history in a more specifically operational way. Moreover, I use a contribution of science research elaborated by Wolfgang Krohn and Günter Küppers in 1989 that demarcates itself from Niklas Luhmann's autonomist system theory while adopting several of its elements. They understand scientific activity as an interaction between a system and its environment, concede equal legitimacy to intrascientific and extrascientific environments, and thus arrive at complex modelings of scientific innovation processes (Krohn / Küppers 1989, 27).

Combining the communicational approach with the system–environment–interaction model, I can specify that the social history of ideas is in its essence culturally shaped: the first interaction of the system of mathematics with its environment occurs primarily in the field of culture. The system of science, and together with it the subsystems of the various disciplines, has in any case been part of the respective system of national education and thus integrated into a defined environment of specific cultural and epistemological shape providing a common system of communication since territorial states emerged in modern times. This primary relation between system and environment makes that studies about professionalization and institutionalization are so significant for clarifying the role of "social"—or better, cultural—elements in producing ideas. In a first stage, communication is not universal, but takes place within determined cultural and social borders; and mathematical production occurs within concrete institutional structures that express historically variable social valorizations of knowledge. It is evident, however, that more precise results require comparative studies of different environments. Since the paucity of such studies constitutes a considerable drawback for context–oriented approaches, a systematic comparison of different cultures or nations will be attempted here.

My approach is specific in that it requires the investigation to be holistic, the term "holistic" being intended in several dimensions:

• Primarily, it has become a truism in the theory of research into the history of mathematics that an investigation into a concept's history may not be confined to the wording of the basic definitions, but must unravel their meaning and scope within the respective author's own context of application. (To what extent this demand is met in practice is another issue). This prelude to a more than pointwise analysis of the concept's use, however, must be generalized even more basically:

- On the one hand, the analysis may not be confined to how an individual definition was applied, but rather must be conducted within the context of an entire concept *field* that determines how the respective concept was justified and applied.
- And on the other hand, the analysis must refrain from isolating individual authors or mathematicians, but rather must understand these as belonging to a cooperative mathematical community. Since mentalities and basic epistemological convictions are among the first things to be shared within a common culture, a concept's understanding and its intended horizon of meaning can be obtained only from a holistic analysis of mathematical production within a given period of the respective national culture.
- In a next step, developments within different cultural contexts must be compared to analyze how differing concepts from other communities and contexts were received, and what influence they may have had.

   Since this approach implies priority for the respective national and cultural context, it assumes that there is no a priori global or uniform development, independent of culture. It remains, therefore, an open question how and when convergence toward an international mathematical community set in.
- Finally, "holistic" character also refers to the texts under scrutiny; sources must be explored in their entirety, including material from archives and from *Nachlässe*. All this must be integrated into the evaluation.

Textbooks are particularly suitable text types for such a program. Since they are directed to a broader public, they yield good indicators as to the intended conceptual horizon of a certain period and culture. In the sense of Thomas Kuhn's "normal" science, textbooks characterize the shared corpus of problems and of methods to solve them. An instructive example is Leo Corry's volume on the origin of abstract algebra (1996); the author uses algebra textbooks as his major sources, these having been innovative by disseminating new developments while at the same time normalizing the respective status of knowledge concerned.

To sum up: this study's conceptual–historical approach may be characterized by the concept pair of *critique of tradition* and *attempt at reconstruction* coined by the historian Ernst Schulin (Schulin 1979, 16): a critical analysis of positions hitherto presented in the historiography of mathematics as the study's object and a hermeneutical reconstruction of how concepts developed as the study's method.

2. Categories for the Analysis of Culturally Shaped Conceptual Developments

What follows will outline the categories along which this study's analysis of culturally shaped processes of generalization and the conflicts fought over these processes primarily proceed.

The first category has already been described in the previous section: within the respective historical context, a concept will never appear isolated—rather, it is an element of a concept field that conveys its meaning to this element. Historical reconstruction thus must grasp how an element evolved together with its surrounding and supporting field.

The second category is that generalization, in many respects, turns out to have been a process of making *implied* meanings explicit. As Lakatos so nicely pointed out in his study on the polyhedra formula, a critical examination of conventional conceptual views leads time and again to uncovering "hidden lemmas" (Lakatos 1976, 43 ff.)—that is, to revealing assumptions implied. It is evident that such implied assumptions may prove to be inconsistent, or valid only for special cases, when we try to make them explicit or consciously reflect on them, thus calling for a conceptual frame modified by extension and generalization.

One of the consequences of this explicatory category is that concepts develop in a process of being differentiated ever further. Emerging concepts are in most cases still very general, in the sense that they can be used in wide fields of application. Bachelard pointed out that such original, unspecific generality, where experimentally testable features are hardly discernible, entail the danger of inducing gross mistakes, a fact that prompted him to assign overly general propositions as "premature" to the prescientific stage in his three-stages model of scientific development (Bachelard 1975, 55–72). In fact, while emerging concepts are being developed further, some of their fields of application become differentiated in subfields where the concept will then be given more specific meanings. Just think of the concept of function and its further differentiation into continuous functions, differentiable functions, etc.

An important further category of the process of explication is the concepts' sign aspect, their symbolism. There is a close connection between the sign and the signified, i.e., the concept. Upon encountering different sign representations, we as a rule cannot assume an identical concept meaning. This linguistic side of mathematical concepts has always had a special importance in periods of restructuring or modernizing the science.

Algebraization is closely linked to the category of sign representation. While the development of algebra has been based on elaborating suitable sign systems since Viète and Descartes, generalization was understood to be growing permeation of mathematics with algebra. Geometry, in contrast, was conceived of as confined to the limited frame of classical mathematics. Algebraization drives explication. This is why algebraization has been also reflected as "calculization"—as in Sybille Krämer's book (2001)—a process that not only makes mental entities graspable by the senses, but also signifies how they are produced.

Algebraization, as one of the central manifestations of generalization thus means a double process here:

- firstly, the process of transforming concepts originating from geometry into an algebraic form,
- and secondly, the process of internally changing algebraic propositions, which since Nesselmann's characterization of 1842 has generally been seen as taking place in three stages. The first stage is "rhetorical algebra," in which all propositions are presented exclusively in verbal form because of the total absence of signs. The second stage is the intermediate one of "syncopic algebra," which, while still operating verbally, introduces abbreviations for terms or operations frequently used. The third and last stage is "symbolic algebra," which relies on a sign language independent of words (Nesselmann 1842, 302).

Both the categories of the extent of sign use and of how far algebraization has progressed are subject to social variables. These scientification processes are dependent in their intensity on the ups and downs the estimation and valuation science as a whole encounters in the societies or nations concerned. These ups and downs find their classical expression in the polar opposition of a dominance either of the *analytic* or the *synthetic* method, and they will be documented as a pervasive category in the developments of concepts treated in this study.

At the same time, extent and modes of algebraization are also directly linked to cultural and epistemological traditions in the societies and professional communities concerned. These were conducive in the then leading countries of France, England, and Germany to variations of their own that remained stable over extended periods. Their differences will make the importance of culturally specific historical reconstructions particularly salient.

3. An unusual Pair: Negative Numbers and Infinitely Small Quantities

> The doctrine of opposite quantities and the calculus of the infinite have both experienced the same fate, inasmuch as one has been formerly happy with their correct application, thereby bringing pure mathematics to a high degree of perfection, while also enriching the mathematical sciences as such with the most important inventions by applying them, before the first concepts and principles on which they are based had been completely brought to light. [...] If one is convinced of a result's correctness beforehand, one will unconsciously ascribe more completeness and evidence to the concepts and principles from which it was derived in a shortcut than these actually have, and only he to whom both truth and its justification are new will feel the defects of the latter.
>
> —*Hecker* 1800 b, 4

The objects of this investigation into concept development shall be two concepts belonging to the foundations of mathematics: that of negative numbers and that of infinitely small quantities. Not only do the debates about clarifying the concepts of negative numbers and of "classical" infinitely small quantities both form typical cases of concept development for themselves, but the development of each of these concepts was by no means unrelated to that of the other.[1] Numerous mathematicians have made contributions to both problem areas. Their respective positions yield telling profiles on foundational issues of mathematics, but far from the claim that they show mechanist correspondences or directly interpretable "scales."

Both problem fields are typical indicators for the respective views on the relations between intuition, rigor, theoreticity (*Theoretizität*), and generality of mathematical concepts. At the same time, they are closely connected to contemporary epistemological views—both to general philosophical orientations

---

[1]   The distinction between "classical" and "modern" infinitely small quantities, introduced here as one of the results of long-term analysis, will be explained later in Chapters V, VI, and VII.

and to specific ones for mathematics and the sciences. Therefore, the development of these concepts did not occur in an isolated way, but as part of more pervasive thought processes belonging to an epoch or a culture. This is why it was suggestive to pursue the very differences that are traceable for cultures existing side by side in Europe.

In traditional historical accounts, the development of the negative numbers appears to be just a small strand that can be presented without connecting it internally to other developments in the mathematical edifice. Actually, developing this very concept forms an essential element in the process of generalization in mathematics: it is pivotal not only for a more abstract understanding of mathematical operations, but also for the emergence of a concept of quantity that facilitates justifying the new, rigorous analysis. As Bos has shown, multiplication, for instance, had been executable only in a restricted manner as long as the concept of geometrical quantities prevailed, and this not only impeded an unrestricted application of the concept of equation, but also the introduction of foundational concepts of analysis in their then insufficiently general form (Bos 1974, 6 ff. and passim). As will be shown, one focus of the efforts at clarifying the concept of negative numbers was on making the operation of multiplication generally executable; in this process of development, the concept of geometrical quantities lost its status of fundamental concept for algebra and analysis.

In contrast to what historiography says, classical authors of function theory, and in particular of complex functions, at the apogee of establishing modern analysis were perfectly aware of the fact that progress in analysis was closely linked to extending the concept of number. Weierstraß, in his lectures on function theory, always presented the underlying concept of number, and was the first to develop a comprehensive and rigorous conceptual system for negative numbers (cf. Spalt 1991). Riemann, in his respective lectures on function theory, always emphasized in detail the introduction of negative numbers as a decisive step in the generalization process of mathematics:

> The original object of mathematics is the integer number; the field of study increases only gradually. This extension does not happen arbitrarily, however; it is always motivated by the fact that the initially restricted view leads toward a need for such an extension. Thus the task of subtraction requires us to seek such quantities, or to extend our concept of quantity in such a way that its execution is always possible, thus guiding us to the concept of the negative (Riemann 1861, 21).

It must be noted that Riemann, in particular contrast to received historiography, emphasized the systematic link between introducing negative numbers and the larger edifice of mathematical concepts: "This extension of the area of quantities incidentally entails that the meaning of the arithmetical operations is modified" (ibid.), quoting the operation of multiplication as an example.

Fontenelle, L. Carnot, and Duhamel shall be mentioned in this introduction as pioneers linking the foundational effort concerning the concepts of negative number and infinitely small quantity as mathematicians exemplary for this study's period of investigation at its beginning, middle, and end.

Carnot, who lived and worked in the middle of the transition from the Enlightenment to modern times, made a strong effort at anchoring mathematical concepts in intuition, and at establishing them rigorously at the same time. Negative numbers and limits, or infinitely small quantities, were his major points of approach toward rigor and generality. The contributions of Fontenelle's early attempts at founding theoretical concepts of mathematics are less familiar.

Bernard le Bovier de Fontenelle (1657–1757), permanent secretary of the Paris Academy of Sciences, is known as the author of *Eléments de la Géométrie de l'Infini* (1727), severely criticized by his contemporaries, in which he boldly developed a calculus for operating with the infinitely large and the infinitely small. At the same time, his volume also presented new contributions toward clarifying the concept of negative quantities: Chapters II to V of the first part dealing with *la grandeur infiniment petite*, and with incommensurable quantities are followed by chapter VI on *Des Grandeurs positives et negatives, réelles et imaginaires*. Based on the concept of quantity thus developed, Fontenelle investigates applications to geometry and to differential calculus. I shall treat in detail the pertinent sections of this in my own next chapter.

At the end of the entire period, there is J.M.C. Duhamel (1797–1872). While he himself was not a "Modern Man," he made an impressive effort to convey the general character of mathematical concepts in his teaching. Again, the two foundational concepts that form the bulk of his presentation of generalization in mathematics were negative numbers and infinitely small quantities.

Particularly instructive are his five volumes *Des Méthodes dans les sciences de raisonnement* (1865–1873), in which Duhamel summarized his own work extending over several decades of reflection and presenting the foundational concepts of the sciences. This work is noteworthy not only as a rare and late approach toward methodologically reflecting on mathematics, sciences, and even on parts of the social sciences, thus avoiding confinement to a narrow, technical specialization, but also because it again tackles the foundational issues, yet unresolved since the beginning of the nineteenth century, trying to solve them methodologically. As Duhamel says in his introduction, he had planned such a methodological work since his youth to assist in clarifying and removing the "obscurities" that he had observed already in the mathematics courses in school. He must thus have been first motivated around 1810, and obviously by doubts concerning the foundational aspects of algebra. These "obscurities" had not been removed at the time of his subsequent studies at the *École polytechnique* (1814 and later); instead, new ones had been added to the old (evidently concerning foundational concepts of analysis), and his fellow students had seen no reason for concern (Duhamel vol. I, 1875, 1). Duhamel's late work thus documents that there had been no decisive further development in the foundational issues in France from 1800/1810 until the century's last third.

At the same time, Duhamel's work underlines how important it was for the methodology of science and epistemology to focus on foundational concepts, an approach typical of the way authors strove for rigor about the year 1800. In his section on the most frequent causes of error in thinking, Duhamel claims that the causes lay less in errors of deduction than in the inexactness of basic assumptions, the easiest way to err being in establishing general propositions (ibid., 21).

Although Duhamel's views belong to the close of my study's period of investigation, they may nevertheless be considered typical of how people strove for rigor across the entire period.

Eventually, the conceptual link between the developments of the concepts of negative numbers and of infinitely small quantities is also confirmed by a partisan opposing the view that algebra's task was to generalize: by d'Alembert, who heatedly argued against admitting the two notions as mathematical concepts, thus indirectly contributing to their further clarification (cf. Chapter II. 2.8.).

Chapter II

Paths Toward Algebraization – Development to the Eighteenth Century. The Number Field

1. An Overview of the History of Key Fundamental Concepts

For the presumably first, albeit rather sketchy, historical study on textbooks about infinitesimal calculus, the conceptual basis had already been that analysis cannot follow up isolated concepts, but must rather pursue the connections within a concept field. The study listed, as elements of this field, "for differential calculus and integral calculus three concepts (are) basic: number, function, and limit" (Bohlmann 1899, 93).

Studying the two concept developments on which we intend to focus will be preceded by an introductory overview about how essential elements of this concept field were developed up to the eighteenth century. The intention is to sketch the conceptual frames, in line with the received literature, sufficiently to prepare for the subsequent in-depth analysis within this concept field. This introduction intends to present only those aspects that are relevant here for concept development. This is why I do not adopt Bohlmann's subdivision of the concept field into elements. Instead, I start from the general position that concepts are subject to continuous differentiation (*Ausdifferenzierung*, cf. Chapter I.). One original notion evolved, by continuous differentiation, into several separate and independent concepts. This kind of continuous differentiation can be established for the foundational concepts relevant here: the concepts of number and of function do not form entirely separate concepts, but have emerged by way of continuous differentiation from the holistic concept of *quantity*. If you take quantity as the original fundamental idea, you will see that yet another concept differentiated from this original makes up an element of this concept field: this is the concept of variable. Three foundational concepts thus must be considered to have been successively differentiated from that of *quantity*: the concepts of *number*, *variable*, and *function*. And what is called limit by Bohlmann constitutes but one element of the comprehensive field of limit processes in mathematics, which was eventually differentiated further into the concepts of *limit*, of *continuity*, and of *convergence*. The concept of the *integral* shall also be included in this overview, because it is of special importance for developing the concept of infinitely small quantities.

1.1. The Concept of Number

The concept of number is of interest here foremost in two of its aspects; firstly, in its capability for conceiving of limit processes, i.e., in particular for investigating intermediate values, completeness, etc. Secondly, the differentiation of the concept of number from that of quantity and from that of magnitude is relevant. The first aspect, which primarily is about how the concept of the real number is formed, has always been intensely investigated and presented in the literature (cf. Gericke 1970). The discovery of incommensurability and of the existence of irrational quantities in Greek mathematics, in particular, belongs to this aspect. The long-debated question whether the Greeks already had a notion of real numbers has meanwhile been decided with negative outcome. Likewise, there is agreement that there were, until the eighteenth century, no efforts by mathematicians to conceptually clarify the real numbers. The term real numbers had been in use since about 1700 to characterize rational and irrational numbers, as opposed to the complex or imaginary numbers. Due to the prevalence of the concept of geometrical quantities, the completeness of the domain of real numbers was implicitly assumed as given; this implicit assumption was also expressed in the terminology used. Quantities were divided into *discrete* and *continuous* quantities. Discrete quantities were understood to be both concrete and abstract, or pure numbers, while continuous quantities were understood to be real numbers given geometrically by segments of straight lines.[1]

How the concept of number became differentiated from the more general concepts of quantity and of magnitude has been less intensely investigated and presented. In the French usage, it is particularly clear that the character of *quantité* is more basic. The concept of *grandeur*, formerly often used synonymously with *quantité*, is now basically limited to meaning concrete numbers. For clarifying the concept of negative numbers, the separation between the concept of numbers and that of magnitudes or *quantités* will prove to have been decisive. In his account of the history of negative numbers, Sesiano chose the differentiation between quantity and number as the basic category for analyzing their development (cf. Sesiano 1985). It is all the more important to emphasize how drawn out and tedious the historical process of differentiation

---

[1]   The characterization discrete-continuous for quantities was established already by Aristotle in his doctrine of categories (cf. Boscovich 1754a, V).

between quantities and numbers was, since scholars are apt to assume that a concept of pure numbers had surfaced already in ancient Greek mathematics, similar to other eminent achievements of the same. What had emerged, however, was a concept of number tied to geometry. Only the integers were understood as numbers (ἀριθμός) at all, while other number areas, in particular fractions, were understood to be quantities; and Euclid understood even the integers geometrically, as segments of straight lines. Arithmetic, at that time, formed an integral part of geometry.[2]

With regard to the possibility of infinitely small quantities being taken to mean non-Archimedean quantities in the eighteenth century, a more special concept development will be discussed at this point. In his Book V, on number theory, Euclid excluded non-Archimedean quantities. Felix Klein has pointed out that Euclid excluded non-Archimedean quantities from his own concept of number to enable himself to found the theory of proportions—and together with it an early concept of irrational numbers (Klein 1925, 221).

The only place where Euclid mentions the admissibility of non-Archimedean quantities is the angle concept. This is where he begins with admitting, besides rectilinear angles, angles whose boundary lines are formed by curves (Book I, Def. 8) as well. A special case of these so-called *hornlike* (or cornicular) angles is that of the mixtilinear angles, in which one of the two boundary lines is a straight line and the other a curve. Klein has shown in detail that the hornlike angles form a model of non-Archimedean quantities (ibid., 221–224).[3] Euclid used the hornlike angles, which are not commensurable with rectilinear angles, only once in his *Elements*, in Book III, 16. This proposition has subsequently provoked an extensive debate, however. It says:

> The straight line drawn at right angles to the diameter of a circle from its extremity will fall outside the circle, and into the space between the straight line and the circumference another straight line cannot be interposed; further the angle of the semicircle is greater, and the remaining angle less, than any acute rectilinear angle (Heath, vol. II, 37).

Proclus, in his commentary on Euclid, rejected this assumed commensurability of mixtilinear or hornlike angles with rectilinear angles. Referring to the so-called Archimedean axiom, he declared that quantities having a ratio to one another, when multiplied, can exceed one another. Accordingly, therefore, the hornlike angle would be able to exceed the rectilinear one as well—this possibility, however, being excluded by it having been shown that the hornlike angle will always be smaller than any rectilinear angle (Proclus 1945, 251).

In modern times, the debate was resumed, due to Euclid's intensified reception prompted by the new printed editions: as debates whether the mixtilinear angles quoted in Euclid's above proposition were admissible and

---

2   Cf. N. Rouche 1992, 169 ff.

3   For a nonstandard interpretation of angles of contingency cf. Laugwitz 1970.

feasible as angles between a circle and a tangent, now called *angulus contactus*, or angle of contingency. The confrontation between Jaques Peletier and Christopher Clavius, in particular, has become well known.[4] In his own edition of Euclid of 1557, Peletier had rejected using heterogeneous, incommensurable angles. The angle of contact, he claimed, was no genuine angle, and in general not even a quantity; rather, the straight line touching the circle coincided with the latter's circumference, the angle of contingency thus being zero, and the angle formed by a semicircle and its diameter being a right angle (M. Cantor, vol. 2, 1900, 559 f.). Against Peletier, Clavius, in the second edition of his edition of Euclid (1589), admitted the angle of contingency to be a genuine angle, that is a quantity infinitely divisible. He agreed with the Euclidean proposition that the angle of contingency will be smaller than any possible rectilinear angle and that at the same time, the angle of the radius will be smaller than right, but greater than any rectilinear acute angle. On the other hand, he conceived angle of contingency and rectilinear angle as heterogeneous quantities that are not comparable with each other (ibid., 560 f.).

The debate after Peletier and Clavius found a first conclusion with John Wallis who—in a treatise of 1656—*De angulo contactus et semi-circuli tractatus*, and in a later one of 1684—adopted Peletier's view according to which the angle of contingency was a "non-angulum" and a "non-quantity." Wallis was the first to introduce, at the same time, however, the term degree of curvature.[5] The curvature behavior of curves had indeed been the mathematical context for which the concept of the angle of contingency was intended. As the concept of curvature was specified and operationalized, in particular by developing differential calculus, the debate about the angle of contingency waned, the geometry textbooks of the eighteenth and the nineteenth centuries touching it only marginally, if at all, mentioning the concept of mixtilinear or that of curvilinear angle, and bare of any further discussion of its foundation or of its mathematical application.

While Proclus's use of the Archimedean axiom still shows an explicit reference to the concept of quantity, the notion of mixtilinear and cornicular angles was used in modern times neither for reflecting on the concept of quantity nor for extending or generalizing the concept of quantity or of number. Typical for this tendency is Tacquet's position concerning the controversy between Peletarius and Clavius, in his own geometry textbook of 1654, in which he completely excluded the angle concept from the concept of quantity proper. In a first hint at differentiating the concept of quantity and particularly at introducing angles as equivalence relations, Tacquet declared that angles were not quantities, but rather *modi* of quantities, comparable to one another only with regard to their congruence or noncongruence (cf. Klügel 1803, 290). Boscovich, one of the

---

[4]   For relevant publications between 1550 and 1650 see Giusti 1994. On the debate between Clavius and Peletier cf. in particular Maierù 1990.

[5]   A new thorough analysis of both studies by Wallis and of their statements about the Peletier/Clavius debate is Maierù (1988).

most prolific authors regarding foundational issues of the eighteenth century, also examined—in his detailed study *De continuitatis lege* (1754)—various arguments as to whether infinitely small quantities objectively exist. He also discussed, as one of these possible cases, the angle of contingency as an angle that is infinitely small when compared to rectilinear angles. Boscovich adopted Tacquet's argument that angle was not a quantity—usually defined via inclination—but rather a *modus* of a quantity. Mixtilinear angles, he said, were in principle incomparable with rectilinear ones, hence had no ratio to a rectilinear angle at all, and for that reason could be neither infinitely large nor infinitely small (Boscovich 1754, XXXVII; cf. also Manara 1987, 179).

1.2. The Concept of Variable

While the question of how the concept of function was formed has always been an essential element of investigations and accounts within the history of analysis, less attention has been given to how the concept of variable emerged, although the concept of function presupposes the concept of variable, and although the latter expresses an equally fundamental change from Greek mathematics. In the latter's prevailing geometrical character, quantities had been understood to be *constant*. The history of the concept of variable has recently been discussed as a constitutive part of the development of the foundational concepts of analysis by E. Giusti (1984).

In modern times, acceptance of the concept of the *unknown* in algebra was the preliminary stage for establishing the concept of variable. In algebra, Descartes placed the unknown in a dualistic opposition to the known quantity (*quantité inconnue* versus *quantité connue*) and this dualism was transferred to the concept of variable. In all the definitions of variable since L'Hospital's textbook of 1696, variables are explained by their opposition to constants:

> One calls *variable* quantities those which increase or decrease continually, and by contrast *constant* quantities those which remain the same while the others change (L'Hospital 1696, 1).

Chr. Wolff's *Mathematisches Lexikon* of 1716 already contains the key terms *Quantitas constans* and *Quantitates variabiles*. The entry on constants says:

> *Quantitas constans*, an invariable quantity, is the name of a quantity which always maintains the same quantity, whereas others increase or decrease (Chr. Wolff in 1716, 1144).

Also, in d'Alembert's entry "variable" in the *Encyclopédie*, the opposition between constant and variable is decisive for the definition, but he relates the variation to "a certain law." As examples he names abscissae and ordinates of curves. Constant quantities "do not change," an example being the diameter of a circle (Encyclop., XVI, 840).

Besides the dichotomy of constant–variable, the frequent explicit demand that it change continuously is noteworthy in these definitions of a variable. Something that might be an additional quality, for ensuring the completeness of the domain of definition, was already integrated, by overgeneralizing, into the basic definition. Possible anomalies in case of limit processes could be thus excluded from the very outset. Some textbook authors even insisted on a dominant character of the demand for continuity in their own definitions, like the Oratorian priest Reyneau in his influential textbook *Analyse demontrée*, who required that the quantities increase or decrease *insensiblement* (Reyneau, vol. 2, 1738, 152).

In Euler, a substantial change in the concept of variable can be noted; he gave up the dichotomy between constant and variable, replacing it by a universality of the variable. For him, the constant presents a special case, since he understands the variable to be an indeterminate that is able to assume certain values:

> *A constant quantity is a determined quantity which always keeps the same value.[...] A variable quantity is one which is not determined or is universal, which can take on any value* (Euler 1988, 2).

> Since all determined values can be expressed as numbers, a variable quantity takes on the totality of all possible numbers.[...] Hence a variable quantity can be determined in infinitely many ways, since absolutely all numbers can be substituted for it (ibid., 2 f.)

We will again find the identical condition, based on geometrical ties, in the concept of function, where the entire domain of the real numbers is implicitly assumed in any case to be the domain of definition and of values.

1.3. The Concept of Function

Youschkevitch's voluminous contribution (1976) is valued as the classical account of the history of the concept of function. Dhombres published some supplements to it, in particular about Euler's concept of function (Dhombres 1986). A summary has been given by Medvedev (1991).

Functional relations were used not only in Greek mathematics—Ptolemy's *Almagest* being a well-known example—but by the Babylonians as well. The concept of "function," however, evolved from the general concept of quantity only in modern times, becoming an independent mathematical object, a concept proper. Issues taken up in physics, and above all the progress in kinematics, proved decisive for this process of ongoing differentiation. This context of application for a long time shaped what the concept of function contained, and how it developed. Another essential factor for this autonomization, however, was the progress in algebraic symbolism, which permitted the representation of even the most intricate equations and formulae by means of a limited number of signs. Because of this context, functions were understood to be *equations* between two variables. The attachment of the concept of function to the formula, i.e., to the "calculation's expression," was to continue determining this concept's form and content for a long time to come.

In Descartes's *Géométrie* (1637), we find an elaborate form of the concept of function in the shape of his formulation of a reciprocal dependency between two quantities given by an equation, both of which can assume an arbitrary number of values. According to this tradition, functions were at first restricted to algebraic functions. Newton and Leibniz, however, extended their investigations to transcendental functions. The problems of how to develop transcendental functions into series and the latter's impact on the meaning of the foundational concepts were to become the principal focus for research into analysis.

In his *Method of fluxions*, Newton introduced the distinction between independent and dependent (variable) quantities, as *quantitas correlata* and *quantitas relata*. In Leibniz's manuscripts, the term function is found for the first time in 1673, for the relation between ordinates and abscissae of a curve given by an equation. In his publications of 1684 and of 1686, he already divided functions into two classes: algebraic and transcendental. The first general definition of the new mathematical object of function was published by Johann Bernoulli in 1718:

> One calls function of a variable magnitude a quantity composed in a certain manner by that variable magnitude and by constants (Joh. Bernoulli *Opera Omnia*, tom. 2, 1968, 241).[6]

This definition gave no attention yet to the distinction between single-valued and multivalued functions.

Essential contributions to further elaborating the concept of function were made by Euler. His definition remained tied to the formula, to the calculation's expression, while he generalized it for analytic functions:

> *A function of a variable quantity is an analytic expression composed in any way whatsoever of the variable quantity and numbers or constant quantities* (Euler 1988, 4).

Euler understood as functions analytically expressible those that can be developed into infinite powers series. He admitted for this not only positive integer exponents, but also arbitrary ones. "Should anyone doubt," Euler argued, "his doubt will be eliminated by the very development of one or another function" into a power series (Youschkevitch 1976, 62). Euler's definition of function as an analytic expression whose most general form is a power series was to remain the predominantly recognized determination during the entire eighteenth century .

Euler developed the concept of function further in his own work mainly in two respects: firstly, in discussing the meaning of the continuity and the discontinuity of functions. For Euler, a function was continuous ("continua") if

---

6   "On appelle fonction d'une grandeur variable une quantité composée de quelque manière que ce soit de cette grandeur variable et de constantes."

its equation, its formula remained unchanged—hence, if the function was describable by just one calculational expression. Thus, the hyperbola with its two branches belonged to the continuous functions.

Conversely, he regarded a function as "discontinua" or "mixta" when it was composed of continuous parts, but subject to several equations—piecewise continuous functions, whose pieces could well be connected and whose graphs could be traced by a free stroke of the hand.[7] This distinction was later supplemented by Arbogast's *fonctions discontigues*, whose various parts were conceived of as unconnected (see below, Section 1.5.). The second further development originated from discussion of the physical problem of the vibrating string: It made clear for the first time that functions are also representable as a superposition of trigonometric functions. The consequences for the concept of function and for analysis in general, however, were drawn only gradually. In his textbook on differential calculus, Euler already made allowance for these extensions by defining the concept of function more generally:

> If some quantities so depend on other quantities that if the latter are changed the former undergo change, then the former are called functions of the latter. This denomination is of broadest nature and comprises every method by means of which one quantity could be determined by others (Euler/Michelsen, Vol. 1, 1790, XLIX; Translation quoted from Youschkevitch 1976, 70).

1.4. The Concept of Limit

In analysis, the limit concept functions in a way analogous to that of the concept of quantity within mathematics as a whole. It forms the essential basic concept, while several other foundational concepts emerged from continuous differentiation, for purposes of studying limit processes, among them that of continuity for functions and that of convergence for sequences. We shall begin here by summarizing the history of the concept of limit.

Among the numerous historical accounts of how this concept was developed, Hankel's contribution of 1871 is notable in that it distinguishes itself by considerable conceptual precision. The so-called method of exhaustion of Greek mathematics again and again provided the key source for all the conceptual developments in early modern times. Hankel pointed out that two different versions, between Euclid's and Archimedes's method of exhaustion, must be distinguished. Proposition 1 of Euclid's Book X forms the common basis for the two methods:

> Two unequal magnitudes being set out, if from the greater there be subtracted a magnitude greater than its half, and from that which is left a magnitude greater than

---

[7]   See Youschkevitch 1976, 64 ff. It seems that Euler understood continuity in the sense of Aristotle as connectivity of adjacent parts; thus his formulation in the treatise *De usu functionum discontinuorum in analysi* (1763), cf. ibid., 67. Cf. Section 1.5.

its half, and if this process be repeated continually, there will be left some magnitude which will be less than the lesser magnitude set out (quoted from Edwards 1979, 16).

This proposition finds its typical application in the inscribing polygons with an ever increasing number of sides into a curvilinear figure, say within a circle: a square inscribed within a circle is larger than half this circle's area. If we continue inscribing polygons having a number of sides 2^n, while applying the above proposition, the polygon's area will approximate the circle's area up to any quantity, however small.

Hankel noted that a general principle for limits of sequences implicitly underlies this method of exhaustion, formulating it as follows:

If the terms of two series of indefinitely increasing quantities:

$$a_1, a_2, a_3, \ldots \text{ and } b_1, b_2, b_3, \ldots$$

are in the ratios:

$$a_1 : b_1 = a_2 : b_2 = a_3 : b_3 = \ldots = \alpha : \beta$$

and if the a_i indefinitely approach a quantity A, the b_i a quantity B then A und B are in the same ratio:

$$A{:}B = \alpha{:}\beta \text{ (Hankel 1871, 186).}$$

Such a general principle is not found, however, in the mathematics of the ancient Greeks, although the frequent application of related conclusions might have us expect them to have become aware of a proposition about limits of series. In every demonstration, all the individual steps of proof are reiterated. Without such a proposition, the method of exhaustion, Hankel says, could not be characterized as scientific (ibid., 187).

The method of exhaustion used by Archimedes proceeds somewhat differently. To prove that a curvilinear figure C has the same area as a rectilinear figure D, he encloses C by inscribed and circumscribed polygon trains that ever more closely approach the value sought. One of the typical proof procedures is the following:

If one has two infinite series of quantities

$$E_1, E_2, E_3, \ldots, \quad U_1, U_2, U_3, \ldots,$$

and two fixed quantities $C, D,$ and if one can prove that the former quantities $E_1, E_2,$ $<D$ and $<C$ and that approximate C arbitrarily, and that moreover the quantities of the second series are $> D$ and $> C$ and that these also arbitrarily approximate C; then there must be $C = D$ (ibid., 188).

Archimedes, too, conducted proofs separately for every case while refraining from reducing them to a general theorem.

As Hankel stressed, these rigorous methods to determine limits were used by Euclid, Archimedes, and other Greek mathematicians, but these authors generally avoided infinite processes, never imagining the transition to the limit as something actually accomplished. Their avoidance of using and accepting the infinite thus already tended toward developing crucial elements for an algebra of inequalities.

In the Christian Middle Ages, however, scholastic philosophy, which dominated scientific debate, saw no epistemological obstacles in the transition to the infinite. In the mathematics of the early modern period in Western Europe, this philosophical acceptance led to establishing the method of the *indivisibles*: the method of exhausting curvilinear figures by an infinitely large number of rectilinear figures. The method's first known eminent representative was Kepler, who, starting from computing the volumes of wine casks, developed stereometrical principles according to which any continuously curved solid can be treated as a polyhedron having an infinite number of infinitely small sides. A typical example worth noting is computing the volume of the sphere. The sphere was understood as composed of infinitely many pyramids, their vertices lying in the sphere's center and their bases touching its surface from within. For F as base area and h as height, the pyramid's volume is

$$\tfrac{1}{3} \, Fh.$$

For arbitrarily small areas of the base, h can be regarded as identical with r (radius of the sphere). Hence, thanks to the already known area of the sphere $4\pi r^2$, the volume of the sphere results as (cf. Kepler 1615 and C.H. Edwards 1979, 102)

$$V = \frac{4\pi r^3}{3}.$$

It is typical for Kepler's method that:

- The indivisibles have the same dimension as the figure which has to be determined.
- The computation is done in each case for a particular geometrical figure, due to an ad hoc subdivision into indivisibles.

In his two volumes *Geometria indivisibilibus* (1635) and *Exercitationes geometricae sex* (1647), Cavalieri developed more general methods for determining volumes; he succeeded in making the indivisibles a widely accepted mathematical concept. In contrast to Kepler's ad hoc methods for the respective figure under scrutiny, Cavalieri established direct correspondences between the indivisible elements of *two* geometrical figures: the area or the volume of one of the two figures being known and the other figure being sought.

Moreover, Cavalieri assumed that the indivisibles of geometrical figures are quantities one dimension *smaller*. He understood, for instance, "all" indivisibles of an area to be the "aggregate" formed by an infinite number of parallel and equidistant segments of straight lines, interpreting the indivisibles of a solid accordingly as the aggregate of parallel and equidistant intersecting planes. On the basis of these two assumptions, Cavalieri's theorem can be understood:

> If two solids have equal altitudes, and if the sections made by planes, parallel to the bases and at equal distances from them are always in a given ratio, then the volumes of the solids are also in this ratio (Edwards 1979, 104).

An example of how to apply this theorem is to determine the volume of a circular cone of height h and base radius r. The cone is thus compared with a pyramid of identical height and with the unit square as base. The indivisibles as sections at the height x are to one another in the ratio πr^2. The result is

$$V(C) = \pi r^2 \cdot V(P) = \pi r^2 \frac{h}{3}.$$

In Cavalieri's work, the method of the indivisibles reached its summit of explicit elaboration and application. Not only did it meet contemporary criticism because of the methodological and epistemological problems it raised, but at the same time it represented a developmental conclusion because of the emergence of a novel fundamental idea, which now began to prevail. Upon summing up the developmental stages of the limit concept hitherto accomplished, it is seen that all these attempts concerned "measuring" given geometrical solids, surfaces, or lines, by means of approximating curvilinear figures by rectilinear ones. The figures given were always fixed and invariable, the applied methods thus corresponded to static conceptions.

The emergence of the concept of function in the seventeenth century therefore marked a fundamental change for the meaning and scope of limit processes. Functions were at first understood primarily as *kinematic* objects—as quantities variable with time. They were functions of only *one* parameter, the name "fluents" being typical for this concept's reach (see Bourbaki 1974, 225 f.).

For Hankel, the "prehistory" of the limit concept ends in the seventeenth century. After summarily listing some mathematicians of the seventeenth century, he immediately switches to the modern concept of limit as established in the nineteenth century. He does not address Newton's essential suggestion toward developing the concept of limit further. This is why these achievements and the developments prompted by them in the eighteenth century will not be discussed in this overview, but instead in Chapter III.

1.5. Continuity

The received view about the concept of continuity in the historiography of mathematics is that Bolzano and Cauchy were the first to define the concept rigorously, and that while there had been some discussion about continuity during the eighteenth century, profound reflections on its meaning nevertheless did not begin before the close of the eighteenth century (cf. Grabiner 1981, 87 ff.). The memoir with which L.F.A. Arbogast (1791) won the prize offered by the St. Petersburg Academy in 1787 is considered to be the principal document of the emerging discussion about the concept. Edwards even considers this treatise to have been the first to clearly elaborate the intermediate value quality (Edwards 1979, 303).

Indubitably, the development of the concept of continuity is tied both to the formation of the concepts of variable and of function, and to advances in the study of sufficiently large classes of curves. Indeed, considerable progress was made only by the middle of the eighteenth century, due to the debates about the equation of the vibrating string, and in particular due to the Euler's contributions. On the other hand, the definitions of continuity, as given by Cauchy and Bolzano, were no sudden innovations without identifiable precursors. I will examine the actually provable contexts of origin in the Chapters III and V.2., in particular Cauchy's immediate context in the *École polytechnique*. The intention here is again to do no more than sketch the respective status of research in the historiography of mathematics.

As is seen from Euler's standard textbook *Introductio in Analysin infinitorum* of 1748, continuity was at first understood as a geometrical quality: as a quality of curves. Continuous curves were characterized by the fact that they could be expressed by an analytic expression. In contrast, discontinuous curves consisted of several segments that belonged to different functions and hence did not correspond to just one analytic expression, but to several. This explains why Euler called the non-continuous curves "discontinuous" or "mixed" curves:

> The idea of curved lines at once leads to their division into continuous and discontinuous or mixed ones. One calls a curved line continuous when its nature is determined by one specific function of *x*; however, it is called discontinuous or mixed and irregular when different parts of it, *BM*, *MD*, *DM* sc., are determined by different functions of *x* (Euler/Michelsen, vol. 2, 1788, 9).

The contemporary idea to call curves continuous when they were representable by a function, that is, by a single-valued analytic expression, is found in d'Alembert (cf. Bottazzini 1986, 23–24). Hence, in contrast to today's understanding, curves could be understood as discontinuous ones generated by an arbitrary movement of the hand, or representable by several analytic expressions (ibid., 25).

In his later treatise of 1763 *De usu functionum discontinuarum in analysi*, also on the problem of the integration of partial differential equations, Euler, specifying the concept of continuity, stressed that it is necessary for continuous curves to obey a single analytic law. A hyperbola's two branches thus form a continuous curve (Youschkevitch 1976, 7 f.).

The historical literature always refers to Arbogast's treatise of 1791 as to that which offered new conceptual proposals. This is said firstly because he explicitly formulated the intermediate value property for continuous functions (Edwards 1979, 303; Grabiner 1981, 92; Bottazzini 1986, 34) and secondly because he introduced a new term: "discontigue." While curves, according to Euler's specification, had been considered to be discontinuous as well if their various parts were attached to one another, provided that these were defined by different "laws," Arbogast now called curves *discontigue* if their various parts were unconnected (ibid.). In all these works, this continued conceptual differentiation is emphasized as an important achievement, because with it, and with the novel term, the discipline had come closer to the meaning of discontinuity as it is understood today.

It must be pointed out, however, that Arbogast's reflections on the meaning of *continue*, *discontinue* and *discontigue* still refer to curves, and that functions, for him, were only of secondary importance for representing particular parts of a curve. Arbogast assumed functions as basic concepts only when reflecting on intermediate values. It must be noted that with Arbogast, just as with many contemporary mathematicians, the concept "loi de continuité" occurs in a twofold meaning: both as the analytic (formulaic) expression of a curve or a part of it, and as the conceptual content of the function's property of continuity. Steps toward conceptually defining the continuity of functions could not be taken before the relationship between these two meanings had been clarified. One of the conditions for this was to abandon the hitherto prevailing epistemology of mathematics by adopting an algebraic–analytic view of mathematical objects, a move that conferred a more fundamental status on functions than on the curves they represent.

This kind of change took place in France after the Revolution of 1789, when the analytic method began to prevail (cf. Chapter IV.2). While the debate on the continuity of functions, which will be reconstructed in Chapter III.9. in more detail, now became more heated, it is not appropriate to consider Cauchy's works as the conclusion of clarifying the concept of continuity. Where not only a concept, but also its negation, has an independent meaning, this negation can be profited from to ascertain how far the original concept reached, and what it meant over a determinate period of time. In our case "discontinuity" may serve to determine the then intended reach of "continuity" more precisely. Cauchy's textbooks contain but few explicit reflections on discontinuity, while Ampère, who held lectures on analysis alternating with those of Cauchy at the *École polytechnique*, explicitly discussed "discontinuity" in these both before and after Cauchy's *Cours d'Analyse Algébrique* was published. Ampère explained discontinuity as "rupture de la continuité." Discontinuity, besides, was only a special form of continuity here: a piecewise continuity.[8] One can thus conclude that continuity, in the French context at the beginning of the nineteenth century, signified continuity in an interval, and not point-wise continuity—in contrast to Bolzano's view. This also implies that the further differentiation of the concept of continuity had not yet progressed far enough to grasp and examine the local behavior of functions in the full scale of its differentiatedness.

1.6. Convergence

How the concept of convergence developed has been studied in detail by Grabiner (1981), in particular how novel Cauchy's works were compared to the mathematics of the eighteenth century. The study of infinite series, above all of

---

[8]   In his later work about "fonctions discontinues" (1849) as well, discontinuous functions are for Cauchy functions that experience an interruption of their continuity at isolated places. This is obviously connected with Cauchy's understanding functions also as equations, as given by terms, and not as a correspondence (Hawkins 1970, 11–12). On Ampère see below, Chapter VI.6.5.

power series, had become a major field of research in mathematics since Newton and Leibniz. On the other hand, the historiography of mathematics has kept wondering at the carefree manner or the lack of rigor in the summation of such series. An example of this phenomenon frequently quoted is the series $1-1+1-1+\ldots$. By 1700, there had already been debates whether the sum $\frac{1}{2}$ obtained from the formal development of $\frac{1}{1+x}$ for $x = 1$ is correct. Leibniz expressly confirmed this value in a letter of 1713 to Christian Wolff. His justification was to consider the two different series $1-1+1-1$ etc. and $1-1+1-1+1$ etc. and to halve the respective values 0 and 1. Although this argument might appear to be rather more metaphysical than mathematical, it was nevertheless justified, Leibniz said (Leibniz 1858, 382 ff.). Euler, too, expressly confirmed the value $\frac{1}{2}$ of the sum by considering the series formally as a development of $f(x) = \frac{1}{1+x}$, obtaining the sum $\frac{1}{2}$ as $f(1)$ (cf. Dieudonné 1985, 23 f.).

The understanding of convergence of series prevailing in the eighteenth century was that a series converges if its terms (seen absolutely) become ever smaller, approximating the value 0. D'Alembert's convergence criterion must be understood in this sense, a criterion that examines the extent to which the respective ratios of successive terms generally become smaller (Grabiner 1981, 60 ff.). At the same time, there was no strict distinction yet between formal and numerical series.[9] Instead, there was the belief that summation was possible in the neighborhood of the point in question.

What eighteenth-century mathematicians understood the meaning of convergence to be was formulated in an exemplary fashion by Klügel in his mathematical dictionary of 1803: "A series is convergent if its terms become successively ever smaller." For Klügel, this explanation was already sufficient for convergence: "The sum of the terms then ever more approaches the value of the quantity which is the sum of the entire series when continued to infinity" (Klügel 1803, 555).

It is one of the standard propositions in the literature on the history of mathematics that clear distinctions between convergence and divergence, and rigorous research into convergence, were not published until Gauß's work of 1812/13 on the hypergeometric series, Bolzano's works of 1816/17 on the binomial theorem and on the intermediate value property, and Cauchy's 1821 *Cours d'Analyse Algébrique*. Grabiner has pointed out, however, that essential elements of Cauchy's innovations are already found in Lacroix, who treated both the definition of convergence and the exclusion of divergent series. The merit of Cauchy was the systematic development of the concept of convergence in analysis (Grabiner 1981, 101 ff.). Given the habit of mathematical

---

[9]   In Germany, a clear distinction had been established by the philosopher Fries, in particular within his philosophy of mathematics of 1822 (cf. Schubring 1990a, 154).

historiography of conferring priority of discovery to a concept's mere mention in passing, or to its rough definition, Grabiner's evaluation clearly intends to relativize Cauchy's pioneer achievements (and hence also those of Gauß). Since Lacroix's *Traité* presents basically a systematization of already available knowledge, it follows, therefore, that essential foundations for the concept of convergence have been elaborated as early as the eighteenth century.

As the first of these elaborations, the discussions about convergence and divergence in Euler's work of 1754/55 about divergent series need to be mentioned. Euler defined convergence just as we have already quoted from Klügel's dictionary, that is, by referring to ever diminishing and eventually disappearing terms. Divergence is determined not by the fact that the terms either do not decrease below a given limit or grow arbitrarily (cf. Bottazzini 1992, XLIX). This is also the same work in which Euler ascribed the value $\frac{1}{2}$ to Grandi's series $1 - 1 + 1 - 1 + \dots$.

By contrast, the literature tends to neglect Louis Antoine de Bougainville (le jeune, 1729–1811). In the latter's two volumes on integral calculus, which were intended to continue L'Hospital's textbook, he not only claimed convergent series to be the only summable ones, but had already developed a convergence criterion as well.

De Bougainville's first volume contains a Chapter "Théorie des Suites." While defining convergence by indefinitely diminishing terms just as his contemporaries did, he added the important remark that convergent series were the only true ones: "We will prove later on that these series are the only true ones" (Bougainville 1754, 302).

The fascinating thing about this chapter is that it contains a detailed discussion of the summability of series—indeed as basis for integrating developments of series—and this in combination with elaborating a criterion of convergence. Actually, Bougainville called a series "true" ("vraie") if its sum was identical with the expression from whose development it had emerged. And he indicated a ratio test as convergence criterion (ibid., 304). Applying his notions, Bougainville showed that Grandi's series was "faux," i.e., did not have the value $\frac{1}{1+1} = \frac{1}{2}$ (ibid., 311).

In their report of 1754 on Bougainville's book for the *Académie des Sciences*, Nicole and d'Alembert placed particular emphasis on his chapter about series and the convergence criterion: "He teaches the manner to form these series, the means to recognize their convergence or their divergence" (ibid., xxij[10]).

José Anastácio da Cunha (1744–1787) made another noteworthy contribution to the concept of convergence. It has repeatedly been discussed during recent years by historians of mathematics, in particular on the occasion of the

---

[10]   If the last digit in the roman pagination number of prefaces of French books was an "i" it was printed as a "j"; Thus an 8 in roman numerals was given as viij.

bicentenary of da Cunha's death. In his textbook *Principios Mathematicos*, published integrally in 1790, three years after his death, he gave a definition of convergence that basically agrees with the later so-called Cauchy criterion: for a series to be convergent, it is necessary and sufficient that the partial sums become arbitrarily small from a sufficiently high index on. Since the French translation of 1811 from the Portuguese errs in precisely this definition, I prefer quoting a modern translation into English:

> Mathematicians call convergent a series whose terms are similarly determined, each one by the number of preceding terms, so that the series can always be continued, and eventually it is indifferent to continue it or not, because one may disregard without notable error the sum of how many terms one would wish to add to those already written or indicated; and the latter are indicated by writing &c. after the first two, or three, or how many one wishes; it is however necessary that the written terms show how the series might be continued, or that this be known through some other way (Queiró 1988, 40).

As Giusti has criticized, da Cunha omitted to reflect on what the sum of an infinite series actually is (Giusti 1990, 105). Not only did da Cunha's concepts suffer from not extending them far enough, but the impact of his work was obviously much dampened as well, from the fact that it originated from Portugal, a nation at the periphery of the contemporary mathematical world. Although it saw a French translation in 1811 (in Bordeaux), no influence on mathematics in other countries has been noted as yet.

Nevertheless, this example of important innovations even at the periphery shows that a broader context of discussion existed for the concept of convergence as well.

The claim that there was no crucial difference between the positive knowledge about convergence in Lacroix's and in Cauchy's textbooks, as stressed by Grabiner, is not confirmed by Lacroix's voluminous *Traité*, to which Grabiner refers. The difference in the concept's status is much more evident upon comparing Lacroix's own two textbooks. While his voluminous edition, aiming at a learned audience, indeed discusses the concept of convergence, this discussion is absent from the concise version intended for a student audience, mainly at the *École polytechnique*. By contrast, Cauchy's later textbook of 1821, directed to the same student body, assigns a central function to the concept of convergence. Only at that time did the concept of convergence attain the status of a fundamental idea in analysis.

1.7. The Integral

The concept of integral differs from the foundational concepts discussed above in that it is not one of the "primary" fundamental ideas. It will receive mention here, however, in contrast to the analogous concept of derivative, because it later had an important role in effecting changes in the system of foundational concepts that will be studied in the chapters to come.

The concept of integral has seen a peculiar change as to its importance for analysis. While determining areas and volumes constituted the principal object for applying infinite processes since their very beginning, problems of this kind were transmuted into simple inversions of differentiation after the differential calculus had become established. The (indefinite) integral as an inverse of the differential was only of derived significance and had no role of its own in foundational studies throughout the eighteenth century. The concept of integral became independent again only subsequent to Cauchy's studies on the definite integral of 1814. Remarkably, this change was linked to a "resurrection" of the infinitely small.

The methods of exhaustion practiced by the Greeks had been taken up again by the methods of the indivisibles in modern times. Developed first mainly by Kepler for determining the volume of barrels, these methods climaxed with Cavalieri's indivisibles. Skillfully used, they permitted calculating volumes of solids, by comparing these to solids already known, and accordingly for areas. The conceptual basis of the methods was the atomistic assumption that any geometrical figure can be understood as composed of "indivisible," arbitrarily small quantities, thus forming a sum of elements (of "slices" having the lower dimension n–1).

Already in Newton's and Leibniz's first works on the new differential calculus, the integral calculus (then called the "inverse tangent problem") was conceived of as the former's inversion (cf. M. Cantor, vol. 3, 1901, 171). Medvedev has shown, however, that integral calculus, in Newton's early works, was not yet based on differential calculus. Rather, it was derived from the method of calculating areas by means of developing functions into infinite series (Medvedev 1974, 100 ff.). Medvedev also criticized the widespread view that credits Newton with the idea of the indefinite integral as a primitive function, and Leibniz with the idea of the definite integral as a limit of approximating sums (ibid., 117). Moreover, he showed that Newton had already introduced the concept of the definite integral as a limit of sums in 1686 (ibid., 120), the integration constant having first been used to solve a concrete problem in one of Leibniz's papers of 1694 (ibid., 121).

Wherever they treated integral calculus, the textbooks of the eighteenth century formally presented the integral as the inverse of differentiation with the task of determining the primitive function. Without discussing questions of existence, rules for determining (indefinite) integrals were examined. The most comprehensive study of how the concept of integral developed is given by Medvedev (1974). Characteristic for the eighteenth century's state of the concept is Euler's textbook on integral calculus (1768–1770). In its first volume, the extensive three-volume work contains a short general section giving definitions and explanations. The very first explanation introduces the integral as a problem of inversion: "The integral calculus is the method to find, from a given relation between differentials, the relation between the quantities themselves" (Euler 1828, 1).

And Euler adds a characterization of differential calculus and the integral calculus as analogously opposite operations, comparing them to the basic operations of arithmetic: "just as in analysis where always two operations are opposed to one another"—subtraction and addition, division and multiplication, extracting roots and exponentiation (ibid).

Medvedev also examined the developments that led to the rise of the definite integral in the second half of the eighteenth century. A major factor in favor of its increasing importance was the investigation of oscillations and their representation by trigonometric series. The research to determine the coefficients of these series showed that these could be most suitably determined by using definite integrals. A further impulse was provided by the problems raised by multidimensional integration; this is where presenting the integral as a sum becomes necessary. Potential theory necessitated the calculation of definite integrals. Lagrange used the concept of the definite integral throughout his *Méchanique analitique* as an important fundamental notion (Medvedev 1974, 154 and 159 ff.).[11]

Cauchy, however, was the first to raise the concept of the definite integral to the rank of a privileged fundamental notion, and the first to comprehensively make the concept and the existence of the integral a subject proper of mathematical research in his textbook of 1823. This is where the definite integral was introduced as a sum of infinitely small quantities.

2. The Development of Negative Numbers

2.1. Introduction

The history of the concept of the negative numbers has not been examined systematically. In M. Cantor's seminal work one finds, in its second volume, for the period 1500 to 1668, several rather dispersed indications (M. Cantor 1900). The fourth volume of 1903 contains a topical account, by Cajori, on the development of algebra between 1750 and 1800; ten pages give an informed account on negative numbers for this period, which otherwise appears in usual historical reports as not having offered conceptual problems. Tropfke's study, in its new version edited by K. Vogel, contains in its part on arithmetic and algebra a brief section describing the development from the Babylonians up to Peacock

---

[11]   Cf. also on the generalization of the concept of function as a premise for the development of the concept of integral: Hawkins 1970, Sections 1.1 and 1.2.

and Hankel (Tropfke 1980, 144–151). In accounts of the history of the concept of number (e.g., Gericke 1970, 51–57) what is held worthy of being reported on negative numbers in general ends with early modern times. The basic assumption implied in these traditional accounts is that the concept of negative numbers had been in essence clarified by the time of Stevin and Viète and no longer offered conceptual problems.[12]

A first approach to studying the gap of centuries—from the seventeenth to the nineteenth—came from an unexpected quarter: from French *didactique*, based on epistemological categories. It was a paper by Georges Glaeser (1981) on the history of negative numbers from the seventeenth to the nineteenth century, but mainly restricted to the rule of signs. The concept of his investigation turned out to be quite unsatisfactory. Glaeser constructed a frame of reference from an epistemological theory, that of "epistemological obstacles" developed by the French philosopher Gaston Bachelard (cf. Bachelard 1975). Bachelard's theory that the mental development of mankind occurs in three stages reveals decisive weaknesses in Glaeser's adaptation to the historical concept field of negative numbers. Glaeser registered six "obstacles" in the course of its evolution that allegedly prevented a full understanding of the concept of negative numbers; he marked with + or − leading mathematicians from the seventeenth to the nineteenth centuries according to whether they had overcome some of these "obstacles" (Glaeser 1981, 309). Earlier views judged by Glaeser as especially "backward" he commented with three exclamation marks (e.g., ibid., 320). Stagnating at the stage of concrete operations rather than progressing to the stage of formal operations is presented as a main obstacle. Even Euler is grouped there (ibid., 308 f.).

By such categorizations, Bachelard's teleological concept of scientific development is saliently revealed: as a progress necessary in the theoretical character of knowledge, as an eventual triumph of reason. In Bachelard, one recognizes the rationalistic vision of increasing mathematization as the necessary core within the process of developing the sciences. Bachelard distinguished three stages in this process: a concrete, prescientific stage in which the phenomena rule; a concrete–abstract stage in which the physical experience is complemented by some abstractions; and eventually the (present) abstract stage, which is determined by theoretical reasoning (cf. Bachelard 1975, 8). This gradation corresponds quite exactly to the three stages of mental development that Piaget ascribed to the individual: the succession of empirical, then of concrete, and eventually of formal operations.

Stagnation at the stage of concrete operations has appeared, since Glaeser's paper, in numerous mainly didactic publications as the "explanation" for a lack of understanding negative numbers in a modern sense. Historical development is again being understood to be linear, as converging to the modern status; moreover, differences—say between different cultures—are just as little

---

[12]   An exception is Vredenduin's study of 1991 (in Dutch), which also considers the eighteenth and the nineteenth centuries.

considered as are embeddings of conceptual views into differing epistemological frames.

If one takes the historical texts seriously, however, and tries to understand their intentions, one will observe that presuming an *obstacle* where an earlier author did not "unify the number ray" (Glaeser 1981, 308) is rather unhistorical. Looking at concept development not under the limited aspect of the rule of signs, but under the more general one of the *existence* of negative numbers, one will observe that a foundational dimension underlies many of the controversies about their existence, which is not decidable by "true" or "false": this is the relation between *quantities* and *numbers*. The rather unspecific concept of quantity does not lend itself to conceptualizing the notion of "negative quantities," whereas the more specific concept of number proves to be more broadly applicable, and better adapted to developing the notion of "negative numbers"— that is, as more general. The decision for "quantity" as the fundamental idea, and for "number" as the derived concept, or for "number" as the fundamental idea, and for "quantities" as derived concept, depends on which architecture of mathematics is chosen or favored, and on the underlying epistemological concepts, and cannot be decided by truth values. The decision for one or the other side can just as little be qualified as a result of mental obstacles.

Moreover, such studies often assume anachronistic views of concept developments. For example, the alleged mental obstacle to unifying the number ray is induced by presenting this idea as having always been self-evident. As Bos has shown, however, geometrical quantities, like lengths or areas, were *not* scaled quantities as long as the requirement of dimensional homogeneity for these quantities was generally shared, so that the introduction of a unit length was unnecessary; hence, these quantities did not represent real numbers, and lines in algebraic geometry did not mean number rays (Bos 1974, 7 f.).

Another methodological problem is exemplified by the historian of mathematics Helena Pycior's new volume (Pycior 1997). In her Ph.D. thesis of 1976, she had examined the development of algebra in Great Britain from 1750 to 1850. In her new book, she studies—just as knowledgeably and carefully— the preceding period from about 1600. The development of the negative and the imaginary numbers constitutes her main focus. For evaluating this development, Pycior chooses a fixed *étalon*, namely "the expanding universe of algebra" (Pycior 1997, 27 and passim). Such a framework for comparison is not only too abstract and makes no allowance for cultural differences, it also assumes at the same time an inevitability in concept development that actually recognizes only cumulative advance. Her analysis therefore shows problems in conceptually grasping the ruptures present in her period of investigation. While she assures us that "British algebra did not develop in a fundamentally linear fashion," she flinches shortly thereafter from the consequences of that assertion, in line with her own model of universal progress, hastening to affirm, "This is not to say that there was no linear development" (ibid., 3). Undoubtedly, the explanatory pattern she applies to the British controversies about the status of negative and imaginary numbers is of particular relevance for this conceptual development. It is the contrast between propagators of the analytic method who are committed to algebraic procedures by means of "symbolical reasoning" and propagators of the synthetic method who accept merely geometrical foundational concepts.

Moreover, Pycior's book provides evidence of the difficulty in analyzing conceptual developments in their contemporary context and understanding. Her analysis refers to terms like *negative **numbers*** or *imaginary **numbers***. However, the majority of the authors examined by her use terms like *negative* or *imaginary **quantities*** or ***magnitudes***. Where she mentions the use of "quantity" or of "magnitude" in a quote from historical authors, she does not, however, systematically inquire into the intended meaning, but subsumes the respective position into her account as a contribution about "numbers" and their development. Not to include the difference between *quantity* (or *magnitude*) and *number* into the basic dimensions of the research design means to exclude from the analysis a majority of the most important contemporary problems in the development of the algebraic concept field, and thus to miss key historical insights, in particular concerning the extension of the multiplication operation.

Eventually, one notes a generally entirely complete account of concept development in the literature. Most authors report only the formulations of the definitions of negative numbers and justifications for the rule of signs. They omit, however, a discussion of how the respective conceptual view was applied in other mathematical concept fields—say, in analytic geometry. An exception is Tropfke's investigation of the history of quadratic equations, which he relates to the context of the history of negative numbers. He says, however, that there was only *one* normal form of quadratic equation since 1659, due to Hudde (Tropfke 1934, 102). As we will see, the question of the normal form was still controversial during the nineteenth century.

2.2 An Overview of the Early History of Negative Numbers

FROM ANTIQUITY TO THE MIDDLE AGES

A careful analysis as to when negative numbers were believed to exist was undertaken by Sesiano (1985). He gives a survey from antiquity to about 1500, intensely discussing what Italian mathematicians contributed during the Middle Ages. He bases his analysis on how the relation between quantities and numbers was viewed. For the overview intended here, I will thus follow his contribution for the era before modern times.

Generally, it can be stated—provided one deals exclusively with *quantities*— that controversies were about whether to admit negative solutions. A lesser issue concerned the problem whether negative quantities were admissible as intermediate quantities during the process of calculation. Where the problems treated became more abstract, intentions can be noted at "reinterpreting" negative solutions so as to transform them into positive ones, thus removing them from their doubtful status.

In *Old Babylonian* times (beginning of the second millennium BCE), a sign from cuneiform texts, translated by "lal," has been interpreted to indicate "being less"—but only in texts about economics, not in mathematical ones. It also occurs in subtractive writing of numbers, e.g. $20 - 1 = 19$ (Neugebauer 1934, 17). Quoting Neugebauer, Tropfke commented, "In serial texts, changing data carried out according to a certain pattern will occasionally lead to negative numbers, but we are ignorant of how that was understood and whether people calculated with such quantities" (Tropfke 1980, 144). Høyrup has taken up this aspect in his systematic re-analysis of the Babylonian texts to say, "The widespread legend that the Babylonians made use of negative numbers comes from misreading Neugebauer's treatment of the topic" (Høyrup 2002, 21). The only thing observable is the use of subtractive quantities/numbers, without, however, performing operations on them (ibid., 296).

It is well known that red rods were used for calculating with positive quantities and black rods for negative quantities in *Chinese mathematics*. Negative values were permissible as intermediate values during calculation, but not as solutions of systems of equations. The literature refers to a typical example, one of the tasks contained in the classical mathematics textbook *Chiu-chang suan shu* (mathematics in nine chapters, approx. 250 BCE). In modern terms, it can be written as the system of equations

$$2x + 5y - 13z = 1000$$

$$3x - 9y + 3z = 0$$

$$-5x + 6y + 8z = -600$$

having the solutions $x = 1200$, $y = 500$, $z = 300$ (Tropfke 1980, 145; Sesiano 1985, 107 f.).[13]

Diophantus's works (approx. 250 CE) form the apogee of algebra in *Greek mathematics*. Since compositions of quantities with additive and subtractive quantities in algebra were subjected to operations of multiplication and of division, Diophantus, in the introductory part of his *Arithmetica*, introduced the *rule of signs* (Diophantus/Czwalina 1952, 6). The rule of signs' function, however, was only to permit operating with so-called "complex" expressions like $(a+b)$ $(c-d)$, no reflection about or acknowledgment of the existence of

---

[13]   Lay-Yong and Tian-Se strove to infer from the practice of such examples that the Chinese were the first to have had "the earliest negative numbers" and "the concept of negative numbers" (Lay-Yong, Tian-Se 1987, p. 222 and 236). They did not see that operating with subtractive quantities as intermediate entries means just one step in a long process of conceptualization.

negative quantities being implied. Rather, Diophantus rejected negative solutions, just as he did irrational or complex ones, avoiding roots of such kind by restricting the equations further, when necessary.

There is only one problem in which Diophantus obtained an equation with a negative solution. After formally operating with a system of nonlinear equations, he ended up with the linear equation

$$4 = 4x + 20,$$

commenting, however, that this this was "absurd" (ἄτοπον)), since "4 units[14] could not be smaller than 20 units." He therefore felt compelled to change his original hypotheses (Sesiano 1985, 106).

Hitherto, just a small part of the surviving texts of *Arabic mathematicians* have been evaluated. Nevertheless, the texts already studied give no indication that mathematicians within this cultural context considered negative solutions acceptable. The six classical types of algebraic equations of first and second degree were always conducive to positive solutions. Mathematicians who treated problems of indeterminate algebra adhered to Diophantus' model, selecting coefficients so as to obtain positive solutions (Sesiano 1985, 108).

In his careful analysis of Ibn al-Ha'im's (1352–1412 CE) commentary on algebra and related Arabic texts, Abdeljaouad has recently shown how their authors explicitly thought about repeated "negations" of quantities, and about operating with "subtracted" (*manfi*) and "confirmed" (*muthabbat*) numbers as intermediate entries (Abdeljaouad 2002).

Indian mathematicians developed mathematics predominantly in connection with astronomy. The textbooks known were mostly introductory parts to works on astronomy.

The view on negative numbers in *Indian mathematics* was not uniform. Brahmagupta (599–approx. 665) treated in an early textbook on astronomy how to calculate with negative quantities in all the elementary arithmetical operations, and in squaring and in extracting square roots, demonstrating the rule of signs.[15] He called positive quantities "property" or "fortune," and a negative quantity "debt" or "loss." Negative quantities were distinguished from positives by a superposed point. Thereby, he already established a general expression for equations of second degree.

Mahavira (around 850) used negative numbers and even discussed the taking of square roots for negative numbers, but declaring them to be impossible because negative numbers could not be squares (*Lexikon* 1990, 304).

Bhaskara II (1114–ca.1191) had a different view. His chapters on arithmetic and on algebra form the introductory part of a textbook on astronomy. In quadratic equations, he sometimes admitted only one solution, even if both

---

[14]   I.e., the $4x$ in the above equation.

[15]   Cf. Colebrooke 1817, 339–343. Algebra was presented in the eighteenth chapter of Brahmagupta's textbook.

solutions would have been positive. For problems with concrete quantities, he rejected negative solutions, in the case of more abstract problems, however, he reinterpreted negative solutions so as to be able to admit them as positive ones.

An example of his former approach is a riddle about monkeys. The fifth part of a troop of monkeys less three, squared, had gone into a cave; only one monkey was still to be seen. What was the monkeys' number? Translated into modern style, the following equation arises:

$$\left(\frac{x}{5} - 3\right)^2 + 1 = x,$$

with the two solutions $x_1 = 50$ and $x_2 = 5$. Since one obtains for x_2 the value $\frac{x_2}{5} < 3$, Bhaskara stated that a double value arises here, but the second is not to be taken, because it is incongruous: "People do not approve a negative absolute number" (quoted from Sesiano 1985, 106).

While negative quantities were rejected here even as intermediate values in calculations, Bhaskara proceeded differently in the case of geometrical problems. While a negative number of monkeys did not appear to be reinterpretable in positive terms, he reinterpreted negative geometrical line segments as having the opposite direction. In a problem about determining the lengths of line segments on sides of triangles he obtained a negative result, to which instead of excluding it, he gave the following interpretation:

> This [i.e., 21] cannot be subtracted from the base [c = 9]. Wherefore the base is subtracted from it. Half the remainder is the segment, 6; and is negative: that is to say, is in the contrary direction.(quoted after ibid., 107).

Bhaskara's later works are thus considerably more reluctant to admit negative quantities than the earlier works by Brahmagupta. This would seem to make evident once again that there need not be continuity in scientific progress even within the very same cultural context.

Moreover, another interpretation seems possible to me. Bhaskara II's chapters on algebra contain a separate section on addition, subtraction, multiplication, and division of positive and negative quantities: "Logistics of Negative and Affirmative Quantities." Not only do the rules they present agree with those established by Brahmagupta, but they are formulated even more extensively and explicitly, and illustrated by examples (Colebrooke 1817, 131–135). The sum of –3 and –4, for instance, is given as –7 (more exactly: as the sum of $\dot{3}$ and $\dot{4}$ with the result $\dot{7}$ (ibid., 131)). Because of Bhaskara's rejection of isolated negative solutions, these rules were apparently intended for intermediate calculations, and hence did not basically differ from Diophantus's approach. Since Bhaskara's rules for calculations are even more explicit than those of Brahmagupta, it may be assumed that Brahmagupta wished only to establish similar rules for operating on subtractive quantities, and therefore also rejected isolated negative solutions. Brahmagupta indeed presents only one root in his own general solution of quadratic equations, giving only one solution, a positive one, for all the problems presented later as examples (ibid., 346 ff.).

EUROPEAN MATHEMATICS IN THE MIDDLE AGES

While the Italian mathematicians began by adopting the traditions of Arabic mathematics, they gradually elaborated methods and approaches of their own. Possibilities of negative solutions arose in particular from the frequent problem type of having to calculate how to distribute goods among n persons under varying conditions. Sesiano examined the systems of linear equations occurring and their solution sets in detail (Sesiano 1985, 108 ff.).

The works on the solution of systems of equations of this epoch culminated in Leonardo Pisano's (Fibonacci) works, ca. 1170–ca. 1250, in particular in his book *Liber abaci* (1202/1228). Negative values appear in many of his problems, partly as solutions and partly as intermediate values. In each case, Leonardo discussed in detail how to treat negative values. While he generally rejected negative solutions *and* intermediate values as *insolubilis* (cf. ibid., 118), he resorted to reformulating the problem to permit a solution whenever a reinterpretation of such a value was possible. This is true for all problems concerning invested capital or monies. He interpreted negative values as debts, as borrowed money, or as capital invested by one participant in addition to monies jointly invested by several persons n. Where it was impossible to reinterpret negative quantities, for instance negative prices (ibid., 131) as positive ones, he rejected the negative solution as *inconveniens*.

A provençal manuscript of about 1430 in Occitan from the *Bibliothèque nationale* in Paris marks a quite revealing rupture with all prior traditions of coping with negative numbers. That this manuscript, *Compendi del art del algorisme* exists had been known for some time already, but exclusively in the history of literature, as one of the documents of Occitan culture. Its mathematical content was not examined until some years ago by Sesiano (Sesiano 1984). The text is the first to contain a negative solution accepted without restriction or reinterpretation.

The problem concerned was about buying a piece of cloth. From a system of five linear equations in six unknowns, and after choosing one indeterminate, the value $-10\frac{3}{4}$ was obtained for the unknown x_1 ("restan 10 et $\frac{3}{4}$ mens de non res," Sesiano 1984, 52). This negative value was accepted without any interpretation whatsoever. The only hint at its particularity is its exceptional verification by inserting the values in all the equations (Sesiano 1985, 133 f.).

This novel way to treat negative solutions did not remain an isolated case, but was continued. This is evidenced by Frances Pellos's *Compendion del abaco* written about 1460, and printed in Nice in 1492. Pellos copied the entire group of problems containing the negative solution from the earlier Provençal manuscript. He also refrained from comment. Pellos's volume is the first printed document to contain a negative solution (ibid., 134).

A step beyond this first instance of acknowledging negative numbers is Nicolas Chuquet's (ca. 1445–ca. 1488) manuscript *Triparty en la science des nombres,* authored 1484 in Lyon. Chuquet solved systems of linear equations in which the unknowns were pure numbers, and no longer quantities. In such systems, negative solutions appear as well. Chuquet accepted such negative values without attempting to reinterpret them, and only under the condition that these values satisfied the equations.

In a problem with five unknowns, Chuquet obtained the solutions: 30, 20, 10, 0, –10. Chuquet continues after this first appearance of zero and of negative numbers by explaining how to add and subtract zero and negative numbers. In an abbreviated form, this expressed the rule of signs. He also made remarkably explicit that the operations of adding and subtracting acquire a novel meaning from the novel negative numbers. First, he declared that adding or subtracting zero ("0") does not change the result of an addition or a subtraction. Then, he observed that when adding a negative number to another number or subtracting it, the addition will result in a decrease and subtraction in an increase: "Et qui adiouste ung moins avec ung aultre nombre, ou qui d'icellui le soustrayt, l'addicion se diminue e la soustraction croist" (ibid., 136).

Chuquet gave examples for these novel meanings; subtracting minus 4 from 10 gives 14 as difference. And he interpreted negative solutions: "minus 4" corresponds to a person who owns nothing but still owes 4. Chuquet's work contains a solution identical to that "negative" problem in the Provençal manuscript and in Pellos's book (ibid., 137).

In an appendix, Chuquet solved application problems for quantities. Here again, Chuquet went considerably further than his precursors. He even accepted negative amounts of commodities and negative prices; the only condition being here as well that the given equations be satisfied (ibid., 140 ff.).

Luca Pacioli's (ca. 1445–1517) approach is less radical, adhering somewhat more to tradition. He wrote an arithmetic in 1470, and his *Summa de arithmetica, geometria, proportioni et proportionalita* in 1494. Pacioli's general stance was to reject negative solutions; he even showed reserve where negative intermediate values were obtained. Nevertheless, he tolerated negative values in special cases, thus for pure numbers, but in one case for a price as well. In an abstract problem, he goes as far as calling a negative solution "un bellissimo chaso" (Sesiano 1985, 142 ff.).

2.3. The Onset of Early Modern Times. The First "Ruptures" in Cardano's Works

While differences of approach can already be noted in Europe at the beginning of early modern times, it is not yet possible to attribute these to differences between established cultural contexts.

Historiography ascribes an important modernizing role to Michael Stifel's textbook on arithmetic and algebra *Arithmetica Integra* (1544) Stifel (1486 or

1487–1567) lived first as a monk, became a pastor after the Reformation, and has been called a peregrine clergyman because of his frequent changes of residence (Cantor, vol. 2, 1900, 430). In the last period of his life, after 1559, he taught mathematics at the University of Jena (Chemnitius 1992, 8 f.). Noteworthy in his treatment of negative numbers is that he clearly states that these are less than zero. While Stifel termed positive numbers *numeri veri* and negative ones *numeri absurdi*, thus using only positive roots of equations (Cantor, vol. 2, 1900, 442), he characterized positive numbers as *supra 0* and negative numbers as *infra 0, id est infra nihil* (Stifel 1544, fol. 249v.). This is why Cantor emphasized, as Stifel's pioneering achievement, "t h e e x p l a n a t i o n   o f   t h e   n e g a t i v e   n u m b e r   a s   b e i n g   s m a l l e r t h a n   z e r o which thus entered into mathematics" (Cantor, vol. 2, 1900, 442).

To corroborate his evaluation, Cantor also noted that Stifel had declared the zero to be the common limit between positive and negative numbers. In fact, Stifel speaks of the zero as: "0, i.[d est] nihil (quod mediat inter numeros veros et numeros absurdos)" (ibid.; Stifel 1544, fol. 249v.). In the first historical account of the development of negative numbers, Karsten ascribed to Stifel that he had understood operating with negative numbers as the reversal of the usual ways of calculating, making that subtraction, for instance, effects an increase. Thus, Stifel is said to have obtained as result of subtracting –5 from –2 the positive result +3, hence a "numerum *supra* nihil, seu numerum verum" (Karsten 1786, 233). Actually, however, Stifel had worked with subtractive numbers in this case. Subtracting 0–5 from 0–2, he obtained 0+3 (Stifel 1544, fol. 249 v.). Stifel always used negative numbers in compositions, in "binomials"; there is no case in which he ended up with a purely negative result. While he explained the rules for operating with such compound expressions in his book's part on algebra, in particular also quite explicitly for multiplication and division (ibid., fol. 248 v–249), this treatment of numbers was accompanied by explicit epistemological reservations. Stifel not only opposed *numeri ueri* to *numeri ficti* (ibid., fol. 48) or "*absurdi*" (ibid., fol. 249v), but beyond that declared positive numbers to be real (*quae sunt*), while ascribing only an imagined existence to negative numbers (*quae finguntur esse*; ibid.). Stifel did not yet acknowledge fundamentally equal status for the two kinds of numbers.

Petrus Ramus's (or: Pierre de la Ramée, 1515–1572) algebra textbook was often reprinted in France after 1560. The systematic importance of this author for the development of generalization will be discussed in Section 2.7. In the two chapters on arithmetic preceding his algebra, Ramus introduced the four basic operations for integer numbers, for fractions, and for proportions. His algebra consists of two parts; the first expands the object—instead of (absolute) numbers there are figurate numbers, up to biquadratic—and operations on such with not exclusively positive quantities. The second treats equations, i.e., solving of problems.[16] For plus and for minus, he used the signs + and –, and instead of

---

16   The first edition was a separate publication of his algebra not preceded by arithmetic. It had merely 34 pages; while L. Schoner's thoroughly revised edition of 1586

a sign of equality Ramus used the term "aequat." Ramus did not explicitly introduce or use negative numbers, nor did he use negative quantities, restricting himself basically to subtractive quantities. He used the opposition of signs, however, and elaborated the meaning of addition and subtraction as reversed: in case of opposite signs, addition effects subtraction, and the remainder receives the sign of the larger quantity:[17] "In contrariis signis additio est subductio et reliquus habet signum majoris" (Ramus 1586, 325).

As an example, he calculated: $(6q + 8l) - (4q - 4l)$ obtaining $10q + 4l$.[18] Analogously, he explained the effect of subtraction for opposite signs; the result receiving the sign of the larger term: "*Subductio in signis contrarijs est additio, cujus totus habet signum superioris*" (ibid., 327).

As an example he showed $(19q - 8) - (14q + 14)$ with the result $5q - 22$. Because 6q, 8l, etc. signify figurate numbers, this implies that Ramus admitted absolutely negative values for some dimensions. Nevertheless, Ramus did not operate with isolated terms, but at least with a binomial expression. For multiplication and division, Ramus, without hesitation and without justification, formulated the rule of signs: "*Multiplicatio et divisio in signis iisdem faciunt plus, in diversis minus*" (ibid., 328).

Multiplying $9q - 4l$ by $9l$ he thus obtained $81c - 36q$, and for $8q - 9$ multiplied by itself $64bq - 144q + 8l$. Here, Ramus gave also the rule, explained by various examples: $(a + b)(a - b) = a^2 - b^2$. He called opposite quantities like $+b$ and $-b$ "heterogeneous." The second part of his algebra, about equations, contained essentially nothing but positive solutions. Only once, as solution of a quadratic equation, he gave a positive and a negative value—revealing, however, an elementary mistake in calculation (ibid., 349).

Ramus's algebra shows that the foundations of operating on negative quantities belonged to the generally shared knowledge of his time.

Algebra saw a culminating point of its early-modern development in Girolamo Cardano's (1501–1576) work. Beyond the already well-established solving of quadratic equations, his book *Ars Magna* (1545) made the solving of equations of third and fourth degree accessible to a larger public. While Cardano systematically developed operating on negative quantities further, he simultaneously was the first to reflect on the foundations of this new field of numbers, reflections that led him to the first ruptures in the development of this concept.

---

also comprises only 43 pages, it is preceded by the arithmetic first published in 1569 but later revised as well. While the basic versions of 1560 and 1586 formulated in the second part correspond, the texts deviate substantially in the first part of the algebra, the meaning of the statements on subtraction, on opposite signs, and on the rule of signs remained unchanged.

[17]   Schoner added a reference to the respective passages in Diophantus in 1586.

[18]   The q-units are quadratic figurate numbers, and the l-units are linear numbers.

Cardano used the rules hitherto developed for operating on negative quantities. He admitted negative numbers as solutions of equations, calling them *radices fictae*—in contrast to *radices verae*, his term for positive solutions—and interpreted them in his *Practica Arithmeticae* of 1539 in the sense of debts, etc. (cf. M. Cantor 1900, 502). Cardano established the multiplicity of roots, e.g., that equations of the second degree have two solutions, and that, e.g. $x^2 = 9$ has both the positive root 3 and the negative root $-,3$, and accordingly for solutions for equations of third and fourth degree. Cardano stressed in his *Ars Magna* that one obtains the number 9 both by squaring 3 and by squaring -3: since "minus in minus ductum producit plus."[19] Furthermore, he explained, in an argument analogous to that of Chuquet, that to add something negative corresponds to subtracting something positive (*Ars Magna*, Cap. XVIII).

In his *Ars Magna*, however, Cardano took further fundamentally new steps. He began to operate on square roots of negative numbers. Thus he obtained as intermediate results $5 + \sqrt{-15}$ and $5 - \sqrt{-15}$, and as their product the value 40 (since "minus -15" is identical with "+15" ; cf. M. Cantor 1900, 509; *Ars Magna* Chap. 37, problem 3). On the other hand, Cardano's thinking remained determined to a considerable degree by the tradition in algebra. As coefficients in equations, he admitted only positive numbers, treating a large number of cubic equations separately (Peiffer/Dahan 1994, 104), as the Arabic mathematicians had done.

While Cardano thus adhered to the mathematical tradition in his algebra, he deviated—in his late work—from knowledge of algebra that had indisputably been approved at least since Diophantus: from the rule of signs. While admitting that it was indeed "opinio communis" that the multiplication of minus by minus gives plus, an opinion formerly also accepted by himself, he now stated that this was false. The result, he said, was not plus, but minus. This later assertion caused bewilderment in the historiography[20]—provided it was registered at all— and will hence be discussed here in more detail.

Cardano performed these conceptual reflections in his book *De Regula Aliza* (1570). By selecting *De operationibus plus et minus, secundum communem usum* as heading of the sixth chapter of this book, he indicated for the first time that he was taking his distance from the established rules of operation, while quoting the familiar rule of signs: "In the multiplication and in the division one obtains plus always from the same signs and minus from opposite signs."[21]

He did not provide an explicit critique of the tradition at this point, but gave it only later in his chapter 22, headed *De contemplatione plus et minus et quod*

---

19   Cardano 1663, 222. "ducere in" means "multiply."
20   The *Mathematiker-Lexikon* speaks of an "error" (Lexikon 1990, 90). A recent Italian article speaks of "incomprehensible" justifications, and says that this error committed by Cardano does not diminish his eminence at all (Dell'Aquila/Ferrari (1994, 348 ff.).
21   "In multiplicatione et divisione plus fit semper ex similibus [et] minus ex contrariis" (Cardano 1663, 384).

minus in minus facit minus et de causis horum iuxta veritatem. This is where Cardano established a first alternative rule of signs: in the case of products and of divisions, in which at least one of the two signs is minus, the result also bears the *minus* sign. To justify this, he gave an example from Euclidean geometry (cf. Figure 1, which reconstructs geometrically Cardano's verbal argument).

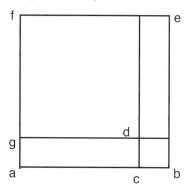

Figure 1, Cardano's refutation of the rule of signs

If a large square with side length $ab = 10$ is given, and the area of the square *df* is sought, with $cb = 2$, one must subtract the area of the two rectangles *gc* and *de*, i.e., 16 and 16, from 100. Since $100 - 2 \times 16 = 68$, one must subtract 4 in order to obtain 64. Thus, one gets −4 instead of the expected +4 (Cardano 1663, 399)! Cardano commented that for this reason minus multiplied by minus gives minus. The traditional error of those stating that minus multiplied by minus gives *plus* was evident, he said. Minus by minus would give no more plus than plus by plus gave minus (ibid.).

Actually, the equation $64 = 100 - 2 \times 16 - 4$ is not about multiplying an expression in parentheses: the latter would be written $(10 - 2)(10 - 2)$, and multiplying it would yield $100 - 2 \times 10 \times 2 + 4$.

While Cardano was well aware of this approach, he interpreted it, however, differently in line with his new view on foundations. Admitting to have hitherto argued along with the "opinio communis," he tried to explain why minus multiplied by minus only *seemed* to give plus and how this illusion might be comprehended (ibid.). He fell back on the same example: After taking away the two rectangles *gb* and *ce* from the large square, the small square *ab* had been subtracted twice instead of only once—it must therefore be added again, Cardano said. While it *seemed* as if minus by minus gave plus, this was false: "Sed non est verum" (ibid.). The +4, he said, was not the result of an operation of multiplication, but of applying Proposition VII of Euclid's second book (ibid., 400).[22] Essential for Cardano's reasoning was the following justification, for it

---

[22]   Cardano's argument shows that he was well aware that rectangle formation in Greek geometry was not to be understood as an equivalent to the algebraic operation of multiplication, as maintained since the end of the nineteenth century by the adherents of a "Greek algebra," in particular in Euclid's Book II.

shows that his decisive reason for rejecting the then already time-honored rule of signs was epistemological. For Cardano, the positives and the negatives constituted two strictly separate areas, as it were two different worlds. It was impossible to mingle them or simply to switch from one to the other, "because nothing can trespass its forces."[23] Upon multiplying plus by plus, he said, one would remain within the plus area. Likewise, multiplying minus by minus could not escape from the minus area—the result would remain minus. If plus was multiplied by something beyond its own area ("*extra ipsum*"), that is, by minus, the result would be minus (ibid.). In Cardano's late work, the overt rupture with the traditionally unproblematic application of the rule of signs was due to the conflict he was the first to realize between the extended rules for the operation of subtraction and their linkage to the operation of multiplication. Cardano conceived of this conflict as an epistemological one—as a separation between two different areas—and refrained from explicitly reflecting upon extending the concept of multiplication.

In *De Regula Aliza*, Cardano discussed no further application of his own alternative rule of signs; likewise, he did not discuss its effects on his earlier results obtained by applying the traditional rule. In a later text, the *Sermo de plus et minus*,[24] he underlined his new view that the multiplication of minus by minus would give *minus*. He offered a differentiated view, however, on the operation of division there. Minus divided by minus could in some cases give plus, and in other cases minus (Cardano 1663, 435), an assertion that did not altogether simplify his argument (cf. Marie 1883, 266).

2.4. Further Developments in Algebra: From Viète to Descartes

In the generation succeeding Cardano, excellent achievements were accomplished in algebra by François Viète (1540–1603) and Simon Stevin (1548–1620). Their work, however, did not advance the concept field of negative numbers along a straight line.

---

23   Ibid., 400: "quia nihil potest ultra vives suas." Tanner analyzed Chapter 22 and the *Sermo de plus et minus* in detail, in particular to judge Cardano's influence on Harriot (cf. below, Section 8.2.); she noted the qualification of minus as "alienum" by Cardano, but did not consider its epistemological function. The separation of the areas of the positives and the negatives and the above justification has therefore been taken into consideration neither by Tanner (Tanner 1980b) nor by Pycior, who relies on her (cf. Pycior in 1997, 24 ff.).

24   This text has been published from Cardano's *Nachlass* (M. Cantor in 1900, 540). Since it refers to Bombelli's algebra of 1572, it was written between 1572 and 1576.

Viète, active as a lawyer in France, published a series of algebraic works after 1591. One of his contributions to the process of generalization was that he introduced the use of letters in algebra—not only for indeterminates, but for coefficients as well. His *In Artem Analyticem Isagoge* (1591) contains an explicit reflection about the conditions of the operation of subtraction. Subtraction, he says, is executable only if the subtrahend B is smaller than the minuend A. For this case, Viète uses as sign of operation "–," However, if it was not known at first which of the two quantities was larger, the sign "=" was to be used for this indeterminate direction of subtraction. Viète thus made clear that he considered possible only subtraction leading to a positive result (cf. J. Klein 1968, 331 ff.). While this was in line with how this operation was viewed traditionally, it had not previously come under debate. Viète's differentiation of signs suggested that there might also be another meaning of this operation.

The rule of signs in its traditional meaning was quite unproblematic for Viète (cf. ibid.). He did not accept negative roots of equations, however (cf. Cantor 1900, 636).

Stevin, active in the Netherlands mainly as an engineer, but also as an educator of Prince Moritz von Nassau, published numerous textbooks, his *L'Arithmétique* (1585) belonging among the more important. Herein, Stevin formulated the rule of signs as a theorem, "proving" it in the traditional way: for "complex" terms $(a–b)$ and $(c–d)$—what was taken away too much by multiplying the middle terms must be restored by the product of the two final terms.[25]

He operated with negative numbers, as had meanwhile become usual, but showed evident reserve in respect to admitting them as roots of equations. While Stevin declared that resolutions by minus (*"solutions par –")* did exist for some problems—e.g., $x^2 = 4x + 21$ is satisfied by $x = –3$—and gave all three roots for cubic equations, even if one of them was negative (cf. M. Cantor 1900, 628 and Stevin 1958, 667 f.), positive roots were his priority of interest, and he considered negative roots only inasmuch he would not risk disregarding any positive solution (cf. Stevin, Werke II B, 642 ff.).[26] Where solving quardratic equations is concerned he considered, however, only three of the four possible cases of combining the coefficients' positive and negative signs. He refrained completely from considering the fourth case $x^2 = –ax – b$, obviously because it gives negative solutions (ibid., 594).

Another generation later, Albert Girard (1595–1632), likewise an engineer in the Netherlands and the first editor of Stevin's works, had no reservations against negative solutions. He was one of the first to formulate the fundamental

---

25   Stevin *Works*, Vol. II B, 560 f. The editors of these collected works canceled as allegedly insignificant the parts on subtraction in the preceding section introducing the arithmetic operations (ibid., 552).

26   Stevin commented on this example of a problem, noting that there were even more solutions than those shown by him. And although they "ne semblent que solutions songeés, toutesfois elles sont utiles, pour venir par les mesmes aux vraies solutions des problemes suivans par +" (ibid., 642).

theorem of algebra and thus granted equal status to all the roots of an equation: positive, negative, and imaginary.

In his *Invention nouvelle en l'algébre* of 1629, Girard expressly stated that negative roots of an equation must not be omitted, naming, e.g., +3 and –3 to be the square roots of +9. He was also the first in early modern times to interpret the negative numbers geometrically: the minus sign, he said, indicated an inverse movement (M. Cantor 1900, 787 f.). Incidentally, Girard belonged to the minority who adopted Viète's sign "=" to indicate the indeterminate difference.

René Descartes (1596–1650), the well-known rationalist philosopher and naturalist, active for some time in the military and otherwise as a private scholar, gave quite decisive impulses to the further generalization of mathematics by establishing algebraic geometry. In particular, the symbols he used, and his notation (he wrote equations with signs for plus and minus and the sign for equality, signs for powers by superscript exponents, used the radical sign, and introduced the practice of using the first letters of the alphabet for constant quantities and of the last letters for variable quantities) made mathematical texts easily readable.

In his famous *Discours de la Méthode* (1637), among whose appendices is *La Géométrie*, Descartes developed principles about how to clearly form concepts. There are no explicit reflections, however, about the foundations of arithmetic and algebra, and in particular none about negative numbers. Operating with negative numbers is nevertheless treated exhaustively in his texts. In the literature, his view on negative numbers, however, is assessed differently. Rouse Ball, for instance, says, "He realised the meaning of negative quantities and used them freely" (Rouse Ball 1908, 276).

A critical view is often deduced from the fact that Descartes called negative solutions "false" roots (cf. Tropfke 1980, 147; Dell'Aquila/Ferrari 1996, 322). Since Descartes at the same time called positive solutions "true roots," the term at first was used only to differentiate between positive and negative solutions (cf. Gericke 1970, 57). What is more revealing is that he characterized negative quantities as "less than nothing." Besides, he used the interesting name "defect of a quantity" for negative quantities:

> It often happens that some of the roots are false, or less than nothing. Thus if we suppose x to stand also for the defect of a quantity, 5 say, we have $x + 5 = 0$. (Descartes, *Œuvres* VI, Géométrie, 445).

On the other hand, Descartes's 1638 short account of his *Géométrie* contains a note that clearly shows the influence of Cardano's epistemological view, so that here as well we have an implicit allusion to the problems of extended multiplication:

> *Note*, that one must take care when multiplying by itself a sum which one knows to be less than zero, or where the greater terms have the sign –; for the product will be the same as if they had the sign +. Thus, $a^2 - 2ab + b^2$ is equally the square of $a - b$ as of $b - a$; so that, if one knows a to be less than b, one cannot multiply $a - b$ by itself, since it will produce a true sum in place of one that is less than nothing: which

will cause an error in the equation. (Descartes, *Œuvres* X, 662; cf. Dell'Aquila/Ferrari 1996, 323).

It has often been said that Descartes shirked considering negative roots; even Charles Adam and Paul Tannery, the editors of his *Collected Works*, commented on the section about solving quadratic equations as follows: "Descartes ne reconnait nullement les racines négatives des équations" (cf., e.g., Descartes *Œuvres* VI, 375). But only a couple of lines later in this text, Descartes gives both the positive and the negative solutions of a quadratic equation (ibid., 376).

Glaeser's and Dell'Aquila/Ferrari's summary criticism that Descartes suggested skillful tricks to eliminate negative solutions for equations is just as unjustified. While Descartes explained how to transform "fausses racines" into "vrayes," without making the "true" at the same time "false" (ibid., 450), this refers to a method for transforming equations, and does not convey an eliminative attitude toward negative solutions. Elsewhere, he even showed how to reduce the number of positive solutions in favor of negative ones (ibid., 448).

There is a problem concerning the conceptual level in Descartes, however, actually in the relation between the concept of number and the concept of quantity, which has not been noted as yet in studies on the history of negative numbers. It appears in his treatment of quadratic equations. While Descartes speaks, upon introducing the algebraic treatment of geometrical problems, of *one* type of quadratic equation,

$$z^2 = -az + bb$$

(ibid., 373), this equation is not intended to be the normal form. Rather, there are different normal forms, depending on the signs with which the coefficients are provided. Actually, Descartes elsewhere presents a general form of the quadratic equation, which,because of the variability of the signs can maximally assume four different forms:

$$xx = + \text{ or } -ax + \text{ or } -bb$$

(ibid., 386). In fact, Descartes considered only three normal types in his account of solving quadratic equations,

$$z^2 = az + bb,$$

$$yy = -ay + bb,$$

$$z^2 = az - bb,$$

whereas the case in which only minus signs appear is not a subject of discussion (ibid., 374–376). Not only is it noteworthy that Descartes does not count the case of two negatives roots among the "normal" cases, but moreover, the implied concept of quantities is of fundamental importance. Contrary to the apparent generality of the domain of values, the coefficients a, b, etc. in the equations are not able to assume arbitrary values in the entire area of the defined numbers; rather, they are evidently understood to be absolute quantities that are able to assume only positive values. While Descartes had declared in the overview already mentioned that the first letters of the alphabet stand for known terms, that is, both for numbers and for quantities ("soit ligne, nombre, superficie, ou corps": Descartes *Œuvres* X, 672), because the coefficients in the equations represent geometrical objects, in particular lines, and hence substances. These quantities are understood as representing absolute values. A separation of the concept of number from the concept of quantity was not realized here, and hence no fusion of positive and negative numbers to the unified area of integers.

Although Descartes extensively operated with negative numbers, these were not completely employed on an equal basis with positive numbers, but with some restrictions due to epistemological reservations and to a specific concept of quantities. This ambivalence at the beginning of the intense modern development of algebra was to essentially affect the later history of the negative numbers, and not only their history. Thus it has already been noted often that Descartes, the founder of analytic geometry, always stayed within the first quadrant with his graphs of figures and curves (cf. M. Kline 1972, 311).

2.5. The Controversy Between Arnauld and Prestet

A New Type of Textbook

The texts mentioned above of authors from early modern times were for the most part research publications directed to a small audience of scholars; they were not objects for systematic teaching, say, at universities. A decisive change in this respect was initiated in France, by one of Antoine Arnauld's (1612–1694) textbooks.

The very title of his book expressed its bold program: *Nouveaux Élémens de Géométrie*—the title "elements of geometry" having hitherto been used in Europe exclusively for Euclid's textbook, "elements" and "Euclid" having become practically synonymous in mathematics. Not only did the title for the first time express the claim to present a textbook superior to Euclid's,[27] but its very concept contained a radical critique claiming to improve the realization of Euclid's intentions. In contrast to criticisms of Euclid aimed at improving the overall presentation or the wording of some propositions, Arnauld rejected Euclid's textbook, traditionally praised as the classical model of a rigorous and methodical presentation, for basic defects in methodical architecture. In doing so, Arnauld continued Petrus Ramus's (1515–1572) critique of Euclid, implementing it by restructuring.[28] In fact, Arnauld realized a new methodological structure. Instead of the pervasive switching between the more geometrical and more arithmetical/algebraic books in Euclid's work, Arnauld began with four books presenting the foundations of operating on quantities in

---

[27]   After Arnauld, this title, "Nouveaux Élémens ..." has been used in France repeatedly by textbook authors (not, however, in Germany and in England).

[28]   For his critique of Euclid cf. Schubring 1978, 40 ff., and here Section 2.7.

general ("la quantité ou grandeur en général"), and only then developed the application of this theory of general quantities to geometry in the books that followed (Arnauld 1667, ii).

Beyond that, Arnauld's textbook had revolutionizing effects, since it adopted the new algebraic notation in equations elaborated by Descartes, addressing a larger public due to his algebraized, more readable style, and thus inaugurating the new textbook format. Subsequent to Arnauld's *Elements*, a large number of mathematics textbooks perfecting this new style were published in increasingly rapid sequence. While Arnauld's textbook had still been addressed to an unspecified general audience, the textbooks that followed were clearly motivated by purposes of university teaching. The publication of Arnauld's textbook may be considered as the onset of modern textbook production in general.[29]

ANTOINE ARNAULD

Antoine Arnauld was an eminent person who had an important impact on French intellectual life. Originally a theologian, he taught at the Sorbonne in Paris until expelled from there by the Jesuits; then he lived in the cloister Port-Royal near Paris; he was also a philosopher and a mathematician. He was one of Descartes's significant dialogue partners in questions concerning the method of cognition; and it was in particular by his own close contact with Arnauld that Leibniz, during his stay in Paris, became intensely acquainted with mathematical research (Bopp 1902, 203). Together with P. Nicole, Arnauld is the author of the famous Logic of Port-Royal: *La Logique ou l'Art de penser* of 1662, which, just as their grammar for linguistics, exerted profound effects on French philosophy. Together with other members of Port-Royal, Arnauld was one of the leading representatives of Jansenism in France. Jansenism, a reform-Catholic current in favor of a French national church, presented an acute challenge to the Jesuits, and they fought it violently. Eventually, the Jesuits succeeded in having the King of France destroy the Cloister of Port-Royal; because of its destruction, Arnauld left France and went into exile in the Netherlands and Flanders in 1679.

Arnauld's geometry textbook is particularly valuable for our context because it gave rise to the first known direct debate and controversy between two mathematicians over how to conceive of negative quantities. This controversy must be understood against the backdrop of a profound theological–philosophical conflict of long standing between Arnauld and Malebranche.

This quarrel is all the more unfortunate, actually, as Arnauld and Malebranche were convinced Cartesians, both actively promoting Cartesianism, in particular in mathematics. Their debate cannot be detailed here (cf. A. Robinet 1966), but must be mentioned briefly as the backdrop for the controversy about negative numbers. Nicole Malebranche (1638–1715), also a theologian and prominent ph

---

[29]   In one respect, Arnauld differs from his successors: he obviously had another view of authorship. All his books appeared anonymously, both his logic and his geometry, the first edition of 1667 and its second of 1683 as well.

ilosopher, promoter of mathematics and the sciences, started a theological career, was ordained a priest, and became a member of the Oratorian Order. In 1699, he was elected to the Paris *Académie des Sciences*. He was the principal representative of a Christian Cartesianism in France. By means of occasionalism, he attempted to solve the soul–body problem caused by the Cartesian dualist doctrine of the two substances. Arnauld replied with his 1683 *Des Vrayes et des Fausses Idées* to Malebranche's two major works, *De la Recherche de la Vérité* (1674/75) and *Traité de la nature et de la grâce* (1680). Besides the theological concept of grace, an epistemological question was at stake as well. Arnauld criticized Malebranche's thesis that the knowledge of objects required ideas that exist independently of the human mind, and were distinct from it and its perception (W. Doney 1972, Mittelstraß 1984, Piclin 1993). The related question as to what extent theoretical concepts have to be representable by empirical knowledge constituted the core of the controversy between Arnauld and Malebranche's disciple Prestet.

In his geometry textbook, Arnauld's introductory chapters on the doctrine of general quantities are on general conditions, or "suppositions." The second of these conditions was the "knowledge" that the multiplication of two numbers is commutative (Arnauld 1667, III). Explanations of general terms like axiom, definition, theorem, followed, and then explanations of signs used, like $+, -, =$, and the proportion sign $::$; he continued with *principes generaux*, on the relation of part and whole, and with axioms on equality and inequality. Based on this, Arnauld introduced the four basic operations. Subtraction was explained quite generally as the remainder $b–c$ of two quantities (ibid., 6). And by introducing letters, Arnauld stressed extensively that upon designating quantities by letters one no longer needed to pay attention to the letters' meaning, since the task now was only to investigate identities and relations. Besides, he added, it was one of his book's particular advantages that it trained the mind to understand things in a "spiritual manner," without the help of any "sensible" images (ibid., 4).

After that, Arnauld began treating "complex," i.e., compound, quantities. Their introduction contains a hint at the concept of opposite quantities. The plus and the minus of the same quantity cancel one another, resulting in zero: "Le plus et le moins d'une même grandeur ou terme sont égaux à rien, ou valent zero. Car l'un ostant ce que l'autre a mis, il ne demeure rien" (ibid., 9).

Using the multiplication of complex quantities $(a–b)$ and $(c–d)$, Arnauld explained and "proved" the rule of signs. The last of its four cases reveals a noteworthy influence of Cardano's last works.[30] Only reluctantly, Arnauld

---

[30]   The Leyden edition of Cardano's works appeared in 1662, so that it might have influenced Arnauld quite directly. On the other hand, Cardano and Arnauld seem isolated with their critique of the rule of signs. However, this can also be an error of our historical perspective: according to Rider's bibliography, between 1570 and 1667, at least 82 algebra textbooks were published, which have hitherto not yet been evaluated with a focus on this question (Rider in 1982, 26–49)!

quoted the traditional rule here that minus by minus makes plus, adding that their product properly gives, due to its inherent nature, *minus*:

> MINUS by *minus* gives *plus*: that is to say that the multiplication of two terms, both of which have the sign *minus*, gives a product which must have the sign plus. [...] This appears rather strange, and in fact it cannot be imagined that this could happen other than by accident. For of themselves, *minus* multiplied by *minus* can only give *minus* (ibid., 13).[31]

Arnauld neither explained the inherent reasons for which the product should yield a minus sign, nor why the plus result should appear only accidentally. The justification for the plus sign's occurrence given by Arnauld following the quotation above corresponded to the traditional justification for compound expressions: because of the terms having a minus sign, one subtracted too much, and this excess had to be compensated by adding the final term. No randomness of the "plus" result was evident from this reasoning, either. Obviously, Arnauld was prompted by profound epistemological concerns, similar to those of Cardano, that operations within the negative area would remain there with their result as well, making mathematical justifications proper marginal.

The statement that a proposition concerning the foundations of mathematical operations was of random validity had to cause a stir. Moreover, it was strange that Arnauld had nevertheless presented this proposition, which was allegedly true only by accident, as a valid rule.[32] Within the very foundations, there existed, hence, a striking contradiction to the otherwise rigorous and methodical structure of his book.

JEAN PRESTET

One of the contemporary reactions to this contradiction has come down to us: that of Jean Prestet (1648–1691).[33] Prestet, of poor origins, was ordained as a priest only late in his life; as adolescent, he had been a servant to the *Oratoire* in Paris. He soon became one of Malebranche's scribes, then his disciple in mathematics; eventually, Prestet, guided by Malebranche, began writing a textbook in Arnauld's new style in 1670. After publication in 1675, the *Oratoire* delegated him to positions for teaching mathematics, the most prominent of these being the new mathematical chair of Angers in 1681. He died soon after having published the second edition of his textbook (Robinet 1960).

The title of the textbook was even more ambitious than Arnauld's: *Elémens des Mathématiques*. As is shown by how the title goes on, Prestet's intention was to extend the usually prefatory general part about quantities to a separate

---

[31]   "Car de soy-même *moins* multiplié par *moins* ne peut donner que *moins*."

[32]   In the other parts of the volume, Arnauld did not return to the validity of the rule of signs, firstly since its main part treated elementary geometry, and secondly because he was entitled to apply the traditional rule due to its attested contrariness.

[33]   P. Schrecker was the first to give access to this debate in 1935 (Schrecker 1935).

book on arithmetic and algebra. It continues, *ou Principes généraux de toutes les Sciences qui ont les Grandeurs pour Objet*. Prestet's volume contained no geometry,.In spite of its apparent claim at totality, the book rather intended a program of algebraization for mathematics, as emphasized in the preface, and as demonstrated in its second edition with challenging innovativeness (see below).

Prestet's textbook indeed contains the first account of the concept of negative numbers in which the negative numbers are presented as having the same status as positive numbers, and in which the rule of signs is "proved" not geometrically, but algebraically.[34] At the very beginning of his presentation of the concept of quantities, Prestet explains that the quantities are a composite of a positive area and of a negative area, each alone being infinitely large. The starting point for this conceptualization was the introduction of the zero, which did not have any absolute character or precarious exceptional status, but was understood to be a relative quantity between the positive area and the negative area that enables a comparison of the relations between quantities from the two areas: "Le rien ou le zero nous sert de milieu[35] pour faire les comparaisons des grandeurs, et pour juger de leurs rapports" (Prestet 1675, 3).

Prestet's concept of quantity, however, is not exempt from contradictions. While he stresses that he is not interested in the nature of quantities and in what they are *in themselves* ("dans elles mêmes"), saying that only the *relations* between them are essential, he characterizes the positive and the negative quantities, despite this relational view, in terms of *existence*, as "being" and "non-being" ["non estre"] respectively, and as bearing an increased or a reduced degree of reality:

> Magnitudes have more or less reality as their being takes them further from zero, and they have less reality when their non-being takes them further from this same zero. It became customary to call <u>positive</u> or <u>true</u> every magnitude which adds to zero, and <u>negative</u> or <u>false</u> every magnitude which takes away from this same zero. (ibid.).

The positive quantities extend to where nothing can be added any more that the quantity does not already have, this being the infinitely true magnitude: "grandeur infiniment vraie." And the negative quantities extend to where nothing can be subtracted any more that has not already been subtracted: this being the infinitely false magnitude, "grandeur infiniment fausse." In defining subtraction, Prestet restricted the quantities to be subtracted to positive quantities, in order to avoid confusing the algebraic sign with the sign of operation, an error committed by many later authors: "The addition of true magnitudes are marked by the sign +, which signifies plus, and the taking away or the subtraction of these same magnitudes are marked by the sign –, which

---

34   Prestet thus continued—after the interim ruptures in the understanding of the concept since Cardano—Chuquet's approach; but what had remained mostly implicit and had become clear only in the application, was developed here to an explicit foundational part of its own.

35   "Milieu," probably to be understood as the "middle;" in the second edition substituted by "terme ou un point fixe" (Prestet 1689, Vol. I, 3).

signifies minus" (ibid.). Prestet was the first to present addition and subtraction very lucidly as the operations inverse to one another:

> The + and the – of equal magnitudes are each mutual takings away, the + takes away from the –, and the – takes away from the +. The position or the possession of a thousand écus takes away the negation or the privation of a thousand écus, and the negation or the privation of a thousand écus takes away the position or the possession of a thousand écus, [...]. Or what is the same thing, +1000 écus –1000 écus are equal to zero (ibid., 10).

Although Prestet did not speak specifically of "opposite quantities" here, his lucid confrontation of *positing* ("*position*") and of *canceling* ("*négation*") refers to this concept as a possible next step of development. As a "clear" conclusion, but without formal proof, Prestet added the rule of signs; we quote here the rule for the equal signs: "D'où il est clair, 1°. que plus plus ou + + est égal à moins moins ou – –, et que – – = + +" (ibid.).

In his discussion of the four basic operations on integers,[36] Prestet took up the rule of signs once more. Thus, he explained it for multiplication by interpreting multiplication as a repeated addition. The product $(-2) \times (-4)$ was to be understood as a twofold negative addition of –4; since a singular negative addition of –4 gave +4, a twofold had to give +8 (ibid., 20).

Of particular interest under the aspect of foundations of arithmetic is that Prestet did not restrict the definition of subtraction to the case of a positive difference. Rather, this definition is entirely general, and this is why Prestet also specifically notes that the difference can be negative: "When the number to be taken away is greater than the number from which one takes it, the difference or the remainder is negative" (ibid., 17).

He exemplified several such negative *remainders*. Moreover, in a comment, Prestet explained for those who might at first be surprised at the result $4 - (-3) = 7$ that such a subtraction must be understood as "descending" from the positive into the negative area: Firstly as descending ("*descendre*") by 4 units from +4 to 0, and then as descending from 0 to –3 by three more units, thus altogether by 7 units (ibid., 18).

Thanks to its extensive, coherent, and consistent presentation, Prestet's textbook of 1675 was the hitherto most advanced presentation of the negative numbers. It represents a significant summit in the history of the negative numbers.

THE CONTROVERSY

Within a period of only eight years, two textbooks had thus been published that transcended by far the traditional level of explicity for the concept of negative numbers while assuming divergent approaches. This alone would have

---

[36]   "grandeurs entières," later in France with a restricted meaning: positive integers, i.e., natural numbers.

suggested a controversy between the two French authors.[37] Moreover the heated theological–philosophical dispute between Arnauld and Prestet's teacher and master Malebranche may have fueled this debate in mathematics. Prestet gave no dates for his correspondences with Arnauld, but wrote in 1689 that the exchange of letters had taken place "more than ten years ago" (Prestet 1689, Vol. 2, 366), that is, before 1679, the year when the conflict between Arnauld and Malebranche broke out. The period of 1681/82 is more probable, however.

Prestet published his debate with Arnauld in his own textbook's second edition. It had begun with a letter that he directed to Arnauld, the content of which he reported only briefly. Arnauld's reply is rendered in more extensive excerpts, and Prestet's answer probably almost completely. Prestet gave as his motive for publishing the correspondence that he had observed what difficulties even intelligent and enlightened people experienced when trying to obtain an unambiguous concept of the nature of roots and of negative quantities for themselves, and all the more so one of imaginary quantities. His answer to the difficulties expressed by Arnauld might be helpful, Prestet argued.

As he explained in an annex to his second edition, he had held Arnauld's geometry textbook in great esteem for its order and method. Because of his wish that the latter's geometry book should contain only completely certain knowledge, Prestet said, he had suggested to Arnauld to modify his paragraph on the rule of signs and had presented him his reasons why, to his mind, minus by minus could not give minus (Prestet 1689, Vol. 2, 366).

Arnauld had answered that he had already made up his mind to change this part in a new edition he was preparing. He said to have himself reflected the matter some time ago and become convinced that one could in fact say that minus by minus makes plus (ibid). In what followed, Arnauld, however, had given four reasons why he himself had conceptual reservations with regard to isolated negative quantities (*ces moins sans rapport à aucun plus*). These four reasons were the following (ibid., 366–367):

1. It was clear that one can subtract two *toises* (fathoms) from five *toises*. But it was impossible to subtract seven *toises* from five *toises*. It was thus incomprehensible for him ("inconceivable"), how one could say "five *toises* minus seven *toises*." Arnauld assigned a meaning to arithmetic operations as far as they were executable with concrete quantities.

2. Arnauld declared it to be likewise incomprehensible why the square of -5 could be the same as the square of $+5$. Obviously, Arnauld was assuming like Cardano that results of operations in the negative area had to remain within it.

3. Moreover, Arnauld had been the first to introduce the so-called proportion argument, which was to develop a high power of persuasion for the future. If $(-5)(-5) = (+5)(+5)$, the proportion

---

[37]   The first edition of Prestet's volume was published anonymously, just as Arnauld's. While it was not difficult to detect Arnauld as the author, many thought Malebranche to be the author of the other volume.

$$1 : -4 :: -5 : 20$$

had to be valid as well. This proportion, Arnauld had said, arose from the "fondement de la multiplication," according to which unity is to a factor as another factor is to the product; this was true for integers and for fractions. In the above proportion, however, the rule ever applicable for all other proportions did not hold: if the proportion's first term exceeded the second, the third must also exceed the fourth.

Arnauld had said he himself had already considered whether the problem could be solved by disregarding the signs + and − in the proportion, that is, by considering absolute numbers (*en faisant abstraction des signes*). He had, however, feared that this might result in undesirable consequences, since negative quantities could actually exist; they should accordingly have to be made allowance for in the proportions.

4. Finally, Arnauld had explained that he rejected isolated negative quantities. A negative quantity like −10.000 écus could be a real thing, such as a man's debts, which could cause him to be thrown into debtor's prison to compel him to find the money to somehow pay his creditors. One could say that a man had −10.000 écus only if one could attribute to him some power (*puissance*) however fictitious to obtain 10.000 écus to satisfy his creditors. Negative quantities, Arnauld had said, were to be credited as possible only in relation to some plus, but without such a relation they were but a fiction, *une chimère, une montagne sans vallée*.

In his response, Prestet had reduced the four problems to two: to whether the subtraction was executable and to the argument of proportion.

With regard to the first set of problems, Prestet referred first to an epistemological issue: whether abstract concepts were admissible. People possessed a notion of plus and minus; this did not comprise quantities that, whether disparate, or not, were certain ways to understand quantities either as added or as subtracted ones. While applying this to the results of operations, he, Prestet, continued to rely, however, on the substantialist concept of "existence," for his own reflections were based not on numbers, but on quantities existing in real life. Only positive quantities existed, he said, they were characterized by the positive idea of plus (Prestet 1689, vol. 2, 368). Conversely, negative quantities did not exist and had better be designated as "rien" or "zero." For to be a negative quantity and to be no quantity at all appeared to be the same thing.

Prestet tried to escape this contradiction by means of an *operational* approach: while −2 *toises* did not exist, one might actually use this expression without relating it to any expression in plus—as an *indication* that 2 *toises* are lacking to return up to zero again. Such an operative meaning indicating the absence of 2 *toises,* Prestet declared, was no chimera or fiction.

Regarding the proportion argument, Prestet did not address Arnauld's problem of contradiction to the definition directly. Instead, he argued by quoting those parts of the rule of signs that had been accepted by Arnauld, i.e., for mixed terms. As Arnauld had pointed out in his own textbook, minus by plus gave minus, just as plus by minus gave minus. Therefore, $+1$ divided by -1 gave -1, just as -1 divided by $+1$. Since the results were identical, the ratios were identical, too; compounded, this yielded the following proportion:

$$-1 : +1 :: +1 : -1.$$

Just as admissible and valid, hence, for him was the proportion

$$1 : -4 :: -5 : 20.$$

Prestet did not recognize the principal importance of the problem of how to define an enlarged multiplication. In the same way, he dealt with Arnauld's question concerning the equality of $(-5)^2$ and $(+5)^2$: equality resulted from the two proportions

$$1 : -5 :: -5 : 25 \text{ and } 1 : +5 :: +5 : 25.$$

As a complement, Prestet demonstrated the equality of the two squares, by means of how he had interpreted the definition of multiplication already in his own textbook: -5 times -5 signified subtracting -5 five times; subtracting once gave $+5$, subtracting five times hence gave 25 (ibid., 370).

THE DEBATE'S EFFECTS ON THEIR TEXTBOOK REEDITIONS

The two opponents each had a second edition of their mathematics textbooks printed—Arnauld in 1683, and Prestet in 1689. Which effects of their debate can be observed in the new editions? Let us begin with Prestet. He changed the title and made it analogous to Arnauld's by preceding it by a *Nouveaux*. Actually, the textbook had been profoundly revised. The one-volume work having 428 pages had become two-volumed with more than 1100 pages; its concept had been changed as well. While the preface of the first edition had favored the analytic method, the second edition's considerably enlarged preface pleaded with fervor for the analytic method and for algebra's superiority over the synthetic method and over geometry (cf. below, Section 2.7).

While Prestet had left the concept of negative quantities virtually unchanged, he now abstained from the qualifications "existence" and "non-existence," and from attributing greater or lesser degrees of reality. The comparison with zero continued to be pivotal:

> We call *positive* or *real* or *true magnitude*, every magnitude which adds to zero, or which is worth more than nothing; and *negative* or *defective* or *false magnitude*, every magnitude which takes away from zero, or which is worth less than nothing. (Prestet 1689, vol. I, 59).

One change, however, was caused by the new structure. While the first edition had been entirely based on the concept of *quantity*, the second began with *number* as the fundamental idea. After an introductory part on methods, a part titled "science générale des nombres" followed; it virtually constituted an independent arithmetic based on *"nombres entiers ou naturels."* The definition of subtraction in this arithmetic was restricted to cases in which the subtrahend was smaller than the minuend (ibid., 23). Algebra as the calculus with letters constituted the work's third part, and it was based on the concept of quantity: *La science générale des grandeurs*. Here, subtraction was again defined without restriction concerning the relations of quantities.

Prestet had entangled himself in considerable technical difficulties, however. To keep the two meanings of the minus sign as algebraic sign and as sign of operation separate, he had implicitly restricted the use of letters to positive quantities alone. He was thus able to characterize positive quantities as those endowed with the sign +, or without sign, and negative quantities as those endowed with a minus sign (ibid., 59). Subtraction, as in the first edition, was thus restricted to positive quantities in both terms: "take away incomplex numbers from each other when they are both true" (ibid., 60). This prompted Prestet to prescribe use of the minus sign for the addition of negative quantities (ibid., 61), and use of the plus sign for their subtraction (ibid., 62). This kind of conceptual differentiation opposing time-honored use proved untenable in practice.

His presentation of the rule of signs had remained basically unchanged (ibid., 62 ff.). Since multiplication was understood as repeated addition, and subtraction of negative quantities as some kind of addition, the rule of signs of multiplication resulted already from the rules for subtracting these quantities (ibid., 62).

In *Arnauld's* second edition, there was a series of changes.[38] It is important to note that he did not use the term "negative numbers" or quantities, just as he had not done in the first edition, but only discussed the use of the minus sign. A relevant change was in how he defined subtraction: it was now confined to cases giving *positive* results: "Soustraire, ou *soustraction*, c'est retrancher une moindre grandeur d'une plus grande" (Arnauld 1683, 7).

Arnauld thus continued to reject isolated negative solutions. Against that, he had changed his mind regarding the rule of signs. The propositions saying that the product of twice minus gives again minus, or that plus arises only accidentally, no longer appeared. He accepted the rule of signs without reserve, undertaking to prove it completely. For this purpose, he elaborated the concept of multiplication, as Prestet had used it, even more systematically.

For the product, Arnauld introduced the distinction between multiplier (*multiplicant*) and multiplicand (*multiplié*). For the proof of the rule of signs, he considered the four cases:

$$1. \quad + \times + \qquad 3. \quad - \times +$$
$$2. \quad + \times - \qquad 4. \quad - \times -$$

In the two first cases having a positive multiplier, in essence the usual meaning of the multiplication is valid: as a reiterated execution—in the first case of addition, in the second case already by an extension of the concept, namely of

---

[38]   Robinet has told without giving evidence that this edition contains a polemic against Prestet (Robinet 1961, 210). Actually, this is not true. The publication of their letters by Prestet six years later certainly occurred to support Malebranche in his controversy with Arnauld and to show the "narrownesses" of Arnauld.

subtraction; hence, the sign of the multiplicand remains the same. In the two last cases, however, a new meaning of the operation of multiplication is revealed, an extended meaning: as a *subtraction* of the multiplicand—it must be subtracted as many times as the multiplier indicates. Hence, in both last cases the sign of the multiplicand must be changed:

> In the 4th case where the multiplicand has a *minus*, the product must have a *plus*; [...] to multiply –3 by –5 is to take away 5 times –3. Now, to take away one times –3, is to set down +3, as has been said on the subject of subtraction; thus to take it away 5 times, is to set down +15; which was what had to be proved (ibid., 18).

Although this approach was more reflected and more explicit than those of previous authors and was even suitable to make the rule of signs plausible, it was as yet no real proof. The last justification already assumed by its $--3 = +3$ what had to be proved. Moreover, Arnauld did not discuss the consequences of the distinction between multiplier and multiplicand for the commutativity of multiplication which he had assumed in his textbook's introduction (ibid., 2).[39]

In spite of this "solution" of his problem with the rule of signs, the proportion argument continued to present a stumbling block for Arnauld. He devoted a note to its discussion in his second edition, which covers more than a page in small type, where he mentioned for the first time that due to the new multiplication concept he had been able to solve "la plus grande difficulté" that had led him earlier to assert the accidental character of the plus result. He then presented the proportion argument, first presented in his answer to Prestet as one still bothering him. Arnauld gave the argument with so many doubts of his own that the reader was unable to decide whether he himself considered it refuted or not. Arnauld had added an observation that shows that he was the first to have become aware of the fact that operating on negative quantities required a new, extended concept of multiplication and that the "usual concept of multiplication" did not suffice for this purpose:

> I see no other answer to this [concerning the proportion argument] than to say that the multiplication of minus by minus is carried out by means of subtraction, whereas all the others are carried out by addition: it is not strange that the notion of ordinary multiplications does not conform to this sort of multiplication, which is of a different kind from the others (ibid., 19).

Later mathematicians, however, for a long time did not take this hint at enlarging the concept of multiplication, but preferred to focus on the proportion argument Arnauld had presented here for the first time, attempting either to refute or to apply it. Thus, the argument began to have an intensive aftereffect: $1 : -5 = -3 : 15$ could not be valid, since the second term was smaller than the

---

[39]   He had presented it indeed as unimportant which of the two factors was chosen for which function (Arnauld 1683, 16). But he had obviously operated on numbers only. The general case is, however, that only the multiplier needs to be a pure number, a scalar, while the multiplicand can be a quantity.

first, requiring that the fourth also ought to be smaller than the third, which actually was larger (ibid).[40]

Summing up the controversy between Arnauld and Prestet, one will observe that Prestet had left his concept of negative quantities practically unchanged. Although he still considers the linkage between the number concept and empiric substances as given, the decisive aspect for him was a criterion of intra-mathematical consistency as to whether quantities are able to satisfy the given equations. The only change in Prestet's concept was that he excluded negative quantities from arithmetic, referring them to the next stage of generality, to algebra. This kind of hierarchization of mathematical knowledge was later to become typical of how the French saw the relationship between arithmetic and algebra.

Two points are noteworthy with regard to Arnauld's positions. Firstly, he changed his mind in the course of the controversy from postulating an only accidental validity of the rule of signs to affirming its full validity. The artful distinction between *multipliant* and *multiplié* he seems to have first introduced illustrates how poorly reflected and explained were the foundational aspects of the mathematical operations hitherto intensely used. It became explicit that the implied premise was that mathematical statements should be always valid over the entire field of numbers and quantities known; no explicit differentiation between ranges of argument and ranges of values had as yet been established. In his closing reflection on the proportion argument, Arnauld for the first time voiced the idea that validity—at this point the validity of the definition of proportions—might be restricted to certain subranges of numbers, i.e., might be invalid for negative numbers.

The second remarkable point appears in the changed definition of subtraction: in spite of its general formulation in the first edition, Arnauld apparently had not intended it to be unrestrictedly executable. His correspondence with Prestet shows his steadfast refusal of isolated negative quantities; the debate had made Arnauld aware of the consequences of his, according to him, own definition, which he himself considered as too generally formulated. In the second edition, he thus explicitly excluded isolated negative numbers.

The reasons advanced by Prestet and Arnauld in their debate "spanned" the largest part of the space from which the arguments on the status of the negative numbers were obtained in the times that were to follow.

The concept of multiplication is a privileged instance illustrating how different views of the same concept could "coexist" within the self–same person and thus provoke specific conceptual ruptures. The algebraic view of multiplication, as an iterated addition and containing the differentiation of

---

[40]   Arnauld had here himself mentioned an extended notion of multiplication, namely the product concept as introduced by Descartes (Descartes 1974, 370): as the fourth proportional to the unity, to the multiplier, and to the multiplicand (Arnauld 1683, 19); however, because of the proportion argument, he saw no possibility to apply this extended notion to negative quantities.

multiplier and multiplicand, a view that incidentally agreed exactly with the view of multiplication in Indian mathematics, had not been assumed to be exclusive by Arnauld. In the second edition, he complemented it by the Cartesian notion of product for geometrical lines: "the most natural notion of multiplication in general [...] is that the unit [...] must be to the multiplier as the multiplicand is to the product" (Arnauld 1683, 19).

Prestet, who had used the algebraic view of the rule of signs for his "proof" of the rule of signs, however, used the geometrical concept as well. A productive use of this geometrical version is his discussion of imaginary quantities. To my knowledge, the first indication of their geometrical interpretation appears in Prestet. He qualified negative quantities to be "linear" and imaginary quantities to be "plane" quantities, in the form of mean proportionals:

> imaginary quantities, originating from the second degree, imply planes and they are complex [compliquées], as when one wishes to take a mean proportional $\sqrt{-ab}$ between a positive magnitude $+a$ and the negative $-a$ (Prestet 1689, II, 371).

2.6. An Insertion: Brief Comparison of the Institutions for Mathematical Teaching in France, Germany, and England

In even another regard, Arnauld's and Prestet's textbooks present a change which was to become constitutive for the future. While the authors discussed before had published their treatises mainly in Latin, thus addressing an international (learned) audience, these new textbooks were written in French and, hence, oriented with priority toward the French education system and the culturally interested French public. Teaching mathematics now became a part of the emerging national education systems. This integration into different systems began to shape mathematics itself in a "style" specific for each case. Hence, it is necessary to compare the educational systems of the three countries with the mathematical cultures best established in a brief overview:[41] France, Germany and England. This comparison should serve as a matrix for a structural analysis of how the concept of negative numbers, and likewise later that of infinitely small quantities, underwent distinctive developments in these communities.

UNIVERSITIES AND FACULTIES OF ARTS AND PHILOSOPHY

A major basis of comparison is the status mathematics had within the universities. How things began in the Middle Ages is relatively uniform for Western Europe. Mathematics, as the quadrivium, then consisting of arithmetic, geometry, astronomy, and music, formed a stable, although marginal,

---

[41]   A more detailed analysis of this institutional development in Europe is given in (Schubring 2002).

component of the propaedeutic teaching provided by the faculty of arts. Teachers were in general not especially qualified for this subject and changed very often, as courses were given by young lecturers who strove to rise to the better and more renowned faculties. The relative uniformity of this structure throughout Europe dissolved into differentiation at the beginning of early modern times. The emergence of national or territorial states and the schism of the Christian faith were conducive to marked differences among the universities. These began to be redefined into components of individual states' sovereignty. The fates suffered by the universities' faculties of arts is an indication of this progressive development.

The position of the traditional faculties of arts was in jeopardy from two sides, firstly from their precarious situation with regard to the three other "superior" or professional faculties. Wherever they were able to attain some degree of independence and a status beyond their propaedeutic functions, the subjects taught there had better opportunities of development toward disciplines of their own. Secondly, they were endangered by their relation to the system of secondary schools, which had expanded since Humanism, since both types of institution competed as to their propaedeutic tasks. Beyond the case in which secondary school and faculty of arts functioned consecutively, there were the two extreme forms under which either the faculty of arts was "soaked up" by the secondary schools or the secondary school became an integral part of the faculty of arts as its preliminary stage.

The decisive factor in the emergence of the various functions and structures was the conflict between Protestantism and Catholicism, which had been ongoing since the Reformation. While the faculty of arts was able to attain a relatively independent position documented in its rise to "faculty of philosophy" in Protestant territories, in particular Lutheran ones, it not only remained confined to subordinate functions in territories of Catholic faith, but was even in essential parts replaced by colleges having the status of secondary schools.

This radical change of function in Catholic territories was basically a consequence of the work of the Jesuit order. It had been intent on establishing a Catholic education system since about 1550, during the first century of the Counter Reformation. The almost total disappearance of the faculty of arts under the Jesuit system was directly connected with the success of a new structural element in the education system, the establishment of secondary schools providing an instruction systematically organized and serving to prepare students for university studies. In their adaptation of this model originating from Humanism, the Jesuits succeeded in transferring practically the entire teaching program of the faculty of arts to their own colleges, named *Kolleg* in Germany and *collège* in France. Their curriculum became restricted to the study of Latin as a transformation of the *trivium*, and to philosophy. The faculty of arts was therefore to all intents and purposes reduced to holding the *examinations* necessary for passing to the superior faculties.

In contrast to the professors at the Protestant faculties of philosophy, the teachers at the Jesuit colleges were ordained priests. Possible personal scientific orientations had to submit to rigid reglementation by the order's superiors and to the predominant task of spiritual education. Starting from the seventeenth century, other orders began to rival the Jesuits in Catholic states with *collèges* of their own, such as the Benedictines and the Oratorians in France, but these remained within the structures established by the Jesuits and did not attempt institutional change.

In comparison, the situation can be summarized as follows: In Germany, conflicting structures existed side by side in Catholic and in Protestant countries. In states with Lutheran denomination, the faculties of philosophy were able to hold their own against the *Gymnasien*, the classical secondary schools, and to develop into nuclei of subsequent growth of academic disciplines. A somewhat fragile division of labor was established between scholastic instruction in the preparatory subjects and a higher, general scientific education. In the Catholic territories of Germany, the Jesuits succeeded in gaining control over the totality of the faculties of arts, and to substitute them by their own *Kollegs* (cf. Hengst 1980, Meuthen 1988, J. Steiner 1989). The German classes possessing *Bildung* were almost exclusively oriented toward the universities. Eminent scholars outside the university system, such as Leibniz, were exceptional.

The French structures were very different from the German. In France, the universities were uniformly Catholic.[42] The teaching tasks of the faculty of arts had been pervasively transferred to *collèges,* reducing these faculties to holding the examinations necessary for the admission to the three professional faculties (Julia/Verger 1986, 141– 152). One of the effects of this development was that in France, these faculties did not foster the emergence of scientific disciplines. Scientific culture in France, on the other hand, was not confined to the universities and their context. In the aristocracy in particular, considerable groups existed as support for scientific activity. The *Académie des Sciences* in Paris, founded in 1666, soon became a crystallizing core for mathematical and scientific research. Beyond that, the universities were confronted with increasingly acute competition, in particular from the first half of the eighteenth century, by the expanding military schools, which became institutions offering comprehensive instruction in mathematics and in the sciences (cf. Taton 1986: Ve Partie). They were also challenged by the *Collège royal*, founded in 1530, which was unique in that it provided lectures and courses in modern science without leading to any examination or degree.

The structures in England were different again, the "Collegiate University" prevailing (McConica 1986a). By the middle of the sixteenth century, the

---

[42]   By the occupation of Alsace in 1681, the Protestant university of Strasbourg became French but remained a special case in France, with an independent faculty of philosophy.

transition occurred to the university exclusively organized into *colleges*.[43] Now not only did the students have to live together in *colleges*, supervised by tutors assigned to each, but teaching was basically done within these *colleges* as well. In contrast to the Catholic model established by the Jesuits, teaching in these English colleges was not restricted to the faculty of arts's subjects proper, but included the subjects offered by all the faculties (cf. McConica 1986b). In Oxford and Cambridge, there was thus a double structure of *colleges* and of faculties. While exams were left to the faculties (Fletcher 1986, 185 ff.; Leader 1988, 102 ff.), there were also salaried professors giving "public" lectures as a part of the activities of faculties side by side with the colleges' *lecturers*. Teaching within the faculty of arts was not only the most extensive part, but the students had to devote most of their studies to its subjects. The dynamism unfolded in the humanist reform phase during the first half of the 16th century had to submit to orthodoxy due to the so-called Elizabethan reform of 1570.

THE STATUS OF MATHEMATICS IN VARIOUS SYSTEMS OF NATIONAL EDUCATION

With regard to mathematics as well, grave differences between European states and their ruling Christian denominations developed at the beginning of modern times. Schöner is the first author to have described the rather marginal function of mathematics at the end of the Middle Ages—a situation rather uniform for the universities in Europe—in his profound study (Schöner 1994).

Philipp Melanchthon, Luther's advisor in all questions of teaching and education, was a fervent advocate of a strong position of mathematical teaching in schools and at universities. In Lutheran territories, it was thus possible for mathematics to undergo a continuous development at the universities. Mathematics was among the first subjects for which tenures for specialized professors were established as part of the process of reorganizing the faculties of arts into faculties of philosophy.

The situation in the Catholic states was considerably different. In the founding document for Jesuit college instruction, the "Ratio atque institutio studiorum Societatis Jesu" of 1599, mathematics was a marginal subject having no relevance for exams. Two things became decisive for mathematics within the Catholic–Jesuit system of education. Mathematics was no longer a part of the teaching of the faculty of arts; instead, lessons in the colleges were organized strictly according to a hierarchy of classes. Teachers were not in general professional specialists, as in the Protestant system, but more or less generalists, their teaching resembling the medieval system of courses held by ever varying lectors. The second structural pattern was that teaching of mathematics was intended only for the end of the curriculum, in the closing two-year philosophy

---

[43]   The transition to the "Collegiate Society" is detailed for Oxford in the third volume of the new university history (McConica 1986a), and for Cambridge in the first volume of the university history (D.R. Leader 1988).

course, namely as part of the physics course held in the second year of the study of philosophy. Accordingly, the subject matter of mathematics was reduced to astronomy and to some basic notions of mathematics required for that.

In France, where the Jesuits' model of instruction had been adopted also by their subsequent competitors, the Benedictines and the Oratorians, teaching mathematics at universities remained in general restricted to that part of the physics class in the final course of the *collèges* attended by only a minority of students. The absence of an academic structure, however, made for creating a number of complementary institutions. The *Collège de France* in Paris had at first two professorships for mathematics, astronomy, and physics in the seventeenth and eighteenth centuries and then five beginning in 1768/69 (cf. Collège de France 1930, 15 f.). During the seventeenth century, mathematical professorships were established at some universities by a number of endowments offered by the king, or by urban foundations, as in Angers. The audience was composed not only of the regular college students, but also of a more general public, for instance of young gentlemen preparing for a military career (Belhoste 1993). Eventually, this professional formation became institutionalized separately from the universities in military schools providing a strong component of mathematics and the sciences. This explains the substantial production of mathematical writings in France beyond the academic context.

Mathematics experienced a quite peculiar development in England. The literature always mentions the extraordinary position of mathematics in Cambridge University: it was the almost exclusive exam subject for the *undergraduate* students. Gascoigne emphasizes the often overlooked development that a basic curriculum reform had been effected in Cambridge about 1700, by accepting Newtonism (Gascoigne 1989, 7). The ruptures caused by Humanism and religious schisms are thus essential for understanding the position of mathematics at the English universities, just as they are for Germany and France. Developments in England after the period of Humanism confined mathematics to marginality for a long time in a way analogous to that in the Catholic states, but also to Calvinist territories like the Netherlands, where mathematics went into decline after the first strong impetus in favor of it during the seventeenth century.

The eminent role that mathematics eventually attained in Cambridge after the mid-eighteenth century due to the *Senate House Examination* was not really conducive to a progressive development of mathematics.[44] While the subject was indeed intended only for that minority of the students who strove for an "honors" degree, it determined the style of the entire university studies. Obviously, mathematics served as a substitute for logic, a subject prescribed since 1570 but meanwhile considered outdated. In line with this one-sided function, the study of mathematics was primarily restricted to geometry, in its Euclidean version (Gascoigne 1989, 270 ff.). This context favored neither an

---

[44]   In the nineteenth century, the name changed to *Mathematical Tripos.*

advancement of mathematics nor a transition toward generalization by means of algebraic methods.

NEW APPROACHES IN THE EIGHTEENTH CENTURY

In most European states, qualitative reforms of the educational systems occurred in the eighteenth century, predominantly during its second half. Among the aspects of these reforms were attempts at specialization within the universities; they were initiated at Northern German universities, and constituted an essential contribution to the institutional development of mathematics.

As is well known, this structural change is related to two universities that were outstanding among the large number of minor universities as new establishments, and were the first to practice a new style: Halle (1694), in the Kingdom of Prussia, and Göttingen (1734/7), in the Dukedom of Hanover. Halle, a center of Pietism and of Enlightenment in Northern Germany, had a great force of attraction for students. The utilitarian orientation of these two ideological currents produced a modernization of teaching. Göttingen had already been established in the wake of the state-promoted concept of mercantilism. To attract a large number of students, professors were obliged for the first time to publish research results in the hope that they would become known and attractive. This is the first form of the "research imperative" to come. Personalities like Christian Wolff, Johann Andreas Segner, Christian Hausen, Abraham Gotthelf Kästner, and Johann Karsten are representatives of this new approach.

For France, it was a new type of school established by the Crown that rivaled the functions of the universities, and eventually made them obsolete: the military schools. For the young noblemen who prepared for a career as a military officer, some mathematics teachers had already been attached to regiments in the seventeenth century. Professional training was first institutionalized for future artillery officers. In 1720, the Crown founded écoles régimentaires d'artillerie for five garrisons of artillery regiments. Each of these schools employed a mathematics teacher. Among these were also known mathematicians like Bernard Forest de Bélidor (1693–1761), Sylvestre François Lacroix (1765–1843), and Louis François Antoine Arbogast (1759–1803). Mathematics formed the central part of theoretical training and shortly afterward as well of the entrance examination the aristocratic candidates had to pass after 1755. For this exam, a new function was created, that of the examinateur permanent, held by members of the Académie. Other military schools for young noblemen, but at a lower level, i.e., for officers' careers in infantry and cavalry, requiring less mathematical knowledge, were likewise founded in this period.

For future naval officers, there were, besides the royal chaires in Jesuit collèges, naval schools proper, employing teachers for training in mathematics and engineering. The formalization of these officers' formation led to establishing private preparatory schools, in particular because of the entrance examinations with their increasing focus on mathematics. All this made for an ever stronger presence of mathematics in the general culture.

The teaching of mathematics undoubtedly attained its highest and most innovative level in this period at the *École du Génie* in Mézières, the school founded in 1748 for training military engineers, in particular in fortification technology, the military formation valued as particularly *savant*. Both the entrance and the final examinations at this school were assigned to the same *examinateurs permanents*: to a mathematician and member of the Paris academy.

The Catholic states had become a privileged region for realizing particularly profound reforms. On the one hand, the time lag to be caught up in order to satisfy the requirements of mathematics and the sciences occasioned by mercantile policy, on the other hand, the dissolution of the Jesuit Order required the Crown to take urgent substitutional action.

The most radical structural reform for mathematics was realized in Catholic Portugal. Among these comprehensive reform measures was the complete reorganization of its only university, in Coimbra, in 1772. The university was now structured into six faculties of essentially equal status. Besides the traditional faculties for theology, for civil law and for canonical law, and for medicine, two new faculties were created, one for philosophy, and one for mathematics. The philosophical faculty was actually a science faculty. The mathematics faculty, like the philosophy faculty, was for the first time entitled to organize its own courses, with graduation rights of its own; in particular, teaching posts were reserved for its graduates. With the establishment of this first mathematical faculty, a specialized scientific education in mathematics was institutionalized for the first time.

2.7. First Foundational Reflections on Generalization

As shown in Section 2.5., Arnauld's and Prestet's textbooks form the starting point for the development of the modern textbook. Their work is a landmark in the process of generalizing mathematics, since it permitted mathematical productions to reach a larger audience. At the same time, however, they oritented the process of generalization in an even more fundamental sense. Their writings heralded the subsequent triumph of the analytic method. With their contributions to the familiar *querelle des anciens et des modernes*, the authors were pioneers in claiming priority for the modern age in mathematics as well.

In mathematics, the debate between tradition and modernity until the nineteenth century was to a large extent a conflict about whether the synthetic or the analytic method should prevail. This pair of opposites had been traditionally known from Pappus's (around 300 CE) *Collectio Mathematica*[45] and was treated as complementarity between the method of "composition" and the method of "resolution" for solving problems without attributing particular priority to one of the two poles. The unproblematic opposition turned, however, into a debated one due to Ramus's sharp critique of Euclid's Elements, and of their underlying synthetic method.

Petrus Ramus may be considered one of the most eminent humanists. A representative of modernization, he was a sharp critic of Aristotelianism and in particular of the Jesuits. Similarly to Melanchthon in Germany, he advocated for an enhanced role for mathematics in schools and at universities. Stinging rhetoric repeatedly made him a victim of persecution who had to flee the country. His conversion to Calvinism in 1562 was logical, but led to new persecution. In 1572, he was murdered in the Saint Bartholomew's Day Massacre.[46]

Ramus, who had himself endowed the *Collège royal* with a chair for mathematics, was intent on anchoring mathematics in society. His critique of Euclid was thus motivated didactically. His own volume *Scholarum Mathematicarum* (1569), the first methodological reflection on mathematics in print, examined why mathematics was held in so little esteem by scholars, and by the public at large. Mathematics, he said, was basically accused of two things (reproaches which were to be voiced in later epochs as well), which constituted the *pestifera duplex opinio*: of *inutilitas* and of *obscuritas* of mathematics (Ramus 1569 [quotations after the edition of 1599], 39). Ramus discussed the two charges in two extensive chapters. Convinced that the utility of mathematics would be admitted by many, he considered *obscuritas* the more serious problem by far. Even those who recognized mathematics' utility, he said, were still persuaded that it was incomprehensible (ibid., 39 and 72). As the cause of this general view, Ramus identified the classical text for mathematics instruction, precisely the [then] 15 books of a text surpassing in obscurity anything else ever written by a human hand: Euclid's *Elements* (ibid., 72). Euclid's work, he said, had nevertheless been estimated to be above criticism, and hallowed all over the world, for almost 2000 years (ibid., 74). Remarkably, Ramus interpreted the familiar anecdote where one of the Ptolemies asked Euclid for a simpler approach to mathematics, getting the answer that there was no "royal road" to mathematics, as proving the unintelligibility of Euclid's *Elements*. That there was no royal road was the guideline for Ramus's critique of the *Elements*.

In contrast to later critiques of Euclid, Ramus did not criticize individual mathematical or methodological problems. He did not hold any mathematical errors against Euclid for the simple reason that the latter, as Ramus repeatedly confirmed, was not known for any mathematical discoveries, but only for giving

---

45   In the preface to Book VII (Pappus/Jones 1986, 82–83).
46   An extensive literature analyzing Ramus's works has been published. With regard to mathematics one should mention M. Cantor 1857, Hooykaas 1958, Verdonk 1966.

proofs and for generally presenting things anew (e.g., ibid., 82). Rather, the reason for Euclid's overall *obscuritas* arose from the *Elements'* matter and form. The first part of Ramus's critique is a methodological inquiry into the subject matter. Thus, he said, some parts were absent that should necessarily be present in any Elements. In particular, Ramus criticized that there was no introduction of the four basic operations for integers, and for fractions. How was some learner, yet ignorant of addition, subtraction, multiplication, and division, to learn mathematics from Euclid's *Elements* without these parts (ibid., 82–83)?

The next aspect of *obscuritas* in Euclid concerned the relationship between logic and grammar to mathematics, which for Ramus was still central. In this relationship, Ramus held "redundantia" to be a major fault: firstly, by the introduction of superfluous terminological differentiations, secondly by unnecessary repetitions (ibid., 84 ff.).

Ramus saw the main cause for *obscuritas*, however, in a methodological and epistemological fault of Euclid's textbook that Ramus was the first to discuss, i.e., the *Elements'* very structure, which went against the most elementary rules of a methodological architecture. For the ideal he strove for, Ramus coined the keyword of "architectura methodica" (ibid ., 94).

The central issue of Ramus's critique was that Euclid's *Elements*— notwithstanding the high esteem in which the work was held, was no model at all for methodological order and logical consistency. For our purposes, it is important to emphasize that Ramus's major criticism stressed that Euclid gave no thought at all to the process of generalizing mathematical knowledge. Ramus understood arithmetic to be the more fundamental ("prius"), more general and simpler discipline of mathematics, whereas he saw geometry "by its nature" as a particular discipline based on arithmetic, while Euclid, by contrast, had begun his *Elements* with geometry (ibid., 97). Even in geometry, Euclid had not adhered to the basic rule of developing the general before developing the particular (ibid., 98–99).

Ramus's critique of method was continued and even intensified by Arnauld and Prestet. It is typical for these two authors, who were to become so eminent for the further development of mathematics in France, to have been the first to take up this issue as well. And it is just as typical for these two advocates of rationalism to have stressed different aspects again. More precisely, it was Prestet again who was more radical in this field. While he had to a great extent already opted for an autonomy of theoretical concepts in mathematics in his textbook, he was also the first in the field of methodology to go beyond criticizing the "Ancients" (i.e., Greek mathematicians) by energetically proclaiming the superiority of the "moderns" and of their analytic method. For all his biting critique of Euclid, Arnauld seems to have been more moderate in generally criticizing "les géomètres" (quoting Euclid merely as a case in point) while refraining from making a distinction between "ancients" and "moderns."

Arnauld's position shall be presented first. He demonstrated his own concept of method in detail in the fourth section of the famous *Logique* of Port-Royal.

He authored this volume jointly with Pierre Nicole, the fourth section being ascribed to Arnauld alone (Brekle et al. 1993, 513). This is where Arnauld introduced the significant innovation of distinguishing between method of research (*méthode de résolution*) and method of instruction or presentation (*méthode de doctrine*) instead of merely reflecting on a single method, as had been traditional, in particular in Descartes and Pascal:

> Thus there are two sorts of methods: The one to discover the truth, which is called *analysis*, or the method of *resolution*, and which can also be called the *method of invention*: and the other to make understood to others what one has found, which is called *synthesis*, or the *method of composition*, and which can also be called the *method of doctrine* (Arnauld, Nicole 1662, 368).

For this method of presentation, Arnauld established specific rules as well, as a part of his eight principles for "la méthode des sciences." According to the seventh rule, the method of presentation has to follow the natural, true order of things—and that means to start from the more general and the simple and to progress to the special (ibid., 408).

Arnauld's critique of Euclid's *Elements* is based on these methodological rules. Not only did he criticize partial aspects of Euclid's textbook by saying that the frequent use of proofs "par l'impossible," which, while persuading the mind, did nothing to elucidate, or saying that propositions were proved that did not need any proof, or that proofs were too far-fetched. What he criticized above all was the pervasive major fault of Euclid's neglecting the "genuine order of nature." The geometers' most important error, Arnauld said, was to confuse everything instead of adhering to natural order:

> It is herein that lies the greatest fault of the Geometers. They are of the opinion that almost no order would be preserved unless first propositions were able to be used to prove subsequent ones. And thus, without taking the trouble to employ the rules of the true method, which is always to start with things that are the most simple and the most general, in order to pass on later to more compound objects and to particular cases, they muddle up everything, and treat pell-mell lines and surfaces, and triangles and squares: proving properties of simple lines from figures, and they make an infinite number of other inversions which disfigure this beautiful science (ibid., 402)..

Arnauld claimed that this fault permeated Euclid's *Elements*; Euclid had begun by treating extension in the first four books, then generally switched to proportions for all types of magnitudes in the fifth—only to return, in his sixth book, to extension, treating numbers from the seventh to the ninth books, speaking of extension again in the tenth. To quote all the examples of this chaos (*désordre*), one would have to transcribe Euclid in its entirety (ibid., pp.402–403). Arnauld made this critique the basis of his own *new* elements, as he emphasized in the preface:

> [...] because the Elements of Euclid were so muddled and confused, that far from being able to provide the mind with the idea and sense of true order, they could only on the contrary accustom the mind to disorder and confusion (Arnauld 1667, X).

And Arnauld stressed that his own new arrangement not only facilitated understanding geometry by finding principles more fruitful and proofs better than those ordinarily used, but at the same time contained proofs self-evident from the principles established, and comprising a large number of new propositions (ibid., xii).

On the basis of this reflection on methodological procedure and structure, Prestet stridently advocated four new aspects:

- priority for the *generality of method;*
- greatest generality attainable only by *algebra* (which he called *analysis*);
- algebra underlying all the other mathematical disciplines;
- *analysis*, as developed by the moderns, thus being infinitely superior to the "geometry of the ancients" (Prestet 1789, v. I, [8]).

Already in the first edition of his textbook of 1675, Prestet had claimed priority of arithmetic and algebra over geometry. His reasoning is not only more extensive in the second edition of 1689—26 pages of the preface now as compared to the former 9—but also more programmatical, carried along by such a rhetorical burst of reformatory optimism that one is led to conclude that Prestet was sure of being backed by an entire group with conceptions of their own. This group may be assumed to have been that which centered around Malebranche, a group that had a key role in modernizing mathematics in France (cf. Robinet 1961 and 1967).

Prestet's leitmotiv was to develop the learner's intellectual abilities such as to enable him to invent new knowledge on his own: "inventer par luy-même." The necessary basis for this, he said, was to establish the general method, and to avoid getting mired in the thousand volumes of possible particular discoveries:

> The general method is what one ought principally to establish, without uselessly taking pains over all the truths one could discover. For in the end, if all the particular discoveries that could be made were to continue to be piled up in a thousand volumes, one would never finish and no lives of men would suffice to read them, since an infinity of them could be found. (Prestet 1689, I, [8]).

As faults of the "geometry of the ancients," he listed their lack of method, their accumulation of detail, and their attachment to sensual intuition. In their disregard of method, the "ancients" had even occasionally placed things upside down, since they sought less the natural sequence of things than how to use some of them to prove others, up to deducing the simplest things from the most complicated. Prestet proudly pointed out that the improved knowledge of the moderns had overcome the almost superstitious admiration of the works of antiquity:

> Besides the fact that their rather difficult and very limited method is not to the taste of a century where one is better taught [...] no longer affecting, as formerly, an almost superstitious admiration to all their works (ibid.).

Prestet's starting point is the "analyse des modernes" (ibid., [5]). For him, all truths are relationships, "rapports." The only science admitting "verités exactes," hence the perfect science, was mathematics. Only in this science were all *rapports* exactly determinable (ibid., [11–12]).[47]

As all relations could be expressed b y *numbers*, he said, the discipline teaching the necessary operations on numbers constituted the basis of all the other sciences. In particular, it is the condition of all application to *quantities*:

> All exactly known relations being therefore expressible by numbers, it is evident that numbers contain all magnitudes in an intelligble way, and thus the science which teaches how to make all the necessary comparisons in numbers, so as to know the relations, is a general science or the principle of all the exact sciences. For it is only necessary to apply to types of magnitudes what one has discovered in general for numbers, to have a knowledge of almost all particular sciences (ibid., [12]).

Arithmetic thus presented itself as a universal science, basic for numerous other sciences: "l'Arithmétique ou la science des nombres [est] une science universelle dont tant d'autres dépendent" (ibid.). There was another science, however, that was even more general and of greater extent, the *analyse* as developed by the moderns:

> This science which I call *Analysis*, and which is normally called *Algebra*, serves marvellously to enlighten, extend and perfect Arithmetic itself and Geometry, and all other parts which mathematics contains (ibid., [13]).

Due to its general character, this *analysis* at the same time already represents, for Prestet, the general method sought. It permitted the ease that allows one to discover concealed truths as well. And due to its short and simple expressions, it was an excellent guide for investigations (ibid).

In a mode without precedent, full of disrespect and reformatory zeal, he argued that the prevalence traditionally claimed for geometry was quite unsubstantiated, and that geometry was unambiguously second to *analysis*. *Analysis* went far beyond what could be inferred from viewing or from simple "imagination"; it was based on methodologically guided operations of the *esprit* (ibid., [11]). Its proofs were more general and simpler than the geometrical ones, and hence the more natural as well (ibid., [14]).

While *analysis* as a fundamental science need not borrow from other areas, geometry was *imparfait*; it could not do without the means and results of arithmetic and *analysis*. It is understandable from that why Prestet was able to call his textbook *Élemens des Mathématiques*, although it treated only arithmetic and *analysis*:

> It is evident that these Elements comprise the general science or the principle and the foundation of all of mathematics, and not geometry, which depends in several places on the knowledge of these Elements and which would undoubtedly be highly imperfect and very limited if it did not borrow support from the sciences we explained (ibid., [15]).

To give an example, Prestet referred to proportions: without them, almost nothing could be discovered in geometry. And to demonstrate the nature and the

---

[47]   The preface is not paginated, the numbers in square brackets give the corresponding numbers.

properties of these proportions, the geometers were obliged to take recourse to the multiples, equimultiples, and aliquota of magnitudes that one compared and could not determine without numbers. Without numbers, Prestet added, the geometers were unable to compare their lines and figures, the commensurables or incommensurables, except very imperfectly.

Arithmetic and analysis were by contrast entirely different. They could be extended to the infinite without any support by lines, figures, or anything. They were above all the other sciences, which had to rely on them for treatment by method, or for perfection. Prestet proclaimed that *analyse* was infinitely more fertile for discovering truths than geometry, and without its support an infinite number of geometrical problems were unsolvable:

> Analysis is infinitely more productive than figures for discovering truth, and it is quite impossible, if one does not engage its help, to solve an infinity of problems in Geometry (ibid., [15–16]).

The methodical and methodological reflection beginning with Ramus led to a first apogee with Arnauld and Prestet, who were the first to formulate the program of generalizing mathematical knowledge as a challenge postulating the preeminence of the analytic method. This is where the subsequent revolution of conceptual rigor was essentially prepared. The object of foundational reflection had to be the very mathematical fields that had already been sufficiently developed by contemporaries. While Ramus's reflections had still been essentially confined to elementary geometry, Arnauld and Prestet were able to include recent progress in algebra.[48]

2.8. Extension of the Concept Field to 1730/40

2.8.1. FRANCE

The conceptual clarifications achieved by Arnauld and Prestet and their presentation of negative numbers gradually spread in France. This dissemination was essentially effected by the work of professors belonging to the order of the Oratorians.[49] The Oratorians, strictly opposed to the Jesuits, can be generally considered to have been the promoters of the processes of algebraization and

---

[48]   Robinet's critique that Prestet did not include the new differential and integral calculus (Robinet 1960, 98 f., 104; Robinet 1961, 232 ff.), is anachronistic: the first research publication of this new calculus had occurred only in 1684, by Leibniz. Prestet died in 1691, while the first textbook appeared in 1696. Prestet's achievements concerning the proof of the uniqueness of the prime factor decomposition is examined by C. Goldstein (Goldstein 1992).

[49]   The role of the Oratorians for French educational history is studied by Lallemand 1888, and their contributions to mathematics by Belhoste 1993. The case of Angers, one of the few universities held by the Oratorians, is studied by Maillard (1975).

generalization within mathematics in France. It was in particular the group of the *Malebranchistes* at the end of the seventeenth and at the beginning of the eighteenth century, Malebranche's adherents who exercised a decisive influence on the Paris *Académie des Sciences*, and together with it on the development of science in France as well (cf. Robinet 1967).

Two members of the Oratorians, in particular, have the merit of having extended Prestet's pioneering program and secured broader acceptance for it: Bernard Lamy (1640–1715) and Charles-René Reyneau (1656–1728). Their highly successful textbooks were used until the second half of the eighteenth century.

Bernard Lamy, a priest of the order of the Oratorians and one of Malebranche's friends, was not only a disputatious theologian, but also an author successful both with grammar and mathematics textbooks. Between 1661 and 1668, he taught poetics and rhetoric. After 1673, he was active at the University of Angers as a philosophy professor. He was soon suspended, however, by intervention of the Jesuits and the Crown for his Cartesian views. Since he was also accused of antimonarchist teaching, he was no longer permitted to teach at a *collège* or at universities, but taught only intermittently in clerical seminaries (cf. Brekle 1993, 808 f.). This did not affect his productiveness as a textbook author. In mathematics, he published two textbooks, which, while applying Arnauld's and Prestet's concepts, were clearly more successful than the two.

The first of these volumes concerned algebra, which was understood as the basis of all mathematics: *Traité de la grandeur en général qui comprend l'arithmétique, l'algèbre, l'analyse et les principes de toutes les sciences qui ont la grandeur pour objet* (1680). The second was devoted to geometry: *Les Elémens de géométrie ou de la mésure du corps* (1685). The success of both textbooks is evident not only from the fact that each of them saw four editions during the author's lifetime (with changes and extensions, and in titles as well), but from the fact that there were further, even revised, posthumous editions for another half century.[50] The two textbooks could be also used independently of each other, because Lamy, in the later editions of the geometry book, had inserted a chapter on arithmetic covering the four basic arithmetical operations and their application to geometrical quantities—in particular to prepare the theory of proportions.

Lamy's textbooks are revealing documents for the process of reflecting the foundations of mathematics and its operations, and for the intention to create the ability to generalize. The enormous impact of Descartes's work on this process

---

[50]   To mention only the major editions: for algebra: 1680[1], 1689[2], 1704[3], 1706[4], 1731[5], 1738[6], 1741[7], 1741[8], 1765[9], and for geometry: 1685[1], 1692[2], 1695[3], 1701[4], 1731[5], 1734[6], 1758[7]. Moreover, there were numerous parallel editions, by other printers, in particular in Amsterdam (s. Brekle 1993, 806 f.). A study of Lamy's geometric concepts is Barbin 1994.

of generalization is very evident. Each of the two textbooks has a voluminous closing chapter *De la Méthode*. The textbooks' sections are for the most constructed as a sequence of rules; noteworthy is the respective major section *Les Principales Regles de la Méthode* which appear, in line with Pascal and Arnauld, as eight rules. These chapters show less fundamental intention than Arnauld's and Prestet's reflections on method; rather, they mark the transition to reflections pertaining more to intramathematical issues. Not only are the methodological parts more concretized on arithmetical/algebraic and on geometrical subjects and rules; at the same time, Lamy had repeatedly revised the respective parts on method in the different editions, substituting, in particular, the basic tenor of the discussion on *analyse* and *synthèse* as general methods present in the first editions by reasoning concerning more concrete mathematical subject matter.

Lamy shared Arnauld's and in particular Prestet's critique of the *anciens*: their works, he said, not only lacked *netteté* and *clarté*, they were also too long and too complicated—and above all they were deficient in methodological order (Lamy 1692, *Préface*, [12]). This is why he had tried to transform their demonstrations into "general" ones to make them prove several truths at once (ibid., [14 f.]). Since he was treating the *grandeur en général* in his algebra textbook, this book provided the foundation for mathematics as a whole and hence the true "Elements" of mathematics. Euclid had considered, Lamy said, only one "particular species" of quantities, i.e., the geometrical. This was didactically particularly dangerous, because such a textbook supported those who were forever in need of pictures and figures for their demonstrations. Imaging, however, was always a considerable cause for errors. His own textbook, against that, did not require "de se representer des corps," i.e. ,no figurative images (ibid., [16 f.]),

In his presentation of negative numbers, Lamy continued the previous stage of operative understanding; in particular, developments in the sense of opposite quantities can be noted; on the other hand, Lamy's concept was at first still characterized by an epistemological reserve similar to that of Descartes and Arnauld, which no longer appeared, however, in later editions.

In algebra, which Lamy understood as *Arithmétique plus parfaite* (ibid., 57), he introduced subtraction without restrictions concerning the relative sizes of subtrahend and minuend (ibid., 61 f.). This was intentional, for Lamy was at ease operating with negative results. He gave no explicit definition of positive and negative results; rather, he used a concept of opposite quantities mutually canceling one another. Thus he posed the problem, "when the magnitudes which must be added are the same and have opposite signs," hence, e.g., $2d$ and $-2d$ give 0, which he explained at length (ibid., 66).

Lamy justified the rule of signs in the section on multiplication of "grandeurs complexes ou composées," but unlike Arnauld, who did this by means of the conceptual differentiation between multiplier and multiplicand, in the traditional way according to which the excess subtracted must be restored (ibid., 72). Lamy had applied this rule already, when subtracting complex quantities like $c+f$ and $b - d$. One did not want to subtract from $c+d$ the entire b, thus $c+f - b$, but somewhat less. One thus had to change the algebraic sign of d from $-$ into $+$, so as to perform the operation $c+f - b+d$ (ibid., 68).

Nevertheless, Lamy still expressed epistemological reserve concerning this rule of signs, which shows that Descartes's and Arnauld's arguments were effective around 1700, a situation evident from Lamy's closing remark:

> It is not necesssary to search for any mystery here: it is not that *minus* is able to produce a *plus* as the rule appears to say, but that it is natural that, when too much has been taken away, one puts back the too much that has been taken away (ibid., 72 f.).

What is salient here is the contradictory situation of an operative practice of handling negative numbers coupled with an epistemological conviction of a strict separation between the "minus" area and the "plus" area, no transition from one to the other being possible. It is thus symptomatic that Lamy gave no definition of negative or opposite quantities, and restricted himself to introducing them only in their operative execution.

In the sixth edition of his geometry textbook (1734), he omitted the epistemological caution, in the part containing the concise version of arithmetic and algebra for the application in geometry. After establishing the rule of signs, he said briefly, without restricting conditions:

> Minus times minus therefore makes plus, that is, that at the end of the product there is the sign $+$, because having taken away too much, the too much that was taken away has to be put back (Lamy 1734, 132).

Evidently, the concept of negative numbers propagated by Prestet had meanwhile met with broader acceptance, the epistemological cautions and contradictions becoming marginal.

Lamy's textbook on *Grandeur en général* contains no account of how to treat equations of the second or higher degree. This is why Lamy's operative use of his concept of negative numbers cannot be investigated in this context.

While Lamy's textbooks can be considered to consolidate the already established practice of operating with negative numbers, Reyneau was able to advance the theoretical understanding of the concept of number significantly. Reyneau continued Prestet's work in several regards. Not only was he Prestet's immediate successor as mathematics professor at the University of Angers (keeping this chair from 1683 to 1705), but he also handled Prestet's scientific *Nachlass*. Moreover, Reyneau, an Oratorian like Prestet and Lamy, was one of Malebranche's close cooperators and friends. After giving up his Angers chair in 1705 because of deafness, he went to Paris, where he was elected an *associé libre* of the *Académie* in 1716.

During his scholarly leisure in Paris, Reyneau made his Angers lectures accessible to a basically universitarian audience in two influential textbooks:

- *Analyse demontrée, ou la Méthode de résoudre les problêmes mathématiques, et d'apprendre facilement ces sciences* (vol. 1: 1708, 2nd edition 1736; volu. 2: 1708, 2nd edition 1738); a textbook on algebra and on differential and integral calculus,

- *La Science du Calcul des Grandeurs en Général, ou les Élémens des Mathématiques* (vol. 1: 1714, 2nd edition 1739; vol. 2 posthumously 1736).

The *Science du Calcul* is a textbook on arithmetic and elementary algebra completely relying on Prestet's concept, as evidenced by its very title. In the preface, Reyneau stresses indeed that arithmetic and algebra form the "general science" of mathematics, and that one learning mathematics must begin with them (Reyneau 1714, xviij).

In his presentation of negative numbers, Reyneau closely adhered to Prestet as well, developing the concept of opposite quantities considerably further in several respects. For Reyneau, positive and negative numbers are not absolutely established as to their meaning, but are rather relative quantities that mutually refer to each other. The positive was no longer privileged from the outset. He motivated positive and negative quantities not only by applications to assets and debts, but illustrated them also by straight line segments of opposite direction ("sens") in geometry. In a figure (Figure 2), Reyneau not only showed horizontal segments as examples of such opposite quantities, but also vertical ones:

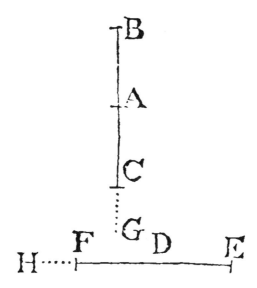

Figure 2, relative position of opposite segments (Reyneau 1714, 14)

Magnitudes can be distinguished as *positive* and *negative*. In commerce, for example, the fortune of a Merchant is a positive magnitude; his debts are negative magnitudes. With lines, and all magnitudes that can be represented by lines, in order to distinguish between the way a line like *CAB* should be understood, in going from bottom to top, and the way of understanding the same line *BAC* taken in the opposite direction, returning from top to bottom, the one taken in the one direction is called *positive* and the one taken in the opposite direction is called *negative*. Thus if we suppose the line *CAB*, going from *C* to *B* to be understood as positive, it will be *negative* taken in the opposite direction descending from *B* to *C*. In the same way, if *FDE* is taken as *positive*, in going from left to right, it will be *negative*, in going from right to left from *E* to *F* (Reyneau 1714, 14).

Reyneau has explicated Prestet's concept of opposite quantities further by a detailed discussion of the internal connection between positive and negative quantities, understood as "retranchements mutuels," as a mutual taking away, where quantities equal in size but of opposite directions canceled another out:

> From which it can be seen that these two sorts of magnitudes, positive and negative, are mutual subtractions of each other: for example, the positive magnitude *CAB*, going from *C* to *B*, being drawn, if the negative smaller [length] *BA*, going back from *B* toward *C*, is superposed, then *BA* is taken away from the positive quantity *BAC*, and all that is left of the positive [length] is *CA*; and if also the negative *AC*, which joined to *BA* is equal to the positive *CB*, is added, it will take away completely the positive *CA,* and zero will be left. If a negative magnitude *BG*, greater than *CB*, is superposed onto the positive *CB*, then the negative *CG* will be left over (ibid.).

With regard to the epistemological objections against connecting the positive area with the negative area, Reyneau's way of taking up and continuing Prestet's de-ontologizing zero must be considered a pioneering achievement. Instead of talking of a metaphysical nothing, Prestet had introduced the zero as an intermediate "term" between the positive and the negative quantities; Reyneau went beyond that in emphasizing that it was by convention that one or the other was called positive or negative:

> It is evident that zero, or nothing, is the term between the positive and negative magnitudes that separates them one from the other. The positives are magnitudes added to zero; the negatives are, as it were, below zero or nothing; or to put it a better way, zero or nothing lies between the positive and negative magnitudes; and it is as the term between the positive and negative magnitudes, where they both begin. For example, with the lines, the point *C* on which lie the negatives *CG*, is the term which separates them, where they begin, and from which they depart toward opposite regions. We call this term the *origin* of the positive and negative magnitudes; and at this term there are neither positive nor negative magnitudes; thus there is zero or nothing. In the same way, *F* is the origin of the positive magnitudes *FD*, *FE* which go to the right, and the negative magnitudes like *FH* which go to the left, and at the point *F* there are neither positive nor negative magnitudes; hence there is zero. It should be noted that, from the opposite directions of the positive and negative magnitudes, it is arbitrary in which way the positive is chosen, the negatives being taken in the other direction; but when one of these two directions has been chosen in a Problem, this should be maintained throughout the Problem (Reyneau 1714, 15).

Finally, it was a quite conceptual clarification as compared to his predecessors that Reyneau was to my knowledge the first to note the two possible functions of the signs + and − , and to reflect on their relation. While he introduced the signs + and − as algebraic signs of numbers and of quantities, however only after introducing the positive and negative quantities (ibid., 16), he explained afterwards, in his next section, that these "also" served as signs of operation, to indicate the operations of addition and subtraction. Reyneau then explained how to proceed if two signs concurred in their different functions of sign of operation and of algebraic sign (ibid.). For this purpose, he explained the conception of opposite quantities still further: indeed, the "−" as sign of operation actually required transition to the opposite quantity:

> From which it can be seen that the sign − before a magnitude simply indicates an opposite. If that magnitude, in front of which is the sign −, is positive or negative, the sign − indicates that one takes the opposite magnitude. Thus $− +a = −a$ and $− −a = +a$ (ibid., 17).

This last determination, however, implied a contradictory definition of negative numbers. While Reyneau had at first given a purely relational definition of positive and negative numbers, he now introduced—in order to explain the signs + and − as algebraic—a requirement already implied in Prestet's conception that presupposed using absolute numbers: One had to put the sign "+" before positive *quantities* and the sign "−" before negative *quantities*. He even added that one had to put the sign "−" before negative quantities in any case (ibid., 16).

Provided that this requirement did not resort presupposing the exclusive existence of absolute numbers, as in traditional mathematics, Reyneau's second definition for negative numbers signified a lack of reflection on the new generality of calculating with letters, as opposed to calculating with numbers. For in the examples that followed, Reyneau used only numbers to explain his definition, e.g., −2 as an example for the algebraic sign of a negative quantity, while in fact general quantities like a, b, can be equally positive or negative, depending of their respective admissible range of values.

Reyneau was virtually the only Frenchman after Arnauld to discuss problems of the concept of multiplication, again prompted by the rule of signs. To prove this rule, he deviated from his previous presentation in introducing the Cartesian concept of proportion for the product, explaining this with the unproblematic example of two positive quantities. He did not attain any conceptual clarification, however, since he relied only on Prestet's conception to derive the extended multiplication from subtraction, although he used Arnauld's differentiation between multiplier and multiplicand. Thus he tried to solve the really difficult problem of establishing a relation between the (positive) unity and a negative multiplier by simply positing that this negative number originated from the unity by subtractions:

> The positive unit +1 can appear, as it were, through subtraction in the multiplier, or rather it may be taken away when the multiplier $−a$ is negative. Therefore if the multiplicand $−b$ is also negative, it must be taken away from the product as many times as the positive unit +1 is taken away from the multiplier $−a$ (ibid., 70).

Reyneau extended operating with negative numbers to exponents as well, forming series with ordered negative exponents (ibid., 128 ff., 135). In the section "Extracting roots from literal numbers," he showed that a^2 has two roots:

$+a$ and $-a$, and that this multiplicity could be expressed by the sign $\pm a$ (ibid., 205).

Reyneau did not treat solving equations in his work *La Science du Calcul*, but did so in the parallel textbook *Analyse demontrée*. This is where he confirmed that he understood quantities at first as absolute, and thus as having positive values: "[…] thus in $x = a$, the root a is positive; but when the value of the unknown is negative, as in $x = -b$, we say the root is negative" (Reyneau 1736, 57).

The volume confirms that Reyneau indeed always understood coefficients as permitting only positive values. When treating equations of second degree, he hence distinguished positive and negative coefficients by different combinations of their algebraic signs; he devised thus six different equation types of second degree: four mixed types (where he accepted negative roots, also admitting the type $x^2 + px + q = 0$), as well as two pure types: $x^2 - p = 0$ and $x^2 + p = 0$ (Reyneau 1736, 58). For equations of third degree, he pioneered a simplification in combining (in his own sense) positive and negative coefficients with a $\pm$ sign into one equation. In spite of this he still obtained four types of equation, since he conceived of cases in which the terms of first and second degree disappear as of separate types (ibid., 59). Reyneau had no problems with operating on imaginary and complex numbers (cf. ibid., 197).

Thanks to a document from Reyneau's *Nachlass* published in Johann Bernoulli's correspondence, we have the fortune rare for this early period of being able to get a glimpse of Reyneau's "workshop." We can thus not only observe his learning process in a relevant field of problems, but also get an insight into the complexity of this field concerning negative numbers, which shows that the new developments of analysis substantially contributed to deepening the understanding of negative numbers. The notes taken by Reyneau document his newly acquired mathematical knowledge during a sojourn in Paris in 1700: "Mémoire de ce que j'ay appris dans mon voyage de Paris en juillet–aoust de 1700." The published note relates the new things in mathematics Reyneau had learned from Pierre Varignon, a leading member of the Malebranche group. It begins with the problem how to invalidate the proportions argument against negative quantities, which must have been relevant still for the group. The "solution" consisted in declaring the minus signs in the proportion to be inessential for the proportion itself, and in separating its qualitative meaning—like indicating an opposite direction—from the concept of ratio:

> On 14 July 1700, I learnt from Mr. Varignon that the ratios of positive and negative magnitudes of the same type are equal to the ratios of the same magnitudes all being taken as positive, the plus and the minus being only signs for calculating the magnitudes, that is to say, for adding or subtracting, and that the magnitudes, supposing they are lines, are on different sides with respect to the point they start from, that is, with respect to the origin. He proves it by the proposition $+2:-4::-4:+8$, which, according to all mathematicians, is true, the product of the extremes being equal to that of the means. Therefore if $\frac{+2}{-4}$ is that by which $+2$ exceeds -4, it must

be that in the equal ratio $\frac{-4}{+8}$, -4 also exceeds $+8$, which cannot be the case. Thus these ratios are the same as those between positive magnitudes, thus $\frac{+a}{-2} = \frac{-a}{+2}$, always half of a. (Joh. Bernoulli 1988, 349).[51]

In his textbook of 1708, Reyneau did not use this argumentation, but rather the more general conclusion that negative solutions indicate a value on the opposite side. This is what Reyneau called an inference not only for the common *calcul*, but also for differential and integral calculus. Indeed, the additional value of this "workshop report" consists in the fact that Reyneau testified how differential calculus and integral calculus had become the new area for applying the concept field of negative numbers and how results of this new *calcul* called for a clarification of hitherto controversial questions about understanding negative quantities. The note continues namely thus:

> Thus when working, either ordinarily or by the integral and differential calculus, a negative solution is found, it only indicates that the magnitude that gives the solution is on the other side of the origin, opposite to the one that was taken to indicate positive magnitudes.

> Thus in the quadrature of hyperbolas, or the area between the hyperbola and one of its asymptotes, if one finds that the sum of the parallel lines that fill that region is negative, this does not mean that the area is infinite, but only that it lies on the other side of the origin, and that it is the quadrature of the region that lies between the other asymptote and the hyperbola. For there are hyperbolas of certain degrees such that the hyperbola approaches one of its asymptotes more closely than it does the other, and that this area is less, with respect to one of the asymptotes, and greater with respect to the other (ibid.).

This interaction between negative numbers and the first results of integral calculus thus boosted the improvement of conceptions of analytic geometry and the representation of curves in the various quadrants.

Reyneau, with his comprehensive operative understanding of negative numbers, was indeed the first to develop a virtually complete analytic geometry in his second textbook. In his preface to the second volume of the *Analyse demontrée*, he had already emphasized "the perfect correspondence of the *analyse* with geometry and even with nature itself," demonstrating how the "lines and figures of geometry are better represented by letters of the alphabet" and how relations between lines and figures are transformable into a *calcul* by means of these letters (Reyneau 1738, iv f.).

This new understanding was based on conceiving the plane as describable by a coordinate system—after having chosen a straight line in the plane as the starting segment for the abscissae *(coupées)*, and of its beginning as the *origine* of the coordinate system. The infinite set of parallel lines intersecting the basic line segment, the *ordonnées,* together with the *coupées,* form the coordinates of

---

51   Such a general acceptance of this proportion had not taken place, as already shown and as will be discussed even more extensively!

the curve to be examined; at the same time, Reyneau adapted the new concept of variable in analysis for analytic geometry, calling it *changeante* (ibid., vj).

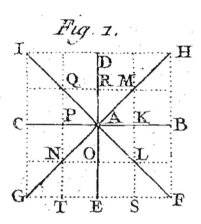

Figure 3, the four quadrants (Reyneau 1736, Planche I)

In a separate section "Sur l'usage des signes + et – par raport à la Geometrie," Reyneau was probably the first to explicitly introduce the four quadrants in the coordinate system of the plane, a novelty that appears self-evident to us today. Using Figure 1 (our Figure 3), he explained, if the straight lines *DAE* and *CAB* intersect at right angles in point *A*, and if one is held to distinguish between parallels to *AB* going right or going left by the problem given, as well as to distinguish between parallels to *DAE* "descending" or "rising," one should call those going right positive, giving them the sign of plus, and call those going left negative, giving them the sign of minus. In contrast to modern convention, he called the descending lines positive, and the rising lines negative (ibid., 9). By means of this, he introduced four quadrants, always with referring to his figure 1. Due to the difference of convention, what is now our fourth quadrant formed the first quadrant for him:

> Calling the angle *EAB* the first, *DAB* the second, *CAE* the third, and *DAC* the fourth, the lines of the first will all be positive; of the lines of the second, those which go to the right are positive and those which rise are negative; in the third, those which go to the left are negative and those which descend are positive; and in the fourth, both are negative (ibid., 10).

As an example of how Reyneau applied this method to representing curves in the four quadrants, we reproduce his figure 22 (our Figure 4) which serves to examine a hyperbola's properties. The axes themselves show no designations, but it may be deduced from the designations of points on the curves that points lying symmetrically to an axis in the positive or negative area are indicated by lower and upper cases of the same letter, to show their correspondence.

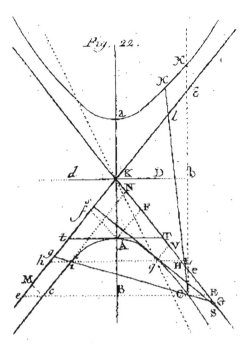

Figure 4, the hyperbola in the quadrants (Reyneau 1736, Planche II)

This introductory part for analytic geometry, however, also hinted at a new conceptual problem. In his "workshop report" of 1700, Reyneau had still interpreted negative areas to express a change of position. In his textbook, however, he now interpreted the product of a line oriented positively and one oriented negatively to be a negative area:

> The rectangle *AH* made by –*AD* and +*AB* will be negative. The rectangle *AG* made by +*AE* and –*AC* will be negative. But the rectangle *AI* made by –*AD* and –*AC* will be positive; and on the opposite side of the negative rectangle *AH* made by –*DA* and +*AB*. […] From which it can be seen that the areas which are on the opposite sides of the line which has been taken as the division between the positive and negative magnitudes are on the one hand positive, and on the other negative (ibid., 10 f.).

Reyneau did not reflect how his inferring a negative area relates to the fact that a positive result is obtained if one of the lines is substituted by its oppositely oriented counterpart. Nor did he argue conceptually or operatively with regard to the meaning of a negative area.[52]

Although Pierre Varignon (1654–1722), essentially a self-educated mathematician and physicist, belonged among the scientifically most productive members of the Malebranche group, his textbook *Élemens de Mathématique*,

---

[52]   In a *mémoire* for the St. Petersburg Academy of 1753, Heinrich Kühn tried to conceptualize negative values of areas as a real substrate for legitimating the construction of imaginary quantities in geometry (Kühn 1753).

published posthumously in 1734, contained no in-depth reflection of the concepts. Varignon was not only active as a scholar in the Paris Academy, but was an enthusiastic teacher as well, after 1688 as a professor of mathematics at the *Collège Mazarin*, one of the *collèges* of the Paris University, and after 1694 in addition to that as professor of the *Collège Royal*. His textbook, before its main part on geometry, contained a short introduction to arithmetic and algebra. Positive and negative quantities were not introduced explicitly; subtraction in the domain of algebra was not unambiguously formulated. At first, he limited subtraction to subtracting a smaller from a larger quantity (Varignon 1734, 12). In his formulation of calculations with letters, however, this restriction was not repeated (ibid., 14). His justification of the rule of signs is remarkable, however. The logical–linguistic argument that a double negation signifies an affirmation appears here for the first time: "The negation of a negation is an affirmation" (ibid., 21).

On the other hand, Varignon did not accept zero as a number; upon introducing the place value system, he explained that the digit 0 signified nothing in itself: "ce chiffre 0 qu'on appelle zero ne signifie rien seul; mais seulement lorsqu'il est mis après les autres dont il augmente la valeur" (ibid., 8).

The next textbook was authored by Dominique-François Rivard (1697–1778). Rivard belonged to no order and was also not a priest; he had just as little connection to the Academy in Paris. Rather, he was—for almost forty years— professor of philosophy and mathematics at one of the *collèges* belonging to the corporation of the Paris University, the *Collège de Beauvais*. Rivard represents the new normality of a university mathematics lecturer; his textbook was destined for his audience. In the preface to his *Elémens des Mathématiques* (1732) he thanked the university for having introduced mathematics into the philosophy course of its *collèges* some years ago.

The textbook, in its fourth edition already in 1744, consists of a first part on arithmetic and algebra, and a second on geometry. Rivard, as the most explicit among the authors hitherto discussed, introduced here negative numbers as quantities opposite to the positives. At the same time, he introduced negative quantities as equivalent to the positives, and as equally legitimate mathematical objects, in the until then most challenging mode:

> It should be remarked that negative quantities are magnitudes opposite to positive quantities. […] With this notion of positive and negative quantities, it follows that both are equally real and that, consequently, negatives are not the negation or absence of positives; but they are certain magnitudes opposite to those which are regarded as positive (Rivard 1744, 66).

And he introduced subtraction within algebra regardless of small/large relations (ibid., 69), while it was defined for arithmetic in the restricted manner (ibid., 17). Just like Reyneau, however, he made parallel use of the definitions of positive and negative quantities by algebraic signs for quantities understood exclusively as absolute numbers: "Non-complex quantities [that is, not composed quantities] which are preceded by the sign + are positive; and those preceded by the sign – are negative" (ibid., 66).

At the level of the general calculus with letters, Rivard used this parallel definition coherently without, however, differentiating between algebraic sign and sign of operation, as Reyneau had done. In particular, he used this definition to explain that a subtraction may have the form of an addition in the domain of algebra (ibid., 69). After having explained the rule of signs by numerical examples, he explained it without any epistemological reservations for isolated negative quantities:

> Suppose it is a question of taking away a negative quantity on its own; it is still clear that the sign must be changed from minus to plus: for example, if one wishes to subtract $-c$ from a, one must write $a + c$. For, to take away a negative quantity is to add a positive one (ibid.).

Although connected with the implied concept of exclusively absolute letter quantities, these reflections became productive for future conceptual clarifications, since they made one aware of the different nature of algebraic quantities as compared to the traditional concepts of arithmetic. In fact, just these reflections were later taken up by Condillac in his *Langue des calculs* to demonstrate the new, theoretical level of the operations in algebra (cf. below, Chapter IV.1.2.).

In Rivard's exceptional methodologically well elaborated and well structured work—thus he discussed for the basic operations in each case in a separate section how to perform the test of operation—there is another noteworthy innovation. Rivard is to my knowledge the first author who explicitly stressed that a^2 arises as a square both from $+a$ and from $-a$. (Rivard 1744, 93). In his third part, about solving equations, Rivard treated negative solutions. For the general method, Rivard also pointed out that there are two roots. From $2ax - xx=b$ one obtains not only the root

$$x = a + \sqrt{a^2 - b}, \text{ but also } x = a - \sqrt{a^2 - b}.$$

There is "également" a "racine positive" and a "racine négative" (ibid., 251 f.).

Rivard's textbook can be considered the presentation of negative numbers most vigorously elaborated among all the texts hitherto discussed in their theoretical and operational respects.

One must be aware that the textbooks analyzed here up to now are not representative, with regard to the degree of generalization they attained in forming the concepts of negative numbers, of the totality of mathematical production in France up to the first third of the eighteenth century. These textbooks, first written by "dissenters" like the Jansenist Arnauld, by the Oratorians, in particular by the Malebranche group, and finally, by the emerging group of mathematics professors at *collèges* of the universities, are more representative of a reform-minded group in France interested in promoting the sciences. Beside these, there were more authors of textbooks among amateurs, private teachers, and lecturers in other educational facilities who, while adopting the innovations, amalgamated them with traditional conceptions without more profound reflection.

Three examples of this broader circle of authors shall be mentioned here: Jacques Ozanam (1640–1717) was educated as a priest, but was predominantly active as a private tutor of mathematics—at first in Lyons, later in Paris. The author of "Dictionnaire Mathématique" (1690), he published a five-volume *Cours de mathématiques* in 1693. The first volume contained an *Abrégé d'Algèbre* (1702) developed more in detail in a separate volume *Nouveaux Elemens d'Algèbre*).

The lack of differentiation between algebraic sign and sign of operation caused Ozanam to generally understand any quantity to be a negative if provided with a minus sign—even for the subtraction of monomials: the second term to be subtracted, for him, was a negative quantity (Ozanam 1697, Vol. 1, 13). Ozanam claimed the subtraction of a larger from a smaller quantity to be absolutely impossible. He explained that only the inverse operation was performable in this case, and that one had to provide the result with the minus sign to indicate that it concerned a quantity that is smaller than nothing, hence a "grandeur fausse" (ibid.).

In contrast to his substantialist rejection of negative quantities, Ozanam justified the rule of signs by the new differentiation between multiplier and multiplicand, adhering in this to Arnauld (Ozanam 1702, 19; according to Dell'Aquila/Ferrari 1996, 330).

The textbooks by the Abbé Deidier (1696–1746) present another example. Trained in a *collège* run by Oratorians, he studied theology with the Jesuits. He became first a priest, then assumed a philosophy professorship in a seminary, teaching primarily mathematics. After that, he accepted the position of a private tutor for the children of a marquis, until he changed to the famous military school of La Fère to teach artillery (Hoefer 1855, Vol. 13, 31). He became the author of numerous mathematics textbooks for military engineers. His textbook on differential and integral calculus will be considered in Chapter III.7. His two-volume textbook *L'Arithmétique des Géométres* (1739) is relevant for this chapter.

In his first volume, Deidier presented the usual arithmetic as well as the commercial arithmetic; in his second volume he covered algebra, *analyse*—i.e., solving equations—and progressions, etc. He himself claimed to have treated everything with the greatest possible "ordre et clarté" (Deidier 1739, Avertissement). Within algebra, for "grandeurs littérales," Deidier devoted a short section to negative and positive quantities. Deidier used the traditional terminology of Descartes's time, calling positive quantities "réelles," and negatives "fausses" or the expression of a "défaut." While introducing both as opposite quantities canceling one another, "s'entredétruisent," he also designated them by the algebraic sign + for positive and by the algebraic sign – for negative quantities (ibid., 8). Deidier's comments reveal that negative quantities were strongly opposed in his time, particularly among the lay public. This opposition prompted Deidier to assert that nothing was simpler than these concepts when one applies his method of motivating them through analogy with debts, or the like:

> Many persons rise up against these negative magnitudes, as if they were objects difficult to conceive, yet there is nothing at the same time more simple nor more natural (ibid.).

He justified the rule of signs, as he said himself, by "la nature même de la multiplication," meaning Prestet's method of pretending to reduce all multiplications to iterated additions (ibid., 12 f.).

In his further algebra and *analyse,* Deidier reverted to pre-Arnauld textbook style as compared to the textbooks analyzed above. His major text was on presenting and solving particular problems, treating solving procedures for isolated cases without explicit theoretical or systematic structure. These parts do not require any negative solutions, and Deidier actually gave rules of thumb on how coefficients have to be chosen to make sure of obtaining only positive solutions (cf. ibid., 115). As coefficients could be only positive for Deidier as well, he presented four types of equations of second degree (ibid., 174). Somewhat later, he systematized the solutions, obtaining eight "general" solutions (ibid., 191).

Deidier's textbook is not only a revealing example for the possibility of different epistemological concepts coexisting in a "superposition" of various historical "layers." It serves at the same time as a first proof that developments of concepts and style need not occur in a "one-directional" and cumulative way, but that ruptures and reversions to forms prevailing at an earlier time may very well occur.

A third example is Bernard Forest de Bélidor (1697–1761), a double-career engineer who lectured at the famous military school *École de la Fère* and attained the rank of field marshal. In 1756, he was elected to the Paris Academy. Bélidor authored numerous textbooks for civil and military engineers, among them of his *Cours de Mathématiques* of 1725, which extensively treated the application of mathematics to the most diverse fields of practice in the manner of an encyclopaedia containing a "crash course" on the algebraic operations in the introductory part on geometry. Positive and negative quantities are introduced rather summarily, again as we already know, as quantities provided with a plus or a minus sign (Bélidor 1757, 11). Bélidor justified the rule of signs by differentiating between multiplier and multiplicand in a mode analogous to Arnauld's, albeit in an abbreviated form (ibid., 14). Similar to Rivard, Bélidor attributed to negative quantities a status equal to that of positive ones: "negative quantities are not less real than positive quantities, but are simply opposite to them: they can therefore by multiplied just like the others" (ibid., 18).

Bélidor applied this view in his discussion of equations of second degree to present an application of negative solutions free of problematizing. Noteworthy and rare among his contemporaries is his explicitness in explaining the existence of two square roots: one with a plus and another with a minus sign (ibid., 159).

In general, he said, the problem presented decided which solution to accept. Bélidor made a point of warning against neglecting or even suppressing negative solutions. Convinced of the analytic method, he pointed out how knowledge profited by these additional solutions, which would never have been found but for the analytic method (ibid., 159).

At the same time, his treatment of equations of second degree is a case in point for the then contradictory generality of mathematical treatment. While Bélidor always presented the general solution with a ± sign preceding the root, he gave like Reyneau six "general" forms for the equation of second degree, because the coefficients are always intended to be only positive numbers.

The development in France up to the first third of the eighteenth century can be summed up as follows: Negative numbers were acknowledged as legitimate mathematical objects; they were more or less explicitly understood to be quantities opposite to positive quantities. Practicing an operative calculus had progressed far; zero was predominantly no longer understood as a metaphysical limit. Conversely, negative quantities were acknowledged exclusively in algebra, while arithmetic, separate from algebra, was understood to be the domain of operating with absolute numbers. While algebra was defined, in contrast to arithmetic, as the domain of general operating on quantities, contemporary mathematicians understood the general letter quantities a, b, c, etc. to be confined to the domain of absolute numbers.

Eventually, the contradiction between the intended generality of algebra and the restriction of its understanding to absolute quantities became an element of the mathematical crisis as to whether negative numbers were mathematically admissible, which became acute in the 1750s, having been triggered, remarkably, by members of the Paris Academy.

Before discussing this critical development for France, developments in England and in Germany shall be examined for resemblances or differences.

2.8.2. DEVELOPMENTS IN ENGLAND AND SCOTLAND

The first English attempts to develop algebra can be dated with two textbooks published simultaneously in 1631: William Oughtred's *Clavis mathematicae* and Thomas Harriot's posthumous *Artis analyticae praxis*, both authors having been strongly influenced by Viète. The two authors are notable for their enthusiastic use of signs; their textbooks contain remarkably little "prose," inventing and applying a great number of signs. This is why Pycior judges them to have been the beginning of symbolic algebra in England.

Oughtred (1574–1660), educated at Cambridge University, was a parson doing mathematics as a private hobby. His textbook, which gave a summary of arithmetic and algebra in a mere 88 pages, had been written for private teaching (Cajori 1916, 17). Indeed, Oughtred distinguished between arithmetic and *specious*, i.e., algebraic, operations, introducing negative quantities in his *specious* subtraction. For practical purposes, however, he confines their use to polynomial expressions, as subtractive quantities. He used only positive quantities for solving quadratic equations. In contrast to that, Oughtred introduced negative exponents, marking them by overlined positive numbers, designating –1, for example, by $\overline{1}$ (Pycior 1997. 49 ff.).

Harriot (ca. 1560–1621), initially occupied as a surveyor after his studies in Oxford, later worked undery the patronage of the Earl of Northumberland, where he was in a position to carry out his mathematical, astronomical, and physical research. His textbook shows no special innovations with regard to the concept of negative numbers; negative equation roots do not appear. He even proves that only positive roots of equations are possible (Cantor II, 1900, 792, Pycior 1997, 58 f.). By contrast, Tanner's analysis of manuscripts from the *Nachlass* yielded that Harriot was the first mathematician to systematically experiment with the rule of signs. His *Nachlass* contains numerous pages with notes in which Harriot tried to establish a consistent connection between the alternative rule of multiplication—*minus* by *minus* gives *minus*—and the other basic operations (Tanner 1980a).

Harriot's approach to that was the "mixed" multiplication of plus by minus, respectively minus by plus. Because first attempts to calculate the result as plus were not satisfactory, Harriot introduced his own sign for mixed multiplication to facilitate his experiments: first a $\vdash$, and later a $\top$. Even then, however, Harriot was unable to attain consistent results: "no interpretation of the intermediary signs gives the desired result in association with the unorthodox 'minus into minus'" (Tanner 1980b, 134).

While Tanner considers this alternative experimentation only as "an entirely unique footnote to the history of mathematical notation in its conceptual aspect" (ibid., 128), the influence exerted by Cardano's *Aliza*, and its having been quoted by Commandino, an influence admitted by Harriot, shows that discussing an alternative concept of multiplication was by no means an isolated instance, but rather regular until about the year 1800.

Two other algebra textbooks that gradually enhanced the acceptance of negative equation solutions in England were also authored by nonprofessional mathematicians. There was the textbook *Teutsche Algebra* published in 1659 by Johann Rahn in Zurich, translated into English and published in 1668 as *An Introduction to Algebra* by John Pell a scholar who had already strongly influenced Rahn's German original, as well as the two-volume textbook *The Elements of That Mathematical Art Commonly Called Algebra* published in 1673–74 by John Kersey (1616–1701) (Pycior 1997, 88 ff.).

A clear break with these hesitant and tentative approaches was made with the ideas conceived by John Wallis (1616–1703), geometry professor in Oxford, one of the most eminent English mathematicians in the mid seventeenth century. While his contribution to the concept of negative number is frequently mentioned in the literature, this is mostly done with little understanding. A view is ascribed to him according to which the negative numbers are simultaneously smaller than nothing, and larger than infinite. Since Cantor (1900, 12), this proposition has virtually been alleged again and again to be the core of his conception (for the more recent literature, cf. Kline 1980, 116, and S. Haegel 1992, 12). What is disregarded here is the specific context of Wallis's considerations, which is that of the transition from $\frac{a}{0}$ to $\frac{a}{-1}$. The more essential thing in this is the context's object proper of integral calculus. In Wallis's time it was already well known that the integral of x^m is given by $\frac{1}{m+1}x^{m+1}$, and the area of ax^m between 0 and 1 by $\frac{a}{m+1}$. Wallis's concern in this part of his own *Arithmetica Infinitorum* of 1655, which has been so frequently reviewed in the literature, was to generalize this solution for negative exponents m. This attempt proves Wallis's ability to make advanced operative use of negative numbers.

To reconstruct Wallis's concept of negative numbers accurately, one must look where he presents the basic concepts of arithmetic and algebraic operations. This he does in his textbook *Mathesis Universalis* of 1656, where Wallis introduces subtraction (*subductio*) separately for arithmetic and for algebraic operations. While he declares it "impossible" (Wallis 1972, vol. 1, 70) to subtract a larger from a smaller number in the field of arithmetic ("subductia numerosa," Wallis 1972, vol. 1, 70), i.e., in the field of operating with positive numbers, he extends the area of numbers in his chapter on "Subductio Algebrica." While he initially introduces positive as well as negative numbers as the *quantitates* beset with plus, respectively minus, he continues by declaring them to be opposite quantities (ibid., 70 f.).

Subsequently, Wallis unfolds an extensive operative application of negative quantities and numbers, in particular also of "isolated" negative solutions, which is free of any epistemological reserve. He justified the rule of signs by an abbreviated version of an argument analogous to that of Arnauld concerning the functioning of multiplier and multiplicand (ibid., 104). Wallis was most probably also the earliest English textbook author to explicitly state that there are *two* solutions for every square number, thus for $r^2 = 27d^2$, equally $r = 3d\sqrt{3}$ as $r = -3d\sqrt{3}$. In solving equations, he thus also extensively used the signs $\pm$ and $\mp$ (cf. ibid., 233). Wallis did not present solving equations of second or higher degree in these instructional texts.

He presented the solving of equations in a later textbook on algebra, in his *Treatise on Algebra* of 1685, and in its extended Latin reedition of 1693.[53] In a

---

[53]   In this historically structured work, Wallis credits his compatriots Oughtred and Harriot with all the essential recent progress, to the detriment of the French mathematicians, in particular of Descartes. Wallis thus seems to have been the first case of nationalist behavior in the history of mathematics (cf. Scott 1981, 133 ff., and Pycior 1997, 128).
One of Wallis's major achievements in this second textbook is that he was the first to have given a geometric interpretation of imaginary quantities (Wallis 1693, 286–287; cf. Scott 1981, 162). In particular, he gave a description of the four quadrants of analytic geometry by means of combining the plus and minus signs of multiplication (Wallis 1693, 287 f.).

mode analogous to his earlier textbook, Wallis introduced negative quantities into the operations of *Arithmetica Speciosa*, explaining the rule of signs just as analogously as something following from "the nature of multiplication" (Wallis 1693, 78). In the solving of equations of second and higher degrees, discussing all the root values is the self-evident basis for him. He has no problem with treating, say, in case of quadratic equations, the case with two negative roots together with the other cases on the same basis (ibid., 141 f.). As he discusses different types of equations according to the respective signs of their coefficients, it is a total of four cases for the quadratic equations (ibid., 143).

Wallis argued often intensely in favor of the superiority of algebra over geometry, considering algebra as the representative of modernity (cf. Pycior 1997, 118 ff.). Quite in contrast to the "modernists" in France, however, Wallis was ferociously attacked for this in his own time; not only by the philosopher Thomas Hobbes (1588–1679), in whose empiricist philosophy geometry represented both the origin of mathematical cognition from the senses and their certainty, but also by his famous mathematical colleague Isaac Barrow(1630–1677), who saw geometry as developed by the "ancients," the Greeks, as the secure fundament of all of mathematics. In the controversies of Hobbes and Barrow with Wallis the negative numbers, however, had no role; they were recognized by both of his opponents (Pycior 1997, 143 and 160).[54]

The next influential English author is Isaac Newton (1643–1727). During his time as professor of mathematics in Oxford (1669–1701), Newton also held lectures on *Arithmetica Universalis*, in which he presented the very foundations with remarkable clarity and extensiveness. *Arithmetica Universalis* was first published as a textbook in 1707.

Before presenting the basic operations, Newton first introduced the various kinds of numbers and the signs. From the very beginning, he introduced positive and negative quantities (*quantitates affirmativae/negativae*), as larger, respectively smaller, than zero (*majores nihilo/nihilo minores*). Upon introducing the signs + and of – , however, he added that common use was to designate the negative quantities by a minus sign placed before them, and the positive ones by a plus sign placed before them, that is, speaking again generally of quantities, and not of numbers. The remarkable thing here is that Newton introduces the indeterminate sign of ± and its counterpart of ∓ (Newton 1964, 3 f.).

In a part separate from the signs, Newton then introduced the operative signs of plus and minus, considering operations like $-5 + 3 = -2$ to be quite self-evident (ibid., 4). He listed the zero without problem among the integers ("integrorum numerorum"). Before, he had stressed that he conceived of numbers not as a plurality of units, but rather more abstractedly as relationships (ibid., 2).

---

54   Cf. Cajori 1929, Scott 1981, Chap. 10; Pycior 1987, 135 ff.; Maierú 1994.

The basic operations subsequently represented are immediately treated across all the kinds of numbers (positive, negative, fractions, powers) introduced beforehand, actually both for numbers and algebraic quantities in the same respective section. For the operation of addition, Newton stresses at once that negative quantities are added just like positive ones; $-2 + -3$ thus giving -5. Algebraic subtraction is defined as unrestricted; only the signs were to be changed. The rule of signs is implicitly used, but not justified.

In the part that follows, *De extractione radicum,* Newton does not treat negative roots. This is done only later, in the parts on solving equations. In its introductory section *De forma aequationis,* Newton first explains the designations used, and for the first time introduces a designation for coefficients (without naming them as such): he was using the letters p, q, r, s, etc.; he said, for "arbitrarily other quantities from which the x [sought] is also determined, if theses quantities are determined and known" (ibid., 54).

It would seem that Newton, with these coefficients, really meant numbers from all the number areas he had introduced, for after this follow normal forms for the equations from first to fourth degrees. For each degree, he is the first to give but a single equation. The general equation of second degree, for instance, is given as

$$xx - px - q = 0 \text{ (ibid.).}$$

In particular, he explains that the general solution for $xx - ax + bb = 0$ is

$$x = \tfrac{1}{2} a \pm \sqrt{\tfrac{1}{4} aa - bb} \text{ (ibid., 58 f.).}$$

The parts following also show that Newton, going beyond Wallis, and in international and intercultural comparison as well, is the author to operate most comprehensively with negative quantities, giving them a legitimate function as mathematical objects.[55]

Another typical textbook is *The Elements of Algebra* (published posthumously in 1740) by the famous blind Cambridge mathematics professor Nicholas Saunderson (1682–1739). He adheres so closely to Newton in his own conception of negative quantities that details are unnecessary here. What is remarkable is the didactical ethic that prompts him to openly discuss all the possible objections against this conception. Saunderson introduces negative quantities as "less than nothing," explaining why this is neither a "very great paradox" nor a "downright absurdity." To substantiate this, Saunderson points to the difference between quantities and numbers; in particular, he shows how "narrow minds" are perplexed by not taking into account that there exist numerous quantities—such as bodies or substances—for which there is no opposite quantity (Saunderson 1740, 50 f.).

---

[55]   Although Pycior devotes an entire chapter to Newton's *Universal Arithmetick,* she did not pursue these concrete forms of Newon's operating in the concept field of negative numbers (cf. Pycior 1997, 192 ff.).

Remarkable is his justification of the rule of signs, which he presents quite extensively, fearing it might be difficult to digest, particularly by people with a weak constitution (ibid., 54 ff.). For this, he draws on arithmetical series and on how to preserve their properties in case of multiplication by a constant factor. In order to show that −3 multiplied by +4 gives −12, one was to consider the part +3, 0, −3 of an arithmetic series, first obtaining +12 and 0 in multiplication, and hence the last term had to give −12, etc. (ibid., 56 ff.).

The clear conceptual penetration of the concept field shows in particular in Saunderson's presentation of the theory of equations. For the quadratic equations, he shows that all equations can be reduced to only one normal form: to $Axx = Bx + C$. Saunderson was the first to explicitly assign both positive and negative values to the coefficients A, B, C (ibid., 172).

The next important work is the *Treatise of Algebra* by Colin MacLaurin (1698–1746), mathematics professor in Aberdeen and later in Edinburgh, the textbook being published posthumously in 1748 from MacLaurin's *Nachlass*. In an excellently explicit way, his textbook develops an operative conception of opposite quantities. He introduces the two basic operations of addition and subtraction as operations having contrary effects. Quantities are either additive, or subtractive:

> Hence it is, that any quantity may be supposed to enter into algebraic computations by two different ways which have contrary effects, either as an *increment* or as a *decrement*; that is as a quantity to be added or to be subtracted (MacLaurin 1748, 4).

And the operations are unrestrictedly executable. If a is smaller than b, "then $a - b$ is itself a decrement" (ibid.). Just as addition and subtraction are opposite, there is an analogous opposition between the quantities' "affections," which are considered in mathematics. MacLaurin calls additive quantities positive, and subtractive quantities negative: "they are equally real, but opposite to each other" (ibid, 6). Negative quantities are thus defined by the operation, not by a sign of an absolute number.[56]

MacLaurin is very precise in describing the significance of the mutual cancellation:

> When two quantities equal in respect of magnitude, but of those opposite kinds, are joined together, and conceived to take place in the same subject, they destroy each other's effect, and their amount is *nothing* (ibid., 5).

MacLaurin also emphasizes the difference between numbers and quantities. While abstract quantities can be both negative and positive, he says, concrete

---

56   Glaeser, who wrongly and with a wholly inadequate approach believed himself justified in attributing to MacLaurin an inability to master two "obstacles" in the concept of negative numbers (Glaeser 1981, 317), drew on MacLaurin's French translation of 1753, which indeed shows remarkable deviations from the original in its re-interpretation to suit the French conception prevailing at the time. The translator, Le Cozic, teacher at military schools and successor to Bélidor and Deidier in La Fère, took liberties with the text, abbreviating some parts and reformulating others to change their meaning, as in inserting a definition of the positive and negative quantities via the signs of absolute quantities (ibid., 316 f.).

quantities are not always capable of assuming the quality of being opposite to each other (ibid., 6). Just as explicitly, MacLaurin elaborated the conceptual distinction between a quantitative aspect and the necessarily associated qualitative aspect of opposite quantities: equality in algebra, he said, required not only equality of quantity, but also of quality (ibid.).

He makes a point of characterizing negative quantities as just as legitimate as positive quantities: "But a negative is to be considered no less a real quantity than the positive" (MacLaurin 1748, 7).

It is also notable that MacLaurin gives a definition of *coefficients*, actually as that number with which a letter quantity appears multiplied. This number states how often the quantity represented by the letter should be taken.

The introductory parts on quantity types are followed by chapters on the basic operations. In treating addition, MacLaurin already presents cases where addition quite self-evidently gives negative results, such as adding a larger negative term, or as an addition of exclusively negative terms. Quite novel, too, is his presentation of the rule of signs. He does not claim to prove the rule of signs, but only to "illustrate" it by arguments. The argument is based on distributivity: it followed from $+a - a = 0$ that $+na - na = 0$. And it resulted for $(+a - a)(-n)$ that the first factor was $-na$, and the second factor need hence be $+na$ (ibid., 12). To my knowledge, MacLaurin also was the first to elaborate the difference between multiplier and multiplicand in the rule of signs: the multiplier, he said, was always regarded only as a number (hence also the letter n in the rule of signs). For quantities of any kind could be multiplied only with numbers, but generally not with quantities (ibid., 13 f.).

For his chapter on powers and roots, MacLaurin clearly took the multiplicity of roots—their having positive, negative, and imaginary values—as his basis. He insists on permitting the possibility of all the roots for an equation being negative (ibid., 144). For equations from first to fifth degree, he explicitly gives the normal form—in each case with one equation only; that is, the coefficients are understood to have a general range of values. MacLaurin even used analytic geometry for applying his concept of negative numbers. In the closing third part of his textbook, about how to apply algebra and geometry to one another, he uses in his figures all four quadrants, while inserting coordinates, even if he does not name the axes. Line segments of equal length, e.g., in the first and third quadrants, are designated by y, respectively by $-y$ (cf. ibid., 318).

MacLaurin's textbook is an excellent example of conceptual clarity and broad operative use of negative numbers and quantities.

At the same time, however, MacLaurin is another example of the fact that a mathematician's work need not be coherent as a whole. In his textbook *A Treatise in Fluxions* published as early as 1742, we find a section about general characters in algebra in which he discusses whether it is possible and whether it makes sense to use generalizing signs in algebra. This discussion was necessary for MacLaurin, since he had attempted to prove the validity of the Newtonian methods of calculus with purely geometrical arguments (MacLaurin 1742, 575) in the book's main part. He began by distancing himself from approaches not specified by himself that had used a complication of symbols to conceal abstruse doctrines that were unable to stand the harsh light of geometrical form.

As a first example, he discussed "the use of the negative sign in algebra." He explained that it was necessary to differentiate between the absolute value of a quantity ("real value of the quantity") and its quality of having the potential to be opposite to another. He thus stressed the achievement of generalization this permitted because it had become possible to group together several cases, and to use their analogy (ibid.). He immediately went on from that to discuss the proportion argument not taken up by English authors as yet. He attempted to declare the problem of negative terms in proportions to be nonexistent by stating that proportions of lines depended only on their absolute magnitude, but not on their quality, i.e., not on their direction. This was true not only for lines, he said, but generally for all quantities. Hence, the proportion of the quantities $-b$ and a was the same as for the quantities b and a. While MacLaurin added a justification of plausibility for the rule of signs, he also made clear that a separate justification for it was not really called for, since the proportion $1 : -n :: -b : nb$ must agree with the proportion $1: n :: b: nb$ because of the fact that the absolute values were the same (ibid., 576 f.). It is very probable that this was where d'Alembert, who had most diligently studied MacLaurin's *Treatise on Fluxions*, found the reasoning most central for his own writings (see below, Section 2.9.3).

It is typical for the close connection between the concept of negative quantities and epistemological foundations that this first sign of a break with traditional views emerges in the context of a foundation of mathematics as purely geometric as possible—and not in the context of an algebraic one.

The four textbooks on which this section focused show over a period of one hundred years in England (and Scotland) a remarkably intense development of the concept of negative numbers in which the legitimacy of this mathematical concept was quite naturally assumed. Since all four authors were influential university professors, a wide dissemination of their views may be taken for granted. Their radical rejection beginning in England about 1750 is thus all the more surprising.

2.8.3. THE BEGINNINGS IN GERMANY

The first textbook author in Germany to adapt the modern style developed by Arnauld and Prestet was Christian Wolff (1679–1754). After his studies in Jena and his *Habilitation* in Leipzig, he became professor for mathematics and natural theory at the University of Halle in 1707. In 1723, he was driven away from there because of theological conflicts, finding sanctuary at the University of Marburg. In 1740, he was reinstated in Halle. His four-volume textbook *Anfangs-Gründe aller mathematischen Wissenschaften* of 1710 became, just as its Latin version *Elementa Matheseas Universae* of 1713, the most successful

German textbook of the entire eighteenth century, in ever new editions. The *Encyclopédie* highly praised this textbook for its quality (*Encyclopédie*, tome V, 1755, 497).

Indeed, the fourth volume, on algebra, and on differential and integral calculus, also shows an effort at conceptual clarity. Thus, Wolff reflected on the relationship between quantities and number, designating quantities as "indeterminate numbers" (Wolff 1750, 1551). Conversely, his book does not contain any independent reflection on the nature of negative quantities, and nothing going beyond the average level of contemporary literature in France. His remarks on that topic are even very brief. In the various reprints edited by him, Wolff did not change these parts.

After introducing the signs of plus and minus, Wolff introduces the negative quantities with the unrestrictedly executable subtraction, without, however, giving them a name of their own. He designates them in a substantialist mode as an expression of "defect," while the positive quantities state an existence:

> All quantities marked with the sign of – are defective, and against that those having the sign of + exist. If I am thus called to add of both kinds, the latter will level out the defect, although the addition has to be converted into a subtraction (ibid., 1557).

Wolff added as a footnote:

> The quantities marked with the sign of – have to be regarded as nothing else but debts, and by contrast the others bearing the sign of + as ready money. And therefore the former are called less than nothing, because one must first give away enough to settle one's debt before having nothing (ibid.).

In the practice of calculation, Wolff had no problem with the admissibility of these quantities. He continues by explaining how subtractions must be executed if "the larger must be subtracted from the smaller" (ibid., 1588).

In multiplication he justified the rule of signs, again very summarily, saying that one obtained the sign of plus in multiplication within a complex in order to level out again the too large "defect" (ibid., 1560). While he did not mention the plurality of roots in his treatment of extracting roots, he pointed out in the solving of equations the existence of two roots, also inserting them in further representation (ibid., 1588).

The Latin version of his algebra—he called it *Arithmetica Speciosa*—is, like other parts of the textbook, more concept-oriented. Here, Wolff uses the terms positive and negative quantities. And the definition he gives is familiar from France: quantities provided with a sign of plus are called *positive*, or *affirmativa*, respectively *nihil major*, and those preceded by a sign of minus are called *privativa*, or *negativa*, or *nihilo minor* (Wolff 1742, 299). Wolff was the first to connect this definition with an explicit mention of absolute quantities. In a corollary to his definition, he said how positive and negative quantities developed from absolute quantities (called *vera*, but not thematized before):

> Quantitas positive prodit, si vera aliqua additur, e.gr. $0 + 3 = +3$, $0 + a = +a$; privativa relinquitur, si quantitas aliqua vera ex nihili subtrahitur, e.gr. $0 - 3 = -3$, $0 - a = -a$ (ibid.).

Wolff preferred the designation *privativa* to that of *negativa*. Indeed, he assigned other qualities of being to the positive quantities than to the negative ones, which expressed a lack (*defectus*). Because of the different qualities, positive and negative quantities were heterogeneous to each other. Positive quantities were able to be in ratio to one another, just as negative quantities could be to another—but there could be no ratio between positive and negative quantities (no. 24, ibid., 300). Wolff justified this view, on the one hand, with his own definition of homogeneous and heterogeneous already presented in his part on arithmetic: homogeneous were only those quantities that, after being multiplied (*aliquoties sumta*), could exceed one another (ibid., 26). Since taking a defect several times, however, made the defect even larger, and could never exceed the positive quantity, positive and privative quantities were heterogeneous (no. 23., ibid., 300). Wolff also tried to justify with this that, e.g., $-3a$ did not relate to $-5a$ as $+3$ related to $+5$, and that 1 did not relate to -1 in the same manner as -1 related to 1 (ibid.). Wolff's intention in this reasoning was to invalidate Arnauld's argument of proportions (also adopted by Leibniz): the proportion $1 : -1 = -1 : 1$, he said, was formed of quantities heterogeneous to each other and hence inadmissible (ibid.). The correct starting point of his ad hoc argumentation was that proportions had originally been formed exclusively for positive quantities, and that nobody had ever tested whether they were applicable to the new number area.

On the other hand, Wolff linked positive and negative quantities in common operations, in contrast to his own view that they were heterogeneous to each other. Thus, he also gave a hint at the concept of opposite quantities: $+a$ and $-a$ mutually canceled out one another ("se mutuo destruunt," ibid.). In his Latin edition, he justified the rule of signs more extensively than in the German version, but only geometrically, by means of a comparison of areas within a parallelogram (ibid., 304). Although Wolff immediately adopted Newton's *Arithmetica universalis*, referring to it in the very first German and Latin editions of his own work, he was still strongly determined by Descartes's concepts. This is particularly evident in his part on solving equations. The plurality of roots is treated consistently, but the positive roots are called *vera*, and the negative roots *falsa*. Coefficients are always understood to be positive so that in quadratic equations, for instance, the cases

$$x^2 + ax = b^2, \quad x^2 - ax = b^2, \quad \text{and } x^2 - ax = -b^2$$

are solved separately. The fourth case, of two negative solutions, is not mentioned at all (ibid., 342 f.).

A deeper reflection of concepts is contained in the textbook *Elementa Matheseos*, by Christian August Hausen (1693-1743), a mathematics professor at the University of Leipzig. In the historiography of mathematics, Hausen is practically unknown today, but during the second half of the eighteenth century, the conceptual achievements of his textbook were pointed out again and again. We shall indeed get to know them not only when treating the concept of continuity, since his presentation concerning the negative numbers had an innovative effect as well. Thus, Wenceslaus J.G. Karsten (1732–1787), himself the author of an important textbook, in 1786 ascribed the prevalence of the modern conception to Hausen's "most excellent book," in which the elementary concepts of positive and negative quantities are from the very beginning presented in the most perfect light" (Karsten 1786, 243). And J.A.C. Michelsen (1749–1797), translator of Euler's textbooks, and intensely involved in clarifying the basic concepts, also underlined Hausen's role in 1789 (Michelsen 1789, 15). Hube, one of Kästner's disciples, even listed Hausen among the top mathematicians in a publication of 1759:

> What else could have moved the greatest measuring artists [i.e., mathematicians] like Newton, de la Hire, de l'Hôpital, Simson, Euler, Hausen, and many others, to place this theory [of the conics] in a better light? (Hube 1759, [v]).

Hausen introduced the subtraction $A - B$ unrestrictedly: for numbers just as for letter quantities. For the case $A = B$, the difference was 0, and for the case $A > B$ he stated with ease that the difference was negative. He represented negative differences as an expression for opposite quantities: opposite to another quantity that, although being of the same kind, was regarded as positive. Hausen tried to grasp the concept of opposition more precisely. It was a case of two contrary determinations, one of which involved the other's absence: "Determinationes oppositas hic quaslibet duas, quarum una involvit absentiam alterius, ut contradictorium sit utramque adesse" (Hausen 1734, 13).

As examples Hausen quoted the sun, which could not at the same time be a certain quantity over the horizon and below it, and the fact that a quantity could not at the same time increase and decrease within the same relation. Opposite equal quantities cancel one another:

> Quoties determinatiom ejusmodi concursus fit, effectibus oppositis se mutuo destruunt, et in statu ejus, quod afficerent solitariae, nihil mutant conjuctae. (ibid., 14).

Hausen supplemented this by an extensive discussion of opposition and mutual cancellation. He also used the term of "absolute quantities" here: positive quantities could also be regarded as "absolute" ones. Subsequent to the discussion of principles of the concept of opposition, Hausen explained the operations of addition and subtraction for opposite quantities.

Also very remarkable is Hausen's extended discussion of the concept of proportions. He used it to define multiplication as a proportion—the product as the fourth proportional to the unit and to the two given factors—in order to arrive thus without further justification at an extension of the admissibility of multiplication for opposite quantities (ibid., 3). In contrast to Wolff, but without mentioning the latter, he declared positive and negative quantities to be of the same genus ("ejusdem est generis," ibid., 14). And without taking recourse to the originally geometric meaning of proportions, he conceived of them exclusively arithmetically, as composed of arithmetic ratios, which he again conceived of, while applying his own concept of opposite quantities, as "rationes oppositae," which led to "numeri negativi" (ibid., 19–21). Hausen concludes that the expression $+1: -1 = -1: +1$ is an admissible proportion, and, more generally,

$$A : -B = -C : +D$$

as well, provided that A, B, C, and D are proportional. Hausen then applies this concept of proportion to justify the rule of signs for multiplication and division.

2.9. The Onset of an Epistemological Rupture

2.9.1. FONTENELLE: SEPARATION OF QUANTITY FROM QUALITY

The above description of how the concept of negative numbers developed in Europe did not offer any clue of an impending crisis, or rupture, in the direction this development was taking. Precisely such a rupture, however, occurred about 1750; it began in France and then spread to England, while there was no influence on Germany for a long time. The crisis originated with scholars belonging exclusively to the Paris *Académie des Sciences*. Their other common feature was that they did not do any teaching. The first theoretical impulse emanated from Bernard le Bouvier de Fontenelle (1657–1757), a further, educationally motivated contribution was made by Alexis-Claude Clairaut (1713–1765), and the decisive formulation, which had real impact, was provided by d'Alembert (1717–1783). Since previous developments give no hint at an imminent crisis, it is suggestive to search for a novel element that may have had a strong effect. This novel element will be found in the concept of *logarithms*. Discovered at the outset of the seventeenth century as an auxiliary computing tool, they had increasingly evolved into an object of mathematical theory in the course of the seventeenth century, having become indispensable for clarifying and making coherent the concept within the field of algebra's and analysis's foundational concepts.

Fontenelle, the Paris Academy's secretary of long standing, published a book in 1727 that was a provocation for many: *Éléments de la géométrie de l'infini*. The most provoking parts were those in which Fontenelle developed an algebraic calculus relying on infinitely large quantities. In spite of its title, the volume was not a textbook for purposes of teaching, but a scholarly contribution reflecting foundational mathematical concepts. That the author was perfectly aware that a real need for clarifying the concept of number had arisen can be seen from his discussing not only the concept of the infinite, but also devoting a separate chapter to the concept of number, and in particular to that of negative quantities.

Fontenelle was the first to examine not only negative numbers, but also the connection to imaginary numbers: *Des Grandeurs Positives et Negatives, Réelles et Imaginaires* is the heading of his book's *Section VI*. At first glance, this seems to be where Fontenelle intends to deepen the concept of opposite quantities. He argues against understanding "the negative" only as a subtracting, as a "retranchement" (Fontenelle 1727, 169). He is the first author to elaborate the differentiation between a *quantitative* aspect and a *qualitative* one, that of oppositeness:

> From this it follows that the idea of positive or negative is added to those magnitudes which are *contrary* in some way. [...] All *contrariness* or opposition suffices for the idea of positive or negative. [...] Thus every positive or negative magnitude does not have just its *numerical* being, by which it is a certain number, a certain quantity, but has in addition its *specific* being, by which it is a certain *Thing* opposite to another. I say *opposite to another*, because it is only by this opposition that it attains a specific being (Fontenelle 1727, 170).

Fontenelle explains oppositeness by the fact that one term cancels the other, negating it, and hence is negative:

> When two magnitudes are opposite, the one excludes or repudiates the other, and consequently is negative with regard to the other, which is positive (ibid., 171).

Upon closer analysis, however, one will note that things positive and negative do not assume symmetrical functions for Fontenelle. Basically, only the negative quantities are endowed just with specific quality, while the only positive quantities have the "privilege" of being endowed with numerical quantity. As is evident from Fontenelle's introductory reflections on the concept of number, there can be no sequential number series

$$..., -4, -3, -2, -1, 0, 1, 2, 3, 4, ...$$

for him, since the zero is not a quantity for him, as he most emphatically proclaims. Before, he had defined as "quantity" what lends itself to an increasing or a decreasing. Fontenelle obviously identifies the zero with the metaphysical nothing; the zero, he says, is no quantity, because it is not capable of being increased or decreased. "A nothing cannot be a greater or a lesser nothing" (ibid., 2).

The zero does not even have a numerical character (ibid., 171). For Fontenelle, the smallest number thus is the one. The one represents the element from which all the other numbers can be generated.

Fontenelle indeed refrains from constructing a general concept of positive and negative quantities as a new common concept of number. An essential condition remaining implicit for this is that he does not distinguish conceptually between numbers and quantities. Rather, he just states that there are no opposites for some quantities (that velocity, for instance, had no *opposé*), and that for quantities having such an *opposé*, the positive part may exist for itself alone—as a numerical quantity. The negative part, against that, necessarily contained the negative quality of oppositeness:

> The negative magnitude, when taken on its own, necessarily includes its specific being in its idea; but the positive, taken in the same way, does not include it necessarily, and so is only positive *improperly*, because it is considered or posited, but not *properly* and with respect to an opposite magnitude (ibid., 171).

Fontenelle's refusal to conceive of negative quantities as of things produced from *retranchements*, i.e., from subtractions, is on the one hand due to his having studied as well whether fractions with integers and powers with roots are able to form opposite quantities with regard to multiplication. In doing so, Fontenelle was the first to implicitly inquire whether there was a second kind of oppositeness—a multiplicative besides the additive one. Since he had neglected to closely inspect the "neutral" element required for this, his result was bound to be negative: neither fractions nor powers could belong to the *grandeurs opposées* (ibid., 172). Although the exponents n and $-n$ were opposed, he said, their powers were due to *retranchements*.

The second reason why Fontenelle assigned to the negative quantities a quality of their own separate from that of positive quantities lay, however, in a conceptual field linked with negative numbers: that of imaginary numbers. Fontenelle tried to grasp the conceptual difference between real and imaginary numbers more exactly. For this purpose, he analyzed the concept of multiplication, or more precisely of multiplication between numbers and quantities, while taking note of the difference between numbers and quantities, multiplication involving, in his own terms, *quantités revêtues d'une idée spécifique* (ibid., 176).

His point of departure was to reflect on the concept of multiplication. The only product he declares to be admissible is that of a scalar—"un pur nombre"—multiplied by a quantity. Conversely, a product of a quantity a with another quantity b made no sense; Fontenelle nevertheless arbitrarily posited that such a product ab could only be a scalar: one would have "necessairement" to take away the *idée spécifique* of quantity; the *idée spécifique* disappeared completely in the product ab, which was thus *purement numérique* (ibid., 175 f.). In case of a negative product $-ab$, such as one consisting of a debt $-a$ and a number b, against that, the specific idea subsisted: "But in the product $-ab$, the idea of specific being remains, and in effect, this idea is properly attached to negative magnitudes" (ibid., 176).

Fontenelle relied on such reasoning to show that a product $-aa$ could not be conceived of as a square: It was not a purely numerical square, for if a is a pure number, $-a$ is not a pure number. At the same time $-aa$ neither could be a square with an *idée spécifique*, because $-a$, understood, say, as debts, could not be detached from its specific idea because of Fontenelle's own willful positing, and hence could not be inserted in a product (ibid., 177).

This chain of reasoning eventually helps Fontenelle to attain his goal. If one takes $-a^2$ as a square, one is taking it for something it cannot be. If one intended, however, to take the square root $\sqrt[2]{-a^2}$, one would have to assume $-a^2$ to be a square. Hence, $\sqrt[2]{-a^2}$ was a completely *imaginary* quantity, which could not be real in any sense.

These concept determinations represent first tentative efforts, and as their author was not called on to make them explicitly coherent in his teaching, he had the privilege of being satisfied with what he himself had posited ad hoc. For later developments, however, it turned out to be crucial that he had determined negativity as a specific quality of oppositeness while neglecting, respectively reducing, the operative side of negative quantities. "Moins que rien," he declared, was neither a mathematical nor a physical concept, but rather "only a moral one" (ibid.).

2. CLAIRAUT: REINTERPRETING THE NEGATIVE AS POSITIVE

The next work in which we similarly find arguments against integrating the negative numbers into an overall conception of real numbers was, in contrast to Fontenelle's a textbook, explicitly addressing laymen. It was likewise authored by a scholar and member of the Paris *Académie des Sciences* who was not involved in any teaching, by Alexis-Claude Clairaut (1713–1765), a scientist mostly known for his research into mathematical physics. Clairaut wrote two elementary textbooks, both intended for a marchioness, and more generally for an elegant public intent on dabbling in leisurely mathematics without having to shoulder any real effort (cf. Glaeser 1983). The two textbooks on geometry (1741) and algebra (1746) attempt to realize a methodological approach according to which the respective mathematical concept field evolves in a seemingly "natural way" from simple inquiries or from useful problems.

While Clairaut himself did not have any problems with operating with negative numbers and quantities, and also quite clearly exposed the plurality of values for roots, such as the two values in equations of second degree (Clairaut 1757, 163), his major concern was to avoid scaring off beginners (*commençants*), and a particular stumbling block ("écueil," Clairaut 1797, 3) in his eyes was multiplication. Since he did not consider isolated negative solutions acceptable for *commençants*, he adopted part of Fontenelle's arguments in favor of separating positive from negative quantities.

For Clairaut, negative numbers did not represent a mathematical problem, but rather a didactical one. Of the fifteen pages of his preface, he devoted more than three and a half to describing his own approach of guiding beginners gradually toward an understanding of the necessity of operating with negative numbers, and of the appropriate rules—in particular the rule of signs. This path led to a discussion of the multiplication of isolated negative quantities: "des quantités purement négatives" (Clairaut 1797, 5).

Operations with negative quantities are indeed extensively presented in an educationally well-considered way. Thus, Clairaut explains the difference between "ajouter" and "augmenter" for addition, and between "soustraire" and "diminuer" for subtraction (ibid., 58 ff.), using more abstract terms that are less dependent on intuition with the intention of leading away from the inability to grasp the fact that adding a negative term will give a smaller result, and subtracting the same will give a larger. On the other hand, Clairaut assumed that the occurrence alone of operations with negative numbers—e.g., divisions $\frac{300}{-10}$

or $\frac{-400}{-10}$ —would involve the learners with metaphysical problems: they might fear executing "mauvais argumens métaphysiques" (ibid., 95). He thus showed other paths to attaining the result without such operations. Clairaut therefore was all the more convinced that his audience intuitively shunned negative solutions. He thus developed a method of interpreting negative solutions away. Already in his introduction, Clairaut had explained that he was liberating operating with negative quantities from everything "shocking," permitting the reader to recognize the nature of negative problem solutions. One should assume the unknown to be of opposite direction:

> When it happens that the unknown in a solution is found to be negative, it must be taken in a sense opposite to that which had been used in expressing the Problem (ibid., 6).

The textbook part did the interpretation by changing direction, but not at all developing this gradually, or "naturally," but rather as an abrupt positing. Clairaut had begun by developing the rule of signs in solving a system of equations concerning the following problem: two sources of different strength fill two different ponds within definite periods. For x and y, the volumes of water provided by the two sources, the results were -30, respectively $+40$ units (ibid., 94 ff.). Subsequent to the equation's "abstract" solution, Clairaut then inquired how the *autre espèce d'embarras* could be solved: the meaning of the negative value of x. For this, he explained, one had to go back to the problem's conditions, and that meant going back to the initial equations.

The remarkable thing, in particular as compared to later developments among French authors, is that Clairaut does not change the equations in order to attain positive values, but instead discusses extensively how the values obtained must be understood to have them satisfy the given equations. He advances as an interpretation posited by him without further explanation that the first source did not pour water into the pond, but instead drew water from it. Clairaut did not understand this interpretation of assuming the unknowns in an opposite sense as a rejection of the method of *analyse*, but rather as the confirmation of the generality of this method, which provided more results than originally intended:

> One sees on this occasion an example of the generality of analysis, which allows cases to be found in a question which one had not first of all anticipated as being able to be included (ibid., 99).

Clairaut thus admitted genuine negative values of unknowns, a fact relieving him from changing the equations, as subsequent authors found themselves compelled to do. He even extended this conception by permitting negative values as well for the coefficients—in contrast to his having defined them in his general definition of negative quantities as quantities preceded by a minus sign, as was the then current practice in France (ibid., 56). He did not substitute according to that, however, as later authors did, say, the coefficient b by $-b$, but inserted for b the negative value -3 (ibid., 100), after having generally explained before that it was permissible to take not only the unknowns, but also the "connues" in the inverted sense (ibid., 99).

It is not only the mathematical recognition of negative quantities that prevails in Clairaut; it is also of prime importance for him that the values satisfy the given equations. He was the first to directly tackle, in a textbook of modern times, the question of how to interpret negative solutions in equations with concrete quantities. Whereas his predecessors had stressed the admissibility of negative solutions, they had not progressed as far as this level of application.

3. D'ALEMBERT: THE GENERALITY OF ALGEBRA: AN INCONVÉNIENT

While the writings of Fontenelle and Clairaut contained new approaches in their thought on negative numbers and quantities, they did not become elements of crisis before d'Alembert's publications. These effected an acute and incisive rupture with everything developed up to the time. D'Alembert wrote articles in the *Encyclopédie* he coedited radically criticizing the then current conception of negative numbers for their false metaphysics, thus attaining a much greater impact than every mathematical author before him. He recognized virtually nothing but positive numbers as admissible mathematical objects. Just as radically, he rejected the generality given by algebra in the solving of equations, labeling it a "disadvantage." Jean le Rond d'Alembert (1717–1783), one of the leading representatives of Enlightenment in France, and at the same time among the most eminent philosophers, mathematicians, and physicists of his time, was a member of the Paris Academy of Sciences. He never accepted any other post, and in particular did not teach. His works on the foundations of negative numbers contradicted other aspects of his work in which he had advocated algebraization and generalization. This raises the question of what the mathematical and/or epistemological reasons led d'Alembert to reject what had been developed up to his time.

D'Alembert did not author a textbook on algebra, or any other integral presentation of his own view of algebra. There are, however, a number of treatises and articles that lend themselves to scrutiny. D'Alembert had no problem with the hitherto established rules for operating with negative quantities and numbers. Thus, he explicitly stressed that these rules were generally recognized to be exact: "the rules of algebraic operations on *negative* quantities are accepted by everybody and generally received as exact," independent of the meaning ascribed to these quantities (d'Alembert, *Négatif*, 1765, 73 left column). He saw the problem exclusively in finding the appropriate "metaphysics" of this concept. If one considered the precision and simplicity of algebraic operations with these quantities, he said, one was tempted to believe that the precise idea one had to ascribe to the negative quantities needs to be a simple idea, and must not be derived from a sophisticated metaphysics (ibid., 72 right column).

D'Alembert's book *Essai sur les Éléments de Philosophie* may serve as an introduction to the metaphysics of negative quantities, a book he wrote in 1758, during a time of crisis when his *Encyclopédie* project was in jeopardy of being wrecked by political opponents. He intended this book to comprehensively present his own views regarding the various fields of knowledge. In his chapter on *Algèbre,* he said that algebra was the leading science in mathematics, but noted at the same time that is was in some aspects not yet free of obscurities—at least in the current textbooks. In a footnote, he quoted as an example that he did not know any textbook in which the theory of negative quantities was *parfaitement éclairci* (d'Alembert 1805/1965, 291). In a subsequent *Eclaircissement sur les élémens d'algèbre,* d'Alembert made this critique more explicit. Again, he highlighted the negative quantities as an example of the lack of exactitude in the textbooks on algebra. He criticized not only the view of negative quantities as smaller than "nothing," and their interpretation as debts as too narrow, but also the view of understanding them as opposite to positive quantities, since geometry also provided examples showing that negative quantities should be taken in the same sense as positive ones:

> Some regard these quantities as *below nothing,* an absurd notion in itself: others, as expressing *debts,* a very restricted notion and for that reason alone hardly exact: others still, as quantities that must be taken in an opposite sense to quantities which are supposed to be positive; an idea for which geometry easily provides examples, but which is subject to frequent exceptions (ibid., 301).

His own answer to the question of what negative quantities really are is more extensively presented in several articles in the *Encyclopédie.* The best known among these is his article *Négatif* of 1765 in the eleventh volume, where he defines these quantities in the very introduction in the ways common in France, that is, as quantities preceded by a minus sign, but at once rejects the view advocated by "several mathematicians" that they are smaller than zero (*plus petites que zéro*)[57]: such an idea, he said, was incorrect (*pas juste*), as would be instantly seen (d'Alembert, *Négatif,* 1765, 72 right col.).

Actually, d'Alembert did not give a straight justification in this article, but tried to refute an argument given by an author not named who had said that 1 was incomparable to −1, and that 1 was in another ratio to −1 than −1 was to 1. The arguments quoted by d'Alembert correspond exactly to those given by Chr. Wolff (see Section 2.8.3), whose primary concern actually had not been to show that the negatives are smaller than zero, but rather that Arnauld's argument of proportions did not make sense because proportions are not defined for negative quantities. D'Alembert's refutation thus did not apply to Wolff's reasoning: While 1 was constantly being divided by −1 in algebraic operations, the proportions argument was about the geometric theory of proportions.[58]

---

[57]   This time thus not below "nothing," but smaller than "zero."

[58]   D'Alembert gave a direct justification in his treatise on the logarithms of negative numbers: while he recognized the proportion $1 : -1 :: -1 : 1$ as correct, on the basis of algebraic calculus, he had no qualms using the traditional geometric proportions argument: if the negative numbers were smaller than zero, there would have to be simultaneous validity of $1 > -1$ and $-1 > 1$ because of the sequence of terms (the text

D'Alembert derived his own determination of negative quantities from geometry. In geometry, negative quantities were often represented by real quantities, which differed "only" by their position with regard to a point they have in common on a line, as he had shown in the article *Courbe*. Consequently, one could infer that the negative quantities ocurring in the *calcul* were indeed real quantities. One had to ascribe to them, however, another idea than that originally assumed: negative quantities *indicated* **positive** quantities—the minus sign simply pointing out an error made in posing the problem and thus a false position requiring correction:

> Thus *negative* quantities really indicate in the calculation a false position. The sign – found in front of a quantity serves to redress and correct an error made in the hypothesis (ibid., 72 left col.).

As an example recurring in variations, d'Alembert quotes the problem of finding a number x that gives 50 when added (*ajouté*) to 100. The rules of algebra yield $x = -50$. This, however, showed that x really was equal to 50, the problem actually having to be formulated as follows: Seek a quantity x and subtract it from 100, a remainder of 50 being left (ibid.). D'Alembert summed this up to say that there are no isolated negative quantities as mathematical objects:

> Thus there is in reality and absolutely no isolated *negative* quantity: –3 taken abstractly offers no idea at all to the mind; but if I say that a man has given another –3 écus, that means, in an intelligible language, that he has taken 3 écus from him (ibid.).

D'Alembert did not reflect on how to maintain his view in the case of quantities for which there is no natural element of opposition as in giving and taking.

He was so radical in his view that he went so far as to declare the long-discussed problem of founding the rule of signs to be obsolete: to multiply $-a$ by $-b$, if one removed the error from the problem and formulated it correctly, was nothing but multiplying $+a$ by $+b$ (ibid.). Whereas later authors adopted many of d'Alemberts approaches and arguments, this argument (of MacLaurin) was never taken up again.

A further typical aspect is contained in the article *Équation* which appeared in the fifth volume in 1755. Its first third has been adapted from the corresponding article of the English encyclopedia, as d'Alembert notes himself, and the latter originated in its essentials from Newton's *Arithmetica Universalis*; it is on methods of solving equations.

After proving the fundamental theorem of algebra, he commented on the significance of the two roots for equations of second degree. While two positive roots solve the equation in the same sense, how should one assess mixed roots? For an example, d'Alembert selects the following problem: Sought is a number

---

shows, due to a printing error, $-1 < 1$); hence, no proportion existed (d'Alembert 1761, 201).

x, x being smaller than 1, for which $(1 - x)^2 = \frac{1}{4}$ holds. Solving first gives $1 - x = \pm\frac{1}{2}$ (this is one of the few times where the sign $\pm$ appears in d'Alembert's writings); hence $x = \frac{1}{2}$ and $x = \frac{3}{2}$. Among these, only $\frac{1}{2}$ solves the roblem, because of the condition $x < 1$. D'Alembert, however, continued the inquiry: why does one get another real, positive root? It was the solution, he said, to the problem, sought is a number x, x larger than 1, for which holds $(x - 1)^2 = \frac{1}{4}$ (d'Alembert, *Équation*, 1755, 849 f.).

The algebraic translation of the former task, he said, was from its nature more general than this former task itself; it contained at the same time the second task as well. D'Alembert now adds a polemic commentary, obviously directed against his fellow academy member Clairaut:

> Many algebraists consider this generality a richness of algebra, which, they say, answers not only what is asked of it, but even more what has not been asked of it, and what one has not thought of asking of it (ibid., 850).

This indeed corresponds to Clairaut's evaluation of the solutions found additionally (see above, Section 2.9.2.). D'Alembert, however, strongly rejected this program of generalization, which had been held in high esteem since Prestet. He was forced to admit, he said, that this richness and this generality presented an inconvenience for him: "For my part, I cannot avoid asserting that this alleged richness appears to me to be an inconvenience" (ibid.).

He considered particularly disruptive the case in which the solutions do not all have the same sign, but in which mixed positive and negative solutions appear. In these cases, he said, interventions into these equations were particularly difficult to carry out: "Negative roots, I repeat, are an inconvenience, above all those mixed up with positive ones" (ibid.).

Let us sum up what has hitherto been obtained with regard to d'Alembert's breaking with the traditional development of the concept of negative numbers:

- For him, *quantities* are the basic and initial concept of mathematics; *numbers* have only a derived status, and not an independent one.
- Since the classical quantities are only positive-valued, the mathematical zero and the metaphysical nothing signify the same; there are no quantities smaller than zero/nothing.
- Where isolated negative quantities appear in a solution, they indicate an error in formulating the problem.[59] In contrast to Clairaut, who was satisfied with a subsequent reinterpretation of the problem conditions, such an error, for d'Alembert, requires one to intervene into the problem formulation and to change the signs of the corresponding equation.

More generally, it can be noted that d'Alembert adopted and radicalized Fontenelle's conception of the two different *êtres* of negative quantities: they

---

[59]   Occasionally, d'Alembert emphasizes, without going into detail, that there is a second meaning of negative quantities still: they could also indicate solutions of the same problem under an aspect slightly different from that assumed in the problem, but nevertheless analogous to the latter's sense (e.g., d'Alembert 1805/1965, 302).

have a numerical quantity and agree in this with the positive quantities; at the same time, they possess quality, an *idée*. This quality, however, has no connection to the first *être*. It designates only a potential difference—one he did not consider relevant—of position with regard to the coordinate axis, as well as a command to remove this as if it were a normal quality by speedily shifting to positive solutions.

Let us return to the inquiry into the causes or motives for this epistemological rupture. D'Alembert himself has extensively described them in his treatise of 1761. The rupture was precipitated by a problem in applying imaginary quantities, a problem in conceiving of the logarithm function concerning negative numbers. D'Alembert wished to demonstrate the real-value character of these logarithms at any cost, and this compelled him to overturn the traditional conception of negative quantities. This reasoning of d'Alembert's developed into a controversy with Euler, and it can be reconstructed from their correspondence.

A controversy about the nature of the logarithms of negative numbers had already been conducted in 1712/13 between Leibniz and Johann I. Bernoulli. Leibniz had declared them to be imaginary-valued, whereas Bernoulli considered them to be real. No agreement was reached, but the public did not learn anything about their controversy. Only the publication of their correspondence in 1745 by G. Cramer made it accessible. D'Alembert's claim in a memoir of 1761 that he had got to know this correspondence only shortly before[60] is difficult to accept, since Cramer's edition certainly must have become rapidly known at the Paris Academy. Soon after this edition appeared at the beginning of December 1746, d'Alembert sent a *mémoire* to the Berlin Academy with results on integral calculus, the first part of which contained a proof of the fundamental theorem of algebra, and at the same time a text in which he explained his own view that the logarithms of negative numbers are real.[61] After Euler had presented his own conception in his reply of the end of December saying that the logarithm of −1 was not real, but imaginary, and even had many imaginary values, a controversy extending over two years ensued, in which no agreement was reached. D'Alembert's arguments in this controversy were virtually never taken up. Youschkevitch and Taton, who comment the Euler–d'Alembert correspondence in detail, note quite generally:

> We should silently pass over the latest arguments of d'Alembert, a certain mixture of reasons which he himself describes as metaphysical, drawn from an improper extension of the properties of ordinary logarithms, and of pseudo-geometric

---

[60]   Actually, d'Alembert wrote this treatise, published in 1761 in the *Opuscules*, already in 1752 (cf. Youschkevitch/Taton, in Euler 1980, 18).

[61]   The pricese wording of this text is not known, since Euler refrained from publishing it at d'Alembert's request, and the original manuscript no longer exists (information from the Archive of the Berlin-Brandenburgische Akademie der Wissenschaften). In his reply of December 29, 1746, however, Euler gave a brief rendering of the argument already used by Bernoulli (Euler *Opera* IV A, vol. V, 252).

considerations, etc. He lacks above all a good definition of logarithm (Youschkevitch, Taton , in Euler 1980, 19).

A similar position was taken by Verley, who briefly remarks in his description of the Euler-d'Alembert controversy that d'Alembert never attained an understanding of multi-valuedness (Verley 1981, 126). The connection to a conception of negative quantities had not been seen, or not been evaluated.

D'Alembert's reasoning in the correspondence extending over two years keeps turning around a point that seems to have been critical for him: whether $\text{Log}(-1)$ is real and whether in particular $\text{Log}(-1) = \text{Log}(1) = 0$, and after Euler's many exhausting mathematical argumentations he withdrew to asking whether one could not at least *assume* $\text{Log}(-1)$ to be real? It is indeed remarkable, on the one hand, how d'Alembert resorts to ad hoc constructions (such as admitting a negative unit for logarithms, as quoted in Euler *Opera* IV A, vol. V, 261) to maintain his own position, or challenges Euler's concrete and extensive reasoning only with some general, unsubstantiated doubts, such as questioning Euler's explicit description of the indefinitely many imaginary values of $\text{Log}(-1)$ as to whether this formula was actually complete and whether a real value did not nevertheless exist after all (ibid., 268 and 290).

On the other hand, Euler, too, had to concede points, and to revoke some things. Thus he was compelled to withdraw (ibid., 270) his own argument saying that e could not be negative for reasons of its expansion into a series alone (ibid., 264) after d'Alembert's objection that expansions into a series will sometimes yield false values, and not the totality of values (ibid., 267). Again and again, reasons went back and forth in their debate whether e^x was a unique or twofold function. First, Euler admitted that the function had two values for $x = \frac{1}{2}$ (ibid.) then d'Alembert conceded that e^0 did not have two, but only one value, adding, however, that one might indeed suppose it to have two (ibid., 273).

Eventually, Euler modified his standpoint with regard to the question so essential for d'Alembert as to whether e^x was multi-valued so extensively that he did not permit his own *mémoire* "sur les logarithmes des nombres négatifs et imaginaires," presented to the Berlin Academy on September 7, 1747, to be printed, but published a different version later in which this point in particular had been changed. As Euler wrote after some pause to d'Alembert on February 2, 1748, he had reflected further on e^x, having come to the conclusion that there were twofold values not only for $x = \frac{1}{2}$, but also generally for $x = \frac{n}{2}$, n being an odd number: there were thus any number of conjugated negative values, these existing, however, as isolated points, and no longer a continuous curve. Such a curve, he said, was present only above the axis (ibid., 280). D'Alembert concluded from that for his own purposes, however, it could therefore be stated that negative quantities could indeed have a real logarithm (ibid., 286).

In August 1747, Euler had referred d'Alembert to his *mémoire* presented to the Berlin Academy, which would soon be printed (ibid., 270). D'Alembert declared himself most willing to submit to these arguments, but after further objections from d'Alembert, Euler wrote back that his forthcoming "piece" would probably not remove all of d'Alembert's doubts (ibid., 275). Euler indeed withdrew this *mémoire* and revised it completely, while he wrote to d'Alembert, obviously at the same time, that he himself was not familiar enough with the issue of the imaginary logarithms any more to give well-founded answers to d'Alembert's new remarks; he thus had to refer him to taking up the matter again some time later (letter dated September 28, 1748; ibid., 293). Euler no longer hoped to be able to convince d'Alembert, and thus ended the controversy. Euler's revised version was published in the Berlin Academy's *Abhandlungen* of 1749 in 1751. In response, d'Alembert sent a *mémoire* to the Berlin Academy in 1752; since it remained unprinted, he published it himself in 1761, in the first volume of his *Opuscules*.

Euler's first version of his *mémoire*, the version he had given in the 1747 lecture in Berlin, already represented an excellent didactical elaboration, which made the then novel matter of distinguishing between areas of definition and infinite complex values comprehensible.[62] It contained, however, a number of allusions to contemporary representatives of Bernoulli's views (Euler *Opera* I, vol. 23, 421 and 425) that would have amounted to Euler making his controversy with d'Alembert public. Euler wished to avoid this, and in his revised version he was indeed in a position, in his review of Bernoulli's arguments, to discuss all of d'Alembert's objections, including the new ones, without naming him. The revised version of 1749/1751 (Nr. 168 on Eneström's list) represents a didactical masterpiece of a technical and unemotional reflection on all the difficulties having arisen as yet in the concept of the logarithmic function, and in its dissolution within the new concept of an infinitely valued complex function. He began with four justifications reflecting J. Bernoulli's position that the logarithms of negative numbers were real and equal to those of positive numbers. Against this, Euler raised six objections. Then he presentend, while quoting three justifications, Leibniz's view that the logarithms of negative numbers were imaginary and differed from those of positive numbers by an imaginary constant. Against Leibniz's views as well, Euler enumerated three objections. Since the respective objections could not be completely refuted, Euler's lucidly conducted demonstration showed that neither side was correct, and that the seemingly clear concept of logarithm must thus be contradictory!

By a subsequent theorem and by three propositions, Euler proved that the contradiction could be removed only by changing the concept of logarithm. To every number now corresponded an arbitrary number of function values, and only in the case of positive numbers was one of these values real (Euler *Opera* I, vol. XVII, 1915). Euler did not explicitly discuss his own concept of negative numbers in either of the two versions of his *mémoire*. He presented them later, in his textbook on algebra (see below, Chapter IV.1.1.).

---

[62]   This *mémoire*, Nr. 807 of Eneström's list, was published in 1862 by Fuss in the *Opera postuma*.

D'Alembert, in his answer of 1752 published in 1761, did not discuss Euler's new conception of logarithms in detail. Rather, he tried to insist by some general remarks that the logarithms of negative numbers *could also* be real, proposing indeterminate definitions of logarithm ("une suite de nombres en progression Arithmétique *quelconque*, répondans à une suite de nombres en progression Géométrique *quelconque*") without developing them operatively, declaring it to be unproblematic to assume $Log(nx) = Log(x)$. D'Alembert concluded that the logarithms of negative numbers could just as well be assumed either to be real or imaginary, the issue being dependent on the choice of system alone (d'Alembert 1761, 181–198).[63]

It is particularly revealing how d'Alembert made the connection between his argumentations on logarithm and his conception of negative quantities explicit here. D'Alembert again confirmed that negative quantities, for him, were essentially identical with the positive, they differed only, he said, by their "opposite" position; a property that he refrained from describing in more exact detail. He inserted it operatively to attain the goal he aspired to with the logarithm: $Log(-x) = Log(x)$. This at the same time yielded the real value of the logarithm of negative quantities: "*lx* and *l.–x* give the logarithmic [curve] two branches, which are equal and similarly placed with regard to the axes" (d'Alembert 1761, 195).

Also important for d'Alembert was Fontenelle's distinction between a quantity's numerical being and its specific, qualitative being, d'Alembert strictly separating the two types of being. He stressed that the negative quantities were "just as real" as the positive (ibid., 202), admitting that they form another progression 0, –1, –2, –3, etc., *qui revient* [...] *en sens contraire* to the progression of positive numbers to which it formed the "complément" (ibid., 187). The minus sign, he said, had no influence on the nature of these quantities:

> The sign – which the algebraic expression carries [...], only indicates its position, and has no effect on its quantity. [...] Negative quantities have no other difference from positive quantities than that they are taken on the opposite side (ibid., 202).

The oppositeness of position was of no consequence for d'Alembert, since he saw relations of quantities determined only via the first, numerical, being: "These magnitudes have no other ratios to each other than that of their quantities" (ibid., 202).

---

63   It has been neglected in the literature on the history of mathematics that all of d'Alembert's arguments against Euler by which he intended to prove his own proposition of $Log(-1) = 0$—even those put forward in obscure texts—have been discussed and refuted in detail, in a treatise by a mathematical outsider of 1801 written quite according to the conception professing dominance of the analytic method. Its author, A. Suremain-Missery, a former artillery officier, most excellently continued Euler's arguments of the study of the complex multiplicity of the logarithm values (in particular A. Suremain-Missery, 1801, 23 ff.). Although Grattan-Guinness mentioned the author's correspondence with Lacroix and with the *Institut* named this treatise, its significance escaped him (Grattan-Guinness 1990, 257).

D'Alembert made quite explicit his own epistemological point of view recognizing only the absolute values of numbers as numbers:

> There can be no ratio between $-a$ and b other than if one compares the magnitude of a with that of b; the sign $-$ is only a denomination. [...] In a word, every quantity on its own has the sign $+$ (ibid., 203).

For this view as well, an epistemological conception can be found to be implicitly determining in d'Alembert. As he had made explicit in his *Encyclopédie* article *Quantité* (*Philosophie*), he assumed the concept of quantity to be bound ontologically. In this article, he had criticized the view of quantity as a substance, since this substance then must also participate in a quantity's changes. D'Alembert was intent on excluding such changeability. He thus underlined that any quantity must be firmly bound to an object, without such a tie, a quantity was a pure abstraction:

> We imagine *quantity*, as an abstract notion, as like a substance, and increases and reductions as modifications, but there is nothing real in this notion. Quantity is not an attribute susceptible of different ways of determining it, some constant, others variable, which characterize substances. *Quantity* requires an attribute in which it resides, and outside of which it is nothing but pure abstraction (d'Alembert, *Quantité* 1765, 653).

In retrospect, we may state that Euler's and d'Alembert's efforts at attaining generality and rigor both had their price. An increase in rigor implied a loss in generality. Euler had to abandon his principle that (real) functions have to be understood as being generally defined, without stating a restricted field of definition. This point of view was a consequence of his own epistemological position of seeing generality represented by algebra. Since algebra assured general validity, deviations were considered to be exceptional values that did not detract from general validity. Fraser has succinctly characterized the identification of algebraization with generality as follows:

> In Euler's or Lagrange's presentation of a theorem of the calculus, no attention is paid to considerations of domain. The idea behind the proof is always algebraic. It is invariably understood that the theorem in question is generally correct, true everywhere except possibly at isolated exceptional values. The failure of the theorem at such values is not considered significant (Fraser 1989, 329).

For the loss of generality, Euler obtained in compensation a rigorous understanding of the logarithm function. In his seminal textbook *Introductio in analysin Infinitorum* (1748), he defined the logarithm only for positive numbers (Euler 1748/1983, §102, 76). D'Alembert intended to maintain the general definition of the logarithm at any cost, but had to accept a restriction of the area of real number in exchange.[64] It is remarkable that a later resistance in Germany against Euler's restriction of the field of definition went along with support for d'Alembert's conception of absolute numbers (cf. below, Chapter II.2.10.4.).

---

[64]   He was unaware of the consequence that he again restricted the area of the logarithm in doing so.

At the same time, d'Alembert is a case in point showing that a mathematician's epistemological views need not comprehend and "direct" his entire field of mathematical activity, but can also affect part of them without entering into conflict with the other parts and their possibly underlying epistemological conceptions. We noted at the beginning of this section that d'Alembert had acknowledged the entirety of rules of operation for negative quantities developed before. An even more far-reaching example is that d'Alembert developed algebraic geometry much more explicitly toward opposite quantities than Descartes did. While Descartes applied his own new method of coordinates virtually only within the first (positive) quadrant, d'Alembert very clearly demonstrated that complete gaphs of curves can be obtained only by taking all four quadrants into consideration. D'Alembert presented this further development of algebraic geometry consistently, in particular in his article *Courbe* in the fourth volume of his *Encyclopédie*, an article he later referred to with great pride as the state of affairs he had achieved. This is where he extensively justifies the necessity of including all positive and all negative axes in determining a curve (d'Alembert, *Courbe* 1754, 379). He stressed that the totality of positive and negative values form the necessary basic unit (ibid.).

D'Alembert's conception of algebraic geometry implicitly and unproblematically assumes that negative values are smaller than positive values, respectively that they lie "below" zero, and below the positive values. The figures illustrating the article *Courbe* in the accompanying *Planches* volumes of the *Encyclopédie* show the entire coordinate surface appropriately situated for curves. The coordinate axes are not separately designated; while points lying symmetrically to the axes are not designated as opposite ones by designations like $M/-M$, their oppositeness is hinted at by designations like M/m.

Generally, this section shows as it were quite incidentally how inappropriate the common notion of mathematics in the eighteenth century is: it is seen as a naïve advance unconcerned with foundations. Frequently, this attitude is ascribed precisely to d'Alembert, whose advice to a beginner, *Allez en avant, et la foi Vous viendra*, is interpreted as unconcern with secure foundations. By contrast, his controversy with Euler shows how intensely he worked on foundational issues. He must thus be taken seriously when he asked Euler not to consider him stubborn when he demanded to be enlightened in order to be convinced:

> You find me, perhaps, very importunate and opinionated to return again to the same things. But the truth can only be found by a great deal of patience and superstition, and I am only seeking to be enlightened so as to give up (Letter of 7.9.1748; Euler *Opera*, IVA, vol. V, 291).

2.10. Aspects of the Crisis to 1800

D'Alembert's break with the conceptual development of negative quantities up to his time had not the same impact all over, but one that varied depending on the respective institutional and cultural contexts of crisis. Notable in France, in particular, is the different effects in the two competing institutional contexts of the universities and the military and engineering schools.

2.10.1. STAGNANT WATERS IN THE FRENCH UNIVERSITY CONTEXT

The textbooks for the *collèges* of the French universities do not show any essential impact of d'Alembert's new conception. They continued to use the traditional concepts, albeit without reflection of their own, and even in a stagnant and bowdlerizing fashion, as compared to the earlier authors. That d'Alembert would encounter resistance here had become clear from the very *Encyclopédie* itself. It contained an article parallel to d'Alembert's own that held a contrary position: *Quantité, en termes d'Algèbre*. It was signed "E," which meant that this article had been written by Abbé Joannes B. de la Chapelle (1710–1792), at the same time author of an influential mathematics textbook for the university context: *Institutions de Géométrie* (1746). De la Chapelle was not himself a teacher of mathematics, but a *censeur royal*—a post salaried by the Crown for the examination of manuscripts submitted to determine whether they were to be authorized for printing with royal *approbation*, or *privilège*. The parallel *Encyclopédie* article classifies the *quantités algébriques* into positive and negative ones, determining their concept as follows: positive quantities shall be called those larger than zero ("au-dessus de zéro"), and negative those that *are* not smaller than nothing (as the positive *are* larger than zero), but *are regarded to be* less than zero—a formulation that most probably was a concession to the Encyclopedia's coeditor d'Alembert (*qui sont regardées comme moindres que rien*: de la Chapelle *Quantité* 1765, 655). The text then identifies positive quantities with additive, and negative quantities with subtractive quantities:

> All that is needed to produce a positive *quantity* is to add a real quantity to nothing; for example $0 + 3 = +3$; and $0 + a = +a$. In the same way, to produce a negative *quantity* all that is needed is to take a real *quantity* away from 0; for example $0 - 3 = -3$; and $0 - a = -a$ (ibid.).

De la Chapelle continued by discussing the view held by "some" authors unnamed according to whom negative quantities were "deficits" (*les défauts*) in positive quantities. Actually, this was but one author, Christian Wolff, whom de la Chapelle discussed just like d'Alembert, quoting him even more extensively than the latter to say that negative quantities were homogeneous among one another, but heterogeneous to the positive ones; one could thus establish, between $-3a$ and $-5a$, the relation $3:5$, but the relation of $1:-1$ was different from the relation of $-1:1$. De la Chapelle countered this Wolffian argument by saying that the proportion $1:-1::-1:1$ was valid, since the product of the outer components was equal to the product of the inner components. De la Chapelle thus added the remark that the concept of negative quantities was not *parfaitement exacte* in (some) authors (ibid.). To close, de la Chapelle listed the most important rules for operating with negative quantities.

In his own mathematics textbook,[65] he said much more, in the part on algebra preceding that on geometry. He did not present a unified conception, however, but merely juxtaposed fragments of different origin. His first determination of concept did not conceive of negative quantities as of subtractive ones, but as a general expression of any kind of subtraction, a conception that disregarded the differentiation between algebraic sign and sign of operation that had been clarified long before:

> Algebraic quantities preceded by the sign $+$ are called *positive*, and those preceded by the sign $-$ are called *negative*. The quantity $a + b$ shows that $+b$ is a positive, and in $p - m$ one can see that $-m$ is a negative (Chapelle 1765, 156).

Somewhat later, Chapelle gave a second determination of concept, which came from another conceptual tradition, defining algebraic quantities as opposite quantities. Here, Chapelle emphasized that positive and negative quantities were similar and hence capable of forming a common concept field:

> Positive quantities are directly opposite to negative quantities to which they are similar, and so these quantities wipe themselves out reciprocally (ibid., 159 f.).

And according to this conceptual tradition, Chapelle stressed that the two kinds of quantities were equally legitimate: "By consequence negative quantities are just as real as positive ones" (ibid., 160). As "proof" for the rule of signs, he used Arnauld's reasoning concerning the function of multiplier and multiplicand (ibid., 165 f.).

Chapelle did not develop anything further on his own. In his part on equations, he described the multiplicity of roots (ibid., 232 ff.). In another of his textbooks, in his *Traité des Sections Coniques* (1750), everything including the figures shows that he used the full system of coordinates and all four quadrants to examine and represent the curves, but points lying symmetrically to the axes are not characterized as correlating points, neither by opposite signs nor by analogous designations (like M/m in d'Alembert).

An even more trivializing conception is found in Abbé Sauri's (1741–1785) textbook *Institutions mathématiques*. This textbook, first published in 1770, saw five reprints—until 1834!—and was widely disseminated in the context of universities as well. Sauri was *professeur de philosophie* at the University of Montpellier, offering in this capacity the mathematics courses within the philosophy class of the *collège*. Sauri's texts no longer contain any hint at the

---

65   The first edition appeared in 1746, the last in 1765.

conception of opposite quantities, and no reflection on concepts either. Only one definition is given at the beginning of the part on algebra:

> Algebraic quantities preceded by the sign + are called *positive*, and those preceded by the sign – are called *negative*. The quantity $a + b$ shows that $+b$ is a positive, and in $p – m$ one can see that $–m$ is a negative (Sauri 1777, 36).

This definition, briefly illustrated, is followed by a naked presentation of the rules for operating with positive and negative quantities without any specifications or clarifications by the author himself. Thus, Sauri notes that it was inconvenient to confuse *augmenter* with *ajouter*: to add to b the term $–d$ gave $b – d$, that is, a diminuition (ibid., 39). He did justify the rule of signs separately for isolated quantities: by means of the differentiation between multiplier and multiplicand in Arnauld's sense (see ibid., 40 f.).

The deficit in determining positive and negative quantities by way of the algebraic signs of numbers implicitly assumed to be absolute has already been mentioned. It is thus no wonder that the wide dissemination of unsatisfactory conceptions eventually led to some kind of fundamentalist reactions (see below, Chapter IV.1.4.).

A first approach at conceptual clarification is finally found in a textbook first published in 1781 for the university context, in the *Éléments de Mathématiques* by Roger Martin. Martin (1741–1811), who obviously did not pursue extended studies, became a professor for philosophy at the *Collège Royal* in Toulouse at the age of less than 20 years. In 1782, he took the college's newly created post for experimental physics. As a freemason, he became at once politically active with the beginning of the Revolution, assuming many political and administrative positions in Toulouse. From 1795 to 1799, he was a member of the first chamber of the French Parliament *(Conseil des Cinq-Cents)*, getting involved mainly in questions of education. After that, he returned to Toulouse, became professor for physics at the *École centrale*, and when courses in natural sciences were no longer being held at the new *Lycées* after the École's dissolution in 1803, the city founded an *École spéciale des Sciences et des Arts*, at which Martin again taught physics. In 1808, when the *Université Impériale* was founded, this institution was integrated as *Faculté des Sciences*, Martin finally becoming a professor of this faculty.[66]

Martin's textbook was reprinted in a somewhat revised version in the year X (1802).[67] Martin was the first author in France to use the term "opposite numbers" (*nombres opposés*) and to make them the basis of his own conception.

Half of the 84-page *Discours Préliminaire* is devoted to discussing the concept field of opposite numbers. Martin was the first of the university textbook authors after d'Alembert to again undertake a comprehensive reflection on foundations. The innovation in his conceptions was first that he made the concept of number the basis of his arithmetic and algebra, the concept of

---

[66]   An exhaustive biography was published by Gros in 1919.

[67]   I was able to study this edition thanks to Colette Laborde (Grenoble), who has this edition in her possession.

quantité no longer being the exclusive basis for the first time. Martin started from the concept of unit (*unité*), establishing that any quantity can be conceived of as a number, with reference to the underlying unit: *Une quantité quelconque peut être conçue comme un nombre* (Martin 1781, viij).

At the same time, Martin's approach shows how multidimensional and complex the transition from the general concept of quantity to the concept of number was. Martin went on to divide numbers into "abstract" and "concrete" according to whether their unit was abstract or concrete, and he declared it to be his position of principle that certain qualities could be connected only with the *concrete* unit—resulting in the fact that only the concrete numbers could form such numbers, like heterogeneous or opposite numbers, because of their unit's different "attributes" (ibid.). As we shall see, this position of principle involved Martin in some conceptual difficulties, but it permits us at the same time to analyze how effective the epistemological beliefs were that were shared by Martin and many French authors.

Since Martin based both his arithmetic and his algebra on the concept of number, he gave in his part on arithmetic a general definition of subtraction not confined to positive remainders. Subtraction, he said, provided the parts the addition of which gave the sum (Martin 1781, 6 ff.). And Martin also was the first to give an exclusively operational definition of opposite numbers—as numbers whose addition signified a subtraction, and whose subtraction gave an addition: *on peut dire en général que leur addition doit se changer en soustraction, et leur soustraction en addition* (ibid., xiij).

Martin stressed that he had for the first time used this property to define opposite numbers, and that it represented an essential concept in number theory despite its simplicity (ibid.). Although formulated quite verbally, it already contains structural analogies to the algebraic definition simultaneously developed in Germany (see below, Chapter VII).

Martin also was the first to criticize the French university tradition for hitherto defining negative numbers virtually exclusively by their being "plus petits que zéro" (ibid., xiv). This, however, was not a critique of the implicit conception of understanding quantities only as absolute, respectively positive, ones; thus a definition of negative quantities via their algebraic sign was considered to be sufficient. For he made a point of stating that one could not imagine any quantity smaller than that which had already attained the term of zero, since zero was the extreme limit of diminishing or decreasing a quantity: *puisque zéro étant la limite de tous les décroissemens possibles d'une quantité* (ibid.).

Martin admitted the notion of "smaller than zero" only as a relative way of speaking: a negative number was relatively farther distant from a positive number than the zero (ibid.).

Martin actually did not adhere consistently to his own conception of taking the number concept as basis. While Martin introduced negative numbers, in contrast to all French authors before him, in the first part of his textbook on arithmetic,

> Two numbers are said to be *opposite kinds*, or simply *opposites*, when the addition of the one reduces the other, and the subtraction of the one increases the other. To distinguish between them, those of the same kind are called *positive* and preceded by the sign +; and those of the opposite kind are called *negative*, and preceded by the sign − (ibid., 10).

Stressing that the two conditions have been chosen at will, he repeated the definition for the representing terms in the second part of his textbook on algebra, which he himself considers to be that science, which investigates the general properties of all kinds of numbers by means of the symbols representing them. Now, however, he again tacitly assumed that the terms merely represented the positive numbers:

> Since there are positive and negative numbers, algebraic terms, which represent them must be distinguished by the opposite signs + and −. Here therefore, as in Arithmetic, terms preceded by the sign + are called *postive*, and those preceded by the sign − are called *negative* (ibid., 44).

Martin later inserted mostly only positive values for coefficients and unknowns, but gave one normal form of $x^2 + px + q = 0$ in the case of quadratic equations, inserting then, after deriving the general solution, a positive value for p, and a negative value for q (ibid., 162). It was undoubtedly one of Martin's decisive steps to choose numbers for his basic concept, and to desist from operating with the concept of quantity in algebra, taking representative terms instead, but the transition to the general concept of the representing sign was not yet complete, because of still prevailing epistemological notions pertaining to substantialist views. While Martin's idea to conceive of oppositeness and negativity as of additional qualities of numbers corresponds to Fontenelle's thoughts (see Section 2.8.1.), the operational novelty of interpreting the additional quality as an "attribute," and hence as constitutive element of a "concrete number," is his own. It guided Martin to reflecting on the concept of multiplication, a reflection that shed sudden light on the fundamental and long unsolved problems of multiplying quantities, and on the immediate connection between extending the area of numbers and defining all the basic operations; at the same time, it led the author into a situation without exit from which he could escape only by willful positing. Martin used the familiar concept of multiplication, according to which the multiplier had to be an "abstract" number, and only the multiplicand was permitted to be a "concrete" number or quantity. Since he himself had postulated, however, that opposite numbers were "concrete" numbers, he saw no possibility of admitting a product from such numbers; products had to agree, accordingly, with the multiplicand in their "dimension." Just as inadmissible for Martin was a product of a quantity of space and one of time, and neither a product of two linear quantities, nor a product of quantity of motion and a quantity of mass (ibid., xv f.).

Martin made his escape from this contradiction by declaring—in a way somewhat comparable to that of Fontenelle and d'Alembert—a potential minus sign preceding the multiplier to be negligible, "it is [done] by abstracting the sign from the multiplier and considering this factor as a pure abstract number" (ibid., xvj).

The concrete method was to be the following: If the task, for instance, was to multiply –6 by –3, the first step was to multiply –6 by 3 as *nombre abstrait*, and after that the result of –18 was to be subtracted from a real term, or an assumed term A. Martin justified this "trick" by explaining that there were sometimes propositions in mathematics that, taken literally, seemed absurd while expressing, if understood correctly and in the "sense of their inventors," precisely a method that could be easily memorized (ibid., xvij).[68]

Such cloaking of inconsistencies could not serve the clarity of concept proclaimed, but was an expression of strong underlying epistemological views. Since these were too strong, it probably was not feasible for Martin to look for an alternative justification of opposite numbers: justifying them not as concrete but as abstract numbers.

The second conceptual problem Martin became entangled in lies in his further basic assumption that positive and negative quantities were *heterogeneous*. By explicitly quoting Christian Wolff ("Wolf"), and by implicitly rejecting Wolff's critique by d'Alembert and de la Chapelle, Martin even extended Wolff's conception. While Wolff had established in a few sentences that positive and negative quantities were heterogeneous, and that no proportion $1 : -1 :: -1 : 1$ could therefore exist (see above, Section 2.8.3.), Martin extended these propositions into a larger system by comprehensively reflecting the concept of ratio (*raison*). His basic assumption here was: a ratio could only exist between two quantities one of which was contained as a part within the other—and which, however, are homogeneous to each other (ibid., xxxiij ff.). There could thus be no ratio between heterogeneous quantities, respectively numbers. In addition to that, he required, for measuring ratios, to use not the arithmetic means, the fractions, but rather to maintain the original concept of measure (ibid., xxvij ff.); since he had postulated positive and negative numbers as heterogeneous before, this made clear that there could be no ratio between these either, and that Arnauld's proportion argument was thus without foundation as well (ibid., xxx ff.).

Here again, Martin had introduced an ad hoc construction in order to avoid another consequence: that of restricting the concept of ratio to positive quantities. Instead of reflecting this concept's original restrictedness in the field of geometric quantities, Martin tried also to maintain the basic epistemological trait of a generality as comprehensive as possible for the concepts. He did not become aware of the fact that he had confined the field of application by his own definition of heterogeneity virtually in the same way.

Martin was also subjected to criticism by one of his teacher colleagues in Toulouse. In his contribution to Cantor's handbook, Cajori mentions a treatise of 1784 by Gratien Olléac, mathematics teacher at the Collège national de

---

68   Martin advanced analogous arguments in favor of division (ibid, xx–xxiv).

Toulouse: *Sur des théories nouvelles des nombres opposés, des imaginaires et des équations du troisieme degré*. Cajori recorded only the explicit critique of Christian Wolff:

> One should give up as absurd Wolff's idea of the heterogeneity of numbers according to which they cannot have any relations with one another, and adopt Descartes's idea of the reality of both negative and positive numbers (Cajori 1908, 86).[69]

Besides openly criticizing Wolff, Olléac actually challenges Martin, without naming him, by quoting the latter's definition of *nombres opposés* (Olléac 1794, 8). Olléac is correct in criticizing Wolff's and Martin's notion of heterogeneity for being inconsistent in merely excluding the existence of ratios between heterogeneous quantities while nevertheless executing arithmetic operations between them (ibid.).

Olléac's own justification of why positive and negative numbers were homogeneous, however, was not suited to solve the conceptual problems. Olléac based his own approach on rejecting Martin's assigning *espèces opposés* to positive, respectively negative, numbers, and virtually assuming conceptual identity for them in implicit adherence to d'Alembert. Quoting the geometric example of two radii lying in opposite directions in a circle, he claimed them to be essentially identical, differing merely in one "aspect," by means of a mental operation that, however, did not change their nature:

> The numbers which are expressions [of the two radii] are called, one positive and the other negative, or opposite numbers of a common number; but these quantities were of the same kind before being envisaged from this point of view, since each one was a radius; thus they must be so after, since the way in which a thing is envisaged is only a purely mental operation which cannot change its nature; [...] so the numbers which express the value of these radii, no matter that they are opposite, are homogeneous (ibid., 11 f.).

After that, Olléac had no difficulty in justifying the rule of signs, since multiplying a negative multiplicand meant nothing but multiplying a positive one, which was being considered only *sous un point de vue contraire*. Otherwise, Olléac adhered to his colleague Martin in understanding a negative product as a subtractive term (ibid., 18 ff.).

Martin remained unimpressed by Olléac's objections to the heterogeneity claimed, and did not revise the second printing of his own textbook of 1802 in this respect.

It can be generally said that Martin's conflict between number concept and proportion concept illustratively demonstrates the sharp contrast between the traditional geometric founding of mathematics and the dimensions of restructuring mathematics under the program of algebraization.

---

[69]   I was unable to unearth this work in any library in France, in particular not in the libraries of Toulouse, and neither in Toulouse's municipal archives. One exemplar, however, is present in the New York Public Library.

It was unfortunately impossible to find further bibliographical data on Olléac. It is only clear that he taught mathematics at least after 1793 at the same institutions where Martin taught physics. I am very grateful to Madame Jocelyne Deschaux, Bibliothèque Municipale de Toulouse, for her extensive research.

2.10.2. THE MILITARY SCHOOLS AS MULTIPLIERS

Shortly after the onset of establishing the military school system in France began the systematic introduction of textbooks, in particular for mathematics. The fact that textbook authors and examiners (*examinateurs permanents*) were in most cases one and the same caused these textbooks to be widely disseminated and applied. Among the three major fields of education—navy, artillery, and engineers—the military engineers (*corps du génie*) were the first to respond. The examiner of the *École de Mézières*, Charles Etienne Camus (1699–1768), was the first to have been officially charged with elaborating a mathematics textbook: the four-volume *Cours de mathématiques* (1749–1752 in first edition). Because artillery was for some time lumped together with the *corps du génie* in 1755, Camus became examiner for artillery as well, which resulted in his textbook being used there as well. After the two *corps* had been separated in 1758, Camus's textbook was criticized by the artillery as too elementary, and too little oriented to practice (Hahn 1986, 529 ff.), yet it remained in use until Camus retired in 1768. With four printings in less than twenty years, it was a quite successful textbook.

The textbook's elementary character is less due to Camus himself, but rather a first typical expression of the institutional "framing" of the subject matter and methods of textbooks. Camus authored his textbook indeed on commission: it was intended for the training of military engineers at the engineering school founded in 1748 by Count d'Argenson, the later famous *École de Mézières*. M.A.R. Paulmy Voyer, Count resp. Marquess d'Argenson (1722–1787) had in 1748, as *Sécrétaire d'État* and war minister, transformed the hitherto informal modes of training for these engineers into a formal, institutionalized education. To ensure the quality of this education, he had at the same time commissioned Camus with authoring a textbook, establishing for this purpose both type and extent of the knowledge required, as well as the method of teaching. It is markedly typical for the sectoring between university and military education that d'Argenson decreed the method for the education of engineers to be the "synthetic" one. The linkage between method of teaching and educational objective had been strictly intentional, as is evident from Camus's introduction to his first volume, on arithmetic:

> M. le Comte d'Argenson [...] has decided on the degree of competence that must be demanded of Candidates: he has even been kind enough to provide all the details of their instruction; [...] he has ordered me to bring together in a single Work, treated synthetically, all the Theory which an Engineer needs to have (Camus 1749, i).

What was understood by the "synthetic" method for engineers can be directly seen from Camus's textbook. It consists of volumes on arithmetic, geometry, statics, and hydraulics, but contains no part on algebra. Camus himself explains this absence of algebra, which might astonish some, as execution of the synthetic method: he had not presented algebra, which he calls *Calcul littéral*, together with the *Calcul numérique*, as other authors had done, because it did not belong among the knowledge mandatory for engineers:

> But having treated the principal parts in which an Engineer needs to be instructed by Synthesis alone, I thought I must reserve literal Calculus for Analysis (ibid., iv).

Camus was so rigorous in his method as to exclude algebra altogether from his textbook for engineers:

> I advise, then, that here I shall not talk about the literal Calculus and Analysis until after having kept my promise and given the Treatises which I have just announced making use only of Synthesis (ibid.).

Camus did not author a textbook of this kind for a more general public. At the same time, it becomes evident that grossly operationalizing the synthetic method in concentrating exclusively on geometry corresponded with a basic conviction widespread among users of mathematics as well.

On the occasion of a navy reform in 1763, Étienne Bézout (1730–1783) not only became examiner for the navy, but was also commissioned with authoring a textbook. His *Cours de mathématiques à l'usage des Gardes du Pavillon et de la Marine* appeared in six volumes starting in 1764. Together with the edition for artillery, where Bézout had assumed the function of examiner in 1768, too, the work led the market;[70] his navy edition saw eleven reprints until 1791, and the artillery version two (1770/72 and 1781) (see Lamandé 1987, 373).

There was one competitor, however, to Bézout's textbook. At the *École de Mézières*, Abbé Charles Bossut (1730–1814), until then mathematics professor of the school of engineers, succeeded Camus as the school's examiner in 1768. In his function of professor, he had striven to raise the level above that of Camus's elementary textbooks. As examiner, he published a multi-volume *Cours de mathématiques à l'usage des éleves du corps royal du Génie*, starting in 1771:

* *Traité élémentaire d'arithmétique* (1772),
* *Traité élémentaire d'algèbre* (1773),
* *Traité- Traité élémentaire de géométrie et d'application de l'algèbre à la géométrie* (1775),
* *Traité élémentaire de mécanique: statique* (1772),
* *Traité élémentaire d'hydrodynamique* (2 volumes, 1771).

Beyond that, Bossut also published revisions of the textbook, such as:

* *Cours de mathématiques à l'usage des écoles royales militaires*, in two volumes (1782),

a volume intended for the formation of infantry and cavalry officers. In his textbooks, Bossut adhered to the algebraization tradition. Thus, he emphasized algebra's high degree of generality, which permitted easy and manifold applications:

> It is evident [...] that by one and the same algebraic calculus one resolves all problems of the same kind, proposed in all the generality of which they are

---

[70]   "Gardes du pavillon" (guardians of the flag) is an ancient name for French navy officers.

susceptible; and that the applications of this calculus to all particular cases are no more than subsequent operations, reduced to their greatest degree of simplicity (Bossut 1781, 286).

Bossut introduced negative quantities via the concept of oppositeness:

[quantities] can be [...] opposite to each other, as to the way they exist; and to mark this oppositeness, these quantities are distinguished in general as *positive* and *negative* quantities (Bossut 1773, 8).

Subsequently, Bossut also quoted the traditional definition by preceding signs according to which a quantity provided with a minus sign is a negative quantity (ibid.).

As in the university tradition, Bossut also emphasized that *soustraire* did not always mean *diminuer*, but rather was an increase in case of negative quantities (ibid., 14). In line with the level of the textbook's audience, no further conceptual reflections followed. Nor was the rule of signs justified in detail.

In the section on powers and roots, Bossut explained the multiplicity of roots (ibid., 74 ff.). He also used the $\pm$ sign, for example for square roots. In representing equations of second degree, Bossut not only explained that there was only one normal form, $x^2 + ax = b$, but also gave as general solution $x = -\frac{a}{2} \pm \sqrt{b + \frac{a^2}{4}}$ (ibid., 184 ff.), introducing, in contrast to his own definition of the signs of positive and negative quantities, the coefficients a and b pointedly as *quantités réelles, positives ou négatives* (ibid., 186). In his examples, however, he inserted only positive values for the coefficients.

Bossut was a quite successful textbook author. The abbreviated version of his textbook of 1782 saw its third printing already in 1788. The version for engineers attained four printings by 1789; a revised edition appeared in 1800.

Bossut had no hope, however, of catching up, in the dissemination of his textbooks, with Bézout, whose textbooks saw a general dissemination and acceptance that no other mathematician before him had attained in France. A comparable success after him was achieved only by Lacroix. Étienne Bézout (1730–1783), member of the Paris Academy since 1758, became mathematics teacher and examiner for navy officers in 1763. About 1800, together with the new educational system's consolidation, we find *four* authors simultaneously publishing reeditions of Bézout: Peyrard, Reynaud, Lacroix, and Garnier! In manifold reeditions, these textbooks appeared until 1868 (cf. Lamandé 1987, 375). In the survey of the ministry of education of the year VII (1799) among the mathematics teachers of the *Écoles Centrales*, 50 of 69 reported that they were relying entirely, or in parts, on Bézout's textbook. L.F.A. Arbogast, who regretted in his answer that he was compelled to use this textbook, of which he was critical, explained its use with the fact that it had already been in the hands of the majority of students (Lamandé 1990, 32).

Bézout based the algebra part of his textbook on d'Alembert's conception of negative quantities, elaborating it so far as to enable it to achieve its true dissemination in practice; in some parts, he adopted d'Alembert's formulations word for word.

For an analysis of Bézout's conception, it is sufficient to examine the volume on algebra in his principal work, his textbook for the navy.[71] This is where Bézout steps forth as the first author who voiced a clear programmatic challenge to the self-understanding then current in French mathematics: to the belief that the analytic method is the best for taking the path toward increasing generalization. This is evident from the fact that in the six-volume work, algebra, as the general science of the quantities, no longer precedes geometry, but forms only the third part, *after* arithmetic and geometry. Beyond that, Bézout explicitly weighed the advantages of the analytic and the synthetic methods for his part on algebra. In contrast to the tradition since Ramus and Arnauld of ascribing better success in learning to the analytic method, Bézout was the first to introduce the distinction between *génie* and the public at large, declaring the analytic method to be suitable only for the "inventeurs," while the synthetic method was more appropriate for the majority: whereas the synthetic method gave the rules to be applied directly, the analytic method led via a complex sequence of operations and *raisonnements* toward the general rules (Bézout 1781, v). While this might raise beginners' curiosity, it entailed the danger that they might feel humiliated if the *raisonnements* did not appear by themselves and from "the depth of their own mind," as desired:

> This last method may seem preferable to the first, in that it appears it must flatter the self-esteem of beginners and stimulate their curiosity. But on reflection, while attention is necessarily divided between three objects, namely the statement of the question, the reasoning needed to express it algebraically, and the operations which need to be carried out with the aid of signs whose significance escapes one, the more easily to the extent that one is less experienced in representing these ideas in an abstract manner, it appears to me that it is doubtful that this method is the best to begin with, for the greater number of readers. Will it not produce, on the contrary, an effect totally opposed to that which some claim for it? The reasonings it demands, however simple to start with, where, doubtless, one treats only simple questions, these reasonings, I say, where the one who uses them is faced with having to draw on his very own resources, will they not humiliate him when they are not apparent to him? The method of invention presupposes always a certain finesse; it is the method inventors ought to follow, and consequently the method of men of genius; now they are certainly not the greater number (ibid., vj).

At the same time, to Bézout is due the merit of having been the first to transform the premise for algebra hitherto implicitly practiced in France into an explicitly formulated requirement. The basic concepts of algebra, the letter

---

[71]   The analogous volume in his later, parallel textbook for artillery is the second of the four volumes. As to their conception, the two versions of Bézout's algebra are identical; they are even widely identical in text. The artillery manual is in parts of somewhat better quality of print, and more extensive in some presentations. It is notable that text problems are differently "cloaked" despite containing the same numeric values: Thus, a problem presented in the navy issue with pure numbers (Bézout 1781, 67) was reformulated for the artillery textbook with a vocational reference: with the corresponding number of "rounds for muskets" (Bézout 1779, 52).

quantities, are presented, and assumed, only as absolute quantities: "Letters represent only the absolute value of the quantity" (ibid., 79).[72]

The fascinating question is always how far an author keeps such a view throughout his entire textbook, or whether he nevertheless inserts negative values. This can be easily tested by how the solution of equations is treated. For Bézout, the test is difficult, since he does not discuss, or represent, any normal forms for equations of various degrees, but only individual concrete cases. And in contrast to all the textbooks analyzed here, he does not give any *formula* for the solutions of mixed equations of second degree, but a verbal formulation consisting of *nine* lines (ibid., 129 f.). From the other examples, however, it can be seen that Bézout remained consistently true to his own conception of absolute quantities. The zero, however, he had already admitted without problem in his arithmetic when introducing digits.

If there are essentially only absolute quantities, what can negative quantities then be? Bézout devoted an entire paragraph of eight pages to this question: *Réfléxions sur les quantités positives et les quantités négative.* These reflections are far more important than those of, say, de La Chapelle or of Sauri. Bézout saw the conceptual solution in the fact that quantities dispose of an *inner* property: they can be regarded under two opposite aspects, that is, as having the capacity of increasing another quantity, or as having the capacity of decreasing another quantity. As long as the quantity is represented only by a letter or a number, one could not make out its capacity aspect. To indicate its effect on other quantities besides the numerical quantity, one made use of appropriate signs:

> The same quantity can be considered from two opposite points of view, as being able to increase [another] quantity or being able to diminish it. While the quantity is represented only by a letter or by a number, nothing shows which of these two aspects is being considered. [...] The most natural way for the difference to be revealed is to show it by a sign which indicates the effect which the [quantities] can have on each other (ibid., 79).

As such signs one used the signs of plus and minus, which Bézout clearly distinguishes from their function as signs of operation. On the basis of this symmetric ascription of function, Bézout was led to establish that negative quantities are just as real as positive quantities: "Negative quantities therefore have an existence just as real as that of positive quantities" (ibid., 80).

He thus had no problem with representing the multiplicity of roots, and with showing, in particular, that both $+a$ and $-a$ are roots of a^2 (ibid., 125 f.). Bézout also acknowledged the rules for operating with negative quantities. He justified the rule of signs only by the application of the distributive rule upon multiplying the two composite terms $(a-b)$ and $(c-d)$ (ibid., 18 f.). On the other hand, Bézout continues in the same quotation as follows: "and they are no different except that they have an opposite sense in calculations."

---

[72]   It is remarkable that the designation "valeur absolue" is used here as something evidently familiar. The history of the concept of absolute value has seen little research as yet. Cf. Duroux 1983, Gagatsis 1995.

There was thus no longer oppositeness of "position," as with other authors of his kind, but oppositeness of "meaning"—a term that was at first indeterminate. Bézout now explained the difference of "meaning" in a way not to be expected by someone familiar with the presentation used up to then. Compared to the hitherto symmetric functions of the positive and the negative, it really represents a rupture that Bézout does not admit negative solutions, designating them as "impossible," or more precisely that he understands them as a demand to "reverse" the problem formulation. In this concrete mode of dealing with negative solutions, one may immediately recognize an elaboration and didactical implementation of d'Alembert's views concerning the epistemological exclusion of negative numbers:

> 70. If then, after having resolved a question, it happens that the value of the unknown found by these methods is negative; for example, if one arrives at a result like this, $x = -3$, it must be concluded that the quantity one has designated by x does not have the properties one had assumed in carrying out the calcuation, but quite contrary properties. For example, if one proposes this question. Find a number which added to 15 gives 10; this question is evidently impossible; if one represents the desired number by x one will have the equation $x + 15 = 10$ and, consequently, by virtue of the rules given above, $x = 10 - 15$ or $x = -5$. This last conclusion makes me see that x which I had considered before to be added to 15 to make 10, by right ought on the contrary to be subtracted. Thus all negative solutions indicate something false in the proposition of the question; but at the same time indicating the correction, in that it shows that the desired quantity ought to be taken in a sense completely opposite to the one that had been chosen.

> 71. Let us conclude then from this that, if having resolved a question in which some of the quantities were taken in a certain sense; if, I say, one wishes to resolve this same question in taking these same quantities in an entirely opposite sense, it suffices to change the signs the quantities carry at the moment (ibid., 81 f.).

It was this very epistemologically justified rejection of negative solutions, which, in their form of reformulating the equations with the objective of securing their positive character, was to have the most profound effects in France.

2.10.3. Violent Reaction in England and Scotland

As described in II.8.2, foundational controversies about the respective role of algebra and geometry had become public during the second half of the seventeenth century, conflicts, in particular debates, about how to justify concepts. Geometrical and empiricist justifications confronted algebraic generalizing systems of signs. While negative quantities had not been affected by the debates of that period, they now became the core of the conflict.

Robert Simson (1687–1768) can be identified as a pioneer opponent of algebraization and negative numbers in the first half of the eighteenth century. Simson, a self-educated mathematician (he had studied Euclid's *Elements*) was a mathematics professor at the Scottish University of Glasgow from 1711 on. An admirer of ancient mathematics, he was active as an editor of the works of Greek geometers. In this capacity, he sharply criticized modern algebraic development, condemning in particular Newton's use of algebraic methods in geometry. Simson did not publish these views, however, discussing them only with his disciples. He voiced his rejection of negative solutions of equations in 1764 in one of his letters (Pycior 1976, 49 ff.).

The first publication of this kind of approach is due to Thomas Simpson. Simpson (1710-1761) was a mathematics teacher in an area of applications and thus quite distinct from the universities, at the English Military Academy of Woolwich. To support his courses, he published his *Treatise of algebra* in 1745, the long-term influence of which is attested by a total of ten reeditions.

Simpson professed in his preface the belief that was to become the leitmotiv for a revisionism stressing the advantages of the synthetic method of the ancients over modern analysis with regard to rigor:

> We will see the Advantages which the ancient synthetic Method of Reasoning has, in many Cases, over the modern Analysis, especially in what regards Neatness and Perspicuity (Simpson 1745, vii).

At first, no divergence is apparent in his textbook as compared to the established view in England; he introduced negative quantities and operating on them in the familiar way, and the rule of signs as well. The difference is revealed only upon closer inspection. The operations and the rule of signs are defined for compound terms ("compound quantities"), in a way analogous to that of Diophantus. In discussing their applicability to isolated negative quantities, however, he made clear that he considered quantities smaller than zero to be absurd: they were just as impossible as imaginary quantities:

> For it ought to be considered that both $-b$ and $-c$, as they stand alone, are, in some Sense, as much impossible Quantities as $\sqrt{-b}$ and $\sqrt{-c}$; since the Sign $-$, according to the established Rules of Notation, shews the Quantity, to which it is prefix'd, is to be subtracted, but to subtract something from nothing is impossible, and the Notion or Supposition of a Quantity actually less than Nothing, absurd and shocking to the Imagination (ibid., 24).

It was thus ridiculous, he said, to pretend one might be able to prove by some kind of reasoning what the product of $-b$ times $-c$ should be, "when we can have no Idea of the Value of the Quantities or Expressions to be multiply'd." Only operations on real, positive quantities were admissible; numbers as algebraic expressions were legitimate only as measures of geometrical quantities:

> All our Reasoning regards real, affirmative Quantities, so the algebraic Expressions whereby the Measures of those Quantities are exhibited, must be likewise real and affirmative (ibid., 24 f.).

This fundamental belief, however, did not keep Simpson from applying virtually unchanged procedures in his concrete operations, such as using both a positive and a negative square root (ibid., 99).

The first radically consistent rejection of algebraization and generalization was published by Francis Maseres (1731–1824) in his treatise *Dissertation on the Use of the Negative Sign in Algebra* in 1758. Maseres had studied in Cambridge and had concluded his studies of mathematics with the good qualification of fourth "Wrangler." In 1755, he became a Fellow of Clare College in Cambridge and competed in 1760 for the Lucasian mathematics professorship. Failing to obtain it, Maseres switched to the profession of lawyer. It is not known who prompted his ideas on mathematics, or who had influenced him; influences from France have not been investigated as yet.

Basically, Maseres confined the meaning of negative numbers to subtractive quantities, defining subtraction exclusively for positive remainders (Maseres 1758, 1). He thus did not admit isolated negative quantities; for him, an expression like $(-5) \times (-5)$ made sense only in the form 5×5, without making allowance for the minus sign (ibid., 2). Maseres thus also denied as a matter of principle that there were two square root values.

His concept of numbers that admitted only absolute numbers becomes particularly clear from how he treated the solving of quadratic and cubic equations. For equations of the second degree, he declared that all such equations are reducible to the following three forms:[73]

$$xx + px = r,$$
$$xx - px = r,$$
$$px - xx = r.$$

p and r being assumed here as strictly positive. The possible negative form of the coefficients must be expressed by the signs of operation. Furthermore, it is typical that the fourth form of combining the signs $+$ and $-$, $xx + px = r$, does not appear, just as it did not in Descartes (cf. above, Section 2.4.) its solutions becoming negative (ibid., 20). Analogously, for equations of third degree, Maseres treated the various cases of allowed combinations of the coefficients' plus and minus signs as separate problems and cases.

In a programmatic manner, Maseres emphasized that the two equations $xx + px = r$ and $xx - px = r$, for instance, by no means expressed the same problem, but rather formed different equations for different problems:

> This method of uniting together two different equations may perhaps have its uses; but I must confess, I cannot see them: on the contrary, it should seem that perspicuity and accuracy require, that two equations, or propositions, that are in their nature different from each other, and are the results of different conditions and suppositions, should be carefully distinguished from each other, and treated of separately, each by itself, as it comes under consideration (ibid., 29).

The main part of Maseres's 300-page treatise is devoted to studying equations of third degree. The author succeeded in distinguishing sixty different cases of respective constellations of coefficients and their algebraic signs, a result inducing him to claim that there were just as many "normal forms" (ibid., 200 ff.).

---

[73]   In mathematical historiography, the underlying number concept has been given little attention. The lack of care in relating Maseres's ideas provides some examples; Pycior reports it thus: "He considered two possible forms of a quadratic equation" (Pycior 1976, 57), while Arcavi/Bruckheimer speak *of four* forms (Arcavi, Bruckheimer 1983, 11).

William Frend (1757–1841) published his anti-algebraic views even more extensively and radically in his voluminous algebra textbook *Principles of algebra* of 1796. Frend had also studied in Cambridge and had passed his exams in 1780 as second "Wrangler." He began working as a tutor at the famous Jesus College in Cambridge, but had to give up that post later because of confessional conflicts. Frend reviewed MacLaurin's algebra textbook incisively for his admission of negative quantities. The legitimate concern of his critique was that MacLaurin introduced his notions not conceptually, but by examples of applications:

> Now, when a person cannot explain the principles of a science without reference to metaphor, the probability is, that he has never thought accurately upon the subject (Frend 1796, x).

Frend made clear his own epistemological principles, saying that only positive numbers were permissible, and that algebraic operations must not change this "nature" in any way. Concerning positive numbers, he proclaimed:

> No art whatever can change their nature. You may put a mark before one, which it will obey: it submits to be taken away from another number greater than itself, but to attempt to take it away from a number less than itself is ridiculous. Yet this is attempted by algebraists, who talk of a number less than nothing, of multiplying a negative number into a negative number and thus producing a positive number, of a number being imaginary. Hence they talk of two roots to every equation of the second order, and the learner is to try which will succeed in a given equation (ibid.).

Frend's algebra textbook is based on rejecting negative numbers, and in particular multiple roots. Since it is not immediately evident in case of algebraic differences which of the two terms constitutes the larger, Frend introduced a sign for general subtraction, in the form of a horizontal S, virtually in the function of the absolute value:

> $\sim$ is the mark of difference: $a \sim b$ means the difference of the two numbers a and b, which will be either $a - b$ or $b - a$, according as a is greater or less than b; and we read $a \sim b$ thus, the difference of a and b (ibid., 4).[74]

Like classical Greek authors, Frend presented the rules of signs in the context of multiplying compound terms. He did this without justification, however, in particular for multiplying two terms having a minus sign (ibid., 19).

---

[74]   Klügel reported in a treatise of 1795 that this sign occurred for the first time in volume 51 of the *Philosophical Transactions* of the Royal Society. That article was pubished in 1761, being the review of an astronomical treatise. The same sign is used there on page 928, but rather marginally and without explanation. Hence, one must suppose that the sign was well known among the contemporaries, at least in England. The sign fulfills the same function as Viète's sign "=" (cf. above Section 2.4.). The author was Henry Pemberton (1694–1771), member of the Royal Society since 1720, physician and professor of medicine at Gresham College in London. He had also published on mechanics and astronomy. Indeed, this sign was known in England, in any case, by Oughtred. It appears first in this function in 1652 in an appendix to his algebra textbook (Pycior 1997, 48).

After having presented the basic operations, Frend explained extensively in an elementary way suited to the addressed level of students how to solve equations of first degree by algebraic manipulation, for natural numbers. He then went on to simple problems, which were solved with linear equations. Only after that, he introduced fractions, ending with powers and roots. The basic method for treating roots was that there were only positive root values in any case (ibid., 89 ff.).

In the subsequent section on equations of second degree, Frend also posited as fundamental premise that only positive solutions were possible (ibid., 104 ff.). Here, he extensively examined four cases as independent problems: the pure quadratic equation $x^2 = b$, and the three forms of mixed quadratic equations Maseres had already used. Frend went so far as to exclude negative terms *within* the process of solving the equation as well. Thus, he performed estimations during this process to determine the positive roots, thereby obtaining only the positive value for mixed roots (ibid., 106). The type of equation that may yield two positive roots raised more difficulties. Calculating an example, Frend surprised the reader with the result that *two* (positive) values satisfied the equation (ibid., 111), without explaining how to handle the two solutions and why there were two solutions in this case instead of only one, as had been shown before. In an analogous mode, Frend exhaustively examined various coefficient constellations for the equations of third degree as separate isolated cases ("forms").

One of the numerous texts in the appendix to Frend's textbook is noteworthy: "A remark on an error in the reasoning of the late learned French mathematician, Monsieur Clairaut, in that part of his *Elements of algebra* in which he endeavours to prove the rules of multiplication laid down by writers on algebra concerning negative quantities" (ibid., 514–518). Frend here indeed revealed an error in the argumentation of many authors who wanted to give a "proof" for the rule of signs. Clairaut had started from the polynomials $(a - b)$ and $(c - d)$, multiplying them as $ac - bc - ad + bd$; claiming on this basis that a and c might be assumed to be zero, thus obtaining $-b\times-d = +bd$. Frend, however, showed that Clairaut had at first assumed $(a - b)$ and $(c - d)$ to be positive, and that it was neither justified nor legitimate to extend the admissible operations at that point to purely negative terms. Frend additionally argued that Clairaut had defined negative quantities only as subtractive quantities and thus had not been entitled to use these as isolated quantities:

> The author on this occasion seems to have forgot his own definition of the sign –, by which he made it to be a mark of the subtraction of the quantity to which it is prefixed from the quantity that goes before it; from which definition it is plain that the said sign always supposes the existence of two different quantities, of which the one is to be subtracted from the other, and consequently that it can have no meaning when applied to a single quantity, as b or d, independently of some other and greater quantity, as a or c, from which it is to be subtracted. There cannot therefore exist any such quantities as $-b$, or $-d$; and consequently no propositions concerning them can be either true or false. Consequently the proposition which the author there

endeavours to demonstrate, to wit, "that $-b\mathbf{x}-d$ is equal to $+bd$; when $-b$ and $-d$ are not preceeded by two greater quantities a and c from which they are substracted, but are considered as single and independent quantities," is so far from being *true* that it is not even *intelligible* according to the only idea of the sign – which the author has given us in all the preceeding parts of the book (ibid., 517).

While it was incorrect of Frend to claim that Clairaut had had only subtractive quantities in mind, Frend's critique clearly showed the deficiencies of defining negative quantities only by their algebraic signs. Frend's critique concerns section 60 in Clairaut's algebra textbook (pp. 96–97 in Lacroix's edition of 1797).[75]

In the same appendix, Frend sharply criticized Euler's algebra textbook for including a multitude of errors, caused by Euler's use of the "perplexing" and absurd doctrine of negative quantities for which Frend attributed an unusual predilection to Euler (ibid., 518). Frend did not detail the alleged errors, however.

An often quoted article by John Playfair (ca.1748–1819) also proves general reservations against the methods of algebra. Playfair had first been active as a priest, then as a private tutor, and eventually as mathematics professor in Edinburgh. Although his treatise read before the *Royal Society* in 1778 was not directly about negative numbers, but rather about imaginary numbers, it reflects the paradoxical situation that operating with meaningless algebraic signs may nevertheless lead to a rigorous result—meaning, for Playfair, a geometrical result (Playfair 1778). It was this very reflection on "artificial symbols" (ibid., 319) and on the possibility of such symbols mutually "compensating" one another with the result of obtaining a "real quantity" that turned out to be so tempting for Carnot (see below, Chapter V.I.3.).

Cajori, who analyzed English publications of the time, asserts a great impact of Maseres's and Frend's concepts in England, for the first half of the nineteenth century as well (Cajori, in: Cantor, Vol. IV, 1908, 85 ff.). While Pycior mentions some voices criticizing that mainstream, too, she has to admit that the concepts presented above dominated debates in England until about 1840 (Pycior 1976, 61 ff.). Pycior attempted to identify an anti-Newtonian attitude as the common feature of these views. This would utterly contradict, however, the traditional interpretation according to which the small progress of mathematics achieved in England is explained by the exclusive partisanship and admiration for Newton's infinitesimal calculus. The authors oppose Newton only insofar as the latter had taught newer algebra as well. A more conclusive approach is to see the motives for Simson, Maseres, and Frend in the absolute model function of Euclid's *Elements*, which still waxed in England in the course of the eighteenth century. Within the frame of this Euclidean epistemology, mathematics basically consisted of investigating isolated cases. A generalizing algebra did not fit

---

[75]   This section followed the deduction of the rule of signs by means of an example: the two springs problem (see above Section 2.9.2).

within this epistemology. Confining oneself to isolated cases without generalizations permitted one to circumvent negative solutions, though at some cost. The model function of Euclid maintained by the British educational system until after 1900 raised this epistemology to the status of a fundamental cultural attitude.

2.10.4. THE CONCEPT OF OPPOSITENESS IN GERMANY

Developments in Germany during the eighteenth century remained to a great extent exempt from being influenced by the radical changes in England and France. Wolff's views found no influential followers, while the now important textbooks were more or less strongly based, in critical dispute with Wolff, on Hausen's views. The quintessential point of the conceptual development was formed by the concept of 'oppositeness,' which was deepened philosophically and historically. In Germany, its linkage to the concept field of the foundations of mathematics was reflected much more explicitly and comprehensively; a challenge producing new solutions repeatedly was constituted by the problem of finding an adequate frame for the operation of multiplication. In this, the tendency toward algebraization came into conflict with the resort to justification by geometrical objects and applications.

Hausen's work, published in Latin, saw only one edition and served primarily mathematics professors themselves. A decidedly more successful author was Johann Andreas Segner (1704–1777), who continued Hausen's views. Segner became the first mathematics professor of the Göttingen reform university in 1735; in 1755 he moved to the university of Halle. In his textbook *Rechenkunst und Geometrie* of 1747, he distinguished, analogously to Hausen, between quantities and numbers, assuming numbers to be the more basic concept. His work was the first to clarify that the signs of the *numbers* represent the criterion for positiveness or negativeness:

> Upon using these signs + and − , the quantities whose numbers are designated by + are regarded as really positing something and are thus also called *positive*. By contrast, those quantities whose numbers have the sign − are regarded, with respect to the former, as *negative* or *privative* because they always negate or destroy from the former considered to be positive just as much as they themselves amount to (Segner 1767, 27).

Segner emphasized that no "internal constitution" made the quantities positive or negative, but rather that the respective attribution was willful positing. Segner had begun his introduction of positive and negative quantities by saying that there were quantities that were diminished by addition, or increased by subtraction (ibid., 25). He also elaborated on the underlying conception of oppositeness (ibid., 27).

Segner discussed the rule of signs in this textbook only in a brief algebraic section—in some kind of elementary algebraic geometry (*Anfangsgründe der Berechnung ausgedehnter Größen*). Although Segner had developed the conception of oppositeness within the frame of arithmetic; for numbers, he fell back for the rule of signs, that is, for the operation of multiplication, on geometrically justified notions, that is on the theory of proportions. For this purpose, he used Hausen's version of the product as the fourth proportional, striving scrupulously toward inferring lucid conclusions, from the ratios between the first two proportional terms to the last two, by the four possible combinations of signs in the multiplication (ibid., 641–645). As he had to admit himself, he fell back for this on "examples," since "otherwise, we might be at a loss for words" (ibid., 643). Since Segner assumed that the algebraic signs of the third and fourth proportional are determined by those of the first and second proportionals, the exemplary basic question for him was to show the ratio $1:-a$ to be possible. For this, he had to clarify how $-a$ can originate from the unit, thus from 1. Segner resorted for this purpose to the visualization of having a person walk a certain distance "toward evening," up to $+a$. Afterwards, the person was to walk back "toward morning," first until the starting point, thus making the unity vanish, and then to continue walking the distance a "toward morning" to make the distance $-a$ originate:

> $-a$ will thus always originate from 1, while the unity is destroyed step by step, and by having the quantity which has made vanish the unit, grow still farther thereafter in the same manner (ibid., 643).

Further illustrations referred to the debt/asset example.

Abraham Gotthelf Kästner (1719–1800) followed Segner as mathematics professor in Göttingen in 1756. By his well-elaborated lectures and by the numerous editions of his textbooks, he had an enormous impact, becoming, in this respect, the genuine successor to Chr. Wolff in Germany. This was caused in particular by the fact that the university of Göttingen became the center of mathematical studies precisely during the time he was active there. Kästner had studied in Leipzig with Hausen, and emphasized later over and over again Hausen's great influence on his own mathematical views (cf. the preface in Kästner 1792, vi). Kästner also exerted an important influence on the development of mathematics in Germany by not only working on foundational questions himself, but by also stimulating his disciples to like research.

Kästner's major work, "*Die mathematischen Anfangsgründe*," appeared in 1758 as a first edition in four volumes, later extended to eight. He developed his basic conception of negative numbers in his first volume, on arithmetic, adopting, just as Hausen and Segner had done, the number concept as basis and developed from that the notion of oppositeness for numbers and for quantities. This concept was to generally understand *quantities* as *numbers*, so that the laws of arithmetic were applicable not only to numbers, but also to quantities:

> When quantities of one kind, i.e., quantities where one can be a part of the other, can be compared in such a way that one examines how many times the one is completely contained in the other, or that one measures both against a part they have in common; then one can consider them once and for all as numbers, and this subjects all quantities which can be measured in such a way, to arithmetic, permitting one to apply the doctrines of numbers to all such quantities (Kästner 1792, 71).

As examples, he quoted concrete numbers and opposite quantities. Kästner based his definition of these quantities on a general concept of oppositeness he posited, but did not justify in detail:

> Quantities of the same kind which are considered under conditions that one diminishes the other shall be called opposite quantities. E.g., assets and debts, walking forward and walking backward. One of these quantities, as one likes, shall be called positive or affirmative, and its opposite negative or denying (ibid.).

Kästner's principle of oppositeness implied from the very outset that positive and negative quantities are "of the same kind," hence homogeneous. Implicitly, he thus criticized Christian Wolff's notion.[76] By quoting further examples, Kästner underlined how general quantities, if capable of oppositeness, can be transformed into one another: "Debts are denied assets, and assets can be regarded as denied debts" (ibid., 72).

Kästner discussed the relation between algebraic sign and sign of operation without, however, reflecting on it extensively (ibid.). Kästner stressed, also in implicit criticism of Wolff, that negative and positive quantities were of equal status: "This negative which remains is a real quantity, only opposite to that considered to be positive" (ibid.).

A new element as compared to the previous discussion in Germany was Kästner's reflection on the "nothing," and his differentiation between an absolute and a relative "nothing." He explained, "in themselves [...] all denied quantities are more than nothing because they are real quantities." But adding a negative quantity to its opposite affirmative quantity yielded zero, or nothing. From this, it followed, on the one hand, that one could call negative quantities, as compared to positive qunatities, to be "*less* than nothing." On the other hand, from this also followed the relative significance of Nothing:

> This expression "less than nothing" assumes a meaning of the term "nothing," which, in a certain way of considering it, relates to the "something" (*nihilum relativum*) and which can be distinguished from an unrelated "nothing" without a relation (*nihilum absolutum*) (ibid., 73).

After these explanations of concept, Kästner introduced the arithmetic operations on opposite quantities. Addition and subtraction were, in the already established way, reversals of the respective operations for natural numbers. He derived the rule of signs for multiplying quantities from Hausen's concept of product, as an arithmetic calculation with ratios, refraining from justifying it in detail, but giving a rather summary explanation, adding, "This is accepted by everybody without proof" (ibid., 79).

---

[76]   In the preface, Kästner explicitly defended Wolff against precipitate critics: His textbook is pioneering the new trend of reflecting on the foundations of mathematics instead of only accumulating scholarship. Hence, the defects of his book are pardonable (Kästner 1792, [ii–iii]). Kästner permitted himself, however, some criticism: thus that it was "a big mistake against method" to found "the doctrine of fractions on that of ratios" (ibid., [iv]).

Kästner treated solving equations in the third volume of his *Anfangsgründe*, the *Analysis endlicher Größen*. The multiplicity of roots was self-evident to him (Kästner 1794, 52). Moreover, he noted with a clarity and explicitness rarely attained before that coefficients are not limited to the positive area, but are capable of values both in the positive and in the negative area (ibid., 55). For equations of second degree, he gave as solution the single general formula bearing the +/–sign.

The elements of analytic geometry contained in the same volume, and presenting the properties of parabola, ellipse, and hyperbola, show quite clearly that Kästner had no problem with using all four quadrants for curves, and operated with both positive and negative coordinates (ibid., 204 ff.). The coordinate axes are not scaled and not presented as number lines.[77]

Analytic geometry can indeed serve as a particularly illustrative indicator for the degree of acknowledgment and operativity of negative numbers. This is particularly evident from a textbook authored by M. Hube, one of Kästner's disciples, and advised by the latter in this endeavor; it was probably the first elementary textbook of analytic geometry ever published (Hube 1759). For the figures, the coordinate axes also are not designated or provided with numbers; the text clearly introduces the coordinate axes as spanning the entire plane: each axis contains both all positive and all negative numerical values, the axes not necessarily being right-angled, but also being skewed-angular (Hube 1759, 3). Kästner extensively stressed in his preface to Hube's textbook the advantages of this first representation of the conic sections by the analytic method, by means of equations, as compared to the traditional synthetic approach (Kästner 1759).

A conception different from that of Segner and Kästner was developed in the third textbook series that became influential during the second half of the eighteenth century in northern Germany: the eight-volume *Lehrbegriff der gesammten Mathematik* (1767–1777) by Karsten. Wenceslaus Johann Gustav Karsten (1732–1787), raised in the small duchy of Mecklenburg Schwerin, began as a lecturer at the duchy's universities of Rostock and Bützow, and became in 1778 Segner's successor as mathematics professor at the prestigious university of Halle.[78]

In contrast to Hausen, Segner, and Kästner, Karsten introduced oppositeness not for numbers, within the frame of arithmetic, but in the second part of his algebra textbook under the heading "general art of arithmetic." For Karsten, the basis of this algebra was newly conceived basic operations, namely, the "four general kinds of operation," generalized from the basic arithmetical operations

---

[77]   In his preface to Hube's textbook on analytic geometry, Kästner, by contrast, sharply crititized Wolff for his concept of negative quantities: The expression "nothing" had induced him, by a wordplay, "to expel the denied quantities [...] from the realm of true quantities." After such examples, one must not be surprised, "when other geometers arrive at deductions, which seem unbelievable to themselves," and in particular did not distinguish between signs and objects (Kästner, in Hube 1759, [xxxi]).

[78]   For Karsten's biography, see below Chapter III.9.

by including opposite quantities. Karsten did not refer the concept of oppositeness to numbers, but rather to quantities, which he obviously considered to be more general. For this purpose, however, he neither defined the concept of "quantity" nor discussed how it related to numbers.[79] For Karsten, the concept of quantity essentially implied the meaning of variable, because he simultaneously ascribed to it the meaning of a continuous movement:

> If two quantities are in such a relation to each other that the one decreases just as much as the other one increases, and vice versa, then they are called *opposite quantities* (Karsten 1768, 64).

With this definition referring to quantities, Karsten implicitly disproved Wolff's view that positive and negative quantities were heterogeneous. Atlthough the positive and the negative part can be considered independently and may thus seem to be of different kind, they can be subsumed in a common "superior" quantity and will then form a homogeneous quantity:

> Such opposite quantities, considered for themselves, are quantities of a different kind, or are to be regarded as having different denominations. However, they are always situated under a common principal concept, and can in so far be considered as quantities of the same kind (ibid., 65).

Karsten introduced the designation of a positive or negative quantity as an abbreviation in order to avoid naming the subconcepts and the principal concepts separately:

> If one assumes that one speaks of the principal concept which the two opposite quantities have in common, one can indicate each of them by negating its opposite (ibid., 66).

Karsten was very cautious in linguistic denomination; instead of *negative*, or *denied* quantity it was more appropriate to speak of a *denyingly expressed* quantity, and accordingly of a *positively expressed* quantity. Karsten has called the new meaning of this basic operation, which arose from adding opposite quantities, a "general addition" (ibid., 68), and their subtraction analogously a "general subtraction."

For the multiplication (and division) of opposite quantities as well, Karsten introduced a more general notion. Due to his preference for the traditional concept of quantity, he also relied on a traditional fundamental concept, on that of ratio, which he understood, however, in a way analogous to Hausen and Segner, largely as a concept to be thus arithmetized. The product results as the fourth proportional to unity and to the two given quantities ($1 : a = b : P$). To obtain this result, he extended the concept of ratio, without discussing whether that was admissible. For the four proportional quantities, he required not only the usual equality $A : B = C : D$, but in addition that if the first two terms were opposite to each other, the last two must be opposite, too (as well as the negated

---

[79]   Only at a later occasion did Karsten give an explanation of his concept of quantity within the "general art of arithmetic," saying that letters "can be understood as entirely general signs, by which one comprehends numbers, lines, and generally every other kind of the quantities" (Karsten 1768, 73).

case); Karsten thus obtained four cases of permissible ratios (ibid., 76 f.), which precisely expressed the rule of signs (ibid., 79).

It shows Karsten's grasp of the consequences of changing the foundations of mathematics by changing one of the basic concepts that he did not only reflect on the transition to more general concepts for operations, but that he noted the necessity to change the concepts of relation as well, implementing this for the concept of equality, and for the relation between larger and smaller. Regarding the former, he stated that it could also be applied to negative quantities. It could be said "with some reason, [...] that every negative quantity is smaller than any positive one" of the same kind, a statement that enabled him to show how to calculate with such relations:

$$-a < +a, \; -2a < +a, \; -3a < +a; \; -2a < -a, \; -3a < -2a.$$

With regard to the respective positive quantity, it could thus also be said that a negative quantity *was smaller than nothing* (ibid., 78).

Of particular interest is Karsten's reflection on the concept of equality, because it already contains a clear formulation of the absolute value. For opposite quantities to be equal, not only equality of their "quantity" (their value) is required, but also equality of their "position" (their direction):

> Two opposite quantities can never be called identical quantities not even if they are equal considered for themselves, without paying attention to their oppositness. By no means $+a = -a$, although a signifies the same line and means the same quantity. Here, however, only those lines are identical which have both the same quantity and the same position; and generally those quantities are identical only if not only their *quantitative* amounts can be substituted, but when they are moreover not opposite to one another (ibid., 77 f.).

At the same time, this quotation shows how helpful a separate term to designate equality of quantities (like "absolute value") would have been.

In the part on solving equations of this textbook, Karsten presented the multiplicities of roots (ibid., 251 ff.). For coefficients in equations, he did not explicitly state that they can likewise be positive and negative, but his practice made clear that he held this to be self-evident.

On the basis of these three important and influential textbooks, the extended number concept, together with an understanding accordingly generalized of the foundations, was well anchored at least in Protestant Northern Germany by the second third of the eighteenth century. The wide acceptance of this generalization and its effect are illustrated by two new developments:

- for the first time, there was historical and comparative reflection on the development of the concept of negative quantities;
- the concept of negative quantities was taken up in philosophy with the intention of philosophically reflecting on the foundations of mathematics, and to apply the concept within philosophy itself as well.

We shall first discuss how historical and critical reflection set in, because this development for the first time raised the problem of how to justify the concept of number beyond the level of its former, more or less implicit, treatment in the elaboration of textbooks.

In the literature on the history of negative numbers, it is quite unknown that a German historical-critical account about the negative numbers was published as early as the eighteenth century. This is indeed impossible to see from the treatise's convoluted heading: *Ueber eine Stelle in Herrn Lamberts Briefwechsel, von verneinten und unmöglichen Wurzelgrößen*. This 80-page essay had been written by the W.J.G. Karsten already mentioned, and was published in 1786! Its key proposition is that the concepts of positive and negative quantities gradually emerged "by abstraction" in the course of generalizing the solving of numerical problems. After giving an account of contemporary positions, Karsten wrote an overview of the concepts of mathematicians before his time, introducing it by saying:

> The way algebraists of old choose to present the nature proper of numbers designated by (–) is indeed not as completely convincing as nowadays expected from an author who wishes to explain the foundations of mathematical science: we must take into account, however, that we now stand on the shoulders of the first inventors, which makes it easy for us to see farther than they did (Karsten 1786, 209).

The principal aspect under which Karsten examines and presents earlier concepts is to what extent negative quantities were accepted as real mathematical objects or the extent to which no status of quantities was attributed to them being understood as defects. For most of the authors who used expressions like *numeros absurdos* (Michael Stifel) or *racines fausses* (Descartes), Karsten concludes that they were using such expressions merely as willful terms (*Kunstwörter*) while actually having the correct thing in mind. Generally, Karsten sees an unequivocal trend toward the state of the art accepted in the mathematics of his time.

Among the numerous authors Karsten studied, beginning with Stifel, Viète, Descartes, Schooten, and Newton, he basically criticized only Christian Wolff, for seeming to have misled "toward thinking [...] as if the negative was to be sought in the matter itself and not alone in the expression" (ibid., 241).

As eminent pioneers of the conception that is relevant for himself, Karsten, conversely, praised Hausen and Segner (ibid., 243). It is fascinating to see which non-German authors of his own eighteenth century Karsten mentions: while lightly praising the Englishman MacLaurin and slightly criticizing the Italian M.G. Agnesi, he makes no reference at all to any French author (with the exception of Claude Rabuel, as a commentator of Descartes). He is all the more severe, by contrast, is his wholesale condemnation of French authors, whom he accuses of adhering to an obsolete lack of rigor:

> Authors from abroad, in particular French, and even recent ones among them, continue to speak in this outdated manner, at least partially, but only when presenting the elements, not as conscientiously observing the necessary rigor as German authors do (ibid., 249 f.).

Undertaking historical research on how concepts have hitherto developed signifies a new level of metareflection that permits one to expect further impulses toward advances. An additional dimension of metareflection was opened some years after Karsten's publication, by an essay in a journal that, again as a first, realized a methodological–critical analysis of various conceptual approaches. Its author, Georg Simon Klügel (1739–1812), was Karsten's successor to the chair at the university of Halle. Klügel pointed out that one of the essential methodological reasons for conceiving of negative numbers was the striving for generalization in mathematics, toward summarizing and concurrently treating and solving related problems, thus identifying the concept of negative numbers as a prime expression of the analytic method. At the same time, he critically examined views rejecting, or restricting, the use of this concept, and was thus able to identify these views as expressions of the synthetic method. Klügel focused on English authors as proponents of this method; he did not quote any French authors at all:

> The analytic differs from the synthetic method particularly in that the former embraces several cases in a single formula, while the synthetic discusses each case separately. The reason for this is that analysis expresses the connection of the quantities by equations, and that it uses the general properties of the equations, as well as the rules for connecting them to give the value of each quantity by those belonging together with it, or to develop their relations. According to the synthetic method, one must seek a separate path for each problem, having no other general formulae for calculating than the proportions together with their modifications, except for the propositions already found. One must therefore always make an effort to discover identical ratios. While the synthetic method avails itself of such propositions which state an equality, it does not use algebraic equations (Klügel 1795, 312 f.).

For Klügel, this striving for generalization was inescapable for mathematics:

> Moreover, it is necessary to present all related cases of a connection between quantities in one calculation, to economize on repetition, and to avoid a too cumbersome set of propositions, as well as in particular to survey all the differences in a formula at a glance (ibid., 313 f.).

In Greek mathematics, Klügel said, such striving for generalization had been just as nonexistent as it was among some of his contemporary "recent" mathematicians (ibid., 311). By these "recent" mathematicians, Klügel meant his English colleagues who were stuck with the Euclidean method, trying to avoid the occurrence of negative quantities by considering separate cases, and thus to prevent any generalization:[80]

> Things here are just as with the geometry of the ancients, and of the Englishmen imitating them, according to whom negative quantities will not occur in any proposition, since it is determined in any case what is a sum, or what is a difference, and since it can never be demanded, for a given difference, to subtract the whole from the part (ibid., 316).

Klügel warned in particular against adopting and using the new sign suggested "by the English" for indeterminate subtractions (precisely the sign ~

---

[80]   Klügel's concept was contradictory, however, because he suggested, at the end of his paper, a method to avoid negative quantities and to operate only by absolute quantities (cf. below).

used later by Frend as well),[81] a sign intended to ensure that the result of subtractions was always positive. Noteworthy and instructive is Karsten's and Klügel's growing awareness of cultural difference, and in particular their tendency toward national distinction. Klügel added this warning: "I [find] it necessary to stand up against this innovation, and to make sure that German analysts will not fall for it" (ibid., 471).

Besides this onset of historical-critical reflection, the conceptual consequences of introducing negative numbers were analyzed and developed further in philosophy. Immanuel Kant pioneered this philosophy of mathematics with his 1763 essay "Attempt to introduce the conception of Negative Quantities into Philosophy." For his mathematical foundations, Kant explicitly drew on Kästner:

> Perhaps no one has determined in a more clear and definite way what is understood by negative quantities than the famous Mr. Professor Kästner, in whose hands everything becomes precise, comprehensible, and enjoyable (Kant 1968, 782).

Kant's contribution to the foundations of mathematics consists in addressing the question always underlying the controversies on the status of negative numbers, the question what "nothing" means, in treating the nothing that occurs when opposite quantities are combined, in stating it as an explicit problem, and even in having proposed a solution for this problem. Kant's solution consists in differentiating between logic, or philosophy, and mathematics.[82] While an absolute nothing emerged in cases of a logical opposition or contradiction, a relative nothing resulted when two quantities mutually "canceled" one another, this relative nothing being the mathematical zero. Kant conceived of this differentiation conceptually as of the difference between a logical repugnancy and a real opposition:

> THINGS contrary to one another means that one cancels what the other effects. This contrariety is twofold, either logical by way of contradiction, or actual [real] *i.e.*, without a *démenti* [contradiction]. The former, the logical, is that which so far alone has claimed attention. It involves both assertion and denial of one and the same thing. The conclusion of this logical relationship is nothing at all (*nihil negativum irrepraesentabile*), as the Principle of Contradiction expresses it. [...] The second, the actual [real], is where two predicates of one thing conflict, but not by the principle of contradiction. Here also the one cancels what the other causes, but the result is something (*cogitabile*). [...] For the future we shall call this actual nothing, Zero = 0.

---

[81]   Klügel did not reproduce the sign, but referred for this to a paper in volume 51 of the *Philosophical Transactions* (i.e., from 1760), without naming its author, cf. Section 2.10.3. above. In addition, Klügel mentioned that the Italian Antonio Cagnoli (1743–1816) had also used this sign in his trigonometry textbook of 1786.

[82]   Kant had been prompted to this by a treatise of the philosopher C.A. Crusius, in which Crusius in 1749—discussing Newton's contrasting pair of attraction and repulsion—had understood negative quantities as negations of quantities; hence, Crusius had claimed attraction like repulsion to be "positive" effects or positive causes (Kant 1968, 781; Crusius 1774, §295, 739 f.).

A detailed analysis of Kant's contribution was published by M. Wolff (M. Wolff 1981, 39–77).

[...] It is easily discoverable that this Zero is a relative nothing, [since, namely, only a certain consequence does not occur] [...]. The *nihil negativum* is not therefore to be expressed by Zero, because Zero involves no *démenti* [contradiction]" (ibid., 783 f.; Engl. transl. by Irvine 1911, 117–119).[83]

By means of this conceptual differentiation, Kant successfully deontologized the "nothing" in mathematics, attributing to it the exclusively relational character of the zero. This relativization at the same time accepted that negative numbers enjoyed the mathematical status of real numbers. The important step away from metaphysics, and toward the program of algebraization of mathematics, this meant was rapidly adopted in Germany, whereas France and England did not follow suit for a long time to come.

In Germany, embedding the process of generalization in mathematics in the entire scientific debate as a whole contributed to a stronger acceptance of abstract concepts within the larger cultural context as well. Moses Mendelssohn's change of view concerning negative numbers on the basis of his scientific correspondence provides an informative example of this, as has been shown by Hans Lausch (Lausch 1993). Mendelssohn (1729–1786), a culturally highly influential philosophical author and significant representative of Enlightenment in Germany, used algebraic modes of expression in one his essays on sensations as well. Comparing this essay's three editions of 1755, 1761, and 1771 reveals differences of essence in the propositions made on negative quantities. While the 1755 text qualified isolated negative quantities as an "absurdity" (quoted from Lausch 1993, 25), obviously under the spell of Wolff's textbooks, Mendelssohn's debate with Thomas Abbt (1738–1766), a philosopher of Enlightenment with an interest in mathematics, made Mendelssohn modify this point in the second edition of 1761. Abbt had tried to convince Mendelssohn of the "reality" of the negative quantities. One might "conceive negative quantities also in abstracto and think of them as entities without positing anything positive, provided only they are based on a relation or a position, or on something similar" (ibid., 28). After this, Mendelssohn deleted the qualification of absurdity, proposing the concept of the two modes of being instead:

> With respect to magnitude, a negative quantity is not distinct from a positive one at all, but it is distinct with respect to the operation which is to be executed with this quantity (ibid., 29).

In 1771, eventually, Mendelssohn referred to Kant's treatise quoted above, adding as an afterthought: "Herr Abbt is thus right in stating that a negative quantity is something just as real as a positive quantity" (ibid., 35).

Due to the wide dissemination of mathematical culture in Germany during the last third of the eighteenth century, more and more mathematical textbooks were published, and it is no longer feasible to discuss all these here. Generally, it can

---

[83] David Irvine did not translate Kant's preface. Some words used by him are peculiar, so that the literal translations are given in square brackets, such as the half sentence, which he omitted to translate.

be said that concepts adhering to, or enhancing, those proposed by Kästner or Karsten prevailed.

One very informative example, however, shall nevertheless be quoted here, since it shows how the philosophical reflection, in particular Kant's, contributed to the emergence of a separate foundational discipline of mathematics. The example is J.G.E. Maaß's (1766–1823) textbook *Grundriß der reinen Mathematik*. Maaß was professor for philosophy at the University of Halle, who lectured on mathematics side by side with Klügel.

In this textbook, Maaß understood *logic* as the basis for the mathematical operations. In particular, the concept of oppositeness arose from logic. Maaß thus presented logic even before arithmetic proper, saying that two other relations with regard to quality could occur with numbers beside quantitative relations: having the same direction (*Einstimmigkeit*) or oppositeness. The quality of oppositeness signifies that numbers "cancel" each other if one wants to unite them. The quantity considered as canceled is called positive, and the canceling quantity negative (Maaß 1796, 18). After having established these foundations, Maaß introduced the basic arithmetic operations (ibid., 23 ff.).

The authors analyzed up to now were all active at Protestant universities in Northern Germany. A last example may show, however, that the Catholic south was not exempt from this process of generalization: a *Handbuch der Elementar-Arithmetik in Verbindung mit der Elementar-Algebra*, published in 1804 by Andreas Metz (1767-1839), first a *Gymnasium* teacher, after 1798 professor for philosophy at the University of Würzburg, and after 1805 for mathematics as well. As Metz explained in its preface, his textbook intended to integrate the modern theories from Segner's, Kästner's, Karsten's, and Schultz's textbooks.[84] Indeed, he inserted a general part on the foundations of arithmetic before his arithmetic proper, introducing, among other notions, the concepts of homogeneity and heterogeneity of quantities. Metz explained the concept of number to be an object of "general pure *Mathesis*" (Metz 1804, 3). According to his view of opposite quantities, these were not to be considered as homogeneous in themselves; they had first to be transformed into "unanimous" quantities: he hence understood assets and debts to be different qualities, but debts, for example, could be regarded as a negative asset and therefore as homogeneous quantities as well (ibid., 49). He also established the purely algebraic proposition

$$-7 < -3 \text{ (ibid., 53)}.$$

He justified the rule of signs by logic; positing the negative negatively meant to posit it positively (ibid., 59).

In Germany, too, however, the concepts did not develop "free of contradictions," nor even continuously, say, in the sense of a progressive algebraization. The unresolved problems concerning the extension of the basic operations, and concerning the relation between the concept of number and of

---

[84]   For Johann Schultz, see below in this Section.

function, also challenged both concepts established and the relation between algebra and geometry.

Johann Andreas Christian Michelsen (1749–1797), mathematics teacher ("professor") at the *Berlinisch-Cölnisches Gymnasium* in Berlin after 1778, provides the first instance of such "diverging" views. He not only authored numerous textbooks on elementary mathematics, but also translated several of Euler's eminent textbooks from the Latin into German: the *Introductio* (1788) and the *Differential Calculus* (1790). Michelsen supplemented these translations by extensive annotations and annexes also criticizing some of Euler's central ideas, attempting to prove alternative concepts. The spirit of this critique and his revisions completely correspond to his own fixation on the idea of striving for rigor in the foundations of mathematics, and of attaining it.

In his textbook *Buchstabenrechnung und Algebra* (1788), he presented opposite quantities so as to show no obvious discrepancy to the now habitual form in Germany (Michelsen 1788a, 58 ff.). In reflections on foundations laid down in his *Gedanken über den gegenwärtigen Zustand der Mathematik* (1789), and in his translations of Euler, however, he presented a divergent concept; his propositions are noteworthy not only as an adaptation of d'Alembert 's view on negative numbers, but also as a response to Euler's view of the logarithm function.

Like d'Alembert, Michelsen postulated the generality of functions, i.e., requiring that all functions should be defined for all number areas and not be subjected to any restrictions in their domain of definition. Hence, in a mode analogous to that of d'Alembert, Michelsen rejected in principle Euler's restriction of the logarithm function to positive real numbers; rather, it should be defined for all real numbers, and for imaginary numbers as well (Michelsen 1788b, 522). In addition, he accepted only *one* value for the function in every case. As the reason for Euler's alleged "difficulties" with the logarithm function, Michelsen attributed to him "not sufficiently clear and determined concepts of positive and negative numbers" (ibid., 525).[85]

His alternative concept was precisely an explication of d'Alembert's ideas, which had partially remained implicit as yet (cf. above, Section 2.9.3). Particularly noteworthy is the fact that Michelsen had been the first to extensively declare absolute numbers to constitute the basic concept and to

---

[85]   Karsten, who in 1786 analyzed the controversy about the logarithms of negative numbers in a detailed treatise, criticized Leibniz, like Bernoulli and Euler, for having "premised the concept of what a negative quantity or a negative number was as being entirely known," while d'Alembert had been the only one to state that the concept of negative number needed to be clarified beforehand. Actually, d'Alembert's explanation of this concept, Karsten said, was "just as little satisfying" (Karsten 1786, 294 f.). Karsten reviewed d'Alembert's arguments against Euler in great detail, essentially rejecting them. In particular, he sharply criticized d'Alembert's definition of the logarithm via general ratios as too unspecific (ibid., 333 ff.). He emphasized that it was a common motivation of numerous such authors to obtain that roots of a negative quantity be negative, and not imaginary (ibid., 328).

propagate the distinction between absolute and positive numbers as necessary (e.g., Michelsen 1789, 20). Michelsen understood numbers as sets of units; absolute numbers formed the basis, as repeatedly taken unit (Michelsen 1788b, 525). For him, the positive and the negative numbers presented "a special kind of concrete numbers" (ibid., 526). He declared the set of their parts, the absolute number, to be the "essential" property, whereas their additional designation as positive or negative numbers constituted "something random" or external (ibid).

Michelsen thus attained, like d'Alembert, the goal set for the context of the logarithm function, of maintaining the logarithm as everywhere defined and of obtaining only real values at the same time. For the negative numbers, in this same context, were virtually identical with the positive, because one just had to modify their "absolute number by adding a random determination" (ibid., 527). Like d'Alembert, Michelsen only posited his view, but did not apply it operatively, nor develop it systematically in a textbook, say, for algebra.

It is telling for Michelsen's awareness of the problems of founding the number concept that he not only repeatedly pointed out the necessity of differentiating between absolute and positive numbers, but also always emphasized, every time he discussed the problem of how to conceptualize opposite quantities, that one of the basic problems consists in how exactly to grasp an adequate conception of multiplication (cf. Michelsen 1789, 18). In his discussion of the conceptual problems, however, Michelsen focused on application to geometrical quantities; for him, the multiplier, if it was a quantity at all, did not become transformed into a scalar by tricks or ad hoc definitions; rather, the product should contain its elements, and in this way he attained, by multiplication, quantities of higher dimension: the product of two lines to yield a rectangle, and that of three lines to yield a parallelepiped.[86] In his algebra textbook, however, Michelsen had not defined multiplication other than as was standard in Germany since Hausen,: as a ratio (Michelsen 1788a, 65), although he had declared the application of this multiplication concept to opposite quantities to be dubious in his critical *Gedanken...* (Michelsen 1789, 37). Michelsen thus remained quite inconsistent in his efforts at establishing rigorous foundations.

An almost immediate response to Michelsen came from Johann Schultz (1739–1805). Schultz, preacher to the Prussian Court and after 1786 mathematics professor at the University of Königsberg, close friend of Kant and the latter's mathematical adviser (cf. Schubring 1982), not only developed noteworthy conceptions on the infinite, but also contributed to the foundations of the number concept.

In his quite influential textbook *Anfangsgründe der reinen Mathesis* (1790), he began with numbers as the fundamental idea. More explicitly than other

---

[86]   Michelsen went as far as to geometrically admit—to my knowledge, as the first since Viète—a "product by more" lines than three, thus admitting higher-dimensional products! He stressed that such a product has to be considered "as a quantity of just so many dimensions, not representable in intuition" (Michelsen 1789, 176).

authors, he stressed that absolute numbers constituted the basic concept: numbers by themselves were neither positive nor negative, and nor were they opposite to one another. Furthermore, he emphasized that only "unanimous" quantities, i.e., quantities of the same kind, were comparable. Opposite quantities were inhomogeneous in themselves and had to be rendered unanimous, of the same kind, before one were able to operate on them (Schultz 1790, 120 ff.).

It is particularly interesting that Schultz inferred consistently from all those manifold debates about how to multiply negative quantities that multiplying quantities by quantities must be systematically *excluded*, and only multiplying quantities with numbers admitted. Schultz presented a systematic deduction of this restriction to multiplication, also visualizing it by differentiating according to signs. For purposes of multiplication, he distinguished between multiplier and multiplicand. Only a number is permissible as multiplier, the latter being expressed by letters m, n, etc., whereas quantities designated by letters a, b, etc. were permissible as multiplicands, in the sense of *quanta* (Schultz 1790, 61). Since Schultz wrote the multiplier on the right, he consistently symbolized multiplication in the following form:

$$a \times n.$$

By means of these forms, Schultz also discussed the rule of signs, by always returning to a multiplication by an absolute number. Thus he first transformed, say, $-a \times +m$ into $-a \times m$ and then interpreted the product $-a \times m$ as a quantity to be subtracted (ibid., 127 ff.). According to Schultz's view, multiplying negative quantities by one another was not possible, the possible operation only being to multiply a negative quantity by a number (ibid., 130). Nevertheless, he attempted a proof of the rule of signs that implied a *petitio principii* (ibid., 131). It was left to Förstemann (1817) to elaborate a really consistent separation of quantities from numbers (cf. Chapter VII).

How d'Alembert's ideas had been taken up in Germany, perhaps conveyed by Michelsen's publications, and how absolute numbers were differentiated from positive numbers, can be elicited from some of Klügel's writings. In an annex to his 1795 essay containing his reflections on how to methodologically justify negative numbers (cf. above), he demonstrated, quite in contrast to the generality of method required in the rest of his text, how to avoid "the concept of opposite quantities," by restricting oneself to "*absolute* quantity," that is by assuming "that the *Subtrahendus* is less than the quantity from which the subtraction is made" (Klügel 1795, 479). If the *Subtrahendus* was larger, however, a "different case" was present "from that for which the calculation was intended" (ibid). In such a "different" case, the conditions and with them the algebraic signs concerned should be changed in the "initial equations," according, in fact, to d'Alembert's and Bézout's conception. It is quite characteristic that applying this approach in the annex, contradicting his general views in the main text and excluding opposite quantities, consisted in declaring the logarithms of negative numbers to be real numbers, by virtually identifying positive numbers with negative ones: "The *Logarithm of a denied number* is identical with the logarithm of the same number regarded as positive; [...]" (ibid., 481).

Actually, Klügel only postulated this view and did not try to present the operative use of logarithms conceived in such a way. In his entry of 1805 on opposite quantities in the highly influential *Mathematisches Wörterbuch* he edited, he did not take up this "French" concept (Klügel 1805, 104 ff.).[87] While agreeing without problem to operative concretizations, he alluded to his persisting epistemological reservations against negative quantities in the entry "equation" of the same encyclopedia. He declared that "the negative roots serve to unite two equations differing only in their algebraic signs, into a single equation" (ibid., 364). This assumes hence an epistemological distinction of the positive from the negative, which are only afterwards composed to an external whole for purposes of operating.

Klügel's "alternative" positions illustrate a conflict that was basic for the stage of the generalization process discussed here. Does the more general meaning constitute the simplest version of the concepts, that which is subject to the least number of conditions, or is the more general meaning that stage of degree of conceptual development which permits the largest number of applications?

This conflict was explicitly reflected in a series of publications in 1799 and 1800 by a mathematician otherwise unknown; these treatises illustrate that it was impossible to solve the problem of how to justify negative numbers within the frame of an understanding of mathematics already directionally developing toward algebraization but still being determined by the traditional concept of quantities. Analyzing these writings will hence conclude our part on the development in Germany—before the genuine transition to algebraization. The author of this treatise in a series comprising three parts was Peter Johann Hecker (1747–1835). Born as a son of a parson, he was a student at the renowned *Realschule* directed by his uncle J.J. Hecker in Berlin. Later he studied theology in Halle, and mathematics with Segner. Beginning as a mathematics teacher in Berlin, he was appointed as Karsten's successor at the tiny University of Bützow in 1778. In 1789, upon the dissolution of this university and its union with that of Rostock, he changed to that university.[88] Published by him as the university rector, as a *Programmschrift* for the three occasions of Christmas 1799, and Easter and Pentecost 1800, his reflections about the operations on opposite quantities were to remain his first and only mathematical publication, but one

---

87   In his textbook on arithmetic and algebra of 1792, Klügel did not introduce opposite quantities at all, mentioning negative quantities only marginally—in his section about progressions—without explaining them operatively, in Wolff's sense, as an indication of a "deficit" that one can represent, say, by a debt (Klügel 1792, 50 f.).

88   I owe these hitherto unknown biographical data to Peter Strassberg's dissertation, *Peter Johann Hecker - Mathematiker und Kalendermacher? [...]*, Universität Rostock 1988, which Wolfgang Engel was so kind to place at my disposal in November 1988.

testifying to his acute penetration of contemporary foundational problems. Noteworthy in Hecker's approach—besides his profound knowledge of the pertinent and recent literature—is his inclusion of a mathematical field that had barely been tackled until then in the debates about negative numbers despite the fact that it constituted one of its major areas of application: trigonometry. Hecker attempted to draw on this *application* for *justifying* the concept of negative numbers.

Hecker criticized the presentation of the concept of opposite numbers in the (German) textbooks, since they did not make clear this elementary topic to the students. His first objection was that many authors subsumed opposite quantities virtually only by accident under a "superior" concept. One had to suppose, he said, that quantities stop being opposite as soon as one no longer considered them under the alignment to this superior respect (Hecker 1799, 6 f.). Hecker did not understand his own critique as general; essentially, however, it aimed at a transition from the substantialist concept of quantities to the abstract concept of number.

His second criticism concerned the fact that the meaning of the basic operations in arithmetic was being treated differently from this meaning in algebra. The most frequent way of presenting them was to define the basic operations "at the beginning of arithmetic" for positive numbers, resulting in addition signifying real increase and subtraction likewise real decrease, but to assign another meaning to the operations later upon treating opposite quantities, resulting in subtraction often being executed by addition. Then, the signs + and – were also attributed new, additional meanings (ibid., 9 f.). Hecker quoted Segner's and Karsten's textbooks as examples of how to avoid this contradiction. These textbooks explicitly distinguished between addition, subtraction, etc. in arithmetic and general addition, subtraction, etc. in the "general art of arithmetic." This approach caused less difficulty for the learner, Hecker claimed, since the "concepts for these kinds of calculating were only extended [...], and not changed" (ibid., 11).

Here, the keyword of the *extension of the number field* was formulated for the first time. Hecker, however, was not satisfied with this solution. Instead of changing, or extending, the concepts later, he said that they could be taught from the very beginning in their more general form. This was not more difficult for the beginner, he claimed, and arithmetic and the general art of arithmetic were not divergent sciences; rather, the latter was a part of the former (ibid., 11 f.). To attain readily teachable presentations of that kind, however, required solving a conceptual problem, that of finding a general notion of the operation of multiplication. As Hecker stated, the common understanding of multiplication in arithmetic as a repeated addition contradicted the practice of multiplying quantities. The explicit conception, he said, was,

> that the multiplier generally must be a number, namely an abstract (*unbenannte*) number, and that multiplication be possible in no other case. [...] It will always remain obscure to the beginner whence it came that while one could add and

subtract, e.g., lines and lines, and divide them by one another, one could not multiply them? (ibid., 14).

Hecker attached the following note to the word "multiply": "At least not without changing the concept of multiplication." Hecker established in clearer and more explicit terms than all his predecessors that the problem of the rigorous justification of negative numbers was linked to finding an adequate definition of multiplication. In particular, he stressed that he did not consider its common definition via reiteration of the unit to be promising, since inferring the multiplier from the unit was not possible if the latter was a negative number (ibid., 16).

In his three essays, Hecker attempted to develop general concepts for the four basic operations, which were be applicable to negative numbers as well. In the end, he failed upon grappling with multiplication, since he fell back on the concept of geometrical quantities for this purpose. Despite his clear diagnosis that the basic operations had hitherto not been introduced in a sufficiently simple and general manner, Hecker considerably complicated the conceptual construction and did not succeed in solving the problem proper.

An additional element of his conceptual construction was that he added, as the "most natural," a "theory of proportions" to those "doctrines which must be taught first in pure mathematics," before he felt able to develop the doctrine of opposite quantities (Hecker 1800a, 12). Hecker needed such a general theory of proportions as foundation, since his goal was not only to introduce multiplication as a proportion, as Segner and Karsten had done, but likewise all the four basic operations. Thus, he introduced addition and subtraction as forms of an arithmetic ratio, and multiplication and division as forms of a geometric ratio (ibid., 13 ff.). Even for this general introduction of the operations, notwithstanding the fact that it relied on quantities, Hecker was unable to avoid assigning an exceptional position to multiplication. Since the multiplier constituted the "exponent" of a geometric ratio, it had inevitably to be "an abstract number," in order to ensure that the product of a line with, say, a line was excluded (ibid., 20 f.). Hecker thus had to note that his intended general multiplication could not be introduced in an elementary way, but only after the general theory of proportions (Hecker 1800b, 29).

For introducing the opposite quantities themselves, Hecker integrated an even more extensive concept field: the concepts of variable and of function (ibid., 8 ff.). His intention in doing so was to conceptually improve Segner's and Karsten's attempts at justifying negative quantities by alluding to a continuous transition of a variable quantity by decrease—across its vanishing at zero—until it assumed the opposite value. For this purpose, Hecker used the application of opposite quantities in trigonometry, with trigonometric functions. Since each of the two arcs a and b could assume, for the trigonometric functions of sine and of cosine, "all possible values, by increase or by decrease," it was legitimate, he said, to consider "them as *variable* quantities, and the quantity sought as a *function* of both" (ibid., 9). For Hecker, the trigonometric functions provided a suitable model for conceptually grasping the notion of opposite quantities, namely as "transitions from decreasing to increasing values via their vanishing" (ibid.).

Rather far from his intention proper of introducing the basic operations in a way so simple and so general that they could be lucidly taught to beginners, Hecker now had even included parts of analysis into the foundations of arithmetic to be able to introduce negative numbers. Indeed, Hecker had not only demanded that opposite quantities be understood in such a manner that "the given quantities of a problem are to be considered as *variable quantities*," and that "the quantities sought [be regarded] as a *function* of them" (ibid., 12); he had moreover premised the validity of the *law of continuity* for securing the effective transition to the opposite value (ibid., 15). Hecker was able to legitimize this further additional assumption only by an ontological appeal to geometrical quantities (ibid).

While Hecker's treatises certainly constitute a culmination point in the reflection on foundations, and in the context of concept fields of mathematics concerned with justifying negative numbers, they also clearly demonstrate that no further substantial progress was achievable within the scope of the traditional concept of quantity, and that a breakthrough would necessitate finding some other way.

2.11. Looking Back

The problem of elaborating a coherent mathematical status for negative numbers developed over long periods of time, assuming at last a quite fundamental character because—or although—negative numbers became an important operative tool in an increasing number of concept fields in mathematics. The problem originated from the generalization of only one operation, subtraction, and it occurred in almost all of the great cultures. It challenged the traditional first understanding of mathematics, its first "paradigm" in Kuhn's terms, its understanding of being a science of quantities: of quantities that, while being abstracted to attain some autonomy from objects of the real world, continued at the same time to be epistemologically legitimized by the latter. The various cultures succeeded over a long time in finding various auxiliary constructions that permitted them to remain within the existing paradigm. Where the geometrical concept of quantity and the synthetic method prevailed, this could be obtained by considering individual cases. This meant that operations had to be limited to subtractive quantities. The fact that this meant that generalization had to be given up in treating problems became itself a problem if the operations of multiplication and division were to be extended to this new kind of quantity.

It is typical that it was still possible, during the period of exclusive dominance of the geometric concept of method and quantity, to legitimize operations by limitation to subtractive quantities, and that insurmountable conceptual conflicts began to become manifest only as the independent function of *algebra* started to grow toward the end of the Middle Ages and the beginning of modern times in Europe. The massive increase of algebraic methods which were relatively independent of geometric methods gave an impetus toward expanding the operations of subtraction and multiplication (division being treated as a corollary to multiplication because it functioned analogously), but this tested the limits of the paradigm of quantity.

In three countries of Europe, different mathematical cultures gradually evolved in modern times, which at the same time developed different approaches in dealing with the problem of generalization. Due to the efforts of the Malebranche group, respectively of Wallis and Newton, the path toward algebraization first seemed to be open both in France and in England. Contributing to the *querelle des anciens et des modernes*, they confidently propagated the superiority of modernity. This development, however, was ended by a rupture, obviously a side effect of Berkeley's attack against Newton's foundation of analysis. In England and Scotland, this rupture led to a return to the traditional concept of quantity and to the synthetic method. The prevailing attitude in France was ambivalent in not denying mathematical legitimacy to negative quantities on the one hand while at the same time suggesting a return to the synthetic method because of the necessary consideration of individual cases. In Germany, the concept of opposite quantities was elaborated for the most part without ruptures. Here again, however, algebraic generalization of operations and maintaining the quantity paradigm became increasingly irreconcilable, leading to eclectic "epicycles."

Acceptance, respectively rejection, of the zero adhered widely to the same patterns as those of the negative quantities, since legitimizing the zero via the concept of quantity was analogously problematical.

Euler's radical algebraization, which was not tied in with national cultures, met with violent resistance in many of them, leading to first cases of fundamentalist exaggerations.

All paths of solution within the prevailing paradigm having been exhausted, it was suggestive to assume that only radical solutions by different means would be able to deal with the "Gordian knot." Carnot's (1801/1803) and Förstemann's (1817) approaches were such radical steps.

The distinctive ways the concept of number was conceived of would take effect on the mathematization of limit concepts in analysis.

Chapter III

Paths toward Algebraization—The Field of Limits: The Development of Infinitely Small Quantities

1. Introduction

After using the case of the concept field of negative numbers to examine how the concept of number developed until about 1800, this chapter will analyze how limit processes were conceptually grasped, and how they evolved by differentiation.

In contrast to the case of negative numbers, in which the concept was developed almost exclusively in practice, that is, mainly in textbooks, and where independent conceptual reflections and treatises bearing on these began to appear, except for some rare exceptions, not before the middle of the eighteenth century, the corresponding development for the concept field of limit processes was characterized from the very beginning by the fact that the problem was reflected theoretically and at the same time presented for practical purposes of teaching.

Another contrast to the history of negative numbers is the fact that the conceptual history of limit processes has been exhaustively studied. Since there is an almost bewildering host of literature in particular concerning the developments in the eighteenth century and their previous history in the seventeenth, this historical field might appear to have been already exhaustively treated. In fact, however, there is something left to be desired in spite of this plenitude of literature. The historical studies have discussed and presented the concept developments predominantly as an intra-*mathematical* one. In the seventeenth and eighteenth centuries, however, the concept field of limit processes formed a broad context not confined to mathematics in which these issues were controversially debated among theology, philosophy, mathematics, and physics, and in physics's subdepartment of mechanics. This concept field had not yet been developed into separate disciplines, not in England, France, or Germany. Concept formation focused on epistemological enquiries into the nature of limit processes and of continuity; such debates were led primarily under aspects of *mechanics*, not under aspects of mathematics.

Since these issues have not been sufficiently researched as yet, our analysis will focus on the relationship between mathematics and mechanics. In a way analogous to our use of negative numbers in the development of the concept of number as red flags to indicate conflict about the tendency toward algebraization, the concept of infinitely small quantities will serve as indicator for the modes of algebraization concerning limit processes.

Here again, the period around 1800 marks an obvious step. In Germany, mathematics had been segregated from the complex field linking it to other disciplines. Theory formation had become increasingly independent, leading eventually to a rigorous foundation of analysis via its arithmetization. In France, an analogous process seemed to prevail toward the end of the eighteenth century: within the comprehensive social process of modernization after the Revolution, the system of education was radically reconstructed on the basis of optimistic views that knowledge was teachable, and the analytic method generally applicable, as Condillac had said. For mathematics, this ensured the massive dominance of the program of algebraization, and the reshaping of mathematics thanks to reflecting on its signs, and on its language. Soon after 1800, however, the pendulum moved the other way in a mode I should like to describe as the "rétour du réfoulé." Side by side with re-establishing pre-Revolutionary structures, the defamed synthetic method was restored as dominant value, and the requirement that concepts be generalizable was replaced by that of their being of easy intuitive grasp. Lazare Carnot, first inclined toward the analytic method, became a propagator of the *rétour* to the dominance of the synthetic, not only for negative numbers, but also for limit processes, replacing the concept of limit by that of the *infiniment petits*.

Since the ruptures in the two concept fields occurred as parts of the same process, this again elucidates the connection between the two basic concepts. We will thus begin by presenting the development up to the establishment of limit processes about 1800, and then examining these ruptures in more detail. Our presentation will take into consideration both the theoretical reflections in essays submitted to academies and in individual publications, and the practical applications in textbooks.

An analysis of how the meaning of infinitely small quantities developed has not been undertaken as yet, except in connection with biographies of some mathematicians. This analysis is all the more called for, since a regular boom in favor of infinitely small quantities has set in since the 1980s as a consequence of establishing non-standard analysis (NSA). Both those in favor of NSA and its practitioners always referred to the earlier mathematical use of infinitely small quantities to legitimize its present development, never omitting to mention the alleged suppression of this promising practice in the wake of the mathematical purism and rigor of the Weierstraß school, just as authors of mathematical-historical works in which actually infinite or infinitely small quantities occur see fit to refer to NSA work in order to show how up to date their own historical work is. The references of present-day NSA theoreticians and practitioners, however, have been confined, since Lakatos's (1976) first attempt of this kind, to claiming that Cauchy was one of the founding fathers of NSA.

The remarkable point is that the emergence of NSA was just as determined by the didactical problems of teaching analysis as returning to the synthetic method of the *infiniment petits* had been in France after 1810/11. The theory of the foundations of differential and integral calculus has always been fraught with substantial problems for beginners of all times, and has led to widely different approaches to reducing their problems of understanding.

The NSA developed by Robinson indeed acquired fame and effect in mathematics only after Keisler's textbook *Elementary Calculus* (1976), which not only was the first to use NSA as an alternative approach for teaching differential and integral calculus, but also claimed that this method was more accessible for students. In a review that has become famous, Errett Bishop interpreted the Robinson/Keisler approach as a response to the quality problems encountered by *undergraduates* studying mathematics at US-American colleges and universities (Bishop 1977). This evaluation as a situation-specific response to problems in US-American higher education has always been most sharply refuted.[1] At the 7th International Congress on Mathematical Education of 1992, Hodgson explained the didactical reasons for the NSA approach, quoting the dominant role of Keisler's textbook as a case in point. At an English university, he said, this textbook had been in use as a basis of teaching analysis during the last decade with the objective of avoiding the "pitfalls" of the traditional approach, and of fostering the development of intuition (Hodgson 1994, 166).

A quite analogous approach, however, was undertaken almost simultaneously in the Federal Republic of Germany: it used the concept of Lipschitz continuity and thus of the "benign" functions developed since the end of the sixties by Hermann Karcher in Bonn, and elaborated and developed further by Herbert Möller in Münster (cf. Möller 1981). Although their approach is likewise oriented toward reducing the problems of understanding in the foundations of analysis by means of intuitively more accessible concepts requiring a smaller formal apparatus, it has been quite unable to win as much celebrity and as many adherents as NSA. There are only isolated cases in which this German approach was adopted in textbooks of secondary education.

2. From Antiquity to Modern Times

Clarifying continuity's conceptual context and status proved to be a foundational conceptual problem for mathematizing limit processes. Since continuity was for

---

[1]   Cf. the contributions to the first year of the *Mathematical Intelligencer.*

a long time understood to be a law inherent to the processes of nature, mathematics was at the same time destined to model this very nature, and a mathematical theorization independent of this ontological tie was inconceivable. Where debates on continuity took place prior to the eighteenth century, these were theological–philosophical, respectively physical-mechanical debates about the general validity of the law of continuity, and about the latter's consequences for the structure of matter and for particular laws and phenomena of physics. The various views on the structure of matter due to different prevailing epistemological positions led to distinctive mathematical modeling, and hence to different concepts of limit processes.

CONCEPTS OF THE GREEK PHILOSOPHERS

It is a remarkable fact that a considerable part of the debates in early modern times corresponds to the discussions already led by the Greek philosophers of antiquity. The theoreticians of early modern times relied to a large part on concepts shaped in antiquity.

The primary theoretical problem tackled by the Greek philosophers of the fifth century BC consisted in the attempt to clarify the constitution of matter. Thus, the Pythagoreans advocated the hypothesis that all bodies were composed of material points, or monads. The Eleats, in contrast, claimed that matter was indivisible and integral. The very purpose of Zeno's famous paradoxes was to prove that the atomistic–mathematical view led to aporias (cf. Vita 1989, 218). According to another interpretation, Zeno's goal was to disprove Anaxagoras's atomism (Mau 1954, 18 f.).

Eventually, two opposed currents formed. The atomists—first Leukipp, then Democritus, and later Epicurus—advocated that there were final elements that were indivisible. Since the properties of geometric figures were modeled according to those of physical bodies, they assumed the existence of indivisible elements for geometry as well. Thus, Democritus adheres to his own atomism not only for the physical division of concrete bodies, but also for mathematical–geometrical division (ibid., 22).

The most prominent representative of the opposition is Aristotle. In a sharp polemic against the atomist current, he presents his own conception of continuity in Book VI of his *Physics*. Continuity, for him, is the basic concept characterizing every quantity divisible into an arbitrary number of elements. The essential thing in his view is that these elements, which are divisible in themselves again, are understood to be homogeneous to the initial quantity (cf. Vita 1989, 219). A typical definition of continuity in Aristotle is the following: "And it is manifest that any continuum is divisible into parts without limit" (Aristotle, *Physics*, VI, Chap.1, 231 b 16). Among these continuous quantities, Aristotle names space, time, and motion.

Beside this determination of divisibility, Aristotle provides a further determination about succession (two elements are successive if no other element homogeneous to them can be interpolated between them) and contiguity (two elements are not only successive, but also touch):

> For I mean by one thing being continuous with another that those limiting extremes of the two things in virtue of which they touch each other become one and the same thing, and [...] are held together, which can only be if the two limits do not remain two but become one and the same (ibid., V, Chap. 3, 227 a 11–14; Transl. Wicksteed).

Quite similarly, Aristotle says in his *Metaphysics*:

> The continuous is a species of the contiguous. I call two things continuous when the limits of each, with which they touch and by which they are kept together, become one and the same, so that plainly the continuous is found in the things out of which a unity naturally arises in virtue of their contact (Aristotle, *Metaphysics*, XI, Chap. 12, 1069 a 4–7. Transl. W.D. Ross).

It is quite obvious that these determinations of concept are modeled after relationships between physical bodies, that mathematics has been formed as an abstraction from physics. For Aristotle, the idea that the elements extend materially is quite central: on the one hand for proving that the elements are homogeneous with the whole, and thus for legitimizing further divisibility, and on the other hand for refuting the view according to which the straight, for instance, is composed of single points.

Such points, accordingly, would have to be conceived of as indivisibles, while it was true for the latter that "the indivisible has no parts" (*Physics*, VI, Chap. 1, 231 b 3). A continuum, however, could not be composed of indivisibles, since the latter formed separate wholes that had no common boundary. Aristotle profoundly reflected this issue, giving an exhaustive justification, which shall be quoted here because of its persistent influence on later developments:

> No continuum can be made up of indivisibles, as for instance a line out of points, granting that the line is continuous and the point indivisible. For two points cannot have identical limits, since in an indivisible there can be no distinction of a limit from some other part than the limit; and (for the same reason) neither can the limits be together, for a thing that has no parts has no limits (ibid., 231 a 21–28).

And points, in particular, taken as examples for indivisibles in general here, were unable to touch:

> Again, one point, so far from being continuous or contiguous with another point, cannot even be the next–in–succession to it, or one "now" to another "now," in such a way as to make up a length or a space of time; for things are "next" to each other when there is nothing of their own sort between them, and two points have always a line (divisible at intermediate points) between them (ibid., 231 b 4–10).

Aristotle thus also stressed that because of extensiveness "nothing which is finite and divisible is bounded by a single limit" (*Physics*, IV, Chap. 10, 218 a 23–24; transl. Wicksteed).

The essential thing for Aristotle's understanding of continuity is that he already postulates the intermediate value property in connection with discussing the continuity of motion:

> The magnitude over which the change takes place is continuous. For suppose a thing has changed from C to D. Then if CD were indivisible, two things which have no parts would be consecutive, and since this is impossible, the space between must be a magnitude and therefore be divisible without limit. So the subject effects innumerable changes before it has effected any given change (*Physics*, VI, Chap. 2, 237 a 31–34).

Epicurus, who developed atomism further after Aristotle's critique of this theory, tried to solve the problem by abandoning Democritus's identification of physical atoms with "abstract mathematical minima" (Mau 1954, 25), and by introducing a differentiation between physical reality and mathematical modeling (see ibid.). For the bodies of the real world, Epicurus assumes the existence of absolute minima, while he assumes the quantities of abstract mathematics to be continuous, to be infinitely divisible (ibid., 47 f.). Because Epicurus was negated, respectively rejected, during the Middle Ages and early modern times, this conceptual differentiation remained without having much effect.

It was of great systematic impact for the development of the entire philosophical–mathematical concept field of limit processes that Aristotle applied the concept pair of the actual and the potential developed in Socrates's school to the analysis of motion (Mau 1954, 25 f.), thus becoming the first to analytically apply the differentiation between actual and potential infinite, which was to determine the debates to come (ibid., 28).

It was this very distinction of Aristotle between the actual and the potential infinite that was heatedly discussed in the European Middle Ages, mainly under theological aspects. Since concepts were not recognized to be field–specific in this time, concepts were not admitted if they seemed to run against Christian doctrines. Remarkable here is the almost general rejection of Aristotle's analysis according to which actually infinite quantities are impossible. To adopt this proposition in the frame of this theologically dominated debate would have amounted to heresy: to denying the ascription of infinite qualities to God. Fascinating and effective contributions to the concept of the infinite were made by the Scholastics, despite their bad reputation in the history of science. Pierre Duhem examined this matter in more detail, in particular in his studies "Infinitely Small and Infinitely Large" and "Infinitely Large" (Duhem 1985).

In this context, the scholastics also took up Zeno's paradoxes, discussing with great intensity whether it was possible to attain an infinite limit by infinite division (see Duhem 1985, 56 ff.).

3. Early Modern Times

In early modern times, too, the debates of principle about divisibility and continuity were continued, becoming detached, however, from the theological context. In contrast to the former abstract theoretical character of the debate, the primarily philosophical concepts changed into operational concepts in mathematics, where ever more fields of application opened, making the concepts themselves develop further. Where Kepler had been concerned with determining volumes that led to mathematizations (see Chapter II.1.4. above), and Cavalieri was initially intent on determining areas of surfaces, a purpose for which he developed his concept of indivisibles, the final concerns were rectifications of curves, and more generally, integrations. Mathematizations were prompted by the latest developments in physics. Recent historiography of mathematics has focused on studying these influences exerted by physics. Mahoney has shown, in particular, the prime importance of the new science of motion, mechanics, for the formation of infinitesimals (Mahoney 1990).

Johannes Kepler (1571–1630) was still firmly anchored in the theological context that decisively determined his reflections on mathematical foundations. In his research into astronomy, he evaluated approximate numerical solutions—of an equation of sixth degree not exactly solvable—as solutions that did not meet the ideal of geometric rigor. Even improved approximations, he said, would not compensate for their inherent fault of being of numerical character. God, he said, was a geometer, not an algebraist, and it was not permissible in geometry to consider any quantity as negligibly small (Field 1994, 22). As Field notes, Kepler thus rejected mathematical atomism, assuming exact congruence between the truths in mathematics and the truths in nature, as both were expressions of God's activity (ibid., 223).

Galileo Galilei (1564–1642), in contrast, proved to be an advocate of atomism, and hence criticized Aristotle's notions. Galileo was intent on proving that matter was composed of atoms, and thus had to challenge Aristotle. Since Aristotle had inferred his rejection of atoms from his own definition of continuity, Galileo rejected that definition in his *Discorsi*. He did not replace it by another definition, however, but opposed it with his own understanding of continuity as a property given from the very outset that is expressed at the same time, say, by the geometrical straight line. In particular, Galileo disagreed with Aristotle's distinction between the actual and the potential: the continuous, he said, is always divisible, it is composed of an infinite number of parts. These parts could not be *quante*, i.e., could not have any quantity, since their composite would otherwise yield an infinite whole. According to Galileo, the continuous is thus composed of an infinite number of *non-quante* parts, or of indivisibles (Giusti 1989, 86 f.). Since Giusti has shown, Galileo had to face a problem because of this view: If the continuous is composed of indivisibles, how is it then possible to compare two continuous quantities with one another, such as a larger line segment with a smaller? Galileo's answer was radical, but of little operational use: since there were no relations of order between infinite quantities, he said, it was impossible to compare them. As justification, he presented the famous example of the relation between natural numbers and their squares. It could not be said of any set that it was larger or smaller than the other (ibid., 87). Giusti thus sums up that the continuous, for Galileo, is composed of indivisibles, and that no relations exist between indivisibles—as infinite quantities (ibid.). Galileo, who was motivated here by physical theory, did not distinguish between physical and mathematical atomism.

Bonaventura Cavalieri (1598–1647), one of Galileo's most gifted disciples, developed the indivisibles further into an operative mathematical concept, in particular in his major work *Geometria indivisibilibus continuorum nova quadam ratione promota* (1635). His basic notion for this was that relations between the indivisibles of two figures can be used to determine the relation between the two figures themselves (see Chapter II.1.4.). In order to be able to apply the classical theory of proportions, Cavalieri had to make the opposite of what had been the result of Galileo's reflection the basis of his own theory: the postulate that the indivisibles of two quantities could be compared to one another. To make the classical theory of proportions applicable to his indivisibles, however, these quantities must not be infinite, but finite ones.

Popular lore about Cavalieri assumes that the (infinite) sum of a quantity's indivisibles is identical with the quantity itself. Recent research, however, has shown that Cavalieri avoided this kind of identification, assuming only that the totality of a quantity's indivisibles was congruent in one sense to the quantity itself, thus sharing the property of finiteness with it (Andersen 1985; Giusti 1989, 88). The principle of applying the rules of proportions consisted in the rule *ut unum ad unum, sic omnia ad omnia*, that is, as one indivisible relates to another, all relate to all others (Giusti 1989, 88).

It has been little noted yet that studies of physics were of great importance for Cavalieri's mathematical works as well (Baroncelli 1992), but how mathematical and physical atomism were related for him was bound to remain unclear because of his own concept of indivisibles.

It was in particular Cavalieri's operation of forming an aggregate of "all lines" that—against Cavalieri's own intentions—was actually understood to be an infinite sum formation and thus provoked much criticism (cf. Giusti 1982, 89 ff.; Malet 1996, 18 f.).

Indeed, Cavalieri's disciple Evangelista Torricelli (1608–1647) contributed to spreading the view that Cavalieri had considered the unification of all the indivisibles of a figure to be identical with the figure itself (de Gandt 1992, 106). It was Torricelli who essentially changed the conception of indivisibles, thus preparing the transition to the new concept of infinitesimals. Cavalieri (see

Chapter II.1.4. above) had conceived of the indivisibles as being one dimension inferior to their figure—indivisibles and their figure were hence "heterogeneous" in order to ensure mathematical "atomism," but this made applying the concept of ratio much more complicated. Torricelli, in fact, was the first in modern times to conceive of the indivisibles as being "homogeneous," of being of the same dimension.

In the brief period from about 1640 to 1670, mathematizing concepts concerning analysis were unfolded on a scale without precedent in many places in Europe, and transmuted into concepts ever richer in applications. For the first time, a large number of scholars in Italy, France, and England, many of them working in close communication on the same problem field, developed the foundations for differential and integral calculus within a short time.

François de Gandt, who has studied Torricelli's works and manuscripts in great detail, was able to show therein a step-by-step transition from Cavalieri's concept of indivisibles toward a new theory of indivisibles in which the indivisibles formed infinitely small sections (*tranches infiniment minces*) having the same dimension as the figure concerned—in case of a surface area hence pieces of parallelograms (de Gandt 1992, 106 ff.).

The basis of this conceptual change was a reinterpretation of Aristotle's major argument against classical atomism: the argument of the infinite divisibility of the geometrical continuum. By accepting—in contrast to Aristotle—*actually* infinite quantities operatively, mathematicians now admitted an actually infinite division, and were thus empowered to argue that a finite line segment was divisible into an infinite number of infinitely small parts (Malet 1996, 19).

This change remained at first primarily one of practice, for the designation "method of indivisibles" remained widely in use during the entire seventeenth century. The new concept, which was only later named *infinitesimals*, or *quantité infiniment petite*, a designation we might translate into "preclassical" infinitely small quantities, shared with the indivisibles of old the property that a finite number of them could never become just as large as, or larger than, a finite quantity. Malet has pointed out that the new infinitesimals differed from the indivisibles of old mainly in three points:

> But Cavalierian indivisibles and Barrow's "indefinitely small linelets" (*indefinite parvis lineolis*) differ in three essential aspects. Infinitesimals are *divisible*; that is, it makes mathematical sense to halve an infinitesimal. They are *homogeneous* with the magnitude they originated from. In particular, by taking together infinitesimal surfaces, a larger surface is gained, which will be infinitesimal itself, if their number be finite, or it will be finite, if their number be infinite. Finally, infinitesimals can actually be reached by a process of infinite division, a process which can never yield a point, say, starting from a linear segment (ibid., 19).

Even Malet's more recent research, however, does not take note of how far the objects to which the concepts were applied were modified simultaneously with these conceptual changes. Torricelli essentially studied types of area and volume determinations analogous to those of Cavalieri, but he also already used the newly conceived indivisibles for studying curves, to solve tangent problems (ibid., 279).

An explicit reflection on the "old" versus the "new" meaning of indivisibles was first undertaken by Blaise Pascal (1623–1662), in a fragment of one of his letters of 1658:

> I have no difficulty in using the expression *the sum of ordinates*, which seems not to be Geometric to those who do not understand the doctrine of indivisibles, and who imagine that it breaks the rules of Geometry to express an area by an indefinite number of lines; this only comes from their lack of understanding, since nothing other is understood by it than the sum of an indefinite number of rectangles made from each ordinate with each of the small portions [which are together] equal to the diameter, whose sum is certainly an area which differs from that of the semicircle only by an amount less than any given quantity (quoted from Malet 1989, 232).

In this fragment, Pascal called the infinitesimal sides of the rectangles homogeneous to the initial area *portions*, or parts. With Pascal, it becomes already clearer that the infinitesimals are no longer applicable to static quantities alone, but also to dynamic quantities. According to Pascal, time and motion belong among the objects that are to be studied by mathematics (Malet 1996, 28).

In a mode quite analogous to that of Pascal, Wallis, in his algebra textbook of 1685, explained the reshaping of the indivisibles into infinitesimals by accepting the actually infinite quantities:

> According to this Method [of Indivisibles], a Line is considered, as consisting of an Innumerable Multitude of Points: A Surface, of Lines [...]: A Solid, of Plains, or other Surfaces [...]. Now this is not to be understood, as if those lines (which have no breadth) could fill up a Surface; or those Plains or Surfaces, (which have no thickness) could compleat a Solid. But by such Lines are to be understood, small Surfaces, (of such a length, but very narrow) (quoted after Malet 1996, 23 f.).

Another quite analogous introduction of infinitesimal quantities was given by Isaac Barrow (1630–1677), a theologian, and as a mathematician the predecessor of Newton as a professor at Cambridge University. In particular with his textbook *Lectiones geometricae* (1670), Barrow made important contributions to the new field of studying curves, and to tangent problems. While Barrow used the terminology of indivisibles according to which all parallel lines form a plane, for instance, he added the qualification that this was a case of an infinite, or indeterminate sum: not a sum of lines, but rather a sum of "parallelograms of a very small and inconsiderable [*inconsiderabilis*] (if I may say so) height." Analogously, he understood planes that were to form a body to be "prisms or cylinders of uncomputable height."[2]

---

[2]   *Lectiones mathematicae* (1664-1666), ChapterIX; quoted after Malet 1996, 33.

Barrow justified these infinitesimals with a critique of classical atomism which shows that mathematical epistemology, for him, is directly coupled to physical ontology. Barrow criticized the atomists' arguments against the infinite divisibility of quantities. The latter had actually intended quantities to mean particles of matter. Barrow rejected the atomist argument according to which matter was only an aggregate of indivisibles—no matter whether of finite or infinite number. The only alternative acceptable for him was that a finite quantity could have infinitely many parts (cf. ibid., 33).

It is generally typical for the conceptual development of mathematizations in the seventeenth century that their ties to theological dogmas and discourses weaken. It is particularly true for the concepts of the infinite that they became embedded into the conceptualization of the physical world (Malet 1996, 149 ff.). The strengthening of the ties of mathematics to physics is particularly clear in Newton's foundation of infinitesimal calculus.

4. The Founders of Infinitesimal Calculus

With Leibniz and Newton, the founders of infinitesimal calculus, the study of limit processes definitively acquires a new character. Its principal objects are no longer static quantities like individual surfaces or volumes, but rather the properties of *curves*.

In a fragment written about 1680, the *Geometria curvilinea*, Newton made explicit the necessity of a new conceptual frame. The traditional Euclidean geometry had assumed straight lines. Since it did not provide the means to study curves, a new conceptual system was now required:

> [Euclid] has delivered the foundation of the geometry of straight lines. [And since Euclid's elements are scarcely adequate for a work dealing, as this, with curves, I have been forced to frame others (Newton 1971, 423).

Newton was indeed the first to introduce a new meaning of the indivisibles since Torricelli, that of "infinitesimals" (cf. Newton 1969, 265), or *quantitates infinite parvae* (Newton 1969, 80), and the first to use *infinitesimals* operatively to a large extent. The new operative function becomes particularly clear from the fact that Newton introduced a separate sign for this kind of quantity: the o.[3]

The analysis of Newton's conception of the infinitesimals and of limit processes, however, is complicated by the fact that this conception was not unified. Recent research has elaborated that there are three or four different conceptions. There is no agreement, however, as to how these differences should be evaluated. While Kitcher ascribes different status to them according to

---

[3]   Cf. for this Whiteside's notes No. 13, p. 262, and No. 33, p. 271, in vol. II of *Newton's Mathematical Papers* (Newton 1968).

whether they are used in a *context of discovery*, or in a *context of justification* (Kitcher 1973), Sageng (1989) and Guicciardini (1989) place more emphasis on temporal succession—a succession that they consider again determined by Newton's persistent efforts at mastering a foundational problem. This problem was precisely to avoid using the *infinitesimals* involving the infinite, and to operate exclusively with finite quantities. In this effort, Newton was confronted with the methodological contradiction of striving, on the one hand, for an efficient application of his own differential and integral calculus according to methods as general as possible, while on the other hand working within a mathematical context that interpreted rigor as something optimally realizable with methods of geometry.

Kitcher distinguished three different methods of Newton for founding the new calculus:

*   The method of the calculus of fluxions,
*   The method of the first and last ratios, and
*   The method of the infinitesimals.

Kitcher's thesis is that these three methods do not represent alternative approaches at foundation, but were used by Newton in a parallel mode, and for distinct purposes:

> Instead of seeing each set of concepts as a candidate for the true foundation of the calculus, I shall contend that we should recognize that each occupied a special place in Newton's total scheme. Putting the matter briefly, the theory of fluxions yielded the heuristic methods of the calculus. Those methods were to be justified rigorously by the theory of ultimate ratios. The theory of infinitesimals was to abbreviate the rigorous proof, and Newton thought that he had shown the abbreviation to be permissible. Rather than competing for the same position, the three theories were designed for quite distinct tasks.(Kitcher 1973, 33 f.).

While this thesis is doubtlessly presented lucidly and convincingly, it relies in its essentials on presentations Newton himself had written against Leibniz in his controversy about priority with the latter. It cannot be excluded that they are a rationalization after the fact.

Sageng indeed criticized Kitcher's thesis in that it neglected how intensely Newton strove to liberate himself from the concept of infinitesimals (Sageng 1989, 164). Guicciardini, too, shows that Newton tried to dissociate himself from the concept of infinitely small quantities by means of a theory of limits. He sums up Newton's diverse approaches as follows:

> Newton left to his followers a wealth of theorems and results achieved with the calculus of fluxions, but he left also a rather confused presentation of the methods and of the concepts which had to be employed (Guicciardini 1989, 6).

Thus, Sageng shows that Newton practically resorted to tricks also in his *Principia mathematica* (1687) which contains the first published form of this calculus of fluxions, with the intention of concealing his factual use of infinitely small quantities. In Lemma II of the *Principia*'s second volume, Newton intends to show that the moment (the momentarily vanishing increment) of the product of two fluents (corresponding to variables) A and B can be calculated as follows: If a is the moment of A, and b is the moment of B, then the moment of the product AB is[4]

$$aB + bA \quad \text{(Newton 1972, 364)}.$$

According to the method using infinitesimals Newton had hitherto applied, he would have had to calculate the moment of the product, that is, of the rectangle AB, like this: A and B would have increased by their moments to $A + a$ and $B + b$, the following product resulting:

$$(A + a) (B + b) = AB + aB + bA + ab.$$

The increment would thus have been

$$aB + bA + ab,$$

ab having been neglected as infinitely small as compared to the other terms. Geometrically, this would have meant disregarding the smaller rectangle ab as compared to the larger rectangles aB and bA.

Actually, however, Newton chose another method, an ad hoc method, by taking as initial product the rectangle formed by the sides enlarged by half a moment, and as the rectangle to be subtracted formed by the sides reduced by half a moment:

$$(A + \tfrac{a}{2})(B + \tfrac{b}{2}) - (A - \tfrac{a}{2})(B - \tfrac{b}{2})$$
$$= AB + \tfrac{a}{2}B + \tfrac{b}{2}A + \tfrac{1}{4}ab - AB + \tfrac{a}{2}B + \tfrac{b}{2}A - \tfrac{1}{4}ab$$
$$= aB + bA$$

This ad hoc method, later sharply criticized by Berkeley, does not only reveal that Newton shirked speaking of infinitesimals of second order, but also shows Newton's discomfort with the method of treating the increments, on the one hand, as non-zero within the calculation, while regarding them in the product, on the other hand, essentially as zero (Sageng 1989, 174 ff.).

This means that Newton continues using essential features of the method of infinitesimals in his method of fluxions as well. He had most explicitly presented this method in his privately disseminated essay *De Analysi Per Aequationes Infinitas* of 1669. Two characteristic features of infinitely small quantities are particularly salient:

- On the one hand, they have, like finite quantities, the property of being able to function as divisors;
- On the other hand, they have, like zero, the property of being negligible if added to a positive quantity x:

$$x + o = x.$$

---

4   As in his earlier writings, Newton uses here lowercase letters for infinitely small quantities, and for the appropriate finite quantities the corresponding uppercase letters. In other writings, he conceives of the moments as products $o\dot{x}, o\dot{y}$ etc., i.e., as products of arbitrarily small intervals of time $o = o(t)$ and of the fluxions $\dot{x}, \dot{y}$ etc.

Thus Newton, in proving the Sluse rule for integrating the curve $z = \frac{2}{3} x^{\frac{3}{2}}$, obtains the equation

$$\frac{4}{9}(3x^2 o + 3xo^2 + o^3) = 2zov + o^2 v^2.$$

First, he divides the equation by o, obtaining

$$\frac{4}{9}(3x^2 + 3xo + o^2) = 2zv + ov^2 ;$$

he continues by assuming o to be infinitely small, and "thus" zero, resulting in the vanishing of the terms multiplied by o. There remains only

$$\frac{4}{9} 3x^2 = 2zv \text{ (cf. Sageng 1989, 166 f.).}$$

Newton called the latter approach he developed the method of the "prime and ultimate ratios." This was the approach at explicitly mathematizing limit processes. By this method, Newton intended to avoid using infinitely small quantities, and to be able to operate exclusively with finite quantities.

In his work *De Quadratura Curvarum*, which he revised again and again after 1691, finally publishing it in 1704, he took a position against any concept of atomism—that is, against the possibility of quantities being composed of smallest particles; he chose by contrast a strictly kinematic view. He conceived of the continuity of quantities as given by continuous motion. He strictly rejected explaining curved lines by lining up points, but understood them as being generated by the movement of points. Newton underlined that mathematizations were ontologically tied to physical reality:

> Mathematical quantities I here consider not as consisting of least possible parts, but as described by a continuous motion. Lines are described and by describing generated not through the apposition of parts but through the continuous motion of points; [...]. These geneses take place in the reality of physical nature and are daily witnessed in the motion of bodies (Newton 1981, 123).[5]

Because of this kinematic view, all variables are dependent on time, and the basic variable is formed by time. Although Newton stresses the continuous generation of curves—"times [are generated] through continuous flux" (tempora per fluxum continuum, ibid.)—it is noteworthy that he occasionally speaks of "parts of time."[6]

Examining such quantities enabled Newton to obtain finite values by relating the velocity of their motions—the "fluxions" to the increments of the "fluents." Thus, he formed the "first ratios": "indeed they [the fluxions] are in the first ratio of the nascent augments" (ibid., 125). Newton uses several examples to demonstrate how to determine the "last," respectively "ultimate ratios." The example quoted here shall be the differentiation of x^n:

---

[5]   In D.T. Whiteside's translation.

[6]   Thus of "temporis particulis quam minimis genita", "very smallest particles of time," Newton 1981, 123.

Let the quantity x flow uniformly and the fluxion of the quantity x^n need to be found. In the time that the quantity x comes in its flux to be $x + o$, the quantity x^n will come to be $(x + o)^n$, that is (when expanded) by the method of infinite series

$$x^n + nox^{n-1} + \tfrac{1}{2}(n^2 - n)o^2 x^{n-2} + \ldots;$$

and so the augments o and $nox^{n-1} + (n^2 - n)o^2 x^{n-2} + \ldots$ are one to the other as 1 and $nx^{n-1} + (n^2 - n)x^{n-2} + \ldots$ Now let those augments come to vanish and their last ratio will be 1 to nx^{n-1}; consequently the fluxion of the quantity x is to the fluxion of the quantity x^n as 1 to nx^{n-1} (ibid., 127–129).

Newton stressed that he used only finite quantities here, and thus had attained agreement with the geometrical methodology of the Greeks:

> In finite quantities [...] to institute analysis in this way and to investigate the first and last ratios of nascent or vanishing finites is in harmony with the geometry of the ancients, and I wanted to show that in the method of fluxions there should be no need to introduce infinitely small figures into geometry (ibid., 129).

Newton reflected the idea contained in the concept of "first and ultimate ratios" of determining, for the case of approximating time-dependent variables to a certain point in time, the limit to which the variable approximates and that it eventually reaches in his view in his *Principia mathematica* (1687) explicitly as a study of limits, introducing the term *limes* for this.

Newton introduced this method as an alternative both to the ancients' geometrical method with the purpose of avoiding their lengthy "ad absurdum" demonstrations, and as an alternative to the method of the indivisibles: the hypothesis of the indivisibles seemed too offensive (*durior*), he said, and the method was deemed not to be sufficiently geometrical. The alternative was to determine the ultimate sums, respectively ratios, of vanishing quantities, respectively of the first quantities generated, and this consisted precisely in determining the limits of the sums, respectively of the ratios: "that is, [to reduce] to the limits of those sums and ratios" (Newton 1969, 38; orig. 1972, 87).

And he emphasized his own concept of continuity, saying that he did not mean indivisibles, but rather conceived of vanishing divisible quantities; and that he did not mean the existence of ultimate parts, but that he rather meant "[not] indivisibles, but evanescent divisible quantities; not the sums and ratios of determined parts, but always the limits of sums and ratios" (ibid.). Here, Newton took pains to stress that his own expression "ultimate ratios" indeed meant *limits* that are more and more approximated by the ratios:

> For those ultimate ratios with which quantities vanish are not truly the ratios of ultimate quantities, but limits toward which the ratios of quantities approach nearer than by any given difference (ibid., 39; orig. 88).

Newton simultaneously declared that the limit exists, and that it is a genuinely geometrical task to examine these values because the limit is fixed and determined: "And since such limits are certain and definite, to determine the same is a problem strictly geometrical" (ibid., 39; orig. 87 f.). Finally, Newton gives a hint at the conception of mathematizing limit processes that was to be formulated later to say that infinitely small quantities are special variables, namely null sequences. Wherever he himself spoke in the book's subsequent parts of smallest quantities, or of vanishing, or of last quantities, these one must never understand to mean fixed quantities having a determinate value, but rather quantities that were indefinitely diminishable:

> Therefore in what follows, for the sake of being more easily understood, I should happen to mention quantities at least, or evanescent, or ultimate, you are not to suppose that quantities of any determinate magnitude, but such as are conceived to be always diminished without end (ibid., 39; orig. 88).

While Newton did not develop an algorithm or calculus for determining limits, he gave a principle for this that was to be comprehensively elaborated by subsequent authors: the principle that the limit could be inferred from the behavior of variables within the finite:

> Quantities, and the ratios of quantities, that in any finite time tend constantly to equality, and before the end of time approach one another more closely than by any given difference, become ultimately equal (Newton 1969a, 29; orig. 1972, 73).

This principle forms the basis for deciding on the identity of expressions in the new calculus.

Guicciardini called this conception an "intuitive theory of limits" (Guicciardini 1989, 5). In any case, it can be stated that Newton did not elaborate an algebraized theory of limit. He presents no operative treatment for limits, no designation for the limit processes' variables and indexes, and in particular no notation for limit. As the above quotations show, description and argumentation are exclusively verbal. Newton hence meets objections to his concept also exclusively in the verbal mode. Thus, he states, for instance, that the term "last velocity" meant neither the velocity of the body before it arrived at its destination, nor a later velocity, but precisely the velocity of arrival (ibid., 87).

His approach involving infinitely small quantities is also remarkable for the fact that explicit operation has a relatively limited scope: in contrast to later authors, he does not present an elaborated calculation, say, for the ratio of infinitely small quantities of different order ("dimension").

A further remarkable thing is that Newton, in his mathematical conceptions, is by no means a "physicalist" dyed in the wool. While he is a strict antiatomist with regard to founding calculus, using the physical categories of time and motion to avoid decomposition into indivisibles, he held pronouncedly atomist views with regard to physics. He understood matter as composed of atoms, conceiving of atoms as of things hard, indivisible, and impenetrable (cf. Scott 1970, 3 ff.). Newton thus presents a case rare in our context, in which mathematical continuity is compatible with physical atomism.

In contrast to Newton, *Leibniz* did not consider the rigor of foundation the methodologically guiding element. For Leibniz, the Greeks' geometrical method was not the goal paradigm, nor did he feel compelled to elaborate other theories because of a desired clarity of methods. While he remained rather more indefinite and undecided with regard to the epistemological foundation in particular of calculating with limit processes, he achieved an exhaustive algorithmic elaboration of the new calculus, developing algebraizing signs and notations for the new concepts that made application attractive and easy, independent of clarifying its "metaphysics."

Gottfried Wilhelm Leibniz (1646–1716) was essentially a self-educated mathematician and universal scholar. While he completed his university studies in Germany with a doctoral degree, his mathematical studies with E. Weigel in Jena remained relatively elementary. Only after he had got in touch with the mathematicians of Paris after 1672 did he acquire knowledge of the new mathematics. Eventually, Leibniz entered the service of the Prince Elector and Duke of Hanover as librarian and legal advisor, traveling in this capacity on diplomatic missions. Leibniz represents the rare case in Germany of a scholar who never taught at a university. It is hence characteristic that he became nevertheless the initiator of the Berlin Academy of Sciences (1700), the place where research was for the first time institutionalized in Germany, even before the modernization of the universities. It is just as characteristic that he did not author any textbooks, in contrast to Newton, but preferred to publish his research results essentially in individual articles in periodicals, if they did not remain manuscripts that are still gradually being edited.

Leibniz was thus not burdened with university traditions of knowledge and their strict rules in favor of emulating the rigid ideals of the geometry of the ancient Greeks. This is why he was able to develop the novel differential and integral calculi as an algebraizing technique, as a calculus. On the other hand, he remained of course attached to his own period's mathematical context. Thus, it can be noted that he assumed *curves* to be the major objects of the new theories, in contrast to the practitioners of the indivisibles, and like Newton, continuing to conceive of curves primarily as geometrical objects.

In his masterful reconstruction of Leibniz's concepts of the differentials, Bos (1974) demonstrated how deeply these concepts were formed by underlying geometrical concepts of quantity. The original object of Leibniz's infinitesimal calculus was the study of curves by means of the algebraic techniques developed by Descartes. Analysis was only a tool for studying geometrical quantities, since analysis had not yet attained a status independent of geometry. To move toward this status implied dissociation from the mathematical object, in which the relations between geometrical quantities and curves were studied, and moving toward a general study of relations between quantities expressed by letters and numbers. This change of standpoint meant at the same time to switch from variable geometrical quantities to functions of one variable:

This change of interest from the curve to the formula induced a change in fundamental concepts of analysis. While in the geometrical phase the fundamental concept in the analytical study of curves was the variable *geometrical quantity*, the separation of analysis from geometry made possible the emergence of the concept of *function of one variable*, which eventually replaced the variable geometrical quantity as the fundamental concept of analysis (Bos 1974, 4).

That there was no concept of functions had a number of fundamental implications. Neither the geometric variables nor the equations formed with them were functions—but with that there was no unequivocal establishment of function, say, by an "independent" variable x and a "dependent" variable $y = f(x)$. Since there was no differentiation between a dependent variable and an independent variable, and since x and y were virtually interchangeable, the differential quotient was also not a reasonable concept. For these reasons, *derivation* could not be a basic concept either—the basic concept being, by contrast, the differential, and this implied ever new repeated reflections of principle concerning the role and legitimacy of infinitely small quantities (ibid., 6 and 8 f.).

At the same time, Bos elaborated that the concept of quantity was still traditionally attached to geometry, and the fact that the number concept had not yet been segregated—that is, the very basic problems that caused the concept field of negative numbers to evolve (cf. Chapter II.)—was a considerable restriction to unfolding Leibniz's creations. Since quantities were always conceived of as equipped with dimension, equations could be constituted only in a limited way; they had to agree with the law of homogeneity. Only quantities having the same dimension could be added, and there were conceptual restrictions to multiplication (as already discussed in Chapter II):

> Only quantities of the same dimension could be added. In certain cases the multiplication of quantities was interpretable, as for instance in the case of two line segments, the product of which would be an area. But multiplication was never a closed operation; that is, the product of two quantities of equal dimension could not have the same dimension. Hence within the set of quantities of the same dimension there was no multiplicative structure and no unit element. A choice of a privileged element in the set of quantities of the same dimension (as a base for measuring, for instance, or as a fundamental constant for certain curves or actually as a unit element) was therefore always arbitrary; the structure of quantity itself did not offer such a privileged element.

> These possibilities of multiplication and addition made possible the algebraic treatment of quantities, although with certain restrictions. The special nature of multiplication induced a law of homogeneity for the equation's occurring in this algebraic treatment: all the terms of an equation had to be of the same dimension (ibid., 6).

This required homogeneity of dimension resulted in an equivocal character of the higher differentials in Leibniz's differential calculus. If a derivative had been conceived within the geometrical context, "then derivation would correlate a ratio (the derivative) to a variable that has the dimension of length" (ibid., 8).

The ratio thus would have had to be interpreted as a line segment, and would thus have required the choice of a unit length. Because of the arbitrary selection of this unit, derivations of higher order would not have been uniquely defined (ibid.). Leibniz's higher differentials indeed had to be abandoned eventually (ibid., 66 ff.).

Leibniz's calculus proper, in contrast to Newton's, started from algebraic concepts, and not from geometrical–physical principles. Leibniz considered curves to be identifiable with polygons composed of arbitrarily small line segments. By considering the sequences of ordinates, abscissae, etc. of the individual finite line segments, he was able to apply the theory of number sequences as well as that of sequences of sums and of differences to these sequences. Since he identified the curve with a polygon, Leibniz assumed the differences to be infinitely small. As Bos stresses, this meant at first that these differences were considered negligible as compared to finite quantities, but also as unequal to zero. Moreover, Leibniz also supposed them to be fixed quantities:

> Therefore, the fundamental concepts of the Leibnizian infinitesimal calculus can best be understood as extrapolations to the actually infinite of the calculus of finite sequences. I use the term "extrapolation" here to preclude any ideas of taking a limit. The differences of the terms of the sequences were not considered each to approach zero. They were supposed fixed, but infinitely small (ibid., 13).

Leibniz, too, developed a criterion of equality for quantities in his new calculus. In his answer to Nieuwentijt of 1695, he declared two homogeneous quantities to be equal that differed only by a quantity that was arbitrarily smaller than a finite quantity:

> And such an increment *(namely the addition of an incomparably smaller line to a finite line)* cannot be exhibited by any construction. For I agree with Euclid Book V Definition 5 that only those homogeneous quantities are comparable, of which the one can become larger than the other if multiplied by a number, that is, a finite number. I assert that entities, whose difference is not such a quantity, are equal. [...] This is precisely what is meant by saying that the difference is smaller than any given quantity.[7]

As fundamental concepts of the new calculus, Leibniz introduced *differential* and *sum*. By means of "extrapolating" to the actually infinite, the finite sequence of differences is transformed into a sequence of an infinite number of infinitely small terms, these terms are called differentials. The differential is an infinitely small variable: the difference operator Δ transforms into the differential operator d, which assigns to a finite variable y the infinitely small variable dy. It is not necessary here to establish which variable is to be the "independent" one (Bos 1974, 16 f.).

In an analogous way, the operation of summation is generated from the finite sum operator Σ. Leibniz, however, had initially conceived of the sum operator

---

[7]   In Bos's translation into English (Bos 1974, 14). The Latin original is in Leibniz 1858, 322. This criterion is quite different from Johann Bernoulli's or L'Hospital's first postulate (see Section 6. below).

analogously to differentiation such as to assign an infinitely large variable $\int y$ to a finite variable y. Only later, under the influence of Johann Bernoulli, did he understand the sum as formed of rectangles with sides y and dx, and he adopted the designation $\int y\,dx$ and eventually the designation of "integral" (ibid., 17 and 20 f.). In this way, it becomes possible to formulate the famous main theorem of differential and integral calculus in signs:

$$\int y\,dx = Q,$$

according to which differentiation and integration are mutually inverse operations (Q stands for *quadratura* here).

Leibniz always avoided raising infinitely small quantities to the status of fundamental concepts of his own theory. In his first publication on this topic of 1684, his definition of the differential did not refer to such quantities. Rather, he relied on fixed and finite parts of line segments (ibid., 65). The explicit calculus of the various orders of the infinite that Bos presents in connection with Leibniz's foundation of differential calculus (Bos 1974, 22f. and 33), which says that differentials of the same order have a finite ratio, that a higher differential is infinitely small with regard to a differential of lower order, that ddy, for instance, the differential of second order, is infinitely small as compared to the first differential dy, and that $d^{k+1}y$ is generally infinitely small as compared to $d^{k}y$, all these propositions refer to Johann Bernoulli's conceptions, who had integrated other influences as well (see below, Section 6.).

Leibniz also designated the homogeneity of dimension just outlined a*lgebraic homogeneity*, supplemented by a "transcendental" law of homogeneity according to which all terms of an equation must be of the same order of infinity—terms of inferior order being negligible (Bos 1974, 33). He did not explicitly introduce this transcendental homogeneity before 1710, however. Actually, he first used transcendental homogeneity in his treatise *Symbolismus memorabilis* of 1710 for proving the product rule (Leibniz 1858, 379). He mentioned having already "presented" the product rule in 1684; the "law of homogeneity" that he briefly mentioned in 1684 was unconnected at this point with the product rule, but was an observation concerning an equation's consistency: an equation was homogeneous, he said, if every summand has as a factor a dx or a dy (Leibniz 1989, 113), no comparison of different "orders" or powers having been implied at this point.[8]

---

[8]   In his commentary to the translation of Leibniz's most important works on infinitesimal calculus into French, Parmentier tried to establish too strong a continuity between the treatises of 1684 and 1710: For the rule of product, he quotes an explicit form containing all the terms and a justification (Leibniz 1989, 103, FN 31) which, however, does not appear before 1710, and he interprets the mention of the "lex homogeneorum" of 1684 (Leibniz 1858, 224) as the transcendental law of homogeneity, and as justification of the rule of product (Leibniz 1989, 113, FN 64) despite the fact that it appears in quite another context. Parmentier's intention of

The general opinion formed by those who applied and practiced the Leibnizian differential calculus ascribed to Leibniz having used the infinitely small quantities himself to found his calculus. Even the first users, and others later, were surprised that Leibniz did *not* adopt such a point of view. Actually, Leibniz repeatedly dissociated himself from such recognition. Some of the many proofs for this shall be quoted here.

Already in his answer of 1695 to the Dutch mathematician Bernard Nieuwentijt (1654–1718), who had criticized Leibniz's methods among other things for the latter's assuming infinitely small quantities to be practically zero, Leibniz argued extensively that for himself quantities were equal not only if their difference was completely zero, but also if their difference was *incomparably small*:

> Caeterum aequalia esse puto, non tantum quorum differentia est omnino nulla, sed et quorum differentia est incomparabiliter parva (Leibniz 1858, 322).

In a letter to des Bosses of 1706, Leibniz stressed that for himself the infinitely small quantities were not really existing mathematical quantities, but only *fictions* that had their uses in the course of calculus:

> Ego philosophice loquendo non magis statuo magnitudines infinite parvas quam infinite magnas, seu non magis infinitesimas quam infinituplas. Utrasque enim per modum loquendi compendiosum pro mentis fictionibus habeo, ad calculum aptis, quales etiam sunt radices imaginariae in Algebra. Interim demonstravi, magnum has expressiones usum habere ad compendium cogitandi adeoque ad inventionem.[9]

Leibniz had written things similar already in the 1690s to Johann Bernoulli, as he communicated to Varignon in 1702, that is:

> Infinities and infinitely small quantities could be taken as fictions, similar to imaginary roots, except that it would make our calculations wrong, these fictions being useful and based in reality (Leibniz 1859, 98).[10]

Pasini, as well as Gerard and Wybe Sierksma, have recently pointed out the deepening dissent between Leibniz and Johann Bernoulli that finally put an end to their communication about this matter. In August 1698, Johann Bernoulli, referring to the infinite divisibility of matter, had claimed the "actual" existence of infinitely small quantities (Pasini 1993, 117; Leibniz 1856, 529). Leibniz retorted that he had hitherto left open the question of their existence; he could

---

predating Leibniz's mode of proof is also evident from his translating Leibniz's expression of 1710 "docuimus" with "montré." In the same vein, he speaks of an analogous "demonstratio" in Leibniz's letter of 1677 to Oldenburg, although this letter contains also only a statement of the rule (Leibniz 1849, 154).

[9]   Philosophische Schriften, Bd. II, 1879, ed. C. J. Gerhardt, 305. Quoted after Bos 1974, 55.

[10]   Costabel interpreted this to have meant Leibniz's letter of June 7, 1698, in which he designated infinitely small quantities as "imaginaria"; cf. Leibniz, III/2, 1856, 409 f. (Costabel 1988, 176).

recognize actual existence, however, only for finite, albeit arbitrarily small, quantities:

> With regard to the expression used by Bernoulli of "infinitesimal," "esistono in atto tutti i termini della serie; tu ricavi da ciò che esiste anche l'infinitesimo, ma io nego che ne segua altro che esiste in atto ogni frazione finita assegnabile, di qualsiasi piccolezza" (Pasini 1993, 117).[11]

In his previous letter of July 29, 1698, Leibniz had underlined that one remained in mathematics always within the field of finite quantities:

> Because even if I recognize that no material particle exists that is not simultaneously actually divided, we nevertheless do not as a result arrive at indivisible elements or smallest things, not even at infinitely small particles, but only at continually smaller ones, albeit actually ordinary quantities, just as when adding up we arrive at ever larger ones (Sierksma 1999, 447; Leibniz 1856, 524).

And while he admitted infinite sets in a further letter of February 21, 1699, he ascribed to them neither any number, nor the character of a coherent whole:

> I will concede the existence of the *infinite multitude*, but this *multitude* is neither a *number* nor a coherent *whole* [unum totum]. It means nothing more than that there are more parts, which could be referred to by any number at all, just as there is a multitude [...] of all numbers; this multitude, however, is itself neither a number nor a coherent whole (Sierksma 1999, 447; Leibniz 1856, 575).

Shortly before, in January 1699, Leibniz had answered to one of Bernoulli's arguments stressing that there could be only finite numbers:

> If one says that an infinite number of parts exist, [...] this does not mean that a certain *number* is being referred to, but only that one wants to say that more exist than any finite number would be able to express (Sierksma 1999, 448; orig.: Leibniz 1856, 536).

Leibniz made his convictions known not only in his private correspondence, but also in public. In a published letter of 1701, Leibniz even stated that the differential as infinitely small quantity could be considered in its relation to the variable like the ratio of Earth's radius to the distance of the fixed stars. Since the diameter of a game ball again related like a point to Earth's radius, the finite quantity of the ball's diameter was something infinitely small of second order as compared to the distance of the fixed stars. The point was only to make the distance smaller than any given quantity; this was methodologically equivalent, he said, to the method of exhaustion:

> For instead of the infinite or infinitely small, one takes quantities as large or as small as is necessary for the error to be less than the given error, such that it differs from the style of Archimedes only in the expressions which are more direct in our Method, and more conforming to the art of invention (Leibniz 1859, 96).

This letter printed in the *Journal de Trevoux*, a scientific periodical published by the Jesuits of France, was felt to be a scandal by the adherents of Leibniz's

---

[11]   Omnesque seriei hujus terminos actu existere; hinc infers dari et infinitesimum, sed ego nihil aliud hinc puto sequi, quam actu dari quamvis fractionem finitam assignabilem cujuscunque parvitatis (Leibniz 1856, 536).

calculus in Paris. In view of their own acceptance of the actually infinite, they considered it incomprehensible "que dans son calcul il ne prenoit point l'infini à la rigueur," that he did not take the infinite rigorously, but only as an element of comparison (Johann Bernoulli 1988, 308). Pierre Varignon, then involved at the Paris Academy in his own conflict with Rolle (see below), consequently wrote a letter to Leibniz in November 1701 urging him to clarify this matter, even urging an answer via Johann Bernoulli.

In his answer of February 2, 1702, Leibniz emphasized that his calculus could exist independently of *controverses metaphysiques*," and that it was not necessary that infinitely small lines existed *dans la nature*" in a strict sense. To avoid such subtleties, it was sufficient to explain the infinite by the incomparable. At the same time, he underlined again that infinitely small quantities were just as useful as ideal quantities, as imaginary roots were in algebra (Leibniz 1859, 91 f.). On the other hand, he also declared that the incomparable quantities need by no means be considered as fixed, determined quantities, but could also be inserted as rigorously infinitely small quantities:

> One must realize [...], that the common incomparable quantities themselves being neither fixed nor determined, and being able to be taken as small as is wished in our Geometric reasoning, make the use of infinitely small quantities rigorous (ibid., 92).

After having learned that Fontenelle was planning a book on the *metaphysique de nostre calcul*, Leibniz stressed a little later in his letter to Varignon of June 20, 1702, his own conception that the infinitely small quantities had the character of ideal quantities, respectively of well–founded fictions:

> I am not entirely persuaded myself that our infinite quantities and infinitely small quantities should be taken to be other than ideal objects or as well founded fictions (ibid., 110).

In the same letter, Leibniz also gave the reasons why he rejected infinitely small quantities in principle. For him, he said, infinitely small quantities were closely connected to the physicalist conception of atomism, and he was rejecting atomism on philosophical grounds. While he recognized ultimate substance, he recognized them only as immaterial principles:

> I do not believe that there are or even that there could be infinitely small quantities, and that is what I believe to be able to prove. It is that simple substances (that is, those that are not aggregated objects) are truly indivisible, but they are not material and are but principles of action (ibid.).

Leibniz's rejection of infinitely small quantities in differential and integral calculus thus is a direct expression of his fundamental philosophical conception, his theory of *monads*. According to this theory, the world is composed of arbitrarily many monads which, while being as simple substances without extension in space and immaterial, at the same time represented active principles. It is not permissible to interpret Leibniz's fundamental conceptions on the basis of how they were taken up by Johann Bernoulli, as the latter, an advocate of the infinite *in actu*, had molded these concepts according to his own preferences. This specific remolding of the Leibnizian theories in France by Bernoulli will have to be discussed below. At this point, we shall analyze whether Leibniz developed a justification that offers an alternative to the infinitesimals, or whether he confined himself to the algorithmic part.

Leibniz indeed gave such a justification: in a manuscript obviously written about 1701 in which the rules of his calculus first published in 1684 were to be demonstrated. This is where Leibniz erected his proofs on the *law of continuity*, which was of general and comprehensive theoretical importance for him. He formulated this law, which appears in various versions in his writings, as follows: "If any continuous transition is proposed terminating in a certain limit, then it is possible to form a general reasoning, which covers also the final limit."[12]

The essential proposition in this formulation of the law of continuity agrees with one of Leibniz's two main versions: that propositions which were valid in the finite maintain their validity for the limit as well in case of continuous processes. For Leibniz, the law of continuity was a law of nature of general validity. What results is quite an analogy between Newton's justifications and Leibniz's theories. Where Newton had justified mathematical theory from the laws of motion in mechanics, and ultimately by the continuity of time, Leibniz founded mathematics on laws of nature, which were general for himself. Mathematics served to mathematize processes of nature, and thus had to keep in harmony with nature.

5. The Law of Continuity: Law of Nature or Mathematical Abstraction?

The concept of continuity was thus integrated into analysis as a premise for mathematizing physics. It was a drawn-out process to transform the concept of continuity from this epistemological function into an operative function as an intramathematical concept.

The remarkable thing is, however, that Leibniz published the law of continuity not only in "metaphysical" form, but also in mathematizable form. Although he published both forms himself, his contemporaries took notice only of the metaphysical form. The conclusion is that to move toward an intramathematical concept at Leibniz's time was beyond the contemporary

---

[12]   In Bos's translation (Bos 1974, 56). The Latin original published in 1846 says, "Proposito quocunque transitu continuo in aliquem terminum desinente, liceat ratiocinationem communem instituere, qua ultimus terminus comprehendatur." The expression "terminus" Leibniz uses does not signify the same as "limit" in Newton (cf. above).

problem scope—in particular because the geometrical concept of quantity that provided an a priori guarantee of continuity was still dominant.

One of the formulations best known to this day is the "metaphysical" saying that nature never proceeds by leaps: *La nature [...] ne fait jamais des sauts*" (Leibniz 1890, 567). This classical formulation can be considered to be the origin of the modern concept of continuity. It was indeed most thoroughly taken up, thus by Boscovich and by Carnot, and it entered Lacroix's eminent presentation via Carnot's propositions. There is, however, a second meaning of Leibniz's law of continuity, which is derived from another formulation of this law, according to which no transition is effected by leaps: "nullam transitionem fieri per saltum" (Leibniz 1879, 168). This results in the second meaning that laws remain valid in the transition from the finite to the infinite. In his solution submitted to obtain a prize of the Berlin Academy, Simon L'Huilier extended this second metaphysical meaning mathematically to the theorem that *lim*, the limit of a variable, had the same property as the *quantité variable* before the passage to the limit. Leibniz gave a formulation suggesting this second meaning in his letter of February 2, 1702, to Varignon: *il se trouve que les regles du fini reussissent dans l'infini* (Leibniz 1859, 93).

Leibniz gave an additional formulation of the principle of continuity that was not only more explicitly mathematized, but also that clearly distinguished in a hitherto unprecedented manner between independent and dependent variables, thus laying the foundation proper for the mathematical concept of continuity. Leibniz first published this "mathematical" version in 1687, in a letter in French addressed to the periodical *Nouvelles de la République des Lettres*, edited by Bayle in connection with his controversy with the Abbé Catelan and Malebranche. In a Latin fragment dated from the same year, Leibniz elaborated on this version (Leibniz 1860, 129–135). In his letter to Varignon of February 1702, he explained this conceptuality again (Leibniz 1959, 91-95). He also unfolded it in his famous letter that underlies his bitter controversy with Maupertuis after 1751, and which Koenig declared to have been addressed to Jacob Herrmann, while Cassirer considered it to have been addressed to Varignon (Leibniz 1904, I, 74 f.). While it is debated today whether this letter is authentic, the point here is merely that it was heatedly debated in the 1750s and that mathematicians must thus have been familiar with it.[13]

---

13   This text is rendered in Cassirer's edition (Leibniz 1904, vol. 1, 74-78) both in German and in the French original (556-559). While researchers into Leibniz are convinced that the letter is genuine, the researchers into Euler are skeptical; a detailed analysis of this problem is given by Pulte (Pulte 1989, 216-225). There is unanimity today that the letter—provided it is genuine—was not addressed to Varignon. In a recent publication, Herbert Breger (Leibniz-Archiv Hanover) has systematically presented the reasons that speak to forgery (Breger 1999). The problem of his analysis, however, is that he understands, in his principal argument, the concept of *limes* exclusively "as technical term of analysis" (respectively of algebra as well) and hence does not include the reference to geometric quantities that is evident from the context (ibid., 379). As we shall see here (Section 8.), however,

In a letter to Bayle of 1687, Leibniz presented the law of continuity as a "principle of general order" that had its origin in the infinite and was of fundamental importance in geometry and in physics:

> When the difference between two instances in a given series or that which is presupposed can be diminished until it becomes smaller than any given quantity whatever, the corresponding difference in what is sought or in their results must of necessity also be diminished or become less than any given quantity whatever. Or to put it more commonly, when two instances or data approach each other continuously, so that one at last passes over into the other, it is necessary for their consequences or results (or the unknown) to do so also This depends on a more general principle,: that, as the data are ordered, so the unknowns are ordered also [Datis ordinatis etiam quaesita sunt ordinata]. (Leibniz 1969, 37; orig.: Leibniz 1887, 52).

In the version of the Latin fragment "On the principle of continuity" the differentiation of the two series into independent ("given") variables, and dependent variables, or variables sought, was formulated even more clearly:

> If in the series of the given quantities, two instances approach each other continuously, so that the one eventually becomes the same as the other, then the same has inevitably to occur in the corresponding series of the derived or dependent quantities which are sought. This is a consequence of the following, even more general principle: a determined order in what is given corresponds to a determined order in what is sought (Leibniz 1904, 84).

The philosopher Cassirer saw in this mathematical version "only another expression" of the modern definition of continuity (ibid., note 56, 84 f.), and the historian of mathematics Demidov attributed to Euler the development of a definition of continuity in Cauchy's and Bolzano's sense on the basis of this Leibnizian law of continuity (Demidov 1990, 37 f.).[14] The two authors, however, make no allowance for a rather essential conceptual difference as compared to the modern concept of continuity. While the modern concept concerns the continuity of *functions*, Leibniz's mathematical version referred to variable geometrical quantities.

Although Leibniz repeatedly returned to this mathematical version of his law of continuity, it was not taken up by his contemporaries, with the exception of Boscovich (see below), and hence was not developed further mathematically for a long time to come. The metaphysical version, in contrast, was accepted much more readily, and saw a much stronger impact. In this form of a natural law, the law of continuity can be considered to be a principle of *conservation*. The reception and effect of this principle was determined in particular by the fact that

---

the concept of *limes* was first formulated in geometry. Another problem is raised by his translating Leibniz's French *continu* into the German *Kontinuum* (continuum) instead of simply *stetig* (continuous) thus misleading by connotation to technically inappropriate concept fields.

[14]   This Russian essay was made accessible to me by a German translation by I. Maschke-Luschberger (Bielefeld).

it challenged another principle of conservation: Descartes's principle of the conservation of motion. The conflict was initiated by more precise physical research into a specific problem of motion: the *impact of hard bodies*.

Descartes obviously was the first to establish the law of conservation for motion. At the same time, he gave the application to the impact of hard bodies. Since the quantities of motion are conserved, motion cannot be lost, not even when particles of matter collide; on the impact of a hard body on another the former was thrown back, he said, but did not lose any of its motion. If it collided with an elastic body, on the other hand, it communicated its motion to the latter (W. Scott 1970, 6). Descartes's application of the laws of conservation to the impact of particles provoked a heated debate, which ended up changing the views on hard and elastic impacts. After 1673, Edme Mariotte formulated the new view that all motion was lost upon the impact of hard bodies, these hard bodies remaining at rest, while motion was conserved only in the case of impact of elastic bodies (ibid., 14).

Leibniz, who had changed his mind from adhering to Gassendi's Epicurus–inspired atomism to the antiatomism propagated by the Cartesians, intended to escape the consequence of nonconservation of motion upon the impact of hard bodies. Leibniz's solution was to introduce the principle of continuity as a central law governing physical facts. Since according to Mariotte the motion of hard bodies would immediately and without transition upon impact pass into rest, Leibniz inferred from the general validity of the law of continuity that there could not be any perfectly hard bodies. Subsequent to Leibniz, an entire school developed this argumentation further, with a large-scale impact. The controversies about the relation between the law of continuity and the impact of hard bodies had the decisive effect for mathematical concept development of discussing, in this debate on physical phenomena and laws (!), the intermediate value property for the very first time.

The priority of the principle of continuity in physics was most extensively and massively proposed by Johann Bernoulli. Bernoulli participated in the contest called by the Paris Academy in 1724 about clarifying the motions of hard bodies upon impact.

In the treatise submitted by Bernoulli, he showed not only that he had been the one to most thoroughly take up Leibniz's law of continuity, but also that he had raised it to the position of central physical principle so far as to examine physical phenomena as to whether they were compatible with the law of continuity. Bernoulli thus went so far as to deny that perfectly hard bodies existed, although the contest task was to explain the motions upon impact of two such hard bodies, declaring the respective physical observations to have been erroneous:

> Partisans [of the theory of] Atoms have attributed a hardness of their nature to their Elementary Corpuscles; an idea which appears to be true when one only considers things superficially; but which one soon realises contains a manifest contradiction when going into it a little deeper.

5. In fact, a similar principle of hardness cannot exist; it is a chimera which offends that general law which nature constantly observes in all its operations; I speak of that immutable and perpetual order, established since the creation of the Universe, that can be called the LAW OF CONTINUITY, by virtue of which everything that takes place, takes place by infinitely small degrees. It seems that common sense dictates that no change can take place at a jump; *natura non operatur per saltion*; nothing can pass from one extreme to the other without passing through all the degrees in between (J. Bernoulli 1727, 9).

As a consequence of this interpretation of the Leibnizian law of continuity, Bernoulli quite explicitly postulated the intermediate value property as a physical, not as a mathematical, concept:

If nature could pass from one extremity to another, for example, from rest to movement, from movement to rest, or from a movement in one direction to a movement in the opposite direction, without passing through all the imperceptible movements that lead from the one to the other; the first state must be destroyed, without nature knowing to which new state it must become; for in the end by what reason should one be chosen for preference, and of which one could not ask why this one and not that one? since having no necessary connection between these two states, no passage from movement to rest, from rest to movement, or from a movement [in one direction] to a movement in an opposite direction; no reason at all will determine producing one thing rather than any other (ibid.).

With this, Bernoulli's argument was at the same time founded in the philosophy of nature, in Leibniz's theory of sufficient reason (cf. W. Scott 1970, 22 ff.). Bernoulli rejected the concept of absolute hardness as contradicting the principle of continuity, admitting only elastic bodies whose deformations made possible the required transitions and intermediate magnitudes in case of motion and collision. Having taken this position, Bernoulli could not be awarded the Academy's prize; but the Academy called in 1726 a contest on the analogous task for elastic bodies, and now Bernoulli was able to attain a prize for his work (ibid., 24).

In his review of W.L. Scott's book, Thomas Hawkins showed his surprise from the perspective of modern science at the fact that the theory of atoms and the principles of conservation were once thought to be contradictory in the history of science (Th. Hawkins 1970, 119 ff.). Actually, this was even a long-lived controversy: the debate on the alleged contradiction between the theory of atoms and the principle of continuity was carried on for the entire eighteenth century.

Contemporary authors frequently referred to Euler as a representative of a position analogous to that held by Bernoulli, creating the impression that the view was backed by Euler's authority as well. Thus Kästner declared in his textbook on mechanics in the chapter "On the Law of Continuity" that "Herr Euler" considered the "contradictions between the law of continuity and perfectly hard bodies [...] to be a proof of the infinite divisibility of matter" (Kästner 1766, 352 ff.).

Gehler's influential encyclopedia of physics drew on Euler even more extensively in its chapter on continuity. It begins by saying that the law of continuity had been regarded "recently," i.e., in the eighteenth century, "by the masters as a natural law established in every rigor." The prime operative purpose is again listed as the rejection of perfectly hard bodies:

> It has been generally claimed accordingly that no change of determinate quantity in nature could happen suddenly, but must always come to pass by infinitely small steps. Authors took the liberty of rejecting everything contradicting this law. Since the velocity must change abruptly upon the moment of the impact of perfectly hard bodies [...], this has been sufficient for some mathematicians, like Johann Bernoulli and Euler, to deny the possibility of perfectly hard bodies in themselves (Gehler, Band 4, 1798, 209 f.).

Gehler then drew the immediate conclusion for the concept of atomism:

> If perfectly hard bodies are hence impossible in themselves, there cannot be any atoms either, and matter must be infinitely divisible. Such was Euler's inference (ibid., 210).

Euler indeed took such a view. It is not found, however, in his textbook on mechanics of 1734, but rather in his treatise *De la force de la percussion et de sa véritable nature* of 1745. He fully adopted Leibniz's law of continuity in this publication, calling it an "incontestably certain law of nature according to which nothing happens by leaps," and thus also rejected the existence of perfectly hard bodies: "It follows therefore that a perfect hardness [...] does not accord with the Laws of nature" (quoted after Pulte 1989, 159).

A particularly exhaustive and profound discussion of the law of continuity was achieved by Boscovich. Roger Joseph Boscovich (1711–1787) belongs among the innovative scientists of the Jesuit order. Educated at the *Collegium Romanum*, the Jesuit's central college in Rome, he became professor for mathematics and philosophy there. He was active mainly in Italy, but for some time also in Paris, and won acclaim in particular in the natural sciences as a universal scholar. The concept of continuity belonged among the foundations of his natural philosophy, to which he devoted a textbook of his own: *Theoria Philosophiae naturalis* (1763). Beyond that, he published a treatise on the law: *De Continuitatis Lege* (1754).

For Boscovich, the law of continuity was the basic concept not only for physics, but at the same time for geometry as well, the latter being understood as the theory of the curves of motion. Boscovich discussed the significance of the law of continuity more extensively and explicitly than all his contemporaries of the eighteenth century. The literature frequently quotes the understanding in Boscovich that conceives of the law of continuity as of an intermediate value property and that shows how strongly Boscovich had been influenced by Bernoulli's reception of the Leibnizian concept of continuity:

> The Law of Continuity, as we here deal with it, consists in the idea that [...] any quantity, in passing from one magnitude to another, must pass through all intermediate magnitudes of the same class. The same notion is also commonly expressed by saying that the passage is made by intermediate stages or steps; [...]

the idea should be interpreted as follows: single states correspond to single instants of time, but increments or decrements only to small areas of continuous time.[15]

In his special treatise on the law of continuity, Boscovich, in a manner not accidental for a Jesuit, adopted the underlying definition of continuity directly from Aristotle: as the connectedness of the respective end of the whole's parts, geometry, as distinct from numbers, serving as model:

> Nam partium numeri nullus est communis terminus, quo eae partes conjungantur... At vero linea est continua; quia communem terminum sumere licet, quo partes ejus conjunguntur, nempe punctum.[16]

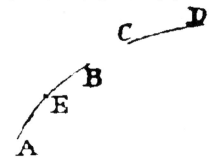

Figure 5, "lesion" of continuity

Boscovich explained this definition using the example of a curve ABCD (Figure 5): In B and C, continuity was injured (*continuitas laeditur*) because of the jump between B and C; AEB, against that, was continuous because the point E was the common endpoint of the parts AE and EB (Boscovich 1754a, V).

Boscovich, who used *limes* synonymously with *terminus*, emphasized that it was necessary in case of every continuous quantity to distinguish between what formed *terminus/limes*, respectively, and the part that was limited by that: while such a part was infinitely divisible, the part representing the *terminus* here had to be *indivisibile* (ibid.).

With these determinations of concept for continuity, Boscovich studied geometrical curves, in particular the conics, summing up as their common property both mathematical and physical that all their changes were continuous and that jumps never occurred: "[...] observatur ex eo, quod in omnibus Geometricis curvis, nihil usquam mutatur per saltum, sed mutationes omnes motu continuo fiunt" (ibid., XLI).

---

[15]   Boscovich, *Theoria Naturalis Philosophiae*, 32; English translation quoted after: G. Brittan 1993, 216.
[16]   Boscovich 1754a, V. Boscovich quotes Aristotle's theory of categories here, chapter 6, according to the Paris edition of 1619.

In case of curves having various branches, like the hyperbola, he conceived of each branch as a continuous geometrical curve of its own (ibid., XXX). Boscovich even studied curves with jump discontinuities as to whether they contradicted the law of continuity.

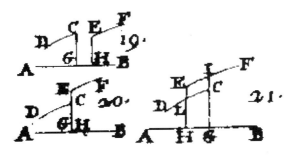

Figure 6, a to c: jump discontinuities and continuity

For this purpose, he constructed three curves: one curve having a jump discontinuity (his Figure 20; here: 6b); a second by pulling the first apart at this point (his Figure 19; 6a), and a third by superposing the two (his Figure. 21; 6c). Boscovich established that these curves signified no contradiction to the law of continuity, since they did not represent any geometrical places. The first curve did not because two ordinates corresponded to a point, and the third did not because two ordinates corresponded to every point in the area of superposition (ibid., XLIIIf).

While Boscovich's justification is based on the classical concept of curve from elementary geometry, it is of great systematic importance in several respects, firstly, because he discussed the possibility of discontinuity in geometry at all; secondly, because he conducted this study as ascription of axes of abscissae and ordinate; and thirdly, because he introduced, in direct connection with this analysis, the mathematical form of the law of continuity as its most general form, by quoting it, in his own, Boscovich's, translation into Latin, from Leibniz's publication of 1687 (ibid., XLIV).

Boscovich used this version containing the correlation of two serial values to study the law of continuity generally for mathematical linkages as well. In the shape of geometrical curves, he formulated how two series of quantities can correspond functionally: "per lineam exprimitur nexus binarum quantitatum, quarum primam exprimit abscissa, secundam ordinata" (ibid., XLVIII), discussing the type of transition from one value to the next. Here again, Boscovich obtained for geometrical quantities generally the result that intermediate values are always assumed without jump:

Hoc ubi generaliter habeatur, habebitur transitus per omnes magnitudines intermedias, et incrementum per omnes gradus infinite parvos sine ullo saltu utcumque exiguo (ibid., LIII f.).

Responding to a large number of counterarguments, Boscovich went on to discuss the law of continuity for physical processes of nature as well, eventually obtaining the general result that the law of continuity was of general validity both in geometry and in mechanics, provided finite quantities were concerned (ibid., LXVIII). A central example for that was for Boscovich the impact of hard bodies. He proposed an original solution to this much-discussed problem. In contrast to the usual partisans defending the law of continuity as a law of nature, Boscovich *did not* deny that perfectly hard bodies existed. Instead of claiming some elasticity for all bodies, as the Leibnizians did, he assumed that direct contact was never established between the two bodies, but rather that a vis repulsiva," a repulsing force, was exerted in case of small distances, a force that made the change of velocity occur not suddenly, but continuously (ibid., LXXIII f.).

In his *Theoria philosophiae naturalis*, Boscovich advocated this view of a transition from attracting to repulsing forces (cf.. Brittan 1993, 223 ff.), but this attempt to harmonize the Newtonian and Leibnizian conceptions of matter did not find a large number of adherents. It must also be noted that despite the fact that Boscovich adopted Leibniz's mathematical version of the law of continuity, and of his extensive discussion of relational dependencies in mathematics, this reflection on the concept of continuity was practically never carried on throughout the eighteenth century. Authors like Kästner who took up Boscovich's work focused only on the metaphysical aspect of excluding jumps.

That Boscovich's approaches were taken up so unwillingly may have been due to the fact that Boscovich did not implement his reflections into a textbook on analysis of his own. In his *De Continuitatis Lege*, he announced a presentation of infinitesimal methods for the fourth volume of his textbook *Elementorum Universae Matheseos* (Boscovich 1754a, xxxv). Acutally, however, only the first three volumes appeared in 1754 (on arithmetic, algebra, elementary geometry, and the conics). He never published a fourth. In the third volume, he again reflected on the law of continuity in connection with the conics, discussing at the same time, however, infinitely small and large quantities as *mysteriis*" (Boscovich 1754b, 297 ff.), thus creating a barrier to integrating them into mathematical operation.

By discussing the incomplete reception of Boscovich here, I am anticipating the development during the second half of the eighteenth century. The debate on the law of continuity's validity continued to be heatedly led over this period. The point of this discussion for our purposes is that it was being conducted *only marginally* for mathematics, Boscovich's reflection not having been taken up, that it concerned rather more physics, and that it for the first time integrated, and seriously accepted, empirical arguments in a quite novel way. Up to that time, debates in mathematics and physics had touched empirical facts only superficially. Just as Boscovich had excluded empirically possible discontinuous curve circuits from geometry by giving priority of definition to an as it were ontological concept of curve, the controversies about perfectly hard bodies in physics had been primarily determined by the debaters' underlying philosophical beliefs. As we shall see, arguments relying on empirics to correct previous fundamental beliefs were still being advanced with utmost reluctance. It is highly typical for the development of the basic concepts of mathematics that views seriously challenging traditional basic positions in order to proceed from the continuity of geometrical curves toward developing a general concept of function did not appear within the horizon of thought before new models of reality no longer exclusively determined by philosophy, but more strongly molded by empirics, had blazed their path in physics.

Kästner discussed the law of continuity in detail in his textbook volumes. The concepts he innovatively introduced in the part on the fundamentals of geometry typically became relevant for applications not in the mathematical part proper, but rather in his volume on mechanics, in his chapter on the impact of perfectly hard bodies. Kästner introduced the law exclusively by way of the intermediate value property:

> It is held because of this law in particular, that no change may occur suddenly, but rather that every change always passes by infinitely small stages, of which the trajectory of a point in a curved line provides a first example (Kästner 1766, 350).

He immediately linked it to his own presentation of the controversy on the existence of perfectly hard bodies:

> This law of continuity does not hold in case of perfectly hard bodies, and this reason was sufficient for some mathematicians to deny that perfectly hard bodies could exist (ibid., 351).

Kästner at first agrees with these mathematicians by pointing to *experience*:

> For the question whether there are such among those [bodies] we know, must be denied on the basis of experience already. Even the most hard matters we can experiment with show some degree of elasticity (ibid.).

One of the examples he quotes from experience is the following:

> If one breathes on a plane of steel polished like a mirror such as to render it matte, and then drops a ball of hard stone on it, the ball will make a round spot where it hit which is the larger the higher it was dropped, as proof that it flattened at the point of contact, and this the more the heavier the impact was. Its jumping back is due to its former shape being restored (ibid.).

In contrast to former participants in the debate, Kästner raised the critical question whether the general validity of the law of continuity had really been proved as yet. He declared that the confirming experience hitherto had to be a strong argument, but stressed that induction from empirics did not ensure general validity:

> I must confess that no stronger proof for this is known to me than the memory that this has been found correct in innumerable cases in nature by experience. Whether one was justified because of this in extending this to everything as well which does not fall under our experience as well, I should like to leave it to everybody's own judgment (ibid., 353).

Kästner quite acutely characterized a second possible justification of the law. It may generally be considered to be typical for epistemological conceptions concerning the mathematization of natural phenomena until the close of the eighteenth century, as deduction from proclaimed *necessities of reasoning*:

> Can something more in this matter be demonstrated by deducing from concepts than by experience? [...] Those who defend the law of continuity claim that without this law it was impossible to comprehend how a subsequent condition arises from a preceding one (ibid., 356).

To illustrate his doubts regarding this approach, Kästner quite excellently made explicit this law's intermediate value significance, which had been formulated mostly by Johann Bernoulli:

> Any change involves at least two conditions, one preceding and one following, which are distinct from one another in such a way that the difference between the former and the latter can be established. Now the *law of continuity* prohibits the thing which is being changed to transcend abruptly from the former to the latter. It must pass through an intermediate condition which is as little distinct from the previous as from the subsequent one. And because the difference between this intermediate condition and the previous condition can be established still, there must be an intermediate condition between these two as well, and this must continue in the same way, until the difference between the previous condition and the one immediately succeeding it vanishes. As long as the set of these intermediate conditions can be established, every difference between one and the next can be established as well: hence their set must become larger than any given set if these differences shall vanish, and thus we imagine infinitely many conditions where one differs from the next to an infinitely small degree (ibid., 354 f.).

It is telling for Kästner's highly critical intelligence[17] that he evaded the suggestibility of his own formulations, stating instead that he could not see any necessity of thought according to which the transition was bound to occur by an infinite number of intermediate conditions. He was well able, he said, to imagine a transition by distinct, "determined differences" (ibid., 358).

In a way quite uncommon for the textbook literature, Kästner allowed the dilemma between the two insufficient approaches to stand for the reader, and avoided giving a glib answer. Rather, he made clear that an unsolved foundational problem was present here.

He merely provided three hints as to the direction he was thinking in to find a solution. The first was as it were "metaphysical" in the classical way: While the law of continuity, as the difference between the *phenomena* that were accessible to us only on the one hand, and reality on the other, was "of great use for calculating the things given by nature,"

> our entire knowledge of nature is actually nothing more than a knowledge of phenomena, which would present something quite different to us if we saw the real in them (ibid., 363).

---

17   And against the widespread opinion concerning the mathematicians of the eighteenth century that these did not care about foundations!

Since the law of continuity could not be considered to be universally valid, one had to have additional knowledge in the individual case in order to be able to decide:

> Whoever wishes to extend this law to the real must justify his inferences by a law other than that, the suspicion remaining that he took images for things (ibid.).

The second hint consisted in differentiating between mathematics and physics. While the mathematician was permitted to abstract from the extension of the particles of matter, thus assuming a continuous transition through all intermediate conditions, the physicist was called upon to make allowance for finite dimensions and boundaries between conditions (ibid., 359 ff.).

His third hint was not quite compatible with the second. It consisted in attesting "large exceptions" from the law of continuity to geometry, legitimizing doubts as to whether is was "quite generally" valid in mechanics (ibid., 354). As much as it makes one sit up to find numerous exceptions from the law of continuity stated for geometry already in the 1760s, this does not correspond to the underlying conceptual horizon which consists in confronting curvilinear figures with polygons: whereas general continuity is ascribed to curvilinear lines, it is denied to polygons (see below, Section 7.):

> The law of continuity in geometry is unbreakably respected in case of curvilinear figures, but can it be maintained for straight-lined figures as well? If it is actually impossible for a point to abruptly change its path, then no point can move around the circumference of a quadrangle or triangle (ibid., 353 f.).

With Kästner—just as with Boscovich—geometrical curves are privileged; they do not represent abstract functions, but are mathematizations of a time-dependent concept of motion, and to see an injury of continuity in the changes of direction made by broken lines or polygons means that Kästner implicitly identified continuity with differentiability.

In the entry on "continuity" in his encyclopedia on physics, Gehler, although publishing thirty years later, completely adopts Kästner's positions, adding nothing about the state of the art to the latter's mechanics. He strictly adheres to Kästner's text in rendering the debate, only leaving out the challenge to metaphysical justification by the necessities of thought.

The article begins by distinguishing between continuous, extended quantities in geometry and successive conditions of physical quantities. In these as well, a change could take place in such a way that it did not happen "leapwise by clearly distinguished steps, but gradually transcended from any condition to the other by way of all kinds of intermediary conditions which can be considered to be infinitely many, but infinitely distinct steps" (Gehler, Bd. 4, 1798, 209).

After rendering the debate, Gehler also inquires whether the empirical proofs were as yet of universal validity: "But has the law of continuity been proved true in such rigor and generality?"

Gehler saw the dilemma's dissolution primarily in the metaphysical distinction between being and appearance. In case of a physical change of condition, continuity

> could be only appearance, and in this case Euler's entire argument against the atoms would disappear; for one would be justified to apply the law of continuity only where experience shows that it agrees with the phenomena. [...]. The law of continuity thus belongs to the clothes of things which we must need rely on wherever reality seems impenetrably cloaked with it, but which we do not consider to be reality itself, and which we may still less cloak with things which do not serve us to see them (ibid., 211 f.).

For all the explicit reflection on intermediate value properties on the one hand, and on the necessity of empirical proofs of continuity in concrete cases on the other, the strong epistemological link to a geometrical–mechanical concept of quantity is so dominant that transforming the epistemological preconditions into an object of a mathematical concepts was impossible.

6. The Concept of Infinitely Small Quantities Emerges

In the previous sections, we have seen how the transition from the concept of indivisibles to infinitesimal concepts took place. Now, however, we encounter a remarkable contradiction, which the literature, in my opinion, has not noted enough. Leibniz as one of the founders of infinitesimal calculus, *did not* introduce the term "infinitely small quantities" as a basic concept of this new theory.[18] Rather, Leibniz used as basic concept first "differences" and later "differentials." Nevertheless, Leibniz's theory, in particular, was understood to be linked to the basic concept of "infinitely small quantities" since the literature of the early eighteenth century. The early propagators of Leibniz's theory, the Malebranche group in France, were firmly convinced that this corresponded to Leibniz's intentions. In the literature on the history of mathematics, this contradiction was noticed only inasmuch as Leibniz's disassociating himself from this interpretation in France was taken note of. In a letter to Dangicourt of 1716, Leibniz summed up in retrospect this dissociation which had shocked his Paris friends in 1702:

> When our friends debated in France with the Abbé Gallois, Father Gouye, and others, I told them that I did not believe at all in the existence of truly infinite magnitudes or truly infinitesimal magnitudes. ... But since the Marquis *de*

---

18   Mancosu mentions that Leibniz explained the differential in 1684 "without any reference to infinitely small quantities," but thinks that Leibniz in most of his other articles had "introduced differentials directly as infinitely small quantities" (Mancosu 1996, 155 f.).

L'Hospital believed that in saying so I betrayed the cause they begged me not to say anything (quoted after Mancosu 1996, 72) .

Conversely, we find the term "infinitely small quantities" with Leibniz's rival Newton. The latter, in his *De quadratura*, speaks of the quantity he designates by "*o*" as of "quantitas infinite parva," calling the appropriate elements $\dot{x}o$ "infinitissime parva."[19]

Leibniz himself, like Newton, did not author any textbook on infinitesimal calculus; while he had the intention to do so, he was never able to realize it. The calculus was disseminated in textbook style by another author who was a member of the Malebranche group—the Marquis Guillaume François Antoine de L'Hospital (also: L'Hôpital; 1661–1704). The text, published anonymously in its first edition of 1696, underlined in its title *Analyse des infiniment petits, pour l'intelligence des lignes courbes* the concept that L'Hospital considered to be fundamental. Since the general public interpreted this textbook to be an explanation of Leibniz's theory, ascribing the infinitely small quantities to the Leibnizian calculus goes back to L'Hospital's textbook.

Actually, only the second of the two postulates with which L'Hospital begins presenting the new theory is contained in Leibniz's treatise of 1684: the identification of a curve with a polygon of infinitely many arbitrarily small line segments (Leibniz 1989, 111). The first postulate is not found in Leibniz.[20] It concerns the extended concept of equality, according to which adding an infinitely small quantity does not change a finite quantity:

> It is required that two quantities should be taken to be the same if they differ from each other only by an infinitely small quantity: or (which is the same thing) that a quantity that is increased or decreased only by a quantity infinitely smaller than itself, may be considered to have stayed the same (L'Hospital 1696, 2 f.).

The direct model for this added postulate is found in Johann Bernoulli's draft of his lectures on differential calculus:

> Postulates: 1. A quantity diminished or enlarged by an infinitely smaller quantity is neither diminished nor enlarged (Joh. Bernoulli 1924, 11).

We have become familiar with Johann Bernoulli as a firm advocate of the existence of actually infinitely small (and infinitely large) quantities, in his discussion with Leibniz toward the end of the 1690s (cf. Section 4. above).[21]

---

[19]   It is difficult to understand why Whiteside translated this expression as "indefinitely small quantity" (Newton 1976, VII, 63/64).

[20]   The extended concept of equality is often directly identified with the concept of the comparison of the orders of infinitely small and infinitely large quantities. In an explicit mode, the latter appears only later, as we shall see. Bos describes this concept, but on the basis of Johann Bernoulli's works (Bos 1974, 22). Leibniz, however, did not explicitly use it in 1684: he gave no justification for the vanishing of $dxdy$ in the product rule (Leibniz 1989, 106); the law of homogeneity mentioned elsewhere in this writing does not necessarily mean the distinction between algebraic and transcendental homogeneity (ibid., 113; see Section 4. above).

[21]   Johann Bernoulli's correspondence with Leibniz began in 1693, after the former's courses for L'Hospital.

How L'Hospital related to Bernoulli is known. In winter 1691/92, Johann Bernoulli gave private lessons to him in the new theory; this is how the Malebranche group gained access to this knowledge. Furthermore, the Swiss mathematician ceded exploitation of his own lectures to the marquis against a yearly life annuity. Only after L'Hospital's early demise did Bernoulli claim priority for L'Hospital's textbook of 1696, but published himself only the subsequent part of his lectures: integral calculus. Whether Bernoulli's text on differential calculus had been conserved was for a long time unknown, until it was discovered in Basel by Paul Schafheitlin, who published it in 1922 in the Latin original, and in 1924 translated into German. This text was copied by Nicholas Bernoulli, one of Johann's nephews, probably in 1705, and it remains open whether it is congruent with Johann Bernoulli's original (Costabel 1992, 14 note 11).

The next question is how Bernoulli attained the additional and explicit postulate. The brothers Jakob and Johann Bernoulli acquired the new theory between 1687 and 1690; which and whose works they studied for this is not known in detail (Bos 1974, 21). It is notable, however, that Johann Bernoulli's first postulate corresponds precisely to Newton's basic equation

$$x + 0 = x \text{ (see section 4. above),}$$

which Newton had formulated already in 1669 in a text privately disseminated.

That the adoption of infinitesimal calculus in France by the Malebranche group was not based exclusively on the Leibnizian concepts, but at the same time took in Newton's work, is confirmed further by the first far–reaching scientific dispute about the new theory, bitterly led in the Paris *Académie des Sciences* from 1700 to 1705.

This controversy was the first on foundations of the new theory, more than thirty years earlier than the much better known debate in England, which emerged after Bishop Berkeley's critique of Newton's concepts. This dispute is also structurally a new element in the history of science. Led twenty years after the controversy between Arnauld and Prestet on the status of negative numbers, this was the first dispute not conducted by letters, or by other forms of indirect communication, but a dispute in direct, oral confrontation. Several mathematicians had acquired membership in the Paris Academy after the 1699 reform. This accumulation of competence created for the first time a mathematical *community*, albeit small, which was able to interact in direct communication. In 1700, it split into two subgroups: into one rejecting the new theory and declaring the traditional methods of Fermat, Hudde, Huygens, and others to be superior, and a second, which defended the concepts challenged. The attacks were advanced by the algebraically oriented Michel Rolle (1656–1719). The major adversary, however, was Abbé Jean Gallois (1632–1707), after whose death the opposition group dissolved; Rolle had defected to the Malebranche group in 1706 already. The attacks assumed a finitist position that rejected infinitely small quantities for reasons of principle. The decisions made in 1706 by a committee of the Academy that caused the adversaries to desist—

albeit only on moral grounds—was for this reason called a *paix des infiniments petits* by the Academy's secretary Fontenelle in 1719. This peace followed after a lengthy and intense debate in which the arguments for and against had been heatedly exchanged for more than a year in the Academy's sessions, after which they had been carried into the general public despite the Academy's warnings against that.

Rolle conducted the controversy on two levels: on a technical level, and on the level of foundations. On the technical level, he argued that the new method led to errors, and that it brought nothing new. Varignon, who took on the role of counterpart to Rolle, succeeded in showing that Rolle, in his examples submitted, had committed errors both in applying the traditional methods and in applying differentials. These analyses of error at the same time contributed to clarifying the nature of maxima and minima in more detail.[22]

The foundational argument, according to which the new methods were allegedly not rigorous, went precisely against using infinitely small quantities— and in particular against L'Hospital's and Johann Bernoulli's first postulate of the equality relation. Since it was an essential element of the alleged lack of rigor of differential calculus, Varignon tackled the critique of L'Hospital's concept of equality:

> 2nd difficulty
>
> If a magnitude plus or minus its differential could be taken as equal to that magnitude (Joh. Bernoulli 1988, 356).

To refute this objection, Varignon not only directly and indirectly adopted arguments from the text of Newton's *Principia*, but also referred several times to this publication of Newton:

> One can refer Mr. R[olle] to the 1st definitions of the analysis of infinitely small quantities [by L'Hospital] and to the scholia of the first § of Mr Newton (ibid.).

Varignon declared, however, that he would treat the objection more extensively yet. In this version, Varignon's foundations relied exclusively on Newton. His foundation is indeed not an independent presentation of the concept of infinitely small quantities, but rather one closely adhering to Newton's *Principia*, in connection with appealing to this scientist, who was obviously thought to be an authority on the matter. Varignon also treated the arguments already advanced by Rolle—before Berkeley—which said that these quantities were simultaneously being considered as *something*, and as nothing:

> Mr. R[olle] is mistaken in taking differentials to be fixed and determined magnitudes, and even more for taking them to be absolute zeros. It is this that has caused him to find contradictions, which dissolve when reflecting that the calculation does not suppose anything of the sort; on the contrary, the nature of differentials consists in their being variable and not fixed and in decreasing

---

22   A major source for accessing the controvery are Reyneau's excerpts. They were published by Peiffer (1988b). Further information is contained in the Academy's records; samples of these were published by Blay (Blay 1986). Summaries of the controversies have been given by Blay (1993) and Mancosu (1996).

continuously toward zero, *influxu continuo*, not considering them even at the point, so to say, of their disappearance. *Evanescentia divisibilia*. It is this that the word *variable* means in definition 1 of l'anal. des inf. pet. and in the 2nd definition. It is a portion of magnitude whose decreasing can proceed as far as zero, and whose increasing can return from zero to something, this is what are called by M. Newton *fluxiones, incrementa vel decrementa momentanea* (ibid.).

The literature recorded with astonishment that Varignon relied on Newton to justifiy the Leibnizian calculus. Thus, Peiffer observes, "on peut s'étonner que Varignon construise sa défense sur les *Principia*" (ibid., 352, n. 2).

And Mancosu likewise remarks, without offering an explanation, "It is interesting that Varignon appealed to Newton's *Principia* as the source of rigorous foundation of the calculus" (Mancosu 1996, 167).

If we assume, however, that Johann Bernoulli could not take—as the basis for L'Hospital in this part—the infinitely small quantities as foundational concept from Leibniz's theory, but that they stem from his adoption of Newton's works, Varignon's exclusive appeal to Newton is readily explained. At the same time, this shows that differential and integral calculus were adopted and further developed in France precisely not directly from Leibniz, thus forming an alleged acute delimitation between "continental" and English mathematics, but rather that the French mode of adopting the two calculi already went by independently processing the contributions of *both* founders of infinitesimal calculus.

Varignon's retort just quoted permits yet another analyis. It has become common to assess L'Hospital's textbook only as a didactically clever elaboration of Bernoulli's ideas, since the latter's lecture drafts have been discovered.[23] Actually, however, L'Hospital did make foundational contributions of his own. While Bernoulli assumes without further explanation the undifferentiated concept of "quantity," L'Hospital begins his own textbook with a reflection on the concept of quantity, basing infinitesimal calculus on the concept of variable (L'Hospital 1696, 1). In his answer, Varignon is thus able to interpret the infinitely small quantities as variables. By this interpretation, he simultaneously explains Newton's concept of limit. Even if this did not immediately clarify the concept of limit, the controversy resulted, among other things, in a more exhaustive presentation of the connections, and thus in preparing further steps toward an algebraized concept of limit.

Upon comparing Varignon's explication with Newton's explanation concerning the concept of *limes* quoted above (4.4), it is evident that Varignon connected Newton's presentations quite clearly to the concept of variable, anticipating the later interpretation of the *infiniment petits* as variables, as null sequences:

Since the nature of differentials [...] consists in their being infinitely small and infinitely changeable up to zero, in being only *quantitates evanescentes*, *evanescentia divisibilia*, they will be always smaller than any given quantity whatsoever. In fact, some difference which one can assign between two

---

23   Cf. Rebel 1934, 14; Schafheitlin 1924, 7 ff.; Costabel 1992, 13 ff.

magnitudes which only differ by a differential, the continuous and imperceptible variability of that infinitely small differential, even at the very point of becoming zero, always allows one to find a quantity less than the proposed difference (Joh. Bernoulli 1988, 357).

With his presentation, Varignon had intended to give a proof for the *infiniment petits* in the mode of *les anciens*, that is to say by the methods traditionally admissible. This intention, however, had been confined to explaining the differentials as quantities that could be viewed as smaller than any given quantity. This, however, did not suffice to "prove" the specific nature of this new type of quantity. At least, however, Varignon was able to point out that it belonged to the generally recognized practice both of *algebristes* and of *géomètres* to neglect infinitely small quantities in establishing the sum of infinite series (ibid.).

Finally, Varignon had to deal with a third foundational argument advanced by Rolle. It is very telling that Rolle "separated out" the central method in the new theory, which had not been formulated as an explicit principle by its founders themselves: the method of comparing infinitely small, respectively infinitely large, quantities according to their orders, respectively powers: "Whether in Geometry there are infinitely great quantities of any sort, and infinitely small quantities of any sort" (Joh. Bernoulli 1988, 353).

Both for Newton and for Leibniz, this method had not been an independent basis; both had used it self-evidently to explain that the product of two differentials, respectively of infinitely small quantities, was negligible, or vanishing. Only Leibniz has explicitly treated the vanishing of higher differentials as opposed to that of lower differentials: with the purpose of ensuring the homogeneity of geometrical equations (Bos 1974, 33). Bernoulli, in his lectures on differential calculus, also did not make the method of comparing the various orders a separate principle. We find it proclaimed as a general method, on the other hand, for the first time in L'Hospital's preface to his *Analyse*:[24]

> [This analysis] does not restrict itself to infinitely small differences; but it reveals the ratios of the differences of these differences, also those of the third, fourth, etc. differences, without ever finding a term where it must stop. So that it does not just encompass the infinite; but the infinite of the infinite, or an infinity of infinites (L'Hospital 1696, Préface [I f.]).

Varignon considered this method to be a cornerstone of the theory. In his letter to Johann Bernoulli of January 20, 1702, in which he bade the latter to intervene with Leibniz in order to receive the latter's rapid answer to his own concerned inquiry as to whether Leibniz really saw the *infiniment petits*, like the Paris group, as a rigorous concept, or possibly only as fictions to express things incomparable, Varignon had declared the method of considering quantities of different orders as mutually vanishing to be indispensable in Leibniz's theory:

---

[24]   There is dissent in the literature as to who should be considered the author of this preface. Traditionally, Fontenelle is named as the author; Costabel has objected to this (Costabel 1992, 14).

[In case Leibniz recognized only the second interpretation:] I no longer see where he will find the different degrees of infinites which he needs for his system: he will only find finite quantities, greater than others, which will no longer be of use to him (Joh. Bernoulli 1988, 310).

In his answer, however, Leibniz did not mention this point claimed to be a foundation of his theoretical edifice. Varignon, on the other hand, had attempted quite extensively to present the method's consistency in his conflicts with Rolle: on the one hand using examples from traditional geometry and mechanics, and on the other hand by referring to analogous uses by Huygens and Pascal—that is to say, by mathematicians who were authoritative for Rolle (ibid., 353–355).

In his own comments concerning L'Hospital's textbook, Varignon did not add the method of comparison as a foundational concept, in the generality defended by him and attacked by Rolle, but only in its weakest form, for the product: "Every product of an infinitely small quantity with another infinitely small quantity is nothing" (Varignon 1725, 2). Only later in France, the method of comparing orders became one of the definitory foundations of the *infiniments petits*.

7. Consolidating the Concept of Infinitely Small Quantities

After we have studied how the concept of infinitely small quantities emerged in mathematics, we intend now to describe how this concept came to be consolidated. This consolidation and dissemination was effected by *textbooks*, and this is why our analysis will focus on them.

The first and at the same time most influential textbook propagating this concept was the *Analyse des Infiniment Petits, Pour l'intelligence des lignes courbes* by the Marquis de L'Hospital of 1696, a work we have had occasion to mention several times already. Deviating from his model of the lectures of Johann Bernoulli, who used a yet unspecified concept of quantity, the marquis introduced, as his basis, the concept of variable. It is characteristic that the continuity of the variables is assumed a priori, and that variables are strictly separated from constants, as he states in his first definition (see the quotation in Chapter II.2).

In the subsequent second definition, the infinitely small quantities are already introduced as a basic concept of the new calculus: as arbitrarily small changes of a variable. In spite of its being embedded into the context of variables, this definition is to a large degree static as yet. The designation of "portion" for the changed part still contains evident connotations to the static context of indivisibles. On the other hand, this was intended to indicate that the part was homogeneous with the whole quantity: "The infinitely small portion, by which a variable quantity inceases or decreases continuously, is called the *Difference*" (ibid., 2).

In the subsequent definitory circumscription, L'Hospital then spoke of a *quantité infiniment petite*: in formulating his first operative fundamental principle, that of the extended relation of equality (as already quoted in the previous section).

As second principle of order, he listed the Leibnizian principle according to which a curve may be considered to be the union of an infinite number of tiny parts of straight lines (ibid., 3). L'Hospital first applied the extended equality in proving the product rule of differentiation, that $dxdy$ is a *quantité infiniment petite* in relation to the two other terms ydx and xdy. He proved this proposition by reducing it to an infinitely small quantity of first order—according to the extended relation of equality he had introduced himself—by a formal division: "for if, for example, one divides ydx and $dxdy$ by dx, one finds firstly y and in the other case dy, which is its difference, and consequently infinitely less than itself" (ibid., 4).

As the textbook goes on, this rule proves to be the basis of operating with infinitely small quantities. L'Hospital did not include integral calculus; he justified this by referring to the book Leibniz was planning on this topic.

The next textbook to be reviewed here is Varignon's supplementary *Eclaircissemens sur L'Analyse des Infiniment Petits*, the text of which, despite the posthumous publication of 1725, had most probably been authored by Varignon in close contact with L'Hospital. Varignon keeps referring to concrete pages and articles of L'Hospital's textbook, his book thus forming a unity with L'Hospital's original text.[25] Varignon's book is of particular interest, since he had been the foremost defender of the new theory in the controversies at the Paris Academy (see Section 6).

His sections on foundations indeed contain some remarkable shifts of emphasis. In his new version of the introductory definition, he already uses "variable" as an independent term, not only within the composite term "variable quantity." Moreover, he explains the term "continuous" by referring to the Leibnizian terminology requiring that there be no jump:

> Every quantity which, keeping the same expression, increases or diminishes continually (*non per saltum*), is called a *variable*, and that which, with the same expression, keeps the same value, is called *fixed* or *constant* (Varignon 1725, 1).

The subsequent, reformulated second definition understands differentials no longer in a purely algebraic way, but documents that Newton's views had been adopted by its kinematical reference to elements of time: "The *difference* or *differential* of a quantity is the instantaneous increase or decrease of its value" (ibid, 1 f.).

The continuity of change is both assumed and presented as ensured by the reference to time. In the following second proposition, infinitely small quantities are again introduced as *portions*, while *infiniment* is substituted by *indéfiniment*: "Thus the difference of a variable quantity will be an indefinitely small portion whose value increases or decreases continuously" (ibid., 2).

Varignon obviously had deemed this indéfiniment" as a term better suited to express the Leibnizian conception as the latter had formulated it himself in his letter of 1702. On the other hand, Varignon used this *indéfiniment* synonymously with infiniment," as can be seen from his exclusively using *infiniment* in his subsequent first fundamental principle by which he considerably abbreviated L'Hospital's formulation:

> Every quantity which is only increased or decreased by an infinitely small part with respect to its total can be taken to be the same as it was before this change (ibid., 2).

As a new basic assumption, a proposition on infinitely small quantities of higher order has been added, as already shown in Section 6, according to which the product of two *indéfiniment petite(s)* quantities was *nul* (ibid.). This additional assumption obviously was a response to Rolle's attacks concerning the issue whether the new calculus was really founded; the surprising element in it, however, is its absolute proposition, not made relative by any formulation of ratio to another quantity, that the product yielded *zero*. In his further presentation of calculus, Varignon designated such terms without further ado as being "nul," and by no means as being, in relation to other quantities, as small as to be vanishing.

A very illustrative work is Reyneau's first textbook of 1708. Because Reyneau, as member of the order of Oratorians, also belonged to the innovative Malebranche group, his presentation of the new theory can also be expected to be a response to the disputes at the Paris Academy between 1701 and 1705. Because of the close cooperation within the Malebranche group, this answer can further be expected not to have been Reyneau's alone, but one jointly elaborated. Indeed, the textbook's title already underlines its character of intentional retort: *Analyse demontrée*, a title implicitly refuting the objection that the new theory was lacking rigorous proof. The textbook is systematically illustrative also because it contained for the first time integral calculus as well, which had still been absent in L'Hospital's work.

This almost unknown two-volume textbook realized a most fascinating structure;[26] it forms the first integral presentation of the field of *analysis* newly created since Descartes. This book, which we have already praised in our Chapter II on negative numbers for its innovative contributions, offers a type of algebra in its first volume: an excellently methodologically structured presentation of solving equations up to equations of fifth and sixth degree, and

---

[26]   Moritz Cantor reports that he "never saw" this work (Cantor, Bd. 3, 1901, 571).

methods of numerical approximation, Reyneau devoting particular attention to incommensurable quantities.

The second volume, bearing the subtitle *Usage de l'Analyse...*, intends to present the new branches of mathematics as a methodologically guided application of an algebra, or *Analysis* understood in a unified way: understanding both analytic geometry and differential and integral calculus as applications of *analysis* to geometry, and to physical–mathematical problems. The distinction made in the title and in the text between *calcul ordinaire de l'Algebre* and *calcul differentiel et calcul integral* underlines the unified character of the classical and the new theories. The subtitle also clearly shows the intention of responding to the attacks at the Academy, since it expressly professes to prove the new methods: "*Ces derniers calculs* [i.e., differential and integral] *y sont aussi expliqués et demontrés.*"

The entire presentation of the extension of the *calcul ordinaire* to the *calcul differentiel et integral* shows an essentially defensive style. The previous programmatic certitude of the superiority of modernity as compared to *les Anciens* has disappeared; what can be recognized is rather an almost anxious effort to prove a factual identity of the new methods with the classical methods of the ancients. Just as telling is the strong appeal to Newton's concepts for introducing and legitimizing the new concepts.

The general foreword of the second volume is concerned with making the new appear basically old:

> [...] it is not a new thing in Geometry to consider parts of magnitudes of such extreme smallness that they do not enter into comparison with the ordinary magnitudes which are being determined. The most ancient of Geometers, as can been in Euclid Book XII and in the works of Archimedes, used these infinitely small parts as the principle of some of their demonstrations (Reyneau 1708, vol. 2, xij f.).

The new thing was essentially only the mode of designation, and the development of appropriated modes of operation:

> This has not been a new discovery in our time, except to employ in Geometry these parts of whole magnitudes, so small that they have no finite ratio with them. What the illustrious Authors of the differential and integral calculus have added, to the suppositions the Ancients made of their nature, is only to provide convenient expressions for these small parts, which are first elements of the magnitudes; and to discover a calculus that was so suitable that the methods of Analysis could be applied, and so that one could go back up from these infinitely small quantities to the whole or integral magnitudes of which they are the first elements (ibid., xiij).

Reyneau underlined in particular that the basic concepts and the certitude of proofs of the *Anciens* and of the moderns agreed:

> The foundation of the differential calculus is common to the ancient and to the new Geometers. The certainty of the demonstrations based on this foundation is the same (ibid.),

the only difference being that the *ancients* conducted their proofs *per absurdum*, while the new calculi proceeded by direct proofs.

In his introduction to the part on differential calculus, Reyneau underlined even more concretely that the basic principle of differential calculus, the extended relation of equality, was not new either:

> C'est une chose ordinaire aux anciens Geometres de regarder deux quantités comme étant égales quand elles different moins entr'elles qu'aucune grandeur finie et déterminée, tant petite qu'elle puisse être, en demeurant finie ou bornée (ibid., 144).

This was the principle on which most of the proofs in Euclids 12th book relied. To legitimize his views further, Reyneau appealed to the principle of continuity as to one anchored in nature, his strong recourse to time becoming evident at this point:

> These calculations follow nature in the resolution of physico–mathematical Problems, which concern only movement and figures, which begins and behaves ordinarily by infinitely small degrees at each moment of time, each of these instants being also infinitely small (ibid., 146).

In formulating the principle of the extended relation of equality, Reyneau at the same time presented a conceptual determination of infinitely small quantities that was equivalent to him: as quantities that were smaller than any given finite quantity — obviously with the intention of elaborating on the identity of the new theories with the geometry of the *ancients* (ibid.). In this introduction, Reyneau already extended the principle of equality by sleight of hand practically to a principle for comparing infinite quantities of different order by declaring it to be valid as well for the relation between finite quantities and infinitely large quantities:

> Just as one can consider magnitudes as infnitely small in comparison with the finite magnitudes of which they are the differences, in the same way one can consider the magnitudes as infinitely large in comparison with other finite magnitudes which become equal to zero in comparison with these infnitely large magnitudes (ibid., 150).

The presentations of the foundations of differential calculus following the introduction markedly replaced L'Hospital's purely algebraic foundations by referring to a geometrical–kinematical foundation of concepts amounting to a direct adoption of Newtonian views. The first fundamental principle (*supposition ou demande*) indeed supposes that all geometrical figures can be conceived of as of being formed by *motion*: lines created by moving a point, angles by revolving a line segment with regard to a second, areas by moving a straight line or a curved line, volumes by moving areas, etc. (ibid., 151). The first definition by introducing the concept of variable—continuity again being assumed—(*qui augmente (diminue) insensiblement*) was followed by a second fundamental principle that has been additionally inserted as compared to former presentations: the introduction of infinitely small units of time called *instants*:

Each portion of finite time, no matter how small, is infinitely divisible just like length, and these portions of infinitely small time [...] are called *instances* [*des instans*] (ibid., 152).

Reyneau also claimed that infinitely small parts existed as well for velocity, motion, and for every imaginable kind of quantity. On this kinematic basis, Reyneau now introduced, in his second definition, the differential as the infinitely small increment in the smallest unit of time:

The infinitely small increase or decrease experienced by a quantity at each instance by any speed whatever, in the making of a line or figure, is what is called a *difference* (ibid.).

While Reyneau introduced an additional postulate concerning infinitely small units of time, the possibility of identifying a curve with a polygon having an infinite number of sides was no longer a basic postulate for him, as it had been for his predecessors, but merely a corollary (ibid., 153). In a second corollary, Reyneau even strengthened the reference of his basic concepts to mechanics, declaring that one could understand every infinite part of a curve also as being formed by the motion of a point on which two forces were acting from different directions (ibid., 154).

In his presentation of the rules for operating with the differential calculus, we find another marked adoption of Newtonian concepts. For the product rule—which increasingly proves to be a decisive test for dealing with infinitely small quantities—Reyneau gave two proofs, the first being an abbreviated form of the proof in L'Hospital. Without giving any justification, as his predecessors used to do, Reyneau simply states that $dxdy$ was an infinitely small quantity in relation to $ydx + xdy$. The second proof, however, represents the Newtonian "trick" using the half–increments $x - \frac{1}{2}dy$, which is calculated exhaustively, $dxdy$ seeming to vanish even without being neglected (cf. Section 5. above; ibid., 156 f.).

After finishing his presentation of the major operations of differential calculus, Reyneau inserted a section in which he reflected anew the "exactness of the proofs" of the new theory. Without naming Rolle, he tackled the latter's objections that while infinitely small quantities were being inserted as real quantities, they were inserted as vanishing quantities in other operations. Again, Reyneau tried to invalidate this objection by appealing to the *ancients*. These, "too," had operated with infinitely small quantities only *during* the process of proving; after establishing the proof, the ancients "too" had assumed that these quantities vanished, since only in this case would the inscribed figure, for instance, become exactly equal to the initial figure, thus attaining the exactness required (ibid., 164 f.).

Only in case of differentials of higher order, Reyneau left his defensive style, adopting, albeit quite cautiously, an argument pattern used by Prestet. The geometry of the ancients had not known these differentials of second, third, etc. order; it had been confined to mathematical problems for which these differentials were unnecessary. In Reyneau's own time, however, much broader problems could be worked on, including those penetrating the infinite. Nevertheless, Reyneau made an effort, even before presenting his set of concepts, to make the *possibilité* of such mathematical concepts intuitively graspable by means of geometrical concepts (ibid., 165 ff.).

With regard to integral calculus, which is presented in an entirely computanional mode, it can be noted that the integral is introduced exclusively as an indefinite integral: in a purely algebraic mode, as an inversion of differentiation (ibid., 159 ff. and 230 ff.).

While the next text to be discussed originated from a Swiss national, Jean-Pierre Crousaz (1663–1750), professor of mathematics in Lausanne, it can legitimately be counted in the French context, since its author resided in France for a long time. The text makes an effort to improve on L'Hospital's textbook in understanding, and it was published in Paris. Crousaz's *Commentaire sur l'Analyse des Infiniment Petits* (1721) is structurally significant, since it is the transition from the first generation of analysis textbooks to the second (Schubring 1994). As Crousaz says himself, the first–generation textbooks had been addressed to the *savants* themselves. Since many conditions and intermediate steps had been self–evident for scholars, he said, the first textbooks had been incomprehensible for the larger public (Crousaz 1721, Preface [2 and 5]). Crousaz thus set himself the goal of making the new theories accessible to an audience larger than that of scholars by commenting L'Hospital's textbook in detail. A large part of Crousaz's explication is assumed by his reflecting on infinitely small quantities. In a 27–page *Discours*, Crousaz undertook to clarify their nature and application more precisely:

> I intend to establish what is meant by the term *infinitely small quantities*, and to dissipate the obscurity and equivocal nature of this term (ibid., Preface [4]).

In the following chapters as well, he repeatedly tried to give grounds for the *Infiniment Petits* being real (cf. ibid., 164 ff.).

The most remarkable result is that the educationally motivated restructuring led for the first time to making explicit the general idea of comparing infinitely small quantities of different order. Crousaz criticized that L'Hospital assumed such a conception, even announcing it in his preface, but then passed over it too quickly (ibid., Preface [5]).

In his discourse on the *Infiniment Petits*, Crousaz declared clarifying the divisibility of quantities to be a necessary basis. The principle to be assumed was that any geometrical quantity was infinitely divisible. There was no ultimate term of division, he said, and hence no ultimate *particule indivisible*. An *infiniment petit* was by no means the ultimate part of a line beyond which division could no longer be executed. Crousaz emphasized that *infiniment petits* were not without dimension, but rather maintained their dimension, and hence the dimension of the original quantity (ibid., 2 f.). With this determination, Crousaz adhered to the Leibnizian concept of incomparability, using it to define infinitely small quantities:

In a word, a part as small as one wishes and meeting the required condition is called *infinitely small*, because the line with which it is compared contains it so many times that the imagination is lost in making this comparison, as it comes to be lost in the infinite. (ibid., 4).

As example, Crousaz quoted, just as Leibniz had done, the ratio of an arbitrarily small, microscopic particle to the earth's diameter. With this, however, Crousaz did not close his introduction of infinitely small quantities, but became the first author to go on to formally introduce infinitely small quantities of second, third, and higher order.

According to his own claim, Crousaz justified these quantities of higher order; actually, however, he explained them by keeping to an *infiniment petit* of a certain order, and then obtaining an *infiniment petit* of the next higher order by arbitrarily dividing it:

But then, Matter is divisible to infinity. In this *Infinitely Small Qmount* which has been determined, one can again suppose that there is a part also small in comparison with it, which is itself small in comparison with the drawn line.

A given line, and in general a visible and imaginable quantity, is called a finite Quantity.

The *Infinitely Small Quantities* of which one imagines it is composed, and of which it contains an infinite multitude, are called *Infinitely Small Quantities of the first degree*. Infinitely Small Quantities of these latter are called Infinitely Small Quantities of the second degree. The divisibility of Matter to infinity forces the recognition of Infinitely Small Quamtities of the third degree, of the fourth degree, etc. (ibid., 4 f.).

After having given geometrical reasons for the existence of such infinitely small quantities of higher order—designating, for example, an infinitely small rectangle by *filet*, and thus as an infinitely small qunatity of second order (ibid., 9)—he developed in detail an entire algorithm for multiplying, and even for dividing, *Infiniment Petits* of various orders. Whereas multiplication by a finite quantity does not change the order of infinity, multiplication by an infinitely small quantity changes the order additively—and division, analogously, diminishes the order subtractively. Crousaz did not present this algorithm with variables, but with concrete numbers: for showing, for instance, that multiplying an *Infiniment Petit du second genre* by an *Infiniment Petit du premier genre* gave an *Infiniment Petit du troisième genre* (ibid., 11 ff.).

The basis of this algorithm was that an infinitely small qunatity of a certain order related to the next higher order just as an infinitely small of the next lower order related to the original quantity (ibid., 8). Moreover, Crousaz assumend that *Infiniment Petits* of the next higher order can be formed from infinitely many *Infiniment Petits* of a certain order, and that for instance, a *ligne finie* could be formed from *une infinité de lignes Infiniment Petites du prémier genre* (ibid., 5). Crousaz concluded by stating that the *Infiniment Petits* did not form any infinitely small parts in an absolute sense, but only in a relative sense (ibid., 15).

In a closing part of his first *Discours*, Crousaz also presented an additive algorithm, for operating with polynomials consisting of finite quantities, and of infinitely small quantities (ibid., 20 ff.).

Since his intention was to comment L'Hospital's textbook, Crousaz also confined his own book to differential calculus. This is perhaps the reason why Crousaz limited his algorithm of comparability to the orders of infinitely small quantities, and did not expand it to the infinitely *large*. Fontenelle, whose articles in the publications of the Paris Academy between 1700 and 1711 Crousaz draws attention to, because Fontenelle had already used *différens ordres d'Infinis* (ibid., 166 f.), collected these dispersed articles himself, finally summing them up into a general theory of the infinite in a book: in the volume of 1727 reviewed already in our Chapter II, on negative numbers. This is where Fontenelle presented an entirely general algebraical–formal conception of arbitrary orders of infinitely small and infinitely large quantities, and of their comparability.[27]

As has been mentioned, this is not a textbook in the narrower sense, but rather a first exhaustive publication of reflections on the foundations of mathematics. The book's intention was to comprehensively clarify and present the theories underlying differential and integral calculus. Fontenelle understood this theory to be the science of the infinite, explicitly subsuming, under this *Infini*, both the *Infiniment Grand* and the *Infiniment Petit*. To justify operating with any kind of infinite, Fontenelle—probably as a pioneer as well—introduced the distinction between the metaphysical infinite and the geometrical infinite. While the *Infini Métaphisique* was absolute (*une grandeur sans bornes en tous sens, qui comprend tout, hors de laquelle il n'y a rien*, the *Infini Géométrique* was a relative quantity with which mathematical operating was possible: "It is simply a magnitude greater than any finite magnitude, but not greater than all magnitudes" (Fontenelle 1727, Préface [12]).

In line with his own algebraical view, Fontenelle chose a *sign* for the quantity thus legitimized—"∞" for an infinitely large quantity—then developed a comprehensive algebraic calculus for the quantities thus designated, as a basis for an application in geometry by means of differential and integral calculus, the latter two being understood as a science of the infinite. As he showed in detail, additive and multiplicative operations with finite quantities do not change an infinite quantity:

$$\infty \pm a = \infty; \quad n\infty = \infty \text{ (ibid., 31)},$$

whereas multiplying an infinite quantity by another changes their type of infinity, with

$$\infty \times \infty = \infty^2$$

---

[27]   An introduction to this book was given by Blay in his preface to his own reedition (Blay, Niderst 1995).

infiniment plus grand then ∞. Fontenelle thus constructs a series of powers of the infinite as a geometrical progression, powers that represent the orders of the (geometrically) infinite:

> So as many possible powers of ∞ as there may be, there are as many *orders* or *genera* of Infinities which always exceed each other. ∞ is of the 1st order, or genus, ∞^2 is of the 2nd, etc. (ibid. 34).

Quite analogous to the assertion that adding a finite quantity does not change an infinite quantity, adding an infinite quantity of lower order does not change an infinite quantity of higher order, e.g.,

$$\infty^2 \pm \infty = \infty^2 \text{ (ibid.)}.$$

For Fontenelle, this vanishing of terms of lower order with regard to terms of higher order is true both for infinitely large and for infinitely small quantities, forming for him—in contrast to the founders of differential and integral calculus—the central conceptual basis for the entire new theory: "The great principle and the richness of the Calculus is to make all the magnitudes of a lower order disappear in the presence of those of a higher order" (ibid., 393).

Fontenelle established the connection between infinitely small and infinitely large quantities also algebraically, by way of the fractional term he explicitly introduced as basic relation:

$$\frac{1}{\infty} \text{ (ibid., 116)}.$$

With this algebraization of the infinitely small, Fontenelle was able to handle the orders of infinite quantities in a unified way as positive and negative powers of the basic term ∞.

This algebraic determination of the infinitely small as the reciprocal of an infinitely large quantity was also the concept underlying his definitional introduction of the *grandeurs infiniment petites*. Every quantity a that was infinitely less large than a quantity b was *infiniment petite* with regard to the latter (ibid., 116).

Fontenelle also implemented his claim to prove the existence of such quantities by reciprocal reference to his method of proceeding in case of infinitely large quantities. Just as the infinitely large quantity was produced from a finite quantity—that is to say by infinite augmentation—the infinitely small quantity must be understood as infinite diminishing of a finite quantity (ibid.). Fontenelle also expressed this diminishing as infinite division:

> An Infinitely Small Quantity is a part of a Finite Quantity arising from division carried on to the infinite, or an *infinitieth* of a Finite Quantity (ibid.).

In this way, Fontenelle obtained the definition of the infinitely small as a fractional quantity:

> Thus, an Infinitely Small Quantity is a fraction, but an infinitely small one or, what comes to the same thing, a fraction whose numerator is finite and whose denominator is infinite (ibid.).

Since such an infinitely small quantity was also a quantity itself, it could again be infinitely diminished, respectively divided, which is how Fontenelle here again justified a geometrical progression of orders of infinitely small quantities, up to $\frac{1}{\infty^\infty}$ (ibid., 117).

In a mode analogous to the character of the infinitely large, Fontenelle also stressed the relative character of the infinitely small. An infinitely small quantity $\frac{1}{\infty}$ could always be taken as zero—but not as "zero *absolu*," because $\frac{1}{\infty}$ in itself always remained a quantity,[28] but rather as "zero *relatif*," that is to say with reference to a finite quantity (ibid., 118).

Fontenelle's notion of a complete sequence of positive and negative orders of the infinite had an acute conceptual consequence. He had to abandon the conceptual basis of Newton and Leibniz according to which the infinitely small quantities are of the same dimension as the respective quantities, having to fall back on Cavalieri's conception. An infinite whole, he said, could not be composed of infinitely many parts of its order:

> A whole cannot have an infinity of infinite parts of its own order, since if it can only be divided into a finite number of parts, they would be infinities, and of its own order, but the smaller they are, the more there would be of them, from which it follows that if there were an infinite number of them, they would have to be infinitely less than they were, and no longer of the same order as the whole. [...] Hence every infinitieth part is of a lower order than the whole (ibid., 52).

Since he implicitly identified dimension and order of the infinite, Fontenelle arived, on the other hand, at a remarkably different conception of the integral, conceiving of it not only as of an inversion of the differential operation, but rather as of a summation of elements of lower dimension. Fontenelle mentioned this briefly in his discussion of "hyperbolic spaces":

> *ydx* is the element or infinitely small quantity of all the curvilinear spaces, such that all the Hyperbolic or Asymptotic Spaces are just the sum of an infinite sequence of *ydx* (ibid., 399).

Fontenelle did not expand this concept to an integral calculus; but it was later used again to found the definite integral.

Fontenelle declared himself to be satisfied with the eventual triumph of the conception of the infinite. At the Paris Academy, there were no loger two factions, he said, and went as far as to dissociate himself even from Leibniz and the latter's explicit conception of the *Infinis de differents ordres* as *Incomparables*. It was sufficient, he said, to rely on the knowledge acquired through Leibniz, instead of appealing to him as an authority: "If M. Leibniz has faltered, one should rely more on the knowledge one has gained from him, than on his authority itself" (ibid., Préface [9]).

Finally, another conceptual achievement of Fontenelle has to be noted. He reflected in detail on how far properties valid within the finite maintained their

---

28   This wording shows not only that the zero continued not to be recognized as a quantity, but also that this special character of the zero was required to legitimize operating with *infiniment petits*. We shall deal with the conceptual consequences from this ambivalent status later.

validity after the transition to the infinite as well. Fontenelle stressed that it did not necessarily follow that if a property emerged and was conserved within the finite then it was always valid in the infinite as well, but rather that this created only the possibility of this being the case. He made a point of requiring a *uniformity* of development to permit inferring from the finite to the infinite:

> If one identifies a property in the Finite [...] and one finds the same property in the Infinite, it is certain that it is constant in all the intermediary stages, provided the transition is *uniform*, that is, that the other properties which increase or decrease are always present (ibid., 37).

Fontenelle, with his realism of concept concerning the infinite, can be considered a prominent propagator of the program of algebraization. As permanent secretary of the Paris Academy's mathematical class, he was certainly an eminent representative of the leading mathematicians in France. The Paris group around Malebranche, however, did not represent the entire mathematical *community* in France. While the resistance of the group around Rolle at the Academy had been removed, programmatic resistance against the program of algebraization now emerged for the first time. The first leader of this programmatic resistance was Abbé Deidier, from the emerging system of military education, whose textbook *L'Arithmétique des Géométres* (1739) has been studied in our Chapter II.2.8.1.

This textbook for arithmetic was the first part of a four-volume work on elementary and higher mathematics, the second part being *La Science du Geometre* (1739). The third and fourth part, both published in 1740, were in contrast presentations of the novel higher mathematics. The third volume, *La Mesure des Surfaces et des Solides par l'Arithmetique des Infinis et les Centres de Gravité*, was an explicit rejection of differential and integral calculus and was presented as a realization of *Synthèse*.

Deidier claimed to have presented this part entirely in a synthetic manner, explaining "synthetic" in a very simple way as dealing with figures, while the counterpart of the analytic was identified as that having to do with calculating, something apt to disorient beginners:

> I have handled this treatise in an entirely synthetic manner, so that those who are Beginners can get used to envisaging figures and discovering their properties, and I have abstained from calculations for fear of losing their attention (Deidier 1740, vol. 1, ix).[29]

From the viewpoint of the synthetic method, Deidier sharply criticized the differential and integral calculus of "the moderns." Although deemed to be the best method by the *savants*, it was *un peu trop abstraite et métaphysique* for those accustomed to *synthèse* (ibid., x). To justify his argument, he again drew on the contrast between calculus and geometry. Although the *calcul* was certain,

---

[29]   At least the issue from the Herzog-August-Bibliothek Braunschweig accessible to me has been incorrectly bound: in the third volume (*La Mesure ...*), title page, and Préface are bound together with the text of the fourth volume. To compensate for that, the fourth volume boasts the title page and Préface of the third.

it always left some *obscurité*, and doubts that could be removed only by geometric proofs. The object of this third volume, he said, was thus higher geometry, elaborated by using the means of Wallis's arithmetic of the infinite, as well as Guldin's and Tacquet's method of the center of gravity. He declared in particular Wallis's method of measuring areas and volumes to be identical with Cavalieri's method of indivisibles. It is very revealing that Deidier names as the essential difference between the modern theories and those of the various predecessors that the smallest particles were of the same, respectively of different, dimensions. Cavalieri's method displeased most of "the moderns," because a line's element was also a line, and not a point, an area's element being likewise an area, and not a line, and a solid's element being a solid as well, and not an area (ibid., xi). Deidier was the first author *after* the genesis of differential and integral calculus who spoke in favor of reapplying Cavalieri's concepts—in the name of a direct geometrical intuitive grasp, and in the interest of the beginners. Deidier affirmed that Cavalieri's method had maintained its full force and vigor (*conserve toute sa vigueur*, ibid.) even now. Nobody could be prevented from calling Cavalieri's points infinitely small lines, his lines "areas of insensible thickness," and his areas "solids having a depth that is smaller than can be expressed." In this way, Cavalieri's notions and those of his antagonists could be made to agree (ibid., xj f.).

Deidier begins his fourth volume, *Le Calcul Differentiel et le Calcul Integral, expliqués et appliqués à la géométrie*, seemingly with an argumentative turnabout. He criticized *Synthèse* for being limited, from its very nature, to detail, for giving access to the general only by permanent detours, as a method in which figures must be multiplied while most carefully observing positions and angles in order to arrive at rather more accidental propositions, before attaining more or less pertinent results. In his preface, Deidier praised the immense progress in mathematics realized by Leibniz's discoveries: *les Routes tatonneuses des Anciens furent abandonnées* [the tentative paths of the acients were abandoned] (Deidier 1740 vol. 2, xj).

Nevertheless, Deidier still succeeded in establishing a bridge to his own synthetic method, which he no longer designated as a general method, but as a didactical method for beginners. The new theories, he said, were too abstract for "beginners," who were too little habituated to things requiring rigorous attention. Analysis was said to be abstract and without imagination, in order to permit emphasizing the easy grasp of detail and visible objects, as in geometry. Upon taking this path, the beginner could be insensibly guided toward abstraction:

> It is not be easy for [Beginners] to apply themselves initially to a Study which imposes an eternal silence on the imagination, the detail and the perceivable objects such as Figures suiting them better, they must be given time to do this, and not be led into the abstraction of general methods except by small degrees which would block them if they were to be aware of the change [to abstract methods] (ibid., xij).

This modified approach to synthesis enabled Deidier to justify his own textbook on this abstract calculus. In his own view, he was continuing Crousaz's argumentation concerning the first textbooks of analysis. These had been written for scholars, he said, who had to learn these new things themselves: "M. le Marquis de l'Hospital wrote at a time when even the Savants had need of instruction" (ibid., xiv); for these scholars, however, it had not been necessary to present all the foundations in detail. In contrast, Deidier said that he had chosen the goal, in his "second generation" textbook, of making analysis accessible to the nonspecialist as well.

His conception of differential and integral calculus in this volume offers no further innovation; it confirms rather more the eclectic character of Deidier's approach. It is essentially a bowdlerized version of Reyneau's textbook on analysis. What is interesting with regard to the architecture of textbook knowledge is how Deidier modified Reyneau's overall structure. While Reyneau still had intended a unified textbook of *analyse* as a whole, comprising both "ordinary" algebra and the new methods of *analyse*, Deidier preceded his chapters on differential and integral calculus by a *Traité préliminaire* in which he presented those parts of algebra and of *Géométrie sublime* that were contained neither in his own textbook on arithmetic, nor in that on geometry, although they were required as foundations of the *calcul*. Deidier obviously was the first to develop an architecture in which some part appeared in the function later fulfilled by algebraic analysis.

Deidier enumerated the concepts of differential calculus by briefly stating six "principles," without bothering with a detailed discussion or justification. They adhered essentially to the foundations in Reyneau. The first principle is the definition of infinitely small quantities. They are being introduced only as willfully small quantities: "1°. Une grandeur qui est moindre que tout ce qu'on peut assigner, s'appelle un *infiniment petit*" (ibid., 227).

Changes of variables are related to the independent variable of *time*, also with *instant* as the smallest unit (ibid., 228). To establish the product rule, he declares that *dxdy* corresponds to a rectangle having infinitely small sides, thus being infinitely small with regard to the rectangles corresponding to *xdy* and *ydx* (ibid., 231).

Upon introducing the "differences" of higher order, Deidier was innovative in differentiating according to whether the differentials concerned were constants or variables, and for the case that a differential itself formed a variable, he involved the next higher order of differentials (ibid., 302). Deidier was thus enabled to conduct concrete detailed studies of curve behavior: establishing turning points, assessing negative values, etc. In this field, he achieved a considerable extension of Reyneau's concepts.

Integral calculus was introduced as the "inverse operation" of differential calculus (ibid., 390 ff.).

8. The Elaboration of the Concept of Limit

8.1. Limits as MacLaurin's Answer to Berkeley

Great Britain, where the *infinitesimals* were created, was the quickest country to give them up again. They were replaced by methods of limit, but always based on geometric–kinematic processes. The first evidence for this change is found in *Methodus Incrementorum Directa et Inversa* (1715) by Brook Taylor (1685–1731), a lawyer and natural scientist educated at Cambridge. A member of the Royal Society since 1712, and the society's secretary since 1714, Taylor gave up his office there in 1718, continuing as a private scholar. The famous theorem bearing his name is found in his book of 1715. In 1717, the author described the book's objective as deriving the ratios of fluxions directly from the ratios of increments (Guicciardini 1989, 32). This approach of trying to obtain the *ultimate ratios* from finite differences made *infinitesimals* obsolete. Taylor rejected infinitely small quantities: neither did such mathematical objects exist ("as if there were any such thing as a real Quantity infinitely little"), nor had Newton used this notion: the quantities he uses were either finite, or zero (ibid., 43). Taylor's method may be understood as an attempt to transmute Leibniz's concept of differences into Newton's concept of fluxions. Taylor's book differs from the French textbooks reviewed above in that he does not explicitly introduce, or examine, his fundamental concepts, but uses them implicitly. The volume indeed begins at once by stating a problem that is then solved by applying rules (Taylor 1715, 3 f.). Taylor described his intentions in great detail in several reviews of his own work (cf. Feigenbaum 1985, 11 f.). These self–presentations show his view that the value for *vanishing* increments is also obtained immediately from the study of finite ratios. But a reflection or discussion of the problem concerning the transition from finite differences to limit is found neither in his book nor in his reviews (cf. ibid., 12 ff. and 38 f.; Guicciardini 1989, 32 f.). In this formal method, Taylor can be considered a predecessor to Euler's method of differences. The virtually total abandonment of the method of infinitely small quantities by the British mathematicians was eventually effected by the much discussed attacks of Bishop George Berkeley (1685–1753), (cf. Guicciardini 1989, 38 ff.). After first attempts by Thomas Bayes (1702–1761) and Benjamin Robins (1707–1751) at establishing a limit method (cf. ibid., 46 and 45), the eminent Scottish mathematician Colin MacLaurin's (1698–1746) voluminous *A Treatise of Fluxions* (1742) formulated the mathematical rejection of the *infinitesimals*, making a geometrical method of limit the basis of *calculus*.

The starting point for MacLaurin's comprehensive approach to justifying Newton's *calculus* was that he rejected—like Newton (cf. Section 5.)—assuming infinitely small quantities, admitting only finite quantities:

> I have always represented Fluxions of all Orders by finite Quantities, the Supposition of an infinitely little Magnitude being too bold a *Postulatum* for such a Science as Geometry (MacLaurin 1742, iv).

Like Newton, MacLaurin advocated atomism; he thus wrote polemics against the law of continuity, arguing that it confined nature, and he also criticized the exclusion of perfectly hard bodies:

> Nature is confined in her operations to act by infinitely small steps. Bodies of a perfect hardness are rejected, and the old doctrine of atoms treated as imaginary, because in their actions and collisions they might pass at once from motion to rest, or from rest to motion, in violation of this law (ibid., 39).

MacLaurin admitted only quantities having a "real existence" (ibid., 3). Infinitely small quantities hence were not admissible quantities; an infinite division was not executable. The key term in MacLaurin's discussion is "assignable." Division into an assignable number of parts is admissible: "but it [a given magnitude] cannot therefore be conceived to be divided into a number of parts greater than what is assignable" (ibid., 43).

In contrast, for instance, to Varignon's argumentation against Rolle, MacLaurin stated that the "ancient geometricians" by no means admitted quantities that could become infinitely large, or infinitely small (ibid., 40). He was just as convinced that the *ancients* as well never replaced curves by polygons (ibid., 3 and 33). MacLaurin thus removed one of the essential pillars of Leibniz's and the French authors' concept of differential calculus, replacing it by another concept, which he claimed to have been the basis of the geometry of antiquity, or at least of Archimedes's geometry, namely by the concept of limit. Archimedes, he said, did not substitute polygons for curves, but had rather refined the circumscribed and inscribed polygons to a degree permitting him to obtain propositions on the inscribed curves as their "limit."

The major part of MacLaurin's substantial *Introduction* was a historical presentation in which the author showed in detail which far-reaching results in higher geometry Archimedes already had obtained with the method of *limit* ascribed to him. Basically, the only element that had escaped him was the study of the "hyperbolic areas" (ibid., 29).

The really surprising thing about MacLaurin's most substantial text is that he used from the very outset (ibid., 10) the concept of *limit* as something quite evident that need not be explicitly introduced or analyzed. Merely in the later supplement to the first volume,[30] there is a brief paragraph "Of the limits of

---

[30]   MacLaurin wrote the main part of this work in 1738.

Ratios" (pp. 420–424), but this does not contain any foundation of the method of limit either, but rather practical hints on how to calculate the limit ratios.

Noteworthy in MacLaurin's practical use of the limit method, however, is that he always includes a reflection on whether the limit sought really exists (cf., for instance, ibid., 271 ff.), something that most of his successors did not do. The constant inquiry into whether there is an *assignable limit*, or whether "it has no limit," is the consequence of MacLaurin's foundational position of exclusively admitting finite, that is to say *assignable*, quantities.

MacLaurin thus directed his main criticism—not only in the historically oriented *Introduction* in which a part on Archimedes is succeeded by a presentation of how infinitesimals came into being—but also pervasively in the volume's main part, against Fontenelle's *Géométrie de l'Infini* and the latter's realism of concept, whose unsophisticated formulation MacLaurin quoted as follows:

> Geometry is entirely intellectual, and independent of the actual description and existence of the figures whose properties it discovers. All that is conceived necessary in it has the reality which it supposes in its object (ibid., 42; orig.: Fontenelle 1727, Préface [XI]).

It has always been noted that MacLaurin's *Treatise* was hard to read (cf. Guicciardini 1989, 50). This ponderousness, however, is not so much a result of his using indirect geometrical proofs. The main characteristic of his style is rather that he formulates purely verbally in the traditional way—without equations, and without using algebraic signs at all, confining himself exclusively to the study of geometric loci. As MacLaurin declared, this style was intentional in order to thwart any critique appealing to the traditional methods:

> In order to obviate any suspicious of this kind,[31] we endeavoured to describe it in a manner that might represent the theorems plainly and fully, without any particular signs or characters, that they might be subjected more easily to a fair examination (ibid., 575).

It is thus understandable that MacLaurin, because of his own geometrical conception of limit, which went against algebraization, made no attempt at all toward algebraizing this concept: neither in his first, exclusively geometrical, main part, nor in his second, more algebraical, part, in which he admitted, for instance, infinitesimal methods as heuristic, and essentially exact, abbreviations. MacLaurin introduced no sign of his own for *limit*, and did not state any rules for operating with it. It is essential for understanding MacLaurin's position with regard to foundations that he assumes as an unquestioned premise that all variables ("fluents") are based on geometrical–kinematical processes—on operationalizations of motion, space, and velocity (cf. ibid., 52 f.). The premises, worded even more acutely than in Newton, cause continuity and

---

[31]   saying that the use of symbols and algebraic calculations "might serve to cover defects in the principles and demonstrations" (ibid., 575).

differentiability to be self–evident fundamental assumptions without further reflection.

8.2. Reception in the Encyclopédie and Its Dissemination

The most intense adoption of the rejection of infinitely small quantities and their replacement by the limit method, however, did not occur on the basis of MacLaurin's work, but rather by a French author, who again had read MacLaurin's book and made the latter's ideas his own. This author was d'Alembert whose impact on the development of this concept field was just as important as his effect on changing the conceptual status of negative numbers, which we have already seen. Here again, his motive was to work in favor of conceptual clarity and rigor. But in contrast to the case of negative numbers, where his effort at rigor led him to reject algebra's potential to generalize, d'Alembert favored a position in this concept field that supported algebraizing the fundamental concepts of analysis.

D'Alembert first published his position concerning the problem of infinitesimals in 1743, in his one and only textbook *Traité de Dynamique*. This was not a separate reflection, but rather a remark following a theorem's proof: on the conservation of the velocity of a body moving along a curve. While he inscribed a polygon to the curve, he did not identify the curve with it, but argued that the *sinus versus* of the exterior angles could be assumed to be smaller than any given quantity (d'Alembert 1743, 35). In his subsequent *Remarque*, d'Alembert considered the problem that this theorem would otherwise be proved by means of using infinitely small quantities of first and second order, an *infiniment petit du premier [genre]* resulting as the total loss of velocity (ibid., 36).

Actually, however, the loss of velocity was exactly zero, and not infinitely small. D'Alembert admitted that the proofs might turn out to be rather long, if one intended to prove the properties of curves "in all rigor," and that *la méthode des infiniment petits* considerably abbreviated the proofs, although the method was not as rigorous: *elle n'est pas si rigoureuse* (ibid., 36).[32]

D'Alembert defended his rejection of this method precisely as an impact of Berkeley's critique, which had been provoked by the very method of infinitely small quantities. He pointed out that this method was to be rejected in particular because it might mislead beginners who, often unable to penetrate a method's spirit, might consider the *infiniment petits* to be real, and misinterpretations of

---

[32]   D'Alembert's rejection of the method of the *infiniment petits* in his textbook on mechanics is all the more remarkable, since the view later so strongly prevailed in France that the method was indispensable in analysis for the reason alone that one was compelled to use it in mechanics; cf. Chapter IV.3.6. below.

that kind had called forth volumes already denying the certitude of mathematics (ibid., 36).

A footnote to this text specifies that these *Livres contre la certitude* meant Berkeley's *Analyst*. In the same note, d'Alembert noted that in his own conception of the method of limit he was relying, in addition to Newton's *Principia*, mainly on MacLaurin's work published the year before. In this relatively brief *remarque*, d'Alembert did not enter upon the method of limit in detail, but drew mainly on Newton's explanation of his method of first and ultimate ratios (ibid., 36 f.; see Newton's quote above, Section 4).

In the *Encyclopédie*, d'Alembert explicitly discussed the concept field of the infinitesimals in several entries, objecting sharply against the admissibility of the *infiniment petits* as mathematical concepts. His reasons for rejecting these quantities were quite analogous to those for his rejecting negative numbers as an independent number area (cf. Chapter II.2.9.3.). He refused them a "reality" in the sense of an ontological reference. In this case, however, he opted in favor of another conceptual view, which proved to lead further, whereas he had not proposed any alternative of this kind in the case of negative numbers. D'Alembert's primary objective in the *Encyclopédie* was to clarify the foundations of the basic concepts of analysis, or to clarify *la métaphysique du calcul différentiel* (d'Alembert, *Différentiel*, 1751, 985).

The precise and most exact *métaphysique* of differential calculus had been applied by Newton, he said, altough the latter had permitted only little insight into it:

> It can be said that the metaphysics of that great geometer [Newton] on the calculus of fluxions is most exact and most illuminating, to the extent that he allows us a glimpse of it.

> He never regarded the differential calculus as the calculus of infinitely small quantities, but as the method of first and last ratios, that is, the method of finding the limits of ratios. [...] The differentiation of equations only consists in finding the limits of the ratios between the finite differences of the two variables that the equation contains (ibid., 985 f.).

In the *calcul différentiel*, there were no *quantités infiniment petites*; this theory was concerned only with *limites* of finite quantities ("il s'agit uniquement de limites de quantités finies"). The *métaphysique* of the infinitely large and the infinitely small quantities was completely unnecessary ("totalement inutile") for differential calculus. The term of *infiniment petit* was being used merely to abbreviate the expressions (ibid., 987). D'Alembert considered the fact that the concept of infinitely small quantities had nevertheless found such a large number of adherents to be, on the one hand, a result of this theory's first adherents' incomprehension, and on the widespread bad habit of the "inventors" of shrouding their discoveries in mystery on the other. D'Alembert condemned such behavior as that of charlatans, professing his own didactical credo that truth must be simple and accessible to all:

> But the inventors have tried to surround their discoveries with as much mystery as possible; and in general men do not despise obscurity, provided

it results in something marvelous. Pure charlatanism! The truth is simple, and can be made accessible to all when one cares to take the trouble (ibid., 988).

D'Alembert, in his entry *Différentiel* of the Encyclopédie, used examples to illustrate that the differential quotient is based neither on a reckoning with zeros, nor on a calculation with infinitely small quantities. To give an example, he differentiated the equation $ax = yy$; from $adx = 2ydy$, he obtained

$$\frac{dy}{dx} = \frac{a}{2y}.$$

According to that, $\frac{dy}{dx}$ was the limit of the ratio y to x One could, however, also set $z = 0$ and $u = 0$ within the fraction

$$\frac{z}{u} = \frac{a}{2y + z} \text{ , thus obtaining } \frac{0}{0} = \frac{a}{2y}.$$

D'Alembert denied that such terms made sense mathematically:

> Qu'est-ce-que cela signifie? Je reponds, 1° qu'il n'y a en cela aucune absurdité; car $\frac{0}{0}$ peut être égal à tout ce qu'on veut: ainsi il peut être $\frac{a}{2y}$. Je réponds, 2° que quoique la limite du rapport de z à u se trouve quand $z = 0$ et $u = 0$, cette limite n'est pas proprement le rapport de $z = 0$ à $u = 0$, car cela ne présente point d'idée nette; on ne sait plus ce que c'est qu'un rapport dont les deux termes sont nuls l'un et l'autre. Cette limite est la quantité dont le rapport $\frac{z}{u}$ approche de plus en plus en supposant z et u tous deux réels et décroissants, et dont ce rapport approche d'aussi près qu'on voudra. Rien n'est plus claire que cette idée; on peut l'appliquer à une infinité d'autres cas (ibid., 986).

D'Alembert also rejected the idea of higher orders of infinitely large and infinitely small quantities: all these were only cases of limits of finite quantities (ibid., 987).

D'Alembert's entry proper *Limite* in the *Encyclopédie*, in contrast, is very short, while he repeats the position of principle that differential calculus can be correctly erected on *limites* only, "La théorie des *limites* est la base de la vraie Métaphysique du calcul différentiel" (d'Alembert, *Limite*, 1765, 542).

He was the first to introduce approaches to a conceptual reflection of the concept of limit at this point, declaring that while the quantity in question could arbitrarily approximate the limit, it could never attain it:

> The *limit* never coincides [with the quantity], or never becomes equal to the quantity of which it is the *limit*; but the former always approaches it closer and closer, and differs from it by as little as one wishes (ibid.).

This determination of concept, he said, at the same time implied—just as in the case of negative quantities—the exceptional position of zero. Besides, the determination made clear that here again the fundamental concept is one of a geometrical quantity, and not at all one of the concept of function. An approach intending to make allowance for the aspect of algebraization is found in the entry *Différentiel*. In this entry, d'Alembert distinguished between the *limite géométrique* of quantities as the determination of a certain line of figure, and the *limite algébrique* as the algebraic term expressed in letter quantities (d'Alembert, *Différentiel* 1751, 986). His primary interest, however, was the geometrical problem. Differential calculus consisted in finding the algebraic expression of the ratio for the lines already known:

> This calculus consists only in determining algebraically the limit of a ratio which has already been expressed in lines, and in making these two limits equal, by which one of the desired lines can be found (ibid.).

In evident agreement with this dominance of the geometrical concept of quantity, no attempts at reflecting, or establishing, operations with limits are found in d'Alembert.

This dominance of the geometrical in d'Alembert is all the more salient, since there is, for the entry of *Limite* in the *Encyclopédie*—just as for the entry of *Négatif*—a parallel, competing entry, and again by the same author, by the Abbé de la Chapelle! In this case, the abbé's parallel entry, while not advocating an essentially different position, clearly tends toward a more strongly algebraic–operative understanding of the limit concept.

De la Chapelle first presented his notion of limit in his textbook *Institutions de Géométrie* 1746, thus being probably the first to explicitly apply limits in elementary geometry rather than in analysis, as had been common. The section concerned treats the *Solidité des Corps*, beginning with a sharp polemic against a traditional view—most remarkably directed not against the method of infinitely small quantities, but rather against the method of indivisibles. De la Chapelle brands the adherents of this method *Indivisibilistes*, talking about them as of members of a sect: *Sectateurs de Cavalieri* (de la Chapelle, vol. 2, 1765, 338). Since the issue is methods for determining surfaces and volumes, it is suggestive that de la Chapelle targeted Deidier as the most recent representative of this sect. Deidier had just propagated, in the third volume of his own textbook, a return to the synthetic method, and to the method of indivisibles (cf. above).

Like MacLaurin (cf. Chapter II.2.8.2.) before him, de la Chapelle did not dare to present the method of limit as something novel. He also strove to enlist the support of the authority of the *anciens*, designating the limit method as the latter's method of exhaustion (ibid., 343). He made a point, however, of declaring that the new aspect he himself had introduced was to add two new *propositions* to this method, which permitted to establish it as *indubitable*:

- The first was the proposition that, if two quantities A and B are the limit of the same *quantité C*, the two quantities A and B are equal to one another. The purpose of this *proposition* was to permit the insertion of a limit already numerically known (ibid., 363).

- The second proposition consisted in transferring the property of limit to the product. If C is the limit of a quantity A, and D the limit of a quantity B, $C \times D$ is the limit of $A \times B$ (ibid., 360 f.). He needed this latter proposition in particular for determining the volume of solids whose surfaces were bounded by curved lines.

Only at this point did de la Chapelle give a definition of *limite*—and this merely in a footnote:

> One says that a magnitude is the limit of another magnitude when the second can approach the first closer than a given amount, as small as one is able to suppose. From which the difference between a quantity and its limit is absolutely indeterminable (ibid., 360).

The remarkable thing in this definition is the indeterminacy of the expressions: *grandeur* and *quantité* are used side by side, without any difference of meaning. Even more revealing is the fact that the term "variable" is not used, and that the term "constant" also does not appear.

In his entry *Limite* in the *Encyclopédie* paralleling d'Alembert's contribution, de la Chapelle repeated this definition verbatim, adding, after giving some examples, his two *propositions* (de la Chapelle, *Limite*, 1765, 542). Despite the indeterminacy of his definition of limit—d'Alembert had followed MacLaurin in giving no definition at all; the second *proposition*, in particular, offers an approach to operating with limits, and thus toward a process of algebraizing this concept that had been used in a purely geometrical way to begin with.

8.3. A Muddling of Uses in French University Textbooks

With regard to the foundations of differential and integral calculus in the context of French universities, loss of reflection and bowdlerization quite analogous to the reduction we have already established for the treatment of negative numbers can be stated, beginning with the second half of the eighteenth century. One of the first cases in point is the exceedingly successful textbook *Leçons élémentaires des mathématiques* by the Abbé Nicolas Louis de La Caille (1713–1762), professor of mathematics at the *Collège Mazarin* of the Paris University, a book that after its first edition of 1741 was republished and translated many times (cf. Boncompagni 1872). At first, the textbook contained merely algebra and geometry as an introduction to the study of mathematics and physics, but it was consistently extended in its subsequent reeditions. In any case, the textbook also contained a section on differential and integral calculus from the fourth edition of 1756 onwards. A mere three quarters of a page sufficed to present the principles (*principes*) of infinitesimal calculus, followed by a presentation of how to execute differentiation and integration that was exclusively in a calculatory mode.

The part on "principles" consisted—besides an explanation of the designations for the new kinds of calculi and terms—of merely one section, stating that any quantity could be imagined as continuously changeable by increments, and that these increments occurred by infinitely small steps. Such infinitely small steps were called differentials, and operating with them was called infinitesimal calculus (La Caille 1758, 203).[33]

---

[33]   I am quoting here after the Latin translation of the 1756 edition, which was improved by the author's own corrections.

Integration is defined formally as the operation inverse to differentiation (ibid.).

La Caille's successful textbook was continued, extended, and reshaped by his successor at the *Collège Mazarin*, Abbé Joseph–François Marie (1738–1801), after 1768. This textbook, too, saw many reprints, after the Revolution as well. Even after Marie's death, the "La Caille–Marie" was separately continued by several authors (cf. Boncompagni 1872). In the part on differential and integral calculus, Marie extended the introduction considerably. Instead of simply stating unequivocal principles, Marie eclectically juxtaposed the positions taken by Newton, Euler and Leibniz (Marie 1778, 428 ff.), albeit with marked sympathy for Newton, whereas he quoted Leibniz's principles as being less "exact." These principles prove to be the French adaptation of Leibniz's concepts (cf. above): conceiving of the *infiniment petits* as of a reduced form of the *élémens* of variable quantities, and operating with different orders of *infiniment petits* were views very welcome to Marie (ibid., 430).

In his own eclectic enumeration, Marie does not omit to point out d'Alembert's view according to which the concept of *limite* contained the correct metaphysics of the new theory (ibid., 428).

The textbook *Institutions Mathématiques* (publication beginning in 1770) authored by the Abbé Sauri also contains an extensive part on differential and integral calculus. This part presents in a mode pertaining only to calculation, and without foundation or reflection, the rules of operation for differentiating and integrating quantities. For the substitutability of $x \pm dx$ by x, Sauri refers only to an earlier section, *De l'Infini*. In this section of a mere three and a half pages, immediately before the conclusion of his first part on arithmetic and algebra, Sauri proves to be a representative of the French reception of Leibniz, that is to say, as an adherent of the *infiniments petits*, in whose use he at the same time integrated the *limes* terminology as well.

A "quantité *infiniment petite*," he said, was a quantity smaller than any given quantity, or one conceived of as being diminished beyond any assignable limit: "A quantity smaller than any given quantity, or a quantity that can be conceived as becoming smaller than any assigned limit" (Sauri 1772, 131).

Sauri designated such quantities, like Fontenelle before him, by $\frac{1}{\infty}$ (and infinitely large quantities analogously by ∞, using for their powers the notion of comparing the various orders of the infinitely small, of the finite, and of the infinitely large (ibid., 131 ff.). Sauri even went as far as considering the zero as *quantité infiniment petite* (ibid., 133).

While all these textbooks present the total of mathematics, differential and integral calculus thus representing a part of the respective textbook, the Abbé Girault de Keroudou's textbook is the first for the university context forming a separate textbook of infinitesimal calculus: *Leçons analytiques du calcul des fluxions et des fluentes ou Calcul différentiel et intégral* (1777).[34] As the title suggests, this work in some way combines Leibniz's and Newton's theories. As the author, a professor at the Paris *Collège de Navarre*, states, his textbook was intended to continue his predecessor Jean-Mathurin Mazéas's (1713–1801) *Éléments d'arithmétique, d'algèbre et de géométrie, avec une introduction aux sections coniques*, which continued to be used for the basics, and saw six reprints between 1758 and 1777.

While Girault de Keroudou's conceptual basis was Newton's theory, he always made an effort to integrate Leibniz as well, at least in his terminology. Thus, he introduced in his own first definition the modification of a fluent, which he then at the same time designated as a variable:

> The Fluxion of a magnitude is the difference between the state of that magnitude before any change and its state after that change. The magnitude that produces a change is called a *changing, variable Fluent*. That which causes no change is called *constant* (Girault de Keroudou 1777, 1).

Remarkably, he explained infinitely small quantities in the terms of the Archimedean axiom, by the comparability of such a multiplied quantity with an arbitrarily large number:

> The change is infinitely small when one cannot assign a sufficiently large quantity so as to express how many times it is contained in the *Fluent*; since when the unit, followed by one hundred thousand zeros, is too small to express the ratio of the flux to the fluent [le rapport du changement à la changeante] (ibid., 1 f.).

The author declared this second kind of fluxions, comprising infinitely small changes, to be the object of his lectures. For his designations, he chose the Leibnizian terminology containing the characteristic "*d*": he designated a fluent's x fluxion as dx (ibid., 2). His further presentation was again in a computational mode. Integral calculus was introduced as "méthode inverse des Fluxions."

Another textbook hitherto unknown in the literature confirms the French penchant for eclectically uniting Newton's and Leibniz's approaches, Newton's concepts being preferred in the presentation—practice, however, being primarily based on Leibniz's concepts. Making appeal to Newton not only resulted in constant reference to the expression of *limite*, but also in an ever broader dissemination of the concept of limit. This textbook is *Leçons élémentaires de Calcul Infinitésimal* (1784) by the Abbé Raymond-Roux, who has not been studied as yet otherwise.[35] According to the title page, he was professor of philosophy at the Collège Grassins of the Paris University. In his preface,

---

[34]   It has proved impossible to establish biographical data on this author. The Bibliothèque Nationale's catalogue, as his last publication, lists one of 1791.

[35]   In Taton's standard work on mathematics education in the eighteenth century, Raymond-Roux is not named, not even in the lists of professors at the Collèges –, and Mazéas is not listed either (Taton 1986, 156). It was not possible to find an indication of life data in any other biographical encyclopaedia, either. The Bibliothèque Nationale's catalogue lists only this single publication.

however, he mentions that his work was addressed to the students of the Collège de Navarre entrusted to his care, and was to be considered to continue Mazéas's textbook. Since Girault de Keroudou had already claimed the same function for his own textbook, it may be supposed that Raymond-Roux did not teach infinitesimal calculus to the philosophy class of his *Collège*, and held these courses additionally at the Collège de Navarre, after Girault de Keroudou had left.

The textbook, which like its predecessors was completely devoted to differential and integral calculus, relies on the *infiniment petits* as its basic concept. The author considered it necessary to motivate this concept in a substantialist vein—by their real existence in nature:

> There certainly exist in nature magnitudes which are negligible and imperceptible with respect to other magnitudes; these quantities are what we understand as *Infinitely small Quantities* (Raymond-Roux 1784, 1).

He made no distinction in calling the method of determining the ratios of these quantities either *Calcul Infinitésimal*, or *Calcul des Différentielles*, or *Calcul des Fluxions* (ibid.). It is remarkable how Raymond-Roux stressed the finiteness of the *infiniment petits*, quite in agreement with Leibniz's original approach. They could be imagined, he said, as the fraction from a quantity divided by the unit followed by many zeros, representing this as follows:

$$\frac{x}{100....0} \text{ (ibid., 2).}$$

Immediately after this first definition, Raymond-Roux introduced the *fluxion seconde*, *fluxion troisième*, and even the *n*th fluxion, as an analogous expression merely modified by raising the denominator to the corresponding power (ibid.). He used the occasion of this generalization to introduce Newton's and Leibniz's terminologies in a parallel mode. These disturbingly complicated expressions had been simplified, he said, Newton by introducing the point notation above the fluents, and Leibniz by means of the characteristic *d*. Since it was almost impossible to avoid confusion in using the points, Raymond-Roux declared that he was going to use Leibniz's *d*–notation. He immediately added, however, that he used Leibniz's terminology "only because of this advantage" (ibid., 3). Indeed, Raymond-Roux claimed a little later, like Marie before him, that Leibniz's principles did not seem to be as exact as Newton's ("ne semblent pas aussi exacts"), without giving reasons for this, just as Marie had given none (ibid., 8). He added that Leibniz's method had the advantage of leading to the same results while relying on only two principles; the author's eclecticism is evident in particular from the fact that he had by no means explicitly discussed analogous principles in Newton.

It is very telling to observe how the choice of principles, and their ascription to the various founders, changed over time. What Raymond-Roux now quoted as Leibniz's two principles agreed only partially with the principles quoted by L'Hospital, and these again were only partially congruent, as we have seen, with Leibniz's own ideas. As first principle, Raymond-Roux introduced the extended relation of equality, in the version saying that infinitely small quantities are negligibly small. As second principle, he named the concept of the comparability of orders (powers) of the *infiniments petits* (ibid., 8). The direct consequence and major application of this second principle was the evanescence of *dxdy* as compared to *ydx* and *xdy*, required to establish the product rule (ibid., 9). The principle pivotal for Leibniz and L'Hospital of the substitutability of a curve section by a polygon was no longer explicitly mentioned as a basic assumption.

Parallel to the method of fluxions and to the method of differentials, Raymond-Roux introduced the method of *limites* as well. Newton, he said, had considered the calculus of fluxions as the method of the first and last ratios, hence as the method to find a ratio's limits. He gave several examples of determining such limits, but did so without systematically developing them, and without presenting a calculus with limits (ibid., 5 ff.). Nevertheless, he concluded from his examples that determining the limit permitted the most precise definition of the calculus of fluxions, adopting literally the same definition as d'Alembert (as quoted above).

This close attachment of algebraic concept formations to geometrical existence is very revealing for the stage of transition within the process of algebraization. Questions of continuity, differentiability, and existence of limits cannot arise here, since geometrical existence is always assumed from the very outset.

Like the textbook authors reviewed above, Raymond-Roux understands integral calculus formally as the inversion of differential calculus: as determining the elements from their infinitely small parts (ibid., 154). There is no longer recourse to Newton in this second part on integral calculus any more, not even terminologically.

It is salient for all these authors within the university context in France that their basis continues to be the geometrical concept of quantity; there was no use of the concept of function.

Corresponding to our section of negative numbers, it will now be revealing to examine how far differential and integral calculus were taught in the competing system of education for military engineers in this period, and according to which conception.

The first official textbook for military schools was published from 1749 to 1752 by Camus (cf. Chapter II.2.10.2.). As we have seen, the four–volume work did not contain any part on differential and integral calculus: this met the author's and the commissioning authority's declared intention of presenting mathematics synthetically (ibid.). It is revealing that Camus resorted to an obsolete theory from the time *before* Leibniz and Newton in connection with applications requiring the use of at least some elements of calculus. In a way similar to how Deidier, to realize his own synthetic approach, had used the indivisibles as an ersatz theory, Camus resorted, in the third part of his work, on statics, to the equally obsolete theory of the *centres de gravité*. These notions had been developed in particular by Guldin (1577–1643) in order to permit the execution of calculations of integration in mechanics, before analysis had been elaborated. Camus's recourse to this obsolete theory can be understood to have been a surrogate for presenting the modern concepts (Camus 1766).

Camus's successor Bézout, on the other hand, devoted half a volume in his textbooks to differential and integral calculus, subsequent to algebra, as an introduction to mechanics. From its size, this half volume corresponds to the volumes proper on differential and integral calculus published by Girault de Keroudou and Raymond-Roux.

The presentation of analysis in this half volume is one of the rare cases in which different conceptions of knowledge for different social user groups become explicit. It is known that Bézout first wrote his textbook for navy engineers, and later authored a parallel work reduced as to volume and content for artillery officers.

While Bézout stressed the central importance of differential and integral calculus for mechanics in his navy edition, he almost made excuses for having this part precede mechanics, in his edition for the artillery, whose engineers were always confronted with the most moderate demands among the diverse *corps* of engineers. As he explained in his *avertissement* for the artillery edition, some had criticized him on the basis of the navy edition for intending to subject the study of mechanics to learning differential and integral calculus. This, however, was a misconception he was warning against here. Everything commonly found explained in textbooks on mechanics without the assistance of these *calculs* was being presented here without these as well. Whoever wished to confine himself to what was usually contained in the textbooks could pass over anything bearing the imprint of this method:

> Those who wish to restrict themselves to what is found to be the most useful in these books, will be satisfied in leaving to one side everything that is based on these methods (Bézout 1772, without page number).[36]

In their presentation of the foundations of differential calculus, the artillery and the navy versions on the whole agree.

In his textbook for navy officers, Bézout justified placing analysis before mechanics with the necessity of abstracting from the particularities of individual techniques, and of using more efficient methods to cope with the large number of pertinent factors in order to make mechanics useful. This purpose was served by presenting the foundations of analysis, which he called an "introduction aux Sciences physico–mathématiques" (Bézout 1770a, iii f.). In a mode analogous to Reyneau's universal conception of analysis, Bézout declared moreover that this *calcul* was by no means a new method: it was rather an application of the *calcul* presented in his volume on algebra, and represented even a simplification of the latter's rules (ibid., 2).

Bézout maintained this segregation of knowledge types in later editions as well. While navy officers continued to go without a presentation slighting

---

[36]   The (anonymous) reedition of the artillery version in the year VII (1799) contains an identical *Avertissement*. The rest of the text is identical, too.

analysis—as in the second edition of 1775—the artillery candidates continued to be met with the confining *Avertissement*.

Upon comparison with university textbooks, it is notable that Bézout does not place several methods side by side, or parallel, but exclusively presents *one* method. Without mentioning Newton's concepts or terminologies, he uses the *d*–characteristic, inserting the *infiniment petits* as the sole principle.[37] In doing so, Bézout assumed a philosophical, or particle conception different from that of, for example, Raymond-Roux. While the given quantities concerned formed the "elements" for the latter, the *infiniment petits* hence being arbitrarily small parts of these elements, the elements, for Bézout, represented precisely the *infiniment petits*. The object of differential calculus, he said, was to determine their elements from given quantities, while the object of integral calculus was to return from the elements to the quantities themselves:

> Firstly, to reducing the quantities to their Elements; and the method of doing this, called *differential* calculus. Secondly, we shall show the way to come back up to the quantities from their Elements, and we shall call this method the *integral* calculus (ibid., 2).

Bézout explicitly defined the elements as the infinitely small increments of variable quantities (ibid., 3). More generally, Bézout defined infinitely large, respectively infinitely small, quantities as such to which no other—sufficiently large or sufficiently small quantity—could be assigned that expressed the former's ratio, that is, that stated how often one quantity was contained in the other (ibid.). To this definition, Bézout added that quantity could lose the ability to be diminished or to be increased without losing the character of quantity. For this reason, there was, for a quantity arbitrarily small or large as compared to another, always a third quantity that was still larger, or smaller.

For operating with these infinitely large, respectively infinitely small, quantities, Bézout developed only a single principle as necessary basis: the principle of the various orders of the infinitely large and of the infinitely small (ibid., 3 ff.). Bézout named this principle *principe sur l'omission des quantités infinies des ordres inférieurs* (ibid., 9). This principle, of course, found its first application in the product rule, for omitting the term *dxdy* (ibid., 15), after that for general powers, etc. The differentials of second, third, etc. order were not a central concern of teaching for Bézout, in contrast to Raymond-Roux, but were only optional as a text in small type.

He introduced integral calculus—in his presentation one and a half times more substantial than his presentation of differential calculus—like all his predecessors formally as an inversion of differentiation (ibid., 97). It is interesting to note that he introduced the concept of *function* at the very beginning of this part, in Euler's sense as an expression of *calcul*: "We shall call the *function* of a quantity any expression of a calculation in which the

---

37   While there is a section on "limites," this concerns the second meaning of the term *limite*, that is, the limits of an interval for which certain properties are valid, here mainly for establishing extreme values (Bézout 1770a, 56 ff.).

quantity features, moreover in whatever way that it enters it" (ibid., 98). This, however, was only a momentary glimpse of an innovation; *quantité* and *variable* were the basic concepts of the further text.

Since the overwhelming majority of the French engineers in most of the *corps* had received their schooling in analysis after 1770 using Bézout's textbooks, and since these were identical in the two different issues for artillery and navy with regard to how their foundations of analysis were conceived of, it is evident that the conception of the *infiniment petits* represented the basic belief concerning analysis most widely socially shared among engineers in France about 1800 (cf. for this Chapter IV).

While Bossut, the third eminent examiner and textbook author for the French military schools (cf. Chapter II.2.10.2.), published a multi-volume *Cours de Mathématiques*, like Camus and Bézout, this publication did not contain any part on differential and integral calculus during the *Ancien Régime*. The textbook comprised, on the one hand, the usual fundamental parts on arithmetic, algebra, and geometry, and additionally special volumes on mechanics, which were for the first time differentiated into statics and dynamics, as well as a volume of its own on hydrodynamics. Only in a volume on mechanics and hydrodynamics published relatively late, did Bossut present, virtually as an annex, some aspects of differential and integral calculus, which he said to be important for discussing the *mouvement varié* (Bossut 1782, 431–452).

I have not been able to find any statement by Bossut concerning this gap in his textbooks. After the Revolution, he published a *Traité de Calcul Différentiel et du Calcul Intégral*, in the year VI (1798). This publication will be treated in the section on the algebraized analysis of the Revolutionary period, since it was written completely in the spirit of this time, implementing Euler's calculus of differences. It is hence also directed to a more general public, not specially to the military schools.

8.4. First Explications of the Limit Approach

As we have seen, the method of *limites* had hitherto rather been proclaimed more as a theoretical approach, but not really elaborated in practice; nor had differential and integral calculus been presented as a whole on this basis.

A first methodological explication of the limit method as such had been given in de la Chapelle's geometry textbook of 1746 (cf. Section 8.2.). The first elaborations for differential and integral calculus going beyond eclectic mention in university textbooks were presented in textbooks addressed to a larger public not directly connected with the school context. A common problem in these two textbooks of 1777 and 1781 by J.A.J. Cousin and R. Martin is that they interpreted the *limites* method, on the basis of the indications given by Newton and MacLaurin, as a conception elaborated already by the *Anciens*, and hence tried to improve its precision by drawing on concepts of elementary geometry.

The first textbook to claim being exclusively founded by means of the *limites* is the work *Leçons de Calcul Différentiel et de Calcul Intégral* (1777) by Cousin. While Jacques-Antoine-Joseph Cousin (1739–1800)—a member of the Paris *Académie des Sciences*—had been a mathematics professor at the *École royale militaire* in Paris from 1770 on, until this institution for the education of young noblemen founded in 1755 was closed in 1776 (cf. R. Hahn 1986), his main position was that of professor for mathematics and experimental physics at the *Collège Royal*, which he held from 1769 until his death. His lectures at the *Collège Royal*, an institution that held no examinations and conferred no degrees, were addressed to a general public. Cousin's textbook shows that it had not been structured for purposes of teaching, but is rather some kind of compendium presenting contemporary knowledge. Its bulk of more than 800 pages confirms that it was not intended for teaching.

In a lengthy *Discours Préliminaire*, Cousin developed the *métaphysique* of the calculus. For this purpose, he criticized all the methods hitherto used: he rejected the doctrine of the *infiniment petits*, saying that one was unable to form an *idée nette et précise* of these quantities. Leibniz's solution of replacing infinitely small quantities by incomparably small quantities, he said, completely destroyed the exactness of method (Cousin 1777, V f.). Newton's method of fluxions, while not contradicting mathematical rigor, had to rely on the concepts of motion and velocity. This, he said, introduced an idea completely alien to mathematics, which was moreover by no means simple. MacLaurin's justification for claiming that the method of fluxions was just as rigorous as the method of the *anciens* relied on the theory of motion (ibid., vj f.).

In contrast, Cousin conferred on d'Alembert the merit of having been the first to show that the real basis of differential and integral calculus was derived from the method of the *anciens* that was known under the name of *Méthode des limites*. Before d'Alembert's publication in the *Encyclopédie*, this method, *la vraie métaphysique*, had been entirely unknown (ibid., vij). He himself (Cousin) intended to present it as clearly as possible.

Cousin explained the limit as a quantity arbitrarily approximated indefinitely by a variable without ever becoming identical with it, as in the example of a circle inscribed or circumscribed by polygons.

> In increasing the number of sides of these polygons, they always approach closer and closer a circle, and may differ from it by as little as one wishes; but rigorously speaking, they never coincide with it (ibid., x).

Not only did this definition permit one to conceive of the infinite as the limit of an indefinitely increasing variable, Cousin also used it to make clear that for him the zero possessed an analogous exceptionality, and did not belong to the normal field of numbers. Diminishing variables could arbitrarily approximate zero, but never reach it: "the idea that we have of zero is that of a limit whose diminishing ratios can approach it continually, without ever meeting it" (ibid., ix). Later, Cousin declared explicitly that neither the zero nor the infinite were quantities, "Neither infinity nor zero are quantities," explaining, "these are limits which quantities can continually approach, without ever meeting them" (ibid., 25).

For operating with limits, Cousin, like de la Chapelle, on whom he obviously relied, listed *two* principles. His first operative principle is identical with de la Chapelle's first (two quantities that are the limit of the same quantity are identical); Cousin replaced de la Chapelle's second principle concerning the product of quantities, however, by a principle concerning the relations between two variables:

> If two magnitudes which increase or decrease continually maintain between themselves the same invariant ratio, this ratio will be that of the limit of the two magnitudes (ibid., x).

Cousin gave no justification for these principles, which he declared to be the basis of the entire method of *limites*; later, however, he gave one example of its application. Upon approximating a circle and an ellipse—the two having a common axis, the ellipse's major one—by inscribed polygons, the two polygons have the same ratio as the ellipse's major axis to its minor axis. This ratio, he said, remained the same for their limits as well, these limits being the circle and the ellipse (ibid., xiij).

These first elaborations of the limit method were thus not yet specific for analysis; they were still based on reflecting concepts of elementary geometry. Consequently, this stage is comparable to that of the indivisibles, during which static objects of geometry were also treated without yet disposing of the concepts of variable, or of function.

This parallel character is also evident from Cousin's general definition of the concept of *limes* in his principal text:

> One says that a magnitude has another magnitude as a *limit*, when one imagines that it can approach it to the point where it differs from it only by a quantity as small as one wishes, without ever coinciding with it (ibid,17).

Here, Cousin adopted the vague concept from de la Chapelle's geometry textbook, and from the latter's *Encyclopédie* entry, without introducing the term "variable," and without differentiating between variable and constant quantities.

On the other hand, a substantial novelty can be noted. Cousin was the first to use the concept of function in a textbook quite comprehensively, inserting, in immediate connection with that, the differential quotient, and not the differential, as the basic concept of differential calculus, enabling him to use his version of interchanging limits for ratios. This seems to yield a clearly systematic structure of his textbook:

- an algebraic introduction using the calculus of finite differences and sums;
- a presentation of the limit method;
- differential calculus by means of limits of finite difference quotients;
- integral calculus.

An analysis of detail, however, shows that the eclectic character is maintained in this textbook as well. The introductory part with his list of methods used to determine series of differences, in particular difference quotients, clearly shows an adoption of Euler's views. This also helps to explain why the basic concept used here is that of *function*. The second part, on the limit method, does not evolve from this first part; it determines the concept of limit—as has been quoted—by the concept of quantity taken from elementary geometry, beginning with discussing limits in determining segments of circles, cones, and pyramids. Geometric curves, and not functions, continue to form the object of this chapter.

In preparation of the next chapter, Cousin increasingly discusses, however, the ratio of the differences between ordinates and abscissae, and the latter's limit (cf., e.g., ibid., 27). In doing so, Cousin at least achieves an important innovation. He had obviously realized that the limit method required the limit to have a sign of its own. He thus declared that using a separate *sign* for the limit was required, while not, however, introducing, a general sign, but only a sign for the limit of the difference quotients:

> We shall use a sign to indicate the limit of the ratio of the differences of two variable quantities; let the two quantities be x and y and $\Delta y : \Delta x$ the ratio of their differences; we shall always use in what follows $dy : dx$ for the limit of this ratio (ibid., 32).

In the third part of his textbook, Cousin again did not use "function" as a basic concept, but rather "quantité variable." In line with that, he defined as the object of differential calculus to find the *limites des rapports entre les différences des quantités variables* (ibid, 72). This chapter, too, contained essentially examples of application, and not conceptual explications. It would have been interesting to know, for instance, that Cousin established the product rule by means of the *limites,* but he disappoints in this basic argument of the *calcul* by not only discussing the differential instead of the differential quotient, but by claiming, on top of that, that the proposition required no justification: "Whatever the ratio of the variables y and x, it is clear that the differential of the product xy equals $ydx + xdy$" (ibid., 76).

Cousin formed the higher differential quotients formally as limits of the corresponding finite series of differences. Integral calculus was for Cousin as well the formal inversion of differentiation, based on the indeterminate integral of a differential. The fact that Cousin gave precedence to treating the integration of differential equations is of systemic character. *Equations*, for him, were the algebraical version of what he considered to be the genuine mathematical substrate: geometrical curves indeed, but not as expression for functions.

The second author to be discussed is Roger Martin, who introduced new approaches here as well as in his treatment of negative quantities (see Chapter II.2.10.1.). It is indeed characteristic that his textbook *Élémens de Mathématiques* (1781) has *two* points of emphasis, reflecting the concept of number, and the *limes* method. While he only elaborated on the conception of Cousin's textbook in some part, he also developed conceptions of his own that were based, in particular, on a different view of the infinite. Martin's textbook, according to its subtitle *à l'usage des écoles de philosophie du Collège Royal de Toulouse*, was clearly intended for the university context. Since he gave, on the other hand, a more detailed presentation of his views on the infinite, and on differential calculus, in two treatises for an academy, his publications can also be seen as having been addressed to a general public. The two publications were indeed noted abroad as well. They were quoted in Klügel's widely disseminated *Mathematisches Wörterbuch* even in 1831 in the entry *Unendlich* in the part on *Gränzen* (Klügel, *Wörterbuch* vol. 5, 1831, 537). Evidence for the reception of his textbook will be given in Section 9.

Like Cousin, Martin was convinced that d'Alembert, with his idea of the method of *limites*, presented as an adaptation of the ancients' method of exhaustion, had been the first to formulate the exact justification of differential calculus (Martin 1781, lvj). He differs from Cousin, however, in his independent definition of limit:

> By the *Limit* of a variable quantity is understood the value or state to which it always tends as it varies, without ever reaching it; but which, however, it can approach so that it differs from it by a quantity less than any given quantity (ibid., 317).

Not only is the condition for the smallness of the difference formulated here almost precisely like that which later became its standard form, but clearly relating the process of limit to variable on one side, and on the variable's (constant) *value* as limit on the other, signifies a genuine innovative achievement. Martin had first introduced the concept of *valeur* in his section on *analyse*, explaining *analyse* as "the means to find in an algebraic expression the value of specific quantities, which are combined with others." Without formally defining *valeur*, Martin's text nevertheless makes clear that by quantities he meant constant quantities that satisfy the respective algebraic conditions, for instance roots of equations (ibid., 132). Martin did not introduce, however, a sign of his own for limit.

A further innovative achievement of Martin consists in his clarifying the following problem, which had not been solvable for a long time: $\frac{1}{\infty} = 0$ had always been accepted as an identity. Likewise, $\frac{a}{\infty} = 0$ and $\frac{b}{\infty} = 0$ held as well, but it was not clarified why it was not permitted to conclude from this that $a = b$, or $a = 1$.

By means of the concept of *limes*, and of his own interpretation of the infinite, Martin was able to give the solution. He said that $\frac{1}{\infty} = 0$ did not designate a genuine division, because ∞ was not a number; rather, it served as designation for a limit, the limit of a sequence of fractions having the constant numerator 1 and an indefinitely growing denominator. The equation $\frac{1}{\infty} = 0$ thus indicated that the sequence's limit was zero. $\frac{a}{\infty}$, with $a \neq 1$, but designated the limit of another sequence; and that the identity of the respective variables did not follow from the identity of the limits (ibid., lix f.).

In the foundations of the limit method he explicitly gave, Martin relied directly on Cousin, and hence on conceptions of the object pertaining to elementary geometry. While he did not adopt de la Chapelle's and Cousin's first principle, he adopted Cousin's second, supplying it as a "theorem" with a proof. The interesting thing, however, is that Martin generalized this principle/theorem, thus establishing the first principle of interchanging limits. As theorem III, he established the following proposition:

If there are two sequences of increasing respectively decreasing, variables, of which the sequence having the elements a, c, e, g, ... leads to the quantity u, the second sequence having the elements b, d, f, h, ... leading to the quantity x, then the sequence of the ratios of the respective terms, i.e., $a : b, c : d, e : f$,.... leads to the ratio of the two limits u and x.[38] He explicitly added a formulation for the principle of interchanging limits, using it repeatedly in what followed: *Ou bien que la limite des raisons sera la raison des limites* (ibid., 321).

It might have been noted that Martin, in this "theorem," designated the limit as "quantité," not as "valeur." This was of systemic significance for him, since he continues to declare that the theorem did not hold in the case in which the limits were zero. Zero, he said, was not a *quantité*, and hence there could not be any ratio between two limits of this kind.

We may consider this to be a basic belief of many mathematicians of the period. The zero is seen as an exception analogous to the infinite, it is not accorded the status of a number. This view had acute and long–term consequences for elaborating the concept of limit (cf. Chapters V, VI).

Martin, who adopted some of Cousin's concepts, criticized the latter, however, for radically excluding infinitely small and infinitely large quantities from mathematics (ibid., lvij). Since Martin did not base his own infinitesimal calculus on the concept of function, but continued to assume the geometrical concept of curve, using differentials, and not differential quotients like Cousin, he was in dire need of the *infiniment petits* for founding differentials of higher order. To do this, he proceeded as follows, declaring the *infiniment petits* to be a special variable: *infinitely small quantities are nothing other than the final values of quantities which decrease to the point where they vanish* (ibid., 328).

Since he had already done for multiplying negative quantities, Martin sought to attain his goal by means of an ad hoc construction. He interpreted the "ultimate values" in a twofold way:

- They could be understood, on the one hand, as limits of these quantities; in this case, the *infiniment petits* were zero, and a comparison of order between them was absurd. They could not become an object of mathematical operations for this reason.
- On the other hand, they could also be conceived of as *termes de leurs dernières raisons*, this obviously being meant as an ultimate element in

---

[38]   It is clear from the notation that indexing terms of sequences was not yet mathematical practice.

the limit process, which, according to definition, could not be identical with the limit. As such, they were *vraies quantités*; they were necessarily smaller than any given quantity.

He then designated the *infiniment petits* thus conceived of as "*dx*" in Leibniz's style, without any regard for his own conception of limit. He added the interesting remark that such *dx* could again be seen as new variables that likewise possessed ultimate values, these were denoted by *ddx*, as *infiniment petits* of second order (ibid., 328). These again could be interpreted as having two meanings, leading to the generation of *infiniment petits* of third order, etc.

After the concepts developed to precede it, his presentation of differential calculus is disappointing. It is done virtually as traditional calculation with differentials, with only occasional verbal references to limits. He does not use differential quotients; while he relies on the concept of variable, he does not distinguish between independent and dependent variables. Hence, he considers it to be the task of differential calculus to determine the ultimate ratio between the finite differences Dx and Dy of the two variables x and y that were connected by a law (*loi*) (ibid., 331).

Martin insisted on his geometric concept of curves so far as to declare that it was impossible to differentiate *simples quantités,* only *equations* could be differentiated, since proportions and ratios could be formed only with equations (ibid., 333).

He justified the product rule in the "Leibnizian" sense by way of comparing the orders of the *infiniment petits* (ibid., 335). Integral calculus, for him as well, was a formal inversion of differential calculus, again worded, however, in the geometrical terminology of ratios: to determine the ultimate ratios of two variables from the ratios of their differences (ibid., 347 f.).

Martin propagated his own view of the *limes* method in two academic treatises. They were read at the *Académie des Sciences, Inscriptions et Belles-Lettres de Toulouse*, the first in 1779 and the second in 1786, and published in this academy's *Mémoires* in 1782, respectively in 1788.[39]

The first treatise is practically a brief version of the appropriate parts from his *Discours Préliminaire*, and the section on the method of *limes* in his textbook of 1781. Martin emphasizes here that his goal was to show that the Leibnizian conception, although erected on the concept of the infinite, of the *infiniment petits*, and of their various orders, and hence based on inadmissible hypotheses, could be rigorously proved, when founding it on the *limites* method practiced by the *Anciens* (Martin 1782, 43 f.).

The second treatise of 1786 does not contain any further elaborations on the concept of *limes*. It is interesting, however, because it quite evidently offers an answer to the famous contest question posed by the Berlin Academy in 1784

---

[39]   The Academy of Toulouse belonged among the numerous provincial academies in France that had an important function in disseminating and developing the sciences under the *Ancien Régime*.

requesting a clear theory of the infinite, including an explanation why it was possible to derive correct results from the current concepts in spite of contradictory assumptions. Martin read this treatise on March 30, 1786, at the Academy of Toulouse, two months before the Berlin Academy made known their decision on the works submitted (see Appendix A).[40]

In his second treatise, Martin concentrated on an analysis of Leibniz's concepts, underlining contradictory elements in the use of infinitely small and infinitely large quantities in the adherents of Leibniz, and declaring that the contradiction existed in not separating the two meanings of the *infiniment petits*; while the *calcul* was based on a method that *redresse les suppositions*, that is, that corrected erroneous assumptions by again separating the two meanings confused (ibid., 56).

In his summary, he thus underlined three results of his *mémoire*, quite in line with the request of the contest task. He had given, he said, a clear and precise theory of the mathematically infinite; he had shown how true theorems resulted from contradictory assumptions, and he had shown that the concept of ultimate ratios—he no longer spoke of the concept of *limes*—was a safe principle suitable for replacing the concept of the infinite (Martin 1788, 71 f.).

Remarkable for the two authors treated in this section, Cousin and Martin, is furthermore that they edited their textbooks a second time, after the French Revolution, and this means during the euphoric stage of algebraizing analysis. It is thus revealing to look for changes in these second editions. Martin's second edition came out in 1802, twenty–one years after his first. As we have seen (cf. Chapter II.2.10.1.), Martin was actively involved in parliamentary work after the Revolution. Conceptually, the second edition is identical with the first, and likewise in the terms and signs employed. Small changes occur only in the wording.

Cousin published his second edition in 1796, hence also about twenty years after his first. Cousin is not known to have taken an active part in the Revolution. He succeeded, however, in becoming a member of the newly established *Institut National des Sciences et des Arts* in 1795, immediately after the *Académie des Sciences* was dissolved in 1793. His textbook remains also unchanged as to its conception in the second edition; for example, his definition of *limes* continues to be as vague as in the first edition; the concept of function is the basis of only some parts. There are considerable didactical changes in the textbook, however, and it is much more clearly structured. Notable also is the comprehensive *Introduction*, which precedes the differential and integral calculus proper, representing some kind of "algebraic analysis." The formerly two introductory chapters have been supplemented by another three: one on the

---

40   The announcement was made in the Academy's session of June 1, 1786. AAdW, *Akten der alten Akademie:* I–IV, 32 Registres de l'Académie, Fol. 418.

application of algebra to geometry, one on the method of the indeterminates (on development into series), and one on the application of the limit method to mechanics.[41]

Particularly remarkable, however, is an innovation regarding signs. In this second edition, Cousin for the first time applies the sign "lim" for the limit, inserting it operatively, for instance, in the equation

$$\text{lim.} \frac{\Delta(E.CK)}{\Delta E} = x \text{ (Cousin 1796, 135).}[42]$$

In his use of signs, however, Cousin was not consistent; he confined himself in this respect to one chapter, to that on mechanics. In the chapter preceding that, he always used a verbalized form, such as

$$limite\ de\ \frac{\Delta LF}{\Delta x} = \frac{a - x}{\sqrt{2ax - x^2}} \text{ (ibid., 122).}$$

8.5. Expansion of the Limit Approach and Beginnings of Its Algebraization

Subsequent to these first attempts at elaborating the concept of limit, a considerable growth in reflection on this concept can be observed. One of the causes for this can be seen in the well–known contest of the Berlin Academy of 1784 already mentioned in the previous section, a competition that gave higher scientific recognition to the rejection of the *infiniment petits*.

The treatise awarded the prize for solving this task by the Berlin Academy in 1786, Simon L'Huilier's *Exposition Élémentaire des Principes des Calculs Supérieurs*, was of considerable impact indeed. L'Huilier (1750–1840), born and educated in Geneva, was at first a private teacher in the household of a Polish prince, and after 1795 professor of mathematics at the Geneva *Académie*, an institution of higher education. His mathematical works focused on topics of elementary geometry, in particular polyhedra, and polygonometry, and on determining geometrical loci (Wolf 1858). L'Huilier was no expert in analysis. In an early work of 1780, he developed methods for isoperimetry to obtain results of differential calculus geometrically (ibid., 405).

In his prize–winning essay, which was published at once in 1786, he had strongly objected to using infinitely small and infinitely large quantities while

---

[41]   Since Cousin, in this introductory part, already treated, among other things, centers of gravity, accelerated motion, and curved trajectories of bodies, he thus adopted the conception that differential and integral calculus were not required for the principles of mechanics. This view represented social consensus even after the revolutionary period.

[42]   Cousin always put a period after "lim," probably to suggest the abbreviation.

elaborating the limit concept further. He, too, understood this concept to be in agreement with the methods of the *Anciens*, albeit appropriately extended:

> The method of the ancients, known by the name of the Method of Exhaustion, properly understood, suffices to establish the principles of the new calculus in a certain manner (L'Huilier 1786, 6).

He was strictly opposed to founding the new calculus via *infiniment petits* and via infinitely large quantities—as had already been implied in the conditions of the Berlin Academy's contest. L'Huilier's justification of this rejection, however, is remarkably simple. For him, their inadmissibility as mathematical concepts followed directly from the definition of *quantité*: quantity was everything that was able to be enlarged or diminished. This definition excluded those states of quantities in which all possible enlargements had already been effected, and likewise all those states in which all possible diminishments had been executed (L'Huilier 1786, 122).

As L'Huilier says himself, he considered these quantities only as completed, absolute ones: *l'infiniment grand et l'infiniment petit absolus* (ibid.).

The second characteristic moment of L'Huilier's view is his conception of the zero, and this again shows the close connection between the conceptions of infinitely small quantities and the concept of number, in particular the status of negative numbers, and thus of zero as well. For L'Huilier, only the *zéro absolu* is smaller than any assignable (*assignable*) quantity (ibid., 137); the zero being, however, the expression of a *privation de toute existence*— the cancellation of any existence (ibid., 126). Numbers are justified only by the ontological character of the existence of quantities. The zero is not admissible as a relational concept, but is only the metaphysical embodiment of the nothing. Negative quantities are still less admissible.

Because of this ontological view, the zero acquires an exceptional status, which is analogous to the infinitely large. The law of continuity, L'Huilier says, is not valid for either. Continuity between "non-being" and existence was just as impossible as between the finite and the infinite: "is there not an immense step between existence and nothingness and between the finite and the infinite?" (ibid., 158).

In his dispute with other "erroneous" conceptions of foundations, L'Huilier concentrated on advocates of the infinitely large, and on Fontenelle among these, despite the latter not having proclaimed his own geometry of the infinite as a basis for analysis (cf. ibid., 127 ff.). L'Huilier's criticism of positions pertaining to the *infiniment petits*, by contrast, was rather brief and cursory. Since he identified the infinitely small quantities with non-existence, he declared that the nothing could not be made a constitutive part of finite quantities (ibid., 126). He criticized L'Hospital, in particular, for the latter's alleged even greater obscurity in introducing various orders of infinitely small quantities (ibid., 138). L'Huilier's position was not consistent, however; in a later chapter on applying differential calculus in physics, he explained that the *infiniment petits* must be admitted in physics because of the *imperfection* of human observation (ibid., 179).

L'Huilier criticized Euler's conception of differential calculus as of reckoning with zeros much more sharply and extensively (ibid., 147 ff.). His own approach, by contrast, was to conceive of differential and integral calculus as of determining the limits of ratios: "the limit of the ratio, which these changes in the variable quantities approach the more closely as they are smaller" (ibid., 148).

L'Huilier gave no source for his concept of limit, but it was quite obviously influenced by both Cousin and Martin. L'Huilier's definition of limit is somewhat more specific than Cousin's; it uses as a basic concept that of *quantité variable:*

> Let there be a variable quantity, always smaller or always greater than a proposed constant quantity; but which can differ from it less than any proposed quantity smaller than itself: this constant *quantity* is said to be the *limit* in greatness or in smallness of the variable quantity (ibid., 7).

While L'Huilier uses Martin's more specific definition of limit, he introduced the novelty of differentiating between *limite en grandeur* and *limite en petitesse.*

By introducing this differentiation, L'Huilier expressed the insight that the limit concepts before his time had made no allowance for the fact that the approximation of the limit can be effected both from a variable with increasing values, and from a variable with diminishing values. Since the concept of the absolute value was not yet at his disposal, he separated the definitions:

• "left-handed" limit—*limite en grandeur*;
• "right-handed" limit—*limite en petitesse.*

He was not able, however, to encompass variables showing alternating behavior by this means.

According to his own mathematical field of work, both the concept of variable and the concept of limit were motivated and applied geometrically. Their primary context was that of geometrical *curves*, not that of functions (cf. ibid., 8 ff.). While L'Huilier did discuss functions and their limits in his introductory part, understanding functions on the basis of Taylor's theorem in a formal algebraic way as power series (ibid., 21 ff.), these formally algebraic parts are placed eclectically beside the dominant view of curves. Typical of that is that he shirks from reflecting on or defining the limit concept for functions, providing instead a supplementary definition for variable ratios on the basis of his own definition of limit for geometrical variables, with the obvious intention of covering the subsequent special case of the ratio between the difference of function values and variable values in differential calculus, a device permitting him to avoid introducing "function" as a fundamental concept:

> 2nd *Definition.* Let a variable ratio be always smaller than a given ratio, but which can be made greater than any assigned ratio less than the latter: the given *ratio* is called the *limit in greatness* of the variable ratio. Ditto; let a variable ratio be always greater than a given ratio; but which can be made smaller than any assigned ratio smaller than the latter: the given ratio is called the *limit in smallness* of the variable ratio (ibid., 7).

L'Huilier used this second definition also to introduce, as one of its special cases, the general concept of limit for functions—as a ratio of equality (*rapport d'égalité*): the limit of the ratio between the function and a constant (ibid., 21).

It is just as typical for his nonsystematic founding of his concept of function that while introducing the concepts of maximum and minimum for functions in his chapter IV, he quotes exclusively examples that assume curves or curve segments, which are not based on the concept of function (ibid., 78 ff.).

In his introductory part, L'Huilier stated as an explicit assumption in most of his propositions about forming limits that the limit exists. One of his typical formulations is, *Soient deux quantités variables susceptibles de limites, ...* (ibid., 11). In his applications, however, the distinction between *susceptible* and *non-susceptible* was not mentioned.

L'Huilier's handling of the limit with regard to signs is also revealing. In his introductory part, he at first does not use a sign proper for limit, expressing all his propositions about limits verbally, e.g., by *le limite de*. Only after twenty pages does the sign of "lim." used also by Cousin appear—also noted in the same function as "Lim."—but only as an unmotivated switch without any explication and justification (ibid., 24). In the subsequent text as well, L'Huilier used the sign only as an occasional abbreviation, without any reflection on the use of sign, or any independent operative application. The only place where he explicitly mentions the sign, he does so to replace it by the differential quotient:

> To shorten the calculation and make it easier by the use of a more convenient notation, it is suitable, instead of using $\lim.\frac{\Delta P}{\Delta x}$ for the ratio of the simultaneous changes of P and x, to use $\frac{dP}{dx}$; so that $\lim.\frac{\Delta P}{\Delta x}$ and $\frac{dP}{dx}$ stand for the same thing (ibid., 31).

In contrast, a new element conducive to progress was that L'Huilier stressed that the expression $\frac{dP}{dx}$, for which he declared that despite appearances it was not a fraction and thus not decomposable (ibid., 32), was itself a function of the variable x with which all the operations possible for functions could be executed (ibid., 33). As examples, he listed only addition, subtraction, multiplication, division, and raising to a power of functions, but not differentiation.

In a later chapter, however, he did treat differential quotients of higher order, expressly naming the functional character of the first derivation as justification for the fact that differential quotients of the second order, as well as analogously even higher orders, could be found according to the same rules applied to the original function. For these definitions, L'Huilier used the limit sign operatively (ibid., 43 ff.).

L'Huilier's outstanding novel achievement was a considerable expansion of the limit concept's applicability. While his predecessors gave hardly more than two basic rules, he provided a systematic elaboration. How this begins provides an additional illustration of how much he was influenced by Martin. L'Huilier began by giving Martin's theorem regarding how two variables conserving a constant ratio behave at the limit (ibid., 11). Further elaboration with such "theorems" concerned refinements of the propositions on ratios, but not algebraic operations with variables or limits; this further underlines that L'Huilier drew on geometry for his concepts. He supplemented these propositions with theorems on the limit behavior of functions, propositions transferring his previous statements on *limites de rapports* to polynomial functions.

A novel approach in this was the attempt at inferring the product rule from deriving two functions P and Q of x having the limits A and A'. For this, he obtained the expression

$$\text{Lim.}\frac{\Delta PQ}{\Delta x} = PA' + QA \quad \text{(ibid., 31)}.$$

For the general progress of the concept of limit, it is important to note that L'Huilier began by adopting Martin's interchangeability proposition concerning limit processes (as one of his *théorèmes*): "The limit ratio of the ratio of two variable quantities which are capable of having limits is equal to the ratio of their limits" (ibid., 24), and continued by generalizing it in the sense of the second metaphysical meaning of Leibniz's law of continuity (cf. Section 5. above): *Si une quantité variable, susceptible de limite, jouit constamment d'une certaine propriété, sa limite jouit de la même propriété* (ibid., 167).

L'Huilier declared this proposition about the continued validity of variables' properties for the limit as well to be the guiding *principe* of his treatise, which at the same time was its decisive core (ibid.)

In his prize–winning treatise, L'Huilier did not name any mathematician on whom he relied in conceiving his solution. In a later footnote for publication he listed as authors relevant for the issue of avoiding infinitely small quantities d'Alembert and Cousin, as well as the Germans Kästner, Karsten, and Tempelhoff (cf. Section 9. below). He omitted to mention Martin, and he included a reference to Robert Simson (cf. Chapter II.10.3.)—one of whose fundamental definitions he had copied almost verbatim—only in the prize–winning treatise's reedition of 1795 (L'Huilier 1795, 1). Simson, whom we have already got to know as an advocate of purely geometrical methods, and as an adversary of approaches favoring algebraization, was the first to devote a separate treatise to the concept of limit. This treatise, which remained fragmentary, was published posthumously in Simson's collected works in 1776, and reedited by Maseres in 1807 under the title *De Limitibus Quantitatum et Rationum*.

Simson's work is exclusively devoted to developing and applying the concept of limit for elementary geometrical quantities. This also explains his focus on application to geometrical ratios (*rationes*). This orientation helps us to understand the eclectic character of L'Huiliers prize–winning essay. L'Huilier had adopted Simson's geometrically founded concepts, and then simply attached some parts on functions without providing a separately developed concept of limit for function. After two introductory definitions, of constants and variable quantities, Simson had indeed given the two definitions of a variable's lower and upper limits, definitions that L'Huilier directly took over as his own basis (see above):

> III. Si quantitas mutabilis semper minor fuerit quantitate data, sed ita augeri poterit, ut major fiat quacunque quantitate data quae minor est prima quantitate data; vel si quantitas mutabilis semper major fuerit quantitate data, sed ita minui poterit, ut minor fiat quacunque quantitate data quae major est prima quantitate data; in utroque casu quantitas prima data dicatur Limes quantitatis mutabilis.

> IV. Si ratio mutabilis semper minor fuerit quam ratio data, sed ita augeri poterit, ut major fiat ratione quacunque data quae minor est ratione prima data; vel si ratio mutabilis semper major fuerit quam ratio data, sed ita minui poterit ut minor fiat ratione quacunque data quae major est ratione prima data; in utroque casu dicatur ratio prima data Limes rationis mutabilis, in primo sciliect casu dicatur Limes crescentis rationis, in altero, Limes decrescentis rationis. (Simson 1776, 3 f.).

We see from this that L'Huilier borrowed the differentiation into upper and lower limits from Simson as well. He drew on Martin only for his clear classification of the limit as a constant.

Simson did not use a sign proper for *limit*, writing his text completely verbally, without any algebraization, and exclusively using ratios. In this vein, he also discussed—circumventing differential calculus in doing so—the ratios of increments, for instance the cases of secants, formation of rectangles, etc. Only at the close of his texts did Simson use x and dx to designate the limits of increments, thus showing that his intention was to clarify foundational problems of analysis (ibid., 26).

L'Huilier's reedition of his essay is a remarkable example for the impact of differing styles in various cultural contexts. He wrote this new version in Latin during a prolonged stay in Tübingen, where he was staying after being invited by his friend Christoph Fr. Pfleiderer (1736–1821) in order to escape political unrest in Switzerland. Pfleiderer, who had first worked in Warsaw, had been professor of mathematics at the University of Tübingen since 1781. While he had induced L'Huilier to come to Warsaw in 1775, they had not cooperated much there. In contrast, he did his reediting of his 1786 essay completely under Pfleiderer's influence. The geometrical context was now considerably reduced, and all the chapters containing geometrical applications were deleted. On the other hand, there was no focus on the concept of function; the basis was rather more the formal approach of understanding functions as power series—making appeal to Taylor's theorem—and of establishing differential calculus on the general validity of the binomial theorem. Just as was typical for the vast majority of mathematicians in Germany, L'Huilier thus understood differential and integral calculus as directly derivable from a finite calculus of differences. Although L'Huilier developed the concept of limit in 1795 as well, this concept has virtually no explorative power or reach, since his now formal–algebraic approach implies no problems at all in the transition to limit, but excludes them from the very outset.

In contrast to this marginalization of his own original approach, L'Huilier considerably increased the number of propositions about limit behavior (cf.

L'Huilier 1795, 12 ff.), also presenting a more strongly operative use of the sign *lim* (cf. ibid., 37 ff.).

One of the possible backgrounds for this strengthening in the presence of a basically inappropriate conception may be that a book was published between L'Huilier's two versions of his essay of 1786 and 1795 that developed the concept of limit in a quite novel way.

This book presents the first tentative approach to an algebraic elaboration of the limit concept and is hence of great systemic interest. It was written at the periphery of the mathematical Europa of the time, in Portugal, and thus represents—after da Cunha's work (cf. Chapter II.1.6.)—the second conceptually eminent work originating from Portugal. The book is at the same time a further instance of the effect that specialized teaching had on strengthening reflection on foundations. Its author was among the most excellent students and graduates of the first specialized studies in mathematics at the Faculty for Mathematics of the University of Coimbra, which had been reformed in 1772 (cf. Chapter II.2.6.). Francisco de Borja Garção Stockler (1759–1829) is well known in the history of mathematics as author of the first presentation of the history of mathematics in Portugal (1819); he is at the same time a most eminent figure in the history of science of Portugal and Brazil (cf. Saraiva 1993, 1997). The importance of this special book has not been recognized and acknowledged as yet.

Stockler was born in Lisbon; his father, Consul Christian Stockler, came from Hamburg; his mother, Margarida Josepha Rita d'Orgiens Garçaõ de Carvalho, was of Portuguese origin. Stockler, destined for a career in the military, took up his studies of mathematics as a cadet of a cavalry regiment at the mathematics faculty in Coimbra, not in 1784, as the biographical literature will have it, but already in 1782, and he spent the obligatory four years of studying mathematics.[43] He graduated in 1785 as bachelor of mathematics. He assumed the professorship for mathematics at the Royal Naval Academy. In 1797, he was elected a member of the Portuguese Academy of Sciences. At the same time, he was involved in several military campaigns. As the Academy's secretary and well–known Jacobite, he was also actively involved in 1807 in having General Junot, Napoleon's commander in chief, elected as member of the Lisbon Academy. After falling into disgrace for this at the royal court, he followed the exiled king of Portugal to Brazil in 1812, developing a considerable activity fostering science in that country. After 1820, he held, as a general, several posts of governor in Portugal under changing political circumstances.[44]

His 100-page volume *Compendio da Theorica dos Limites, ou Introducçaõ ao Methodo das Fluxões* was published by the Academy of Sciences in Lisbon

---

[43]   Information graciously given by the Archives of the University of Coimbra.

[44]   Information given by the Academia das Ciências de Lisboa; as well as the entry "Francisco de Borja Garção Stockler" in Innocencio Francisco da Silva, *Diccionario Bibliographico Portuguez.* Tomo (Imprensa Nacional: Lisboa, 1859), 354–358.

in 1794, but had already been submitted to this body in 1791 by Stockler (Stockler 1794, X). Stockler followed many of his predecessors in arguing that the limit method had already been the basis for the Greeks, and in particular for Archimedes. It had, however, not been made explicit by them. Modern geometers had generalized the idea of limit, but had at the same time introduced the imperfect and repellent (*imperfeitas, repugnantes*) concepts of *Infinitos* and *Infinitesimos* (ibid., IV) in order to avoid the laborious methods of the ancients. For Stockler, extensive debate on the concept of infinitely small quantities was superfluous; he just briefly mentioned the lack of rigor in their principles (*a inexactidaõ dos principios*) before presenting his own view. Stockler drew on six sources for his own development: Section I of the first volume of Newton's *Principles of Natural Philosophy*; MacLaurin's textbook; d'Alembert's entries *Limite* and *Differentiel* in the *Encyclopédie*; the second chapter of Cousin's textbook; the part concerning the principles of infinitesimal calculus in the Abbé Martin's textbook, and finally L'Huilier's prize–winning essay. Stockler noted proudly, however, that almost all of his own results were "completely new" (ibid., VIII). He also observed that he had elaborated the foundations that had hitherto remained implicit for his own lectures at the Royal Naval Academy in Lisbon (ibid., XI).

Not only is Stockler's precise knowledge of the international literature remarkable, his list of sources being identical with our own reconstruction of the concept's development, but he also confirms that Martin's textbook of 1781 was not only regionally significant, but had been noted internationally. Martin's reception indeed forms one of the most important bases for Stockler. Stockler widely surpasses his sources, however, by separating the concept of limit from the geometrical concept and by algebraizing it as an operative explication for variables, and finally for functions as well. Furthermore, the beginning of operating with inequalities can be found with him.

On the basis of Martin's concepts, Stockler adopted L'Huilier's differentiation into left- and right-hand limit, introducing, however, because of the characteristic exceptional status of the zero, a further differentiation, practically raising the null sequence to the status of basic concept of the limit theory.

Stockler began his own presentation, like almost all authors before him, by introducing the concepts of constant and variable, but as in Martin by using the value concept, with constants being capable of only one value, while variables could assume various values (ibid., 1). In a manner general for all authors in the eighteenth century, he did not explicate the range. Stockler now used these determinations to define the general concept of limit in close agreement with Martin, but with stronger emphasis on the fact that the limit forms a constant value, the arbitrary approximation being formulated identically with Martin:

> A constant quantity is called "Limit" of a variable if the latter is capable of increasing, or of diminishing—even if the variable's value can never become equal to that of the constant—in such a way that it can approach the constant so far that the difference becomes smaller than any given quantity, however small this may have been chosen (ibid., 2).

Stockler hardly used the concept of limit in this general form, but rather in specified forms, which at first, as in l'Huilier, distinguished between left- and right-hand limit because the absolute value was lacking. What is called *limite en grandeur* in l'Huilier, Stokcler names *limite em augmento*, and L'Huilier's *limite en petitesse* becomes Stockler's *limite em deminuiçaõ* (ibid., 2). The conceptual innovation surpassing Martin consists in Stockler's opposing to these two concepts of limit seemingly complementary concepts of unlimitedness: variables that have no limit of diminishability (*que naõ tem limite em deminuiçaõ*), respectively no limit of augmentability (*que naõ tem limite em augmento*).

Actually, however, only the variable having no limit of augmentability proves to be unlimited and thus as not generally operatively applicable. The variable without limit of diminishability, in contrast, constitutes despite its name not only a limited quantity, but also a new basic concept of the limit theory. The reason for that is the exceptional status of the zero. As we have seen above, Martin had not acknowledged the zero as *quantité*, and it was hence not a possible value for variables either. For this reason, Martin had introduced the *infiniment petits* as a special, that is finally vanishing, variable (cf. Section 8.4. above). For Stockler, too, the zero had an exceptional position, even if he did not explicitly comment on that matter; his entire practice shows clearly that the zero could not be reached as a limit. In contrast to Martin, however, Stockler did not introduce *infiniment petits* as auxiliary concept, but defined another concept of limit: that of the variable diminishing without limit. Since this did not mean a variable dropping into the negative "infinite," but rather a variable approaching zero as a barrier, this shows not only the factual limitedness of this variable, but also the exceptional character of the zero. The inconsistency in Stockler's conception was his refraining from discussing the possibility of negative limits.

Even his first uses of the concept of "variable without limit of diminution," he took this to mean variables that we might most simply designate as null sequences here. His first theorem concerning these quantities said that the sum of an arbitrary number of such variables is again such a variable. The proof for that showed the operationalization that any such variable could become smaller than any quantity being assumed ever so small—Stockler not assuming this quantity to be positive; typical for him is consistent use of an algebra of inequalities that Grabiner establishes as an innovation only with Cauchy (Grabiner 1981, 54):

> Let n be the number of the variables z, y, x, etc. and k a quantitiy which is so small as one wishes to assume it. One may thus assume
>
> $$z < \frac{k}{n}, \ y < \frac{k}{n}, \ x < \frac{k}{n},$$
>
> and correspondingly for all further variables, and one obtains
>
> $z+y+x+$ etc. $<k$. QED (ibid., 4).

Another proposition, Stockler's third theorem, further illustrates the basic characteristic of this only seemingly unlimited concept of limit as null sequence. If two different variables x and y have the same limit a, their difference $x - y$ has no limit of diminution (ibid., 7).

The subsequent development of the properties of these null sequences leads Stockler eventually to formulating a *Principio Fundamental* of limit theory, which shows the null sequences to be the very basic concept of this theory. Any variable possessing a limit can be conceived of as sum or difference from a constant—i.e., the limit—and a null sequence:

> Each quantity capable of a limit has necessarily, in whatever state of greatness it might be, to be equal to its limit plus or minus a variable quantity which has no limit of diminution (ibid., VI).[45]

With these conceptual foundations, Stockler developed a purely algebraical conception of operating with the various limits. In contrast to his predecessors who had focused on operating with geometrical ratios, Stockler conceived of the behavior of limit algebraically, and systematically developed operating with it: for all applicable operations like addition, subtraction, multiplication, division and raising to a power, and also for such linkages between two or more variables, between a constant and a variable, etc. Proofs for the individual theorems are given consistently by algebraic operations with inequalities. We shall quote some examples of such theorems here:

- If two variables are null sequences, their sum, their difference, and their product are the same; the analogous holds for the product of a variable and a constant, and for the division of a variable by a constant (ibid., 3–10).

- A power a^x, where $a<1$ is a constant and x a variable with positive values and without limit of augmentation, forms a null sequence (ibid., 22 f.).

- The limit, both *em augmento* and *em deminuiçaõ*, of the sum—resp. the difference, resp. the product—from a constant a and a variable x with limit b is equal to the sum/ difference /product of a and b. He showed the analogue for two or more variables (ibid., 28 ff.).

These theorems were immediately conducive to propositions of inter-changeability for limits, and were additionally supported and generalized by creating the sign for the concept of limit.

They were indeed immediately followed, in the section about the two concepts of limit proper succeeding the presentation of the fundamental principle (see above), by an explanation of the convention concerning the sign for limits, containing the *Lim*, which had meanwhile become standard, Stockler refraining from differentiating between left- and right-hand limit for this sign:

> If we desire to express a variable's limit in calculation, and if we have not yet established a letter of the alphabet for this, we shall write the first three letters Lim. of the word *Limite* before the term or the expression represented by the variable. To express the Limit of x, we shall hence write Lim. x; to express the Limit of xy, we shall write Lim. (xy); to express the Limit of y^x, we shall write Lim. (y^x) ; and thus analogously (ibid., 28 f.).

Both in introducing and using signs, Stockler was again more explicit and

---

45   This principle is presented in more detail in the second section (28 ff.).

precise than his predecessors. While Stockler formulated his own propositions ("theorems") without using the sign of limit, he always used this sign operatively in his proofs. This use at the same time shows the interchangeability of algebraical operation and limit process. Thus, the result for Lim. $x = a$ and Lim. $y = b$ is that $a + b = $ Lim. $(x + y)$, and $ab = $ Lim. (xy) (ibid., 30 ff.).

Stockler stated the interchangeability also for transcendental functions, in particular for the logarithm: For $b = $ Lim. x and a a constant, Lim. $(a^x) = a^b$, and analogously Lim. $(y^x) = a^b$ (ibid., 55 ff.).

The new thing in Stockler's approach was that he did not confine himself to the limits of variables, but also explicitly introduced and discussed limits of functions (ibid., 66 ff.). In particular, he explicitly stated the substitutablility of limit processes for functions, as his Theorem XXII: "The Limit of any function Fx of a variable x which is capable of a limit is equal to the homologous function of its limit" (ibid., 68). In sign form, Stockler expressed this theorem as follows: For $a = $ Lim. x , it follows that Lim. $Fx = Fa$.

In the conditions for his theorems, he introduced, in case of general propositions, the condition that the variable(s) concerned are capable of a limit: *capaz de limite* (ibid., 11 ff.). He did not make this condition, however, an element of its own theory for the purpose of studying the behavior of functions. There is one exception that is particularly interesting, since Stockler was familiar with da Cunha's works, whose innovative concept of convergence we have already mentioned above. Stockler used the example of the sum $\frac{a}{1-r}$ of the geometric series ar^n to show in detail that the limit exists only if the series is convergent, i.e., if $r<1$ holds in this case (ibid., 46 f.). He went on to discuss that a series' convergence is a condition for the sum's existence. At the same time, he did not exclude the nonconvergent series as mathematical objects, but rather admitted them as formal series—with the differentiating designation of *Limites de expressaõ* instead of the designation *Limites* (ibid., 47 ff.).

In his closing chapter, Stockler studied applications for the development of transcendental functions into series. He announced continuations of his textbook, which were never published, however.

Stocker's reception of Martin's ideas is of particular systematic interest, since the transition from Martin's *infiniment petits* to Stockler's variables as null sequences proves the contemporarily intended "classical" meaning of the infinitely small quantities, and since it simultaneously refutes the projection into history practiced by the adherents of non-standard analysis.

9. Operationalizations of the Concept of Continuity

In Section 5., we had seen that a reflection on continuity going further than Leibniz's mathematical version of his own law of continuity had been pursued only occasionally in the eighteenth century: by Boscovich in Italy after 1745, the latter's postulating continuous curves, which precluded his ideas from becoming operative, and by Kästner in Germany after 1750, an author who already argued empirically and referring to mechanics, and no longer excluded discontinuous quantities on epistemological grounds, but admitted a differentiation into continuous and discontinuous.

Since the reflections in Germany concerning the concept of continuity represent an essential contribution to the development of the fundamental concepts of analysis, we shall briefly treat here how the concepts of analysis developed in Germany in connection with the concept of continuity, such as can be reconstructed from textbooks.

The new differential and integral calculus was presented in Germany for the first time in a work already reviewed here for its contribution to the development of negative numbers in Chapter II.2.8.3., the voluminous and influential textbooks authored by Christian Wolff, printed in the Latin version within the first volume of the *Elementa Matheseos Universae* first edition of 1713 as part of algebra, in the German version, by contrast, in the last (fourth) volume of the first edition of 1710, likewise as part of algebra.

Wolff introduced the new theory without any profound philosophical reflection, and differentiated his concepts less than contemporary French works. In contrast to his part on negative numbers, the Latin and the German versions did not differ as to their conception.

Wolff's introduction was based neither on the concept of variable, nor on that of function, but in an overly general manner on "quantities." By using this undifferentiated concept of quantity as a fundamental concept, Wolff had implicitly chosen one of the traditional philosophical assumptions concerning the infinite, according to which the infinite sum of infinitely small quantities yielded a finite quantity:

> Differential calculus is a science of finding, from a given quantity, an infinitely small one, of which an infinite number, taken together, become equal to the given quantity (Wolff 1750, 1799; in Latin: Wolff 1742, 545).

To define the infinitely small quantities, Wolff relied on Leibniz's specification of incomparability: "An infinitely small quantity is that which is thus a part of the other so small that it cannot be compared to it" (Wolff 1750, 1799).

While Wolff's Latin version is more precise in using the expression "smaller than any assignable quantity," it refers, by using the term *particula*, to physicalist notions of particles:

> *Infinitesima* seu *quantitas infinite parva* est particula quantitatis adeo exigua, ut eidem incomparabilis existat, seu quae omnis assignabili minor (Wolff 1742, 545).

In spite of this basing on incomparability, Wolff drew from his definition of infinitely small quantities the same conclusion L'Hospital and his successors had drawn in France. Infinitely small quantities were to be considered, "in view of that quantity to which they cannot be compared, for nothing"; moreover, they could neither increase nor decrease finite quantities (Wolff 1750, 1800); and two quantities differing only by an infinitesimal quantity were "equal" (*aequales sunt*, Wolff 1742, 545).

Wolff introduced a distinction between "variable" and "invariable" quantities only after having concluded his explanation of infinitely small quantities. He assumed variable quantities to be continuous (Wolff 1750, 1802; Wolff 1742, 546). Immediately afterwards followed the presentation of the usual rules of differential calculus, and their applications. In this, a difference between the German and the Latin versions is remarkable: in the proof of the product rule. The German version used Newton's "trick" involving half increments of the sides of a rectangle xy:

$$x - \tfrac{1}{2}dx, \quad y - \tfrac{1}{2}dy, \quad x + \tfrac{1}{2}dx, \quad y + \tfrac{1}{2}dy \quad \text{(Wolff 1750, 1803)},$$

in which seemingly no term is neglected, while in the Latin version the product term $dxdy$ is quoted as *habetur pro nulla* against ydx and xdy (Wolff 1742, 547).

Integral calculus, again introduced formally as the inversion of differential calculus, was still strongly determined by the traditional geometrical methods (squaring, cubing, inversion of the tangent method, etc.) by using geometrical loci, and without use of more general methods, or the concept of function.

Wolff's textbooks remained for fifty years almost the only ones in Germany presenting analysis. Reflection of the foundations of analysis hence occurred elsewhere, in what was obviously the first explicit treatment of the concept of continuity in a mathematical textbook after Leibniz's introduction of the law of continuity. Christian Hausen,[46] mathematics professor in Leipzig, in his textbook *Elementa Matheseos* (1734), not only achieved the necessary extension of the concept relation for negative numbers by a new definition of the concept of multiplication (cf. Chapter II.2.8.3), but also introduced the concept of continuity as a further fundamental concept. Hausen's effort is thus a further example of the parallel character of, and connection between, reflections on the foundational concepts in the area of numbers, and in that of functions.

Hausen began the part of his textbook on geometry by defining continuity, at the very outset of introducing concepts like point, straight line, area, space, angle, circle, etc., even preceding the traditionally introductory postulates and axioms. The first intended application of Hausen's concept of continuity was to determine the nature of *extensive* quantities. To introduce this concept, Hausen drew, just like Boscovich eleven years later (see Section 4. above), on Aristotle's definitions of continuity.

The definition of continuity proper was preceded by introducing *terminus* (boundary) in order to be able to express the "beginning" and "end" of a quantity element:

---

[46]   Christian August Hausen (1693–1743) was the son of a theologian, and had studied from 1710 to 1712 at the University of Wittenberg (ADB [Allgemeine Deutsche Biographie], vol. 15, 440).

[1.] *Terminus* est, quod alicujus extremum est, seu ubi id quod hactenus ponebatur cessat et definit, vel incipit, quod antea non erat (Hausen 1734, 87).

With this, Hausen was able to define a continuous quantity as one whose totality of parts, or elements, were connected, i.e., possessed the same boundaries, or were superposed: "2. *Continuum* est, cujus partes quaevis vel conterminae sunt, vel interjectas habent ipsis et inter se conterminas" (ibid.).

The essential thing in this definition of continuity was the notion of parts, or elements, of the quantity concerned. Hausen himself indicated this by his remark that all quantities possessing a common boundary were homogeneous with regard to one another, and that a continuous quantity hence consisted of elements homogeneous to it—thus providing a decisive answer to the question as to the nature of infinitely small quantities: *Quare continuum ex homogeneis constat* (ibid.).

A first application was to introduce the concept of extension. Extension, he said, was the continuity of space and of the bodies that fill part of space.[47] Hausen's concept of space was tied in a substantialist way to physical space; he discussed this definition via physical concepts: quantity, distance, position, and mass of bodies in space (ibid., 87 f.). Hausen related the concept of continuity of extension to all geometrical dimensions: to solids, areas, and lines. The point, in contrast, as boundary of any linear extension, was nonextensive, and thus not capable of quantity, nor of further division (ibid., 88).

In his preface to the first volume, Hausen had announced two additional volumes, the second, on *geometriam generalem* and on mechanics and optics, the third on analysis. He died before being able to publish them.

Hausen's work was continued by his disciple Abraham Gotthelf Kästner (1719–1800). Since Kästner's definition of continuity, for instance, was practically the verbatim translation of Hausen's Latin text, it may be assumed that the latter's basic concepts of analysis can be recognized in Kästner. In his autobiography, Kästner described how Hausen had convinced him to study to become a teacher of mathematics, and how much he owed to Hausen's training. Kästner's father had intended him for a lawyer's vocation, and at the time he met Hausen he had not only studied law, but had passed all the necessary examinations for that profession (Eckart 1909, 6 ff.). Kästner also described how mathematics within the Faculty of Philosophy was at the time generally only taught to boys at a stage preceding their serious professional studies,[48] a reason why Hausen's students at his advanced level were scarce and the latter had no grounds for permitting more than occasional glimpses into analysis in his teaching:

About the calculus of the infinite I had heard nothing but what had been explai-

---

[47]   "3. *Extensio* est Continuitatis Spatii et Corporum Spatii partes replentium."
[48]   Kästner began his studies after having been accepted at university at the age of twelve.

ned by Hausen in connection with his *Elementorum*, and that was very little. I had read Wolff's *Analysin infinitorum* on my own, and that was not much more (ibid., 9).

At the same time, Kästner claimed to have attained a higher degree of rigor than that which he attested to Hausen:

> Whoever has read his [Hausen's] elements will know that he often was satisfied with inductions, and I was never satisfied myself before I had shown the general correctness of the same (ibid., 11).

As has already been shown in Section 5., Kästner discussed the law of continuity extensively in his part on mechanics, conceiving of it as of an intermediate value property, while challenging the latter's general validity. In this text, he referred to his remarks on continuity in his textbook's part on geometry, where he indeed had gone beyond Hausen in publishing a detailed reflection on continuity in mathematics. Like Hausen, he began the textbook on geometry in his own work—even before the postulates—with a section on continuity. His introductory definitions are without exception verbatim translations of the corresponding parts in Hausen:

> 1. We call a thing's *limit* its extreme, or where it ends.
>
> 2. We call continuous *quantity* (continuum), the parts of which are linked in such a way that where one ends, the other begins at once, and where there is nothing not belonging to this quantity between the former's end and the latter's beginning.
>
> 3. *Geometrical extension* is a space filled by a continuous quantity (Kästner 1792, 177).

Kästner's definitions of solid, area, line, and point were likewise adopted from Hausen. New in this is, on the one hand, that Kästner refrained from discussing the filling out of space with physical particles, and on the other a note attached to the definition of extension calling for a separation between mathematical concepts and their ontological interpretation:

> It is not necessary to get involved here with metaphysical studies of space and of continuity. The concept of geometrical extension is an abstract one, however one may otherwise imagine these things to be (ibid.).

Kästner likewise adopted Hausen's definition of straight and curved line, while using them for his own attempt at elaborating the concept of continuity into an operative mathematical concept. Both Hausen and Kästner tried to define the difference between straight and curved line by means of the absence, respectively presence, of curvedness:

> A *straight* line is one whose points lie all in one direction; AB (Fig.1). A *curved* line $ABCDEF$ (Fig.2) is one in which there lie points between any pair of points f, c, however near the latter are assumed to be, which are not situated in a straight line with the pair of points assumed (ibid., 181).

Kästner himself was not quite satisfied with this definition of a straight line, for he added as an observation, "Nobody will get to know the straight line from an explanation, and nobody is in need of that, either" (ibid.). Nevertheless, this distinction is the central means for Kästner in permitting operative use of the concept of continuity:

> In the case of the curved line, the point which moves starting from *A* does not at once move into the direction where *B* lies in view of *A* [...]. Hence, a curved line is described by a point which continually deviates from the direction in which it presently moves in the very next moment, or which continually changes its path (ibid.).

Subsequent to these definitions, Kästner presented in detail his conception (briefly mentioned above in our Section 5.) of viewing curves as continuous lines, polygons, in contrast, as discontinuous ones. The condition for his argument was that he conceived of curves as variables having time as a parameter. Because this notion was highly unusual at the time, and has fallen into oblivion today, we shall quote it at length here, using Kästner's own figures to illustrate it:

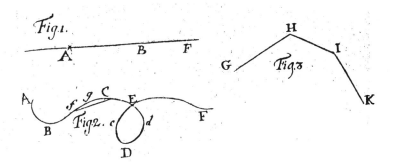

Fig. 7, continuous and "discontinuous" lines: Kästner 1792, table I

The geometers assume here besides that this change occurs in agreement with the law of continuity. The point has considerably changed the direction of its path from *A* to *B*: Before this change could become as large as it is in *B*, all small changes falling between had to occur. The point describing a curved line never changes its paths suddenly, as in fig. 3 from *GH* to *HI*, from *HI* to *IK*, but insensibly in such a way that it cannot be said how much it has deviated from at any moment from its path of the moment before. Thus it will not pass in fig. 2 from *CE* immediately to *EF*, but describe the loop *EdDeE*, which turns it by insensible changes of the path *E* into the direction from which it deviates further, to go on to *EF*. Hence, the curved line in fig. 2 differs from the composition of straight lines in fig. 3 in two things: 1st in that no straight part can be defined there like *GH*, *HI*, *IK* here; 2nd in that one cannot determine there how far the direction of the describing point differs in any moment from the directly preceding one (ibid., 182).

In his geometry textbook, Kästner showed, as an application of the concept of continuity, that continuous variables have limits:

> XII. (one) says: an *infinitely small* chord be equal to its arc, was losing itself in the latter, and was distant from its center by the radius. One calls it *infinitely*

small, a status which [...] the chord continually approaches, not having reached it yet as long as it has a defined quantity, but getting as near to it as one desires, because it can diminish, from a certain quantity on, by all the smaller until nothing.

XIII. This *infinitely small* thus means that a quantity can diminish by way of all defined values until nothing, and consequently become smaller than any quantity, however small it is given. Just as the *infinitely large* indicates an augmentation of the very same kind (ibid., 293 f.).

These reflections, first published in 1758, are remarkable not only for their operationalization of the concept of continuity within mathematics, but also for their clear definition of infinitely small quantities as variables having the limit null. We have not found a comparable explicitness in the contemporary reflections in France. Just as clearly, Kästner presented the apparent omission of infinitely small quantities as finding limits. In one of his lectures at the Göttingen Academy of 1759, he used a concrete example to present how a limit process is executed, characterizing the alleged omission of infinitely small terms as an abbreviated way of speaking:

Let be $q = \frac{ax+b}{\alpha x+\beta}$ and one posits $q = \frac{a}{\alpha} + r$; one obtains $r = q - \frac{a}{\alpha} = \frac{\alpha b - a\beta}{\alpha(ax+b)}$, where the divisor grows indefinitely if x grows, the *dividendus* remaining unchanged, hence x be always being assumed so large that r becomes smaller than any given quantity; and consequently q is less different from $\frac{a}{\alpha}$ than any given quantity. If x thus grows indefinitely, the variable value of q approaches the limit $\frac{a}{\alpha}$ indefinitely, this being the true concept and the geometrical proof of the proposition that one may consider, in the value of q, the finite quantities b; β in comparison to the infinite x, to be nothing. It elucidates that this proposition is only an abbreviated expression of the former, and that b and β in comparison to ax, αx are not really omitted, but that they do are not included in that quantity which q approaches indefinitely. One grasps how this can be applied to more composite expressions (Kästner 1759a, 498 f.).

In his own textbook on the *Analysis des Unendlichen* (first edition of 1761), Kästner introduced infinitely small and infinitely large quantities as fundamental concepts, however, using the existence of different orders of such quantities to justify the "vanishing" for the corresponding terms. Here again, he stressed, albeit only in a footnote, that the infinitely small quantities were intended only as an attempt to express the limits of ratios for finite quantities (Kästner 1770, 12). This work by Kästner is an impressive first example of a comprehensive reception of Euler, who had had virtually no impact on the authors hitherto quoted. Kästner not only studied Euler's eminent textbooks on algebraic analysis concerning differential and integral calculus, making allowance for new books by Euler in the various reprints of his own works, but also carefully assessed Euler's writings in the St. Petersburg Academy's publications, integrating the results from these in his own books. Kästner's reception referred to individual results, however, not to Euler's entire conception and methodological approach. While Euler's concept of function was used in

Kästner's textbook on analysis, for example, it did not become a methodological moment of restructuring. In general terms, this part of the work is less a textbook to be studied in connection with lectures; but since there was little demand for such high–ranking subject matter, it is rather more a handbook concerned with the contemporary status of knowledge, quite comparable to the later three-volume handbook published by Lacroix. To inform the reader, for instance, Kästner also presented Newton's method of fluxions.

With his textbook, Kästner showed that he was at the level of the international development of analysis, being able to include even the most recent results of research. In particular, he stressed on several occasions in his book the fundamental importance of MacLaurin's *Treatise* for clarifying the foundational concepts (e.g., in his preface, ibid.). The fact that he nevertheless did not take Euler's approach featuring the theory of functions as his own basis agrees with the general situation then attained in the development of mathematics. How relatively isolated Euler's approaches were as compared to the contemporary trend is illustrated by the lectures *Principj di Analise Sublime* that the young Lagrange (1736–1813) gave in 1759 at the artillery school of the Kingdom of Piedmont in Turin.[49] Although Lagrange already entertained a close correspondence with Euler (see Borgato, Pepe 1987, 14 ff.) and despite the fact that he had studied the most recent contemporary literature, he mentioned the concept of function only briefly, but did not make it the basis of his own presentation. His basis, by contrast, was the concept of variable quantity, still dominantly motivated and applied in geometrical studies of curves.

After introducing the differentials as algebraic quantities, he translated them into geometrical quantities, since these were allegedly better represented in geometry; here, Lagrange still understood geometry to be the dominant methodology; on this basis, he also assumed the continuity of the changes of variables:

> Le differenze, che abbiamo sin qui esaminate appartengono alle quantità algebraiche; nella Geometria esse si determinano molto più facilmente; imperciocché basta supporre che ciascheduna delle linee, che hanno fra di loro un dato rapporto si muti continuamente di posizione (Borgato, Pepe 1987, 129).

It is not correct, either, that one can establish an "acceptance of the limit approach" by Lagrange (Grabiner 1997, 403). While Lagrange used the term *limite* in the manuscript of his lectures, he did so in the sense of Newton's ultimate ratios, and not as an algebraic instrument of its own. Nor did he apply the concept of limit operatively; it appears only for purposes of verbal description. He uses it, moreover, to proceed to transposing algebraic processes into geometrical ones, and to justify them from geometry:

> Ora siccome abbiamo ritrovato nelle espressioni delle differenze algebraiche certi limiti ne' loro rapporti; tali limiti dovranno anche esistere ne' rapporti delle differenze geometriche, che alle algebraiche corispondono; [...] si supporrà

---

[49]   The manuscript of these lectures was edited in 1987 by M.T. Borgato and L. Pepe (Borgato, Pepe 1987).

primieramente che le differenze siano prodotte, e si ricercheranno i loro rapporti supponendo, che diminuiscano sino a svanire del tutto, cioè che linee variabili ritornino nella loro prima situazione (Borgato, Pepe 1987, 129).

Lagrange, too, presented in this early text a mixture of different foundational approaches, discussing Euler's interpretation of the differential quotient as a ratio of 0:0 while at the same time quite clearly disassociating himself from it, presenting the conception of the infinitely small quantities together with their various orders as a better comprehensible theory—a position quite surprising for an author who later in his life became an advocate for a foundation doing completely without infinitely small quantities (ibid., 161 ff.).

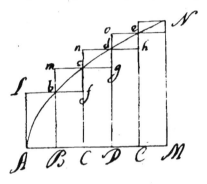

Fig. 8, Lagrange on areas under curves (Borgato, Pepe 1987, 184)

To this text by Lagrange, Borgato and Pepe ascribed the merit of having given a geometrical interpretation of the integrals that was similar to the later, so-called Cauchy integral (ibid., 37). There is indeed, in the part on integral calculus, after integration is introduced as a formal inversion of differentiation, a section on the geometrical determination of integrals. At first glance, this seems to be about finding definite integrals by means of refining lower and upper sums by finite differences (cf. Figure 8; ibid., 184). While the problem is indeed to approximate the area content below a curve by means of rectangles with equidistant, decreasing distances dx, Lagrange does not use any interval nesting for refining lower and upper sums: instead, he uses a method applied already by Newton in his *Principia* (cf. Sageng 1989, 173). The small rectangles overlying the curve are in their sum precisely identical to the last rectangle *CNM*. If dx continues to diminish and finally becomes zero, the differences formed by the small rectangles will vanish, and the integral becomes equal to the entire area below the curve (ibid., 184). In this geometrical presentation, only individual points could be designated; for arguing with intervals, and with upper and lower sums, the forms of designation for indexed sequences like $x_1, x_2, ..., x_n$, etc. are missing here.

An attitude toward Euler's concepts similarly ambivalent to Lagrange's can be found in W.J.G. Karsten (1732–1787), the second eminent textbook author of the second half of the eighteenth century in Germany besides Kästner. While he shows an intense reception of Euler's writings, he adopted only some results, but not the latter's fundamental conceptions, Euler's conception of the differential quotient as the ratio of two actual zeros in particular having met with widespread rejection. In his curriculum vitae of 1766, written for the occasion of his election to the Bavarian Academy of Sciences in Munich, Karsten described his own path of development in mathematics.[50] After studies in Rostock and Jena, he became lecturer for mathematics in Rostock in 1755, and started to work his way into modern mathematics—primarily under the guidance of Franz U. Th. Aepinus (1724–1802), who worked at the St. Petersburg Academy after 1757. Karsten reports how intensively his teacher guided him:

> He acquainted me with the best mathematical authors, in particular with Euler's works, and helped me to overcome the first difficulties which I necessarily had to encounter with this author, as I had as yet not read any other analytic writing but Wolff's *Elementa Algebrae* (C.V., 13 f.).

It is remarkable that Hausen—in addition to Segner and Euler—is received here as a foundational author:

> First of all, I worked, besides Euler's works, through Hausen's *Elementa Matheseos*, and through Segner's lectures on the art of calculation and on geometry" (ibid., 14).

He had been unable, however, to accept the two approaches to founding analysis known to him. The difficulties he had initially encountered with differential calculus, he said, had been less concerned with its rules, but "rather with the theory." He quoted a critique of Aepinus from 1753 pertaining to the concept of infinitely small quantities that had become formative for him:

> On the whole, we believe that the modern *Mathematici* who have brought the infinitely small quantities into geometry do not merit much gratitude. All the proofs founded on it lack by far the clearness, care, and what is convincing in the other geometrical demonstrations. One does not need them either, as one may in all the cases where the infinitely small things occur make use of the manner of demonstration serving Archimedes in his books *de circuli dimensione* and *de figuro et cylindro*, attaining a much firmer conviction by this (ibid., 16).

Karsten thus was unable to accept L'Hospital's and Wolff's presentations, but neither Euler's approach, which he had at first studied with suspense:

> I was very glad to obtain, just at this time in 1756, Herr Euler's differential calculus; from this excellent work I got to know the entire scope of this science for the first time. Now I freely confess that I felt something to be missing in my firm belief in the theory's correctness. But I dismissed my doubts for the time being, and only made an effort at getting quite familiar with the rules (ibid., 17).

---

50   For knowledge of this c.v. I am indebted to Karsten's descendant Johannes Karsten (Wismar), who placed a typewritten transcript at my disposal:
   "W. J. G. Karstens kurzer Entwurf seiner Lebensgeschichte von ihm selbst aufgesezt, und der Churfürstlichen Akademie vermöge des §. XXXVI ihrer Gesezze übergeben im Jahre 1766."

Karsten had in his own words obtained access to a satisfactory foundation, however, only after having studied MacLaurin's *Treatise*. While the latter's presentation appeared to him "rather far-fetched," he had been prompted, by a hint in the *Treatise*, to study Newton's natural philosophy, and had become convinced by the latter's section *de methodo rationum primarum et ultimarum* that the limit method was the appropriate basis. In the volume on analysis in his textbook, Karsten had indeed used the limit method as basis of his own presentation. In this book of 1786, he was thus able to declare with regard to the Berlin Academy's contest that he had always avoided the concept of the infinitely small:[51]

> that I did not find it necessary at all to use the term of infinitely small. Everything has been reduced, as in my other writings, to the concepts of the limits of the ratios and sums (Karsten 1786a, xiv).

In the second chapter of this textbook, "zu den ersten Gründen der Differentialrechnung mit einigen Anwendungen," Karsten indeed used the limit method for presentation. The method, however, had been little elaborated by him. He used the concept of *Gränze* (limit) without defining it, and he applied it to geometrical quantities or, more precisely, to ratios (ibid., 36). His basic concepts here were invariable and variable ratios. In some places, however, Karsten also formulated applications for functions. Although he had introduced differential calculus using the concept of geometrical ratios, Karsten was able to formulate the dependence between the dependent and the independent variable for attaining limits quite precisely. Thus, he explained for two "variable quantities" z and y, whose ratio was given by

$$\frac{y}{z} = \frac{\beta + \gamma x + \delta x^2 + \dots}{p + qx + rx^2 + \dots},$$

that if x became [absolutely] ever smaller, "the exponent[52] of the ratio $\frac{y}{z}$ simultaneously approaches the limit $\frac{\beta}{p}$ if x approaches the limit 0" (ibid., 36).

He obtained the product rule, by contrast, without explication of limit processes, by formally transposing the binomial theorem for $(a + b)^n$ to $(x + dx)^n$, by simply omitting the higher terms without giving a reason (ibid., 41). Karsten used the concept of limit only in this introductory part, and only verbally, without operationalizing limit processes.

For our context, the development of the concept of continuity in the frame of the basic concepts of analysis, it is of particular interest that Karsten continued Hausen's and Kästner's reflections on foundations without interruption. It is remarkable that while Karsten uses the concept operatively in the same field as

---

[51]   The first printing of his seven–volume textbook of 1762 contained no part on differential and integral calculus. The book *Mathesis Theoretica*, separately published in 1760, served as a substitute. Karsten integrated the part on analysis only into the second edition of his *Lehrbegriff*.

[52]   "Exponent," in the theory of proportions, meant the value of ratios (as division of the "antecedent" by the "consequent").

Kästner in mechanics, in discussing hard–body impact, his epistemological choice was different from that of Kästner. The latter had shown three alternatives to the dilemma that while general validity of the law of continuity was philosophically required, considerable objections spoke against general validity on the level of individual sciences like mechanics (cf. Section 5. above). In his own mathematical practice, by contrast, Kästner had admitted both continuous and discontinuous quantities in mathematics itself as well—albeit not with sufficient reasons (see above).

Karsten, in contrast, adhered to the second alternative sketched by Kästner, according to which the mathematician was permitted to abstract from the physical properties of the bodies and to assume continuous transitions (cf. Section 5.). He made a distinction between the principles in "theoretical mathematics" and those in the natural sciences. Just like Kästner, and referring to him, Karsten discussed the validity of the law of continuity in mechanics, in his section on the impact of bodies, and like the former with regard to the question whether perfectly hard bodies existed. He quoted the law of continuity at this point as an intermediate value axiom, his presentation being somewhat simplified as compared to that given by Kästner:

> 235 §. One imagines, actually, in theoretical mathematics (41 §. A. R.), all the variations happening to a quantity, by its increasing or diminishing, as such which do not happen abruptly. [...] Briefly, before $\frac{1}{2}A$ can become of A, gradually all the values falling between A and $\frac{1}{2}A$ must have become of A. One calls this rule the *law of continuity*, and several naturalists have assumed this rule to be a general law of nature. By force of this law, no change in nature shall occur abruptly (Karsten 1769, 223).

The "A. R." is a reference to the volume *Ausführung der Rechenkunst*" (continuing the art of calculation), the second part of his textbook series. Paragraph 41, in the section *Zu den Ausdrücken des Unendlichen*, indeed states that the law of continuity generally holds as intermediate value law in theoretical mathematics, in Karsten's *Allgemeiner Rechenkunst* (general art of calculation):

> 41 §. If a quantity, e.g., a line AD, is considered as a variable quantity, and if one posits it has changed by a certain segment, e.g., by Dd, one assumes as a condition in the general art of calculation that this change does not happen abruptly; but rather that AD, during this change, passes through all values falling between AD and Ad. It follows from this that if AD was positive, but has become gradually smaller, and later negative, that there must have been $AD=0$ during this time. Upon returning, the point D cannot come to ∂ if it has not passed through A, and by D falling on A, there is $AD=0$. Similarly, the negative value $A\partial$, if the same gradually diminishes, cannot become positive, if it has not at one time been $-A\partial = 0$. (Karsten 1768, 95).

A   d   D   δ

Fig. 9: taken from Figure 178, Table I (intermediate values)

The parts relevant for § 41 are here extracted from his Figure 178, to which Karsten refers for this purpose.

While Karsten thus claimed general validity of continuity for pure mathematics, he did not consider this generally given in physics. What is new is mainly his proposition that a certain occurrence in nature need not be generally continuous or discontinuous, but that a process may show different aspects of which some can be continuous, while others are discontinuous. This differentiation according to particular sciences—albeit confined to physics— represents an important step beyond Kästner. Karsten indeed continued his § 235 of mechanics quoted above as follows:

> 236 §. Thus a body never changes its place abruptly. From that, however, it can hardly be concluded that the other circumstances one distinguishes in motion in mechanics could not change abruptly (Karsten 1769, 224).

As example, however, he quoted here that already drawn upon by Kästner. A point moving along a straight–lined figure changed its direction abruptly at the angle points (ibid.). In contrast to Kästner, Karsten had assigned this case not to geometry, but to mechanics. Karsten rejected inferring propositions in mechanics only from philosophical principles, requiring empirical justifications:

> 237 §. One can thus not prove the impossibility of perfectly hard bodies for the reason that the change of velocity does not occur according to the law of continuity upon the impact of such masses (234 §.); and as long as its impossibility has not been proven from other reasons, a special study is required which investigates according to which laws the motions of perfectly hard bodies are changed by the impact (ibid., 225).

The textbooks authored by Kästner and Karsten represented for an extended period the most advanced reflections about the concept of continuity in Germany. Neither Johann Schultz nor Klügel continued this reflection. In his influential *Mathematisches Wörterbuch*, which was typical for the contemporary conception of mathematics, Klügel devoted only a single sentence to continuity in the first volume of 1803, an entry moreover still evoking the concept at Hausen's level:

> CONTINUUM, the continuous, or the immediately connected. A quantity is called a continuous one, a CONTINUUM, if its parts are all connected in such a way that where one ends, the other immediately begins (Klügel 1803, 553 f.).

Only the fourth volume, published by Mollweide as new editor in 1823, contained a rather modern formulation, explained as a supplement to the entry of old:

> *Stetig [continuous]* s. CONTINUUM. To what is observed there should be added: The continuity of a function consists in that its values will change infinitely little for infinitely small changes of the variable quantity. Otherwise, the continuity is interrupted. The function tang x, for example, adheres from $x = 0$ to $x = 90° - \omega$ to the law of continuity, ω being as small as one may wish, for $x=90°$, however, continuity is interrupted (Klügel 1823, 550).

Klügel's entry "Gränze" [limit] within the *Wörterbuch*, too, continues to show a strong confinement to geometrical ties. He defines the concept exclusively verbally, without algebraization, and without separate signs:

> *Gränze [limit]* of a quantity is that quantity which the latter, considered as a variable one, may approach ever more closely in such a way that the difference may become smaller than any quantitiy, however small it may be assumed (Klügel, vol. II, 1805, 646).

As examples, he quotes primarily geometrical ones: the circle as limit of inscribed or circumscribed polygons, the cylinder as limit of prisms, etc. Only after that does he name the sum of an infinite series as limit of the arithmetic sum of the series' terms. Johann August Grunert, Klügel's second successor as editor of the *Mathematisches Wörterbuch*, continued this geometrical tradition in the subentry *Gränzen der Verhältnisse* [limits of ratios] of the entry *Verhältnisse* in 1831 still; "variable ratios" and their limits are presented and discussed exclusively in proportional notation (Klügel, vol. 5, 1831).

Quite frequently, the contemporary literature mentions a textbook on analysis for artillery engineers authored by George Friedrich Tempelhoff (1737–1807), a Prussian lieutenant of artillery: *Anfangsgründe der Analysis des Unendlichen* of 1770. It begins with a relatively naïve part in which Tempelhoff bases his textbook on the concepts of infinitely small and infinitely large quantities—which he conceived to be reciprocal quantities—as well as on their various orders:

> I am basing the theory on the proposition that $\frac{1}{0}$ means and *infinite*, and $\frac{0}{1}$ an *infinitely small* quantity (Tempelhoff 1770, iv),

claiming to be able of rigorously proving this, while he actually only calculated some examples as illustration (ibid., 123 f.). In subsequent parts, however, he not only relied on the concept of function, using it operatively, but also introduced and applied the differential quotient as *Gränze der Verhältnisse* [limit of the ratios], for the purpose of thus "developing the true principles of differential calculus" and escaping the difficulties allegedly associated with the concept of the infinitely small quantities (ibid., v f.):

> One sees indeed from the solution [of the previous problem] that one must not interpret *dy, dx* as *quantities*, but rather as *mere signs* serving *to indicate a certain operation* (ibid., 240),

that is, the operation of finding the limit of the ratio of differences of Δy to Δx (cf. ibid., 259). Tempelhoff's textbook was the first attempt at realizing an algebraization of the fundamental concepts of analysis in textbooks for a public of engineers.

A reception of the works of Boscovich on the one hand, and of the German authors Hausen, Kästner, and Karsten on the other, concerning the concept of continuity can be found neither in England nor in France. What d'Alembert considered generally relevant in the concept of continuity can be seen from his entry *Continu* in the *Encyclopédie*. For mathematics, he noted only that it considered discrete and continuous quantities, and that continuous quantities were given by extension, of lines, areas, solids, and that they form the object of geometry. A definition of continuity, by contrast, is found only in the subentry on physics, and d'Alembert had borrowed the latter from Formey.[53] The latter again defines continuity in Aristotle's (and Hausen's) sense:

> Thus shall we call that which has parts ranged one after the other, such that it is impossible to insert other parts between them in another order; and generally *continuity* is understood wherever nothing can be placed between two parts (Encyclopédie t. IV, 1754, 115).

Besides the adjective *continu*, d'Alembert also included the noun *continuité*. He made a point of saying that this entry belonged only to physics; this may be the reason why the later excerpt of the mathematical entries from the *Encyclopédie*, the *Encyclopédie Méthodique*, contains no entry on continuity at all! For the entry *continuité*, d'Alembert did not write a text of his own, but adopted a foreign one from E. Chamber's encyclopedia of 1728. The text reports on discussions about continuity among the Scholastics, and is thus based on the concept of the connectedness of neighboring parts (ibid., 116). Finally, d'Alembert did insert a separate entry on the law of continuity, printing, with regard to this *loi de continuité*, a text by Formey (ibid., 116 f.). The law of continuity was presented as a principle founded by Leibniz, according to which nothing in nature occurred *par saut* [by leaps], but that all the intermediate stages had to be passed. Formey quoted Leibniz's deduction of this principle from the axiom of sufficient reason. He then went on to explain that the principle was most strictly observed in mathematics: "Cette loi s'observe dans la Géométrie avec une extrême exactitude." It was not violated by the points of inflection of curves, which seemed to contradict the law—this again refers to differentiability—since the points of inflection represented infinitely small knots. The principle applied also to physics: "La même chose arrive dans la nature." Everything was dominated by *gradation*, and all transitions went by *nuances*. Formey presented the arguments of the Leibnizians against Descartes with regard to the impact of hard bodies in a more or less neutral tone, but on the whole his entry, cosigned by d'Alembert, did not contain any argument against the *loi de continuité*. He even closed his presentation by assuring "[la] *loi de continuité* que la nature ne viole jamais" (ibid., 117).

As we had to state already in our part on the development of negative numbers, Euler's contributions to the fundamental concepts of analysis were rapidly and profoundly received, but remained of limited effect for a long time. Euler's concept of function was not accepted as a basis, nor was his introduction of the distinction between "continuous" and "discontinuous" taken up and deepened. In our introductory section in Chapter II.1.5., we had already quoted

---

[53]   Jean-Henri-Samuel Formey (1711–1797), author and philosopher, worked in Berlin, first as a teacher at the Collège Français, and after 1744 as secretary of the philosophical class of the Berlin Academy. He became prominent, among other things, with works on Wolff's philosophy, and in defense of Leibniz's theory of monads against Euler.

Euler's introduction of the term "discontinuous" in the second volume of his *Introductio in Analysin Infinitorum* (1748). In contrast to views before him, which were all determined by Aristotle's notion of the composition of parts, Euler relied exclusively on internal properties of curves (Euler 1945, 11).

It must be noted, however, that Euler himself did not assign much importance to the distinction between continuous and discontinuous, since he went on to state in the next paragraph, "in geometry we are especially concerned with continuous curves" (Euler 1988, vol. 2, 6),[54] all essential curves lending themselves to mechanical description and being expressible by a single function. Since the irregular or discontinuous curves were continuous merely in segments, and the single–valued functions had already been assumed to be continuous, his contemporaries were unable to recognize an essential deviation from the traditional construct of the dominance of geometrical, continuous curves.

Debates beyond that on the concept of continuity were carried on in academy treatises, and remained confined to a small circle of scholars without changing the generally shared notions about foundations. The best-known and most heated debate of that kind concerned a problem from mathematical physics: the problem of the vibrating string (cf. Youschkevitch 1976, 64 ff. and Truesdell 1960, 237 ff.). The principal opponents in this debate were again d'Alembert and Euler, who remained just as basically irreconcilable in this controversy as well. The problem of finding the equation for a vibrating string firmly fixed at its two end points was about solving the partial differential equation

$$\frac{\partial^2 y}{\partial x^2} = \frac{1}{c^2} \frac{\partial^2 y}{\partial t^2} \ .$$

D'Alembert gave as solution of the differential equation, in 1743 already,

$$y = f(ct + x) + g(ct - x),$$

for the case in which c is a constant function.

The controversy now was about which functions f and g were possible, considering the initial conditions. While Euler declared perfectly arbitrary functions to be admissible, in particular "discontinuous" functions, which, for him, were functions with jump discontinuities, d'Alembert admitted only "continuous" functions, and this practically meant that the functions had to be at least twice differentiable.

Many eminent mathematicians supplied further research treatises on this problem: Daniel Bernoulli (1753), Lagrange (1759), Condorcet (1771), Laplace (1779). An important systematization of these works, above all with regard to the concept of continuity, was achieved by L.F.A. Arbogast (1759–1803)[55]

---

[54]   "De curvis autem continuis in geometria potissimum est sermo" (Euler 1945, 12).

[55]   Arbogast's biography has not been exactly researched as yet. To begin with, about 1787, he was *professeur des mathématiques* at the *Collège* of Colmar, later at the *École d'artillerie* in Strasbourg. He then changed over to the University

1791 in a treatise written as answer to a contest question of the St. Petersburg Academy, an achievement that has been briefly pointed out in Chapter II.1.5. already. Besides his own concept of "continuous" and discontinuous (for a curve with corners, resp. for a function composed of several analytic expressions), Euler had introduced the concept of *discontinue*, [non-coherent]. According to his illustrative figures, these were functions with jump discontinuities. Arbogast made a point of deriving the introduction of this third term from Euler's comment on what the latter understood the "perfectly arbitrary" solution functions *f* and *g* for the partial differential equation of second order to be: separate coherent curve segments that were not interrupted by any jump discontinuity:

> Curvae ex portionibus variis diversarum curvarum utcumque conflatae, atque adeo curvae libero manu ductu utcumque formatae, hic locum inveniunt; dummodo omnes partes inter se cohaerent et nusquam hiatu abrumpantur (Euler 1763, quoted after Arbogast 1791, 70).

Arbogast, in contrast, systematically examined the solutions of ordinary and partial differential equations, showing that jump functions are possible as solutions, too. Arbogast named curves containing jump discontinuities *discontigue*, non-coherent, and their functions accordingly to be *discontigue*. Actually, Arbogast uses "continuous"—just like Kästner and Karsten—to mean "differentiable." Arbogast's merit was in particular in how he explicated the concept of continuity. He did so, on the one hand, by connecting it to the axiom of the intermediate value, as we know it already from the German discussion within mechanics, and on the other hand by defining continuity—for the first time since Boscovich—for functions, and not only for variables:

> The *law of continuity* consists in [stating] that a quantity cannot pass from one state to another except having to pass through all the intermediary states subject to the same law. Algebraic functions are regarded as continuous, since the different values of these functions depend in the same way on the variable; and in supposing that the variable increases continually, the function will receive corresponding variations; but it will not pass from one value to another without also passing through all the intermediary values (Arbogast 1791, 9).

The connections between the dependent and the independent variables of the function are very explicitly formulated here, albeit in a still completely verbal manner, and not yet algebraized with inequalities and signs proper. Grattan-Guinness's reference to Arbogast's closing statement (ibid., 96) saying that arbitrary functions in general need not be continuous, nor coherent (Grattan-Guinness 1980, 104), however, suggests an incorrect scope of the concept: actually, Arbogast's statement refers to the three types of functions he was able to imagine: *continu*, *contigue* and *discontigue*—in our understanding of the

---

of Strasbourg , until the latter's dissolution, and was after his political activity in Paris teacher at the *École Centrale* in Strasbourg (DSB, vol. 1). While Friedelmeyer's doctoral thesis contained a chapter on his biography, this has unforunately not been included in the printed version (Friedelmeyer 1994).

concepts hence to "differentiable," "continuous," and "piecewise continuous"—
and does not refer to a general notion of discontinuity.

What was the impact of these reflections? The debates on the problems of the
vibrating string remained confined to academy treatises, and did not reach the
textbook presentations. Arbogast himself made his differentiation of the concept
of function one of the starting points of his own exhaustive treatise of 1789 about
the foundations of differential and integral calculus. The treatise remained in
manuscript status, and was not yet fit for print (it contains many work notes),
but it may be assumed that Arbogast used the draft in his mathematics lectures.
The following excerpt shows how he applied his own concept of *discontigue*—
as the most general form of functions: to determining the differentials of functions
as well, as piecewise continuity and as piecewise differentiability:

> § 25. We derive from this and from what precedes it this fundamental rule of the
> differential calculus: *Given any function of x, it can be regarded as being
> composed of many parts, thus on taking separately the differentials with
> respect to each of these parts, as if the others were constant, and re-combining
> these particular differentials, their sum will be the differential of the proposed
> function* (Arbogast 1789, fol. 22v.–23; original: underlines instead of italics).[56]

Arbogast's works could not take immediate effect in France, since the sudden
dominance of the algebraization of analysis in the first years after the French
Revolution did not take note of these approaches. The reflections on continuity
were carried on, after this period of exaltation, in two surprising and hitherto
unobserved instances. It was Prony who in teaching his mechanics at the *École
polytechnique* received Kästner's reflections on intermediate values, thus
introducing the concept of continuity into the instructional context of the famous
École, and on the other hand, one of Kästner's immediate disciples, J.G.
Tralles, was the first to elaborate the epsilon–delta concept of the continuity of
functions in Berlin, also in the new nineteenth century. The two instances of
development will not be treated here, but in the following Chapters examining
the turn at the *École polytechnique*.

10. A Survey

Whereas it was the emergence and extension of algebra that served to give
impulses for developing the concept field of negative numbers, it was primarily
mechanics that promoted the mathematization of limit processes for differential
and integral calculus. And while it was for the negative numbers the concept of
quantity whose epistemological confinement impeded the generalization of

---

[56]   Friedelmeyer, in his study on Arbogast, did not include the author's
reflections on continuity, since he focused on the latter's later elaboration of the
calculus of derivations in which continuity – as in all formally algebraizing concepts
– no longer has any role (Friedelmeyer 1994).

operations, it was the concept of the infinite for analysis that raised problems of mathematization. An additional factor was the dominant philosophical belief that all processes in nature, and in particular in mechanics, were continuous from principle. A mathematization of continuity in the sense of an analytically applicable operative concept was hence not in view for a longer time.

Here again, we were able to establish different national cultures, at first only in two countries: in France and in England. While in England, subsequent to Newton, the mathematical concepts were closely coupled to mechanics, there existed in France—on the basis of an only partially legitimized appeal to Leibniz—a more strongly algebraizing mathematization of limit processes. In Germany, an independent development set in only with the second half of the eighteenth century, remarkably concurrent with the first empirical doubts in mechanics concerning the general validity of the principle of continuity.

The work to make continuity more precise mathematically was thus begun only at a late point. One of the essential preliminary steps for this was to elaborate the concept of limit. This elaboration proved to be tied to the respective conception of the number concepts, and in particular of the status of the zero. This is where the classical conception of infinitely small quantities originated.

While it had already been true in the case of negative numbers that a considerable amount of productive contributions was achieved by persons hitherto considered to have been rather marginal in the historiography of mathematics, it can also be established for the concept of limit that innovations came from the "countryside" (from Martin in Toulouse) or from the "periphery" (from Stockler in Portugal).

Whereas for negative numbers at the end of the eighteenth century, dissolving the "Gordian knot" by algebraizing the concept of number was still below the horizon, something analogous to that appeared to have happened already for analysis—in the wake of an incisive political event: of the French Revolution. Lagrange's program served to cut the ties to mechanics, to establish the independence of analysis, and to eliminate the concept of the infinite. A further conceptual specification of limit processes—be it via the limit method or by the method of *infiniment petits*—was no longer deemed necessary. The development into series seemed to be tractable without problem in a purely algebraic mode, just as results of operating with finite differences seemed to be transferable to differential calculus.

The rollback after the effervescent optimism occurring precisely at its core institution—at the *École polytechnique*—in favor of a renewed dominance of synthetic methods justified by the requirements of mechanics, and the struggle between the more algebraic limit approaches and the more synthetic–intuitive *infiniment petits*, leading to the full elaboration of the classical conception of infinitely small quantities, will be presented in Chapters IV, V, and VI.

Chapter IV

Culmination of Algebraization and Retour du Refoulé

1. The Number Field: Additional Approaches Toward Algebraization in Europe and Countercurrents

We have now analyzed how the concept of negative numbers in three European countries especially important for mathematics developed until about 1800, and we have seen that the concepts in these countries developed not only in different ways, but also within these countries mostly not cumulatively, but with discontinuities. Furthermore, it becomes clear that there was no homogeneous or uniform mathematical community, but one that was clearly defined by culture and nationality.

Additional limitations of mathematical communication become clear in three approaches to a further algebraization of the concept of negative numbers, which were, however, only insignificantly recognized in their time, although they could have exerted an important impact if there had been a common structure of communication. In fact, all three remained with their approaches on the margins of the mathematical communities, though in different ways. This holds true even for the most famous author among the three, Leonhart Euler, with his textbook on algebra. But it also applies to Condillac and to A.-Q. Buée.

1.1. Euler: The Basis of Mathematics Is Numbers, not Quantities

Leonhard Euler (1707–1783) wrote several mathematically and didactically excellent textbooks, although after finishing his studies (in Basel) he worked exclusively at academies (in St. Petersburg and Berlin) and never at a university. While his textbooks on algebraic analysis, and on differential and integral calculus were recognized by scholars relatively quickly, his textbook on algebra did not have a large positive impact on science for quite a long time, although it was translated into several languages (cf. Cajori 1908, 75) soon after its publication in 1770. Because of his algebraization of negative numbers his work continued to be strongly rejected over and over again.[1]

Euler was the first to consistently present algebra as a science of numbers and to conceptually separate numbers clearly from quantities in algebra. Right in the introduction Euler stated that "all magnitudes may be expressed by numbers." A clear concept of numbers and their operations, therefore, is the basis of mathematics as a whole:

> The foundation of all the Mathematical Sciences must be laid in a complete treatise on the science of numbers, and in an accurate examination of the different possible methods of calculations. This fundamental part of algebra is called Analysis, or Algebra (Euler 1984, 2).

Euler put the emphasis explicitly on that *numbers* are the exclusive subject of algebra: "In Algebra, then, we consider only numbers, which represent quantities, without regarding the different kinds of quantity" (ibid.).

Only "in the remaining parts of mathematics" is the application of the science of numbers to the respective quantities to be discussed. Euler was also the first to consistently stress that whenever letters are used to represent numbers generally, the letters stand in for the numbers, so that he, as had been customary before, did not speak of "quantities" whenever letters were used in algebra:

> It will not be attended with any more difficulty if, in order to generalise these operations, we make use of letters instead of real numbers. It is evident, for example, that $a - b - c + d - e$ means that we have numbers expressed by a and d and that from these numbers, or from their sum, we must subtract the numbers expressed by $b, c, e,$ which have before them the sign $-$ (ibid., 4).

Because Euler did not, however, describe explicitly and in detail the relation of the numbers to the quantities, especially in the applications, it seems that this innovation was not recognized in the eighteenth century. Right at the beginning, in the introduction of the elementary operations, Euler also explained negative numbers, after the explanation of addition and the minus sign. Probably as an adjustment to the usual way of speech, Euler defined "numbers with their preceding signs" as "simple quantities," and was thus able to define, quite consequently, "positive quantities" as numbers "before which the sign + is found," and "negative quantities" as those "which are affected by the sign –" (ibid., 19).

Euler's approach of using "quantities" here for specific numbers may have been too subtle to be generally understood then. For illustration only, Euler cited assets and debts as examples of positive and negative quantities. Euler subsequently stated without any reservations that positive numbers were greater

---

[1]   See above for the criticism by W. Frend (Chapter II.2.10.3.) and by Karsten (1786, 240).

and negative numbers less than nothing; later this was to provoke sharp reactions in France:

> In the same manner, therefore, as positive numbers are incontestably greater than nothing, negative numbers are less than nothing. Now, we obtain positive numbers by adding 1 to 0, that is to say, 1 to nothing; and by continuing always to increase thus from unity. This is the origin of the series of numbers called *natural numbers*; the following being the leading terms of this series:
>
> $$0, +1, +2, +3, +4, +5, +6, +7, +8, +9, +10,$$
>
> and so on to infinity.
>
> But if, instead of continuing this series by successive additions, we continued it in the opposite direction, by perpetually subtracting unity, we should have the following series of negative numbers:
>
> $$0, -1, -2, -3, -4, -5, -6, -7, -8, -9, -10,$$
>
> and so on to infinity (ibid., 5).

Although Euler says that this terminology is known, so far it had not been expressed so clearly:

> All these numbers, whether positive or negative, have the known appellation whole numbers, or *integers*, which consequently are either greater or less than nothing. We call them *integers* to distinguish them from fractions, and from several other kinds of numbers of which we shall hereafter speak (ibid.).

At the same time, Euler stressed that "a precise idea [...] of those negative quantities" was "of the utmost importance through the whole of Algebra" (ibid., 5). A first application arose in multiplication. Euler defined it as iterated addition and clearly limited it to numbers (ibid., 6 ff., 79). Euler, however, explained the rule of signs only summarily: if one multiplies a negative number with a positive multiplier, one "apparently" gets the respectively multiplied negative number. The inverse case of a positive number and a negative multiplier is settled with a short "just as"—as the preceding case (Euler 1940, 21). Euler spent a few more words on the case minus times minus, e.g., $-a$ times $-b$, but he was content with a plausibility argument. First Euler stated that it was clear that the product had to be ab, taken absolutely. Since the product of $-a$ and b is already known as $-ab$, the opposite, therefore, "must produce a contrary result," i.e., $+ab$ (Euler 1984, 8).

Also in the second part of his textbook on algebra, on solving equations, Euler consistently kept quantities and numbers separate. He defined the subject of solving equations as follows:

> All those questions were reduced to finding, by the aid of some given numbers, a new number, which should have a certain connection with them (ibid., 186).

Using the coefficient of the unknown number in equations, Euler, however, apparently felt to be compelled to make a concession to dominant epistemological ideas of positive/ absolute numbers. On the one hand, he assumed that the coefficients could take on any values; on the other, he put the $\pm$ sign on the coefficients in their normal form; he had only one normal form:

> The following general formula represents all equations of the second degree:

$$ax^2 \pm bx \pm c = 0,$$

in which the sign $\pm$ means that such terms can be positive or negative (ibid., 217; Euler 1940, 248).[2]

The concession becomes even more obvious later when he explains the normal form with the highest term without a coefficient:

> By these means our equation will assume the form of $x^2 \pm px = \pm q$, in which p and q represent known numbers, positive or negative (ibid., 255).

Euler also presented cubic equations with one normal form with apparently only positive coefficients:

$$ax^3 \pm bx^2 \pm cx \pm d = 0 \quad \text{(ibid., 253).}$$

It was only the biquadratic equations that he established in a truly general normal form:

$$x^4 + ax^3 + bx^2 + cx + d = 0 \quad \text{(ibid., 272).}$$

1.2. Condillac: Genetic Reconstruction of the Extension of the Number Field

The second contribution is by Etienne Bonnot de Condillac (1715–1780). It is highly characteristic that Condillac, the philosopher of the enlightenment, who with his program of the *analytic method*—understood by him in a novel way—influenced so decisively the concepts of education and science in the first phases of the French Revolution, at the same time also wrote a work on the foundations of mathematics, which was to transform his epistemological program into a revision of the basic concepts of mathematics. Unfortunately, Condillac was not able to finish this work, *La Langue des Calculs*. In 1798 the unfinished manuscript was published—practically at the end of those phases of the Revolution, in which the analytical method was dominant. Afterwards his reception was largely negative—almost simultaneously after Condillac's philosophy was held responsible for the Revolution during the restoration and his philosophy was virtually banned. Most of the time only a popular prejudice remained, with which Condillac was labeled a sensualist and epigone of English philosophers.

His major work, *L'essai sur l'origine des connaissances humaines* (1746–1754), is, in fact, a discussion about the increasing abstraction of human knowledge. Especially the abstract science of mathematics was the central methodological key function for Condillac. He assumed the necessary superiority of the analytic method—represented by algebra. In Condillac's reflections on the relation of thought and language, the key requirement for exact and methodical thinking

---

[2]   The English translation in Euler 1984 differed too much here ("may be sometimes positive, and sometimes negative").

was a methodically organized language—"une langue bien faite." Mathematics especially was of an exemplary character for Condillac's program of a *langue bien faite*. The principles of *raisonnement* had to be looked for in the language of mathematics in order to make all sciences just as exact. And within mathematics in turn, algebra represents the most methodical of all languages. Condillac's program, therefore, was even more radical and more comprehensive than the program that Prestet had formulated earlier, which contained a first apotheosis of the analytic method.

It is also very characteristic that Condillac was—as far as I know—the only author in France in the second half of the eighteenth century who opposed d'Alembert's criticism of the concepts of negative quantities. Indeed, a scientifically interesting point of the *Langue des Calculs* is in addition that Condillac reconstructed in it the development of science as a sequence of different phases with an increasing abstraction of concepts, whose single phases cannot be reduced to one another.[3]

In the *Langue des Calculs*—written in about 1778 (Auroux/Chouillet 1981, xxxvii f.)—Condillac sets the following point of departure: "Mathematics is a science that is handled well and its language is algebra." And, "Algebra is a well constructed language, and the only one that is so" (Condillac 1981, 6 f.). On the other hand, he stated that this language of mathematics does not yet have a grammar and that it is the task of philosophy to develop such a grammar: "this language [algebra] still lacks its grammar, and [...] only metaphysics can provide it with a grammar" (ibid., 47 f.).

The first signs of a language are chosen by convention; further signs are created from them by analogy. It is a sign of a good "grammar" to use such analogies that the process of thinking is helped and supported by the development of the signs and that it does not lead to *obscurités* (ibid., 2 ff.).[4] The discussion of appropriate analogies is, therefore, a major issue in the *Langue des Calculs* (ibid., 37); and indeed, the analogies worked out by Condillac of the different phases in the development of the general concept of number can be understood as a first formulation of the principle of permanence, which later was founded by H. Hankel.

Condillac's presentation of how the concept of number developed was not built on historical or didactic studies, it was rather a "rational" reconstruction. A major methodological progress Condillac made is that he analyzed the transition from the concept of quantity to the concept of number as the decisive step of abstraction.

---

3   The only German edition of this work in hand shows how little this concept has been understood even in the present: Georg Klaus, a Marxist philosopher, who translated it in 1959, published only the first of its two parts because he thought that the second part, in which the transition to calculating with letters is described, is only an uninteresting treatment of Euler's algebra (Condillac 1959, LXXVI).

4   For the concept of analogy in Condillac compare Auroux 1981, 11 f.

Condillac reconstructed the genesis of how the concept of number developed as a sequence of four phases:

1. The first phase is completely defined as empirical; in it, the first counting is realized as a calculation with *fingers*. One calculates with fingers to represent a series of units and quantities for oneself. From this first empirical calculation the four elementary operations developed; thus subtraction developed from the inversion of addition: "l'opération qui defait ce que l'addition a fait, est ce qu'on nomme soustraction" (ibid., 14). In this first sense, subtraction has only a limited range.

2. The second phase is marked by the transition to *names*. It would be difficult to make calculations with larger units with the help of fingers only. Using signs, therefore, opens up new ways for the calculation. It is easier to calculate with names instead of fingers; ranges of quantities are grasped, which had been out of reach before. The names for the units are originally given by the ten fingers; on the basis of ten, bigger names can be constructed by analogy. At this level, analogy itself still has an empirical character.

The transition from fingers to a first form of signs for names already represents a necessary condition for the development of algebra, since this transition will eventually lead to the transition from quantities to abstract numbers. Condillac explains this transition in great detail as deontologizing, as abstraction of every empirical meaning:

> Now, since we have become used to using our fingers to represent a sequence of numbers, whether increasing or decreasing, we can represent this same sequence in any other way, by pebbles, by trees, by people, etc., etc.; that is, we can enumerate and denumerate pebbles, trees, people, etc., just as we do with fingers (ibid., 47).

> These ideas that we explained using fingers we can therefore apply by analogy to pebbles, to trees, and to people; and because we can *apply them to all objects in the universe, we say that they are general, that is, applicable to everything.* But when we are content to consider them applicable to everything, we do *not apply them to any object in particular*, we consider them as beings in themselves, and we separate them from all the objects to which they could be applied (ibid., 48 f.; my emphasis, G.S.)

The development of abstract ideas goes back to this separation of the empirical objects:

> Considering numbers in a general way, or as applicable to all the objects in the universe, is [...] the same thing as abstracting or separating them from these objects, to consider them separately; and so we say that the general ideas of numbers are abstract ideas.

> But when these ideas of numbers, firstly perceived in fingers, then in objects to which they are applied, become general or abstract, and we no longer perceive them in fingers nor in the objects to which we have ceased to apply them, where then do we perceive them? *in the names which have become the signs of the numbers* (ibid., 49; my emphasis, G.S.).[5]

---

[5]   Condillac's philosophical position is not, therefore, nominalistic, as Auroux thinks (Auroux 1981, 17), but one of a first theorist of signs.

3. The third phase consists in the development of the signs. Operating with big names for numbers—as Condillac has used them so far—makes a transition to simpler signs necessary. The invention of such signs is prepared by the usage of *cailloux*, of tokens (as calculational devices). They form different signs that are among themselves, however, uniform, to shorten calculating with fingers and with big number–names. According to Condillac, the word *cailloux* is actually the origin of the word *calcul*. This phase finally leads to the invention of numbers and thus to the origin of actual arithmetic.

Condillac differentiates explicitly between operations with quantities and operations with numbers: both follow different analogies (ibid., 223).

4. The final phase eventually consists in the establishment of algebra, of the transition to operating with *quantités littérales*. For Condillac, the transition from numbers to letters means the definitive step for the development of the concept of abstract numbers. It is central for Condillac that this theoretical concept requires a redefinition of the operations; he says that the character of the new concept calls for a revision of the grammar:

> This dialect [algebra] has rules which one must get to know, and a new grammar that must be learned. It is a matter of discovering the use of these general terms, their different meanings and their syntax (ibid., 275).

It is just in this context indeed that Condillac introduces the term negative numbers, when he extends the former meaning of subtraction. He redefines it as an extension of addition:

> A letter preceded by the sign + indicates an added quantity, and addition, and I call it a plus quantity (*quantité en plus*): when it is preceded by the sign –, I call it a minus quantity (*quantité en moins*), since it is a subtracted quantity, a subtraction (ibid., 277).

Condillac was very explicit about this aspect of extension:

> To the quantity a one wishes to add the quantity b? Write $a + b$, and if one wishes to add $-b$, write $a - b$. Note that this latter operation is properly, by the result it gives, a subtraction: but by extension we retain the name addition (ibid., 278).

Condillac did not try to reduce the theoretical terms to empirical ones. On the contrary, he insisted on the new character of these terms and the operations with them. He established a neat distinction between operations with ab solute quantities and with abstract numbers:

> *Adding a subtraction is to subtract, and subtracting a subtraction is to add*: why are we forced to use this contradictory language? It is because we speak two languages, which are too different to be always analogous to each other. [...] Thus, properly speaking, there are no minus quantities in ordinary languages, nor even in arithmetic. But in algebra, where the signs are indeterminate, one would not know how to pronounce the difference: one can say only that either $a - b$ or $b - a$ gives the unique answer to some one who asks what is between a and b (ibid., 294 f.).

The clarity with which the philosopher Condillac—in contrast to the French writers of textbooks—elaborates the epistemological differences of arithmetic and algebra is remarkable:

> There is no contradiction except in the words sum and remainder, which are not words of algebra: but what is an addition in algebra is called subtraction in arithmetic [...] when these two dialects are used together it is then not possible to avoid contradictory expressions (ibid., 295 f.).

It is, however, inevitable to connect the two dialects, since the *langue du calcul* is formed out of the calculation with numbers and the one with letters. It is difficult, therefore, to "speak algebra" already at the beginning, since every transfer of a *quantité en moins* to arithmetic or to every-day language would lead to an inconsistent language. Condillac uses the instructive picture of a learner of a foreign language to solve the problem: the capacity to *parler algèbre en français* would have to be acquired (ibid., 296 f.).

Condillac's main emphasis on the analysis of operations—corresponding to his program of elaborating a coherent "grammar" and his concept according to which negative remainders can only be "indicated," but cannot be shown in real empirical termini, underline Condillac's already modern operationalistic approach. Condillac, however, did not draw the logical conclusion from this approach, as did Martin Ohm later (see below, Chapter VII.3.), to accept only absolute numbers as real terms and to understand all other kinds of numbers only as "indicated" operations; he rather gave the results of operations the equal status of mathematical objects:

> When I say that quantities are added or subtracted, and then I make a distinction between plus quantities and minus quantities, I am not confusing them with the operation that adds them or subtracts them (ibid., 298).

Condillac criticized and rejected, therefore, the identification of negative quantities with subtractive quantities:

> When the addition of a quantity is called a *positive quantity*, and the subtraction of a quantity is called a *negative quantity*, there is a confusion between the expression for the quantities and the expression for the operation which adds or subtracts them, and such language does not throw light [on the matter] (ibid., 298 f.).

Condillac explained the rule of signs analogously to Prestet. I quote his argumentation of the case minus times minus, which he discusses using the example $-2ax-4a$. He separates the case according to factors, declares $axa = aa = a^2$ as known, and turns to $-2x-4$, in which $+2x+4 = 8$ is known.

> Thus to multiply -2 by -4 will be to make the subtraction minus -2 as many times as there are units in 4: but making a minus subtraction is to take away, and taking away a subtraction is to add. Since, in this multiplication, -2 and -4 are subtracted subtractions, it is evident that -2 changes into $+2$ and -4 changes into $+4$. It is not therefore surprising that the product of -2 and -4 is $+8$, just like that of $+2$ and $+4$ (ibid., 286).

As already mentioned, Condillac rejected d'Alembert's criticism of the concept of negative numbers. He cited for this both d'Alembert's statement from the *Eléments de Philosophie*, according to which the theory of the negative numbers was not yet *parfaitement éclairci* (cf. Chapter II.2.9.3), and the further part with the criticism of the previous conceptions (cf. ibid.).

Condillac commented on this quotation by saying that he does not regard these points of criticism as "decisive" as d'Alembert (Condillac 1981, 300).[6] Condillac saw a part of the problems discussed by d'Alembert in the selection of misleading starting terms. If you begin badly, you cannot well explain what you actually mean. Condillac explained at this point why he had chosen the terms *quantité en plus*, and *quantité en moins*. For the traditional terms, positive and negative quantities, had the connotation that some are not quantities whereas only the others were true quantities—and this was not, however, what one wanted to say:

> The designation, for example, of negative quantities as opposed to positive quantities seems to imply that there are quantities which are not quantities, and quantities which really are quantities. Since this stated absurdity is not what one wishes to say, we do not hear what has not been said (ibid.).

Condillac intended to discuss the theory of negative numbers in more detail, that is, while treating equations of second degree, which is where it becomes most visible, which conception of negative numbers the respective author was defending: "we shall achieve to clarify this theory when we deal with equations of the second degree" (ibid.).

The meaning of this last sentence had never been clear in the previous editions of *Langue des Calculs* since the first publisher, Laromiguière, had exchanged the word "équations" with "opérations," which does not make any sense. Only since the 1981 critical edition by S. Auroux and A.-M. Chouillet has the original text been made accessible. Condillac did not mention a conception of oppositeness.

1.3. Buée: Application of Algebra as Language

In Chapter V, we shall discuss the—mostly negative—reception of Condillac's manuscript, published in 1798, in the context of the general break with the analytic method. The only positive—though indirect—reception of Condillac's conception of mathematics comes from an author who not only cannot be placed into a certain national context—like Euler— but who also was—like Condillac—outside the usual institutional structures: Adrien-Quentin Buée (1748–1826).[7] Buée, an amateur, remained also outside the *scientific community*; born in Paris, he was ordained as a priest and worked as the organist in the church of S. Martin de Tours. Since the beginning of the Revolution he published religious–political tracts. Because he did not want to take the oath, required of clerics, to the new constitution, he emigrated on August 10, 1792—

---

[6]   In his manuscript Condillac had written "tranchante;" the editor Laromiguière replaced this word with "décisive" (Condillac 1981, 300 and 504).
[7]   Although the work came out after 1800 and already represents a reply to Carnot, it is discussed here because it coninues Condillac's conception.

the day of the storming of the Tuileries and the dismissal of Louis XVI—and lived in exile in England, from where he returned to Paris only in 1813 (Dict. Biogr. Franç. VII, 615).

Besides his religious–political writing, there is, except for a work on crystallography, only one publication on mathematics: *Mémoire sur les Quantités imaginaires*. Not only was this work of an outsider presented by a member of the Royal Society, William Morgan, on June 20, 1805, and published in their *Philosophical Transactions* in 1806, but it also contains, for the first time—if one does not take into account the 1797 treatise by Caspar Wessel, which was not discussed by his contemporaries—the solution to the problem of the graphical representation of imaginary numbers.

Buée's fundamental achievement regarding the concept of negative numbers is that he consequently thought through the distinction of a quantitative and qualitative aspect of numbers, first introduced by Fontenelle, and clarified it conceptually. Buée was the first to explicitly establish the idea that the signs plus and minus have two different meanings—as an operational sign and as an algebraic sign—and to further interpret these meanings. He defined the first meaning as an arithmetical operation that—when applied to lines—defines *length* (the numeric aspect in Fontenelle), whereas he characterized the meaning as an algebraic sign as a geometrical operation that defines *direction* (in Fontenelle it is the quality of the specific being).

Concerning conceptual development, it had an especially far–reaching effect that Buée did not separate the two meanings—as had been done in the discussion of the two aspects in France so far, since Fontenelle and d'Alembert—but unified them on a higher plane: "When, therefore, these two operations are reunited, what is carried out is in reality an arithmetico–geometrical operation" (ibid.).

This enabled Buée to solve a central problem that Carnot had raised (see Chapter V.1.6.). Carnot, like Frend, took into account only the first, arithmetical, meaning of the signs + and –. In this case you are on the first level of algebra, which Buée called universal arithmetic.

> In ordinary algebra, that is, in algebra considered as universal arithmetic, where an abstraction is made of all types of quality, the signs + and – can have only the first of these meanings (ibid., 24).

But if not only the arithmetical, but also the geometrical meaning is required, e.g., in operations with isolated negative quantities, then you operate on a higher level of algebra, which Buée called—clearly referring to Condillac—the "mathematical language":

> By consequence, whenever the result of an operation is preceded by the sign –, for this result to have a meaning one must consider here some quality. Therefore algebra ought not any longer be regarded simply as universal arithmetic, but as a mathematical language (ibid., 25).

With this definition, Buée rejected d'Alembert's criticism that it is not allowed to regard quantities as less than nothing: not that the quantity is less than zero, but in the sense of the second meaning, the quality is inferior to *nullité* (ibid.).

Buée rejected the convention, common in eighteenth-century France, to understand negative quantities as being characterized by a minus sign, and rather took a minus sign as a sign of oppositeness. If, therefore, $+t$ means a past time, then $-t$ means a future time of the same amount (ibid., 24). Buée was thus able to understand the two meanings of the signs plus and minus more generally as the expression of two different types of operations, whereas he characterized the first type as arithmetical:

1. Placed in front of a quantity q, [the signs + and –] are able to indicate, as I said, two opposite arithmetic operations of which that quantity is the subject.

2. In front of this same quantity, they are able to indicate two opposite qualities, having as subject the units of which that quantity is composed (ibid.).

In "ordinary algebra" only the first of these two meanings is valid, since there you abstract the quantities from each quality. This means that Buée took the arithmetical meaning as operating with absolute numbers. As far as this is concerned, Carnot's and Frend's view is correct (ibid., 24 f.).

Buée, finally, took Condillac's conception of understanding algebra as language a more concrete step further in that he characterized absolute numbers and quantities as *nouns* and the preceding signs as *adjectives* expressing their qualities:

According to the second meaning given to the signs + and –, they indicate two opposite qualities having for subject the units of which a quantity is composed, so since a quality cannot be separated from its subject, the signs + and – cannot be separated from their units. In the language of algebra, these units are substantives and the signs + and – are adjectives (ibid., 26 f.).

Through Buée, therefore, there has been a significant indirect effect of Condillac, for Buée's grammatical approach of interpreting positive and negative qualities was used by Cauchy, in the appendix of his landmark textbook on *Analyse algébrique*, as explanation for operating with positive and negative numbers (Canchy 1821, 403).[8]

In the further part of the text Bueé introduced $\sqrt{-1}$ as the result of an arithmetic–geometrical operation and presented his own solutions to problems posed by Carnot.[9]

---

8   "Les signes + et – placés devant les nombres peuvent être comparés suivant la remarque qui on a été faite [in a footnote Cauchy refers here to the *Transactions* without naming the author], à des adjectifs placés auprès de leurs substantifs."

9   In the context of the reflection on language in France, a work by an anonymous author was published in 1802, which dealt with the usage of *expressions négatives* in the French language. In this book also negative quantities in mathematics are explained as analogy to negations in language and the rules of signs are presented as rules of grammar (De l'usage 1802, 5 ff.).

1.4. Fundamentalist Countercurrents

It has already been mentioned in Chapter II.2.10.1 that the manifest contradictions in the conceptions of negative quantities especially in France led us to expect the rise of specific counterreactions. In fact, even positions that can be called fundamentalist were propagated. By these, for the time being, we shall understand those concepts that on the one hand justly address contradictions within the foundations and suggest solutions to them, but on the other hand, propagate these solutions so one-sidedly and absolutely that their supporters thought they could reject the whole former mathematics and solely develop the right mathematics. They especially rely on procedures of elementary geometry one-sidedly and represent thus an orientation toward the synthetic method. Two such positions will be presented here, which are marked by rejecting the common rule of signs and by claiming that they are able to prove that minus times minus is not plus, but minus. In contrast to the authors of the sixteenth and seventeenth centuries, who had already proposed such a notion, these texts are of a far different format and quality. These were no longer single short passages, but detailed discussions and explanations.

I became aware of the French supporter mostly because of my search for Lazare Carnot's *Nachlass*. One of the various family branches owns the castle of Presles, where Carnot lived between 1807 and 1815. The current owner of Presles, Thierry Carnot, invited me to Presles in 1987 and allowed me to examine the present materials of the *Nachlass* and to see Carnot's library there. I could especially take a look at the catalog of Carnot's library written in 1815. Among the titles listed in the catalog, I was drawn to special attention by the following entry (in the part on the softcover and hardcover books in octavo on *Sciences*): *Du calcul des négatives, par Dom Bidona*. This author could not be found, but since the catalog had been written not by Carnot, but by a different person, typos could not be excluded. A list of anonymous French authors by A.E. Barbier finally brought the solution: The complete title is *Exposition du calcul des quantités négatives*, Avignon (Besançon) 1784, in -8, and as author was mentioned dom Donat Porro.

The writer of the catalog, therefore, had transcribed a handwritten note not correctly. The book in Carnot's library turned out to be identical with a book that I had found in a bibliography mentioned also as anonymous with the title *Exposition du calcul des quantités négatives* and that had turned out to be extremely rare, which I had been finally able to examine in 1986. About the author Porro, who is otherwise unknown, we read in Hoefer's *Nouvelle Biographie Générale* that his common name was François-Daniel Porro, but he had received the cognomen *Donat* after entering the Order of St. Benedict. In the "congrégation de Saint-Vanne" he was released from all religious exercises and duties so that he could pursue his propensity for the abstract sciences. Apart from that, only the dates of birth and death are known: 1729 until January 26, 1795, both in Besançon.

In Hoefer he is called an *algébriste français*, but that is too much of an honor for this obviously self–taught person. This author also attracted Cajori's attention, but he mentions only the work of 1784, apparently without having had access to it, whereas he quotes a few main assertions from Porro's second work of 1789, also very rare, *L'Algèbre selon ses vrais principes*, which correspond to the earlier work. From his book of 1784 it can be gathered that he probably was a member of two provincial academies: Besançon and Angers, since he refers to interna of their meetings (Porro 1784, 276 and 145 f.). It is not clear why he did not want to come forth as the direct author of the book: it was published anonymously. In addition to this, he also wanted to erase his traces, since he named the then papal Avignon, which did not belong to France, as the location of printing for his first book, and London for the second, although both were actually printed in Besançon.

The fact that Carnot possessed Porro's book attributes it a particular importance, so that it should be rewarding to look for similarities in the argumentation.

Porro surely was a man of learning and knew the contemporary literature on mathematics. His approach is fundamentalist in that he wants to base the whole of mathematics on just one principle—i.e., rejecting isolated negative quantities—and in that he does not develop mathematical topics systematically and in that he sees himself as a lonely advocate in the search of the truth (*l'amour de la vérité*) against the blind clinging to the authority of scholars (*l'adhésion aveugle à l'autorité des savans*, ibid., 5). From his narrow point of view he criticized many writers of textbooks that were being used in France then; he disapproved the most of Euler's algebra und Ch. Reyneau's *Science du calcul*.[10] He did not have a good word to say about any of them; the only mathematician who is quoted with approval is d'Alembert, on the one hand, anonymously as author of the article *Équation* in the *Encyclopédie* and of the statement that the multiple solutions, that algebra produces are an *inconvénient* (ibid., 2), and on the other, as author of treatises in which logarithms are claimed not to have imaginary values (ibid., 144).

Porro exposed many weak points in the French conceptions of negative quantities. It is especially characteristic that the point of departure is a criticism of the flaws in the concept of multiplication—the insufficient definition of the multiplier is the pivotal point of his whole argumentation.

Porro, therefore, criticizes right at the beginning that the beginner learns in arithmetic that the multiplier is an abstract positive quantity, but that in algebra

---

[10]   Reyneau is the only one among the critics who is never mentioned by his name, but always only as "the author of *Science du calcul*." The reason for this could be that Reyneau in their first edition is not mentioned by his name, but appears as "l'Auteur de l'Analyse démontrée" on the title page.

he is suddenly confronted with the possibility of a negative multiplier. Porro did not, however, consider changing the concept of multiplication, but stated that because of the false assumption of negative multipliers the rule of signs had to be wrong (ibid., 3 f.).

Porro could rightly disapprove of Euler's explanation of the rule of signs, according to which it had to be $-a\mathsf{x}-b = +ab$ because it was the opposite of $-a\mathsf{x}b = -ab$ (see above). He argued that the concept of opposition could not be applied to $-a\mathsf{x}b$, but that $+a\mathsf{x}+b = +ab$ was the meaningful "opposition" counterpart to $-a\mathsf{x}-b$, so that $-ab$ is the result of this product (ibid., 64).

Furthermore, Porro conducted a remarkably precise analysis of the function of the signs. Not only did he make clear that the signs + and − have a double meaning (*double sens*) in algebra, as operational signs and as algebraic signs (*tantôt signe d'opération, tantôt signe de quantité*; ibid., 18); he also was the first to suggest that because of the ambiguity of these signs a distinction of the signs be made and that the algebraic sign be designated something different (ibid.). His suggestion, however, was not helpful. He proposed placing this sign *behind*. That is, instead of writing

$$+ -b, \text{ he suggested } + b-.$$

Porro himself did not, however, use his suggestion in compound terms; there would have again been a confusion of the algebraic–"post–sign" with the operational sign. Porro could not constructively use his symbolic distinction of concepts and signs for algebraic sign and sign of operation either because he recognized negative quantities only as subtractive quantities, but rejected pure or isolated negative quantities.

He defined positive quantities as those that increase any value, i.e., *par voie d'addition*, and negative quantities as those that decrease any value, i.e., *par voie de soustraction* (ibid., 13). He used the expression $900 - 700$ as an example: the first term shows positive quantities and the second negative (ibid., 14). He also identified negative quantities with *quantités soustraites* (ibid., 15). His rejection of negative remainders becomes clear in that he allows only subtractions with a truly positive remainder as subtraction proper, as *soustraction rigoureuse* (ibid., 24 ff.). Porro summarized his concept of negative quantities in five characteristics. Three of them seem to correspond to the usual concept:

1. Negative quantities are just as *réelles* as the positive ones.

3. As indication of oppositeness: the negative and positive quantities *se détruisent mutuellement*.

4. The same operations can be done with negative quantities as with positive ones (ibid., 14).

The other two characteristics show a differing conception. It is not that conspicuous at first:

2. The negative quantities cannot be less than zero, neither absolutely nor relatively.

It is new, however, that Porro here, for the first time, calls negative quantities *corrélatives* to the positive ones, a term that is to play an important role in

Carnot. Porro repeatedly underlined this rejection of negative quantities as a central point:

> If positive units are taken away from a positive whole, the subtracted units are called negatives because they compose a part [of the whole] (ibid., 21).

From this narrow view, Porro then deduced his conception of allowing subtraction only for homogeneous quantities and of excluding it for heterogeneous quantities. One can subtract positive quantities only from positive ones and from negative quantities only negative ones, i.e., homogeneous quantities respectively, according to him:

> Only positive parts can be taken away from a positive whole, and negative parts from a negative whole, because the parts that are subtracted must be of the same kind, and homogeneous to the whole (ibid., 22).

Positive and negative quantities, however, are heterogeneous: *d'espèce diamétralement opposée.* Porro allowed the subtraction of only homogeneous quantities as the true one, the *rigoureuse*:

$$+12 - +4 = +8 , \quad -12 - -4 = -12 + 4 = -8 .$$

He called subtractions of homogeneous quantities, which go beyond their own "totality," *soustraction par emprunt*:

$$+4 - +12 = +4 - 12 = -8,$$
$$-4 - -12 = -4 + 12 = +8.$$

He understood the improper subtraction, the *différence additionelle*, hence of heterogeneous quantities, which actually is an addition, as the difference, however,

> of $+12$ to -4, which can be expressed as $+12 + 4 = +16$,
> of -12 to $+4$, which can be expressed as $-12 - 4 = -16$ (ibid., 23 ff.).

Porro did not use these distinctions introduced by himself, but limited his usage to the "rigorous" subtraction.

The fifth point in his list of characteristics of "his" negative quantities finally leads us to Porro's specific solution to the multiplication problem:

5. Negative quantities can never take on the function of multipliers, since this is a *quantité abstraite* and has to be positive (ibid., 14).

It becomes clear that with this interpretation Porro took up Fontenelle's conception and made it more explicit, that he led it, as it were, to its consequence. Porro strictly based his interpretation on the initial definition of multiplication, according to which the multiplier has to be a pure and absolute number. Thus, he rejected the manner of speaking used in geometry textbooks, according to which the product of a line and a line had to result in an area, or the product of an area and a line in a solid, etc (ibid., 7). The product rather has to be of the same kind as the multiplicand—the multiplier, therefore, a *quantité abstraite*; Porro illustrated this demand for homogeneity in the representation of the product as a ratio:

$$1^{\text{fois}} : cd^{\text{fois}} :: b^{\text{ligne}} : bcd^{\text{ligne}} \text{ (ibid., 11 f.)}.$$

From this, it followed for Porro that an algebraic product could only be the result of the combination plus and plus or plus and minus, but not of the other two combinations that are standard in the rule of signs (ibid., 4). At first, however, the text seems to allow the rule "minus times minus is plus." Using the example of the product $-4x-3$, Porro explains that it is possible to transport the negative sign in front of the negative sign of the first factor so that $--4x3$ is to be multiplied. This means, however, that this is the subtraction of a negative quantity, so that it is a product of the positive factors $+4$ and $+3$ (ibid., 19). Later he also made explicit the same procedure of "turning over" a negative sign onto another (ibid., 36 f.). Although he did not explain the justification of the "over-turning," it can still be concluded that Porro thought a positive product of two negative factors possible.

But Porro decided actually on a different conception in the remaining work—without making the inconsistency with the first a subject of discussion. Because he held onto the initial definition of multiplication as a positive number of repetitions and because he took over Fontenelle's conception, according to which numerical and qualitative *être* had to be distinguished from each other and the qualitative aspect had be left out in multiplication, Porro explained that with a negative multiplier only the number of units mattered, but that the qualitative aspect was irrelevant: "In the multiplier one has to have regard only for the quantity of the units which compose it, and not at all for the quality" (ibid., 37). Thus the original factors—*les quantités proposées pour en faire le produit*—have to be distinguished from the factors made by transformations, the *vrais multiplicandes*, which always have to be positive numbers (ibid., 36).

From this Porro deduced that there are only two rules of signs, namely one for plus times plus and one for minus times plus (he places the multiplier second), but not the rules for plus times minus and minus times minus. An apparent product of minus times minus has to be transformed resulting in minus, namely as a multiplied subtracting of the multiplicand—as often as the multiplier indicates.

With this "modified" rule of signs, Porro thought that he could solve the problem, which had been discussed since Cardano and Arnauld, of why $(+5)^2$ is the same as $(-5)^2$. He illustrated it using the example of

$3 = 5 - 2$: $(5 - 2)^2$ was, if the subtraction is done first, 9 and if multiplying out the parentheses, $+25 - 20 + 4$, i.e., 9.

$(-5 + 2)^2$ was, however, on the one hand $(-3)^2$, and this was, because of the modified rule of signs, -9, and on the other, when multiplying the terms in the parentheses,

$-25 + 20 - 4 = -9$ (ibid., 43).

Already at this point, Porro was tangled up in inconsistencies without realizing it. That $-5x-5 = -25$ was still consistent, but that both $-5x+2$ and $2x-5$ were to be $+10$ cannot be deduced from his premises. Not only did he omit discussing the commutativity of multiplication with a negative term, even though he transforms the negative multiplication factor into a positive one in one of the products and thus obtains the result of $+10$, the other product is necessarily -10. And that $+2x+2 = -4$ also contradicts his premises. Porro formulated his rule of sign as his own corollary: "For obtaining the product of two negative quantities, one has to give the sign + to one of them and the product will have the sign –" (ibid., 54).[11]

With this incoherent basis, Porro took not only to criticizing the then current textbook authors in France—Reyneau, Deidier, Bélidor, Clairaut, and Rivard— as well as Euler for their description of negative quantities and of the rule of signs, but also to claiming quantities that had been understood as imaginary so far to be real.

The starting point for him was the potentially productive reflection according to which the conventional form of multiplication of a *quantité* with itself is not allowed, since the *quantité* had to be then concrete and abstract at the same time. Porro, however, did not draw the conclusion from this to look only at operations between abstract quantities, that is, numbers, but he used it to revise the concept of exponentiating and extracting roots. The traditional statement according to which −36 is not a square and thus $\sqrt{-36}$ is imaginary is wrong: in fact, −36 is a square, namely that of −6. One can, therefore, extract an even root out of a negative number. For him, imaginary quantities represented *absurdités* (ibid., 80 ff.). The so-called *racines imaginaires* are not *impossibles* for him, but are even roots of negative powers (ibid., 106). He explained it to be unintelligible that impossible quantities be the solution of so many problems. In a typical fundamentalist manner he exclaimed that for a product of impossible quantities times themselves to be real contradicts the evidence and the clarity of mathematics (ibid., 109). For Porro, an application of this conception was the fact that curves are always continuous and do not suffer a disruption because of imaginary values (ibid., 157 ff.).

His discussion of the signs for root extractions is even more sectarian and hermetic. He rejected negative exponents in principle; he replaced them by a minus sign before the base. Thus it is not allowed for Porro to write $a^3 \mathrm{x} a^{-3} = 1$. The correct way is: $a^3 - a^3 = 1$ (ibid., 96).

His treatise of equations of the second degree or higher was similarly peculiar. Every product can have any number of factors, but only one can be designated as the root; all the others function as indicators of the number of iterations in the product (ibid., 49). Therefore, the quadratic equation has only one solution.

Porro also commented on the controversies about the logarithms of negative numbers, and not only did he support Bernoulli and d'Alembert in their rejection of the fact that logarithms of negative numbers are imaginary, but he also wanted to convert negative logarithms into positive logarithms (ibid., 140 ff.).

The combination, typical of Porro, of a good eye for terminological problems and one-sided propositions whose consequences he does not think through,

---

11   He stated similarily peculiar rules of signs for division: for him, only divisions of dividend and divisor of the same kind were allowed, but not those of mixed positive and negative terms (Porro 1784, 66 ff.).

becomes obvious in algebraic geometry, which he also treats in this book, in order to discuss this important field of application.

At first, another inconsistency is conspicuous in that, already in the introduction of the book, he allows negative axes—contrary to his usual rejection of non-subtractive negative quantities—and that he takes all four quadrants into consideration for the discussion of curves:

> In order to represent the oppositeness between positive and negative quantities, geometers choose a point of origin which serves to divide into two, lines which cut each other at right angles, which results in four opposite lines and four opposite angles. The determination of one of the axes or of these angles, as positive or negative, is arbitrary; but once done, it determines the others (ibid., 15).

Porro understood not only the coordinates as divided in negative and positive, but also the curves: in *la branche positive* and in *la branche négative* (ibid., 16). His deviant conception of mathematics can be seen in the next discussion. Porro's general *Proposition* is the claim that the negative ordinates of a curve always correspond to the negative axes (ibid., 163 ff.). Because he cites only conic sections for this and no curves of a higher degree, it is obvious that this *Proposition* is an implementation of his conception, according to which the square of a negative quantity—here of a negative ordinate—has to be negative again.

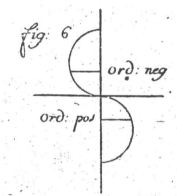
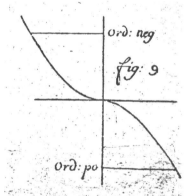

Figure 10a (Porro 1784, Planche)   Figure 10b (ibid.)

Porro also backs it with another geometrical principle: the part of the curve belonging to the negative ordinates does not go through the quadrant neighboring the positive part of the curve, but through the diametrically opposed quadrant (ibid., 164). Porro formulated this principle for all curves in general, but he gave examples only for conic sections. Thus he indeed made the distribution of positive and negative coordinates throughout the four quadrants clearer and more explicit than the previous authors, but his deviant definition of multiplication results in inconsistent graphs without Porro mentioning the contradiction with the definition of curves or commenting on it. The parabola, therefore, has got the form of a curve of the third degree (Figure 9/10b) and the circle does not form a closed line, but two cut halves with S–form (Figure 6/ 10a). For the hyperbola, Porro actually manages to unite the two different branches of the curves into one coherent graph (Figure 10/10c).

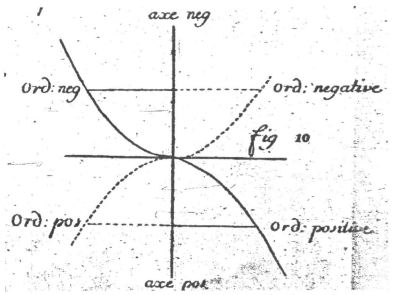

Figure 10c Porro's graphs (Porro 1784, Planche)

For Porro, this algebraic–geometrical principle was also of strong epistemological relevance. Curves are not allowed to be of an amphibious nature, to be half positive and half negative, but they have to be continuously positive or negative:

> Since negative ordinates have to be attached to a positive axis to make an amphibian curve, half positive, half negative, it is something that cannot easily be agreed on (ibid., 194).

Here we still see the epistemological principle operating that we have been able to unravel since Cardano: the positive and the negative fields are two separate entities, between which there normally are no transitions.

Porro agreed with the statement in the *Encyclopédie*—as we know, it was by d'Alembert—that the generality of algebra was an *inconvénient* (ibid., 186). In his criticism of Euler, which surpasses that of Reyneau in its stridency, he stated categorically not only that there was nothing below zero (*il n'y a rien au-dessous de rien*), but also that it was not allowed to look at quantities in isolation. At this point he also used the cue for Carnot again: they are "correlatives" to the positive quantities (ibid., 234).

Porro discredited Euler's explanation of operating with negative numbers as the reason for lacking progress in algebra, for paradoxes and for mistakes in the development: "These are the principles which stopped the progress of algebra, contaminated calculus and occasioned so many paradoxes" (ibid., 238).[12]

KLOSTERMANN: ELEMENTARY GEOMETRY VERSUS ALGEBRA

The second author arguing in a fundamentalist manner against the rule of signs did not investigate the concept field of algebra in a similar way, but limited himself to a small aspect of elementary geometry. The author, Johann Hermann Joseph Klostermann (1730–1810), is not known in the history of science so far. He is listed neither in Poggendorf's *Handbuch* nor in other biographical works. From his writings, I knew only his own note that since 1805 he was *Associé correspondant* of the Göttingen Society of Sciences (Klostermann 1805, 14) and that he worked in St. Petersburg. The Academy of Sciences of Göttingen, the successor to the Society of Sciences, was able to find out his full name and his dates of living for me. From their records, it followed that he became a corresponding member on Kästner's suggestion in 1785. The three works he wrote for the Göttingen society all deal with problems of measuring the length of an arc of a meridian (1785, 1786, 1789). In two small works of 1804 and 1805, he tried to show that minus times minus is not plus, but minus.

Klostermann attempted to represent his new rule of signs as a direct consequence of Euclid's propositions in his second book on squares and rectangles, opposing with the impetus of a fundamentalist all *auteurs*, all *Algébristes* since Cardano, who distorted the clarity of mathematics by claiming that minus times minus was plus. The only modern author mentioned positively is d'Alembert with his polemic against negative quantities being less than nothing (ibid., 28). It is noteworthy that the main object of his devastating criticism is again Euler.

---

[12]   Kästner published a review of the anonymous book in 1784. He gave, however, priority to criticizing a note on the infinite and dismissed the analysis of Porro's operating with negative quantities saying that the "objections are based on incomplete and wrong ideas of algebraic numbers," so that this "(stopped) the interest in going further in the book" (Kästner, "Avignon." Göttinger gelehrte Anzeigen, Der zweyte Band auf das Jahr 1784, 134. Stück, 21. August 1784, 1339–1341).

Porro's writing was received by Jean de Castillon (1701–1791), since 1763 professor of mathematics at the Artillery Corps in Berlin and since 1787 director of the mathematical class of the Berlin Academy. In two writings of 1790 and 1791, he rejected, on the one hand, an extension of the meanings of arithmetical operations and, on the other, proposed different operational meanings for his three kinds of quantities: absolute, positive, and negative. The traditional rule of signs holds only for absolute quantities. Contrary to all the earlier authors, he postulated that there is not just one unity for the positive and the negative quantities, but rather three: one for each type of his three different kinds of quantities. Consequently, he had to reject common rules for operating with quantities. He seems to have been primarily motivated not by wanting to infer imaginary quantities from negative square roots, but by excluding imaginary quantities. The central position of absolute quantities in Castillon points to Carnot (Castillon 1790, 1791).

Even though Klostermann sees the beginning of all evil with Cardano, he practically uses the same argumentation as Cardano to demonstrate that the product of negative factors is again negative (cf. above, Chapter II.2.3). Klostermann does not state a definition of negative quantities, nor does he reflect upon them. He especially lacks any distinction between operational sign and algebraic sign.

His argumentation gains some plausibility only because he visualizes neither his geometric propositions nor their version with general signs and because he works only with a few numerical examples. Klostermann takes Euclid's Proposition II, 4 as the starting point: if a straight line FL is divided at some point O, then the square over FL equals the square over FO and OL plus the doubled rectangle formed by both sections. Klostermann's only visualization is the starting line:

F   O   L

Figure 11, refutation of the rule of signs (Klostermann 1804, 1)

Using this he calculates a numerical example. For the sake of simplicity we shall use letters to designate the single lines. Klostermann chose
$$a = FO = 256; \quad b = OL = 44.$$
Thus the "racine" of the total square is $c = FL = 300$. With $a^2 = 65536$ and $b^2 = 1936$ and $ab = 11264$, the result is $a^2 + 2ab + b^2 = 90000$, in agreement with $c^2 = 90000$.

After this Klostermann wants to determine the smaller a^2 out of the total square c^2. Here he already commits the first *petitio principii*, since he does not differentiate between operational and algebraic signs. The root a, determined out of c, is $300 - 44$. Thus the root of the square over OL is -44, so it is negative! This square is also negative, because it is subtracted from the total square. Confusing subtractive and negative quantities, Klostermann thought he had already proved that a negative quantity also yields a negative square and that a negative square results in a negative quantity, which means that there are no imaginary quantities (Klostermann 1804, 2).

Even though this had already provided sufficient evidence for him, he numerically elaborated on it. The smaller square is the result of gradual subtractions:
$$90000 - 1936 - 11264 - 11264 = 65536.$$
One does not get, therefore, as is claimed by *tous nos Elemens d'Algèbre*,
$$(c - b)^2 = cc - 2cb + bb,$$
for his calculation did not result in $bb = 1936$, but -1936, so the product is $-b \times -b = -bb$ (ibid., 2 ff.). In fact, however, Klostermann, lacking sign symbols,

obscured the fact that he substituted a by c. His numerical calculation is actually, putting in signs,

$$c^2 - b^2 - ab - ab.$$

If the distributive law is applied to the square of the compound quantity $(c - b)$, $cb = 13200$ has to be deducted twice. Since, therefore, the small square b^2 is deducted once too many, it has to be added once. This, however, is a totally different calculation as if—as Klostermann did—one calculates with all three parts.

In 1805, in a review in the *Göttinger gelehrten Anzeigen*, B.F. Thibaut (1775–1832) criticized this method of confusion, since Klostermann's negative terms appear in a transformed equation and not in the "direct multiplication of two equal factors" (Thibaut 1805, 168). Thibaut also strictly opposed the approach of deciding algebraic problems with methods of geometry:

> Basing arithmetic theories on geometric approaches, which also led to this misunderstanding, has to be altogether rejected as completely unscientific. We should not and we must not take refuge in the construction of squares and rectangles in order to learn what sign the product of two numbers has to receive (ibid.).

In spite of this refutation, Klostermann published yet another, longer, work in the same year to defend his theory. In it, however, he produced only further varieties for this single argument. He rebukes all supporters of the common rule of signs because they violate the rules of the "bon raisonnement," lack metaphysics, and confuse the clear thinking of mathematics (ibid., 27 f.). Klostermann failed to demonstrate that his new rule of signs can be applied, and he was content with stating the rule and with discussing the proof thereof.

The positions of Porro und Klostermann surely are extreme. But because they are an exaggeration, they clearly show that the previous paradigm of the dominating concept of quantities did not allow for a constructive development. In the course of the eighteenth century reflections on the foundations had become quite intensive, the concept of negative quantities were ever more completely treated and its ramifications had been followed up, and extremely differing ways had been examined in order to arrive at a satisfying solution. So far, however, a commonly shared and coherent conception had not been found. It had become ever clearer in the numerous research works that the main obstacle lay in the concept of multiplication.[13]

---

[13]   Works by Leonardo Salimbeni (1752–1823), lecturer of mathematics at a *Collegio Militare* in Verona and prominent member of the *Società Italiana*, the first society of mathematicians in Italy, founded in 1782 (cf. Grattan–Guinness 1986a), are illuminating pieces of evidence of the decisive role that the definition of multiplication played. Salimbeni had rightly seen that operations for numbers and quantities have to be defined differently. Regarding the multplication of quantities, he tried to solve this, however, with a noncommutativity of the product, which was probably formulated intentionally for the first time (cf. Schubring 2004a).

The situation that had emerged can be compared to the development of the world view at the beginning of modern times. The Ptolemaic system of the world was exhausted regarding its capacity; because there was a lack of a new conception, ad hoc quantities, the so-called epicycles, were introduced to have the theoretical system correspond again to the data made by observation. Using analogous "epicycles," one tried to readjust the concept of multiplication to the whole field of operating with negative quantities, but instead of a solution, the entire structure became even more complicated. We have already seen this tendency in Hecker's works, who had incorporated further additional and more implicative concepts while trying to establish simple bases. It happened even to mathematicians who were more professional than Porro and Klostermann that while striving for rigor, they got stuck in epicycles and suggested one-sided solutions, which allowed no coherent conceptions or ways of operating with them.

2. The Limit Field: Dominance of the Analytic Method in France After 1789

2.1. Apotheosis of the Analytic Method

In the first years of the French Revolution there was—besides the political changes—a radical structural and conceptual reshaping of the entire system of education and science. Science, which had been of high social prestige already during the Enlightenment, but had been disputed over by politics and had been of limited effect on society, now became the dominating authority in society. Under the banner of the *analytic method*, especially mathematics and chemistry became leading disciplines, which were also to realize the program of disseminating rationality.

A great number of first-class mathematicians—their number could not have been matched internationally—lived in France, none of whom was actively teaching before the Revolution: Condorcet, Lagrange, Laplace, Legendre, Monge—besides an additional number of qualified mathematicians.[14] The institutional teaching, however, was in a desolate state. At the universities mathematics was taught only in *Collèges*, which were similar to secondary

---

[14]   Monge, who had to withdraw from the *École du Génie* in Mézières because of his engagement in the Paris Academy, worked only as an *examinateur permanent* for military engineers, as did Laplace.

schools. The low standards of their textbooks have already been described in Chapters II.2.10.1. and III.8.3. A large part of their mathematics teachers were so insignificant that today one does not know for many of the Paris *Collèges* who was teaching in the 1780s. The system of military schools was also in a crisis: the *École royale militaire* had been closed down, and the *École du Génie* was collapsing because, among other things, their students were reduced to "true" nobles.

After these ailing structures had been entirely broken down during the Revolution until 1792/93, it took another two years until completely new structures could be established in 1794/95. These, however, embodied—driven by a unique social optimism—structures of a public school system and a standardized conception of the curriculum for the first time: the concept of the "analytic method," with mathematics and chemistry as its central elements. There are two roots for this dominant conception:

- On the one hand, it was the tradition of rationalism in France, for which the philosopher Condillac had worked out the analytical method.
- Within mathematics, the analytic–algebraic approaches of Euler and Lagrange could break through against the "synthetic" concepts of the *infiniment petits*.

Since Descartes and Arnauld, it had been a main characteristic of the philosophy of the Enlightenment in France to put the emphasis on the *method*: *l'art de bien raisonner*. During the French Revolution this orientation on methods received highest priority for the entire system of education and science: the aim of the Enlightenment, the elimination of *préjugés* in the thought of man, was now to be realized throughout the entire new Republic by reorganizing the educational system and the sciences.

During the French Revolution and especially after the Thermidor, the overthrow of Robespierre in July 1794, it was the *Idéologues*, a group of philosophers, who determined the politics of education and sciences and successfully propagated the method of *analyse*. A hundred years after Prestet's offensive attack against the method of *synthèse* and his optimistic plea for the *analyse*, it was now the *Idéologues* who resumed the program of the Malebranche group and pushed through the analytic method as dominant orientation—at least for the time of the revolutionary upheavals. In 1794, the *Décade*, the journal of the *Idéologues*, could therefore rightly declare the *analyse* as the fundamental method for education and science:

> This method will be, without doubt, founded on analysis. Locke, Helvetius, and Condillac have sufficiently shown that it is uniquely by means of analysis that we can penetrate with assurance into the sanctuary of science (*La Décade*, 10. Frimaire an III, v. 3, 462).

It was this view of the method, modeled on Condillac's, upon which basis D.-J. Garat, leading *Ideologue* and instigator of the first *École Normale* of 1795 and, since Robespierres's overthrow, commissioner of the Republic for public education (in fact, minister of education), could exclaim enthusiastically that one of the main problems of modern philosophy and science had been solved: "There is no longer a need to search for the best method; it has been found" (D.-J. Garat 1795, 147).

Condillac's interpretation of this analytic method corresponds to the rationality ideal of the Enlightenment. *Analyse* to him means to elaborate how ideas emerge. It is strictly differentiated from the method of synthesis, which is characterized by its use in mathematics:

> With this method the truth is apparently allowed to show itself only if a great number of axioms, definitions, and other supposedly fruitful sentences preceded (Condillac 1977, 109).

The synthetic method is a method that obscures the natural way of the developing ideas on purpose:

> Those who discovered new truths thought that they had to make a secret out of the method they had applied in order to give a greater impression of their genius (Condillac 1977, 108).

Condillac, therefore, established a direct connection between the synthetic method and the dissemination of *préjugés*; he called it a *méthode ténébreuse*, that spreads darkness instead of light (*lumières* = enlightenment) (Condillac 1948, 405).

Later authors, e.g., Lacroix, could no longer understand this emphasis on the analytic method without knowing the special situation of the time. They interpreted analysis versus synthesis in the original sense of Pappus again, i.e., as the dissolving of a whole into its elements and the joining of the parts, respectively. Condillac, however, had a consistent concept of methods. The traditionally separated methods are applied in *analyse* as well as in *synthèse*. For Condillac the difference lay somewhere else. Only *analyse* guides the development of ideas and a real language and thus of methodical thinking (ibid.). As is the case with Prestet, *analyse* means algebraization, with *synthèse* remaining in the geometry of the *Anciens*. The novelty in Condillac is the introduction of *signs* as an independent level in the development of concepts.

Corresponding to Condillac's sensualism, the origin of ideas is seen in the perception of the senses, the *sensations*, which process the external objects. The decisive new step is that Condillac unambiguously assigns *signs* to the ideas: the first signs are pictures of external objects, the additional signs are formed by the human mind through operating with these signs. Condillac and the other sensualists

> found, so to speak, a genealogy in which the first ideas, which generate all the others, are the pictures of external objects and in which the last, the most ingenious concepts, are either a disassembling or an assembling of these pictures (Garat 1795, 146).

Thus interpreting concepts as signs, Condillac not only had worked out a conception of how concepts developed, but also at the same time had developed the concept of method further by reflecting on conceptual operating: he understood the signs as "instruments" that the human mind uses in its "operations" (Condillac, ibid., 309). It was important to correctly operate with signs to methodically think correctly. The analogy with the system of language rules led Condillac to the identification of speaking and thinking. Garat described the discovery by the sensualist philosophers: while so far, languages had been thought of as necessary instruments for the communication of thoughts, they had revealed that languages are necessary to have thoughts: one thinks only when one is talking. The required method of correct thinking could, therefore, be interpreted as a method of exact speaking: *L'art de penser avec justesse est inséparable de l'art de parler avec exactitude* (Garat ibid., 147).

For this new interpretation of the method of operating with signs, of calculation, the model is mathematics, but now no longer geometry, but arithmetic or algebra, which represents itself as the model of calculation with signs: "If the ideas of mathematics are exact, then it is because they are the work of algebra and analysis" (Condillac 1977, 109).

The geometry of the Greeks, the embodiment of the synthetic method, seemed to be limited, static knowledge, which remains attached to details. Jean-Baptiste Biot (1774–1862), educated as a mathematician and physicist during the time of the Revolution, sharply accentuated this criticism in his famous *Essai sur l'Histoire des sciences pendant la Révolution Française*, which originally was planned to be the preface to the new edition of the *Séances*, the lecture notes of the *École Normale* (Guillaume, vol. 5, xvi), and which was written in the spirit of the founding fathers of the *École Normale*:

> The treatises which have appeared up to now show us the ancient geometers as limited by simple elements; their genius is as if confined within a circle of narrow compass from which they cannot escape. If one searches for the reason why these powerful minds were retained by such detail, one will soon see that it is the method [they employed]. Synthesis, which they used, proceeds from known truths to those which are to be proved: and since all truths do not have the one without the other, a mutually intimate liaison, it is only by the use of a sort of tact that one can divine what leads to the goal; one cannot even hope to achieve this unless the goal is very close: the march of science, by this method, is therefore long and difficult (Biot 1803, 23 f.).

Since in mathematics "the method is everything" (ibid., 23), the perfection of philosophy made great progress possible for it: indeed, says Biot, mathematics owes "the largest part of the discoveries made in the last time" to the analytic method developed by Condillac's philosophy (ibid., 24).

Condillac's interpretation of the scientific method found its direct application in the correct operating with signs in chemistry. Lavoisier, who had substituted the phlogiston hypothesis with oxidation theory, developed his *Méthode de Nomenclature chimique* on the basis of Condillac's methodology as operating with chemical signs, as language. The great number of new discoveries in chemistry was understood as proof that the new philosophy opened up the way to great scientific progress (Garat 1795, 147). Biot emphatically highlighted the progress in the sciences made possible by the analytic method:

We, by the bright light of philosophy, therefore also forget those chimerical fears of a return to ignorance, and march on with a firm step into the broad highway henceforth open to the human spirit (Biot 1803, 32).

2.2. Euler's Reception

The dominance of the analytic method in society and especially its operationalizing in mathematics in preference to algebraic approaches allowed a broad reception of Euler's conception of analysis in France for the first time and was the inspiration behind Lagrange's own, even more radical, conception.

Euler's reception took not only a contradictory course, but also one completely lacking in the high esteem in which Euler's works are held in mathematics and in mathematical historiography today. His reception was contradictory in so far as a relevant number of mathematicians in numerous countries read his books and treatises, but mostly only single results were adopted, and not, however, foundational conceptions. The difference to the current understanding of Euler's importance is, on the one hand, that Euler as an academic did not stand in a direct relation to a single educational system and that he thus wrote his textbooks for a general, predominantly scholarly, audience and not for direct use in a certain institutional context. In addition, the specific difference is that especially his basic concepts were met with a strong rejection and opposition not only in algebra, as we have already seen, and that his basic concepts of differential and integral calculus were also rejected, especially by leading mathematicians of the different countries—and not only by "normal" scholars and practicioners!

At this point a description of these foundational concepts by Euler is necessary to be able to differentiate between the rejecting and accepting receptions in the following. In Euler *two* approaches can be distinguished: the first is his conception of sequences of finite differences. By choosing Leibniz's concept of finite differences and its continuation by Taylor (cf. Chapter III, Sections 4. and 8.1.) as his basis, Euler introduced a comprehensive algebraic operationalizing of an independent calculus of differences. The first two chapters of his textbook on differential calculus are dedicated to the presentation of this part of algebraic theory. The operationalizing also included series of differences of a higher degree, so that in this way the basis for iterated differentiation was already prepared. As inverse operation, he developed the formation of series of sums and series of higher order, as algebraic operations.

Euler based the difference concept on the changes of a function y by finite increases ω of the independent variable x and constructed with equidistant intervals the series of x-values

$$x, x + \omega, x + 2\omega, x + 3\omega,\ x + 4\omega, \text{ etc.,}$$

and the series of functional values

$$y, y^{\text{I}}, y^{\text{II}}, y^{\text{III}}, y^{\text{IV}}, \text{etc. (Euler 2000, 2)}.$$

Euler introduced his own operational sign Δ for series of finite differences, which could be made thus:

$$y^{\text{I}} - y = \Delta y; \quad y^{\text{II}} - y^{\text{I}} = \Delta y^{\text{I}}; \quad y^{\text{III}} - y^{\text{II}} = \Delta y^{\text{II}}; \text{etc. (ibid., 2)}.$$

The operative capacity of this sign could be instantly seen for the higher differences. For the second difference, Euler wrote two such signs next to each other: $\Delta\Delta$. For the higher differences he used a power notation: Δ^3, Δ^4, etc (ibid., 3). Euler was especially interested in expressing higher differences by terms of the first series, e.g.,

$$\Delta^5 y = y^{\text{V}} - 5y^{\text{IV}} + 10y^{\text{III}} - 10y^{\text{II}} + 5y^{\text{I}} - y \quad \text{(ibid., 5)}.$$

A main application of this method of insertion was the determination of series of differences of powers and, more generally, of rational functions in terms of the first series of differences (ibid., 6 ff.). By analogy, Euler introduced the sign $\sum$ for series of sums and determined in this way series of sums for rational functions (ibid., 17 ff.).

Euler's second approach consisted in his peculiar approach toward the transition from the calculus of differences to the differential calculus. Euler introduced this approach with an extensive discussion of the concept of the infinitely large. He supported a conceptual realism here that was to allow mathematicians to operate with quantities considered infinitely large. It is remarkable that Euler discussed this by using a philosophical problem of mechanics: the permissibility of the infinite divisibility, of matter. Euler disputed such an infinite divisibility since he was of the opinion that this position was practically identical with the one that "the existence of simple beings that make up a body is completely refuted" (ibid., 50). He himself, however, thought that "the material is made up of simple parts" (ibid., 48). The application of this conceptual realism was to give the infinitely small a real meaning, i.e., as an actual vanishing, as zero. A requirement remaining implicit was that the limit— in contrast to the conception of limit in all the other mathematicians—of a variable be understood as really attained. Euler explained that there was no doubt that every quantity could be diminished in such a way that it disappears entirely and becomes zero:

> But an infinitely small quantity is nothing but a vanishing quantity, and so it is really equal to 0. There is also a definition of the infinitely small quantity as that which is less than any assignable quantity. If a quantity is so small that it is less than any assignable quantity, then it cannot not be 0, since unless it is equal to 0 a quantity can be assigned equal to it, and this contradicts our hypothesis. To anyone who asks what an infinitely small quantity in mathematics is, we can respond that it really is equal to 0. There is really not such a great mystery lurking in this idea as some commonly think (ibid., 51).

For Euler, therefore, the differential calculus was calculating with zeros. He had to explain, however, why he did "not always use the same symbol 0 for infinitely small quantities," but with dx, for instance. For that Euler developed an ad hoc conception: on the one hand, all zeros are the same in relation to each other, but this holds true only in the arithmetical ratio. Instead, he introduced their geometrical ratio as something new, and for these a ratio between zeros could have a finite value:

> We can easily see this from the geometric proportion $2 : 1 = 0 : 0$, in which the fourth term is equal to 0, as is the third. From the nature of the proportion, since the first term be twice the second, it is necessary that the third be twice the fourth (ibid., 51).

Euler did not see any difficulties in justifying the conventional rules of operating with infinitely small quantities with this definition of zero: "since the infinitely small is actually nothing, it is clear that a finite quantity can neither be increased nor decreased by adding or subtracting an infinitely small quantity" (ibid., 52). In the same manner he deduced the legitimacy of operating with different orders of the infinite (small and large, respectively) (ibid., 52 ff.).

To Euler the transition from finite differences to the limit was really unproblematic; he deemed its existence secured by the "law of continuity," which for him was represented by the continuity of the real numbers:

> Both infinitely small and infinitely large quantities often occur in series of numbers. Since there are finite numbers mixed in these series, it is clearer than daylight how, according to the laws of continuity, one passes from finite quantities to infinitely small and to infinitely large quantities (ibid., 90).

As a proof, Euler referred to a series of natural numbers, which he continuously wrote from positive to negative:

$$..., -4, -3, -2, -1, +0, +1, +2, +3, +4, ...$$

As the numbers constantly decrease, they "approach 0, that is, the infinitely small" and finally become negative (ibid., 56). That is why it was no longer a problem for Euler to represent the "analysis of the infinite" and especially the differential calculus as "nothing but a special case of the method of differences" (ibid., 64).

In spite of his approach with finite differences, Euler introduced the integral algebraically as indefinite integral and not geometrically by way of finite surface patches as a definite integral. Euler even took the algebraic interpretation of the operation of integration so far that he equated differentiation and integration with the basic arithmetic operations:

> Since in analysis always two calculating operations are opposed to each other, as subtraction and addition, division and multiplication, extracting roots and exponentiating, so differential calculus is opposite to integral calculus in a similar way (Euler 1828, 1).

It is a widespread opinion to think of eighteenth-century mathematics as unconcerned with the foundations and as interested only in the further development of analysis. As we have seen with the concepts of negative numbers as well as with the infinitely small quantities, the mathematicians were, in contrast, very anxious to clarify basic concepts. The opinion mentioned above is, therefore, rather an incorrect generalization of some of Euler's concepts, which later were rightly called "naive" concerning foundations by mathematicians (cf. Bohlmann 1899). Whereas Bishop Berkeley had criticized Newton's partially nonchalant operating with infinitely small quantities in a strong and dissecting manner and set off an intense debate, there was no such fundamental attack in Euler's case. Sometimes his approaches were indeed criticized, in the prefaces of textbooks for instance, but Euler apparently did not have a key orientating function, so that criticizing him would have caused a deep feeling of unease.

One of Euler's important requirements was therefore generally agreed on and did not need clarification: the existence of limits and the unproblematic transition to the infinite. Euler's approach of understanding the differential quotient as a *ratio* of zeros was not agreed on, however. Since Euler with this approach had excluded a discussion of limit processes and therefore did not contribute anything to an algebraization of limit theory, he understood the differential quotient as the actually obtained ratio of zero to zero. This contradicted all conceptual clarifications obtained so far and was—with a few exceptions—completely rejected by contemporary mathematicians without their involvement in extensive debates—as was the case with Newton and Berkeley.

It is remarkable, however, that no mathematician opposed Euler's idea according to which infinitely small quantities *per definitionem* are to be understood as zeros. This shows that at least in the eighteenth century, infinitely small quantities were not understood as quantities of a completely different kind than the ones introduced so far.

Regarding Euler's second approach, there was no positive contemporary reception. The first approach to algebraization via difference calculus did, however, during the period of dominance of the analytic method, during the revolutionary years, experience a first intense reception in France.

2.3. Algebraic Approaches at the École Polytechnique

This first significant reception of Euler's approach of difference calculus happened in the first regular course on analysis at the newly founded *École polytechnique* in 1795, by de Prony. The lecturer as well as the course and the entire context were extremely exceptional and corresponded to the revolutionary state of affairs.

De Prony was unusual in that not only did he as a representative of the engineer *corps* work as a real scholar on mathematics and physics, but he also propagated pure science and supported the analytic method. For France de Prony forms, therefore, a counterpart to August Leopold Crelle in Germany, who also was as a technician one of the strongest advocates of a pure mathematics (cf. Schubring 1981). Gaspard Riche de Prony (1755–1839) was trained as an engineer, at the *École des Ponts et Chaussées*, not, however, as a military engineer, but for civilian purposes, for the construction of bridges and roads, from 1776 until 1780. He quickly reached leading positions at this school. From 1794 on, he worked at the *École des Ponts et Chaussées* and at the *École polytechnique* at the same time: at the former he even functioned as its director from 1798 on—until the end of his life—and at the new central school he worked as a lecturer (*instituteur*): nominally for mechanics, but actually—at least at the beginning—for analysis. Later, at the *École polytechnique* he transferred to its board of supervisors.[15]

The unusualness of the context will be explained in more detail in Chapter IV.3: the later *École polytechnique*, founded as *École Centrale des Travaux Publics* in 1794, functioned not only as a school ambitiously intended for the entire education of engineers—military and civil—but also for the education of scientists.

After a three-month-long *cours révolutionnaires*, methodically guided intensive courses that ensured the prerequisites for the training (cf. Langins 1987), the regular courses, which aimed at a three–year course of studies, began in May 1795. In this first curriculum there were no lectures proper on analysis in the sense of differential and integral calculus; rather, they were courses—corresponding to the dominant concept of the analytic method—on *analyse*, applied to geometry, mechanics, and technology. De Prony taught the *Cours d'analyse appliquée à la mécanique* here. His first such course showed, however, neither an orientation toward application nor the subject matter of mechanics. The intention of his course, which was published in four parts under the course title in the *Journal de l'école polytechnique*, can be easily gathered from the title, under which these parts were later published together: *Méthode directe et inverse des différences* (1796).

Indeed, the main part of the summaries published in the *Journal* afterwards is a presentation of calculus of differences and sums, following closely the first and second chapter of Euler's differential calculus, and especially adopting his notation. As a consequence, de Prony made the concept of function the starting point and foundational concept of his entire treatise. His text was, therefore, the first treatise in France that had used the concept of function throughout as a basis. This underlines both the modernity of the teaching at the *École polytechnique* and its break with the institutions existing thus far.

In the introduction to his lecture, de Prony declared differential and integral calculus—differing from Bézout, for instance—as the indispensable basis of mathematical physics, which meant that he began his course on mechanics by presenting these foundations:

> The study of the differential and integral calculus has become indispensable, in the present state of knowledge, to all those who wish to concern themselves with the physico–mathematical sciences: I thought therefore that I must begin the course of

---

[15]   More in-depth presentations on his biography can be found in Bradley 1984, 1985, and 1998 and Grattan-Guinness 1990b.

mechanics with an exposition of the principles of the calculus (De Prony 1795a, 92 f.).

The published text, although it summarizes at least forty hours of lectures from March till December, does not, however, discuss any issues from mechanics. In most parts of the lessons, he instead explicates in great detail the calculus of differences (*méthode directe*) and the calculus of sums as its inversion (*méthode inverse*).

After introductory sections on variables and functions, de Prony moves on to discussing the *Différences des Fonctions en général*. He immediately treated these functions as functions of several variables:

$$z = f(x, y, t, \text{etc.}).$$

If every variable respectively gets an increment a', a'', a''', etc. then the function z will be changed to a variation that can be called ω:

$$z + \omega = f\{(x + a'),(y + a''),(t + a''), \text{etc.}\}.$$

The aim of the *méthode directe des différences* is the search for procedures with which the first difference ω and Δz respectively or higher differences can be found as easily as possible (ibid., 107). De Prony explicitly remarks that he means the word increase (*accroissement*) to be "generic," so that both positive and negative differences are included (ibid., 108).

De Prony's approach moving to algebra is underlined by the fact that he dedicated a large part of his lectures to the abstract discussion of differences: "independent of any special relation between the variables" (de Prony 1795 b, 1 ff.). This is why he also discussed in great detail the so-called partial differences, as a finite complement of partial differential equations. Another main part discussed, also mostly formally, the *méthode inverse*, the determination of series of differences of a lesser order and of series of sums of various orders (de Prony 1795 c). Although de Prony explicitly pointed to the difference between *Sommation* and *Intégration* (ibid., 215), he spoke of *Intégration* in the following when he discussed series of differences to simplify things. A last main part, finally, was the presentation of new theories of his own on recurrent series: the development of rational functions into series and its connection to integral calculus (de Prony 1796).

Occasional remarks showed how de Prony understood the transition from finite differences to differentials: in Euler's sense, i.e., differences become zeros in their ratio (1795 a, 111). De Prony explicitly discussed this transition only in a few lessons at the end of his lecture, as *passage de la méthode des différences à celle du calcul différentiel* (de Prony 1796, 543). Unfortunately, he did not give any further details in the print version, since at the same time Lagrange in his lecture had described the foundations of differential calculus according to his new method, and its text was going to be printed here (ibid.). On the one hand, he mentioned only that he had discussed "foundations and practice" of differentiating. On the other, he mentioned two consequences of his discussion. The first is especially important in terms of methodology: he gave the law of continuity as the reason for the transition (*le passage de l'analyse des quantités finies à celle appelée des quantités infiniment petites*) and introduced thus the reflection on *continuity* in the French teaching of analysis for the first time:

> I explained to the students how, with a character distinctive to the differential calculus, one could determine the introduction of the law of continuity between those variables the method considers, and in what sense the word continuity should be taken when it is applied to a similar question (ibid., 544).

Of course, it would have been exciting to learn the precise meaning that de Prony gave to continuity here, since his explanation implies that there can be different meanings.

As a second consequence of his view of the foundations, he explained that it was Taylor's theorem with whose help he determined the transition of the values of the difference quotient to the differential quotient (ibid., 544 ff.).

De Prony's high degree of methodological reflection can also be seen because he took up Arbogast's[16] Petersburg treatise on the nature of arbitrary functions (1791), introducing *discontiguës* curves, and inserted the distinction between continuous, discontinuous, and *discontiguës* functions into his course: under the heading *la fonction arbitraire K peut être continue, discontinue ou discontiguë* he discussed an integral curve, of which he stated before that to it applies *la possibilité d'être discontinue ou discontiguë* (ibid., 511 f.). De Prony did not explain the exact meaning of these notions here.

De Prony's methodological reflection led him not only, in accordance with the analytic method, to constantly making the notation explicit and emphasizing how important signs were for understanding, but also to additional didactic innovations: he was the first to systematically use *Tableaux*: whole-page tables of two or more columns, in which the relation of certain variables is visualized (cf. de Prony 1795 b, 18; 1795 c, 222 f. and 248; 1796, 535). Later, the *Tableaux* were an important means of illustration for Carnot in his *Géométrie de Position* (cf. below, Chapter V).

De Prony's approach to algebraization in analysis also proves the connection to conceptual clarifications for negative numbers: at the beginning of his text there are some remarks on analytic geometry, in which for the first time in France the four quadrants of a coordinate system are explicitly assigned to the respective positive and negative values of the $x-$ and $y-$axes in the plane (cf. Figure 12).

The break with previous tradition, which was probably the most visible to his contemporaries, was the exclusion and rejection of *infiniment petits* by the analytic method. In de Prony the *infiniment petits* were excluded from the foundational concepts of his teaching by simply not being mentioned; only in a heading did they appear in a quotation, as "so-called analysis of the infinitely small quantities" (see above). In Lagrange's famous and epoch–making lectures

---

[16]   Arbogast was originally offered the position of *instituteur* at the *École polytechnique*, but did not take it.

they were, however, explicitly "banned," so to speak, as inadequate foundational concepts.

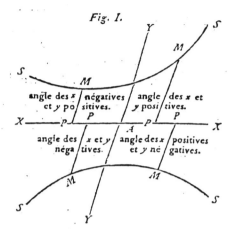

Figure 12, the four quadrants (de Prony 1795a, 96)

Joseph-Louis Lagrange (1736–1813) practically experienced a second youth because of the qualitatively new and different requirements for teaching during the time of the Revolution. Already as a young man he became a teacher of mathematics in Turin, from 1755 on as a nineteen-year-old, first as an assistant and apparently from 1758 on as independent lecturer (Borgato, Pepe 1987, 14 ff.). A lecture of 1759, whose manuscript has been preserved, is still based on the concept of infinitely small quantities (cf. Chapter III.9.). Called to the Berlin Academy in 1766, as Euler's successor, he was exclusively involved in research afterwards, and no longer in teaching. After the death of Fredrick II of Prussia, the situation in Berlin became unattractive for him, and in 1787 he finally accepted the invitation of the king of France to Paris, where he now was a full member of the Paris Academy and where he lived in the Louvre, the king's palace. His first biographer, the astronomer Delambre, reports that in Paris, however, he had lost any stimulus to do research in mathematics. Even when meeting with other scientists, he remained passive:

> I often saw him in companies that had to be of his taste, among the scholars, whom he had visited from so far away, among the most distinguished men of all countries, who each week came to the famous *Lavoisier*, standing, as if he was dreaming, at the window, where there was nothing that could catch his gaze; he did not partake in any of what was spoken around him; he himself confessed that his enthusiasm was extinguished, that he had lost the taste for mathematical studies. If he heard that a mathematician was dealing with this or that work, he said "the better; I had started with it and I do not bother you now with the completion." But this thinking head could only change the subject of his research. Metaphysics, the history of the human

mind and the history of the different religions, the general theory of the languages, the media, botany were all part of his leisure.[17]

His appointment to the *École Normale* and the *École polytechnique* during the Revolution revived his former passion for teaching, and he continued with a project of 1772 dealing with the redefinition of the foundations of analysis on a new basis. This treatise for the Berlin Academy already included the concept of developing functions into series; it was also based on Euler's concept of the calculus of differences and sums, which Lagrange himself had fruitfully used already in 1759 while working on the problem of the vibrating string. It was especially this part of his approach, however, that he was still dissatisfied with in 1772:

> Though the operation by which we have gone from the difference Δu to the difference $\Delta^\lambda u$ and to the sum $\Sigma^\lambda u$ was not based on clear and rigorous principles, it is none the less exact, as can be confirmed a posteriori; but it would be very difficult perhaps to provide a direct and analytical proof (Lagrange 1772/1774; Œuvres, tome III, 451).[18]

In his new approach of 1795-1796, in his lecture on analysis in the first regular year of courses at the *École polytechnique*, Lagrange radicalized his method and completely algebraized it. There no longer was a transition from finite series of differences or sums to differentials or integrals. He now based his approach completely on the method of developing functions in series. Lagrange's approach of developing the increment $f(x + i)$ for analytic functions in power series

$$f(x) + pi + qi^2 + ri^3 + ...,$$

where p as the new function of x is identical with the derived function $f'(x)$, is far too well known to present it here again.[19] His basic position, which he reached and from now on always explained anew, is, however, relevant for our context.

The subtitle of his *Théorie des Fonctions Analytiques* shows the programmatics of his methodology and had an immediately strong effect upon contemporary mathematics:

> Les Principes du Calcul Différentiel, dégagés de toute considération d'infiniment petits, d'évanouissants, de limites et de fluxions, et réduits à l'analyse algébrique des quantités finies.

Because we know the debates on foundations that took place in the eighteenth century, it is not surprising to us that Lagrange rejected the method of fluxions and the *infiniment petits*: the former because Newton had based it on the concept

---

[17]   Here quoted according to Crelle in his translation of Lagrange's theory of analytic functions (Lagrange/Crelle 1823, XLVI f.).

[18]   Lagrange's treatise of 1772 inspired Lacroix to write a handbook with a coherent and modern presentation of differential and integral calculus including more recent research results.

[19]   A more recent summary is in Grattan-Guinness 1990, Chapter 3 (127 ff.); a more recent analysis of details can be found in Grabiner 1990.

of motion. This brings, however, an *idée étrangère* into a theory, which deals with algebraic quantities (Lagrange 1881, 17).[20]   For the rejection of the *infiniment petits* he did not give a detailed explanation in the *Théorie des fonctions*; he expressed that he had identified it with the foundations in Leibniz, the brothers Bernoulli, and L'Hospital and that these had not given any reasons for these basic concepts (ibid., 16). What is surprising, however, is that he also rejected the method of the *limites*. In the *Théorie des Fonctions*, Lagrange gave two explanations, which are both based on the rejection of operating with vanishing quantities or those that have been declared zero: he ranked Euler's approach to zeros next to d'Alembert's approach to *limite* and declared their basic thoughts to be substantially and didactically inappropriate to serve as a *principe* for a science whose certainty had to be based on evidence. He explained their common approach thus:

> Since the differences which are supposed to be infinitely small must become absolutely nothing, and their ratios, the only quantities that actually enter into the calculations, are none other than the limits of the ratios of differences, finite or indefinite (ibid., 16).

The second reason explained the *limite* method as an algebraic form of Newton's concept of vanishing quantities and argued especially against Euler's basic concept of reckoning with zeros: the method had the big disadvantage

> of considering quantities in the state where they cease, so to speak, to be quantities, for although one can always well understand the ratio of two quantities while they remain finite, this ratio no longer offers to the mind a clear and precise idea as soon as both the terms become equal to nothing at the same time (ibid., 18).

In terms of didactics, Lagrange's lecture was a failure in the first year (Langins 1987, 76; Grattan-Guinness 1990a, 108). It was not repeated for the time being. Only in the year VII (1798/99) did Lagrange take on a lecture on analysis again—but this time in a redefined, exceptional framework, not for the normal training of engineers, but for a complementary education of scientists— and even then only as a nonobligatory offering (de Prony 1799, 217). In this lecture, entitled *Leçons sur le Calcul des Fonctions,* he gave a revised and shortened version of his first lecture, as its *commentaire et supplément.*

In his introductory *Discours*, immediately published in the *Journal* of the school in 1799 and identical with the versions taken over into the later print versions of the *Leçons*, Lagrange now differentiates between d'Alembert und Euler. Again, Euler is sharply criticized because of his reckoning with zeros:

> *Euler* regards the differentials as being nothing, which reduced their ratio to the vague and unintelligible expression of zero divided by zero (Lagrange 1799, 232).

D'Alembert, on the other hand, is now criticized, together with MacLaurin, for the *limite*-method. The criticism is mainly aimed at a technical aspect of the definition of limits, because of the lacking of the concept of the absolute value even before a completion of an algebra of inequalities so that it seems as if the variable goes beyond the limit; the criticism is also aimed at the problem, which

---

20   This edition of the work follows the revised second edition of 1813.

has always remained controversial, whether a variable can definitely reach the limit or is only allowed to come close to it at any rate:

> MacLaurin and d'Alembert used the idea of limits; but one can observe that the subtangent is not strictly the limit of subsecants, because there is nothing to prevent the subsecant from further increasing when it has become a subtangent. True limits, following the ideas of the Ancients, are quantities which one cannot go beyond, although they can be approached as close as one wishes (ibid., 232 f.).

Lagrange conceded, however, that one could prove the foundations of differential calculus *rigoureusement* as MacLaurin, d'Alembert, and others had done, with the limit method, if defined in a specific way. But he insisted that the type of *métaphysique* that had to be used for it was contrary or at least foreign to the *esprit de l'analyse* (ibid., 233).

The strong impact Lagrange's two textbooks had consisted mainly in disseminating the message that differential and integral calculus was nothing other than a theory of functions, which could be developed wholly within algebra and free of metaphysics. This directly met the intention of the analytic method in the sense of Condillac and the *Idéologues*: through a clear language, through a good method to easily make understandable even apparently difficult theories and to lead it out of small circles of experts and to make it into a general instrument of application to sciences.

There was a third work, which—although not connected to the *École polytechnique* by institution—demonstrates the break with the traditional presentations of infinitesimal calculus in France and the tendency toward algebraization: it is the two volume textbook *Traités de Calcul Différentiel et de Calcul Intégral* (an VI/1798); by Charles Bossut, then 68 years old and free of his decade–long ties to the system of military engineers of the *Ancien Régime*, during which he did not publish a textbook on infinitesimal calculus despite having written numerous others.

In contrast to the dominance of the synthetic method in the old system of military schools, Bossut also presented a contribution on algebraization with this textbook. Using Newton's *Principia* as an example, he criticized the synthetic method and emphasized the advantage of the analytic method, especially for understanding great discoveries:

> One finds here obscurity, demonstrations drawn from convoluted sources, a usage very affected by the synthetic method of the Ancients, while analysis would have made the spirit and progress of invention much better understood (Bossut 1798, vj).

The methodological optimism concerning the analytic method even leads Bossut to—this explicitly for the first time, as far as I know—substitute the traditional concept of limited human knowledge with the research imperative of the unbounded task of science:

> It often happens that an idea is confined apparently within a fixed and determined space, it grows upon reflection, and becomes the nucleus of a corpus of science which has no bounds (ibid., ij).

Bossut presented his algebraization of the infinitesimal calculus as an adaptation of Euler's concept of calculating with differences and zeros:

> I suppose differentials to be zeros or quantities which are evanescent and indeterminate, between which there can exist certain ratios which are the same as those between finite quantities (Bossut 1798, lxxx).

For Bossut also, the concept of function was the starting point and foundational concept. The calculus of differences was a simplified version of the presentations in Euler and de Prony. Bossut formulated the transition from finite differences to differentials without reflecting on limit processes or the existence of limits. For this, he continued basing himself on the *infiniment petits*, which he, however, explicitly called a *hypothèse*, which led to important shortcuts (ibid., 91). He also added that no definite ontological status had to be given to these quantities; it was enough to consider them as incomparably small:

> It can be seen that it is not necessary to attribute a real and physical existence to an infinitely large or infinitely small quantity: it suffices to imagine that if in a question one attributes to a certain quantity a value, as great or as small as one wishes it to be, there is always, beyond that limit, a quantity of the same type which is even greater or smaller, which is, in this sense, an infinite quantity, or an infinitely small quantity, compared with every finite magnitude of the same type (ibid., 92).

Except for the principal legitimizing of the transition from finite differences to the differentials, Bossut only used the "hypothesis" of the infinitely small quantities to explain the product rule, here also carefully called a means of a "shortcut" (ibid., 99).

Integral calculus is, however, introduced as formal inversion of differentiation, as based on the indefinite integral—despite the presentation of calculating finite sums by the means of equidistant values.

Even though the transition from finite differences to differentials was unproblematic for Bossut, he also strongly emphasized the principle of continuity as a regulating principle in his *Discours préliminaire*:

> Everything is subject to the law of continuity, in the world of the intellect just as in the succession of physical beings. Human knowledge is formed, develops, and propagates itself by gradations which can escape the eye of the common man, but not that of the attentive philosopher (ibid., ij).

With this he confirms that the reflection on continuity had now become a key element in the new French textbooks.

Despite his connecting Euler's approach with older approaches, which had room for improvement, Bossut's textbook confirms the intensity and the extension of the break with the tradition after 1789.

3. Le Retour du Refoulé: The Renaissance of the Synthetic Method at the École Polytechnique

3.1. The Original Conception

As has already become clear in the previous Sections, the *École polytechnique*, a school famous for its role–model function in science, was an impressive realization of the algebraizing program of the analytic method since it was founded in 1794. In contrast to this well-known function, it is less known that in 1811 there was a definite break with this conception, which meant a return to the synthetic method previously pushed successfully into the background, to a new dominance of geometry, in the name of intuitiveness. This drastic turn is addressed here with an expression commonly used in France for such processes: *le retour du refoulé*. The process of reversion will be described here in a condensed summary; a detailed presentation is given in (Schubring 2004).

The starting point is an outline of the original curricular concept from when the school was founded in 1794. The basis of the curriculum of the college, which was called *École centrale des travaux publics* at first, was Condillac's understanding of method and that of the philosophical school of the *Idéologues*, the uniformly understood method of *analyse*. The conceptions of knowledge are clearly formulated in the first curriculum of the new institution. The curriculum was published by the *Comité du Salut Public* after the founding resolutions of the third and seventh *vendémiaire* of the year II: the *Développements sur l'enseignement adopté pour l'École centrale des travaux publics*. At the new school, especially two fields of study were to be taught: mathematics and chemistry/physics. In the understanding of the method of *analyse*, both areas of knowledge are distinguished in that they are languages of a high degree of *perfection*, languages that therefore are an especially suitable means for finding new knowledge and for the presentation and teaching of knowledge. Besides this, the *géométrie descriptive* had a particularly distinguished function because it embodied social unity in the division of work between theory and practice. In the first curriculum, separated subjects were not planned, but integrated concrete realizations of the concept of *analyse*. The central content of teaching, *analyse*, did not mean, therefore, analysis in the sense of differential and integral calculus, but the *mathematization* of the understanding of nature regarding the *movement* of bodies. Thus, analysis was a basic subject related to application, which was to be taught in three different areas of application:

- *analyse appliquée à la géométrie,*

- *analyse appliquée à la mécanique,*
- *analyse appliquée aux calculs des effets des machines.*

The general goal of education at the engineering school was oriented on the role model of scientific generality and rigor.

3.2. Changes in the Structure of Organization

Furthermore, this extraordinary concept is so novel because theory and applications were combined at this new institution and because the basic education was integrated in the education for the different engineering professions in the state sector, both the military and the civil. This institutional unity was, however, broken up after only one year. The *École centrale des travaux publics* became a preparatory school for the *écoles d'application*. As an institution that was to teach the fundamentals of knowledge common to the different areas of application, it changed its name to *École polytechnique* on the 15th Fructidor of the year III (September 1, 1795). The previous three–year-long education remained only as a general framework, but for most of the engineering majors it was shortened to two years, after which there was a one- to two-year period of studying at a school of application. In contrast to the original integrated conception of the education, for every state profession in engineering a separate school of application was founded with the law of the 30th Vendémiaire of the year IV (October 22, 1795):

- for military professions:
  - ➤ *École d'Artillerie,*
  - ➤ *École des Ingénieurs militaires*
- for civil professions:
  - ➤ *École des Ponts et Chaussées,*
  - ➤ *École des Mines,*
  - ➤ *École des Géographes,*

plus three types of naval schools for both military and civil purposes.

The main structural point of realizing the character of a preparatory school was that only those were admitted as students to the schools of application who had successfully finished their studies at the *École polytechnique.*

The reasons for these radical changes, which had far-reaching effects on the curriculum of the *École polytechnique*, have not been the object of more in-depth research, despite the numerous studies of its history. Similarly, there has been traditionally less research on the schools of application, and only very recently were more detailed studies done (Bret 1991, Picon 1992, Belhoste/Picon 1996); there is a special lack of studies dealing with the development of the institutional and curricular relationship of these different *écoles d'application* with the *École polytechnique.*

Preceding all schools of application there were institutions in the *Ancien Régime*, however, in a different way. Military schools for the training and education of officers were of the greatest number; admittance was generally restricted to the nobility. Among the five armies the *corps du Génie*, the artillery and the navy were considered the "corps savants" in the second part of the eighteenth century, because their officers needed theoretical knowledge, in contrast to infantry and cavalry (R. Hahn 1986, 520; cf. above, Chapter II.2.6). The *École du Génie,* for the education of military engineers, enjoyed the highest status. After the Revolution, however, it fell into a crisis, because of its strong ties with the nobility, and practically closed down in 1793. At the beginning of 1794 the parliament, the *Convention*, had ordered that the school be immediately moved to Metz, but reduced in form and function. The school had to give away a major part of its library and laboratories to the new central school, and only the practical military training remained there (cf. Taton 1986, 609).

In 1720 regiment schools in garrisons close to the borders had been founded for the artillery. Since 1779 there was a system of central entrance tests for regimental schools (R. Hahn 1986, 516 ff.). An *école d'artillerie* proper was founded only in 1791, in Chalons, as a consequence of the interventional wars against France, and was to be broken up—as Fourcroy emphasized in his justification of the law in 1795—when the war was over in favor of the practical regimental schools that were continually functioning.

Among the schools for civil engineers, the *École des Ponts et Chaussées*, founded in Paris in 1747, was the most famous (cf. Serbos 1986 and Picon 1992). In 1791 it was confirmed as *école gratuite et nationale des ponts et chaussées* by the parliament,[21] but fell into a crisis shortly afterwards and had almost no students. After the foundation of the new school it was only provisionally maintained by the *Comité du Salut public* (Bradley 1984, 41).

On paper, the system of the schools of application looks so extensive and influential that one can assume a direct change of curriculum. The fact is, however, that the original curriculum was essentially confirmed, so that one has to conclude that the *écoles d'application*, on the one hand, were still being planned and, on the other, that they had not yet become the counterpart of the "users."

It seems that in the literature on the early history of the new school the intended unity of theory and applications has been taken too literally in an idealizing way. Since no year of students finished the education according to the original concept, it cannot be deduced in which way the connection to the practice had been planned. But it is obvious from the existing concepts that the new central school was not to manage the whole practical education for the single *corps*, which were still kept and could not do it either. The end of the practical training was to take place at the engineering schools that continued to be run by practitioners who worked in their respective jobs The serious change

---

21   Procès-Verbal de l'Assemblée Nationale, tome 61, Séance du 1 juillet 1791, 20.

decreed by the structure law of 1795, therefore, meant that instead of founding *écoles particulières* for short practical phases of training, schools were to be set up with their own educational assignments, which included significant parts of the applicational knowledge besides the practice period and lasted at least one to two years.

3.3. The First Crisis: Pressure from the Corps du Génie and the Artillery

Only two years after its foundation, the school fell into a massive crisis; it originated in the *corps du Génie*.

Not only were the single schools of application connected to the *École polytechnique* to form one system of education, they also each represented one element in their own social system: the respective engineering profession, each of which developed its own intense way of life with its career paths, types of canonized knowledge and hierarchies thereof, with a network of inspectors and specific connections to the state administration, a way of life that could easily turn into a practice of caste spirit. Among the subsystems connected to the schools of application, two had already existed in the *Ancien Régime*, represented by influential corporative *corps*: for the *Génie* and for the *ponts et chaussées*. To both schools admission was socially restrictive. The older *corps* of the military engineers, which went back to the *ingénieurs du roy* of the sixteenth century, had acquired its own administration because of the foundation of the *Département des Fortifications* in 1691 and the change to *officiers du Génie*. During the French Revolution all corporations had indeed been dissolved (cf. Schubring 1991a, 285 ff.), but these social subsystems, which were interwoven with the state administration, were only partially affected. The privileges of the nobility were terminated, and the institutions of education were generally opened.

Only shortly after the new structures and the new curriculum had been introduced, massive opposition against their permanent establishment began to manifest itself, partly openly, partly concealed. This opposition came from parts of the military, but not from the civil corps. The opposition began in the artillery and led at the beginning of 1797—at the same time when royalist forces openly emerged on the political scene ever more strongly and became more and more influential in France—to a frontal attack of the *corps du Génie* on the *École polytechnique*, because of which it fell into an existential crisis that lasted two years, which could be ended only by Napoleon's coup d'état. An analysis of this crisis and a comparison to the field of civil engineering leads to another characteristic of the military schools of the *Ancien Régime*, i.e., the *examinateurs permanents*. Despite the clear and obvious privileged position of these examiners, the French Revolution did not infringe on this structural characteristic. For each of the three *corps savants*—the navy, the *Génie*, and the artillery, there was an external examiner who, appointed for life, tested the candidates and graduates of the respective schools.

The examiner for the artillery was Laplace before the Revolution as well as after, with only a short break from 1793 to 1795. He used this position to work against the integrated teaching conception of the *École polytechnique*. Because graduates that had been examined by him and were to be admitted to the artillery school allegedly knew too little, he criticized the teaching method in December 1796 and called for a standardization of the teaching content and especially for a significant increase in courses on analysis and mechanics.

These attacks alone would not have caused a big crisis, because of the particular status of the artillery corps. The school, however, was hit hard by a new frontal attack at the beginning of 1797, which came from a part that was much more central to the concept of education: from the *Génie* corps. At the *École polytechnique* it was thought that Laplace was also the instigator behind this attack (Pinet 1887, 395). The reason had been an order of the minister of war addressed to the *Comité central des Fortifications* to prepare a reorganization of the *École du Génie* in Metz and by doing so to establish an effective relation between the *École polytechnique* and this school of application, thus implementing the law of the 30th Vendémiaire of the year IV. Instead of following this order, however, and making the still disorganized school ready for its new function, the *Comité* declared on the 6th Pluviôse of the year V (January 25, 1797) that several decisions on the *École Polytechnique* in the law mentioned obstructed the implementation of the order; these regulations had to be changed first. The list of nineteen such demands for change in the *avis* of the *Comité* called into question the substance of the school. That students of the schools of application had to finish their basic studies at the *École polytechnique* was written off with the politically charged word "privilege," and the *Comité* demanded the abolition of this privilege. One of the motives was that the military was in competition with the interior department and especially with the *corps* of the *Ponts et Chaussées*. They alleged that because of better living and working conditions, qualified graduates primarily chose a career as civil engineers, and for artillery and *Génie* only the less qualified were left. To prove this, they referred to the bad results of the examinations administered by Laplace (Fourcy 1828, 113). Further drastic demands included not only such points as barracking the students, changes of the examinations and a strictly limited two-year duration of studying, but also radical interventions in the teaching program. All courses were to be canceled, for instance, that could touch a field of study of a school of application.

To look at the reasons for the crisis, the *Comité central des Fortifications* has to be analyzed. The National Assembly formed the committee in 1791 to counsel the minister of war, and it was especially responsible for all plans of fortifications, for the budget, and for the teaching at the *École du Génie*. Basically, it was a self–administration of the *corps du Génie*, since its members were leading officers of this *corps*. A prosopographical study of its members from 1795 to 1812 found out that all were members of the *corps du Génie* since the time before the Revolution and were graduates of the school of Mézières. Only little by little, from about 1800 on, did it open up. In the decisive year V, the committee consisted of fourteen members. Almost all of them had finished their education at the *École du Génie* in Mézières and were tied to the *Ancien Régime*; the membership of Carnot-Feulins is conspicuous, the younger brother of Lazare Carnot, who at the same time was in the *Comité de Salut public* and in the *Directoire Exécutif* thereafter, the center of power and decision–making in the French state. It has to be assumed that Carnot-Feulins fully shared the *avis*; there are even signs that he belonged among its active supporters (Schubring 2004, Chapter 4). This also makes it doubtful whether Carnot himself—he, too, was a member of the *corps du Génie*—belonged to the supporters of the *École polytechnique*, as is assumed as self-evident in the literature.

The principal attitude of the *Comité* was to not recognize the task and structure of the *École polytechnique* and to demand the restoration of the former *École du Génie*. At the most, basic education in mathematics and physics was granted to the school, but fortification was claimed to be the *Génie* school's own assignment. The *Comité* especially took over Laplace's demands (Schubring 2004, Chapter 4). The *École polytechnique* was to hand over all application courses to the schools of application and to restrict itself to the general education of engineers. Any other purpose of education that went beyond this, for instance a scientific education, was rejected.

Looking at the way the *Comité central des Fortifications* dealt with the matter, one is especially surprised by the carefree attitude with which it ignored the legal regulations, despite the desolate state of the *Génie* school in Metz, and called for a radical reorganization that would have led to a restoration of the structures of the *Ancien Régime*. The Revolution had apparently not infringed on the self–confidence of the engineering *corps* at all. In addition, it clearly relied upon numerous supporting structures in the government and the administration.

The situation for the school became even more complicated from October 1796 on, because on the one hand, the teaching program was sharply criticized and reductions were demanded in parliament, in the *Conseil des Cinq-Cents*, and because Laplace put the ministry of war under pressure by suspending the artillery exams in order to keep the artillery school independent of the *École polytechnique*. The interior ministry felt compelled to have a new organizational plan worked out for the school with a reduced budget. The new plan, which the ministry finally approved in January 1797, whose text has not yet been found and whose precise content is, therefore, unknown (cf. Schubring 2004, Chapter 5), had to be presented to the government, the *Directoire*. The *Directoire* discussed the plan at the end of January, but then tabled it.

This delay gave the fortification committee the chance to speed up the planned extension of the *Génie* school. Indeed, in intensive discussions and meetings from January till May, it set up an organizational plan for this school in

Metz, which had it function practically independently of the central school in Paris. It provided not only a complete education in fortification, but also an independent education in mathematics and physics, for which two lecturers were intended. In addition, the "reintegration" of the teaching materials of the former school in Mézières was called for, which mostly had been given to the *École polytechnique*. The minister of war, Pétiet, accepted this plan and in April presented it to the *Directoire* for approval, also referring again to Laplace's criticism, and demanded a quick implementation. These tactics were successful. The *Directoire* gave priority to the decisions about the *Génie* school, took over the bill of the minister of war, allowed the extension of the school in Metz, and, using this plan as a base, restructured the *École polytechnique*. While it was proposing several changes to the parliament, since these implied altering earlier laws, it immediately put some of them into action as decrees. Thus the Paris school had to give the teaching materials from Mézières to Metz. In addition, it was forced to immediately cancel the courses on fortification and to let their lecturers transfer to the school in Metz (ibid., Chapters 5 and 6). The chemistry laboratories were drastically reduced.

Since Lazare Carnot was responsible for all questions concerning the military in the *Directoire* and since the republican majority of the *Directoire* forced Carnot to fire Pétiet in July 1797 (Amson 1992, 196), it is obvious that Carnot also supported the *corps du Génie*. That Carnot acted more in favor of his *corps* than of the new *École polytechnique* can also be seen from the fact that after his removal from office in September 1797 the school was able to work again according to the conception of 1795: the changed *Directoire* allowed the application courses to continue and even increased the budget (Fourcy 1828, 151 ff.).

The parliament, however, was still working on the assignment of the old *Directoire* to make fundamental changes in the structure. Both chambers of the parliament blocked each other for several years, until 1799. Whereas the first chamber even planned far-reaching improvements, e.g., continuing the application courses, the second called for the abolition of the "privilege," i.e., that only those students could be granted admission to the schools of applications that had finished their education at the central school beforehand. Shortly before both chambers settled on such a law, Napoleon's coup d'état on November 9, 1799, saved the Paris school. Laplace, above all people, now interior minister appointed by Napoleon, changed the bill and presented it to the provisional legislature. There the law was passed on the 25 Frimaire of the year VIII (December 16, 1799). After almost three years of uncertainty there were now solid foundations.

3.4. The New Teaching Concept of 1800

The law of 1799 is regarded as an important success for the *École polytechnique* in the literature. The fact is that the main point of conflict was removed. The so-called privilege, whose abolition already seemed certain, was now definitely confirmed; the candidates for the schools of application had to study at the preparatory school in Paris. Furthermore, now—in contrast to the law of October 1795—an effective link between the *École polytechnique* and the *écoles d'application* was established, for the schools were required to coordinate their curricula. The *École polytechnique* was responsible for the coordination of this, through a newly founded committee, the *Conseil de Perfectionnement*, which represented itself as the new network between foundations and applications in civil and military state offices. In addition to representatives of the leading board of the school, the *Conseil d'instruction et d'administration*, the functional "environment" was represented by permanent examiners and members of the engineer *corps*, who were connected to the schools of application. There was no limitation to the *services publics*, as the fortification committee had demanded, but also an admission of further students for a scientific education. A large part of the application courses were kept, that is, under cover of "applications of descriptive geometry," namely, fortification as well as the *travaux civils*, architecture, mining, and naval constructions, which were related to the construction of roads and bridges (Fourcy 1828, 193 ff.).

Considering the now definitive anchoring of the *École polytechnique* into the state system of education, it has not been taken into account that main goals of the demands by Laplace and the *Comité central des Fortifications* of 1796/97 had been able to prevail and that the original conception of the curriculum had to be abandoned.

A specialization in the respective fields had taken place; leaving the overall concept of *analyse* behind—originally in the sense of the analytic method conception—the mathematical subject "analysis" was established, as differential and integral calculus and theory of curves with preceding algebraic foundations. In the same manner, mechanics had been established as a subject proper. The strengthening and specialization of analysis and mechanics since 1797 are for a significant part the result of pressure exerted by Laplace, which he could effectively exercise because of his extraordinarily strong position as examiner. He did not support descriptive geometry connected with the name of Monge. It is actually a paradoxical result of the pressure from the field of applied engineering to state that a process of specialization took place that increased the theoretical parts of the education and enforced the importance of the foundations, that is, those of mathematics, since chemistry was forced to play only a minor role.

The pressure from the military *corps* to further reduce the application courses was continued also after 1799 and even was, which is remarkable, instrumentalized in the *Conseil de Perfectionnement* to strengthen the foundational education even more. Reacting to demands voiced since 1803 to cut all *cours d'application*—and especially the course on fortification (Fourcy 1828, 265)—in 1805 the council finally decided to completely restructure the courses on descriptive geometry. From this course, in which was to be included everything that was of common interest for the different engineering professions, everything was to be removed that belonged to the details of one profession without disadvantage. From the year 1806 on, the previous application courses, as part of *géométrie descriptive*, were substituted by three new courses. By this the impression of a competition with the "special" schools had indeed been removed, but at the same time this abandoning of the original conception of the curriculum meant an enormous enhancement of both analysis and mechanics. The basic structures of the curriculum of 1806 remained constant for the next decades. Since 1806 at the latest, one can, therefore, rightly speak of the *École polytechnique* as an *École Laplace*, in contrast to an initial *École Monge*.

3.5. Modernization of the Corps du Génie and Extension of the School in Metz

From the beginning of the Napoleonic period, the *Comité des fortifications* changed significantly in terms of its members and therefore also in terms of its politics. The modernization of the *corps du Génie*, which took place at the time, becomes especially clear in the person of Pierre-Alexandre-Joseph Allent (1772–1837), who then was of central importance and function. I became aware of Allent, who has not been noticed in historiography so far, while studying the organizational and curricular relations between the school in Paris and the schools of application. Allent took a "revolutionary" path in his career as an officer: he joined the army as a volunteer in 1792, qualified as an officer by himself and was admitted to the *Génie* corps in 1793. After a long time as a Carnot's secretary, he took over the administration of the corps in 1799 and became the representative of the corps in the *Conseil de perfectionnement* of the Paris school in 1802. He practically was a member of all committees that reformed the curriculum of the school in Metz. He worked out and pushed through the conceptions that defined the definitive relation of the basic education in Paris and the education in the applications for the engineering professions (Schubring 2004, Chapter 7).

The *École du Génie* in Metz remained a nuisance factor in the network of *École polytechnique* and schools of application, but no longer as an exponent and expression of the politics of the entire *corps du Génie*, but on its "own account," so to speak, that is, out of the interest to become an independent institution. The fortification committee no longer let itself be made into the mouthpiece of the school in Metz—in sharp contrast to the previous unquestioning support of their interests—and considered carefully their special interests and the overall aims of the education network for engineering professions. In this, Allent held a decisive and promoting position.

In 1802 the *Conseil de Perfectionnement* had been able to solve one of the basic problems that had existed up till then: The *École d'Artillerie* in Chalons finally was not only definitely included in the network of schools of application, but was even united with the *Génie* school in Metz. The coexistence of two such different fields of education and *corps*, however, did not make the integration into the network any easier. Even though the school, now operating under the name *École d'artillerie et du Génie*, still kept trying to evade integration into the network by obstructing the process, the *Conseil* in Paris, thanks to the intervention by Allent in 1806, finally managed to move ahead with the integration. The conflicts that occurred during this process illustrate the intention of obstruction.

Indeed, it runs like a central thread through the behavior of the school since its transfer to Metz in 1794, not to recognize the character of a—subordinate—school of application and to demand rather the status of an independent "special" school. The main expression of these politics was that the foundations of mathematics and physics were to be taught at their school, and their own lecturers were to be appointed. The ambitions of the school are illustrated by the fact that they wanted to win over Fourier for it in 1797. The fact is that since 1798 the school had been able to have a lecturer for mathematics and physics—whose position was played down on the outside as one of a "répétiteur." When in June 1806 the school went so far as to plan on installing yet another position for the teaching of foundations, namely for descriptive natural sciences, and on top of this, from the money for the application courses, which was to be the actual teaching assignment, the minister of war finally felt compelled to intervene. Together with the *Conseil* of the *École polytechnique* and under the direction of Allent, a detailed curriculum for the school in Metz was worked out and decided on in March 1807, which was to prevent any spreading of the schools of application to the field of foundations and which was to perfect the training in applications. In 1803 Allent developed the conception of integrating the curriculum as a compromise formula. The underlying principle of the relation of *École polytechnique* and the schools of application was that the teaching programs had to fit together, that is, without any gaps and without any repetitions. The characteristic, therefore, was a separation of *theory*, i.e., mathematics and natural sciences, from *applications*, i.e., military disciplines. Thus, the education of engineers was generally structured in two parts: an *instruction preliminaire sur les sciences et les arts* and an *instruction spéciale sur l'artillerie et la fortification*. In contrast to the previous practice in Metz, the first part was to be taught completely at the *École Polytechnique*, at the expense, however, of the application courses there.

Before the final curriculum was laid down, the school in Metz had brusquely and intransigently rejected any coordination with the *École polytechnique* and had even tried to justify its own basic education with a flawed education in Paris. After an inquiry about them, the flaws could not be verified. It was clear to all that this attempt was just an offensive that was to bring relief for the declared intention to get their own curriculum and choice of teaching staff.

With the so-called *plaintes de Metz* the school took the offensive again in 1810. They have always been described in great detail in the literature on the history of the *École polytechnique*. There the complaints appear as a sudden and serious break in an otherwise continuous development, as a drastic threat to the education at the *École polytechnique*, since it seemed, for the first time, that the complaints were based on a reliable evaluation. A thorough analysis of these events, however, forces a different conclusion.

In a report from November 5, 1810, to the minister of war, the *Conseil d'instruction* of the *Génie* school in Metz asserted a number of flaws in the preparatory education in Paris. The report,[22] which declared the flaws as the result of "experiences of several years," claimed them to be a threat to the success of the teaching of applications in Metz:

- Almost all students could draw only very little; there was no precision in the execution of their sketches (*levers*) of the mining industry, buildings, and fortifications.
- The students did not bring with them any memories of the courses taken by them on machines, constructions, architecture, and the military.
- The reason for this was that these courses were of too little importance at the *École polytechnique*.

Since such problems had not been discussed in previous meetings in Metz, and since in 1806 the position of the drawing teacher was to be canceled because of the good education the Paris graduates already possessed, other motives can be assumed here again. Indeed, at the same time there was under a new *sous-inspecteur* a new extension of the school on its way: a resumption of courses in mathematics and physics and a building up of a chemical laboratory. From October 1810 on there were indeed meetings about learning problems students had, but not as a consequence of their education in Paris, but for completely different reasons. There were complaints about the students of the artillery school not working hard enough. This was attributed to the way exams had been administered so far, which made the students believe that they would be assured of a job even without taking any pains. Indeed, this was a main problem of the artillery education, since graduates of the *École polytechnique* with the least capabilities took up this career path. Because all candidates were actually hired due to the needs of war, studying carelessly was all the more understandable. This behavior was changed for a short time by a sudden order of the minister of war in July 1810: the artillery students were to be examined by September 1, and

---

22   The entire text is printed in Appendix B (Schubring 2004).

the forty best were to join the artillery *corps* immediately. This order resulted in hard work and great efforts by the students until the exam, but it had negative consequences for the artillery students who remained at the school afterwards. They thought the lessons, additional from their perspective, to be completely useless. Their teachers complained that it was difficult to receive any drawings from them at all. The *Conseil d'instruction* in Metz could, therefore, be sure that—if its complaints about the *École polytechnique* were going to be checked within the next months—there were indeed students with bad learning results, not, however, caused by the education at the *École polytechnique*.

3.6. The Crisis at the École Polytechnique in 1810/11

The *École polytechnique* had to take these complaints seriously, since after the militarization of 1804, major parts of it were under the supervision of the same minister of war, who was also responsible for the school in Metz, and since this school was the main "customer" of their graduates due to the war, only a few transfers to careers of civil engineers took place.

If one looks at the points of criticism mentioned in Metz, practically only the weakness of the graduates in graphic works remains as a strong point. This weakness can be seen as the natural consequence of the fundamental change in the conception of the curriculum of the year 1800, since from this time on descriptive geometry had been pushed into the background. The *Conseil d'instruction* and the *Conseil de perfectionnement* had indeed recognized this weakness; after 1809 it was discussed several times, and measures of remedy were decided. At the same time, in the fall of 1810, it became clear at the *École polytechnique* that the dominance of analytic methods could no longer be maintained. The *Conseil d'instruction*, therefore, demanded on October 19, 1810, that it be added to the requirements in mechanics that proofs of statics were to be done "synthetically." A subsequent correction explains what was meant here by this ambiguous term: "geometrically." The emerging reorientation concerning the method can also be seen from the fact that the *Conseil de perfectionnement* approved of the preeminence of the synthetic over the analytic proof method.

The first meetings at the Paris school after the *plaintes* became known did not show a basic call for action. The included *Comité des Fortifications* did not take the complaints from Metz seriously either. It made, however, some suggestions to the *École polytechnique*—based on a few key words, which went far beyond the complaints, in an evaluation by Bossut—regarding a reorientation in terms of contents—, which only had the motive with the initial demands from Metz in common. The *Comité* had developed into a competent authority on the whole educational system and convinced, therefore, again under the direction and initiative of Allent, who was committed to the subject, but not partisan, the *École polytechnique* to change its educational concept according to the logic of the engineering education. The *Comité* and Allent managed all the more to have this whole logic prevail, since at the same time the complementary goal of the education of teachers and scientists had been given to an independent institutional system, the *Université impériale*. It now criticized, practically "from the inside," in well-informed details the curricular conception in the fields of analysis and mechanics. For the revision of their teaching programs it demanded especially the following:

- The courses were to be limited to such theoretical questions that could be applied to the *services publics*; exclusion of those questions that do not offer direct applications there (yet).
- Instead of the analytic solutions, "synthetic or approximative" solutions were always to be taught. Such solutions were qualified as "less rigorous or less elegant," but they were to be preferred by the engineer for the sake of shortening or simplification.

At the beginning of 1811 the committees of the *École polytechnique* were able to ward off these demands. Even the repetition of the exams, instigated by the ministry of war, for students in Metz did not lead to comprehensive steps despite the catastrophic results (Fourcy 1828, 300 f.). Only when Etienne Malus noted that students of the second year no longer had knowledge of the first year, an internal change was instituted. At the end of April 1811 it was decided to test students of the second year on topics from the first year. Even though the results were not as bad as had been feared, the *Conseil* was shocked by the forgetting of the "elements." Although the phenomenon of the forgetting of learned foundations is known to us today as a didactic phenomenon at schools and universities that does not allow direct conclusions about bad teaching success, this assessment was shocking for the *Conseil* back then not only because of the lack of a didactic tradition at universities, but also because of the underlying conception of knowledge that was based on the assumption that all knowledge was interwoven and connected, which was built up methodically from a few basic elements and which could be reconstructed by attentiveness and consciousness. The failure of students in the elements was interpreted, therefore, as a failure of the methodological concept of the *École polytechnique*. The conviction of such a failure now caused the breakthrough. For there now prevailed what the school had resisted in the last six months: the abandonment of an independent theoretical perspective and task, and the integration of the school into the exclusive perspective of the education of engineers.

On May 19, 1811, it was decided then to remove from the teaching program all knowledge that was not essentially useful to professional reality. The subsequent general revision of the courses had the effect that in mechanics significant shortenings and exclusion of theoretical parts were decided on. The most drastic change, regarding the self–image of the institution, in the conceptual fundamentals concerned analysis. In June, the preparing commission had proposed a principal change concerning method and content, which the *Conseil* passed on July 13, 1811. The main consequence of the return to the prerevolutionary preeminence of the synthetic method was that the limit method was to be substituted by the method of the *infiniment petits*. In contrast to the original evaluation by the fortification committee, which at the time was based on the assumption that the synthetic method lacked rigor, in the decision of the *Conseil* the method of infinitesimals was characterized to be as analytic as the limit method, which meant that it was comparable in terms of rigor.

That this return to a concept of the synthetic method was the main change in the curriculum is underlined by the annual report of the school for 1811 to Napoleon from May 1812, in which this change was explicated as the most important part of the intended simplification and orientation to real life:

> In the mathematical sciences they [these changes] are aimed at simplifying the studies and orienting toward real life: 1. For the presentation of differential calculus, the method of limits was substituted by that of the *infiniment petits*, which is simpler and to which one has to come back in mechanics.[23]

---

[23]   The precise sequence of meetings is described in Chapter 10 of (Schubring 2004); it is also discussed how the decision on the revsion was achieved.

Chapter V

Le Retour du Refoulé: From the Perspective of Mathematical Concepts

1. The Role of Lazare Carnot and His Conceptions

1.1. Structures and Personalities

> Généraliser, c'est simplifier.
> —Carnot 1785, fol. 85, no. 100

Lazare Carnot epitomizes the movement toward rigor introduced in France by d'Alembert, and the problems and contradictions that this movement encountered in trying to perform generalization while simultaneously retaining intuition. In line with his goal of rigor, Carnot concentrated his work in mathematics on its foundations. All his mathematical texts contain either *principes* or *metaphysique* in their title or aim to develop foundational concepts. It is not just his major impact on general ideas about mathematics that makes Carnot so important, but, in the present context, the fact that his work on foundations deals with precisely the two fundamental aspects addressed in this book: negative numbers and infinitely small quantities, both of which he wished to define precisely on the basis of one global concept.

History as a science reveals a traditional contradiction between accounts oriented toward the analysis of structures and those oriented toward the actions of persons. This results in contradictory depictions of history: structural versus biographical. However, there are also rare personalities whose biographies are so exceptional that they describe and explain the history of general structures as well. One of these exceptional personalities is Lazare Carnot. He embodies the scientific orientation of the Enlightenment just as much as the attempts to put it into practice in the French Revolution. He represents the assumption of leading state functions by the intellectual bourgeoisie just as much as the conflicts with Napoleon's struggle for power and the opposition to the Restoration. This is why Carnot, a leading figure in this period of radical change, continues to fascinate up to the present day, as confirmed by five major biographies in the last 15 years alone. The latest, from Jean and Nicole Dhombres (1997), is simultaneously the most comprehensive attempt to grasp this exceptional personality in terms of his three main fields of activity: science, the military, and politics.

Moreover, one aspect of Carnot that has not received enough attention up to now is that he did not just live through but was also, in part, responsible for the radical change in epistemology in France during the period extending from the beginning of the Revolution across the Napoleonic period and up to the Restoration—from the pinnacle of analytical methods with his scientific orientation and Condillac's algebraization program up to his renunciation of scientific values in favor of those of classical literature and the revival of the synthetic method.

Because I wanted to take a closer look at this conceptual reorientation, I spent a long time trying to obtain access to Carnot's *Nachlass*. This proved to be divided into several parts, mostly corresponding to the different places where Carnot and his family had resided.

Only one part of the *Nachlass* is available to the general public, in the national archives at Paris, an endowment by Paul Carnot in 1955. Although this contains a number of important personal documents, these are not directly relevant to his scientific career. Nonetheless, this is where I found Dégérando's obituary for A. Allent, which made me aware of Allent and his importance (see Chapter IV.3.5.). In addition, there are family archives at two former family homes at the country seat of Presles at La Ferté-Alais to the south of Paris and at Nolay.

Carnot purchased Presles with its small castle in 1807. Although forced to sell it in 1815 when he fled into exile, it remained in the possession of the family: It first went to an in-law before being repurchased by his son Hippolyte. The main building contains not only a museum dedicated to Carnot and the Revolution but also an extensive collection of Carnot's books. Major structural damage to the building prevented access for a long time. After restoration was complete, I was able to spend some time there through the kind permission of the present owner, Thierry Carnot. The actual scientific part of the *Nachlass* is not there, except for a detailed correspondence between the Abbé Grégoire and Carnot. Nonetheless, the library forms an important and interesting collection. It also contains a catalogue of Carnot's library listing a total of 6,519 books from all areas of knowledge including poetry and literature. Looking through this catalogue, I noticed the work of Porro on negative numbers (see Chapter IV.1.4.). However, I was surprised to find that the catalogue and library failed to correspond in two ways—at least for the mathematical literature in which I was interested. First, the Carnot family had always assumed that the catalogue listed those books to be found in the present library. In fact, a large number of the mathematical books in the catalogue are not present, and the family has no explanation for this. I did manage to find a few individual books from the catalogue in the adjacent tower that Lazare's grandson, President Sadi Carnot, had converted into his own library. Second, the library also contains books that are not indexed in the catalogue, which, for all intents and purposes, only goes up to 1815. These include, for example, the German books on mathematics that Carnot acquired during his exile in Magdeburg.

It is conspicuous that a large proportion of the mathematical books date from the period after 1800. A number of these contain dedications to Carnot (e.g., from Lacroix and Legendre). However, there are also earlier books, from the eighteenth century. Some of the mathematical books even contain bookmarks. One can still find, for example, the original edition of Condillac's *Langue des Calculs* (1798): The sewn copy has had its pages cut and has obviously been read.

The second family archive is at the original family seat of Nolay in the Côte-d'Or. Charles C. Gillispie was the first researcher to explore its relevance for the history of science in 1964. One of his finds was the draft of a letter from Carnot to J.K.F. Hauff, the German translator of his 1797 book on the foundations of infinitesimal calculus. This revealed that Carnot had written the first version of this text in 1785 for the prize competition at the Berlin Academy (Gillispie 1971, vii). It was after this that A.P. Youschkevitch and K.R. Biermann discovered the manuscript in the Berlin Academy Archive that was published subsequently by Gillispie and Youschkevitch in 1971. A later visitor to the library was the military historian Charney. In his two-volume work *Révolution et Mathématiques*, he edited some text fragments from Carnot's *Nachlass* on the foundations of mathematics (Charney 1984/85), although he did not catalogue or analyze them.

My contacts with Marie-Sylvie Carnot, a cousin of Thierry Carnot, and her mother, the owners of this part of the family legacy, revealed that "Nolay" actually consisted of two different properties: the manor house in the middle of Nolay in which Lazare Carnot was born and a castle from the Middle Ages, La Rochepot, approximately five kilometers outside the town that the family acquired at the end of the nineteenth century. Although Lazare's *Nachlass* is located in the castle, the house in which he was born also contains an extensive collection of books. While none of its contents can be attributed directly to Lazare's library, this collection does contain old books, some dating back to the middle of the eighteenth century, that had most certainly belonged to his parents and been read by Lazare during his school days (see the next section).

Lazare Carnot's *Nachlass* at La Rochepot is very broad in extent and content. The mathematics collection contains a large number of manuscripts and drafts in which Carnot repeatedly used a variety of different approaches to tackle not only the relation between analysis and synthesis but also its application to both the concept of negative numbers and the concept of *infiniment petits*. Unfortunately, none of these manuscripts are dated. However, they can be brought into some temporal order by comparing them with Carnot's publications. I shall deal with this in more detail in the following sections.

1.2. A Short Biography

Because there are so many publications dealing with Lazare Carnot's life, this section will give only enough information on his biography to understand its relation to his scientific work along with the impact of his personality.

Lazare was born on May 13, 1753, in Nolay as the son of Claude Carnot and Marguerite Pothier. Although their eighth child, he was only the fourth to survive. Lazare was particularly close to the youngest of his three brothers, Claude-Marie, who later also became a *Génie* officer under the name of Carnot-Feulins. As a notary public, lawyer, and *bailli*, his father was a member of the nonhereditary nobility, the *noblesse de robe*. This would prove to be an impediment to Lazare's subsequent career as an officer in the *Ancien Régime*, because of the need to prove true noble descent.

The father first ensured a basic education in grammar and Latin at the school for the children of local notables in Nolay. Lazare attended this school until 1767, when he was 14 years old. He then went to Autan, the nearest larger city, with his elder brother Joseph. Whereas Joseph was intended for a career in the Church and studied at the *séminaire* of the diocese, Lazare attended the *collège* of the Oratorians. Instruction was based exclusively on classical values and the study of classical literature. However, he changed schools again after only one year. Joseph had to return to the parental home because of illness, and Lazare took over his place at the *séminaire*. Although not planning a career in the Church, he was still a devout Catholic like most of his contemporaries. The paradox is that it was within this clerical institution, run by the order of *Saint Sulpice*, that he developed his interest in the exact sciences. He graduated from the final class, that of philosophy, in which he received his first instruction in mathematics and physics. Although they do not give any sources, Dhombres and Dhombres (1997) report that this is where his original disposition toward *calcul* changed into a true enthusiasm for the exact sciences. Instruction in mathematics focused particularly on applied fields, with a strong emphasis on fortification. This meant that he had to teach himself pure mathematics (Dhombres & Dhombres 1997, 27). Although the authors suggest that he did this with the textbooks of LaCaille and Camus, there are, in fact, no indications that Carnot used Camus's textbook, which was specially written for training *Génie* officers. It is also not to be found in the library at Nolay. However, one book that is there is a work that rivaled the usual *collège* textbook from LaCaille and was particularly appropriate for instruction in physics, namely, that of Mazéas, professor of the philosophy class at the Collège de Navarre, University of Paris. This was published in 1761, the same year as the shorter version of Rivard's mathematics textbook, also present in the library. This latter book reports that it was published in this abbreviated form because the philosophy class assigned only four months to the teaching of mathematics (see Chapter II.2.8.1; Mazéas 1761; Rivard 1761). It seems very probable that Carnot gained his basic grasp of mathematics—arithmetic, algebra, geometry, and conic sections—from these textbooks written for university *collèges*.

Carnot spent only one year at the *séminaire* in Autun, graduating in the spring of 1769. It was decided that he should pursue a career as a military engineer. Thanks to the patronage of the Duke of Nolay, who employed Carnot's father as *bailli*, as well as a generous interpretation of the military rank of some of his relatives, Lazare was able to surmount the hurdle of having to prove that his family lived *noblement*. In November 1769, he was granted permission to take the entrance examinations for the *École du Génie* at Mézières. He had already started preparation for this entrance examination in May 1769, but through self-instruction, unlike the majority of candidates, who were "primed" at special institutions in Paris or at the *École royale militaire*. This was his first exposure to the textbooks of Camus: these four volumes—along with some drawing—formed the exclusive content of the entrance examination (Reinhard 1950, Vol. I, 19 ff.).

Carnot failed this examination. The examiner was the recently appointed Bossut. He reported that Carnot's knowledge of Camus's work was merely superficial, although he did possess high intellectual abilities. He also pointed out that Carnot was still very young, indeed, he was only 16.5 years old at the time (ibid., 25). His prior education was certainly insufficient to master a four-volume work completely by himself in such a short period. Another aspect, not discussed in the literature, is that Bossut had previously served as a mathematics teacher in the school at Mézières, where he had had to teach with the work of Camus that was contrary to his own methods. As an examiner, he was now in an independent position and publishing his own rival textbooks. He may well have been biased in his examination by reservations about this work of Camus that continued to form the content of the examination.

However, failing this examination proved very productive for Carnot. He did not give up. Through the continued patronage of the Duc d'Aumont, a place was found for him at the *Pension Longpré*, just about the best private preparatory school for the entrance examination. Although his father was in no way wealthy, he managed to pay the considerably high fees. The advantages of this school were considerable: the director was able to coordinate its program through his personal connections with the examiner Bossut. In addition, it gave Carnot his first opportunity to enjoy a systematic introduction to the study of mathematics and physics. And perhaps more decisive for his future development, one frequent guest at the *Pension* was d'Alembert. He followed the students' exercises, and was occasionally also willing to engage in talks and discussions. Carnot soon came to the notice of this leading scientist through his dogged discussions on foundational issues (Dhombres & Dhombres 1997, 39 f.). It was during this time that Carnot drifted away from Catholicism and started to embrace the ideas of the Enlightenment (Reinhard 1950, Vol. I, 29).

What did they teach in the mathematical courses at the private school of Longpré? It is clear that the focus was on Camus's textbook, which continued to be the official textbook in Mézières. However, with Bossut as examiner, it seems likely that further textbooks were introduced. The Dhombres assume that Carnot also studied Bézout's textbooks (Dhombres & Dhombres, 1997, 41), but this seems rather improbable: Why should Bossut have encouraged the use of a rival textbook within his own sphere of influence? Moreover, the earliest edition of Bézout in the Carnot libraries is the marine version of 1775 in the family library at Nolay. It is far more likely that the students at Longpré worked with texts recommended by the examiner Bossut, particularly ones on those topics not covered in Camus: algebra and hydrodynamics. Although Bossut's own book on hydrodynamics was not published until the following year 1771, he may well have already made individual chapters available at Longpré. His algebra book did not appear until 1773. However, it can be assumed that Bossut used a conceptually similar book in the university tradition during his many years of teaching at Mézières and also went on to recommend such a work in the private preparatory classes.

Carnot entered the private school at Paris at the beginning of 1770 and studied there for almost a complete year up until his second examination by Bossut in December 1770. This time, he passed with flying colors: he gained third place, thereby qualifying for one of the few highly sought after places at the *École du Génie* in Mézières. Carnot entered this school in February 1771 (Amson, 1992, 26 f.) and completed the two-year training as a *Génie* officer.

Carnot's professor for mathematics and the natural sciences was Monge. However, the Dhombres are right to emphasize that no scientific relationship grew up between the two. Although Carnot completed the exercises in stereotomy as a foundation for the practice of drawing ground plans so central for engineers, he seems to have shown no interest in the theory that Monge developed from this: his *géometrie descriptive*. The Dhombres judge him, in contrast, as more of a *calculateur* who acquired differential and integral calculus at Mézières (ibid., 82 f.).

After completing his course at Mézières without any hitches at the end of 1772, Carnot began his career as a *Génie* engineer. Following the customary practices of the time, this led him to a succession of garrisons over the years: Calais, Cherbourg, Béthune, Arras, and Saint Omer. Although receiving numerous commendations for the quality of his engineering works, the final years of the *Ancien Régime* confronted him with increasingly inflexible social barriers that handicapped further advances in his career and even his marriage plans. Carnot countered this by devoting what was clearly sufficient free time to scientific projects. All his writings up to 1789 were submissions to academies in response to open prize competitions. In fact, one of the positive characteristics of France under the *Ancien régime* was to encourage a scientific culture in which broad sectors of society could participate actively by entering open prize competitions set by the Paris Academy as well as academies in the provinces.

Carnot's first submission was a memoir applying the theory of simple machines to friction and to the tension of ropes submitted to the Paris Academy in 1777. Because the Academy was not satisfied with any of the submissions, the competition was relaunched in 1779. Carnot revised his submission in 1780 and was awarded an honorable mention the prize itself went to Coulomb. The theoretical part of his submission was then first published in 1782 under the title *Essai sur les machines en général*.[1] In its definitive form of 1803, *Principes fondamentaux de l'équilibre et du movement*, it became the foundation of the science of machines and in particular of the pioneering work of his son Sadi Carnot.

In 1784, Carnot tackled two completely different competitions. The first was a call by the Dijon Academy for appraisals of Vauban's contribution to fortification (see Schubring 2004, Chapter 8). Carnot won this prize with his *Éloge de Vauban*. This did not just make him well known in France; it also drew him into a seemingly never–ending pamphleteering discourse with opposing appraisals of Vauban, particularly those of exponents of other parts of the army. The second prize question posed by the Berlin Academy addressed the concept of the mathematical infinite. Carnot submitted his memoir in 1785; the prize was awarded to L'Huilier. Carnot read L'Huilier's work and revised his paper, but did not publish it immediately.

Two developments marked Carnot's path into political life: In 1787, like so many intellectuals during the age of the Enlightenment, he joined the Freemasons' lodge at Arras; the same lodge in which Robespierre was an active member. The second development was initially purely private: At the beginning of 1789, Carnot found out merely by chance that Ursule de Bouillet, who had been practically his fiancée for many years—although her father had not given permission for the marriage because of Carnot's lower social status—was to marry an officer of higher social rank on the following day. Although the scandal that Carnot then unleashed prevented the marriage, Ursule's father countered by using his connections at the ministry in Paris to get Carnot dismissed from the army. The minister also sentenced Carnot to two months in the military prison at Béthune. This adjudication and incarceration without prior judicial trial and the behavior of the noble de Bouillet family brought about Carnot's final rejection of the arbitrariness and social injustice of the *Ancien Régime*. Carnot took the side of the *tiers état* (i.e., the commoners), and

---

[1]   Although the Dhombres and Taton assign publication to the year 1783, they do mention the possibility of an earlier printing in 1782, because this publication year is to be found in either one (Dhombres & Dhombres 1997, 19) or two (Taton 1990, 461) North American library catalogues. In fact, I have held a copy of the first edition of 1782 in my own hands in the family library at Nolay.

The commemorative volume embellished with many documents provided by Carnot's descendants, the *Centenaire de Lazare* (ed. by La Sabretache 1923), does not contain any details on his education or his scientific activities.

campaigned for a radical military reform as part of the general calls for reform in France in mid 1789.

Carnot did not enter politics immediately after the revolution, but only in 1791 when elected to the *Assemblée Législative*. Political events radicalized his originally moderate ideas on reform. He became a Republican, one of the *régicide* that voted in favor of the execution of Louis XVI, and, finally, the military expert of the Republic when the traditional military leaders betrayed their country and the European royal houses invaded France. Elected as such to the *Comité de Salut Public*, he successfully checked the invading troops under the conditions of the *terreur* and finally repelled them. It is these achievements linked to the *levée en masse* for which he continues to be remembered in France as *le grand Carnot* or *le Général Carnot*. Although still a relative moderate, he managed not only to survive the heights of the Jacobin Reign of Terror but also to remain in the government after the overthrow of Robespierre in July 1794. In 1795, following the adoption of the new constitution and parliamentary elections, Carnot even became a member of the five-person *Directoire*, the highest executive body of the Republic. Here as well, he was particularly responsible for all military concerns. In this phase, marking the onset of state consolidation, it was Carnot's conciliatory attitude that led to his downfall. There was a deep-running political division among the French at the time: Monarchists were making increasing gains, particularly in Parliament and the administration. However, Carnot was not prepared to counter antirepublican activities by resorting to unconstitutional measures. When, in September 1797, the majority in the *Directoire* executed a coup d'état to save the Republic and suppress the monarchists, by, among other means, revoking their parliamentary mandates, they also decided to arrest and imprison the minority in the *Directoire*. Carnot managed to flee just in time, first to Switzerland and then to southern Germany.

His military and political activities from 1791 onward left Carnot with no time to continue his scientific activities. Nonetheless, in 1796, he was elected to the newly established *Institut National*, the successor to the *Académie des sciences* disbanded during the *terreur* in 1793, as a member of the section for the *arts méchaniques*. Although officially elected for his *Essai sur les machines*, his political functions had exerted a major influence (see Dhombres & Dhombres, 1997, 438 ff.). Hence, the unprecedented political intervention of striking him off the list of *Institut* members after the 1797 coup d'état should also be viewed in light of the political aspects motivating his initial election. Although Carnot was an infrequent participant in the scientific sessions during this politically turbulent period, one important effect of his membership can be ascertained: discussions on basic principles at the Institute led him to publish the revision of his "Berlin" memoir. As mentioned above, Carnot had revised this text thoroughly after reading the winning memoir submitted by L'Huilier (1786) and certainly before embarking on his political career in 1791–1792, but he had then shelved the manuscript.[2] Perhaps encouraged by Lacroix's manuscript for his major textbook on analysis submitted to the institute for evaluation in 1796, Carnot gave Lacroix his manuscript to read. In the printed edition of his 1797 textbook, Lacroix emphasized the importance of this text and encouraged Carnot to publish it:

> Carnot had the goodness to inform me about his memoir analyzing the principles of differential calculus with great care in which he notes that it is due to the *law of continuity* that evanescent quantities maintain the same ratio that they successively approach while vanishing.
>
> This memoir, and it would be desirable for the author to publish it, confirms that if one had only formulated words in the right way, one would also have obtained clearer concepts (Carnot & Hauff 1800a, II).

In fact, it was published in the same year, in June, shortly before the coup d'état, as *Réflexions sur la métaphysique du calcul infinitesimal*. Still in the same year of 1797, but already in exile, Carnot published his *Oeuvres Mathématiques* in Basel, Switzerland, clearly in order to better his precarious finances. This book contained reprints of his *Essai sur les machines* (1782) and *Réflexions*. Through the forced inspiration of exile, Carnot devoted himself completely to his research on the foundations of mathematics. Quite remarkably, he produced a new concept of analytical geometry that he transformed, on the basis of newly formulated conceptions on the relation between analytic and synthetic methods, into a new type of geometry. Carnot's starting point was reflections on trigonometry. He first published this approach in 1800 as a letter to Bossut, his old examiner in mathematics with whom he had always retained a close relationship (Carnot 1800b).

Napoleon's coup d'état in November 1799 led to the overthrow of the *Directoire* and the constitution of 1795. Carnot was finally able to return to Paris in January 1800. He did not just regain his seat in the *Institut*; Napoleon also appointed him minister of war. However, it took only a few months for it to become apparent that such a freethinking personality as Carnot did not fit into the perfect administrative machinery of Napoleon. He resigned, and in the following years, concentrated almost exclusively on the foundations of geometry. In 1801, he published his first version, *De la correlation des figures de géometrie*, and, in 1803, his main work, *Géometrie de position*. This was followed, also in 1803, by the revision of his work on mechanics, *Principes fondamentaux de l'équilibre et du movement*, and in 1806, by a further

---

2   In fact, Carnot submitted a revision of his Berlin text to the Dijon Academy in April 1788 under the title "New ideas on the metaphysics of infinitesimal calculus." On December 10, 1791, he asked for the manuscript to be returned to him in Paris. He received a copy in January 1792 (Taton, 1990, 464), but it is unlikely that he found time to work on it further. Hence, the text published in 1797 is essentially identical to that submitted to Dijon in 1788. In the 1797 preface, Carnot himself says that the text in its current published form had been around for "several years" (see also Tisserand 1936).

composition on geometry. In 1813, he published a new edition of his *Réflexions* on infinitesimal calculus—the third version of a text first produced in 1785.

Whereas Carnot had no problems in surviving the First Restoration of the Bourbon monarchs in 1814, in the Second Restoration, he became a victim of vindictive persecution after having sided with Napoleon in his Hundred Days. Once again, he had to go into exile: After first fleeing to Poland, he finally found asylum in Prussia, the country whose troops had been so decisive for Napoleon's defeat. He died in Magdeburg on August 2, 1823. His remains were returned to Paris and reinterred in the Pantheon during the presidency of his grandson, Sadi Carnot.

1.3. The Development of Carnot's Ideas on Foundations: Analyse and Synthèse

Carnot's writings are characterized throughout by an intensive and continuous effort to explain fundamental issues. He was never satisfied with any position once reached, but was always rethinking it. Hence, before analyzing his views on single concepts and concept fields, it is worth studying the underlying epistemological concepts, particularly to see which elements remain constant and which ones change.

Because the evolution of the analytical method during the second half of the eighteenth century, which peaked during the French Revolution, also had its roots in the philosophy of Condillac, it seems a good idea to start studying Carnot's epistemology of mathematics by looking at his attitudes toward Condillac's philosophy and the analytical method. Indeed, the analytical method and its relation to the synthetic method increasingly became the central object of Carnot's reflections. The change in the way he conceived this relation bears witness to the change in Carnot's epistemology. In contrast, this change did not have any decisive impact on his attitude toward Condillac.

J. and N. Dhombres exhibit a general tendency to play down the importance of Condillac, which is why they also try to minimize the influence of his philosophy on Carnot (Dhombres & Dhombres 1997, 48, 167 ff.). However, Carnot's writings continually confirm his agreement with the fundamental ideas in Condillac's philosophy. In the theoretical conclusion to his treatment on the mathematical infinite from 1785, that is, during his first mathematical period, Carnot refers in great detail to Condillac and his concept of science as language: analysis has to be conceived as a language in which the ability to express oneself more easily, clearly, and precisely depends on the availability of appropriate words and signs (Gillispie 1971, 253).

Carnot emphasizes not only the role of the sign but also the need to perfect language in order for science to advance (ibid., 254). He also refers to Condillac's ideas when presenting the abstraction process in mathematics: from manipulating concrete numbers, across operating with abstract numbers, up to the stage of universal arithmetic (ibid., 255 f.).

In the final revision of his mechanics prize submission, published in 1803 when he was at the peak of his scientific creativity, Carnot bases himself quite explicitly on the philosophy of Condillac and Locke. He declares that the axiom according to which all ideas are derived from sensory experiences is no longer controversial: "[...] axiome que *toutes nos idées viennent des sens*: et cette grande vérité n'est plus aujourd'hui un sujet de contestation" (Carnot 1803 b, 2).

And he even refers specifically to Locke in order to emphasize experience and the *sensations* as the initial form of knowledge (ibid). Carnot stresses that this derivation from sensory experience also holds for mathematics: *Cela est vrai pour toutes les sciences même les plus abstraites, pour les mathématiques pures en particulier* (ibid., 3).

However, his argumentation reveals one major divergence from Condillac. Whereas Condillac thought that a science could develop from empirical origins over a variety of stages into a rational science, Carnot thinks that mathematics always remains bound to the empirical concept of *quantité*:

> Since these sciences are nothing other than a series of reasonings based on quantity, the idea one forms is always the result of experience, since it is only our sensations that can give that to us (ibid.).

This ontological constraint would become decisive in Carnot's changing concept of the analytical method.

A further continuous characteristic of Carnot's ideas is his striving toward generalization. His submission to the Berlin Academy of 1785 emphasized the importance of generalization for the dissemination of knowledge and hence for scientific progress. He remarked that when one starts to reflect on the nature of mathematics, one soon notices that it uses only one procedure to fulfill its sole purpose of simplifying everything, namely, that of generalizing everything. One generalizes in order to achieve a greater simplification, and in particular, in that one "substitutes several particuliar propositions, or those which can be applied to only a limited class of quantities, by general propositions" (Gillispie 1971, 260 ff.).

This last aspect of generalization forms one of Carnot's main arguments in favor of theorizing knowledge. Particularly in relation to mechanics, he stressed repeatedly the need to overcome the restricted practice of focusing on single machines, and that an understanding of machine processes could be attained only through knowledge of their general laws (see Carnot 1797b 2 ff.). In both the 1782–1783 and the 1803 texts on mechanics, Carnot gives detailed and emphatic justifications for the priority of the general. He uses this to criticize the numerous and repeated attempts in his time to construct a *perpetuum mobile*: he viewed such ventures as characterized by a desire to make the special absolute. The usual proofs with which one tries to convince expert constructors of the impossibility of such a project refer to simple machines or to analogies with simple systems of forces. This is what made the constructers believe that it must be possible to construct a special machine with an as yet unknown drive to which such standard arguments would not apply, with the result that it would no longer be subject to the same laws:

> The way to eliminate this error is, without doubt, to show that, not only for all known machines, but even for all possible machines, there is an inevitable law that *what is lost in time or speed is gained in force*; [...] to do this it is necessary to ascend to the greatest possible generality, not to stop with any particular machine, not to rely on any analogy; what is required, finally, is a general proof.

> [...] my object [here] has been simply to present the most general principles as exactly as it was possible for me to do so (Carnot 1803b, xviij ff.).[3]

Even in his final mathematical publication, the third revision of his work on the mathematical infinite (1813), Carnot continues to stand by the priority of generalization: When having to choose between several methods, he would prefer the one that treats the problem more generally (Carnot, 1921, vol. 2, 54).

In contrast to these essentially constant positions on foundations, Carnot's ideas on the analytical method went through major changes. The main expression of this change is found in his discussions over the relative status of algebra and geometry.

In the first version of his submission to the Berlin Academy of 1785, Carnot conceives algebra unreservedly in the sense of the analytical method. In Newtonian terminology, he describes algebra as universal arithmetic, as a further development of arithmetic (Gillispie, 1971, 256). And he argues expressly that algebra is both more general and simpler than arithmetic when presenting his concept of the development of mathematics as generalization *and* simplification:

> The arithmetic of abstract numbers is more general than that of concrete numbers; because each is here being compared with a class of numbers less determined, and so more extended, and which is at the same time more simple, because the combinations are not encumbered by the consideration of the particular properties characteristic of concrete numbers (Gillispie 1971, 258 f.).

In algebra, each quantity is compared to a class even less determined and so more extended; also the principles are more simple than those of arithmetic. In the last revision of 1813, in contrast, Carnot takes a completely different position in which the foundations of algebra are in no way certain and clear: "les principes de l'algèbre ordinaire sont beaucoup moins clairs et moins bien établis que ceux de l'analyse infinitésimale" (Carnot 1921, vol. 2, 62).

Carnot's radical downgrading of algebra is an expression of a more general epistemological transformation in France after the first years of the Revolution combined with a reappraisal of geometry, a revived dominance of the synthetic method, and an increased shift toward empiricism in epistemology in general.

The change in the fundamental epistemological orientation can be demonstrated through Carnot's earlier writings on mechanics. In the concluding section of his 1782–1783 version, he presents two contradictory methods from philosophers for studying the laws of motion. The former philosophers make

---

3   The corresponding passage in the 1783 text can be found in the *Préface*, xj–xiv.

mechanics seem to be an experimental science and the latter a purely rational science. The former compare the phenomena of nature and break them down in order to ascertain what they have in common and to reduce them to a small number of basic facts. The latter, in contrast, begin with hypotheses, then reason about the assumptions, and move to the discovery of laws that the bodies follow in their motions should the hypotheses conform to nature. Finally, they draw conclusions on the exactness of hypotheses that are initially only assumed (Carnot 1797b, 120 f.).

Carnot remains relatively ambivalent about these two alternatives. His further argumentation reveals that the former also do not just proceed from experience, but from *idées primitives* as well. However, his comparative discussion refers to the reliability of such *idées primitives* and, in particular, to the status of definitions of fundamental concepts in each of the two alternatives (ibid., 121).

Although Carnot tends to favor the approach of rational science here, he also sees that its difficulty is to establish appropriate and accessible definitions of fundamental concepts. In his 1803 version, Carnot also presents two contradictory fundamental positions in mechanics, but the meaning has changed completely. He now goes on to explain that the first concept is based on an *idée primitive*, namely, to view forces as causes, but that this is a "metaphysical and obscure notion." With the second concept, in contrast, one drops the metaphysical distinction between causes and effects, and conceives mechanics far more as the theory on the laws of the reciprocal effects of motions (*lois de la communication des mouvemens*). Because this means that the *principes fondamentaux* can no longer be assumed as axioms as was the case in the first concept, it is necessary to derive and prove these *principes*, which can no longer be taken as self–evident propositions, from *experience* (Carnot 1803b, xj ff.).

No reflections on the role of geometry or discussion on the synthetic method can be found in Carnot's writings or manuscripts until the French Revolution. *Synthèse* and thereby its relation to *analyse* becomes a problem for Carnot only when he starts to think about the justification for negative quantities and the rule of signs. The starting point for the subsequent restrictions to the function of *analyse* is a definition of the concept of algebra that can be found very early in Carnot's work. This does not view algebra as being independent, but as being based on arithmetic and serving only to *indicate* its operations: "L'algebre n'est que l'art d'indiquer les opérations de l'arithmétique."[4] This instrumental definition contains the starting point from which he later went on to distinguish operations that can be performed in arithmetic but cannot be performed in algebra.

Carnot's *Nachlass* contains a number of drafts entitled *Réflexions sur la différence caractéristique de la synthèse et de l'analyse en mathématiques*

---

[4]   Carnot's *Nachlass*, Nolay: Carton 28, Chemise 5 (Mathématiques, brouillons). Because the context refers to ideas on the classification of knowledge, the text can be dated to approximately 1780 (see Dhombres & Dhombres, 1997, 70).

(Reflections on the characteristic difference between synthesis and analysis in mathematics). They form prior deliberations leading up to his book published in 1801 on the correlation of figures in geometry. All of the drafts already assume the distinction between executable versus nonexecutable operations in arithmetic, also including the exclusion of negative values as legitimate objects of mathematics.

One seven-page fragment with such a title can be dated to the period before 1800, approximately 1799, because it does not yet contain the terms *directe* versus *inverse*.[5] Here, Carnot works out that the main difference between analytical and synthetic methods—according to his view—is that the latter permit only absolute quantities and thus only executable operations, whereas the former also permit isolated negative quantities. The difference between the methods is revealed in the solution of equations, that is, in the transformations performed to obtain the value sought:

> Analysis allowing itself in this respect a much greater latitude than that which is contained in synthesis; seeing that the latter never allows in its combinations [any quantities] except absolute magnitudes related by signs which can only indicate executable operations; whereas in analysis there is no difficulty in separating out quantities carrying a minus sign and in operating on them as if they were ordinary quantities, conforming to the general rule of signs.[6]

For *synthèse*, he considers the rule of signs to be clearly proven; but this is not the case for *analyse*. For Carnot, the assumption of general validity proves to be the central criterion for the difference between analysis and synthesis:

> But analysis gives, by analogy, a much greater extension to this rule; it allows it for all imaginable cases [...]. It is just this extension which, I repeat, forms the true character distinctive of the analytic method.[7]

At this stage in his ideas, Carnot treats the values that can be obtained by *both* methods as being equally acceptable solutions by distinguishing them with the labels "analytical value" and "synthetic value":

> This is why one must distinguish these values one from the other: I shall thus call *synthetic values* those unknowns which can be obtained only by synthesis and *analytic values* those which can be obtained by analysis.[8]

In a later fragment bearing the same title, Carnot also introduces the labels *quantités directes* and *quantités inverses*.[9]

Carnot goes into more detail regarding his new concept of the relation between arithmetic and algebra from the perspective of the synthetic method in a manuscript entitled *Réflexions sur les principes généraux de l'analyse finie* (Reflections on the general principles of finite analysis). A few pages from the

---

5   Carnot adopted these terms in the spring of 1800 from a book published in Geneva by E. Develey (Schubring 1997, 8 f.).

6   Carnot's *Nachlass* at Nolay: Carton 26, Chemise 4 (Notes diverses).

7   Ibid.

8   Ibid.

9   Carnot's *Nachlass* at Nolay: Carton 28, Chemise 10.

introductory part of this manuscript were edited by Charnay (1985).[10] In this work, which can be dated to approximately 1800, Carnot expressly emphasizes that algebra—assuming the rule of signs to be valid—extends the domain of algebraic language that would be too limited if constrained to the usage of *synthèse* alone: "l'analyse étend par son moyen [i.e., la règle des signes] le domaine de la langue algébrique, dont l'utilité serait fort bornée, si elle était restreinte aux usages de la synthèse" (Charnay 1985, 433).

Here as well, Carnot stresses the unity of arithmetic and algebra as developed by the Oratorians: "L'arithmétique et l'algèbre forment ensemble ce qu'on nomme la science du calcul" (ibid). On the other hand, he also declares arithmetic here to be the only body that can and may perform operations with numbers, whereas algebra only indicates to arithmetic which operations are feasible: "L'arithmétique nous enseigne à exécuter sur les nombres toutes les opérations dont ils sont susceptibles [...]."

He considers that being able to perform operations is not enough. One also has to discover which operations actually can occur and report their nature. That is the task of algebra:

> Algebra is nothing other than an abbreviated form of writing which is used for expressing with conventional signs the diverse relations that can exist between quantities, so as to indicate to arithmetic the operations it may carry out (Charnay 1985, 433)

Although Carnot still continues to assign to algebra the function of deciding which operations are permissible, in fact, this decision has already been assigned to *synthèse*. As before, Carnot formulates here as well that the only values *synthèse* recognizes as solutions are those that are *reélle et positive, c'est-à-dire une valeur absolue* (real and positive, so to speak, an absolute value; ibid., 436). And *synthèse*—as well as determining the beginning of a problem solution—is also exclusively responsible for the solution to a problem; *analyse* can be applied only for the intermediate steps: "It is always by the latter [synthesis] that every problem must be begun and finished, and analysis just fills the gap in between" (ibid., 436).

Here, Carnot gives a breakdown of the steps in solving a mathematical problem that confirms his strong disposition toward methodological reflection. He explains this in detail as a sequence of determination of the premises of the problem, solution of the equations, and application of the formulas. Interestingly, he breaks application down into interpretation and generalization. However, in this context, the meaning of "generalization" has already been transformed: it stands for the search for more or less closely related cases (ibid.,

---

[10]   However, Charnay fails to report two things: First, that he left out the first, highly informative, paragraph of the manuscript (see Chapter I.5); and second, that it is part of a very comprehensive text—of about 110 pages—forming a first draft of the subsequent publication of the 1801 version of the *corrélation* (Carnot's *Nachlass* at Nolay: Carton 28, Chemise 3).

434 f.). This meant, for example, that the equations in the different quadrants may bear different signs and hence represent different, though related, cases.

In the book version from 1801, *corrélation des figures de géometrie*, Carnot addresses only the relation between *analyse* and *synthèse* explicitly in passing, although their limitation forms the main object of the text. In an explicit comment, he reiterates that algebra serves only to "indicate" the operations that have to be made, and that arithmetic has to perform them in order to obtain the result sought (Carnot 1801, 27). He conceives his earlier discrimination between *valeur analytique* and *valeur synthétique* here as *valeur* versus *qualité*: *Valeur* now stands for the algebraic forms, for example, negative or imaginary, whereas *qualité* means exclusively the absolute value independent of the sign of the quantity: *valeur absolue, c'est-à-dire, abstraction faite du signe* (absolute value, so to speak, abstraction made of the sign; ibid., 2). What is remarkable is the expressly empiricist linking of the quantity to a real object:

> I shall use the name *quantity* for the thing itself whose properties are being sought […]. Every *quantity* is a real object, sensible to the mind, or at least its *representation* in calculation in an absolute manner [is so] (ibid.).

In contrast, in the *Géométrie de Position* of 1803, a thorough reworking of the *correlation*, Carnot describes his rejection of the analytic method in detail. This rejection forms a programmatic creed. It is all the more characteristic for the *retour du refoulé*, for the reappraisal of the synthetic method, in that Carnot announces it in direct parallel to the classic topos of the Ancients versus the Moderns, and now revokes the consensus over the superiority of the modern dominant in France since Arnauld and the Oratorians.

In this, his major work, Carnot presents *synthèse* as the generally appropriate method of thinking, not just in mathematics but also in all fields: "Synthesis does not apply exclusively to mathematics: it is in general the art of reasoning justly, whatever may be the subject of the argument" (Carnot 1803a, 15). Although he sees *synthèse* as being limited through its procedure, this means that it never loses sight of its object, and it does not introduce any nonexecutable operations or reason over absurd quantities.

The *analyse*, in contrast, permits nonexistent objects in its operations; it mixes real objects with imaginary ones. However, through certain transformations, it succeeds in eliminating the *êtres de raison* (the imaginary objects), and arriving at results more quickly than *synthèse*. The *synthèse* could also attain the same results, but by taking a slower and more problematic path. This shortcut represents the advantage of the moderns compared with the ancients: "Tel est l'avantage de l'analyse sur la synthèse, et par conséquent, celui des modernes sur les anciens" (ibid., 9 f.).

Because in the heyday of the analytical method it was identified with the use of signs, Carnot concentrates on the role of signs for both methods. He admits that the "ancients" had not used any signs, but that this is not characteristic of *synthèse*. It is far more the case that had the ancients used signs, they would have made far greater advances with *synthèse*. In a clear critique of contemporary civilization, Carnot notes that admitting infeasible operations has enabled the modern age to achieve a wealth of results (ibid., 10).

In order to discredit the enthusiasm for signs in the analytical method, Carnot goes so far as to compare the way it uses signs with the use of hieroglyphs, the Egyptian hieroglyphs were still indecipherable at this time. He points out that whereas *synthèse* proceeds methodologically, according to the conditions of existence and executability, *analyse* enters another path using hieroglyphs that too frequently describe only imaginary objects:

> [...] analysis, on the other hand, arrives there first of all by a swift route characteristic of itself, in forming along its route another chain, not of real objects as before, but of hieroglyphs which, most often, stand only for [mere] objects of reason (ibid., 12).

He considers the path of *synthèse* to be in no way less certain or less instructive (*ni moins sure ni moins lumineuse*). The hieroglyph-like use of signs in *analyse* will become a merely abbreviated form of notation only when these symbols are changed into real objects that are also accessible to reason:

> La route hiéroglyphique [...] demeure toujours analyse, tant que les objets désignés par ces symboles ne sont point devenus réels et appréhensibles par l'esprit (ibid., 13).

By radically reevaluating the synthetic method while simultaneously devaluing the analytic method and its use of signs, Carnot does not just perform a reappraisal of the classic topos of the superiority of the modern age over the ancient. He also reevaluates the previously widespread criticism of Newton's style of presentation into praise for his synthetic method. Whereas previously, Newton had been criticized, and particularly by d'Alembert, for using a hermetic presentation that did not allow any insight into the way he had made his discoveries, Carnot, in contrast, praises him for having justifiable doubts regarding the correctness of the analytical method, and for always testing the results obtained through it with the synthetic method, thereby replacing possibly meaningless hieroglyphs with graspable truths (ibid., 14).

1.4. The Change in Carnot's Ideas

It is natural to ask why Carnot's ideas on foundations changed; whether we can ascertain causes for his rejection of the analytical method. Although Carnot himself tells us nothing about this, we now know that this change occurred round about 1797.

Indeed, there is one strong indication that Carnot had not been able to devote much time to mathematics before his flight into exile in that year: the contents of the library collected while he was living at Paris since 1792 (he had left his earlier books at St. Omer). The Paris library is highly analyzable, because Carnot's official residence at the Palais du Luxembourg was sealed up after his flight. All its contents were seized, confiscated, and carefully inventoried in the search for evidence that he was conspiring with the monarchists. The catalogue of this library in 1797 lists 348 titles as either hardcover books (1–193) or softcover books. Only one of the hardcover books is a mathematical work; and that, an old one: Ozanam's mathematical dictionary. Among the softcover books, there are only a few mathematical works, all published during his years at Paris. Some of these books were presents from the authors. These include de Prony's course in analysis, *Méthode directe et inverse*; Bossut's new textbooks; the *Système du Monde* from Laplace; the translation of Euler's *Introductio*; and Carnot's own *Réflexions*. However, the majority of the titles listed are military, political, and geographical. Philosophy, literature, and poetry are also well represented.[11] This paucity of books on mathematics would indicate that Carnot was not actively engaged in this field in the years up to 1797.

On the one hand, we can assume biographical reasons for the change after 1797: Only just avoiding attempts on his life during the coup d'état and having to associate with the other French refugees in asylum—nearly all monarchists and nobles—may have led to a reappraisal of his earlier values. On the other hand, we can also assume that a closer inspection of the literature on the foundations of negative numbers led him to question his earlier, rather unquestioning, acceptance of their legitimacy adopted from the textbooks of his student days. He was certainly familiar with d'Alembert's criticism of this concept. However, because this was composed of only short statements of principles and did not give a comprehensive presentation of its concept field, its impact was not necessarily immediate. Accordingly, Carnot may have felt the need to examine it only after reading more detailed treatments such as Porro's book (see Chapter IV.1.4.) and the work of British authors (see Chapter II.2.10.3.). A note to be found in the *Nachlass* provides direct confirmation that Carnot actually did read such British works. He comments:

> Dans les transactions philosophiques pour l'année 1778 on trouve un mémoire intéressant de M. John Playfair sur le calcul des quantités imaginaires: on the arithmetik of impossible quantities.[12]

This note is all the more significant, because in this, his first publication, Playfair favors a relation between algebra and geometry that shows many similarities to Carnot's new outlook (see Chapter II.2.10.3.).

Nonetheless, it has to be recognized that Carnot did not just change his ideas on negative qualities, but also on infinitesimal calculus (see below) as well as on the epistemology of mathematics in general. This makes it necessary to look at developments that go beyond the personal level of a single individual.

Indeed, it can be seen that France went through a more general epistemological shift after the first years of the Revolution, and particularly after

---

[11]   Archives Nationales, AF/III/463.
[12]   Carnot's *Nachlass*, Nolay, Carton 28, Chemise 10. This note is part of a set of systematic literature excerpts on recent work in geometry compiled sometime during the second half of the 1790s.

the end of the *Directoire* and the beginning of Napoleonic rule: a turning away from the analytical method and a stronger shift toward empiricist concepts.

This shift can be traced in the concepts of the *Idéologues*, a group of philosophers—influenced strongly by the philosophy of Condillac—who dominated the French education and science system after the fall of the Jacobins. Under Napoleon, however, their influence was increasingly curtailed in a systematic way (Moravia 1968). How far the situation had already changed by the end of the 1790s can be seen in the *Idéologues'* reception of Condillac's *Langue des Calculs*. They gave only a guarded welcome to this unfinished work, published directly from his *Nachlass*, accusing Condillac of believing that algebra was "the language" and thereby the model and goal of all sciences. They most emphatically denied that algebra could take such a general methodological role (see Auroux 1982). Condillac's idea of a genetic evolution of scientific concepts developing through various stages until they become abstract terms was not taken up and developed further by any philosophers or mathematicians. At the end of the line, the *Idéologues* identified the concepts—what they call ideas—on only one level of theory: predominantly in close relation to the sensations, and understanding them as being equipped with an empiricist substance.

A change in the concept of mathematics among the *Idéologues* about 1800 can be confirmed in the work of Destutt de Tracy (1754–1836), one of its leading representatives. His main work is his *Élémens d'Idéologie*, a multivolume publication containing a comprehensive account of the concept of a "science of ideas." In the first edition of his work, Destutt de Tracy still characterizes mathematics as drawing equally on two underlying ideas: the *quantité*—understood as a numerical quantity—and the *grandeur*—understood as the geometric magnitude of figures: "(les) quantités (et les) grandeurs [...] composent ensemble tout ce qu'on appelle les mathématiques pures" (Destutt de Tracy 1798, 398). Both make it possible to recognize the consequences that can be deduced from a single idea, from that of the number unit or from that of the figure:

> They consist uniquely, in each case, in recognising all the consequences that can be drawn from a single idea; for quantities, the idea of unity; and for magnitudes, the idea of a figure (ibid., 389 f.).

At this stage, Destutt de Tracy still emphasizes the nonempiricist character of the two ideas: "Now, the idea of unity and that of the figure are two ideas which are abstract, two ideas of objects that do not exist of themselves in nature" (ibid., 390). Hence, they are not *vérité de fait*, but *de déduction*.

However, in the second edition of *Élémens d'Idéologie*, published in 1804, Destutt de Tracy no longer assumes two foundational ideas but also traces the number back to measurement, and thus to a geometric foundation. He now answers the question regarding the degree of certainty that can be attained in the various sciences by referring to how far they can be measured exactly, and he states that measurability is represented most precisely in the quantities of extension:

> All sciences are more or less certain in the proportion that the objects that concern them are more or less reducible to quantities capable of being perceived by measures that are perfectly exact, and [...] of all types of quantities, extension is that which possesses this precious character most emminently (Destutt de Tracy 1804, 203).

In a comprehensive commentary, he gives a detailed proof that the unit of abstract numbers acquires a real value only when it is related to the measure of an extension (ibid., 213 f.). The applicability of *calcul* to the objects of a science would therefore depend on how far it is accessible to exact measurement. This is primarily the case for geometry followed by all domains that can be expressed in measures of extension (ibid., 215). Destutt de Tracy completes his commentary by reaffirming the new conception that quantities can be an object of meaningful computational operations only when they can be traced back to extensive quantities:

> Any quantity is thus calculable in the proportion that it is reducible directly or indirectly to measures of extension; for it is [extension which is] the property of objects that is the most eminently measurable (ibid., 216).

Because of its new central and universal status, Destutt de Tracy even proposes that geometry should be renamed as the general science of exact measurement without any special restriction to the planet Earth or the cosmos:

> If the words had been used correctly, the science of extension would not have been called *geometry*, which is to say the measurement of the Earth, which is merely surveying, but rather *cosmometry*, since it serves to measure the entire world, or even better simply *metry* (ibid., 217).

Destutt de Tracy's presentation of algebra, in contrast, is accompanied by several critical reservations. In a comparable way to Carnot, he presents the course of an algebraic operation process as an incomprehensible sequence of signs:

> Provided one observes scrupulously the rules of syntax of this language [algebra], which are nothing other than rules of calculation, one is certain to arrive at a valid conclusion, which is, to put it exactly, that one has no need to know what one is saying all the time one is reasoning: also that one never knows it. An algebraic calculation resembles, perfectly and rigorously, the discourse of a man who begins with a true proposition and finishes with another true proposition, who in between talks continuously in a manner unintelligble to others and to himself, and without any linguistic error (ibid., 364 f.).

As well as rejecting the possibility of gaining a better insight, still proclaimed by the analytical method, Destutt de Tracy also states—in a clear criticism of Condillac—that algebraic language is incomplete, because parts of normal language are also mixed into its usage. In the end, the only utility and service that he still assigns to it is its exceptional ability to abbreviate, thus allowing reason to advance further (ibid., 369).

The *Ideologues*' return to geometry as the primary science can also be seen in a text written in 1802 by Marie Fr.P.G. Maine de Biran (1766–1824), in some ways a second–generation *Idéologue*. Unlike many of the *Idéologues* themselves, he possessed a sound mathematical knowledge and could therefore engage in this discourse with greater authority. In his *Mémoire sur les rapports de l'idéologie et des mathématiques*, he claims that the *Idéologie* would "orient" the sciences toward greater conceptual clarity through its ability to perform a clear penetration of their basic terms, whereas the single sciences tend to refuse to consider this and its consequences. He thinks that the *idéologie*'s particular task is to *nettoye[r] le champs de l'évidence*, that is, to "clean" mathematics of inconsistent or misleading concept formations (Maine de Biran 1924, 15). Maine de Biran is the first in France to attack supporters of algebra with polemics; he calls them *algébriers* who are not only unwilling to initiate reforms themselves but would also resist them:

> One may wait in vain for these reforms from mathematicians themselves: *algebrayers* will always use algebra, but will not set themselves aright, will not by themselves keep to the right path (ibid., 23).

He likewise disputes the *clarté* previously conferred on the "algebraic language" (ibid., 19). To counter the "blind and mechanical" practice of all its rules (ibid., 20), he orients himself toward the model of geometry (ibid., 24 f.). The only conceptual field in mathematics that Maine de Biran mentions as being inconsistent and requiring reconstruction is, remarkably, the concept of *quantités negatives*. He states that in the "field of evidence," it is particularly necessary to clean the fundamental section on the negative quantities. One would then recognize that the paradoxical principle of the rule of signs is meaningless. He does not explain what was paradoxical in these rules, but only notes that the signs *plus* and *minus* are not themselves the objects of the operations, but only indicate operations (as in Carnot's terminology):

> (The ideologist) will be able to clarify the essential point about negative quantities, [...] he will recognise the triviality of the statement of the paradoxical precept that – *multiplied* by – gives +, and – *divided* by – gives – [*sic*!], for the signs + and – are not themselves the *objects* of operations, but only indicate the operations of addition and subtraction, and that's all (ibid., 21 f.).

Maine de Biran first comments on the rule of signs within the context of the language discourse: As soon as one permits inexact expressions, one has to face the consequences sooner or later. However, he also gives an explanation of the problems with this conceptual field. This shows, on the one hand, that he proceeds from a strict epistemological separation between geometry and algebra: between geometric constructibility and arithmetic numerical value, and on the other hand, that he has already conceived a separation between the geometric concept of quantity (*quantité*) and the arithmetic number *(nombre)*. He considers that there are no negative numbers in the real world; certain algebraic formulas can be constructed only with lines, but have no arithmetic meaning or value:

> The ideologist will prove that there are actually no negative numbers; he will claim to show that certain results or algebraic formulas, uniquely those that can be constructed by lines, or changed into lines, have no meaning, no arithmetic value (ibid., 22).

He points out that the reason for this difference is the insufficient attention paid to the different nature of continuous and discrete quantities (ibid).

Hence, it can be seen that Maine de Biran adopts epistemological positions that are completely analogous to those of Carnot. Since even the *Idéologues*, the main advocates of the analytical method in France, had transformed themselves into supporters of a new supremacy of geometric concepts, the strength of the epistemological break, the *retour du refoulé*, becomes very clear.

1.5. Relations Between Mechanics and the Foundations of Mathematics

As seen in the biographical section, Carnot did not just work on the foundations of mathematics; he also addressed the principles of mechanics and the theory of machines. He revised his main work on mechanics far more intensively than, for example, his work on the mathematical infinite (four times: in 1778, 1780, 1782, and 1803). One could even say that Carnot had a stronger impact on the formulation of theories in mechanics than in mathematics. In light of the strong controversies among his contemporaries regarding the concept of force and conservation, particularly the *vis viva* controversy, Carnot achieved major advances, beyond those of d'Alembert and Euler. He deduced rules of conservation for mechanics that were to form essential elements in the later derivation of the general law of energy conservation. They simultaneously provided the direct precondition for his son Sadi's second law of thermodynamics proposed in 1824. However, this is not the place to discuss Carnot's concepts of mechanics or their impact in detail. Readers are referred to the recent studies on this topic published by Wilson L. Scott (1970: Chapters V, VI, VII, & VIII) and Charles C. Gillispie (1971: Chapters II, III, & IV). A brief overview can also be found in Gillispie's (1971a) article on Carnot in the *Dictionary of Scientific Biography*. In general, however, it can be noted that it was problems in hydraulics that provided one of the main motivations for Carnot's work in mechanics (see Gillispie 1971a, 73). This was also Bossut's main field of interest. Indeed, Carnot repeatedly referred to Bossut positively in his scientific work; his former examiner continued to exercise a lasting influence.

What should be discussed here, in contrast, is the relations between Carnot's theoretical achievements in mechanics and his concepts in mathematics. As will be shown, it was particularly his innovations in mechanics that led him to the theoretical reflections on analysis that finally helped him to achieve new theoretical clarity and generalization.

What is surprising about these achievements is that Carnot started out by deliberating on hard–body impact, selecting precisely that development of ideas that this book has already identified as being essential for a mathematization of the concept of continuity (see Chapter III). Carnot became involved in this controversy at just about the time when it had attained its peak in Germany. In 1751, Pierre Maupertuis (1698–1759), then president of the Berlin Academy of Sciences, had proposed a principle of least action that he considered to be also applicable to hard–body impact. This was opposed by the Leibnizians at the Academy, and S. Koenig drew on a letter from Leibniz to engage him in a bitter dispute over plagiarism (see Chapter III.5.). When Voltaire and King Frederick II also became involved, this broadened into a nationwide affair in Prussia (see Pulte 1989, 216 ff.).

Even in his very first work on mechanics, Carnot radicalized the approach of Maupertuis, and transformed a principle of conservation that contemporaries had tended to view as "metaphysical" into an operational principle of mechanics. The basic idea in Carnot's discussions on the motions of bodies was to permit the impact of not just elastic bodies but hard ones as well. Right from the very start, he formulates his laws on the motions of bodies for impacts between hard bodies (*les lois du choc des corps durs*; Carnot 1783, 16). He even goes so far as to assume the impact of hard bodies to be the general and simplest foundation for the laws of mechanics and to treat the laws for elastic bodies as mere applications (ibid., 17). Carnot attained this deduction by conceiving elastic bodies as being composed of an arbitrarily large number of hard corpuscles separated by compressible small rods.

For the mathematization of mechanics, the central consequence of this concept that departed from the previous consensus over the exclusion of jumps in nature was to permit not only "continuous" motions (to use the previous terminology) but also "discontinuous" ones. Carnot formulates this consequence for mechanics quite explicitly: "to encompass the question in all its generality, let us suppose that the movement can change suddenly, or vary by imperceptible amounts" (ibid., 18).

In the last version of his book on mechanics of 1803, Carnot reaffirms these preconditions and consequences and formulates them even more impressively. He openly criticizes Maupertuis for the vagueness of his principle of least action derived from the philosophical principle of sufficient reason. He says that although Maupertuis distinguished between continuous changes in motion (*par degrés insensibles*) and those occurring suddenly, he had failed to discriminate between the two cases in his actual principle (Carnot 1803b, vj). Carnot gives Lagrange the credit for having transformed the principle of least action into a truly physical and mechanical law (ibid., vij f.). Looking at his own first work, he is now able to ascertain with some satisfaction that nobody before him had formulated precise laws for the case of the impact of hard bodies, let alone proved them:

> In so far as [the proposition of Maupertuis] relates to the impact of bodies or to any sudden changes, I do not know how anyone could have attempted to express it in a precise manner, nor yet more to have proved it, before the first edition of this work (ibid., vijj).

Carnot reaffirms that he is studying the principles of motion for both hard bodies and elastic ones, and that he proceeds from the impact of hard bodies as the general case within which he treats elastic impacts as special cases:

> Moreover, the general method which I followed in the first edition of this work differed principally from the one usually followed in that I related everything to the impact of bodies or to sudden changes, and I regarded the simple pressure which gives rise to changes taking effect by imperceptible degrees as a particular case of the general problem (ibid., ix).

This quotation simultaneously confirms that Carnot also considers the admission of hard and elastic bodies here to be synonymous with the admission of "discontinuous" (*changemens brusques*) and "continuous" (*changemens par degrés insensibles*) motions.

Numerous historians have analyzed Carnot's decisive impact on the further development of the theory of mechanics. For example, W.L. Scott (1970, 104 ff.) has examined his influence on the concept of work and reversible conversion. He has emphasized the importance of Carnot's introduction of hard–body impact into concepts of conservation for Sadi Carnot's achievements (ibid., 146; see, also, Gillispie 1971, 101ff.; Bailhache 1990, 544). Within his theory of mechanics, Carnot showed that the "continuous" motions stand out from the "noncontinuous" ones. Machines attain their highest efficiency only when no impacts occur and changes in motions occur only continuously—"in inconspicuous degrees" (Carnot 1803b, xxj).

What did Carnot adopt from these ideas in mechanics and transfer to mathematics? Amazingly, none of the extensive historical research on Carnot seems to have examined this relation.[13] The next two sections will address this question in more detail. However, we can already state that although Carnot did not address discontinuous processes in mathematics, he repeatedly emphasized the need to pay attention to the properties of continuity. Carnot was responsible for drawing explicit attention to continuity in the French mathematical community, which, as Chapter III shows, took continuity for granted as an unquestioned precondition.

A further methodological relation between mechanics and mathematics is Carnot's development of a characteristic research approach to mechanics that he later transferred and applied to problems in infinitesimal calculus and in the concept field of the negative numbers. In general, he does not proceed from a single element, but from a "system"—for example, a system of bodies—in other words, from a set of elements attributed to or classified with a property. Another characteristic is that he does not view such a system in isolation, but typically compares two different system states: one as an assumed fixed and given system in an initial state and the other as the system after a certain transition (see, e.g., Carnot 1783, 14, 35; 1803b, 150 ff.).

---

[13]   With the exception of Gillispie's (1971, 121) sweeping classification as "an engineering justification of algebra and the calculus."

Carnot takes this comparison of two system states as a model for the analysis and transfers it to his study of the foundations of mathematics. When studying infinitesimal calculus in his Berlin memoir of 1785, he compares a system of *quantités designées* possessing fixed quantities with a system of *quantités auxiliaries* possessing changing auxiliary quantities (Carnot 1785, 177; see the Section 1.6.1.). To clarify the concept field of negative numbers, Carnot also compares two system states—particularly the concept of the correlation of figures (see Section 1.6.).

A third element making Carnot's concepts in mechanics productive in studies on the foundations of mathematics is practically a consequence of the approach of comparing two system states. The concepts used in his essay on mechanics of 1782/83 already contain the core of his concept of infinitely small quantities: as variables that can be conceived as null sequences. Carnot's basic concept and important conceptual achievement is to introduce the concept of geometrical motions (*mouvements géométriques*) for those changes of position that depend only on the geometry of the system and not on the laws of dynamics (Carnot 1783, 26)—; Lagrange's concept of virtual changes corresponded to one aspect of this concept. He supplements this with the discrimination into proper or absolute geometric motions and geometric motions *par supposition*. By the latter, he understands such motions that are not true geometric motions themselves, but become such when a part of the system can be removed without this influencing the system state:

> It seems therefore fitting to extend the name *geometric* to all movements which, without actually being so, become so, on eliminating some machine or part of a machine which has no influence on the state of the system (ibid., 34).

Carnot's later conception of infinitely small quantities reveals many analogies to the formation of these assumed geometric motions. One system turns into another, because one element can be either replaced by another or removed.

The present analyses show that the transfer of conceptual reflections in mechanics to the same in mathematics is not due to the acquired disposition or thinking style of an engineer. It is far more a sign of a rare striving toward theoretical clarification in mechanics and the effort to mathematize this field. At the same time, Carnot's work on the foundations of mathematics forms an autonomous, inherently mathematically motivated epistemological conception for which a practical, engineer–like orientation would be a completely misleading label.

1.6. Infinitesimal Calculus: Carnot's Shift from the Concept of Limit to the Infinitely Small

1.6.1. THE MEMOIR FOR THE BERLIN ACADEMY

Carnot's concepts on infinitesimal calculus started to be analyzed more closely only after his memoir for the Berlin competition was discovered and published by Gillispie and Youschkevitch in 1971. The first such analysis was that by Youschkevitch himself in the same volume. This focused on the 1785 memoir, and formed the basis for Youschkevitch's analysis of Carnot's contribution to the development of limit theory. However, the 1797 and 1813 versions receive only a brief comparative discussion at the end of his analysis, where he notes a tendency toward a "deemphasis of limit–theory" (Youschkevitch 1971, 168).

Although a French edition of this English–language work of 1971 appeared in 1979, knowledge of the original memoir has spread only slowly (see Barreau 1987, 9, note 1). Dhombres and Dhombres (1997, Chapter VI and annex) follow Youschkevitch's approach and analyze the first memoir in detail, but, yet again, mention the further two versions only in passing (ibid., 455 ff.). The one study that has compared the three versions most precisely up to now comes from Angelo Guerraggio and Marco Panza (1985). Hence, they have also analyzed the transition from the limit concept to the infinitesimals. However, this is predominantly an "internal" analysis, and not embedded in the general development of conceptions on analysis.[14] As a result, their criticisms of Carnot's mathematical concepts are ahistorically strict, for example, when they censure his failure to formalize the limit concept (ibid., 8) or his assumption that functions are continuous (ibid., 10).

Although Carnot's third version presents the relations between his conceptions of analysis and negative numbers, this relation has not been examined up to now—just like relations to his mechanics, to the general development of ideas on analysis, and to the epistemological change in post-Revolutionary France. The goal of the present section is to fill this gap.

What is surprising and new about Carnot's memoir on the theory of the infinite is its theoretical approach to the problem. Completely in contrast to the engineer–like approach assumed by Gillispie, Carnot does not pass over the foundational questions in favor of practical solutions. Indeed, he is the first author in France to consistently avoid presenting practical procedures and

---

[14]   Apart from referring to the main protagonists such as d'Alembert, L'Huilier, and Lagrange as well as the critic Wronski.

choose a theoretical presentation. His memoir starts with an explicit introduction of his fundamental terms and a reflection on their definition combined with a criticism of other authors who use such terms without defining them and reflecting on them. Whereas his treatise on mechanics had already been characterized by a rejection of the traditional stringing together of single cases in favor of a discussion of underlying principles, here Carnot was the first to produce a theoretical model for the analysis that would not become generalized to mathematics until the work of Cauchy.

As already mentioned in the previous section, "system" forms the basic concept for Carnot in the analysis just as in mechanics: as "system of quantities." What is characteristic is that such a system never appears in isolation, but always in a pair: one system of fixed quantities and a corresponding system of quantities going through change. Just as a system is made up of a constellation of bodies or particles of a body in mechanics, it is now the totality of the quantities occurring in a geometric figure that forms the specific "system." He ascertains relations between quantities through comparisons with the quantities in the system in a changed state. Hence, system is the basic concept of analysis for Carnot that serves factually as a substitute for the concept of function, which was still not applied generally as a basic concept. Although Carnot uses the concept of variables, this is secondary to that of the system. Hence, Carnot's understanding of mathematics is still dominated by the traditional model of geometry with its (static) figures as the main object of study.

Indeed, Carnot's memoir starts by introducing and explaining his concept of systems. To convey the *rapports et relations* between the quantities of a system, he introduces a second system that, through continuous changes (*changer par degrés insensibles*) according to a "law," can be brought either arbitrarily close to the first system or into complete agreement with it (Carnot 1785, Nr. 1, 177).

Carnot's basic principle consists in the following assumption: Instead of ascertaining the relations sought directly within the first system, one starts off by comparing the relations between the respective quantities of the first and the second system. Carnot's assumption is that the quantities of the second system can be eliminated through the approximation process, thereby making it possible to obtain the relations sought in the first system (ibid).

The emphasis on the given, constant, system and the merely random character of the variable system is expressed by calling the former the *système désigné*, and the latter, the *système auxiliare*. He correspondingly calls the quantities of the main system *quantités désignées* and the quantities of the auxiliary system *quantités auxiliaries* (ibid., Nr. 3, 178). The terms constant and variable, the initial terms in prior presentations of infinitesimal calculus, are for Carnot, in contrast, derived terms (compared with system): the quantities of the second system that do not change in relation to the corresponding *quantités désignées*, he calls *quantités constantes*, and those that change, *variables* (*auxiliares*) (ibid., Nr. 10, 180).

In the first version of his *Réflexions* on infinitesimal calculus, Carnot does not just rely on the limit method in order to prove infinitesimal calculus. He also develops it further, particularly through a precise specification of the relation to continuity. Carnot explicates this as follows: The *système désigné* is also called the limit of the auxiliary system, because it denotes the end of the change to which the auxiliary system is subject. The auxiliary system comes closer and closer to the main system:

> The designated system being no other than the final state of the auxiliary system, the one related to it and which it approaches more and more closely (ibid., Nr. 4, 179).

Here as well, single quantities are specified only after the entire system has been defined. The limit of an (auxiliary) quantity is the term that this quantity increasingly approximates—interestingly with the following additional remark: to the extent to which the auxiliary system approximates the main system (ibid., Nr. 5). Indeed, Carnot assumes expressly—and in line with his approach to mechanics—that the approximation is continuous or that it follows the law of continuity:

> Thus in general, one calls last values and last ratios of the quantities, those which the law of continuity assigns to these quantities and their ratios, when one imagines the auxilliary system as approaching the designated system continuously (ibid.).

Carnot did not publish anything about what happens if this precondition is not met.

The other surprising element in Carnot's conception is that he does not just subordinate and rank the concept of infinitely small quantities to the concept of limit, but also—once again, the first person to do this compared with his many predecessors—makes detailed attempts to provide a definitional introduction to these quantities. However, this attempt to "normalize" the infinite contradicted the problem set by the Berlin Academy.

Whereas the Academy had described the concept of the infinite as "contradictory," Carnot—in an introductory footnote—assigns this character only to the first imprecise ideas of novices. "In reality, however," he continues, "nothing is simpler than an exact concept of the infinite" (ibid., 174).[15] He justifies this statement—which was completely counter to the traditional view—by claiming that the concept of the infinite relates directly to the limit or to Newton's prime and ultimate ratios, to which nobody had ever objected and that also generally remain undefined because of their intelligibility (ibid).

Carnot gave numerous definitions of infinitely small quantities; these are addressed later. In his Berlin memoir, he formulates all these definitions in the form of limits. The initial definition, given in the first footnote, is analogous to de la Chapelle's formulation (see Chapter III.8.2), but given without the use of variables and only with the unspecific term *grandeur*. What he means by this is the difference between two variables: "Two quantities are said to differ infinitely

---

[15]   Perhaps this acceptance of the infinite should be seen as one reason why the Academy did not award first prize to Carnot's treatise.

little or to be infinitely little different from each other when their last ratio is a ratio of equality" (ibid.).

It is already clear in this first definition that this basic concept of the analysis is linked directly with a specific conception of the number concept. Carnot explains that he would allow only "effective" quantities as values for *grandeurs*, and these exclude not only infinites but also zero: "By this expression magnitude, I understand an effective quantity here. This means a quantity that is neither 0 nor $\frac{1}{0}$" (ibid.).

Even before this, he had not only identified the zero with nothing, but also "quantity" with an ontological existence concept. During the first introduction of infinitely small quantities, he expressly rules out the existence of something between "quantity" and zero, denying "that [something] forms a middle (*milieu*) between one of the quantities and the zero, between *existence* and nothingness (*le rien*)" (ibid., 174).

Excluding the zero from the domain of permissible quantities and tying the number concept to ontological ideas corresponds, as Chapter III has shown, to the broadly dominant concepts of quantity in French mathematics; as representatives contemporary to Carnot, we have met Cousin and Martin (see Chapter III.8.4). This exclusion simultaneously meant that definitions had to be given twofold: here, for the case of a limit unequal to zero and for the case of a limit equal to zero—in other words, completely analogous to Stockler (see Chapter III.8.5).

The definition of infinitely small quantities, which Carnot accordingly gives for the second case, is, on the other hand, one of his conceptual achievements, because this is the first time he produces a precise formulation of something that had been expressed only implicitly and covertly since Varignon. This simultaneously gives the definition its convincing simplicity: Infinitely small quantities are nothing other than special variables, namely, variables with a limit of zero!

For this definition, Carnot starts by differentiating between finite and infinite quantities. He calls all quantities whose limits form an effective quantity finite; the others—that is, those with limits 0 or $\frac{0}{0}$—in contrast, infinitesimal (ibid., Nr. 11, 181). This then enables him to make the following definition: "Among the infinitesimal quantities, those whose limit or last value is zero are called *infiniment petits* (ibid., Nr. 12, 182). And then he finishes off by summarizing it again: "Infinitesimal quantities are simple variable quantities whose limit is 0 or $\frac{0}{0}$" (ibid., Nr. 15, 182).

Carnot addresses potential objections to his definitions, thereby working out the variable character even more explicitly. One objection was that infinitely small quantities are generally conceived as being smaller than any given quantity. However, according to his definition, although they attain zero at their limit, they can otherwise be larger than this or any given quantity. Carnot's answer to this is that an idea of this smallness is necessary only for soothing one's imagination, but not for the computing operations. This requires only the final limit as well as the compensation of the errors (ibid., 175).[16]

This brought Carnot into conflict with La Chapelle, who had labeled the differences as *inassignables* in his article for the *Encyclopédie* (see Chapter III.8.2). However, Carnot went on also to reject the labeling of quantities as *inassignables* for epistemological reasons. Any quantity, as long as it exists, can also be assigned: "this inassignable means nothing, because it is evident that any quantity can be assigned as long as it exists" (ibid., 175).

This clear epistemological specification and restriction to existing and assignable quantities makes it completely clear—along with the equally clear definition of the *infiniment petits*—that Carnot conceives the *infiniment petits* within the conceptual bounds of contemporary mathematics. Hence, it would be completely wrong to interpret him as a precursor of nonstandard analysis. In contrast, this definition formulates a direct and effective precursor for Cauchy's definition of the *infiniment petits*.

Astonishingly, the existing literature on Carnot's Berlin memoir fails to discuss how far he openly contradicts the fundamental conception of the prize question. Using exactly the same terminology with which the Academy had called for a replacement of the infinite, Carnot explains that he can give a clear, exact, and precise concept of it, though admittedly within the framework of limit theory: "Hence, assuming the limit concept, here is a clear, exact, and precise solution" (ibid., 174).

Just as decisively, Carnot does not see anything contradictory in the concept of the infinite as long as it is conceived within the framework of limit theory: "Hence, the infinitesimal quantities are not fanciful beings, but simple variable quantities characterized by the nature of their limits" (ibid., Nr. 13, 182). This remarkable independence of Carnot is also revealed in his distance to d'Alembert, who had wanted, in the *Encyclopédie,* to exclude the *infiniment petits* completely from legitimate mathematical discourse (see Chapter III.8.2). This makes it almost self–evident that Carnot's conception has absolutely no connection to that of Bézout, the author whose textbooks for military schools were increasingly dominating the market.

The best place to find any direct influence on Carnot, in contrast, is in the work of Roger Martin. Up to then, Martin had been the only person to have based infinitesimal calculus on the limit concept and to have introduced the *infiniment petits* with the help of this concept, nonetheless, as we have seen earlier (see Chapter III.8.4), not in a conceptually logical way. A further influence of Martin can be seen in Carnot's separation of the infinitely small

---

16   At the end of the second definition of the *infiniment petits*, that for the limit zero, Carnot also emphasizes that the values for the variables do not need to be arbitrarily small. Nothing forces one to assign more specific values to them rather than to other variables. The character of these quantities is determined far more by the nature of their limits: "le caractère de ces quantités [...] consiste uniquement [...] dans la nature de leurs limites" (Carnot 1785, Nr. 17, 186).

quantities into two differently labeled types. He is the only other contemporary scientist to do this. Because he had not analyzed the variable character of the *infiniment petits* in a clear way, Martin had regarded them from two different perspectives:

1. Either as a term for the last ratios in the limit process and as such *vraies quantités*, that is, objects of operations.

2. Or as the limit value, that is, zero, and therefore unable to form any ratio or be *l'objet du calcul* (Martin 1781, 328).

In a completely analogous way, Carnot distinguishes between whether the limit zero is already assumed or not assumed:

1. If not, then the variables can be assigned a certain value. Carnot then calls such a value *infiniment petite* **sensible** *ou* **assignable**. At the same time, he explains that he prefers to use the expression *indéfinement petite* in order to avoid the misunderstandings generated by the expression *infiniment*.

2. In the case of the limit value zero, in contrast, the quantity is "absolute zero." He then labels this the *quantité évanouissante* (Carnot 1785, Nr. 15, 183).

Carnot uses this distinction to characterize the existing, apparently disparate, methods of infinitesimal calculus and then propose a way of uniting them.

In the Berlin memoir, Carnot applies the concept of infinitely small quantities particularly to establish an operation with limits. In this first version, he works out theorems over limit behavior. He does this by introducing the property of "being infinitely close." He needs this property to assess the approximation process in his two basic systems, a *système désigné* and a *système auxiliare*. Carnot declares two quantities to be infinitely close when the limit of their ratio possesses a ratio of equality, that is, is equal to one: "Two quantities are said to differ infinitely little or to be infinitely little different from each other when their last ratio is a ratio of equality" (ibid., Nr. 18, 187).

Correspondingly, he assumes that in the two systems, the same property should hold for all corresponding pairs of quantities. He thereby introduces the expression of an *équation imparfaite:* "False" equations between the quantities of two systems can become exact ones in the passage to the limit when their quantities are infinitely close: "If A and B differ infinitely little, the equation $A = B$ will be a false equation, but its limit will be none the less exact, and the last ratio of A and B will be a ratio of equality" (ibid., Nr. 19, 187).

Gueraggio and Panza (1985, 9) wanted to show a mathematical weakness in Carnot here, because they considered that he had confused "infinitely close" with the definition of asymptotic equality. However, first of all, asymptotic equality is a recent definition that has emerged only in modern mathematics, and second, there is no reason why Carnot should not have phrased the meaning he intended in such a definition, as long as he applies it consistently (see below). In addition, Carnot evidently models his definition here on the central Lemma I in Newton's *Principia*:

> Quantities, and also ratios of quantities, which in any finite time constantly tend to equality, and which before the end of that time approach so close to equality that

their difference is less than any given quantity, become ultimately equal (Newton 1969a, 29).

Pourciau (2001, 24) has recently defended this definition against traditional criticism: Newton was naturally aware of the conditions under which $\frac{f}{g} \to 1$ and $f - g \to 1$ are equivalent, namely, for infinite and nonvanishing limits of f and g.[17]

The productive aspect is that Carnot not only introduces a limit sign equivalent to that of L'Huilier, but he also applies it operatively, among others, to express the property of being infinitely close. The introduction of the sign is not covert, but is performed explicitly: "I point out that I label the limit or the last value of an arbitrary quantity through this same quantity preceded by the sign L" (ibid., Nr. 36, 199).

He is then able to formulate the above property in the short form $\mathcal{L}\frac{A}{B} = 1$ (ibid., Nr. 51, 214). Carnot applies this limit sign extensively, particularly in proofs (see ibid.). He also uses it in the proof of his theorem that—continuing the work of Martin—now became part of the standard repertoire of limit theory, namely, the exchangeability of the limits with mathematical operations, in this case the formation of quotients:

> The last ratio of any two quantities is always equal to the ratio of their last values, or, which is the same thing, the limit of the ratio of two quantities is equal to the ratio of their limits (ibid., Nr. 88, 242).

The universality of the statement for *quelquonques* quantities continued to be found for a long time among later authors. Carnot also expresses this statement in symbols: "Hence, $\mathcal{L}\frac{Y}{Z} = 1$ is the same as $\frac{\mathcal{L}Y}{\mathcal{L}Z} = 1$" (ibid).

However, it has to be admitted that Carnot is not completely consistent when constructing infinitesimal calculus on the limit concept. Independent of this, he also develops and presents the traditional concept of comparing different orders of the infinite or the infinitely small (ibid., Nr. 21–26, 188 ff.).

To answer the main question posed by the Academy, Carnot developed a comprehensive, systematically linked conception on how consistent propositions can be derived from contradictory concepts. This is the conception of the compensation of error based on the previously developed foundations of the two systems and their relations through *équations imparfaites*. Because this conception of error compensation, which also remains fundamental to the two later versions, has been reported frequently in the literature, it only needs to be sketched here.

---

[17]   Nonetheless, Pourciau reinterpreted Newton's definitions: When Newton talks about "differences," he means the difference between the limit "quantity" and the *ratio* "quantity"—and not between the two "quantities" in the ratio. The problem arose for the author because he was too quick to identify "quantity" with "function," and, hence, no longer saw that the constant limit is also one of the quantities implied.

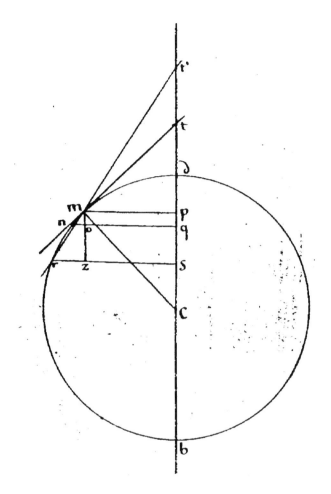

Figure 13: Reproduction of Carnot's figure in his 1785 memoir

However, Gillispie and Youschkevitch's (1971) edition of the Berlin memoir is incomplete insofar as the one figure in the entire text is missing. Although Section 39 clearly states *voyez la figure*, no comment has been made on its nonappearance up to now. Dhombres and Dhombres (1997, 178 f.) constructed three figures specially to overcome this deficit. Because the figure explains the concept of error compensation, it is published here for the first time (Figure 13).[18]

---

[18]   AAdW I–M 842, Fol. 91. The Gillispie and Youschkevitch (1971) edition publishes only the first 90 folios. In the original, the figure on Folio 91 is hidden behind a fold in the paper. A comment in the preface points out that the figure addressed in Nr. 39

In all three versions, the mathematical problem that Carnot uses to explain the concept of error compensation is that characterized as the initial problem in the analysis of drawing a tangent to a circle at a point m. The main system, the *système designé*, particularly includes the circle. To solve the problem, a *système auxiliaire* is formed by making the circle a polygon with a very large number of sides. Increasing or reducing the number of sides leads to the approximation of the auxiliary system to the main system.

If mn represents one of these sides in the auxiliary system, then the rules for geometric proportions produce

$$mo : no :: tp : mp, \text{ or } \frac{mo}{no} = \frac{tp}{mp}.$$

And by entering this in the initial equation of the circle for m, hence $yy = 2ax - xx$, and for n, hence $(y + no)^2 = 2a(x + mo) - (x + mo)^2$, one obtains

$$\frac{mo}{no} = \frac{2y + no}{2a - 2x - mo}.$$

If no and mo are now dismissed as very small quantities, the result for the sought after (sub) tangent is

$$tp = \frac{y^2}{a - x} \text{ (Carnot 1785, Nr. 39, 203 f.).}$$

Although this initially appears to be only an approximation, it actually proves to be the exact result.

Replacing the circle with a polygon already introduces one error. But this error in the equations is compensated directly by the second error introduced by neglecting the arbitrarily small quantities no and mo (ibid., Nr. 46, 209). The solution of the problem is to introduce a few indeterminate quantities, *indéterminées*. This allows a few errors to be committed as long as they are kept arbitrarily small. Once these indeterminates have been eliminated during the course of the operations, one can be certain that the errors committed have been compensated, and the exact result has been attained (ibid., Nr. 48, 211). Carnot generalizes the example to state that the theory of infinites should be a *calcul* of error compensation: "La théorie de l'infini n'est autre chose qu'un calcul d'erreurs compensées" (ibid., Nr. 57, 217).

This is his general answer to the Berlin competition. His concept of the *équations imparfaites* and five "theorems" over their limit properties form its mathematical scaffolding. Berkeley (1734) was the first to propose using the concept of error compensation as a method for explaining the correctness of infinitesimal calculus in his work *The Analyst*. Carnot could also feel that the legitimacy of his approach was bolstered particularly by a comment from Lagrange. In an essay in the second volume of the Turin Academy of 1760–1761

---

is identical with Figure 1 in the 1797 edition except for the capitalization of the letters marking the points. This difference in connotation is correct, but some of the proportions also differ.

entitled *Note sur la Métaphysique du Calcul Infinitésimal*, Lagrange also discussed the reciprocal elimination of error:

> The error of the hypothesis completely destroys what has been done in the calculation. [...] Here it is just as in the method of the infinitely small quantities, where the calculation itself corrects the false hypotheses which have been made (Lagrange, Oeuvres, tome I, 597–599).

However, Carnot does not look any deeper to see what justification underlies his claim for a compensation of error. Indeed, Dhombres and Dhombres (1997, 177–184) present a detailed analysis to show not only that Berkeley's claim to have demonstrated a compensation between two contradictory errors is unjustified but also that Carnot's claim to have demonstrated such a compensation is, despite all its appeal, nothing but a bluff, because no different errors were involved and, as a result, no compensation either. However, they also admitted that the subsequent formulation of a theory of *équations imparfaites* motivated by the compensation concept marked an important step in the algebraization of the analysis as the limit theory of error compensation (ibid., 184).

Finally, Carnot also undertakes a systematic study of the relation between the different methods of infinitesimal calculus in his Berlin memoir. He bases this on the previously described discrimination between a "perceptible" and an "absolute" infinite, that is, between indeterminate (*indéterminées*) and vanishing (*évanouissantes*) quantities.

The first method operates with infinitely small quantities, in other words, with variables to which "effective" values are assigned. This is simultaneously the method in which error compensation is effective and the *équations imparfaites* are applied (Carnot 1785, Nr. 58, 219). The second method, in contrast, assigns the value zero to the infinitely small quantities. This is the *calcul des évanouissantes*. What this actually means, although not named as such, is Euler's reckoning with zeros applied to $\frac{0}{0}$ for the differential quotients. Carnot explains that in this method, it is also the *loi de continuité* that ensures a finite, effective value for the apparently indeterminate expression, thus permitting a mathematical operation (ibid., Nr. 61, 220). For the apparently indeterminate term $\frac{0}{0}$, one could not enter any arbitrary value. Instead, the value is determined by the continuous limit process for the quantities in the numerator and the denominator:

> In the case treated above, for example, as long as mz does not vanish, $\frac{mz}{rz}$ is greater than $\frac{tp}{mp}$: these two quantities do not become equal until the moment when mz reduces to zero. It is true, however, that $\frac{mz}{rz}$ is just as equal to any other quantity as it is to $\frac{tp}{y}$, since $\frac{0}{0}$ is an arbitrary quantity; but among these diverse quantities, $\frac{tp}{y}$ is the only one which is subject to the law of continuity and determined by it (ibid., Nr. 62, 221).

In summary, Carnot states that the difference between the two methods is as follows: Although the equations are always exact in the *calcul des évanouissantes*, one operates with quantities that are *êtres de raison*. In the *calcul des erreurs compensées*, in contrast, one operates with *véritables quantités*, but the equations are false or *imparfaites* (ibid., Nr. 68, 225). Hence, the true method of the *analyse infinitésimale* is to link the two methods together by using, in part, the *quantités évanouissantes* and, in part, the *quantités indéfinement petites*, the choice depending on which is practically feasible (ibid., Nr. 72, 229).

Although this already seems to be his conclusion regarding the methods of infinitesimal calculus, and although both the two single methods and the combined method are all based on the limit concept, Carnot goes on to discuss— initially rather astonishingly—a separate, "true" *méthode des limites*, that then proves to be the superior general method.

As such an independent method, Carnot presents the determination of the limit for the difference quotients:

$$\mathcal{L} \frac{\Delta x}{\Delta y} \ , \ \mathcal{L} \frac{\Delta z}{\Delta x} \ , \ \mathcal{L} \frac{\Delta y}{\Delta z} \text{ (ibid., Nr. 84, 239 f.).}[19]$$

For Carnot, forging a link between determining the differentials and the differential quotients is not mathematically trivial. He therefore stresses that the differential quantities do not occur separately in the actual limit method: "In the latter [method], one cannot enter separately the quantities we have called infinitely small, nor their ratios, but only the limits [...] of these ratios" (ibid.).

He calls this inherent tie a "difficulty" that does not occur in what he calls the normal *analyse infinitésimale*: One cannot separate the variables and perform the transformation operations individually:

> It is that one cannot here, as in the latter method, separate the infinitely small quantities from each other; they must always occur together, which prevents the equations where they are met from being subject to all the transformations which would serve to eliminate them (ibid., Nr. 85, 241).

This "difficulty" confirms once again, for both the mathematical–conceptual and the mathematical–operational sides, that the basic concept in the analysis for Carnot is still the geometric curve concept and not yet the function concept (see Chapter III.4.).

It is only when he reaches this form of the limit method that Carnot emphasizes that it operates with finite quantities alone, and therefore restricts itself to the means of normal algebra. This makes infinitesimal calculus "a simple application of ordinary algebraic calculus." Since a pure algebraic calculation with finite "perceptible" quantities that does not take detours through the "land of errors," this limit method is clearly the one that is general and to be preferred.

---

[19]   Carnot writes ∂x instead of Δx, and so forth.

However, because of the difficulties he describes in separating the variables, Carnot does not want to grant it this status in general. Nonetheless, after developing several theorems on limits—particularly on their substitution (see above)—he himself reports a way to overcome this difficulty: by changing the differential basis. The reproduction of this line of thought in Figure 14 presents the operative application of the limit symbol L, for separating the variables.

Figure 14, Carnot's operating with the limit-symbol (Carnot 1785, fol. 71 f.)

In a final evaluation, Carnot summarizes his opinions on the significance and status of the various methods of infinitesimal calculus he has been discussing. In light of the later editions, it is particularly interesting to see that the Berlin memoir describes all methods as precise and lucid and that none should be assigned an exclusive rank:

> If one were now to ask which, of all the methods that have just been explained, is the most preferable, I would reply that they are all good, exact, and lucid. It would be proper, I think, not to exclude any (ibid., Nr. 98, 252).

However, Dhombres and Dhombres (1997, 190) have found an error in Carnot's first theorem on limit theory: that on the *équations imparfaites* (Théorème I; Carnot 1785, Nr. 51, 214). Because the error is also so elementary, the judges at the Berlin Academy could not have missed it, and this may well be one of the reasons why they did not award him first prize. Put briefly, the theorem states than an exact or *imparfaite* equation does not change its character into an effectively false equation when one replaces one of the quantities with another that is infinitely close to it. As a counterexample, Dhombres and Dhombres (1997, 190) report the equation

$$x^3 = 6(x - \sin x).$$

This is *imparfaite*, because it becomes exact for x approaching zero. However, if $\sin x$ is replaced by an infinitely close quantity, for example, $x - x^3$, one obtains the "exact" equation $1 = 6$. However, only a few lines above, they expressly emphasize that Carnot—following the general practice among all mathematicians in the eighteenth century—implicitly assumed polynomial equations. In fact, Carnot's proof for his Theorem I shows that he assumes both sides of the equation to be polynomial.

The problem with the theorem's validity does not lie in a confusion of quantities with variables (ibid., 189)—for Carnot, variables are special quantities—but in the theorem's implication of two different limit processes without considering whether their sequence could be changed. However, this was not an elementary problem, but at the time, a far-reaching one. Nobody had analyzed relations between multiple limit processes yet, and there was a total lack of symbols for characterizing such multiplicity. The depth of this problem will be seen later, in that it also underlies Cauchy's theorem on the continuity of the limit function, a problem that was solved only when Dirksen formulated an appropriate symbolic characterization (see below, Chapter VI.6.6)..

The further Theorems IV and V particularly contain propositions on substitution: conditions for limits to remain unchanged when variables are replaced by arbitrarily close quantities (Carnot 1785, Nr. 65 & 67). The substitution theorems are no longer presented so explicitly in the later versions. Nonetheless, the substitution concept was to become one of the main methods of infinitesimal calculus in France during the first half of the nineteenth century (see below, Chapter VIII).

1.6.2. THE 1797 VERSION

The first published version of Carnot's memoir on the foundations of analysis appeared in 1797, only a few weeks after Lagrange's *Théorie des fonctions analytiques* and after the first volume of Lacroix's major *handbook*. It thus formed a counterpoint particularly to the work of Lagrange, who wanted to found analysis without the concept of the limit and the *infiniments petits*. We cannot say how far this published version—under the title *Réflexions sur la métaphysique du calcul infinitésimal*—is identical to the memoir submitted to the Dijon Academy in April 1788. The title is slightly different and less modest: *Nouvelles idées sur la métaphysique du calcul infinitésimal* (Tisserand 1936, 399). The Academy did not assess the submitted manuscript, most probably because none of the members felt competent enough (see ibid., 400). As a result, it remained unpublished. In the 1797 preface, Carnot points out that the text in its current form has been around for "several years" (Carnot 1797, [iii]).

On December 10, 1791, Carnot asked the Academy to return the manuscript to him at Paris. However, what he actually received in January 1792 was a copy. Up to now, no trace has been found of the 1788 manuscript in the Academy's archive (Taton 1990, 464). The Berlin memoir was revised either in 1787 or at the beginning of 1788 (ibid.).

Comparing the 1797 version with the original from 1785, the first thing one notices is that there has been hardly any further practical or conceptual development. Basically, it is only an abbreviated version of the original text. The only notable expansion is the addition of two chapters: one on the foundations of differential and integral calculus, and the other containing some examples of its application. Both chapters, however, are computational and not conceptual presentations. Furthermore, the main text has not been abbreviated consistently throughout; there are indications of shifts in emphasis and conceptual changes.

The most conspicuous difference is that the 1797 version lacks the framing concept of a comparison of two systems: the fixed *système désigné* and the variable, approximating *système auxiliare*.[20] Hence, the variable concept is no longer embedded in the system concept. This had previously ensured—through continuous approximation—the categorization of his concept of the *infiniments petits* within limit theory. It is also conspicuous that the text no longer begins with a systematic introduction to and definition of the various basic concepts. Correspondingly, the limit concept is no longer one of the initial concepts; it is defined only much later (Carnot 1797, 24). As a result, the limit concept no longer represents the general conceptual framework for the individual methods as in the text of 1785. The *méthode des indéterminées*, in which the *quantités infiniments petits* are assigned finite values, now appears as the main method of the *analyse infinitésimale* (ibid., 38). The method of reckoning with zeros applied to the *quantités évanouissantes* takes only a minor role. The limit method in a stricter sense—that is, operating with difference quotients—is, in contrast, clearly downgraded. After presenting this method using the means of *algèbre ordinaire*,[21] Carnot determines, as in the first version, that it is preferable to the *analyse infinitésimale*, as long as it is just as easy to apply:

> If this method were always just as easy to apply as ordinary infinitesimal analysis, it would appear preferable; for it has the advantage of leading to the same results by a direct and always lucid route, whereas the latter leads to the true [result] only after having made a detour into, if one may be permitted to say so, the land of errors (ibid., 44).

To confirm that the limit method is not easier, he once again names the difficulty of not being able to separate the variables, however, without mentioning the alternative sketched in the first version of substituting them. Nevertheless, he does add that this difficulty is to be found more in the preparatory considerations than in the operations themselves (ibid.).

---

20   The term *système* is mentioned only once, in relation to the "law of continuity" (Carnot 1797, 50).
21   The presentation of this limit method marks Carnot's first use of the symbol "L" for the limit, now as a Roman letter rather than in script form (ibid., 43).

The reversal revealed here is also illustrated by a remark directed toward critics of the *analyse infinitésimale* who considered its principle to be false and likely to lead to misconceptions:

> The metaphysic that has just been explained easily provides answers to all the objections which have been made against infinitesimal analysis of which many Geometers have believed the principle to be false and capable of leading into error (ibid., 54).

In the 1788 text and the possible revision of 1792, the only major new reference is to the winning memoir published by L'Huilier in 1786. He may have been the target of this criticism, though Carnot may also have been thinking of the mathematics class at the Berlin Academy.

Finally, it is impressive to see how fascinated Carnot is with the concept of infinitely small quantities. Even in the first version, he makes a conspicuous effort to define the basic concepts as clearly as possible. In the new 1797 version, he concentrates this effort on the *infiniments petits*: Whereas the first version had given two definitions for them—objectively agreeing and differing only according to whether the limit was unequal or equal to zero (see Appendix B, Part I)—the new version gives six definitions for this concept!

These six definitions are documented in Appendix B. In conceptual breadth, they do not differ from the definitions in the 1785 version. However, we see that Carnot now carefully avoids using the expression "variable," replacing this specific expression with the unspecific and overly general *quantité*. For example, an infinitely small quantity is no longer a variable with a limit of zero, but a "quantity" with limit zero (see Appendix B, Part II, Numbers 2 and 6). One cause for this change is the decision to drop the two reference systems on which Carnot had based his definition of the concept "variable" (see above).

The 1797 version places more emphasis on defining differences of other quantities, that is, for a limit deviating from zero: as the difference of an auxiliary quantity to its limit (Number 1), as the difference between two auxiliary quantities (Numbers 3 and 4), and as the ratio of the difference of two quantities, whose final ratio is a ratio of equality, to either one of these quantities (Number 5). Despite this extension of the foundation for the *infiniment petits*, Carnot drops the further theorems that added the substitution of quantities by others to which infinitely small quantities have been added. There is no apparent reason for this.

Finally, it is notable that the sweeping statements on generalization and simplification—particularly in the final section—have been dropped.

1.6.3. THE 1813 VERSION

Whereas the 1797 version presents a shortened edition of the original memoir in which the conceptual about-face is not yet complete, the last version of 1813 presents a much-extended text with a radically changed conception.

The first 11 introductory paragraphs are identical in both versions. However, what follows is an almost complete revision. As in 1785, a section of definitions comes next, but this is restricted to the *infiniments petits*, without addressing or introducing the limit concept. Then comes a reintroduction of the system conception, a presentation of the central principle of the *analyse infinitésimale*, and the theorems from the prior version in a somewhat reformulated form. The conception of the *équations imparfaites* is given as an equivalent approach. This completes the presentation of the foundations of the *analyse infinitésimale*. It is now based exclusively on the concept of infinitely small quantities and the comparison of their different orders. The section on how to perform differential and integral calculus first appended in 1797 now follows, strongly extended as a presentation of the algorithm and further augmented by calculus of variations.

For the first time, the work is divided into two volumes. The second volume contains a historical and systematic discussion of earlier and alternative methods of infinitesimal calculus—integrating passages of text from the earlier versions. This new section was certainly encouraged by Hauff's translation of 1800, which had appended a long historical and critical essay. This last version is characterized by a more or less strongly implicit criticism of the Berlin competition. Leibniz's conception is now described as "exact," as can be confirmed through the method of compensation for error (Carnot 1813, 35). Following the French tradition, this implied approval for the Leibnizian method of drawing on the concept field of the *infiniments petits*. Carnot's continued fascination with these quantities is characteristic, and it leads him to engage in a literal fury of definitions.

The 1813 text gives four further definitions of the *infiniments petits*. These differ from the earlier sets (see Appendix B, Part IV). I also found a sheet of paper in his *Nachlass* on which he had jotted down a set of nine further approaches to a definition (ibid., Part III). These notes can be dated clearly to the years between publication of the 1797 and 1813 versions. They are evidently part of his preparations for the last version. They use characteristic expressions found only in this version and do not fit the earlier versions conceptually. Whereas the first two sets (1785, 1797) are based completely on the limit concept,[22] the definitions in the last two sets (*Nachlass*, 1813) are built precisely on its exclusion.[23] The approach of replacing the variable concept by the unspecific *quantité* concept already present in the second set of 1797 is strengthened. The third set goes beyond this and marks the transition to the traditional understanding of infinitely small quantities within the framework of the geometric method of exhaustion: omission of the algebraic definition based on the link to a specific limit, omission of the statement on the actual attainment

---

[22]   With one exception in Definition II.6.

[23]   An exception being the last definition but one in 1813 (Part IV.4), in which a definition from the second set of 1797 is repeated within the framework of the presentation of Newton's limit method.

of the limit, exclusive use of the quantity term, and restriction to statements on an arbitrary diminuition of these quantities.

Typical for the strong change in the understanding of the infinitely small quantities is the first definition in the text of 1813 (Appendix B, IV.1):

> We will call every quantity, which is considered as continually decreasing, (so that it may be made as small as we please, without being at the same time obliged to make those quantities vary the ratio of which is our object to determine), an Infinitely small Quantity (Carnot 1832, 17).

The only remaining difference compared with the traditional geometric quantity concept is given in the last part of the sentence: the condition that as they continuously diminish, the quantities do not induce change in those quantities whose relation one is seeking. This points to Carnot's framing concept of linking together the main system and auxiliary system. Indeed, although he dropped this framing concept in the short version of 1797, Carnot already returns to the definitions on the linkage in his *Nachlass* manuscript. Here is the continuously retained condition with only slight changes in its formulation: "without changing anything in the value of the other—either given or variable— quantities of the system" (see Appendix B, Part III).

This condition corresponds to a basic understanding of infinitesimal motions in both mechanics and analysis that has remained unchanged since Carnot's first writings on mechanics (see above, Section 1.5).

The definitions in the third part, in the *Nachlass* manuscript, reveal yet another remarkable idiosyncracy. All nine definitions do not just relate a *quantité infiniment petite* to the embedded system of quantities, but further assume that operations are performed with several infinitely small quantities in the single system under examination. Hence, all these definitions represent a demand for uniformity in the limit process (even though the limit process is no longer formulated in terms of the limit of a variable). The quantities should *simultaneously* be arbitrarily small or zero (see Appendix B, Part III). The demand that several quantities attain the limit *simultanément* or *en même temps* would have offered interesting possibilities for developing limit theory— particularly through reflections on properties of uniformity. Because these approaches remained unpublished, there is also no evidence regarding how Carnot intended to use them or what had prevented him from doing so. Although the first volume of the 1813 version applies his instruments to more "problems" than just the tangent problem of 1785, including ones on cycloids, pyramids, and spheres or cylinders, the mathematical instruments needed for these traditional geometrical problems are not very demanding. However, there is one suggestion for an application: When describing alternative methods in the second volume, Carnot uses much of the terminology from his manuscripts to describe Lagrange's approach to function theory. He argues that Lagrange's differential terms operate in a way analogous to his own *infiniments petits*:

> Although Lagrange takes his differentials as if they were finite differences, they have a character which essentially distinguishes them from these last: it is this, that they

continue always indeterminate, during the whole operation, make them as small as we please without at all changing the value of the sought quantities. This furnishes means of elimination which never belong to the ordinary calculus of finite differences, in which these differences are fixed. (Carnot 1832, 111 f.).

Hence, it is possible that at the time when he was working on his definitions, Carnot was trying to match his concept more exactly to that of Lagrange.

In the "algorithmic" section following the foundational section, Carnot presents the *calcul* for differentials *dx*. Hence, he continues to proceed from the basis of the geometric concept of curves and not the function concept. He only occasionally mentions functions and uses differential quotients (see Carnot 1813, 77). In the second volume, Carnot discusses further historical methods that can be used to "replace" the one single remaining main method: the *analyse infinitésimale*, that is, the method of infinitely small quantities. The alternatives he suggests are the exhaustion method of the *anciens*, the method of indivisibles, and the method of *indéterminées* (a method of indeterminate coefficients traced back to Descartes, as an extension of the method of *indéterminées* presented in the 1785 text: operating with variables taking "perceivable" values with a limit of zero). Moreover, he also presents some of Newton's different methods: the methods of first and last ratios or the *limites*. Here Carnot draws on earlier versions of his text, but without applying an explicit concept of the variable. He uses the symbol L as an "expression of the limit," without applying it operatively (ibid., Vol. 2, 38). He also presents the fluxion method and the *calcul* of evanescent quantities. Finally, in a major revision compared with the 1797 version, Carnot addresses Lagrange's concept of the analytical functions. His commentary is very brief, with the justification that his textbooks on function theory are widely available. Carnot sees a major analogy between Lagrange's function approach and his own method of the indeterminates. Although Lagrange does not drop any quantities in his approach—all steps continuously remain exact—this is analogous to the method of indeterminates, because, practically speaking, it is only the first term that is important in the expansion of a series. Although the further terms are not dropped there, it is implied (*sous-entendu*) that they cancel each other out in further operations. Carnot declares it to be a true obstacle to the general acceptance of Lagrange's concept that it is so new as to make it necessary to abandon all the long entrenched methods and textbooks (ibid., Vol. 2, 58).

The final comprehensive conclusion practically forms a rejection of the *méthode des limites*. Carnot presents a general comparison with the method of *infiniments petits* that is now understood as its opposite. He tries to present the latter as superior in every way. Hence, the 1813 version is also a legitimization of the radical change at the *École polytechnique* in 1811: the return to the synthetic method (see Chapter IV.3.6.).

Already in the prior section, when presenting the *méthode des limites* as one of Newton's methods, Carnot explains—fully embedded within the context of the debates at the *École polytechnique*—that the concept of infinitely small quantities is in no way "less lucid" than the concept of the limit. Furthermore, he points to two disadvantages with the limit method: It has to dispense with the inclusion of auxiliary quantities, and is therefore unable to vary and transform these expressions. And, as in the two prior versions, he repeats the disadvantage that it is impossible to separate the differentials (Carnot 1813, Vol. 2, 33 ff.). Moreover, as in 1797, he no longer mentions the method for avoiding this problem presented in 1785 (see Section 6.1. above and Figure 14).

In the final section, Carnot expands this devaluation further. First, he explains that the various methods basically form just one single method, namely, an extension of the original synthetic method of exhaustion:

> The different methods which we have described in this work are, properly speaking, one and the same method, presented under different points of view: it is always the method of exhaustion of the ancients more or less simplified, more or less happily applied to the wants of the calculus, and at last reduced to a regular notation [sic!: should be "algorithm"] (Carnot 1832, 116).

Nonetheless, despite the assurance that no approach is exclusive, he presents a criterion for selecting among the methods. One should select those methods that are the simplest and, should two have equal rank, then the one that is the most convincing. Carnot now postulates that such a method exists: the Leibnizian. This means not only that he explicitly rejects the legitimacy of the Berlin submission of 1784 but also that his conception of the *infiniments petits* has swung back to the traditional French one:

> But amongst all these methods deriving their origin in common from the method of exhaustion of the ancients, which is that which offers most advantages for habitual employment? It seems generally agreed that it is the Analysis of Leibnitz (ibid., 121).

He then compiles arguments to counter the limit method. First, he notes that in the second edition of his *Mécanique analytique*, Lagrange himself took the *infiniments petits* as a foundation (ibid., II, 58, 66). Lagrange actually did make this about-face in 1811—but not out of sincere conviction, rather as a reaction to the *École polytechnique*'s ruling in favor of the *infiniments petits*. Lagrange's own commentary confirms his internal resistance (Lagrange 1811, Avertissement, and see Section 2.4.5. below).

Second, Carnot argues that the *limites* method does not dispense with the need to distinguish between *quantités désignées* and *non désignées*, that is, between fixed and variable quantities (ibid., Vol. 2, 69). Surely, nobody had claimed this. The third argument refers directly to the controversies at the *École polytechnique*. It would be a mistake to believe that the limit method is any more rigorous than that of Leibniz:

> Hence we may observe that the expression of limit is neither more or less difficult to define than that of an infinitely small quantity; and that consequently it is erroneous to think that the method of limits is more rigorous than that of the Infinitesimal Analysis (Carnot 1832, 126).

The justification for this is only that the definition of limits is linked, in his opinion, to that of the *infiniments petits* (see Appendix B, Part IV.5). Hence, if, as indubitably the case, the limit method is exact, then the infinitesimal method must be exact as well (ibid.). He once again repeats the formulation of the "disadvantage" that the differentials in the limit method are linked together in differential quotients, whereas in the *analyse infinitesimal*, one can operate with auxiliary quantities in isolation. This leads him to his final conclusion that the *infiniments petits* even form the superior method: a development of the limit method with extended and operative applications:

> The Infinitesimal Analysis is therefore an improvement on the method of limits; it is a more general application of the latter, and such as nevertheless is neither less exact nor less clear (Carnot 1832, 127).

Carnot further supplements this apotheosis by saying that the statement applies for devising not only the foundations but also applications, for devising the equations, and for solving the problems (ibid.). Finally, he also introduces the argument of intuitiveness: It is generally the case in mathematical research that the imagination is directed toward the *infiniments petits* and not to the limits of their ratios (Carnot 1813, Vol. 2, 72).

Whereas Carnot started his memoirs on analysis by espousing limit theory and developing it as an algebraic theory, he had now, after the revolution and in the Napoleonic age, reverted to the synthetic infinitesimal method that continued to be taught in the military corps. He became a supporter and legitimizer of the decisively antialgebraic radical change at the *École polytechnique*.

On the other hand, Carnot clearly made a lasting contribution by making the property of continuity one of the major elements in the basic concepts of analysis. Even in the last version of 1813, he continually emphasized that when the limit of the quotients of *quantités évanouissantes* possesses a certain effective value, this is due to a law of continuity.[24]

1.7. Substituting Negative Numbers with Geometric Terms

1.7.1. CARNOT'S WRITINGS ON NEGATIVE QUANTITIES: AN OVERVIEW

Already in his first mathematical work, the Berlin memoir of 1785, Carnot addresses not just the mathematical infinite, but negative quantities as well, although only marginally in a footnote and placing little emphasis on them. However, this already presents an essential element of his later basic concept. In contrast to the French textbook tradition, particularly that of the Oratorians, he permits negative quantities only as subtractive quantities. In general, the only permissible objects are positive quantities:

---

[24]   Of course, at this point, one can still not expect differentiability to be taken into account.

At each step negative quantities are met which, it must be understood, can represent true quantities only when they are preceded by positive quantities bigger than themselves (Carnot 1785, Nr. 83, 239).

As we have seen in Chapter II, such conceptions that recognized exclusively positive numbers as a foundation for mathematics originated in England, as part of the priority given to synthetic methods by Barrow. MacLaurin took them up again in his *Treatise of Fluxions*, where they also came to the attention of d'Alembert (see Chapter II.8.2). For Carnot, distinguishing the absolute, positive numbers was logically consistent, because of his exclusion of zero from the domain of quantities. This made him more consistent than other French mathematicians such as Cousin and Martin, who excluded the zero while retaining the negative quantities.

After his Berlin memoir, Carnot did not address this conceptual field for a long time, and particularly not in the short version of his *Metaphysics of Infinitesimal Calculus* of 1797. It was only after 1797 that he returned to this field and started to address it intensively, evidently in relation to his work on trigonometry. The first programmatic presentation was his work on the correlation of figures in 1801. The *Géometrie de Position* of 1803, initially conceived as a revision of the *Corrélation*, developed into an extensive work presenting a comprehensive account of his alternative conception of geometry.

After this, Carnot went on to discuss negative quantities in two further publications. An Appendix to a work on geometry in 1806 presented his conception of the algebraic operation with negative quantities. This was the first time he addressed this concept positively. Previously, like d'Alembert before him, he had discussed it only negatively by rejecting it. In an Appendix to the 1813 version of the *Reflections on Infinitesimal Calculus*, Carnot reported how close he found the relation between the concepts of infinitely small quantities and negative numbers. As also formulated in the manuscripts to be found in his *Nachlass*, he considered the principle of error compensation to be the regulating element that corrected originally false assumptions and led to correct results.

1.7.2. CARNOT'S BASIC CONCEPTS AND ACHIEVEMENTS

Carnot's ideas on the concept of negative quantities not only show that he had analyzed foundational issues more consistently than other mathematicians, but also confirm that he did not take an engineering approach to mathematics. Instead of skimming over foundational issues in favor of easily manageable calculations, he preferred to put up with inconveniences in the concrete operations if this meant that he could maintain his basic principles.

It would also be wrong to interpret his stance as indicating a lack of knowledge over alternative positions and approaches. Carnot was well aware of the difference between quantities and numbers. In an extensive footnote to his work on mechanics, published practically simultaneously with the *Géometrie de Position*, he uses the difference between quantities and numbers to explain the old problem of how to divide and multiply inhomogeneous quantities. He takes the example of speed and dividing a linear quantity by a temporal quantity to explain that one does not divide these two quantities here. The problem lies in the ratios of these quantities to their respective units. In other words, one operates with the "abstract numbers" that form these quotients, but not with the quantities themselves (Carnot 1803 b, 11).

In one central problem of negative numbers, Carnot even reaches a conceptual clarity that none had achieved before and many would continue to dispute until much later: in the proof of the rule of signs. Until this time, it had always been assumed that the rule of signs could be proved for multiplication (and for division). Many authors were also convinced that they had provided a proof—even those proposing an alternative rule of signs. Carnot, in contrast— and most explicitly in 1813—considers not only that prior proofs are inadequate but also that, evidently, no proof is possible: "this rule has never been proved in a satisfactory manner; it does not even appear susceptible of being so" (Carnot 1813, Vol. 2, 62).

Carnot does not use the impossibility of proving the rule of signs to conclude that concepts of mathematics need to be extended, but that the traditional quantity concept should be retained while dismissing the concept of negative quantities established in mathematics since the sixteenth and seventeenth centuries. This makes Carnot more consistent and more productive than the British authors who only rejected it, because he works out the geometry of position as an alternative theory.

In the concept field of negative numbers, it is particularly clear that for Carnot, it is his foundational stance, his epistemological conception, that has priority, and that he makes no compromises in favor of, for example, ease of practical application.

1.7.3. THE DEVELOPMENT OF CARNOT'S CONCEPT OF NEGATIVE QUANTITIES

The rudiments of Carnot's epistemology of mathematics have already been reported in Section 1.3. For him, permissible objects of mathematics are exclusively absolute quantities. The nonabsolute quantities occurring in algebraic operations only possess the function of signs or computational quantities that have to be eliminated from a result before it can have an arithmetic meaning. Algebra possesses no independent function in mathematics, but only one that is subordinate to arithmetic.

As his writings on negative quantities show, arithmetic also has a subordinate function: it is subordinate to geometry, and only geometry can guarantee the existence of mathematical objects.

In his 1801 work, the *Corrélation des Figures de Géométrie*, Carnot addresses the twin category of the two modes of being proposed by Fontenelle: the quantitative and the qualitative. Carnot calls these the quantitative relation and the positional relation. The former expresses the absolute value of the quantities; the latter, their position in space, whether they are above or below a straight line and so forth:

> There exist between different parts of every geometric figure two sorts of relationships, namely, the relations of magnitudes, and the relations of position. The first are those between the absolute values of the quantities; the others are those which express their respective positions, showing whether a certain point is placed above or below a certain line, to the right or to the left of a certain plane, inside or outside a certain circumference or curve, etc. (Carnot 1801, 1).

Although in this work, Carnot concentrates on the second mode of being, the positional relation, he starts off by presenting his new ideas on the first, quantitative, mode and on the quantity concept. What he used to conceive in earlier manuscripts as the relation between analytical and synthetic values, he now understands as the relation between value (*valeur*) and quantity (*quantité*).

Whereas *quantité* describes a quantity a itself as its absolute value, with abstraction of its sign, and is identical with its positive value $+a$, $-a$ forms its negative value and $\sqrt{-a}$ its imaginary value. Hence, neither positive quantities nor negative and imaginary quantities exist. Every *quantité* is a real, conceivable quantity, whereas its values are algebraic forms that often indicate impossible operations. Although algebraic forms such as the isolated negative quantities of $-a = 0 - a$ can be used during calculus, they lead to comprehensible results only when transformed into what are really executable operations (ibid., 2 f.).

What is fundamental for Carnot's approach to the algebraic, unreal values is, once again, the relation between two systems: an original system and a transformed or correlative system. Whereas he assumes the operations in the *système primitif* to be executable for the transformed system, he distinguishes between quantities *en sense directe* or simply *directe* and those *en sens inverse* or simply *inverse*. Direct is what he calls those quantities in which the difference between a greater and a lesser quantity is retained, and inverse, those in which the smaller has become the larger and the larger the smaller (ibid., 3 ff.).

Hence, it becomes clear for this concept field as well that Carnot's system concept serves as a substitute for a general function concept. It even serves, in the sense of the synthetic methods of the "ancients," as the restriction to the solution of single cases, but simultaneously as a degree of generalization as well by establishing relations between the single cases. In the intended main application, of trigonometric functions, the one concern is not only to view the equations of a function in the different quadrants as single cases, but also, at the same time, to show that they are systems that have been transformed from *one* original system.

Therefore, Carnot presents the following main principle: in order to ensure that effective operations can be performed in the transformed or correlative system, one has to change the sign in the equation for inverse quantities while leaving it unchanged for direct quantities (ibid., 11 f.).

Here, Carnot always speaks about the *signe de la valeur de la qualité*, clearly meaning the algebraic sign. He likewise explains that an originally absolute value p takes the correlative value of $-p$. However, he never reflects on the relation between sign of operation and algebraic sign, and also gives no rules for entering a quantity with a changed sign into the formulas. To that extent, he relies implicitly on the traditional theory he has rejected.

Regarding his transformation concept, Carnot ascertains that he presents the true theory on those quantities that are falsely called positive and negative in analysis, falsely because neither positive nor negative quantities really exist, nor quantities that are negative in themselves. He believes only in absolute quantities that can be added to others, or subtracted, as long as they are smaller than those from which they are subtracted. This is because nothing exists that is less than zero. When one says that a quantity becomes negative, one means only that the system to which it belongs changes, and that the correlative sign of the value of this quantity in the transformed system is the minus sign (ibid., 23).

In 1801, Carnot even sees a relation between his approaches for the *analyse infinitésimale* and the *quantités directes et inverses*: they show "much analogy." In the former case, for an invariant system of quantities, one observes a second comparison system whose quantities transform into the quantities of the initial system through a continuous approximation to the limit. In the latter case, one likewise compares the quantities of two systems, and these are the absolute values in both systems that are equal to each other at the limit. Depending on whether the difference between two quantities in the auxiliary system continues to have a similar orientation or not, these are either direct or inverse quantities (ibid., 28 ff.). Elsewhere, Carnot phrases this more formal analogy in more absolute terms (see below). He also works out his rejection of negative numbers more extensively and fundamentally in his next work, the *Géometrie de Position*.

Indeed, the 1803 version is not just edited more carefully than that of 1801. In contrast to the first version, Carnot names or implicitly cites a few authors. His main reference is to d'Alembert, on whom he bases much of his work, although also criticizing him. Carnot introduces the *Géometrie de Position* with a 37–page *Dissertation Préliminaire* that focuses on his conception of negative quantities. Right at the beginning, he explains that his *Géometrie de Position* forms a theory to replace that of positive and negative quantities, which he rejects completely:

> The geometry of position is therefore, properly speaking, the doctrine of quantities called positive or negative, or rather the means of providing them; for that doctrine is entirely rejected here (Carnot 1803 a, ij).

He allows, in turn, only subtractive quantities, saying that nothing is easier than negative quantities that are preceded by larger positive ones. However, in algebra, one is continuously finding isolated negative forms whose meaning is impossible to ascertain precisely, because they occur in unexecutable operations. Carnot stresses that the algebraic operations have been proved only for executable quantities, but not for others. Hence, the distributive law of $(a - b)c = ac - bc$ is valid only for $a > b$. It is impossible to prove it for $a < b$. To truly obtain an isolated negative quantity, one has to subtract an effective quantity from zero. However, this means to take *something* away from *nothing*—an impossible operation (ibid., ij f.).

With this criticism, Carnot draws on d'Alembert, on both his article *Negatif* in the *Encyclopédie* and his *mémoire* in the eighth volume of the *Opuscules Mathématiques*, which he quotes in detail. Nonetheless, he censures d'Alembert by saying that although his critique is highly appropriate, he does not offer any replacement for the construction he criticizes, making some of his suggestions for improvements seem to function more as a reinforcement of the rejected theory.

One such imprecision in d'Alembert that Carnot particularly discusses is his argument that the negative quantities possess the same nature as the positive ones, but have to be taken in a *sens contraire*, an opposing direction. Naturally, d'Alembert had already pointed out that positive and negative solutions do not need to be symmetrical to each other. Carnot now claims that such symmetry would actually be the exception. Hence, it seems as if the emergence of his new geometric theory was due to a misunderstanding: the sweeping statement that negative quantities have to be assumed to take the opposite direction to positive ones gives no explanation regarding whether this means the initial quantities or the results. Indeed, the originators of this statement particularly meant the inversion of initial quantities. However, Carnot concentrates on the relation of the positions of the multiple solutions and thereby develops his new theory.

In his introduction, Carnot assembles a number of further arguments against the permissibility of negative quantities, although not all of them are mathematically sound. One even has the impression that he has read Arnauld's objections in Prestet's second edition. For example, he uses the proportions argument, but not in d'Alembert's way that had considered it resolved through MacLaurin's proof that proportions are permissible only for absolute quantities, but fully convinced that the reciprocal larger–smaller relation between the former and latter pair of proportions $1 : -1 :: -1 : 1$ is a decisive rejection of the existence of quantities that should be smaller than zero (ibid., vj f.). Likewise, Carnot considers it to be a decisive argument that—if one views -3 as being smaller than 2, then their squares, that is, $(-3)^2 = 9$ and $2^2 = 4$ have an inverse relationship—in his opinion, in shocking contrast to a clear quantity concept that he considers to imply a retention of the relations between quantities: "which offends all clear ideas one is able to form of quantity" (ibid., ix).

The next argument is even less well founded: if there are quantities that are smaller than zero, then one is even more justified in neglecting them because one can already neglect zero quantities. However, this neglect is certainly not permissible, and, therefore, there are no quantities smaller than zero (ibid.).

The argument after that is of a similar quality: Consider $\sqrt{A-a}$, in which A is invariant and a is initially smaller than A but growing in size. Hence, $\sqrt{A-a}$ will become increasingly smaller and will finally equal zero when a equals A. Hence, if a continues to grow, $\sqrt{A-a}$ has to become smaller than zero and not imaginary. However, this is false; in other words, there are no quantities smaller than zero (ibid., xii). There is even the further reference to Arnauld that the traditional theory of negative quantities may be "accidentally" correct under special circumstances (ibid., xv).

Because his actual intention in this work is geometric, Carnot reports only a short, theoretical indication of an algebraic realizability of his new conception of negative quantities: They can now appear in the course of computing as an algebraic form without any independent meaning. These forms gain mathematical meaning only when they are viewed without their signs. They then appear as the difference of two absolute quantities, the first of which is larger, but, in the intended case of application, smaller than the second:

> From which I conclude, 1^0. that every isolated negative quantity is a being of the mind, and those met in calculations are just simple algebraic forms incapable of representing any real and actual quantity. 2^0. That each of these algebraic forms being taken, abstracting it from its sign, is no other than the difference of two other absolute quantities, of which the one which was the larger when the reasoning was established is found to be the smaller when the results of the calculation are applied (ibid., xviij f.).

And Carnot expressly emphasizes that the sign in front of the quantities does not indicate any specific kind of quantity, but exclusively indicates an operation: "Thus these signs, taken together with these same quantities, do not make new quantities, but complex algebraic forms" (ibid., xix).

Carnot discusses only two prior conceptions of negative quantities, and these in a rather global way: (a) as quantities smaller than zero, and (b) as quantities that have to be substituted in the inverse direction. Whereas he believes the first conception to be completely disproved, he sees a grain of truth in the second that he intends to develop with his new geometry. Further on in the text, as a necessary foundation for this development, he presents the introduction of his likewise epistemologically justified conception that there is no general case, no general equation.

He introduces this by discussing the addition theorem for the cosine:

$$\cos(a + b) = \cos a \cos b - \sin a \sin b.$$

However, this equation becomes incorrect as soon as the angles a, b, and $a + b$ no longer lie in the first quadrant. The supporters of negative quantities nonetheless explained this equation as general and applicable to all cases. However, Carnot declares, without any further justification, that this assumption is false, and he adds, also without any corroboration, that the same supporters— in order to correct their error—change the sign from plus into minus, thereby obtaining the equation

$$\cos(a + b) = \cos a \cos b + \sin a \sin b,$$

an equation that is correct, but reveals that the assumption of negative cosine values contains further errors:

> However those who admit the notion of negative quantities regard this formula as general and actually applicable to all cases; but since this supposition is not true,

they correct their error in saying that, in that case, cos a and cos $(a + b)$ become negative one and the other; that it becomes necessary in consequence to change their signs from $+$ to $-$, which results in

$$\cos (a + b) = \cos a \cos b - \sin a \sin b;$$

a true result but which, by the very fact that it is true, proves that the supposition that cos a and cos $(a + b)$ were negative is a new error; but if there had not been an error, nothing would have compensated the result of the false hypothesis that had been made in the beginning, namely that the formula was applicable to all these cases (ibid., xvj).[25]

In an analogous way, Carnot questions the validity of the formulas

$$\cos(\pi + a) = -\cos a, \quad \sin(\pi + a) = -\sin a.$$

These are *équations fausses*, because the formulas have been found only for the first quadrants and cannot be applied directly and without modification to other cases. Here, once again, is a vivid illustration that the function concept does not form an effective basic concept for trigonometry either.

Carnot summarizes the discussion on the trigonometric example into the general statement that it is a fundamental error to assume that an equation can be valid without modification for all areas of space:

> It is in following the same error that one supposes gratuitously that the same equation can be applied, without modification, to all regions of space; that, for example, an equation given in two variables can relate, without distinction, to points to the left just as those to the right, and those above just as those below the axes of the curve whose nature the equation expresses. This is very convenient, but it is not true: to give this as a principle is precisely to beg the question (ibid., xxj f.).

This methodological and conceptual approach simultaneously explains the continuously emphasized characteristic feature of Carnot's *Géometrie de Position* as a presentation of tableaus—or, as Carnot puts it, *faire la géometrie par tableaux* (ibid., xxxv). Indeed, the idea in Carnot's program of the correlation of related cases is to make their correspondences in the table forms accessible and understandable. His particular concern is to illustrate that one can obtain positive values in all quadrants by performing corresponding transformations.

Otherwise, the 1801 and 1803 versions basically follow the same structure. Carnot simply copies many passages of text directly. Even the analogy between *analyse infinitésimale* and *quantités dites negatives* is formulated in the same terms (ibid., 75 f.).

---

[25]   The text contains a number of printing errors here. The first equation, for example, contains $-\sin a$ instead of $-\cos a$ and sin $(2\pi + a)$ instead of sin $(\pi + a)$.

1.7.4. LATER WORK: L'ANALYSE: LA SCIENCE DE LA COMPENSATION DES ERREURS

Later, in 1806 and 1813, Carnot published two further short versions of his concept for substituting negative quantities. These focused more on its application in algebra, whereas the 1801 and 1803 works had been used to present the geometric alternative.

Both works reveal some variation in individual aspects of Carnot's conception, evidently as a result of discussions with other mathematicians. In 1806, he published the 17–page *Digression sur la nature des quantités dites négatives* as an appendix to a *mémoire* on the relation of the distances between five points. In 1808, Schumacher appended this text to his German translation of the first part of the *Géométrie de la position*. The 1813 work forms an expanded 30–page "footnote" to a comment in the last version of his *Reflections on Infinitesimal Calculus*. This final mathematical publication presents his work on the foundations of two different concept fields as being directed toward a common approach, a unifying principle, an explication already prepared in several manuscripts still to be found in his *Nachlass*.

Primarily, Carnot's argument in 1806 agrees with that in earlier writings, particularly in rejecting the ideas that negative numbers are smaller than zero or symmetrical to positive numbers (Carnot 1806, 97ff.; 1808, 354 f.). As in his earlier work, he also rejects both negative solutions and negative roots. I shall quote his precise summary:

> One never believes that one has obtained a true solution when one has failed to find a positive value for the unknown, because these positive values are the only truly comprehensible ones, and all others necessarily reveal an incompatibility of the given conditions and the assumptions on which the calculation is based. Hence, one has to change these assumptions or carry on modifying the conditions of the problem until one obtains a positive value for the unknown. If the equation has positive and negative roots, only the former provide true solutions (Carnot 1808, 361).

What is new, in contrast, is that Carnot now starts to assign two different meanings to the expression "negative quantities":

1. First, he talks about "every [quantity] that is affected with the minus sign," hence, subtractive, but, of course, still absolute quantities.
2. Second, he talks about quantities that possess a sign that is opposite to that which they should have.

Carnot explains this being opposite to the "inherent" sign of the quantity with the example of the addition theorem for the cosine in the first and second quadrants. Once again, this is quoted in detail in order to reveal how he manipulates the sign while keeping the main term positive:

> For example, in the well known formula

$$\cos(a + b) = \cos a \cos b - \sin a \sin b \quad (A)$$

the last term is negative according to the first meaning. As long as a, b, and $a + b$ are each $< \pi$,[26] the term in question remains positive according to the second meaning, because this sign $-$, with which it is affected, is the true sign that it has to possess. However, if $a + b > \pi$, the formula is false, and one obtains directly through synthesis

$$\cos(a + b) = \cos a \cos b + \sin a \sin b . \text{ (B)}$$

Hence, should the first formula still be applicable to this case, one has to view the first term $\cos(a + b)$ as if it were affected with a sign that is opposite to that which it should possess, because, in fact, by changing the sign of this term

$$- \cos(a + b) = \cos a \cos b - \sin a \sin b$$

one obtains a formula that becomes (B) through permuting the terms. Hence, if one wants to view (A) as also being applicable when $a + b > \pi$, $\cos(a + b)$ is then affected with a sign that is opposite to that which it should have, and is subsequently a negative quantity according to the second meaning although it is positive according to the first (ibid., 356).

The same quantity can be positive according to the first meaning and negative according to the second. To avoid confusion, Carnot uses the expression "negative quantity" only for those affected with a minus sign, that is, for subtractive quantities (ibid., 357). In the 1806 text, he even calls these "properly negative quantities" (*quantités négatives proprement dites*; Carnot 1806, 110). He calls the second–meaning quantities, in contrast, "inverse quantities," in contrast to "direct" quantities, which always have the sign "that they should have" (ibid., 357). The inverse quantities had falsely been labeled "negative," falsely, because the inverse quantities have an alternating sign: plus or minus depending on their position (ibid., 362).

At the same time, Carnot uses this text to illustrate his basic principle that every inverse quantity should be presented as the difference of two absolute quantities: "that every inverse quantity is the difference of two direct ones in which one viewed the larger of the two as the smaller and vice versa when forming the equation" (Carnot 1808, 365, Carnot 1806, 111). Hence any inverse quantity p can be presented as $m - n$ with $m > n$, so that $-p = n - m$. Carnot does this to ensure that no isolated negative quantity emerges in any transformation.

This "pair" concept for inverse quantities makes it possible to understand the twofold attraction of the additional trigonometric functions *sin vers* and *cos vers* for Carnot. Versed sine is a trigonometric function that is no longer common in today's elementary mathematics, although still used in a few professional applications such as navigation:

$$\text{sin vers } a = 1 - \cos a,$$
$$\cos \text{vers } a = 1 - \sin \text{vers } a.$$

---

[26]   Whereas the original always correctly reports smaller or larger than one quarter of the circumference, Schumacher always gives $<$ or $> \pi$ instead of $\frac{\pi}{2}$. Because this change is consistent throughout the German translation, it has to be assumed that Schumacher had not understood that Carnot wants to test for each quadrant separately.

On the one hand, these functions can be used to attain positive values when the sine or cosine function is in the domain of negative values. For Carnot, however, the versed forms have the additional attraction that he can use them to decompose the trigonometric functions into pairs containing a larger and a smaller direct quantity. Whereas this is relatively simple to do for the cosine,

$$\cos a = 1 - \sin \text{vers} \, a,$$

or changed in sign for other quadrants, the decomposition is possible only "by force" for the tangent:

$$\tan g \, a = \frac{\sin a}{1 - 2\sin \text{ver} \, a + \sin \text{ver}^2 \, a} - \frac{\sin a \sin \text{ver} \, a}{1 - 2\sin \text{ver} \, a + \sin \text{ver}^2 \, a} \quad \text{(ibid., 102)}.$$

Finally, there is one new element in the argument. Beforehand, Carnot had emphasized that *one* equation could not be valid for all areas of space. He now stakes out what could be called a "third path": between the complete generality of an equation of the curve (and with it, the function concept) and the decomposition into separate single cases following the synthetic, British style (though without mentioning this). Carnot emphasizes the "advantage" of the inverse quantities:

> [They] make it possible to present a variable system in all of its states through one single formula. If one did not want to apply it, for example, one would need four different equations to represent the four areas of a plane curve divided by its two axes, eight equations for a curved surface, and so forth. However, if one uses inverse quantities, one equation covers the previous eight, when, at the end of the computation, one simply changes the sign of the inverse quantity in the results (Carnot 1808, 362).

In the final 12–point summary of his concept, Carnot declares the distinction between "properly" negative quantities and inverse quantities to be essential, along with the fact that the "properly" negative quantities are impossible quantities only when examined in isolation, and are therefore exclusively computational quantities (ibid., 364).

It may well be that this first explicit discrimination into two different kinds of negative quantity is a reaction to the strong criticism of Carnot's concept of negative quantities published in 1804 by the German mathematician Busse. Busse particularly criticized Carnot's failure to distinguish two meanings of the plus and minus signs: as algebraic signs and signs of operation (see Chapter VII.3.1.). Buée, a French author living in England, also propounded the need for this discrimination in a memoir presented to the Royal Society and published in 1806 (see above, Chapter II).

In his 1813 work, Carnot no longer addresses this discrimination explicitly and concentrates on presenting the calculus for inverse quantities. In particular, he applies the versed sine and cosine functions more extensively in order to obtain a flexible transition between the various positional "cases" (Carnot 1813, vol. 2, 78 ff.). He once again emphasizes that he recognizes the plus and minus signs exclusively as signs of operation, and therefore continues to reject any second function as a sign (ibid., 85).

What is new is the first mention of the concept of opposite quantity (*quantité opposée*)—although only to be rejected out of hand. Nonetheless, this rejection shows his awareness that this concept calls for additional assumptions. He conceives these assumptions as a demand for additional metaphysical ideas: *nouvelles idées premières*. However, he rejects such extensions here. Additional basic assumptions may help to overcome difficulties, but they are rather unphilosophical when they are not absolutely necessary: "The possibilities offered by these first ideas are without doubt useful for avoiding difficulties, but they are hardly philosophical, since they are not indispensable" (ibid., 76).

In this new work, Carnot focuses on two goals. He names one of them expressly, stating that his concepts on the negative have been well received by other scientists. The only objection of which he is aware is that his conception appears to be more difficult than the generally accepted one. He wants to show that this is not the case (ibid., 77).

Indeed, after presenting how to "correct" the equations in each positional situation, Carnot sums up by saying that this solves all difficulties, and nothing is changed compared with how things have always been done before. As usual, one turns each problem into an equation. One assumes that the equation can be applied to all states of the basic system, with the only reservation being to perform the necessary modifications for the specific system states in the result (ibid., 101).

Carnot does not formulate his second goal expressly. However, it is evident from the entire concept of this final text. The inverse quantities are now presented in such a way that their function is to overcome an initial error. This initial error lies in the use of an equation that is directly applicable for only one case. For example, a trigonometric equation that is directly applicable only for the first quadrant is considered to be directly applicable for all other quadrants. This initial error has to be corrected through a second error, in other words, compensated. The second error is to replace the absolute quantity y with the negative algebraic expression $-y$ in order to correct the error due to the false assumption (ibid., 97 f.).

Hence, the second goal of the 1813 text is to conceive error compensation as a general principle for solving the two foundational problems: those on the negative numbers and the infinitely small quantities. Indeed, several preparatory manuscripts in Carnot's *Nachlass* show that—compared with the short version of his *Reflections* of 1797—he was searching for a standardized approach for his concepts. One element of this is his understanding of the analysis as a comprehensive discipline: as containing the finite or common *and* the infinitesimal. In one *Nachlass* text, he formulates what he considers to be the common conceptual structure in both concept fields:

> Analysis is the art of simplifying algebraic calculations by the use of false propositions such that the resulting errors compensate each other.

In ordinary analysis, it is gratuitously and falsely supposed that negative quantities exist, in infinitesimal analysis it is supposed gratuitously and falsely that infinitely small quantities exist.

Thus analysis is literally nothing other than the science of compensation of errors.[27]

There can be no doubt that Carnot was still not satisfied with this draft, because it presents the assumption of infinitely small quantities as false, whereas his publications always define them as real quantities with a limit of zero. The comprehensive concept that satisfied him was to view negative and infinitely small quantities as auxiliary quantities that have to be eliminated at the end of a solution. They have to be eliminated completely from the calculus in order to ensure precise results.[28] Indeed, he states at the beginning of the 1813 text:

> There is a remarkable analogy between the theory of isolated negative quantities and that of infinitesimal quantities, in that both are employed only in an auxiliary way and they must necessarily disappear from the results of the calculations for these results to become perfectly exact and intelligible (ibid., 75).

The concept of error compensation, first developed by Carnot as a strict justification for infinitesimal calculus, is now presented as a general principle designed to establish sound foundations in the "normal" analysis just as much as in the infinitesimal.

2. Carnot's Impact: Rejecting the Algebraization Program

2.1. Initial Adoption

In a paradigmatic way, Carnot ties together his clarification of the foundations of two concept fields in analysis and drives them forward hand in hand. Because these works of Carnot were also adopted—at least in France—hand in hand, I shall try to follow the two conceptual developments in parallel when discussing them. The starting point for both fields was the mathematical lectures at the *École normale* in the year III of the French Republican Calendar (1795). The best French scientists taught the knowledge of the time in a systematized and elementarized way by analyzing the fundamental concepts that gave it its

---

[27]   A note on a small sheet of paper highlighted through the notation "B.B.B." (Carnot's *Nachlass*, Nolay, Carton 28, chemise 10).

[28]   A six–page draft entitled *Essai sur la théorie des quantités dite négatives en algèbre* (ibid.).

structure. This formed the basis for modernizing scientific progress. One pinnacle in this adoption was the works of Cauchy, who adopted Carnot's ideas on both concept fields in a remarkable way.

Because German interest in Carnot has been essentially limited to his work on negative numbers, the controversies over this topic will be presented in a separate chapter.

The *École normale* in the year III was founded on the initiative of the *Idéologues* as the first scientific institute of higher education for training teachers. The plan was to train teachers in so-called revolutionary courses based on the analytical method so that they could go on to disseminate this knowledge in the new primary schools to be set up throughout France. To get the schools up and running as quickly as possible, the institution was given only four months to do its work. This is what gave the intensively compressed courses their "revolutionary" character. The scientific evaluation of this institution is controversial. Although the goal of primary school teacher training was certainly unattainable, there can be no doubt that there were significant achievements in the sense of a secondary school teacher training (see Schubring 1982b).[29]

2.2. Limits and Negative Numbers in the Lectures at the École Normale

Whereas Monge was assigned the courses on descriptive geometry at the *École normale*, Lagrange was asked to give the courses on mathematics. However, Lagrange did not do this alone, but let Laplace assist him. The practical division of labor was that Laplace gave a basic course in arithmetic, algebra, and geometry, with previews on the application of algebra in geometry, on the new system of weights and measures, and probability theory. Lagrange, in contrast, restricted himself to a few specialized courses in arithmetic, algebra, and analytical geometry.

Quite simply in line with the goal of training at this institution, the mathematics courses contained no lectures on infinitesimal calculus. Nonetheless, Laplace gave a short sketch on the *méthode des limites*, characteristically, in his lectures on elementary geometry as a method for solving problems in calculating segments, surface areas, and the volumes of solids.

Laplace does not perform an algebraization of the limit concept, and also does not use any special sign for the limit, but remains completely on the level of rhetorical style. He presents the limit method as consisting of two principles. The first is the extended equality concept, which he applies to the example of the

---

[29]   The production of an annotated new edition of all the courses was initiated to mark the 200th anniversary of the French Revolution. The volume on the mathematical course is available (Dhombres 1992).

extension of a statement on the proportionality of segments on the sides of a triangle to the case of their incommensurability: a line drawn parallel to the base of a triangle will divide both sides of the triangle into proportional parts. To prove this, he calls the two sides A and B, and their parts between the base and the parallel, a and b. If A is divided into n equal parts, and one such part is marked off to B, this part will contain a certain number with a remainder R. If one draws a further line through the end of $B - R$ parallel to the base, this cuts off a part $b - r$ from the segment a. As a result,

$$\frac{b-r}{a} = \frac{B-R}{A}, \text{ or } \frac{B}{A} - \frac{b}{a} = \frac{R}{A} - \frac{r}{a}.$$

If one now arbitrarily increases the size of n, the remainders R and r can be diminished indefinitely. Laplace concludes that the difference $\frac{B}{A} - \frac{b}{a}$ "is" therefore smaller than any given quantity. He therefore formulates as *le premier principe de la méthode des limites* that "two quantities for which it can be proved that the difference is less than any given magnitude are evidently equal to each other" (Dhombres 1992, 89).

It is notable that Laplace avoids the variable concept here and uses only the unspecific quantity concept. Indeed, nothing in the signs used in his proof indicates any functional dependence.

He derives his second principle from the example of determining the area of a circle. He compares the value of the product for a circumscribed polygon with increasing refinement to those for the circle as limit—once again, completely verbally without using any signs. However, for Laplace, it is not enough to confirm that the area of the polygons has the area of the circle as its limit. He generalizes the statement far beyond that which can be proved, and turns it into the general principle that the commutativity of the limits holds for every conceivable expression: "the limit of the expression of a sequence of magnitudes is the expression of the limit of these magnitudes" (ibid., 90).

This takes Laplace far beyond the statements on commutativity that mathematicians had formulated up to this time within the framework of limit theories. Typical until then had been to give the commutativity statement for *one* concrete type of *expression* since de la Chapelle, for a product or a ratio, particularly in the work of Cousin, Martin, L'Huilier, and Carnot. At the time, Stockler was the only mathematician to have addressed a more general commutativity of limits, but he had also proved it in each case (see Chapter III.8.). Laplace, in contrast, had performed an overgeneralization here that has to be conceived as an epistemological evaluation, analogous to the basic beliefs of Leibniz and Fontenelle (see Chapters III.5. and III.7.). The consequences of such epistemological overgeneralizations will be discussed in the section on Cauchy.

In these fundamental lectures on arithmetic and algebra, Laplace also explains the use of negative quantities. However, he does not give an extensive or theoretical explanation, but sketches the practice of dealing with an unrestricted subtraction. Some of the epistemological reflections on this are not explicated; Laplace simply assumes the existence of negative quantities.

The framework that Laplace uses to introduce the corresponding operations is the beginnings of algebra. He says that it starts by taking numbers as being independent of their value and of every number system. After briefly sketching potential types of problems in algebra, Laplace presents the operation rules for algebraic quantities. He introduces subtraction without any restrictions at all. One has only to write the quantity b that one is subtracting behind the one from which it is being subtracted, and change the signs for all the terms in the former: *en changeant les signes de tous ses termes*. This rule is considered to be clear when the number to be subtracted bears the plus "sign." In the case of a minus sign, Laplace gives his own justification. He says that it is evident that the number a can also be rewritten as follows:

$$a = a + b - b.$$

If one wants to take away $-b$ from a in the right-hand side of the equation, then it is clear that this makes the last term disappear. We are left with

$$a + b \quad \text{(Dhombres 1992, 61)}.$$

After addition and subtraction, Laplace goes on to present multiplication. For him, the general case is formed by complex terms; thus

$$(a + b) \times (c - d).$$

However, he formulates the rule of signs with no reference to whether the terms are simple or complex. In other words, he does not restrict the rule of signs to subtractive quantities, but relates it generally to negative quantities and also to negative numbers. Laplace comments that the rule of signs creates "difficulties," because it is hard to understand that the product of $-a$ and $-b$ is equal to the product of a and b (ibid., 62). However, he is the first mathematician who does not initially try to provide a proof for the rule of signs. He wishes only to make it "perceptible" and to do this through distributivity:

> To make this identity clear, we observe that the multiplication of $-a$ by b is $-ab$, since this product is just $-a$ repeated as many times as there are units in b. We observe next that the multiplication of $-a$ by $+b - b$ is zero, since the multiplier is zero; therefore, the multiplication of of $-a$ by $+b$ being $-ab$, the multiplication of $-a$ by $-b$ must have the opposite sign, so equal to $+ab$, in order to cancel it (ibid.).

As Carnot later criticized (see above, Section 1.7.3), this assumes that the distributive law is also valid when one of the factors is zero. It is interesting to see how Laplace falls back on Arnauld's argument on iterated addition (see Chapter II.2.5) when proving the rule that the product of various signs is negative.

In his lectures at the *École normale*, Laplace says that one of the major applications of his concept of negative numbers is in the trigonometric functions. He also emphasizes that the results of studying the values of these functions are a highly appropriate aid in understanding the nature and the application of negative quantities (ibid., 92). By this, Laplace means, on the one hand, the transformation of the sine or cosine from positive to negative values and then back to positive ones again—transformations that Carnot categorically rejected—and, on the other hand, the addition theorem of the sine function that is transformed through a change of sign into the analogous theorem for subtraction. For Laplace, it is self-evident that this involves, in each case, a universal and general theorem, without distinctions of case (ibid.).

2.3. Dissemination of the Algebraic Conception of Analysis

In France, the years after 1794 not only formed the pinnacle of the analytic method but were also a period of previously unmatched productivity in the writing of textbooks on differential and integral calculus. The dissemination of this knowledge reached previously uninformed social classes; knowledge on the analytical method actually became public. One outcome of this process of dissemination was that it was no longer just members of the traditional classes of academic scientists who appeared as the authors of textbooks but also teachers at secondary schools. Finally, the appearance of "hermetic" idiosyncratic conceptions was also symptomatic for this process.

One main characteristic of all writings on analysis in France from 1794 onward is that none of them recommend the traditional conception of the *infiniment petits*—for example, in the sense of L'Hospital—and all propose a more or less strongly algebraic conception. A number of these works have already been analyzed or mentioned:

1. The revised versions of the textbooks from Cousin (1796) and Martin (1802); see Chapter III.8.4.

2. De Prony's lectures at the *École polytechnique* starting in 1795; see Chapter IV.2.

3. Lagrange's lectures at the *École polytechnique* starting in 1795 and his textbooks on function theory (1797, 1806); see Chapter IV.2.

4. Carnot's *Reflections on Infinitesimal Calculus* of 1797; see Chapter V.1.6.

5. Bossut's late work on differential and integral calculus (1798); see Chapter IV.2.

Before considering the development of ideas about analysis at the *École polytechnique* up to the reversal in 1811—and hence in particular the work of the "great" author S.F. Lacroix—I would first like to present two rather marginal authors from the first decade of the nineteenth century: Nicolas Desponts (1749–after 1812), formerly mathematics teacher in the military school at Brienne—where he also taught Bonaparte—and J.B.E. du Bourguet, former naval officer and, about 1810, teacher of the class on *mathématiques spéciales* in the *Lycée Impériale* at Paris. Neither of these authors was a well-known member of the French mathematics community. As a result, their publications on analysis are even more representative of mainstream developments in France.

Desponts had already proposed his own terminology for avoiding infinitely large and infinitely small quantities—*illimitables* and *évanouissantes*—in his first treatise, published in a regional journal in 1802. This was most probably intended as an introduction to infinitesimal calculus for beginners. In practice, he applies limit theory, because he lets these quantities arbitrarily increase or decrease to zero and bases his arguments on the results obtained. However, he neither reflects on the limit processes nor uses a sign for them (Desponts 1802). In a second work published in the same regional journal in 1804 and conceived as a guideline for beginners, Desponts tries—completely in the sense of Lagrange—to present the "true principles" of the calculus called infinitesimal analysis as "emancipated from every metaphysical hypothesis."

Desponts declares the *principe fondamental* of differential and integral calculus to be to reduce expressions to their simplest possible form (Desponts 1804, 4). Although this "principle" is formulated only in everyday terms and contains absolutely no mathematical specifics, he believes that he is able to derive the customary analysis from it. Moreover, although constantly inveighing against the method of the *infiniment petits* and limits, he draws on these just as continuously in order to attain the desired results. Furthermore, he repeatedly criticizes Lagrange's approach of developing functions into series. He accuses him of allowing the actually infinite quantity (ibid., 38). It is notable that Desponts adopts many lines of argumentation from Carnot's *Réflexions* of 1797 (see ibid., 7), particularly also the conception of infinitely small quantities as variables with limit zero (ibid., 104).

Du Bourguet published a textbook on differential and integral calculus in 1810, that is, shortly before the reversal at the *École polytechnique*. This was the first book to be written for students at the new secondary schools, now called *Lycées*. He is also very emphatic in his rejection of infinitely small quantities. His subtitle already demonstrates the reference to Lagrange's program: *Indépendans de toutes notions de quantités infinitesimales et de limites*. He also declares categorically that infinitesimal quantities do not exist and that the so-called fundamental equation $x \pm dx = x$ is "false" (du Bourguet 1810, vij). He rejects all "metaphysical" justifications and statements, and claims that differential analysis is based on pure algebraic principles (ibid., 1). Proceeding from the, indeed, algebraic difference calculus, he completes the transition to the differentials through mere definitional positing—without justification, which, of course, made it easy for him to avoid "metaphysics" (ibid., 12 ff.)! The textbook itself is a simple mechanical presentation and application of rules without any theoretical approaches. The only notable aspect is that du Bourguet addresses the definite integral in a final section within integral calculus. He describes it as *intégrale prise entre les limites* (ibid., 457 ff.), after he has basically introduced the integral as formal inversion.

Finally, the introductory *Discours Préliminaire* contains a characteristic indicator for the pending reversal. Whereas the infinitesimals need to be "excluded completely from pure mathematics," they are even "necessary" in the *sciences physico–mathématiques* (ibid., X). His justification for this interesting asymmetry is that the laws hold only "approximately" in the domain of physical phenomena. Mechanics, which, abstractly speaking, is subject to geometric rigor, loses this rigor as soon as it is applied to real, physical bodies.

2.4. Analysis Concepts at the École Polytechnique to 1811

Chapter IV has shown that initial ideas on how to teach analysis at the *École polytechnique* were based completely on the analytical method, and that the bastion of this method conceived and taught analysis as a specific form of algebra. At the same time, we have seen that de Prony—the exceptional engineer and physicist who was also the *instituteur* for the *cours d'analyse*—had introduced continuity as a central concept for the transition from (finite) differences to differential calculus.

2.4.1. THE REORGANIZATION ABOUT 1799

In the years following its establishment, the algebraization concept remained decisive at the *École polytechnique*. De Prony continued to be the *instituteur* or analysis and taught it in his adaptation of Euler's algebraization, whereas Lagrange taught analysis as algebraized function theory, although only in occasional courses and to particularly gifted and advanced students. Both the teaching program and personnel were expanded in 1796 by introducing a course on *analyse algébrique* to bring students up to the necessary standard to start the course on analysis. This course was assigned to Jean-Baptiste Fourier.[30]

Although the year 1799–1780 brought a degree of change, the major development was an institutional and programmatic consolidation of the institution and its teaching program. The decisive impact on the teaching program of the analysis course came from what was actually a chance event. In 1799, Lagrange announced his resignation from his teaching post at the institution for health reasons. Sylvestre-François Lacroix (1765–1843) was elected to succeed him. At the same time, a fourth *instituteur* post was established for analysis and filled by Jean-Baptiste Labey (1752–1825). Labey, formerly a mathematics teacher in the *École militaire* at Paris and in an *École centrale*, was known for his French translation of Euler's *Introductio in analysin infinitorum* (1797).

Although Lagrange proposed that his vacated post should be devoted to regular course teaching (Fourcy 1828, 191), the institution decided that both Lacroix's and de Prony's courses should be reserved for particularly gifted students and held only once a week.[31] The regular courses—named the first or second *divisions* for each of the two years of the course—were transferred to Labey and Garnier. Jean-Guillaume Garnier (1766–1812), examiner for the regional entrance examinations to the institution in the years 3 to 7 (1794–1799),

---

30   AEP, Conseil d'instruction et d'administration, Vol. 1. Resolution of 8th Frimaire in Year 4: fol. 55.

31   AEP, Conseil d'instruction, Vol. 3, fol. 53–54.

took over Fourier's post as the teacher for analysis in 1798 while he was away on the Egyptian expedition. He remained at this post until 1802. Although criticized for not maintaining enough authority,[32] his teaching texts offer interesting insights into the further development of concepts (see below).

The programmatic consolidation was tackled systematically by the newly formed *Conseil de perfectionnement* after its first session in October 1800. It appointed individual commissions for each subject to work out a detailed course program. For the *analyse*, the commission was composed of Laplace, Bossut, and Legendre. The mechanics commission had exactly the same membership plus de Prony.

The commission took only one week to present its program, and the *Conseil* approved it right away. It proposed a division of the course into three sections: *Analyse pure* (or *Analyse algébrique*), *Calcul différentiel*, and *Calcul intégral*. In the section on differential calculus, the plan was to start the course by introducing the differentials on the basis of the limit method: "establish the notions of the differentials by the *théorie des limites*."[33]

2.4.2. LACROIX: PROPAGATOR OF THE MÉTHODE DES LIMITES

Although Lacroix had not been a member of the program commission, he very clearly influenced this first formal decision on methods. His monumental textbook on analysis was not only the most modern and comprehensive account of the state of the art in analysis but also the most consistent presentation of the limit method. Moreover, this work could be viewed as an explication of the *méthode des limites*, as presented by Laplace at the *École normale* in 1795. At this time, Laplace was the *éminence grise* of the *École polytechnique*.

At present, no detailed biography is available on Sylvestre-François Lacroix (1765–1843), although René Taton (1953a, 1953b, 1959) has published a number of studies on biographical aspects. The one area about which we know least is his education. Taton (1953a, 588) suggests that he may have attended the course given by the Abbé Marie, teacher in the *Collège Mazarin* at Paris.[34] However, we can assume that he basically taught himself at Paris. Indeed, an older biographical handbook mentions that he was self-taught because he was too poor to pay for systematic studies and attend Mauduit's course at the *Collège Royale*.[35] His early intensive astronomical calculations soon brought him into contact with scientists at Paris. While first under the patronage of Monge, he

---

[32]   AEP, personal files of J.-G. Garnier, VI 1 b2.

[33]   AEP, Procès Verbaux du Conseil de perfectionnement, Vol. 1; Première Session, Brumaire–Frimaire in year 9: fol. 7v.

[34]   This seems to be the basis for Itard's claim in his article in the Dictionary of Scientific Biography (Vol. 7, 549) that Lacroix had studied at this *collège*. Because the mathematics courses were also open to guests, this is certainly an unjustified conclusion.

[35]   Nouvelle Biographie Générale, Vol. 28 (Paris 1859), 593–595.

then also gained the patronage of Condorcet. It was also Monge who found him his first teaching post although he was only 17 years old at the time.

During the *Ancien Régime*, Lacroix held only modest posts as a mathematics teacher, almost exclusively at military schools. The first post was at Rochefort starting in 1782 at the *École des gardes de la marine*. He would have used Bézout's textbooks there. In 1786, he moved to Paris, where on the initiative of Monge and Condorcet, he took on the post of mathematics teacher at the so-called *Lycée*, an independent teaching institution.[36] In February 1787, he took a further post in the military system: the *École Militaire* at Paris (see above, Chapter II.2.6). Due to insufficient attendance, the mathematics post at the *Lycée* was discontinued in August 1787. When the *École Militaire* also closed in March 1788, Lacroix had no more employment in Paris. He had to move to another military school, at Besançon, the *École Royale d'Artillerie*, where he taught mathematics, physics, and chemistry. Here as well, teaching was based on Bézout's *Cours*. Through his connections with Parisian scientists, he already became a corresponding member of the Paris *Académie* in 1789. During the revolution, he was close to the group around the *Idéologues*. In 1793, he took over Laplace's post as examiner for the artillery, permitting his return to Paris. After 1794, he held a leading post in the instruction department of the Ministry of the Interior, where he organized the first *concours* for the production of good elementary textbooks (see Schubring 1982b, 116). In 1795, he started off as Monge's assistant in his lectures on descriptive geometry before becoming a mathematics teacher at one of the large Parisian *Écoles Centrales*, the new type of secondary school set up by the *Idéologues*. He kept this job until 1815, even after these schools were converted into *Lycées* in 1803. The year 1799 brought a rapid rise in his career: He became not only *instituteur* for analysis at the *École polytechnique* but also a full member of the *Institut*. After resigning as *instituteur* in 1808, he took on the even more influential post of *examinateur permanent* as Bossut's successor.

Lacroix provides an impressive example of the French practice known as *cumul*, the amassment of several official posts. In 1808, he was, in addition, called to the newly founded *Faculté des sciences*, where he also took over the office of dean. The crowning of his career came with his appointment to the *Collège de France* in 1815.[37]

Lacroix was not just a committed teacher but also the prototype for the modern author of major textbooks (see Schubring 1987). His first textbook appeared in 1795 and tackled descriptive geometry. It was followed by a scarcely interrupted series of textbooks or revised editions and new printings—

---

[36]   In an account of his teaching posts written at Besançon on October 3, 1791 (and still signed in the *ancien* manner of "DeLaCroix," Lacroix does not mention his activity at the *Lycée* (autograph in the possession of the author).

[37]   Lacroix's personal records at the *École polytechnique* reveal the impressive size of his income thanks to the practice of *cumul*. It was also augmented by royalties from his textbooks.

covering all of pure mathematics and probability theory as well. One peak in his productivity was in 1800, when he published not only his monumental work on analysis but also a series of school mathematics textbooks. His textbooks were not only reprinted continuously in France, but were also translated into several languages throughout Europe (sometimes more than once in the same language!) and even used in North and South America.

Despite the range of his textbook production, one major work was particularly productive and innovative: the three-volume *Traité du Calcul Différentiel et du Calcul Intégral*. The first volume appeared in 1797, followed by the second and third in 1798 and 1800. In this work, Lacroix pursued the concept of elementarizing existing knowledge on analysis. In the programmatic sense of the *Encyclopédie*, he wanted to deliver a systematic presentation of principles corresponding to the state of the art in this science. In the second half of the 1780s, Lacroix started to collect the new research findings scattered across many academy treatises, to communicate with leading researchers, and, indeed, to restructure analysis. When he submitted the first volume to the *Institut*, it was well received: the report presented by Legendre assured him not only that he had attained his goal of elementarization but also that his textbook activities were valued just as highly as research (Institut de France, I, 1910, 155).

One frequent criticism of Lacroix's handbook is that it presents three different methods of analysis practically side by side: Lagrange's method of developing functions into series, the limits method, and the method of infinitely small quantities (see, e.g., Grabiner 1981, 79 f.). However, this is actually a misunderstanding, brought about by a comment Lacroix makes in his preface. Although he talks about a presentation from "several perspectives"—these three methods—he actually carries out this intention for only one application of analysis, namely, that of the "theory of curves and areas" (Lacroix 1799-1800, Vol. 1, XLIV).

The total structure of the work does not take the "encyclopedic" form suggested. Lacroix's original incentive and his starting point was, as he emphasized repeatedly, Lagrange's Berlin treatise of 1772 (see Chapter III.9) that had formulated analysis in purely analytical terms, without any consideration of the infinite. Lacroix considered the advantage of Lagrange's approach of expressing changes in a function through developing into series to be, first, that it frees the equations from infinitely small quantities, and, second, that it permits a generalization: in the "detachment" of differential calculus from its application to curves (ibid., xxxvii ff.). Lacroix repeatedly emphasizes his principal rejection of the method of infinitely small quantities, and particularly the concept of higher-order infinitely small quantities (ibid., xxxv). This is how he also declares his hope, "that the novices will know how to thank me for the care I have taken to make them independent from the concepts of the infinite" (ibid., XL).

The major handbook actually does present the treatment on the basis of one single method: the limit method. Using this basis, Lacroix goes on to present

Lagrange's approach of development into series. The introductory section, which is practically an abbreviated algebraic analysis in preparation for the true differential and integral calculus, develops the limit method as a precondition and basis for applying the development into series. This enables him to explicate a precondition that had remained implicit in Lagrange, thus bringing further clarification. However, it is notable that Lacroix's presentation of the limit method is based exclusively on the version in Laplace and in the *Encyclopédie* article, that is, it uses only verbal formulations, avoiding any use of signs and thereby also any independent algebraization and the development of a calculus with limits. Hence, the form of the limit method in Lacroix represents a step back compared with the work of Martin, L'Huilier, Carnot, and, above all, Stockler (with whose work he probably remained unfamiliar). As a result, we have to characterize Lacroix as a conservative modernizer.

Nonetheless, he was indubitably a modernizer. His handbook is the first general textbook since Euler's still isolated work to be based consistently on the function concept. Lacroix even introduces Euler's most general definition of the function concept as a basic definition in the first paragraphs of his work, and he does not tie it to a specific algebraic expression (Lacroix 1797/99, 1 f.).

Lacroix uses the fraction $\frac{a}{a-x}$ to show that the study of limits forms the basis for applying the development into series. He argues that one has to distinguish the notion "development" from that of "value," because a series does not always result in the appropriate value for the function. It may even diverge ever further from the value as the number of terms increases (ibid., 6). Hence, Lacroix justifies the need for the limit method through the necessity to clarify the convergence or divergence of the series development.

After detailed calculations on the convergences or divergences for several values of x in his example, Lacroix formulates the following, purely verbal, definition of *limite*:

> From now on we shall call a limit any quantity that a magnitude cannot go beyond when increasing or decreasing, or even one that it cannot reach, but which it can approach as closely as is wished. (Lacroix 1797, 6; author's emphasis)

It is notable to see the conservative attitude of talking about "quantity" in an undifferentiated way without distinguishing between the constants and the variables.

For Lacroix, the limit method is not only the necessary foundation for analysis, but at the same time, the means to dispense with the use of infinite quantities. Therefore, after reflecting on the contradictions inherent in the concept of the infinite, he determines that:

> Everything that one can prove with the aid of the consideration of the infinite, can be deduced from the notion we have given above of the word *limit* and the two following propositions, which are so clear as to be incontestable (Lacroix 1797, 10).

The two propositions are aimed toward operations with inequalities: any quantity, regardless of how large, can be surpassed by another by as much as one wants. Likewise, for any quantity, regardless of how small, one can conceive one that is even smaller (ibid., 15 f.).

What can be regarded as modernizing is Lacroix's emphasis in the foundations section that not every quantity or function has to allow a limit. Repeatedly—as only L'Huilier did before him (cf. Chapter III.8.5.)—he formulates the precondition that the respective function "is capable of having a certain limit" (ibid., 28). However, in the first edition, he does not explicitly test the pertinence of the precondition in his applications.

After explaining the *limites*, he ascertains that these create a sufficient basis to study limit processes, and he adds two supplementary propositions:

> The principles which we have just employed are sufficient for finding the limits of those quantities, which are susceptible [of a limit], and we shall conclude [...] with the two following propositions, which will be very useful to us (Lacroix 1797, 18).

Both propositions agree in content with the two principles given by the Abbé de la Chapelle in his first explication of the limit method (see Chapter III.8.2.). They do not take into account the further developments by other authors:

> 1. Two quantities that are limits of the same function are equal to each other [...].
>
> 2. When two quantities maintain an unchanging ratio to each other, this is [also] the ratio of their limits. This is self-evident (ibid.).

It is only after this explanation of the limit method that Lacroix goes on to present how functions can be developed into series. This forms the remaining part of the foundations before he moves on to the proper sections on differential and integral calculus.

He introduces the first main part of his handbook on differential calculus with a chapter entitled *Exposition analytique des principes du calcul différentiel*. In line with Lagrange's concept, he shows how the development of a function into series proceeds when the variable x changes into $x + k$ or $x + dx$. Lacroix does not address the convergence of the development into series in this section, because he has already dealt with this as a foundation in the previous section. Although he has not introduced or used any sign for the limit, it becomes evident that he has thought about the role of signs. For example, he notes that he uses a general sign—namely "f"—for functions that do not already possess their own names and no signs but are dealt with generally (Lacroix 1797, 87).

The differential $df(x)$ results by way of the coefficients of the first power of k in the development into series of the function $df(x) = f'(x)k$. One central result for Lacroix is that $f'(x)$ as a "derived function" is, in turn, itself a function, and it can be used in an analogous way to obtain $f''(x)$ as a second derived function, and so forth (ibid., 93 ff.).

The further presentation of the rules and operations of differential calculus starts with the differentials. What was new is his particular use of the differentials in the proof of the product rule. Up to that time, only Lagrange had taken an analogous approach in his almost simultaneously published lectures on function theory:

Developing two functions u and v to $u + pdx + \ldots$ or to $v + qdx + \ldots$ leads to the product

$$\left.\begin{array}{l} uv + uqdx + \ldots \\ + vpdx + \ldots \end{array}\right\} - uv.$$

However, because $qdx = dv$ and $pdx = du$, one obtains $d(uv) = udv + vdu$ (ibid., 102).

Whereas Lagrange presented this argumentation in only a very condensed form (see Lagrange 1881, 40), Lacroix justifies dropping the higher-order terms in the presentation of the product given above. It is impermissible because of the definition of the differential (Lacroix 1797, 102). Hence, only linear terms appear.

One characteristic feature of Lacroix's approach can be seen in his introductory section on foundations: he uses both the differential $df(x)$ and the differential coefficient $\frac{df(x)}{dx}$. He handles the transition between both forms as a *division* (ibid., 95): from $df(x) = f'(x)k$, he forms $f'(x) = \frac{df(x)}{k}$, and then, because "the signs have the same form," he replaces the letter k by a sign indicating the variable, by dx. This gives him $f'(x) = \dfrac{df(x)}{k}$ (ibid.).

This approach confirms that Lacroix conceives differentials as finite quantities. Hence, he is already using the conception of differentials as linear functionals that Spalt (1996, 123 f.) considers to have been introduced by Cauchy. Because of the finiteness of the quantities k or dx, Lacroix switches to and fro between the two forms of representation. For example, he considers $f''(x)$ $dx = df'(x)$ to be equal in value to $f''(x) = \frac{df'(x)}{dx}$ (ibid., 96).

It seems that Grattan-Guinness (1990a) may have misunderstood the new conception of the differentials in Lacroix, and have tried to view his use of both Lagrange's approach with derivation functions and Euler's approach with differentials as confirmation of his "encyclopedism." Although admitting that Lacroix views differentials as being something other than Euler's reckoning with zeros, he still accused Lacroix of wanting to present all available methods (ibid., 141).

In fact, however, this is a unified concept for Lacroix: He obtains the differentials by developing the function into a series, thereby establishing his foundations. For the applications, in contrast, he reshapes the differentials into the differential coefficients with which he goes on to perform his further operations. Indeed, the application chapters following the foundations section use almost exclusively differential coefficients.

Otherwise, Lacroix and Lagrange do not contradict each other in their use of both forms. Lagrange also starts off with a "differential" form of presentation for the development into series with, for example, $f(x+k) = f(x) + kP$. And, after division, $P = \dfrac{f(x+k) - f(x)}{k}$ (Lagrange 1881, 24; Lagrange writes i instead of

k). He even starts off by using the "differential" form for more complicated developments as well:

$$k[f'(x) - F'(x)] + \frac{k^2}{2}[f''(x) - F''(x)] + \frac{k^3}{2 \cdot 3}[f'''(x) - F'''(x)] + \dots \quad \text{(ibid.,}$$

185), but he does not do this in such a detailed way as Lacroix.

At the end of the section on the foundations of differential calculus, Lacroix shows that the differential coefficients in the development of functions into series are identical with the limits of the relations between function values and variable values (ibid., 189 ff.). Therefore, when summarizing his foundations and methods, Lacroix ascertains that differential calculus can be conceived more generally as a calculus of limits: "The differential calculus can thus be envisaged as having the purpose of *finding the limits of ratios that the increments have to each other*" (ibid., 193).

It is only here, at the end of the section on foundations, that Lacroix addresses the method of the *infiniment petits*: first in a footnote showing how easily this method leads to errors. He gives the differentiation of the sine function as an example: for $u = \sin(x)$, he writes $u + k = \sin(x + h)$ and obtains after transformations $k = \sin x(\cos h - 1) + \cos x \sin h$.

Lacroix works out the limits of the terms and then notes that he deviates from other authors here who follow a different argument: they say that if the curve k becomes infinitely small, then $\cos h$ will differ only infinitely slightly from 1, so that $\cos h - 1 = 0$, and finally, only $\cos x$ will be left for the derivation. However, because $\sin h$ itself also becomes infinitely small, it is not clear how the two products relate to each other quantitatively (ibid., 191).

Lacroix then goes on to generalize this footnote comment into a paragraph commenting on the method of the *infiniment petits*. He identifies it—in line with the traditional approach in France—with the Leibnizian method, expressly declaring this to be less rigorous (*moins rigoreuse*), but worth knowing, because it is easier to apply (*plus commode*). He describes its main feature to be the extended concept of equality, in other words, that one can exchange two quantities for each other as long as they differ only in terms of an infinitely small quantity. One necessary conclusion from this principle is the homogeneity of the differential expressions, implying the principle that each higher-ranking infinitely small term vanishes (ibid., 193 f.), that is, the comparison of the orders of the infinitely small (and infinitely large) quantities.

Initially, Lacroix's assessment of these methods as lacking in rigor seems to contradict the explanation in his preface that he follows on from the *méthode des limites* with an inspection of the infinitely small quantities when applying differential calculus to the theory of curves and surfaces: "Finally, I made use of the consideration of infinitely small [quantities]" (Lacroix 1797, xxvij). However, despite the impression this creates, Lacroix does not use the *infiniment petits* in his presentation of these chapters either. What he had meant by this comment is as follows: In the fourth chapter on the theory of curves, he starts off by introducing algebraic geometry, in other words, how curves can be represented by equations. He follows this, in line with the Lagrange program, by considering how the development of functions into series can be applied to these equations—in order to ascertain the values by expressing their coordinates through convergent series (ibid., 362 ff.). The main part is the application of differential calculus to the study of curves, that is, for example, the use of the first and second derivatives to determine maxima, minima, and so forth. The chapter closes with a section working out how the *méthode des limites* can be used to tackle the traditionally difficult study of contact curves in a rigorous and transparent way (ibid., 419 ff.). Within this section, Lacroix adds a few paragraphs discussing the applicability of the "Leibnizian" method of conceiving curves as polygons with an infinite number of sides. Lacroix shows that when applied carefully, the method practically amounts to a calculation of limits and, in this sense, has strong possibilities for applications in curve theory (ibid., 422 f.). In the fifth and final chapter as well, on curved surfaces and curves of double curvature, there are only occasional discussions on how to proceed when one conceives of curves of double curvature as polygons (ibid., 507 ff.).[38]

Hence, it cannot be said that Lacroix's handbook presents an eclectic collection of all the usual methods of his time. Nonetheless, when one considers the breadth of contemporary advances, it would be a completely meaningful approach for the most comprehensive handbook of its time to provide an overview.

However, although Lacroix's handbook makes a major contribution to modernization, his conservative tendencies cannot be overlooked. For example, even in 1797, that is, during the period dominated by the *méthode analytique*, he stresses his sympathy for the *synthèse*. Characteristically, he links this explanation to the classic contradiction between the *modernes* and the *anciens*. The advantages of the *analyse* listed in his preface are in no way a negative judgment of the *synthèse*, it is far more the case that he thinks insufficient attention has been paid to the study of the works of the *anciens*:

> Let it not be thought that, in insisting on the advantages of algebraic Analysis, I should wish to attack Synthesis and geometric Analysis. I think, on the contrary, that the study of the Ancients is neglected too much today (ibid., xxvj).

Chapter V.3., on the impact of Carnot's concepts of the negative numbers in France, discusses how directly Lacroix took up this shift toward the *synthèse*.

The foundations of integral calculus, to which Lacroix assigns the entire second volume of his handbook, contain no conceptual innovations. Taking the

---

38   In the comment in the preface cited above on using the *infiniment petits*, the only concrete context named by Lacroix is the theory of curves of double curvature. He states that if these are well explained, they may be helpful in solving new problems in this context. Lacroix is referring to Monge here. However, Monge's textbook, entitled *Applications de l'analyse à la Géométrie* and based on his lectures at the *École polytechnique*, discusses curves of double curvature in terms of differential calculus with absolutely no reference to the *infiniment petits* (see Monge 1809, 407 ff.).

traditional approach, he conceives integral calculus as a reversal of differential calculus, starting with the differential coefficients and determining the functions from which they are derived. Accordingly, the basic concept is the indefinite integral (see Lacroix 1798, 1 ff.).

Just like Lagrange in his function theory, Lacroix deals extensively with functions of several variables. Because the relations between continuity and differentiability were still below the conceptual horizon of his times, Lacroix assumes that no further preconditions are necessary for the formation of partial derivatives. There is only a multiplication of the limits (Lacroix 1797, 191). He also declares the permutability of the higher-order partial derivatives to be "evident" (ibid., 118), just as Lagrange did during the same period (Lagrange 1881, 144).

The intensive teaching of analysis at the *École polytechnique* triggered a stronger reflection within its entire conceptual field. An increasing rigor in concept formation becomes particularly evident in the teaching notes of the lectures on analysis. An initial example is the "elementary" edition of his handbook that Lacroix first published in 1802 as a text to accompany his lectures. This rapidly became the general textbook at the *École polytechnique*.

2.4.3. FURTHER CONCEPT DEVELOPMENT: LACROIX, GARNIER, AND AMPÈRE

So that it could serve as a textbook (a *traité élémentaire*), Lacroix revised his handbook completely; not just in quantitative terms—the three large volumes became one normal-sized book—but also qualitatively: a handbook aiming to give the most up-to-date presentation of the state of research became a book oriented precisely toward being teachable.

For the textbook version, Lacroix strongly reduced the section on foundations. From the detailed presentation on developing functions into series, he basically left only a few pages on Taylor's theorem. To some extent, this is countered by a more thorough and clearer presentation of the limit concept right from the very first pages. After introducing basic concepts such as constant and variable, the term *limite* is already addressed on the third page; and the first example in the book is used to explain the derivative as a limit. Limits are now applied right from the beginning to determine differentials and differential quotients (Lacroix 1802, 7). There is also a stronger emphasis on ways of operating with limits. We now find statements in a modernized form—no longer in the terminology of ratios—that the limits of products and of the quotients of variables are interchangeable (ibid., 8). However, Lacroix retains his purely verbal use of the limit concept without switching to any sign, thus limiting the operations he can perform.

The importance of continuity is also presented more clearly. It is no longer just considered in the *Préface*, but discussed in the main text—particularly in a section preceding the application of differential calculus to the theory of curves.

Lacroix wants to ensure that the calculus of limits is used to study curves. Whereas the general sections in the first edition of the handbook named "if the limit exists" as a precondition (see above), but did not examine this precondition any further in the applications, this textbook raises the precondition to a postulate. In practical terms, he follows the common practice up to that time of identifying continuity with differentiability. He says that differential calculus is based on the prior fact that "all functions" possess a limit for the ratio between the increase in the values of the function and of the variable. He understands this fact as an expression of the validity of the law of continuity:

> This [differential] Calculus [...] is immediately based on an *analytic fact*, pre-existent to every hypothesis, like the fall of heavy bodies to the surface of the earth, and preexistent in all the explanations that have been given for it; and this fact is precisely the property enjoyed by all functions possessing a limit of the ratio of that their increases bear to their dependent variable (Lacroix 1802, 75).

Lacroix follows this up directly with the first detailed characterization in a French mathematics textbook of the continuity concept as a "law." He does this with the image of a geometric line representing a continuum characterized by completeness without any disruptions:

> By the law of continuity is to be understood what is observed in lines being described [drawn] by movement and by which consecutive points of the same line succeed each other without any gap [between them]. The way in which magnitudes are envisaged in the calculus does not appear to admit of this law when one supposes there is always an interval between two consecutive values of the same quantity; but the smaller that interval is, the more one approaches the law of continuity, which is perfectly satisfied at the limit (ibid., 75).

Hence, Lacroix interprets continuity as a guarantor for differentiability in terms analogous to those used by Carnot (ibid.).

This period is characterized by a rapid development of concepts through an institutionalized discussion context. This is certainly due not only to this prominent explication but also to the opportunity provided by the context of the *École polytechnique* for mathematicians to communicate intensively over the conceptions underlying their lectures.

Before going on to discuss how further mathematicians contributed to this development and the discussion, I shall take a single example to illustrate how Lacroix himself revised one of his conceptions within this discussion context. The topic is the key proposition on the differentiation of the product of two functions.

In the first edition of the textbook version of 1802, Lacroix simply transcribes the proof of the product rule from the first edition of 1797 (Lacroix 1802, 9). For the second edition of 1806, the need to drop the higher terms because only terms of the same order can be present in a differential by definition alone must have made this proof seem circular to him. As a result, he looks for a real proof, and he does this with limits. Instead of developing the function into a series, Lacroix tries to obtain the proof directly for two functions u and v. Their product changes when the variables increase, and, hence, the function value into

$$uv + u\beta + v\alpha + \alpha\beta,$$

making the increase $u\beta + v\alpha + \alpha\beta$.

By "comparing" (i.e., forming a ratio) with dx, he obtains the expression

$$u\frac{\beta}{dx} + v\frac{\alpha}{dx} + \frac{\alpha}{dx}\beta.$$

He labels the limits of the ratios $\frac{\alpha}{dx}$ and $\frac{\beta}{dx}$ with p or q. Nonetheless, he tries to follow a completely different argumentation for the factor $\frac{\alpha}{dx}$ when it occurs for a second time in the last summand: he says that one has to consider that the increase β disappears "at the same time" as dx (from which the quantities u and v are independent), and then recognize that the limit of the term $\frac{\alpha}{dx}\beta$ equals zero. Evidently, he means that the limit of the ratio of β to dx is finite and the limit of the factor α is zero. The limit of the remaining terms is accordingly $uq + vp$. To obtain the differential from uv, he "multiplies" this sum by dx:

$$d(uv) = uqdx + vpdx = udv + vdu \text{ (ibid., 12).}$$

In the third edition of 1820, Lacroix changes his argument once again, however, without improving it. His explanation continues to declare that β and dx vanish simultaneously. However, he does not apply this, but is now more consistent in that he also sets the last summand as p in the limit $\frac{\alpha}{dx}$. Hence, he explains that the limit of the term $\frac{\alpha}{dx}\beta$ is $p\cdot 0$, that is, a total of zero (Lacroix 1820, 14).[39]

We shall now turn to the contributions of other mathematicians to the further development of basic concepts in analysis within the context of the *École polytechnique*. One notable contributor was J.G. Garnier (see Chapter IV.2.). Garnier was the first—very soon after taking over the normal analysis course—to print his own abridged versions of his lectures to students. The archive of the *École polytechnique* contains several of these texts that were evidently printed purely as course material and not for broader distribution. One of them, entitled *Leçons d'Analyse Algébrique, Différentiale et Intégrale*,[40] for the first student admission in year 9 (1800/1801), also contains an abridged version of a lecture headed *Cours d'Analyse Différentielle, fait en l'an 9*. After introducing conceptual terms and presenting the foundations of differential calculus, Garnier goes on in the third section to dwell explicitly on the transition from the differences to the differentials and apply the *méthode des limites* as a means of performing it. This is the first time such an approach was taken at the *École polytechnique*, where differential calculus had always been based on the calculus of differences (see Chapter IV.2.). This section is headed *Notion de la limite et passage du calcul aux différences finies au calcul différentiel*. Garnier does not just motivate and present the concept of the limit in detail (although only

---

[39]   Afterwards, his argument remains unchanged in both the fourth (1828) and fifth (1837) editions.
[40]   AEP, Bibliothèque, A III a 51. The lack of pagination also suggests that these prints were only for internal use.

verbally and without any sign-like form and hence without any more far-reaching operations with limits). He goes beyond this—in contrast to Lacroix's almost simultaneous textbook version—and determines that not all functions assume a finite limit:

> Among functions to be considered, some may have zero or another quantity for a limit: others may become greater than any given quantity; others, finally, may decrease indefinitely.

A further remarkable feature of this discrimination between finite limits and points at which functions become positively or negatively infinite is that the zero is not given any special status, but is classified with the finite values.

This extension of the limit concept was in no way an isolated event: it was already part of the final examinations. One document has been preserved from 1799 in which Garnier wrote down the main parts of his teaching program on differential and integral calculus as a guideline for the two examiners Laplace and Bonne. This reveals that one of the examination topics was that not all functions possess a limit everywhere. From the 19 topics on the foundations of differential calculus, the first three address the introduction of the function concept. Several of the subsequent topics deal with the limit concept. Examples are:

4. What one understands by the limit of a function of one variable.

5. Not all functions are capable of a limit.

12. Limits of the ratios

$$\frac{\Delta y}{\Delta x}; \frac{\Delta^2 y}{\Delta x^2}, \cdots \frac{\Delta^{n-1} y}{\Delta x^{n-1}}, \frac{\Delta^n y}{\Delta x^n}$$

and notation of these limits.[41]

Because Garnier had already left the *École polytechnique* in 1802, after which Lacroix's *Traité elémentaire* became the standard textbook, it was some time before this work on foundations was continued directly. It was taken up again and radicalized by André-Marie Ampère (1775–1836), who took his first post as *répétiteur* in 1804 before becoming a full-time teacher of analysis in 1808. Although Ampère is known mostly as a physicist, he was additionally not only an active mathematician and chemist but also stood out through his strong interest in philosophy. Ampère worked closely with several of the *Idéologues*, most intensively with Maine de Biran, who tended to represent their conservative wing (see Chapter V.1.4.). He is well known for the resulting work on the classification of the sciences. Less well known, and indeed underrated, is Ampère's contribution to basic concepts in mathematics. Even his first scientific biographer declares Ampère's final contribution to mathematics to be minimal (J.R. Hofmann 1996, 59).

---

[41]   AEP, Bibliothèque, III 3b, J.G. Garnier, Programm de Calcul Intégral et de Calcul differentiel, remis le 16 fructidor aux c.en LaPlace et Bonne pour servir aux examens ouverts le 2 Complementaire, an 7.

His contributions to foundations are evidently an outcome of his philosophical reflections. They all emerged within the context of the *École polytechnique*. Usually, only Cauchy's conceptual innovations are perceived as the beginning of the new rigor in mathematics. However, the fact is that Ampère had already achieved important reforms that initially impressed Cauchy, who was his student. Later, Cauchy and Ampère exerted a reciprocal influence on each other in their respective courses. However, Ampère's work on foundations (once again, remarkably, on both negative numbers and the concepts of analysis) is concentrated mostly in lectures at the *École polytechnique* that nearly all remained unpublished. His most well known mathematical work is his *mémoire* on the theory of derived functions published in the Institute's *Journal* in 1806. In the nineth century, this work was attributed with having very far-reaching intentions to achieve conceptual rigor, and with having proved that continuous functions are differentiable except for occasional singularities. However, such a statement is not to be found in Ampère's *mémoire*. Although Volkert (1989) followed up and analyzed the reception given to Ampère's publication, he missed the opportunity to trace the cause for this misattribution.[42] Nonetheless, several recent studies have analyzed the actual intentions and achievements of this *mémoire* (Grabiner 1978; Grattan-Guinness 1990a, 197 ff.; Guitard 1986).

Ampère's (1806a, 149) intention was to prove the *existence* of the derived function $f'(x)$. Declaring this existence to require proof and addressing a comprehensive treatment to it was a decisive innovative achievement. In 1806, Ampère also presented the basic belief in "generality": that a statement still holds for "all" functions even when it does not hold for single values. Moreover, he did not explicitly assume continuity. Evidently, he also assumed this to be "general." Ampère's *mémoire* is important not just as the first work to present the need to prove the existence of the derivative, but also for two particular technical advances:

1. The first was the introduction of an equidistant nest of intervals with increasingly diminishing intervals and the use of inequalities to test the accuracy of the property claimed for each interval. However, this technique is still far from perfect: mostly verbal, without indices, and without the possibility of a sign-like symbolization of the diminishing of the intervals. The operations with inequalities are also verbal, without signs. This restricts Ampère to the illustration of what is still a very rough nest of intervals. The most elaborate presentation attains eight subintervals; further refinement could only be suggested verbally:

One can show in the same way that among the eight fractions

$$\frac{K-H}{k-h}, \frac{H-G}{h-g}, \frac{G-F}{g-f}, \frac{F-E}{f-e}, \frac{E-D}{e-d}, \frac{D-C}{d-c}, \frac{C-B}{c-b}, \frac{B-A}{b-a},$$

---

[42]   This is because Volkert (1989, 62) considered Lacroix's (1810) paper on Binet's work of 1808 to be a paper on Ampère's publication of 1806, and failed to notice that these are two different works.

there is necessarily one greater and one smaller than $\frac{K-A}{k-a}$. Continuing in the same way to intercalate values of x between the preceding ones, a new sequence of fractions will be formed, among which there will always be one greater and one smaller than $\frac{K-A}{k-a}$ (Ampère 1806a, 152 f.).

Likewise, he can still only use verbal terms to express that the length of the interval can be arbitrarily diminished (ibid., 153).

2. The second technical innovation is the very precise claim regarding the validity of statements on specific intervals—and no longer diffusely for all only potential values of a variable. For example, Ampère talks about values "*depuis x jusqu'à x+1*" (ibid., 156) or about values for $f(x)$ "*depuis x = x jusqu'à x = z*" (ibid., 169 f.).

Ampère clearly expresses his rejection of infinitely small quantities in this text. His approach allows one to "to be detached from considering *infiniment petits* (ibid., 162). Although he always draws implicitly on the limit method, he never uses a sign for limits, not even the term *limite*.

Binet's memoir differs markedly from that of Ampère. Paul René Binet (born 1779), one of the mathematicians trained at the *École polytechnique*, revised some aspects of Ampère's "proof" and rendered it more precise in a memoir presented to the *Société Philomathique de Paris* on November 12, 1808.[43] Although the memoir itself was not published, it was reported in detail in the *Bulletin de la Société Philomat[h]ique* in 1809. Ampère wrote this very precise report,[44] leading Grattan-Guinness (1990a, 198) to even assume that Ampère "wrote" Binet's ideas "up for him" so that they could be published. However, this was not the case. Of the numerous independent scientific societies established at the end of the *Ancien Régime*, the *Société Philomathique de Paris* had a special status. Particularly because it had taken in numerous former members of the *Académie des Sciences* after its closure in 1793, it had the closest contacts with leading mathematicians and scientists and functioned as an "antechamber" to the new *Institut National* after 1795 (Mandelbaum 1980, 65 ff.). Important *mémoires* continued to be read at its weekly sessions, by young but also "established" scientists. The *Bulletin de la Société* published reports on its lectures, mostly written by the secretary of the *Société*. From January 1808 to January 1810, Ampère was vice-secretary and edited a number of *procès-verbaux* in this office (ibid., 156).

Basically, Binet wanted to prove the same proposition as Ampère's from 1806. However, he chooses a modified path, thereby making his methodological instruments more precise. He explains more explicitly than Ampère that the proposition should hold for any arbitrary function (*fonction quelquonque de x*;

---

43   Mandelbaum (1980, 176).
44   Ampère's authorship is documented by the presence of the manuscript of this report in his *Nachlass* (Grattan-Guinness 1990a, 198, Fn. 1) as well as the monogram "A" at the end of the report (Mandelbaum 1980, 176). Grattan-Guinness was not familiar with Mandelbaum's study.

Binet 1809, 275 f.). One of his modifications is to assume a local monotony of the function, either increasing or decreasing in the interval under observation (ibid., 276). This demand for monotony was later attributed to Ampère. Binet also expresses his statements in terms of limits. Finally, it is significant to see that he generalizes and algebraizes the method for nesting intervals. Entering the number of subintervals n as a variable permits a general presentation of the interval nesting for each selected interval $[a, b]$ by appropriate signs, enabling him to go on to examine all possible refinements of the subdivision operatively:

$$x = a, x = a + \frac{b}{n}, x = a + \frac{2b}{n}, \ldots, x = a + \frac{n-1}{n} b, x = a + \frac{n}{n} b \quad \text{(ibid., 277)}.$$

Because Binet also presents the values for the functions as a corresponding sequence:

$$\frac{f(a + \frac{b}{n}) - f(a)}{\frac{b}{n}}, \frac{f(a + \frac{2b}{n}) - f(a + \frac{b}{n})}{\frac{b}{n}}, \ldots, \frac{f(a + \frac{nb}{n}) - f(a + \frac{n-1}{n} b)}{\frac{b}{n}} \quad \text{(ibid.)},$$

one potential technical consequence is operations with indexed subsequences.[45]

The discussion on the existence of the differential quotient provides a first impressive view of the density and speed with which a concept can develop when several mathematicians work together in the same institutional context. Ampère's and Binet's findings became disseminated effectively only after Lacroix included them in the second edition of his handbook. Lacroix did not just publish his textbook version, the *Traité élémentaire* after 1802, but went on to prepare a second edition of his handbook, the first volume of which was published in 1810. The self-confident presentation of the unproblematic existence of the derivative still to be found in the textbook version disappeared as a result of the work of mathematicians at the *École polytechnique* in the intervening years.

A significant conceptual innovation has occurred when Lacroix discusses *l'existence de cette valeur*, the *coefficient différentiel*, as a *difficulté* in the second edition of 1810. And Lacroix points out specifically that earlier

---

[45]   Standard French texts up to about 1800 do not give sequences of quantities or variables with a notation identifying the single term of a sequence as a part of a generally labeled sequence, for example, a_3 as part of a sequence (a_n) with the general term a_n. Lagrange used letters in alphabetic order to label elements as part of a sequence, for example, the function terms in developing it into a series as P, Q, R, and so forth or coefficients with A, B, C, and so forth. With such an unspecific approach, he was not able to label the last term of a sequence or a general term. It is notable that Crelle shifted to indexed series in the sections he added to his translation of the *Théorie des fonctions analytiques*, for example, $B_1, B_2 \ldots, B_n$ or P_1, P_2, P_3 with P_n as general term (Lagrange 1823, Vol. 2, 332 ff.). Lacroix had already used general indexed quantities $a_1, a_2 \ldots, a_n$ in both 1798 and 1802, but only in a narrowly restricted field of calculus: within integral calculus to operate with the sequence of approximate values in using approximation to determine integral values (Lacroix 1798, 135 ff.; 1802, 285 ff.). Lacroix, who had studied the contemporary literature intensively, may have been encouraged to introduce this usage—even though very partial—by the publications of the German school of combinatorics, which used indexed quantities as one of their everyday tools.

mathematicians such as Landen or d'Alembert had not confronted this difficulty, because of the historical limitation of their conceptual horizons. This had restricted them to "familiar," that is, simple, nontranscendental functions (Lacroix 1810, 240) and precluded any doubts about the existence of the derivatives (ibid., 240 f.). The justification for doubting the differentiability expressed here represents an erratic element in Lacroix's handbook. In the applied sections, he does not go on to consider whether this doubt is justified.

However, Lacroix could feel justified in not addressing such a research question in the textbook version, because he sketches a proof for the existence of the limit of the ratio $\frac{f(x')-f(x)}{x'-x}$ in a detailed footnote (following his reported "doubts") in the 1810 edition of the handbook (ibid., 241 f.). He attributes this proof to Paul Binet. Lacroix was also a very active member of the *Société Philomathique* over many years and had often taken leading posts there (Mandelbaum 1980, 279 ff.). Hence, he was certainly familiar with Binet's original *mémoire* and actually presents Binet's algebraic method of nesting intervals and its refinements in detail. However, he mixes Binet's and Ampère's approaches together. He replaces Binet's indirect proof that the limit cannot assume infinite values with Ampère's approach of nesting the limit with upper and lower bounds. This modified version of Binet's and Ampère's "proofs" is Lacroix's own contribution: It makes him the first to introduce continuity into the argument as an operative criterion. He concludes from the "law of continuity" that the limit lying between the two bounds can actually be assumed (Lacroix 1810, 241). The precondition of continuity, which later authors attribute to Ampère's "theorem" (see Volkert 1989, 62 ff.), was, in fact, first formulated by Lacroix. Hence, one should really talk about the Ampère-Binet-Lacroix theorem. In this innovation of the operative application of continuity, it has to be stressed that Lacroix mentions it only within the proof procedure; in general, the functions are still assumed to be continuous. For Lacroix, continuity in this context means the intermediate value property.

Although the "doubts" regarding differentiability remained an erratic moment in Lacroix's work, the broad international dissemination, particularly of the second edition of the handbook, led this explication to serve as an impetus for subsequent conceptual extensions.

Although Ampère used the limit method only implicitly in his 1806 memoir and did not apply any signs for limits, he did not just continue his work on the foundations of analysis. He also went on to adopt the limit method explicitly and propose and prove new theorems for its operative use. He took a different direction to that of Stockler and proposed theorems for operating with limits of inequalities.

These new advances on foundations are documented in Ampère's preparatory notes for the 1808/09 course on analysis that he took over formally from Lacroix after the latter's resignation in favor of his new post at the *Faculté des Sciences*. These notes are particularly informative, because they confirm Ampère's independence from Lacroix. Claude Gardeur-LeBrun, *inspecteur des élèves*, had sent him Lacroix's curriculum from 1806 before the course started.[46] Its structure precisely followed the single paragraphs in his textbook and illustrates

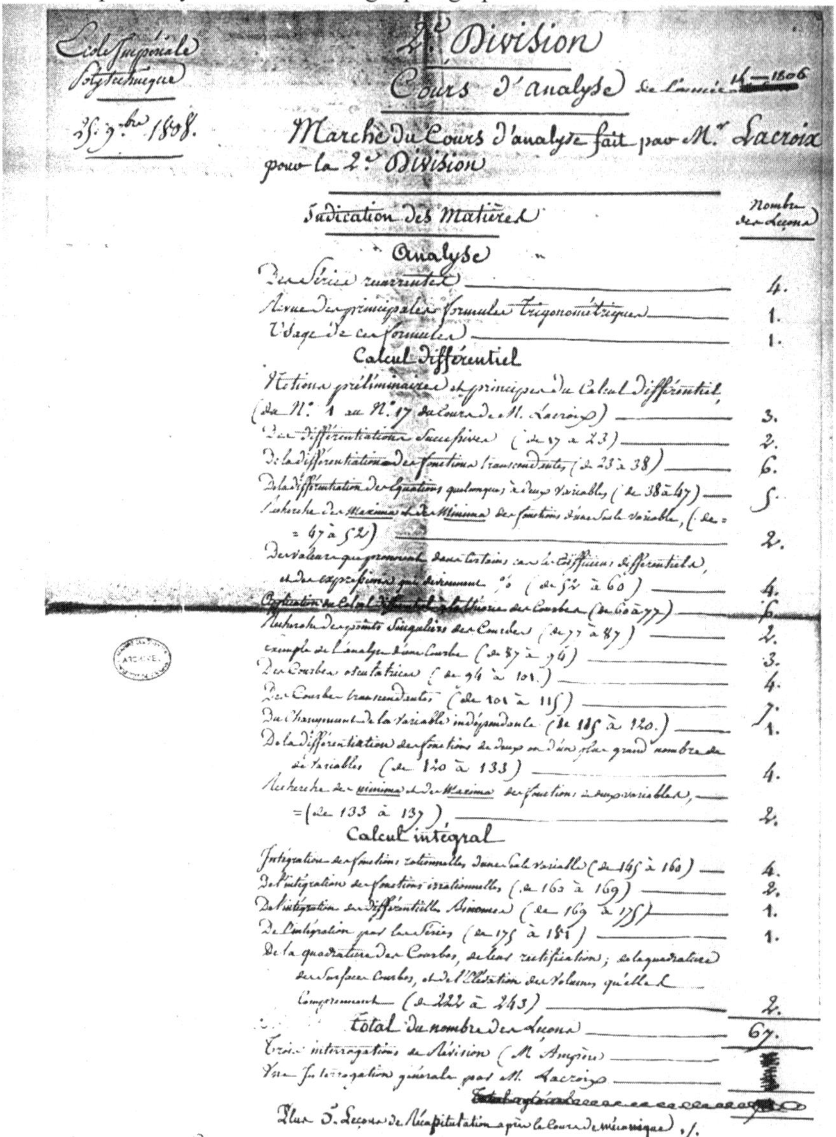

École Impériale Polytechnique

25 7bre 1808.

2e Division

Cours d'analyse de l'année 1806

Marche du Cours d'analyse fait par Mr. Lacroix pour la 2e Division

Indication des Matières	Nombre des Leçons
Analyse	
Des Séries recurrentes	4.
Revue des principales formules trigonométriques	1.
Usage de ces formules	1.
Calcul différentiel	
Notions préliminaires et principes du Calcul différentiel (de N° 1 au N° 17 du cours de M. Lacroix)	3.
Des différentiations successives (de 17 à 23)	2.
De la différentiation des fonctions transcendantes (de 23 à 38)	6.
De la différentiation des Équations quelconques à deux variables (de 38 à 47)	5.
Recherche des Maxima et des Minima des fonctions d'une seule variable, (de 47 à 52)	2.
Des valeurs que prennent dans certains cas les différentielles, et des expressions qui deviennent 0/0 (de 52 à 60)	4.
Application du Calcul différentiel à la théorie des Courbes (de 60 à 77)	6.
Recherche des points singuliers des Courbes (de 77 à 87)	2.
exemple de l'analyse d'une Courbe (de 87 à 94)	3.
Des Courbes osculatrices (de 94 à 101)	4.
Des Courbes transcendantes (de 101 à 115)	7.
Du changement de la variable indépendante (de 115 à 120.)	1.
De la différentiation des fonctions de deux ou d'un plus grand nombre de variables (de 120 à 133)	4.
Recherche des minima et des maxima des fonctions à deux variables, (de 133 à 137)	2.
Calcul intégral	
Intégration des fonctions rationnelles d'une seule variable (de 145 à 160)	4.
De l'intégration des fonctions irrationnelles (de 160 à 169)	2.
De l'intégration des différentielles binomes (de 169 à 175)	1.
De l'intégration par les Séries (de 175 à 181)	1.
De la quadrature des Courbes, de leur rectification; de la quadrature des Surfaces Courbes, et de l'évaluation des volumes qu'elles comprennent (de 222 à 243)	2.
Total du nombre des leçons	67.
Trois interrogations de Révision (M. Ampère)	
Une interrogation générale par M. Lacroix	

Plus 5. Leçons de Récapitulation après le cours de mécanique. /.

Paris le 25 7bre 1808. LeBrun C. S. V. P.

Figure 15: Lacroix's 1806 curriculum as a guideline for Ampère in 1808.

46 Archives de l'Académie des Sciences, Paris, *Nachlass* Ampère, carton 5, chap. 4, chemise 100.

quite impressively how canonically analysis was taught at the institute (see the reproduction of LeBrun's course program in Figure 15). The only changes LeBrun requested were to teach the course in only 55 lecture hours compared with the 67 that had been available to Lacroix. Ampère used this curriculum to jot down his own notes listing his priorities and approaches. The first note states that the course was constructed on the basis of the limit method in line with the general program. Ampère comments that this method is rigorous if one provides a good proof for how close one can get to the limit, showing how he has been thinking about foundations:

> 1°. based on the theory of limits by the program. This method is rigorous when one well proves the proximity [?][47]

The innovation is to be found in Ampère's second comment. Apart from the familiar theorems on limits, which he therefore mentions in only an abbreviated form, he proposes and proves a further one. Both the formulation and the proof of this theorem now use the *lim* sign.[48] He also introduces a sign for "between" or "close to between":

> 2°. besides the known theorems about limits: 2 limits of the same quantity are equal, etc., establish this:
>
> If X always between V_1 and V_2 which have as limit $= V$
>
> one also has $\lim X = V$.
>
> Since X between V_1 and V_2 this yields $X - V$ between $V_1 - V$ and $V_2 - V$, these two become as small as one wishes upon approaching the limit, $X - V_1$ will become it, too, hence Lim. $X = V$. q.e.d.[49]

Ampère follows this predominantly verbal proof by sketching a further, algebraized version, introducing a new sign for the "between" relation:

> write thus

$$X \propto \begin{matrix} V_1 \\ V_2 \end{matrix}$$

> therefore

$$X - V \propto \begin{matrix} V_1 - V \\ V_2 - V \end{matrix}$$

hence Lim X = V.[50]

---

47   Ibid. As this comment was entered afterwards, Ampère squeezed the last word into the margin, making it very hard to read. There was not enough room for him to explain what relation he meant.

48   The first author to mention this inclusion rule in Ampère is Grattan-Guinness (1990a, 199).

49   Ibid. It can be seen that the proof starts off by being predominantly verbal; it is in no way as algebraized and furnished with sequence indices as Grattan-Guinness suggests (1990a, 199).

50   Ampère's *Nachlass*, as above. In the central part, Ampère forgot to insert the "V" on the right-hand side.

The majority of the remaining preparatory notes address concepts on integral calculus. Once again, we find new approaches. One of these presents and discusses the definite integral for the first time in an independent and prominent way.

Lacroix, in contrast, assigns only a few lines to the definite integral in both his 1798 handbook and his 1802 textbook. This is at the end of his introductory chapter on integral calculus, and he does not give it any systematic status (Lacroix 1798, 142; 1802, 294). Ampère, in contrast, introduces the integral straight away by contrasting the two forms *intégrale indéfinie* and *intégrale définie*.

As a *règle générale*, Ampère proposes that for the integral of a function from $x = a$ to $x = b$, one always has to subtract the first value from the second. Evidently, a detailed presentation of the definite integral as part of the foundations was rather uncommon at this time. In any case, Ampère feels obliged to point out expressly that the integral from a to b is not the same as that from b to a, but that they exhibit *signe contraire*.

New elements can also be found in the way he applies Binet's method of nesting intervals to determine integral values. He starts by discussing how to proceed when no closed formula can be found directly for the integration of a function. As an approximation method, he follows the approach in Lacroix's textbook (and, thereby, Euler's integral calculus) of integrating the terms of the development of the function into series as long as the series are *assez convergente*. If this is not the case, Ampère proposes performing an interval nesting in the interval $[a, b]$ with $h = na$ and $b = a + h$. Then, the definite integrals should be determined for the equidistant subintervals with a distance of a: from a to $a + a$, from $a + a$ to $a + 2a$, ..., from $a + (n - 1)a$ to $a + na = b$.

Ampère now points out that the limit has to be determined for two limit processes: for arbitrarily diminishing a and for n becoming arbitrarily large. Although Ampère generally applies the approach presented in the second edition of Lacroix's textbook for increasing the convergence of the series by diminishing the breadth of the intervals—in his notes, Ampère even gives the page and paragraph number from this edition—he diverges from Lacroix in that he does not present an indeterminate sequence a to $a + a$, $a + a$ to $a + 2a$, and so forth (Lacroix 1806, 303), but demonstrates in the above way—following Binet—a clear operationalization of interval divisions.[51]

2.4.4. PRONY: AN ENGINEER AS A WORKER ON FOUNDATIONS

Impulses for the development of concepts in analysis at the *École polytechnique* also came from physics. Remarkably, it was once more reflections over hard–body impact that led physics to focus on the continuity concept as well, making

---

[51]   From his many instructive contributions to foundations, I shall also address his reflections on negative numbers in Chapter V.3.2.

ideas on continuity a common concern in both the analysis and mechanics courses. The main conceptual contributor was, once again, Prony, who has already appeared in this book as an enthusiastic early campaigner for the analytic method (Chapter IV.2.1.). In 1810, Prony published the major work of this program that continued to be oriented toward the clarification of basic concepts: his *Mécanique Philosophique*. Based on his mechanics courses taught since the *École* was founded, his goal was to present a *tableau méthodique* of the main developments, dropping the distracting parts of proofs and intermediate calculations, so that *l'esprit des methodes* could stand out clearly. Prony viewed this as a part of the analytic program, and set this concept of a "philosophical mechanics" alongside the program of mathematics as a language and chemistry as a language (Prony 1800, i).

The methodological orientation and reforming goal can already be seen in his innovative presentation: One-half of the text is organized in *tableaux*, in tables on each right-hand page; whereas the left-hand page gives the continuous explanatory text to these tables (see Figure 16). Each tableau contains four columns covering:

1. An interpretation of the signs in the formulas.
2. A list of the definitions for the concepts used in the text.
3. A presentation of the accompanying theorems.
4. The problems to which they are applied.

It is very likely that it was this methodological arrangement in *tableaux* that encouraged Carnot to develop an adaptation of this presentation for his *Géométrie de Position* published in 1803.

This is not the place to analyze Prony's concepts in mechanics, and I shall limit myself to his introduction of continuity. He considered that it is not necessary to know the nature of the forces in mechanics. Instead, one uses *calcul* alone to represent, measure, and introduce them by the effects they elicit. These effects always reduce themselves to the velocities that the forces are trying to generate or that they give to actually determined masses (ibid., 18). Prony declares that the law of continuity possesses universal validity in mechanics; and he explains this in terms of the intermediate value property: "The law of *continuity* is constantly observed here. A body does not pass from one speed to another without having [passed through] all the intermediate speeds" (ibid., 20).

The surprising aspect is that Prony—almost even more dogmatically than the traditional Leibnizians and without any consideration of the earlier differentiations introduced by Kästner and Karsten—claims an absolute validity for the law of continuity in nature and dismisses all contrary statements as being a consequence of limited human perceptibility:

> However the transition is often so rapid that it seems to be discontinuous, and even in mechanics one considers *abrupt* changes, or variations in speed which are *instantaneous* and *finite*: but these are the limits which the mind conceives and to which the real phenomena approach ever more closely as the time corresponding to an increase or decrease of speed becomes ever shorter (ibid.).

Although Prony was an engineer, he does not assume that the applications—and here he means, above all, the natural processes—are more varied than their modelings in theory. Instead, he reduces the applications to the breadth of the

NOTATION.	DÉFINITIONS.	THÉORÈMES.	PROBLÈMES.
$\varphi =$ la force accélératrice. $v =$ la vitesse. e, τ et φ ont la même signification qu'aux art. 22, 23 et 24.		**11.** La force accélératrice, pour un instant quelconque, est égale à la vitesse que le mobile acquerrait pendant l'unité de temps, si le rapport $\frac{dv}{dt}$ demeurait constant pendant cette unité. **12.** Si dans l'équation $e = f(t)$ on suppose que t augmente de τ, et qu'on développe $f(t+\tau)$, ... les coefficiens de τ, et ... et le double du coefficient de τ^2 seront respectivement l'espace parcouru, la vitesse, et la force accélératrice au bout du temps t.	**19.** Trouver à quelque instant la force accélératrice du mouvement qui a $v = f(t)$ pour équation, ... égale à la force de la pesanteur; problème qui est résolu par l'équation $f'(t) = g.$
10. De ce qu'on entend par force ou puissance, et de ce qui doit la représenter dans le calcul d'après les divers effets de cette cause inconnue que nous observons dans la nature.		**13.** Un corps tombant dans le vide, se meut d'un mouvement uniformément accéléré, dont la force accélératrice est ... 9,809, ou 7,322 mètres, suivant qu'on prend pour unité de temps la seconde sexagésimale, ou ... la 100000.ᵉ partie du jour.	**20.** Trouver les équations des mouvemens, dans lesquels les différences 3.ᵉ, &c. des espaces parcourus sont constantes, ... mouvemens de même espèce, qui, parmi tous les mouvemens de ... $e = f(t)$.
La force accélératrice de la pesanteur terrestre, au niveau de la mer, sera constamment représentée par la lettre g dans tout le cours de cet ouvrage.	**14.** De la pression et de son évaluation.		

. C 2

est uniforme, et changeant d'un moment à l'autre, lorsque la variation est quelconque.

25. LES formules fondamentales d'un mouvement quelconque, considéré uniquement quant à la relation entre les temps et les longueurs des espaces parcourus, sont, en observant que $\frac{d^2e}{dt^2} = \frac{dv}{dt}$,

$$e = f(t); \quad v = \frac{de}{dt}; \quad \varphi = \frac{d^2e}{dt^2}; \quad \varphi = \frac{dv}{dt}.$$

26. LE mouvement uniforme sert de mesure au mouvement uniformément varié, art. 19; et comme ce dernier sert de mesure au mouvement varié en général, art. 24, c'est, ultérieurement, le mouvement uniforme qui fournit le moyen de comparer tous les autres mouvemens entre eux.

27. SI l'on fait entrer dans les valeurs de e, données art. 23, les 4.ᵉ, 5.ᵉ, &c. termes du développement de $f(t + \tau)$, on aura les mouvemens à différences 3.ᵉ, 4.ᵉ, &c. constantes, qui parmi tous ceux de leurs espèces respectives, et pendant un temps fini, après ou avant un instant donné, approchent le plus du mouvement qui a $e = f(t)$ pour équation.

Toute la théorie précédente n'est que celle des osculations de différens ordres, appliquées à des considérations relatives au mouvement.

28. NOUS pouvons maintenant ajouter aux idées abstraites du temps et de la mobilité, celles de la *force* ou *puissance*, de la *masse*, de l'*impénétrabilité* et de l'*inertie*, en faisant encore abstraction de l'étendue et de la figure.

La nature de cette cause de mouvement, nommée *force* ou *puissance*, nous est tout-à-fait inconnue: l'homme appelle *force* la faculté organique qu'il a de se mouvoir, de s'arrêter, de produire ou de faire cesser le mouvement des corps qui l'environnent; et sans savoir en quoi consiste cette difficulté, il a supposé qu'il existait quelque chose de semblable dans les agens physiques qui sont ou qu'il croit être, sur le globe terrestre et dans l'univers, les causes du mouvement de différens corps.

Mais nous n'avons, en mécanique, aucun besoin de connaître la nature de la *force* ou *puissance* qui est représentée, mesurée et introduite, dans le calcul, uniquement par les *effets* qu'elle produit. Ces effets se réduisent toujours à des vitesses

Figure 16: A typical *tableaux* double page from Prony's (1800) *Mécanique Philosophique*.

theory known at that time. He avoids any consideration of not "continuous," that is, not differentiable functions, by claiming a priori that all processes are inherently continuous.

Prony's identification of application and theory was in line with the contemporary epistemological consensus on the principal "continuity," that is, differentiability, of the functions known and used at the time. In 1815, Prony published a revised edition of his 1800 work—now without the *tableaux*—presenting his conception of continuity even more explicitly. This revised edition not only delivers his conception in a more radical form but also introduces an innovative change. While reaffirming that discontinuous processes in nature only appear to be so but are really continuous, he simultaneously discusses with increasing intensity the "hypothetical" possibility of discontinuous processes, and uses this to prepare the reader for a consideration of both continuous and discontinuous functions. Prony's basic problem continues to be determined by the discussion on the impact of hard bodies. The instantaneous generation of a finite velocity observed in this field also does not exist in nature, where all motions in general are continuous. However, seemingly contradictory phenomena do not violate the law of continuity; it is merely impossible for us to detect the continuous transitions with our human senses.

Whereas Prony starts off by reaffirming the epistemological basic consensus on general continuity as in the 1800 edition, he then goes one step further. He declares it to be legitimate to work with the violation of continuity as a *hypothesis*, and to treat it mathematically as an extreme even when it is not confirmed in nature:

> I shall no less resolve important questions of dynamics in the hypothesis generated in such a way, and it is not the only circumstance where, in order to prepare oneself with advantage for applications to physics, one should first treat cases which are *extreme* or abstract, which never rigorously occur in nature (ibid., 30).

This ontological reversal of recognizing discontinuities as a hypothesis, as a modeling in mathematics but not in nature, suggests that Prony's *Mécanique* of 1815 is the first work to consider both continuous and discontinuous motions, based on this hypothetical assumption. Prony goes on to repeatedly present his new concept that it is necessary to address discontinuous actions even when such forces do not exist in nature: "it is necessary even to deal with the hypothesis of the nonexistence of such forces, with *discontinuous* forces, in nature" (ibid., 97)

He considers that although this *discontinuité* is apparent in the phenomena only because of the imperfection of our senses, one can view continuous motion as *le cas limite* of discontinuous actions (ibid.). He illustrates this transition from "discontinuous" to "continuous" actions with the transition from the *polygone funiculaire* to the limit case of a "continuous" curve (Prony 1815, Vol. 1, 184 ff.). This seems to be a clear adoption of the reflections and conceptions on continuity that Kästner developed for the impact of hard bodies.

Prony's book stands out not only through its explicit introduction of *discontinuité* but also through the way it repeatedly discusses continuous and discontinuous situations throughout the text. He finally systematizes this discussion by summarizing it into three options for arbitrary functions: they can be *continues*, *discontinues*, or also *discontigues* (Prony 1815, Vol. 2, 490). This terminology is not new; Prony has drawn on Arbogast's concepts of 1791 (see Chapter III.9). However, he has gone on to actually apply them, and, now, in 1815, to no longer rely exclusively on continuity as being self-evident.

2.4.5. LAGRANGE'S CONVERSION TO THE INFINIMENT PETITS?

We have now outlined the background to the increasingly broad reflection on basic concepts in analysis and the development of increasingly new and more precise contributions within the context of the *École polytechnique*. Before proceeding to the study of the effects of the radical resolution to return to the *infiniment petits* in 1811, we first have to clarify something that seems rather erratic: Is Lagrange's program of rejecting all forms of the infinite in itself contradictory? Or did he personally change or drop his program? It is a known fact that Lagrange exercised a broad international and intensive influence through his algebraization program: the subtitle of his famous function theory, *dégagés de toute considération d'infiniment petits, d'évanouissantes, de limites et de fluxions, et réduits à l'analyse algébrique des quantités finies*, was highly attractive far beyond the world of professional mathematicians. Even Lacroix, in the first volume of the second edition of his handbook in 1810, had still pointed out that Lagrange wanted to exclude the *infiniment petits* not from differential calculus alone but particularly from applications in geometry and mechanics as well:

> In his Theory of analytic functions [...] he deals with it for the first time and endeavors to remove from not only the differential calculus itself, but also its application to geometry and mechanics, all ideas of infinitely small [quantities] and of limits (Lacroix 1810, 242).

However, in the second, 1811, edition of his *Mécanique analytique*, Lagrange (1811, IV) himself declares that the *système des infiniment petits* had been "adopted" into this *Traité*. How can we explain this contradiction, a contradiction that does not seem to have been discussed up till now.[52]

One might initially assume that Lagrange restricts his program to differential and integral calculus in a narrower sense and is not as consistent or comprehensive as Lacroix (1810) claims. However, both the *Théorie des fonctions analytiques* and the *Leçons sur le calcul des fonctions* show that Lagrange worked out a comprehensive and consistent opus. Both books present the application of his conception of analysis to geometry and mechanics in a

---

[52]   Although Bottazzini (1992, XLV f.) mentions this contradiction, he does not study it further because he views infinitesimals and differentials as an expression of one and the same conception.

comprehensive way. For example, his function theory consists of three parts: The first presents the theory with *ses principaux usages dans l'analyse*, the second applies function theory to geometry, and the third applies it to mechanics. At the end of this third part, Lagrange declares with some satisfaction that with the help of his function theory, he has managed to formulate those principles and basic equations of mechanics whose proofs normally require infinitely small quantities:

> Since my intention is not to provide a Treatise of that science, I shall content myself with having established, by the theory of functions, the principles of the fundamental equations of Mechanics which can only be normally demonstrated by consideration of infinitely small [quantities] (Lagrange 1881, 411).

Although his later lectures on the *calcul des fonctions* concentrate on extending function theory itself, his ninth lecture deals with applications to geometry and mechanics. In this lecture, Lagrange clearly emphasizes that a rigorous application of the theory of derivatives is possible for those parts of geometry and mechanics that use differential calculus: "It is on these principles that is founded the rigorous application of the theory of derived functions to Geometry and Mechanics, for which one employs the differential calculus" (Lagrange 1884, 101).

Because Lagrange wanted to establish a program within mathematics encompassing not only analysis but also geometry and mechanics, another possible explanation for the contradiction is that the procedure he follows in mechanics is not coordinated or consistent with that in mathematics. This idea would make it particularly necessary to examine how far the mathematical conception of differential calculus agrees in the two editions of the *Mécanique*. Such a comparison needs to focus only on the first volume of the second edition, because the second volume was revised only in part by Lagrange himself, and more than half of it was simply taken over from the first edition.

The first edition of Lagrange's *Mécanique* appeared in 1788, that is, before the Revolution and hence also before the period dominated by algebraization in the *méthode d'analyse*. Nonetheless, even in this first edition, Lagrange speaks out against the use of approximations via infinitely small distances and in favor of the direct use of the modern methods of differential calculus. He points out that the traditional method of solving mechanical problems addressing bodies of finite mass is to start off with the laws of equilibrium or motion for a definite number of points at finite distances, and then to extend the study to an infinite number of points. Because this renders the distances between the points infinitely small, he argues, one has to perform the reduction and modification required by the transition from the finite to the infinite. This procedure in mechanics is analogous to the geometric and computational methods available prior to differential calculus. However, the latter calculus is so simple because it views curves directly as curves and not initially as polygons that then go on to be treated as curves. By using similarly direct methods, mechanics would also attain simplicity and generality (Lagrange 1788, 50 f.).

In his preface to this work, Lagrange does not just make the famous announcement that he will present mechanics without any need to rely on the usual figures. He also declares programmatically that his methods require neither geometrical nor mechanical justifications but exclusively algebraic ones. This turns mechanics into a new domain of the *analyse*, and in contemporary terms, this means algebra, and he points out that its disciples will be grateful to him for the extension of their domain: "Those who love Analysis will be pleased to see Mechanics becoming a new branch, and know my delight in having thus extended its domain" (ibid., vj).

Indeed, mechanics is then presented in purely algebraic terms, although analysis is applied with differential quotients and not yet with development into series. It is not based, in contrast, on infinitely small quantities and the comparison of such quantities of different order.

It is only at the end of the work, when discussing motions in vessels in the chapter on hydrodynamics, that Lagrange mentions the concept of comparing quantities of different order. But he does this with the express indication that dropping higher-order quantities leads only to an initial approximation, and that such neglected quantities will have to be considered in further approximations (ibid., 487). Outside the context of differential calculus, Lagrange occasionally uses the terminology of infinitely small quantities of different order, for example, when determining the order of magnitude of angles for curves of double curvature (ibid., 101 f.).

In the second edition, which goes into more detail, Lagrange maintains his algebraic claim (Lagrange 1869, III f.) along with the same basic concepts in the use of differential and integral calculus. He does not use the conception of infinitely small numbers as a base here either. He repeats his statement on the advantage of viewing the curves themselves instead of approximating polygons (ibid., 74 ff.). At one point in the second edition where he discusses the neglect of the higher-order *infiniment petits* in order to attain a certain value—when discussing the equilibrium of incompressible fluids—Lagrange points out immediately that although such a hypothesis is legitimate, one should not accept it without proof, in order to avoid introducing deficits in the rigor of the formulas (ibid., 181).

The remark in the preface mentioned at the beginning of this section on being based on the system of the *infiniment petits* should therefore not be taken literally. Initially, it provides a justification for a completely different content. The first statement addresses the notation used: "The ordinary notation of differential Calculus has been maintained because it corresponds to the system of infinitely small [quantities] adopted in this Treatise" (ibid., IV).

Indeed, Lagrange had to explain that he did not base his mechanics on the same conception of analysis as function theory that he had used in his mathematical works from 1795 onward. This was not just because a revision of mechanics would have been a lot of work. What was more decisive was that the new form of notation he had introduced in his function theory was almost completely rejected. Lacroix did not adopt it in either his handbook or his textbook, and even Crelle, a passionate advocate of Lagrange's algebraization program, transcribed Lagrange's notation into the Leibnizian form when translating his lectures and function theory.

Hence, when Lagrange continues to apply *notation ordinaire* in the second edition as he had in the first, in other words, the established notation for differentials, differential quotients, and differential equations, his textbook practice shows that this does not in any way imply that he bases his work on the concept of the *infiniment petits*—just like Lacroix, who also used differentials in another sense while rejecting the *infiniment petits*. Indeed, a glance at the continuation of the sentence quoted above shows that Lagrange has in no way become a supporter of the *infiniment petits*, and still does not rate them as a rigorous method but as a shortcut—whose precision has to be proved in other ways. He names his own method as one such way, and introduces, for the first time, the limit method, but in its original geometric form, in Newtonian terms:

> When one has well understood the spirit of this system, and one is convinced of the accuracy of its results by the geometrical method of first and last ratios, or by the analytical method of derived functions, infinitely small quantities can be employed as a sure and convenient instrument to shorten and simplify proofs. Thus can the demonstrations of the Ancients be shortened by the method of indivisibles (ibid.).

In summary, it can be seen that Lagrange's statement on "adopting" the concept of the *infiniment petits* applies to neither the second nor the first edition of his *Mécanique*. With marked reservations, he seems to have attributed something to his work in response to the pressure of external expectations. We can identify this external pressure through the present study, particularly in Chapter IV, of developments at the leading French teaching institution for mathematics, the *École polytechnique*. The publication date of the new edition of Lagrange's *Mécanique* also suggests such a connection. It was a common practice at the *Institut* for members to present their new publications to their classes. The session protocols document that Lagrange presented the first volume of his revision on September 16, 1811: "Lagrange presents to the class the first volume of the new edition of his revised and augmented analytical Mechanics" (Institut de France 1913, Vol. 4, 534).

The printing of the new edition was finished two months after the fundamental reversal at the *École polytechnique*: the resolution by the *Conseil de perfectionnement* on July 13, 1811, declaring that the analysis course should now be based on the method of the *infiniment petits*—a trend already evident in the commission's deliberations in June (see Chapter IV.3.6.). As the report on this resolution to the Emperor Napoleon (see ibid.) shows, just as much as the subsequent debates within and outside the *École* (see the next section), there was strong public pressure to return to the synthetic method and the "intuitive" nonalgebraic concept of the *infiniment petits*. Lagrange, himself a member of the *Conseil de perfectionnement*, was unable to evade this pressure, the *École polytechnique* formed the main public for his textbooks. He accordingly altered the foreword during printing and introduced a number of explicit usages of the terminology of the *infiniment petits* in the text in order to express a compatibility between his textbook and the teaching program at the *École*.

There can be no doubt that Lagrange's declaration in the preface of the new *Mécanique* served as an important public and lasting legitimization of the resolution on the change of methods. Indeed, the statement that it is impossible to avoid using the *infiniment petits* in mechanics formed the main justification for the reversal (see Chapter IV. 3.). This argument was also used in the report to the Emperor of May 1812 (ibid.). And in the last, 1813, edition of his *Reflections on Infinitesimal Calculus*, Carnot also cites Lagrange's "adoption" of the *infiniment petits* in mechanics in order to back up his own new preference for them (see above).

2.5. The Impact of the Return to the Infiniment Petits

2.5.1. IMPACT INSIDE THE ÉCOLE: THE "DUALISM" COMPROMISE

For the *Conseil de Perfectionnement* at the *École polytechnique*, reversing the prior algebraization program and declaring the *infiniment petits* to be the foundation for the analysis seemed to be a blow for freedom. However, simply passing this resolution in July 1811 was not the same as implementing it, particularly among a staff largely committed to teaching the algebraization concept. Indeed, the council resolution was followed by considerable "grass-roots" resistance. One effective focus was the textbook issue, because it soon became apparent that none were available for the new method, whereas Lacroix's book was firmly anchored in everyday teaching.

Although Lacroix had not been an *instituteur* since 1808, an exceptional ruling in 1810 had formally confirmed his *Traité élémentaire* as an *École* textbook. The reason for this ruling was that the *Conseil de Perfectionnement* had been continuously urging instructors to publish their lectures—or, at least, a *sommaire*—so that students would have reliable course literature at their disposal; in 1810, they wanted to see this finally put into practice. A commission had been set up to select *livres classiques* for each subject. *Livre classique* was the proposed title for the teaching materials underlying a specific teaching course; the idea being that each student should be handed a copy upon admittance to the *École*.[53]

---

[53]   In 1806, in contrast, the *Conseil*—despite urgently calling for the publication of teaching texts—had still ruled against the appendage *livre classique* for such printed versions. It considered this formulation to be acceptable for the *lycées*, but not for an École oriented toward scientific progress (Gilain 1989, 6).

As the commission's report to the *Conseil* on November 29, 1811, revealed, no printed texts from current instructors were available for, among others, *analyse algébrique*, physics, or chemistry. Therefore, instructors were urgently requested to fill this gap. In the meantime it should be filled by books from other authors, however, not for general distribution, but as two copies set out for common reference. For analysis, the commission suggested setting out Lacroix's *Traité élémentaire*. This suggestion triggered a long discussion, with one member proposing that Lacroix's textbook should be designated not only for public use as a common reference but also as a general "classic" textbook for analysis. Eventually, this proposal was approved as an exception to the rule under the condition that the analysis instructors accept this book as the text for their lectures. Further exceptions for other subjects were expressly rejected.[54]

For the new academic year 1811/12, Lacroix's textbook could no longer be retained as a "classic," because its fundamental principles—the *méthode des limites*—ran counter to the new resolution. Resistance within the teaching staff took the form of agitation in favor of its retention. An internal committee, the *Conseil d'Instruction*, met after the vacations on November 15, 1811, and proposed keeping Lacroix's book as a common reference in the public domain because of the lack of textbooks teaching infinitesimal calculus with the *infiniment petits* and the fact that the book provided a number of good exercises for students. However, due to external pressure, the *Conseil d'Instruction* was reluctant to propose this resolution alone, and referred it to the *Conseil de Perfectionnement*.[55] In its next session, on November 22, the *directeur des études* presented a detailed plea. This revealed that in common sense or in a surrogate understanding, the method of the *infiniment petits* means quite simply the use of differentials. Hence, there was no problem in using Lacroix's textbook for integral calculus (which is thus viewed as an application of differentials and not of *limites*). The seriousness of this issue is revealed in the fact that the *Conseil* postponed any immediate decision and referred the issue to the mathematics commission for consultation.[56]

The speaker of the mathematics program commission was no less a person than Laplace. At the following *Conseil* session on December 6, he presented the outcome of the consultations on this central textbook issue: until a good textbook teaching differential calculus according to the *infiniment petits* method became available, two copies of Lacroix's textbook should be laid out for public reference. After such intensive consultation, the *Conseil* then approved this limited retention of Lacroix's textbook.[57]

This was followed immediately by a personal statement from Ampère, himself a member of the *Conseil* and still an *instituteur* for analysis. It shows that he was in no way an advocate of the *infiniment petits*. He announced that in

---

54   AEP Procès-Verbaux du Conseil de perfectionnement, Vol. 3. fol. 103.
55   AEP Procès Verbaux du Conseil d'instruction, Vol. 6. fol. 2.
56   AEP Procès-Verbaux du Conseil de perfectionnement,Vol. 3. fol. 133 f.
57   Ibid., fol. 134.

compliance with the *Conseil* ruling of 1810, he had produced a draft of his lectures. However, he had had to cancel its publication because of the resolution on the change in methods. He intended to revise his text in line with the approved method and hoped to send this to the printers within the next 12 months. However, no such text was published, and I was unable to find any manuscript for a revised version in his *Nachlass*.[58]

None of the analysis instructors at the *École* published a treatment of differential calculus based on the *infiniment petits* during this period. Only one of the physics instructors, S.-D. Poisson, later advocated an "imperialism" of the *infiniment petits* (see Chapter VIII 2.1.); and only one graduate from this period (of 1814), J.M.C. Duhamel, eventually went on to publish a textbook based consistently on the *infiniment petits* in 1856. It was exclusively mathematicians from outside the *École* who published such adapted textbooks directly following the resolution of July 1811 (see below). One reason for the lack of an immediate internal adaptation was that internal resistance led to a compromise solution, a kind of dualism, whereas external mathematicians took the resolution seriously, anticipating that they could increase their potential influence on the *École polytechnique*.

It is not surprising that the resistance from the instructors was manifested in the internal advisory board of the *École*, the *Conseil d'Instruction*. It held its next meeting on August 9, 1811, shortly after the resolution by the main *Conseil de Perfectionnement* on July 13. This session succeeded in finding a way of "interpreting" the resolution. It concluded that the *Conseil de Perfectionnement* had resolved on changes in the teaching program for analysis, but had not presented a definitive version of the new program. Indeed, the teaching program for each single course was specified annually in a formal procedure. To correct this deficit, the *Conseil d'Instruction* set up its own commission composed of the two current analysis instructors Ampère and Poinsot; the long-time analysis instructor, committed algebraizer, and current instructor of mechanics Prony; and, finally, Poisson.[59]

The outcome of this commission was published as the new *Programme d'analyse*, and contradicted the intended exclusiveness of the *infiniment petits*. It proposed a compromise permitting parallel presentations of the limit method and Lagrange's method of developing functions into series. The teaching task was formulated as follows:

---

[58]   I had thought that the following long-missing document entered in the *Nachlass* catalogue (Académie des sciences, Archives: Papiers d'Ampère; IV, 76 Ancien cours d'analyse algébrique à l'ècole Polytechnique) might have been this revision. However, when I eventually traced it, this proved to be a text from 1817/18 (see below, Chapter VI. 6.6.).

[59]   AEP Procès-Verbaux des séances du Conseil d'instruction, Vol. 5., August 9, 1811.

To present the principles of the differential calculus by consideration of *les infiniment petits*; to show, for the simplest cases, the agreement of this method with those of limits or series expansions.[60]

The eclecticism unjustly attributed to Lacroix's major handbook had now become institutionalized. The compromise of a dualism of methods would now determine the development of analysis at the *École polytechnique* and beyond its walls throughout France.

A characteristic insight into this practice is given by the *Registres d'instruction* for the analysis course. These registers, kept by the *École* since 1800, are an informative source first noted and analyzed by Gilian (1984, 1989, 32) that confirm the school-like style of instruction. Every instructor had to enter a short summary of the contents of each lecture.

Analysis was taught in two successive two-year courses given alternately by two *instituteurs* in succession: the first course for the "freshman" year (known as the "second division"), and the second, for the advanced year (the "first division"). The "compromise" was first implemented in the second *division* of 1812/13 by A.A.L. Reynaud (1777–1844), himself a graduate of the *École* in 1796, standing in temporarily for Labey. Differential calculus was first introduced in the fifteenth lecture, in determining the tangents to a curve in geometry. Then, a special problem (tangents to the circle) was used to present a total of *four* methods in parallel, and in a further step, this was even followed by a fifth method as well:

16. To determine the tangent to a circle, 1.) by the method of limits, 2.) by the means of functions that reduce to $\frac{0}{0}$, 3.) by the method of indeterminates, 4.) by the method of infinitely small quantities.

17. Notation adopted in the differential calculus: differentiation of several functions using series.

The general treatment of the differentiation applied two methods, with the first remaining unnamed and the *infiniment petits* once more being presented last:

18. Differentiation of algebraic functions.

19. The same matter treated by the method of infinitely small quantities.

However, this multiplication of methods must have seemed impractical. The next course of 1814/15, given partly by Reynaud and partly by Louis Poinsot (1777-1859), appears to have dropped the geometric introduction and reduced the number of methods:

11. The principles of the differential Calculus.

12. Completion of the preceding lesson and application to simple cases.

13. Application of the given principles to the differentiation of algebraic and transcendental functions.

14. Differentiation of functions of x by the means of series and by the method of infinitely small quantities.[61]

---

60   AEP Bibliothèque, $X^{2b}10$: Programmes d'enseignement, Box 1810–11.
61   Ibid., 1814–15.

The following lecture reintroduced a geometric method in the Leibnizian style. Its content was given as "differentiation of the trigonometric functions by means of the *triangle différentiel*.

These examples of the eclecticism of methods are taken from just one of the analysis courses. What happened in the second cycle of lectures offset by one year? The corresponding part of this course in which the resolution on foundations should have been practiced for the first time was given in the year 1811/12. Its *instituteur* was Ampère. As the entries on his lectures reveal, he made no external concessions in favor of the *infiniment petits*.

With hardly any structural changes compared with his prior course in 1809–10, Ampère started the 1811–12 part of the course on differential calculus with *Notions générales des fonctions et du Calcul différentiel*, without emphasizing one method either at this point or in the following text. The rapid introduction of Taylor's theorem and its application in the development into series shows, however, that he proceeded predominantly from Lagrange's perspective. In the next lecture cycle of 1813/14 as well, Ampère expressed himself only generally, starting with an *Exposition des principes du Calcul différentiel*.[62]

Indeed, Ampère formed the most stubborn focus of resistance to the introduction of the *infiniment petits*. Specifically for his analysis lectures of 1811/12, we possess one reliable document: notes taken by the student Théodore Olivier, who later went on to found the *École Centrale des Arts et Manufactures*. One reason why these can be considered more reliable that many other notes is that Ampère dictated particularly important sections to his students, and Olivier marked these sections with *sous la dictée de Mr Ampère*. One of these marked sections is the important introduction on the *Calcul différentiel* with the *Notions sur les fonctions* (cf. Section 2.4.3.). The important thing to note here is that although Ampère no longer introduces and applies the *limites* explicitly, he also does not introduce the *infiniment petits*.[63]

2.5.2. THE IMPACT OUTSIDE THE ÉCOLE

It is not just the reversal brought about by the resolution of July 1811 that is documented only marginally in the historiography of mathematics. Even less research has been carried out on the reaction in the broader French mathematical community. I shall restrict myself to two of the first textbooks to implement the resolution—those by Garnier and Boucharlat—as well as an extremely informative article written by Servois in 1814.

Garnier was the first to produce a textbook adapted to the resolution, and he published it already in 1811. He was himself a former analysis instructor at the

---

62   Ibid., 1809–10, 1811–12, 1813–14.

63   AEP Bibliothèque, A II a, 56: Ampère, Cours de calcul différentiel et intégral données à l'École Polytechnique, années 1811–12–13. Notes prises par L.-Th. Olivier: 13e leçon.

École, where he had been actively involved in developing the limit concept. Now, however, he was probably working as a private teacher. His book, entitled *Leçons de Calcul Différentiel*, was announced as a third edition to follow his revised edition of Bézout in 1800 and the version for the internal use of students at the *École polytechnique* printed in 1801/02 (see above, Section 2.4.3.). The 1811 version differed completely from this second edition that had been based on the limit method. It was not only a large work (474 pages) but also presented *three* methods of differential calculus. The main method followed Lagrange by means of series development. Indeed, Garnier closely follows Lagrange's text on the theory of analytical functions, although he transcribes Lagrange's notation for derivative functions into the traditional notation of differentials or differential quotients, just as Crelle did in his later Lagrange translations.

At the end of his preface, Garnier declares that he also considers it to be meaningful to present the *méthode infinitésimale* as well as the method of *limites*: the former—completely in line with the phrasing of the resolution of the *Conseil de Perfectionnement*—because it is used pervasively in mechanics and all applied fields. He remarks that he does not need to justify this in detail, because of the admirable—and from him, much praised—*Réflexions* from Carnot. The limit method, in contrast, is the one with which the true differential calculus is normally presented (Garnier 1811, xix f.).

These two supplements to the methods are presented as two appendices, each approximately 20 pages long. Because this text was produced so quickly, they are not well written. However, the text on the *infiniment petits* is highly informative in terms of a reconceptualization in France following the dominance of the analytic method. The first appendix on methods is entitled "Abridgement through the *infiniment petits*. Garnier declares that he has presented the foundations of differential calculus in a purely analytic way in the main text, independent from any hypothesis over the value and nature of increases in the functions. He attributes the method of the *infiniment petits* to Leibniz, again following the French tradition. Continuing to use the phrasing of the resolution of 1811, he declares it to only seem to be less rigorous, but, in fact, very useful in applications. He lists the basic principles of these methods: the increases are infinitely small; they can be neglected compared with finite quantities, thus validating not only the generalized relation of equality but also the principle of homogeneity from which the rules for comparing the orders of the infinitesimals follow (ibid., 439 f.).

These few principles lead to abridgments and simplifications in the procedures for proofs. His first example is a proof for Taylor's theorem: by assuming dx to be infinitely small and setting $n = \infty$, he views ndx as a finite interval. Through corresponding operations, he derives the desired theorem in only one page (ibid., 440 f.). It is notable that in this application, Garnier views the *infiniment petits* as constant quantities and not as variables. Accordingly, he manages to determine the derivatives of the trigonometric functions in only two lines, and so on.

However, Garnier takes care to check each single chapter in order to ensure whether it was capable of being abridged through the *infiniment petits* or not. Remarkably, in the chapter dealing with those exceptional cases in which the development into series does not lead to the goal, he sees no possibilities for applications or abridgments. However, he does see abridgments, in contrast, when deriving functions for several variables: For infinitely small and independent increments dx and dy, the differentials exist without further assumptions (ibid., 448 f.). He particularly discusses abridgments for tangents to curves. For the radius of curvature, however, he does not want to show any abridgment, but the "agreement" between the two methods (ibid., 449 f.). In general, he notes, geometry offers a number of possibilities in determining lengths, areas, and volumes for obtaining complicated formulas *sur-le-champ*, that is, directly. The precondition in each case is to drop curved segments and replace them with straight lines, and so on (ibid., 452 ff.). What is conspicuous about Garnier's approach is that he links up again with the concept of indivisibles in these "abridgments." For example, he labels an infinitely small trapezoid on the increment of a curve as an *élement* of an area (ibid., 452).

Garnier introduces his appendix on the limit method with a lengthy quotation of Lagrange's negative evaluation of the method extending over more than one page (ibid., 457 f.). In the further text, he does not present, for example, an abridged version of his so-called second edition of these lectures. He also drops the algebraizing use of a sign for the limit. The appendix confirms in general that the *méthode des limites* was relatively underdeveloped in France at that time. It had not been able to realize its operative potential because of the dominance of Lagrange's approach of developing into series. What is more remarkable is that properties of continuity are discussed almost exclusively within the framework of this method. Indeed, Garnier emphasizes here—with yet another reverent mention of Carnot's *Réflexions*—that due to the law of continuity, vanishing quantities continue to retain a finite ratio (ibid., 465).

Finally, what is new in this section is that Garnier considers whether Taylor's theorem can lose its central justificatory status through the *limites* approach and the precondition of continuity. If one wants to start by developing the basic rules for differentiation and only then use Taylor's theorem to develop series with preordained differential coefficients, then one first has to demonstrate the existence of these differential coefficients themselves. This is given by Ampère's theorem as simplified by Binet. Garnier sketches the course of the proof—basically following Lacroix (1810)—but in his summary, he drops the idea of such an alternative path, saying that Ampère and Binet's proof is too difficult (ibid., 465 ff.).

In fact, Garnier could have gone on to offer a good way to continue and extend operating with limits, because he presents an approach to inclusion theorems using inequalities to determine the ratio between arc and sine. He includes the arc between the sine and the tangent—while leaving out all labels for variables—and because the ratio of tangent and sine is the unit at the limit, he finally obtains[64]

$$\text{limite}\left(\frac{\text{arc}}{\sin}\right) < 1 \quad \text{and} \quad > 1.$$

Hence, the limit of the ratio is included between 1 and 1, and therefore, it "becomes" a ratio of equality (ibid., 459 f.). However, he does not now go beyond predominantly verbal, nonsymbolic approaches.

Garnier's biography and works have not yet been examined in any more detail, despite his breadth of textbook publications. Together with Prony, he prepared the second edition of Part 2 of Lagrange's *Mécanique* for publication. After leaving the *École polytechnique*, he did not hold any influential posts. In 1817, his influence in France waned almost completely when he became mathematics professor at universities in southern Holland (later Belgium).

The second textbook written in response to the *retour* resolution of 1811 had an amazingly long printing history: First published in 1813, it was reprinted five times during the author's lifetime and went on to reach a ninth edition in 1891. Finally, it was even revised again more than 100 years later, in 1926! Furthermore, it was translated into German, English, Dutch, and Spanish. This repeatedly published work was the *Éléments de calcul différentiel et intégral* by Jean-Louis Boucharlat (1775–1848). Like Garnier, Boucharlat was a rather marginal figure in Parisian mathematics, although still affiliated with the *École polytechnique*. One of the students attending the first course in 1795, he also worked as a répétiteur in the analysis course from 1807 to 1810 before moving on to take a post as *professeur* at the *École militaire de la Flèche*.[65]

Boucharlat's textbook embodies the eclecticism engendered by the dualism compromise on methods at the *École polytechnique*. He presents three foundational methods, claiming that they supplement each other. These are the limit method, the theory of infinitely small quantities, and Lagrange's method. He motivates the theory of infinitely small quantities according to the *retour* resolution of 1811, declaring it to be a means of accelerating the determination of differentials. He writes that it lodges the differentials in memory by means of geometric figures of the greatest simplicity, the figures stimulate the imagination more than abstract terms. Furthermore, this theory is indispensable in the more advanced areas of mechanics and in astronomy. Some problems would be too complicated to solve without it. He claims to provide a proof for one fundamental principle of this theory from which its theorems follow, saying that previous authors had assumed it to be an axiom (Boucharlat 1823, IX).[66]

---

[64]   In this period, the symbol > is also used to mean ≥ (and < to mean ≤).

[65]   For more information on Boucharlat, see Zerner (1986, 1994).

[66]   I refer here to the German translation—poor though it is—of 1820b because I was unable to gain access to the first edition.

The main method in Boucharlat's book is the *méthode des limites*. He bases his approach to it on the text of Lacroix's *Traité élémentaire*, with which he was already familiar from his work as *répétiteur*.[67] He modifies Lacroix's text in a characteristic way. He expands on some details while dropping all the more conceptual elements; Lacroix's approaches for defining limits as well as his general reflections and principles are lacking. Like Lacroix, Boucharlat also does not introduce signs for the limit.

The subsequent section on the *infiniment petits* reveals which fundamental principle that Boucharlat believes he has proved "in a satisfactory way": the principle of extended equality $x + a = x$. To characterize the quality of his work, I shall sketch his "proof" here. Without any further preconditions or explanations, Boucharlat sets as his initial equation

$$\frac{1}{a} + \frac{1}{x} = M$$

and multiplies it by ax. This gives him $x + a = Max$.

He now assumes that x becomes infinite; the fraction $\frac{1}{x}$ will then finally diminish to zero. This transforms the initial equation into $M = \frac{1}{a}$. He enters this "value" into the second equation and obtains

$$x + a = x \quad \text{(ibid., 157)}.$$

It should be noted that Boucharlat not only calls his *infiniment petits* quite unspecifically *quantité*, but by labeling them "*a*," also suggests that they are constants.

These two textbooks published outside the *École polytechnique* are clear evidence that the resolution of the *Conseil de Perfectionnement* actually did trigger a *retour du refoulé*. Calling for a return to synthesis, to geometric intuition, led to the reemergence of really old, traditional ideas on infinitely small quantities that then proved to be amazingly powerful and persistent—as confirmed by the longevity of Boucharlat's textbook.

At the same time, these explications of concepts on what could be called the highest possible level of mathematical discourse show that the entire conceptual range of the period reveals no intention to interpret infinitely small quantities in terms coming even close to a nonstandard analysis. Within the entire context of the mathematics of the time, there was nothing even vaguely comparable with the infinitesimals in this modern field.

The first criticism of the *retour* resolution and its consequences came, in 1814, from a mathematician outside the Parisian mathematical circle but linked to Arbogast and other mathematicians in Alsace, with their ties to mathematics in Germany. This criticism was accordingly published in the one journal that had only slight contact with Parisian mathematicians: the *Annales de mathématiques pures et appliqués*, edited by J. D. Gergonne. Its author was François-Joseph Servois (1767–1847). Originally ordained as a priest, Servois joined the army in

---

[67]   He did not base it on Lagrange as claimed by Dhombres (1987, 150).

1793, where he taught himself mathematics so successfully that he became a mathematics teacher at military schools, starting at Besançon in 1801, then at Metz in 1802, and from 1808 onward, at La Fère.

His extensive article of 1814 is highly interesting for other reasons as well. It starts off with an approach to developing operations with an algebra of functions. Through a composition of applications, he produces an extensive calculus with the inverse element and is the first to apply distributivity and commutativity of operations. An informative example is his general equation

$$f^n f^{-n} z = f^{-n} f^n z = z \text{ (Servois 1814, 94)}.$$

The final section of the article is the first public criticism of Hoëne de Wronski's memoir of 1812. Wronski was the first to question the general validity of Lagrange's claim that all functions could be developed into series. However, he did this in a spiteful, aggressive style. Interestingly, Servois's criticism of Wronski is to a major extent only a pretext for a harsh and fundamental criticism of the resolution of the *École polytechnique*, however, without saying this explicitly.

It must have felt like a fresh breeze to hear the independent and open voice of Servois after experiencing the zealous adoptions and the only subversive resistance within the *École*. Servois countered the shift toward a return to a static and unspecific concept of quantity by pointing out that in principle, the basic concept in the *analyse algébrique* and the *calcul différentiel* is the variable, and that the object of research is given through variations. This, however, makes it impossible to circumvent an intrinsic relation between algebra and geometry. Algebraic calculus deals with discrete quantities, namely, numbers, thereby initially remaining distinct from the extension: the geometric quantities. However, when the variations to be determined numerically become arbitrarily small, the only means of creating a unity between *calcul* and geometry is, of necessity, the *méthode des limites* (Servois 1814, 141 ff.).

Servois was the first mathematician within the long tradition of the French methods discourse to openly declare that Leibniz had not been a supporter of the *infiniment petits*. He uses quotations to point out that Leibniz had considered there was a need for rigorous proofs of Archimedes' method, and thereby also for the method of the *limites* (ibid., 143). Hence, for Servois, the true scandal is the devious way in which the supporters of the *infiniment petits* had managed to get their method recognized and viewed as a synonym for the (Leibnizian) *méthode différentielle*. He even infers that Lagrange declared his use of differentials to be an adoption of the *infiniment petits*:

> How has it come about that this strange method of *infinitely small quantities* should have acquired, at least on the continent, such celebrity; and even to the extent that its name should be found among the synonyms for the *differential method*? (Ibid., 144).

Servois passes harsh judgment on the justifications for this method given in the *École polytechnique*. He argues that they do not abridge the presentation in any way. Even just proving that the so-called orders of an infinitesimal can be applied operatively would require a major analytical effort based on, for example, Taylor series. The same deficit holds for the applications. For example, the hypothesis that one can replace a curve by a polygon and thereby neglect segments is inadmissible and also foreign to Euclid's elements (ibid., 146). It is more the case that enough historical examples show how many misconceptions and obscurities have arisen from manipulations with *infiniment petits* (ibid., 147 f.).

Servois concludes his discussion with a fundamental verdict unheard of in France up to that time. He declares that this method has no theoretical foundation, and that it is didactically dangerous, particularly for novices, because it encourages clumsiness and faintheartedness:

> It is a dangerous instrument in the hands of beginners, who impart, necessarily and for a long time, a character of awkwardness and tentativeness to their investigations in the area of applications (ibid., 148).

Resuming, Servois issues a prophetic warning against the damaging consequences for the progress of mathematics:

> Finally, in my view, *to anticipate the judgment of posterity*, I am bold enough to predict that this method will one day be accused, with reason, of having retarded the progress of the mathematical sciences (ibid.).

As far as France is concerned, the future would show some justification for this judgment. I shall return to this later.

2.5.3. THE FIRST OVERT CRITICISM BACK AT THE ÉCOLE: POINSOT IN 1815

The first published, though rather covert, criticism of the method of the *infiniment petits* by a member of the *École polytechnique* was written by Poinsot in 1815. It may well have been a reaction to Servois's fundamental criticism. Louis Poinsot (1777–1859) was also a student from the early days of the *École*, and, after 1804, a mathematics teacher at one of the major Parisian *lycées*. Although his main scientific interest was in mechanics, he was appointed to teach analysis at the *École polytechnique* in 1809. His article from 1815 in the *Correspondance sur l'École polytechnique*, the *École*'s second journal, published by Hachette since 1804, described itself as an "excerpt from the lectures on *analyse* given during the course he taught to the second *division* in 1814/15. A footnote explains that these are *notes rapides* thought to be of use for students. The title, *Des principes fondamentaux [...] du calcul différentiel*, bore a further footnote immediately locating the article in our familiar context: it recalled that the teaching program now required the foundations of differential calculus to be taught *par la considération des infiniment petits*, while simultaneously showing how this method agrees with that of the *limites* or development into series (Poinsot 1815, 111).

It is interesting to see that the main object of Poinsot's reflections on foundations is to clarify the meaning of *elements*. This casts a light on the conceptual context of the debate on methods at the *École*. Poinsot considers this clarification to be essential for defining the nature of differentials. He assumes that all quantities are formed through the successive addition of their parts. Although this statement may seem trivial, the emphasis is on *homogeneous* parts. He describes these parts that can be used to reassemble the quantities as their elements (ibid.).

This can also be seen as a first notable acknowledgment of Carnot and his principle of substitution (see above, Section 1.6.1.). Poinsot argues that because the elements are no easier to examine than the quantities themselves, one replaces them with quantities that are close to them. These close quantities are called differentials, and the closer these come to the elements, the closer the outcome to the truth. If the differentials are selected so that their final relation with the elements is of the unit ratio—once again, "Carnotian" terminology— then what seems to be an approximation method will become a calculus that is just as rigorous as algebra (ibid., 111 f.). The property of being substitutable with appropriate differentials forms Poinsot's first main principle.

His second main principle is an explication of his central demand for *homogeneity*. His argumentation against operating with inhomogeneous elements is very detailed, showing how the return to static ideas on indivisibles really had developed a considerable momentum. His principle of homogeneity states that all differentials, however small one may assume them to be, must always possess the same nature as the quantity under observation itself. Poinsot points out that even though this principle is actually self-evident, it is often forgotten in the applications. He therefore goes into great detail when explaining that *tranches* of a body, regardless how small, always maintain the same dimension and continue to have their physical form as prisms. Likewise, the differential of a surface is also a surface, and so forth. It would, in contrast, violate the principle of homogeneity if one were to view a solid object as being composed of an infinite number of surfaces; a surface, of an infinite number of lines; and lines, correspondingly, of points. This is a clear criticism of the indivisible method of Cavalieris that avoids systematically imposed, roughest error only through additional implicit assumptions (ibid., 112 f.). As we have seen while discussing reactions to the *retour* resolution outside the *École*, a subsequent revival of the indivisibles was a highly topical issue.

Poinsot's entire argumentation reveals that he considers the method of *limites* to be the obvious foundation. He also declares expressly that the goal of the *calcul différentiel* is to determine *limites*, namely, those of special ratios. At the same time, he is obliged to render lip service to the *infiniment petits*: On the one hand, the limits of ratios are retained as real quantities—whereas the differentials vanish—and thus ensure the retention of the rigor of the conceptions. Both Newton's calculus of fluxions and Lagrange's theory of derivative functions can be understood as expressions of the method of the *limites*. However, the *artifice* of treating the differentials like real quantities is just as sound as these methods and, moreover, more practical in applications to geometry and mechanics (ibid., 114). This definition of *infiniment petits* as variable differentials represents a striking acknowledgment of Carnot's idea of infinitely small quantities and a rejection of the revival of ideas on indivisibles. For *infiniment petits* that are emphasized in such a positive way, Poinsot also sees a concordance with the *limites*, because, in the end, the only thing one may observe when they are used is the limits of their ratios (ibid.).

It is interesting to see that it was a physicist who brought mathematics back to reflections on the elements of a whole. Indeed, Poinsot also explicitly considers the inverse operation of being able to reassemble the whole from its—homogeneous—parts. This motivation in physics paves the way for his new approach to the integral concept. The text on foundations already justifies the integral as a converging sum of the limits of the differentials, and no longer as a formal inversion of the differentiation (ibid.).

After this section on principles, Poinsot goes on to treat the rules of differential calculus as a mixture of Lagrange's method and the *limites* method. He reports that his goal is to find the derivative function (ibid., 115). Another characteristic aspect is that his basic precondition is motivated by physics. Not only are functions assumed to be continuous—as a condition for the existence of the limit of the ratio of the increments—but also the function concept is linked categorically to the continuity concept:

> Consider [...] a quantity y which is continuously dependent on another quantity x, according to some law; y will be what is called a function of x, and this function will vary in a continuous manner at the same time as x; otherwise, it would not be continuously dependent on x as was supposed, and would not be a function of it (ibid.).

Poinsot's rather tautological definition did not just exclude discontinuous functions from mathematics; it also contrasted with Ampère's more far-reaching reflections (see Section 3.2.).

3. Retour to Synthèse for the Negative Numbers

3. 1. Lacroix Propagates the Absurdité of the Negative

In the years just before and after 1800, France went through a transition from the dominance of the analytic to a preference for the synthetic method. This is exemplified by the decline in the *Idéologues*' epistemological and intellectual leadership (see above, Section 1.3.). The person most responsible for this change in the basic principles of mathematics was Carnot. This section examines how Carnot's radical reversal in the approach to negative numbers was received.

One central figure in this reception is, once again, Lacroix, although his involvement took a different pattern to that in the analytic methods. In the latter, he opposed the return of the *infiniment petits* without compromise, continuously propagating his preference for the *méthode des limites*, even though he practiced it in a scarcely algebraized form. With the negative numbers, in contrast, he became the main propagator of a reinterpretation *à la synthèse*.

3.1.1. ...IN ALGEBRA

Carnot and Lacroix had a positive and close relationship. In a letter to Lacroix at Christmas 1802, Carnot tells him that the new edition of his *Corrélation des figures*, the *Géometrie de Position*, will soon be finished, and lets him know that he will be asking him to check the manuscript—particularly the algebraic parts and that on the negative quantities—before it goes to press:

> When I come to Paris I shall prepare a new edition, which will be much enlarged and more correct. I have added a chapter on the application of algebra to elementary geometry, and another on the theory of curves, which are not follow-up treatises, but are applications of my theory on positive and negative quantities and on transversals. I will ask you to give me your advice before printing.[68]

The following year saw the publication of not only Carnot's work containing his most detailed presentation of the conception of negative quantities but also Lacroix's thorough revisions of the two of his textbooks with the strongest relevance for this conceptual field, those on algebra and analytic geometry.[69]

Because the conception of negative numbers is so central to the changes in the algebra textbook, this will be analyzed first. Within the space of only a few editions, Lacroix's algebra textbook went through quite unique qualitative transformations: It presents a rare example of an—in some ways, "ontogenetic"—reflection of the "phylogenetic" course of mathematics.

I shall call the first version his "zero" edition. In 1797, to meet his needs as a mathematics teacher at a Parisian *École Centrale*, what Lacroix published was not his own textbook, but a fifth edition of Clairaut's "genetic" algebra of 1746. He plays down his own role in this publication: His name appears only once, in coded form as "L.C." at the end of the introductory *Avertissement de l'éditeur*. Although Lacroix builds in a whole series of new developments in algebra through comments and additions to the text, he makes no claims to authorship, but attributes them to leading mathematicians such as Euler, Lagrange, and Laplace. His choice of Clairaut corresponds with the still dominant French

---

[68]   Bibliothèque de l'Institut, Papiers de Lacroix: ms, 2396. Carnot's letter, dated 4 Nivôse an 10 (December 25, 1802), was sent from St. Omer, the family home of his wife.

[69]   These were third editions for both textbooks. In my earlier publications, I had assumed that Lacroix did not modify his algebra textbook in the Carnotian sense until the seventh edition (Schubring 1986, 14). In an as yet unpublished M.Phil. dissertation from 1998 on the history of the negative numbers, A.Th. Henley has pointed out that this change is already present in the fifth edition of 1804.

tradition of attributing the leading role in the composition of *livres élementaires* to inventors (see Schubring 1999, 50 ff.). In fact, Clairaut's approach to the negative numbers still shows a clear affinity to the analytic method. He follows the goal of generality, and although rejecting negative solutions, he only reinterprets the results and does not call for a reformulation of the problem (see Chapter III.9.2.).

The first version of the algebra textbook for which Lacroix acknowledges authorship, which I shall therefore call his first edition, was published in 1799. This was also still not all his own work, the largest part was taken from Bézout's textbook on (marine) algebra. Lacroix explains that he did not wish to republish Clairaut's text because of its length, but he did not have the time to put together a complete text of his own. Therefore, he has added his own articles to his *notes* and *additions* to the Clairaut edition of 1797 plus some "pieces" from Bézout's algebra textbook. Lacroix now emphasizes his authorship of the earlier additions and the new articles while minimizing the importance of the parts of the text taken from Bézout. He remarks that there have been some accusations that Bézout lacked precision, and that he has taken this into account in the changes made here (ibid., 74).

In this first "own" version, Lacroix adopts the conception and the texts on negative quantities from Bézout's work. Hence, starting out with Clairaut's continuing acknowledgment of the analytic method, Lacroix takes a qualitative step toward the synthetic method, because Bézout's approach was already dominated by the single-case approach, despite a fundamental recognition of the existence of negative quantities (see Chapter III.10.2.). Hence, Lacroix takes over from Bézout not only the distinction between the sign of the operation and the sign of the quantity (Lacroix 1800, 31), but also the explanation that negative quantities have just as real an existence as positive ones (ibid., 32). He likewise rejects negative solutions, saying that they indicate that a problem has been posed incorrectly. However, they simultaneously show what needs to be done, namely, to replace the sign of the quantity in the first equation with its opposite (ibid., 33).

Nonetheless, Lacroix does not remain consistently "synthetic" or "Bézout-like" here, but mixes the Bézout text with Condillac's conception of the *analogy* between arithmetic and algebraic operations. Even the first edition contains a paragraph not to be found in the original Bézout edition that considers both how the two types of operation relate and how they differ. He states that the algebraic operations gained their name because of the analogy with arithmetic ones, but that they possess an extended meaning because of the generality with which quantities are regarded in algebra. For example, subtracting a negative quantity actually leads to an increase (Lacroix 1799, 39). In a detailed footnote to his second edition of 1800, Lacroix extends this conception of generalization that departs from Bézout's original algebra by paraphrasing Condillac's (1798) posthumously published *Langue des Calculs* (see Chapter IV.1.2.): The change from a minus to a plus sign in subtraction is disturbing only for novices, because they want to trace the concepts they are being taught in algebra back to the ones they already know from arithmetic. However, the extended meaning of general signs in algebra no longer permits any direct comparison with arithmetic operations. Here, Lacroix, along with Condillac, expressly permits the subtraction of a larger from a smaller number: "The subtraction $b - a$ indicated algebraically does not imply necessarily the idea that b is greater than a" (Lacroix 1800, 41 f.).

Lacroix then even goes on to argue that in a certain sense, one may view the negative quantities as "being below zero" (ibid., 43 f.). The third version appeared only one year later, in 1800, as a second edition of Bézout-Lacroix. Its text remained largely unchanged (Schubring 1987, 45).

In the fourth version and third edition, the first algebra textbook "of his own," Lacroix pursues what is now the "latest" state of the art in the synthetic method in algebra. After the prior stages of Clairaut, then of Bézout—though supplemented eclectically with a Condillac who was still just about "modern" in 1799—Lacroix now switches to Carnot's *retour*. In the *avertissement* for this edition, not only does he report that the treatment of the negative quantities has been the main reason for his extensive revision but he also justifies this in light of his intervening experiences with a general school education system. Because the book is not just intended specifically for those interested in mathematics, metaphysical considerations cannot be addressed right at the beginning. He argues that the introduction of isolated negative quantities at too early a stage, as found in most textbooks, will lead to confusion in the novice. Lacroix claims that he has proved the rule of signs a priori for isolated negative quantities (1804, V ff.).[70]

The didactic justification given in the preface is that he has shifted the treatment of negative quantities to a later position, where it will be motivated by the solution of problems. However, this is a pretext, because he—or Bézout—had also introduced this concept in relation to a problem and not dogmatically in the earlier editions. The true methodological reversal, in contrast, is not explicated.

Indeed, the new edition, the first that is all his own work without excerpts from Bézout, presents a rejection of the analytic method. There are no more statements on the equal status of negative quantities, no reflection on the relation between arithmetic and algebraic operations, and no differentiation between signs for operations and signs for quantities. In contrast, Lacroix now describes isolated negative solutions as an *absurdité*. He starts to develop his claimed a priori proof for the rule of signs together with the explanation of the basic operations for letter quantities in a completely traditional way based on

---

[70]   Here I am quoting the fifth edition of 1804 that contains a reprint of the preface to the third edition. Copies of the third edition are very rare. There is one in the *Bibliothèque (Nationale) de France* at Paris. The title page gives the year of publication as "*An XI = 1803.*"

polynomes (ibid., 44 ff.). Negative quantities, in contrast, are treated quite late, when linear equations with two unknowns are addressed.

He presents a very long and elaborate discussion on a system of two such equations given by one problem:

$$12x + 7y = 46 \text{ and } 8x + 5y = 30.$$

Whereas he obtains a value of 5 for x, the result for the other unknown is $y = \frac{-14}{7}$. This leads him to ask in surprise what the minus sign in front of the isolated quantity 14 may mean. He argues that its meaning is familiar when it is located after a larger quantity from which something should be subtracted. But what is subtracting without the presence of a term from which something can be subtracted? To explain this paradox, he says, one has to return to the conditions of the original equations (ibid., 89 ff.). So, he enters the value $x = 5$ into the second equation, obtaining: $60 + 7 \, y = 46$. Lacroix proclaims that the mere appearance of this equation discloses an *absurdité*, because there is no way in which one can obtain the number 46 by adding a quantity to the number 60 (ibid., 86). After a prolonged discussion over "what the absurdity consists of" and many transformations—apparently with numbers—"so that the given problem becomes possible," Lacroix not only reformulates the problem (replacing inputs through outputs), but also depicts the solution in another way, as

$$x = 5^{fr}, \quad y = 2^{fr}.$$

In other words, Lacroix is not operating with "numbers" at all, but with quantities, and assumes implicitly that opposite quantities are not homogeneous and therefore do not admit any common operations. This simultaneously means—just as in Carnot—that numbers are subordinate to quantities as the primary mathematical objects, and that the only permissible operations are those with absolute numbers.

Lacroix does not give any definition for *quantités negatives*, but it is no longer surprising that since his change of position, he continues to admit them only as *subtractive* quantities (ibid., 91). This return to apparently long vanquished, restrictive conceptions can also be seen in the way he labels quantities with a minus sign as negative: *la quantité negative –c* (ibid.).

Lacroix goes on to perform his proof of the rule of signs announced before by nonetheless stating that up to now, it has not been proved for isolated quantities. However, the purported proof contains nothing new. By noting—consistent with his new approach and in agreement with Carnot—that for a quantity b, the largest possible subtraction is $b – b$ with an outcome of zero, he argues—like Euler and Laplace before him (see Chapter V.2.2.)—with the products of a (or $–a$) times $b – b$ (ibid., 91 ff.).

On the other hand, even though this is the first edition that is all his "own work," Lacroix does not remain consistent in his conception. After a detailed account of how one needs to "rectify" the first equations in order to avoid negative results, the solution he proposes is also to perform formal operations with the negative "expressions" that are, according to him, mathematically meaningless. This is the first time he names the criterion of internal consistency for the executability of operations:

> But Algebra dispenses with any research in this regard [correcting the statement of the question], since it can operate appropriately on expressions carrying the sign –; for these expressions, being deduced from the equations of the problem, must *satisfy* the equations: that is, when subjecting them to the operations indicated in the equation one must find, for the left-hand side of the equation, a value equal to the right-hand side (ibid., 88).

In great detail, with all transformations of terms, Lacroix shows how the "expression" $\frac{-14}{7}$—he carefully avoids giving this term a mathematical label such as "quantity" or the like—derived from the two original equations "satisfies" both equations simultaneously with the value $x = 5$.

However, when it comes to both linear and quadratic equations, Lacroix sees no need to apply this epistemological alternative. The reason for this lies in the interaction between form of representation and *degree of generalization* in his book. Lacroix does not give a systematic theoretical exposition of algebra. Instead, he still largely follows the more traditional style of presenting a sequence of practical problems with an extensive discussion of their concrete solutions integrated only loosely into a thematic framework: linear equations with one, two, and more unknowns; quadratic equations with one, two unknowns; and so forth. Hence, the focus is on discussing specific solutions and not on a more universal body of algebra.

As a result, Lacroix's contributions to a systematic approach added, so to speak, *en passant*, do not fit that well into the underlying conception. Whereas he initially uses several pages to present the numerical procedure for computing the root in the context of quadratic equations and then proceeds to operate over several pages with the square root as only single–valued (ibid., 134 ff.), he finally goes on to describe and discuss their double–valuedness, thereby labeling the absolute value of the root as the *valeur numérique* that can be provisioned with the plus or the minus sign (ibid., 155). Finally, Lacroix even presents the general quadratic equation as: $x^2 + px = q$, and explains that p and q may be either positive or negative quantities. His discussion on problems and their solutions nonetheless shows that he understands *quantités negatives* as subtractive quantities here as well. Negative solutions are also still not admitted. They make it necessary to *modifier* the *question* in a way analogous to that for linear equations (ibid., 162). Lacroix's understanding of generality is to view the two—for him—different problems as being linked into the same formula through an analogy: "One sees by this, in the second degree, Algebra combines in the same formula two questions which bear to each other a certain analogy" (ibid., 162 f.). Hence, Lacroix adopts the reduced conception of generality in which the problems solved in the different cases are viewed as being at least related or analogous as advocated by Klügel (Chapter II.10.4.) and Carnot (V.1.6.3).

In his section on linear equations, Lacroix reveals even more reservations regarding the alternative approach of internal coherence. Instead of "expressions," he even talks only about "symbols" for the values of the unknowns used to fulfill the equations in formal operations:

> It is an essential property of algebra that the symbols of the values of the unknowns, whatever they may be, being subject to the operations indicated for the unknowns, will satisfy the equations of the problem (ibid., 101).

In the last part of his algebra textbook, addressing the general theory of equations, which makes up about one-third of the book, Lacroix no longer complies with the epistemological restrictions and *retours* that he had chosen himself for this edition. Here, where he presents more abstract mathematics, namely, cubic equations, quartic equations, and higher, Lacroix limits himself to general procedures and no longer uses a style of presentation based on discussing single problems. In line with this, the defining criterion in this final section is that of internal coherence: whether the conditions of the equation are satisfied. Therefore, Lacroix proceeds as a matter of course from the assumption that both positive and negative roots may be solutions of equations (Lacroix 1808, 304 f.).[71] What has to be taken into account here is that because the presentation is general and does not discuss concrete problems, roots now represent numbers and not quantities.

3.1.2. ...IN THE APPLICATION OF ALGEBRA TO GEOMETRY

In line with the underlying concept in this book, the understanding of negative numbers will be studied not just in algebra but also in further conceptual fields, particularly in analytic geometry. Indeed, when discussing his reconceptualization of algebra in the third edition of his textbook, Lacroix himself points out that a complete understanding of negative numbers may be deduced only in the application of algebra to geometry. He also produced a textbook on this topic that included the two trigonometries as well: his *Traité élémentaire de Trignonometrie rectiligne et sphérique, et d'Application de l'Algèbre à la Géométrie*.

This textbook, whose first edition was published in Year VII (1798/99) and the second edition in Year VIII (1799/1800), also received a major revision for its third edition in 1803—the same year as the algebra textbook. The changes in this revision were analogous to those in the algebra textbook—brought about, once again, by the change in the negative quantities!

The first two editions deal with negative numbers as if they were completely unproblematic. For example, the section on trigonometry treats the known formulas in the traditional way as being uniform for all quadrants—without any differentiation according to whether values are positive or negative. The section on analytic geometry contains a systematic exposition based on types of

---

71   Ampère also noted and criticized the lack of consistency between Lacroix's "lower" and "higher" algebra (see Section 3.2 below).

equation. This does not consider negative quantities in any way—not even in the Bézoutian form of simultaneous algebra problems, that is, without assigning them an indication of an error in the posing of the problem.

Section Number 71 is a characteristic example in which Lacroix sets himself the problem of subsequently explaining "what the signs of the quantities express that represent the lines" (Lacroix 1799/1800, 83). He makes it very clear that the position of the curves in the coordinate system is determined with the help of the plus and minus signs. *Valeurs positives* and *valeurs negatives* are standard expressions here (ibid., 83 ff.).

On the other hand, the textbook contains no major innovations compared with textbooks from the 18th century. Although curves are presented using all four quadrants, the book continues not to indicate units of length on the axes of coordinates, and the semi-axes are not presented as the positive or negative parts. Correspondence is suggested with symmetrically located points marked with signs such as K and K'.

The new third edition, like the algebra textbook, is furnished with a new preface. However despite all its detail, it does not mention the conceptual change in the negative quantities. In line with the previous discourse of the *modernes*, Lacroix emphasizes the incomparable advances in analytic geometry compared with the *anciens*. He says that the still unanswered questions cannot be mastered with *geometric* approaches (Lacroix 1803b, vj ff.). Lacroix even takes up the relation between the synthetic and analytic methods again. On the one hand, he says, it is clear that the only way to deal with the object appropriately is to apply a combination of both methods. On the other hand, he also explains—surprising in light of the change in fashion so far—that the synthetic method has to be subordinated to the analytic in order to gain a true understanding of terms (ibid., xv). According to Lacroix's preface, analytic geometry is not suitable for all students, but for those preparing themselves for an application-oriented course of study: the study of the *sciences physico-mathématiques*, for example, for those preparing for entrance to the *École polytechnique* (ibid., xj).

The section on trigonometry is unchanged in the new edition. Lacroix does not adopt Carnot's fundamentalism in the trigonometric formulas of recognizing only positive values in all four quadrants. In contrast, the section on analytic geometry has been rewritten completely. One characteristic change can already be seen in the introductory section: the paragraph is missing that presents the goal of mastering the, at first glance, seeming difficulty in expressing relative situations (positions) of lines and surfaces through algebraic terms (Lacroix 1799/1800, 70 f.).

The structure of the text—and also the figures—is completely different: The systematic treatment according to types of equation is replaced by a return to a problem-oriented treatment of the correspondence between equations and constructions. This is the reason for a new paragraph here in which Lacroix explains how the algebraic minus sign is to be understood in geometric constructions.

He does this by introducing the same conception as that in the revised algebra textbook. The minus sign should be interpreted as in numbers as an inversion of the terms of the problem or by taking the lines in a direction opposite to the one assumed originally (Lacroix 1803b, 89). Lacroix recalls that the *quantités négatives* have their origin in nonexecutable subtractions, and, in agreement with both Bézout and Carnot, such subtractions indicate errors in the initial conditions. By changing these, one moves from a previously nonexecutable subtraction to a positive result:

> There was an error in the statement of the question, or at least in its application to the particular case being considered; and in redressing this error, that is in modifying the statement in a way so as to render possible the subtraction which could not have been done, one arrives at a positive result (ibid.).

For Lacroix, however, this conception of the application of algebra in geometry leads to a far less decisive change than it does in algebra, or than it does for Carnot, whose restrictive fundamentalism led him to a new kind of geometry. As the main application, Lacroix is satisfied with explaining how minus signs should be applied to mark off segments in the opposite direction (ibid., 89 ff.). Hence, he has no problem in saying that the values of the sine and cosine change their sign when entering the corresponding quadrants (ibid., 92 f.).

Hence, Lacroix's new conception of analytic geometry presents a case in which the change in epistemological perspective has not led to any major change in the practice of mathematics. The most notable effect is found in the form of representation: in a reduction of the degree of generalization of the theories formulated because of the shift back to a problem-oriented discussion of single cases. A broader effect was that no incentives for the further development of the concept of negative numbers came from analytic geometry either, and it continued to give it the character of a theory wrapped in a somewhat mystical ambiguity.

Both Lacroix's algebra textbook and his textbook on trigonometry and analytic geometry were reprinted several times after these third editions—up to the 17th and 9th editions respectively in his lifetime—but no more changes were made to the conception.[72]

The revisions to Lacroix's textbooks had an almost incalculably broad impact in France, which is why they are also analyzed in detail here. These new versions enabled Lacroix to profit from the fundamental restructuring of the French school system. The *École Centrales* set up under the influence of the *Idéologues* in 1795 were abolished by Napoleon in 1803 and replaced by the *Lycées*. These contained not only a clear structure of consecutive classes and a set curriculum but were also subject to strict and centralized control (Schubring 1984, 369 ff.). One aspect of the centralization policy was to permit only

---

[72]   It was not until the 23rd edition in 1871 that the editor Prouhet cautiously suggested, while assuring all reverence to the great master, that one could also operate with negative numbers.

textbooks approved by a ministerial commission. This change is even expressed in what these books were called: Whereas previous textbooks had been called *livres elémentaires*—in line with the idea in the Enlightenment of basing general education on the elementarized sciences—they were now called *livres classiques*—and oriented toward a classical, invariable general education rather than any participation in scientific progress.

In 1799, when specifying *livres elémentaires* for the courses at the *Écoles Centrales*, Lagrange had wanted to approve only one textbook for mathematics—that of Bézout. However, the commission—also supported by Lacroix's wire pulling at the ministry—disapproved of the idea of a single exclusive textbook and permitted a choice between several competing works (including that by Lacroix; see Schubring 1999, 57). By 1803, in contrast, as a result of strict centralization, no selection was permitted. And it was now Lacroix's textbooks that were imposed on the whole of France, and specifically those teaching the revised, "synthetic" conception. His books were selected for practically all parts of the mathematics curriculum: for arithmetic, geometry, algebra, trigonometry and applications of algebra to geometry, his *Complément des élémens d'algèbre*, and even his textbook on differential calculus. Mechanics, which was still regarded as part of mathematics at that time, was the only field that he failed to dominate: Francœur's textbook was chosen here (Belhoste 1995, 79 ff.).[73] The large number of new editions of these elementary textbooks in the following years confirms the enormous influence exercised by Lacroix, particularly through the new version of his algebra textbook.

Although a new list of textbooks was negotiated when the *lycées* were restructured in 1809, and this moved away from a monopoly of only one textbook and permitted a degree of choice,[74] the authorized works of Bézout, Bossut, and Marie on arithmetic and algebra all dated back to the 1770s. Because a current new edition was available only for Bézout—last edited by Peyrard in 1803 and 1808—and this taught the sign inversion method for negative solutions, the rejection of an independent theory on negative numbers continued to dominate France after 1809. The strong influence of Lacroix's textbooks in Europe and the Americas, particularly documented by the numerous translations of his algebra textbook, extended this domination far beyond France. One of these translations—that by M. Metternich, who had to teach the obligatory Lacroix—nonetheless also documents German resistance to the rejection of the negative numbers (see Schubring 1995; and, also, Chapter

---

[73]   It is no surprise to see that Lacroix was a member of the commission responsible for these decisions—together with Monge, his promoter for many years, and Laplace.

[74]   The proposal for the mathematics textbooks in 1809 came from Legendre (AN, F^{17} 12818). After competing with Lacroix since 1799 (Schubring 1999, 71), he had now finally managed to ensure that his *Éléments de géométrie* was authorized alongside the work of Lacroix.

VII.3.3., below). A similar resistance also arose in the Netherlands (Beckers 2000).

3.2. The Criticisms of Gergonne and Ampère

Taken in combination, the power Carnot and Lacroix exerted in France was so strong that it long prevented the publication of any other stance, any opposition to the *retour* to the "synthetic" approach in the negative numbers, or any challenge that they required further proof. The first critical voice was not heard until more than 10 years later, in 1814. Remarkably, this was not only almost in the same year as the first overt criticism of the return to the *infiniment petits* but also in the same place: in the *Annales de mathématiques pures et appliquées*. This time, it was written by the editor of the journal himself, Joseph-Diez Gergonne (1771–1859). Gergonne was a self-taught man who originally made his career as an officer in the Revolutionary army. Here, like many of his contemporaries, he became attracted to the sciences. He taught mathematics at an *École Centrale* before becoming mathematics professor and dean of the *Faculté des sciences* at Montpellier. As the editor of the *Annales* from 1810 to 1831, he was responsible for publishing the first modern scientific journal for mathematics.

What triggered Gergonne's criticism was an article published in his journal by a certain Cach, teacher in a secondary school at Tours. Cach, following Carnot and Lacroix, operated with subtractive quantities in a complicated way, modifying each original equation in order to obtain positive results (Cach 1813).

This led Gergonne to compose an admirably systematic memoir in which—actually, in a thoroughly defensive way—he picked apart the arguments of the "new" theory and "went to bat" for the—as he called it—"old" theory. His memoir clearly reveals the total dominance of the Carnot-Lacroix conception. He talks about the *idées qui sont aujourd'hui généralement en vogue* (Gergonne 1813, 6) and says that the new theory *a substituée* the old theory *depuis quelques années* (ibid., 19).

Gergonne's own approach starts (practically following on from Fontenelle) by attributing to quantities not only their *valeur absolue*, but also a qualitative aspect: a *mode d'existence* that can take the form of an *opposition* between quantities of the same kind (*de même nature*). For such opposite but, at the same time, homogeneous quantities, Gergonne distinguishes between signs of operation and signs of quality. He asks himself how he can define *modes d'existence opposés*. His initial answer is somewhat polemic: he argues that he could just as much expect a definition of space or angle or even of the (Carnotian) *quantités directes et inverses*. However, he then goes on to point to the easily recognizable property that two quantities with the same *valeur absolue* cancel each other: *s'anéantissent par leur reunion* (ibid., 9). Gergonne expresses all this only verbally, without algebraization or signs.

He does not accept that expressions such as $-a$ should be qualified as *soustraction impossible à effectuer* or as *forme algébrique inintelligible*. Instead, he interprets them simply and naturally as opposite quantities. He frequently uses the term *convention* for this, but only to emphasize that the attribution of "positive" or "negative" is arbitrary within a pair of opposites.

Gergonne focuses particularly on Carnot's arguments, engaging in a detailed consideration of the statement that quantities smaller than zero cannot exist. This makes him the first person in France—after more than a century—not only to take up Reyneau's explanation of the relative position of the zero but also to draw attention in France to Kant's discrimination between absolute and relative nothing (see Chapter II.10.4.).[75] Gergonne explains:

> I distinguish first between two sorts of zeros: namely, an *absolute zero*, a symbol for pure nothingness […] and a *limit zero*, or point of departure, which is pure convention, and to which quantities considered as being able to be positive or negative are related (ibid., 10).

He goes on to say that one could also replace above (*au-dessus*) and below (*au-dessous*) just as easily with "to the left of" and "to the right of" or "in front of" and "behind."

However, Gergonne introduces a completely new argument here by pointing to the analogy with the operation of multiplication. His argument also shows remarkable structural insights and a new way of applying structural and axiomatic arguments. He starts off by explaining that the relative zero simultaneously forms the additive unit (*unité*). He then goes on to say that alongside this additive unit, there is also the multiplicative unit: the *one*. Gergonne points out that according to the original epistemological conception, nothing can be smaller than this unit. However, if one simultaneously accepts fractions such as $\frac{1}{4}$ as legitimate mathematical objects, and one accepts that these are smaller than this unit, and if one also accepts that $\frac{1}{4}$ is smaller than $\frac{1}{3}$, then, by means of analogy, one is also justified in conceiving -4 as being smaller than the additive unit and that -4 is smaller than -3 (ibid., 11).

Gergonne is also the first no longer to be impressed by the proportions argument that had not been countered successfully up to that time. He argues that one should apply only those criteria that belong to the core of the concept to its definition. However, the exclusive essence of a geometric proportion lies in the equality of the two ratios of division. The inversion of smaller or larger terms is a nonessential, an arbitrary consequence that, moreover, is valid only when all terms possess the same sign (ibid., 17). Indeed, he continues, the field

---

75   Although he does not mention Kant by name. There was a very strong interest in Kant at that time in France, particularly through Madame de Staël. It can be assumed that Gergonne acquired the notion of opposite quantities from contacts with German mathematicians.

for the application of proportions has been extended to negative numbers. From an axiomatic standpoint, one would have to study how far the concept of proportion changes when the number domain is extended.

It is also interesting to see his criticism of many teaching approaches that try to "avoid" mathematical difficulties. He counters this inadequacy with the maxim that one should face difficulties directly and show how they may be overcome (ibid., 19).

The only weak point in Gergonne's otherwise so well constructed memoir is his conceptualization of the relation between quantities and numbers. His conceptualization of *opposition* is actually founded completely on the quantity concept. He also notes the difference between quantities and numbers, but asks himself how far his justification for negative quantities also holds for numbers. This leads him to stumble over the obstacle set by the dominant belief that "abstract" numbers exist only as absolute and thereby positive ones: *principe, que tout nombre absrait est essentiellement positif* (ibid., 12).

The continued lack of a developed axiomatic approach to numbers and their operations means that Gergonne also possesses no effective argument for proving opposite *modes d'existence* for what are accordingly only positive pure numbers. This is the consequence of a concept of opposites that is bound to the quantity concept. Gergonne tries to justify corresponding operations for pure numbers from the concept of opposites for quantities (through numbers as statistics for ratios), but fails to find a convincing solution, particularly in multiplication (ibid., 12 ff.).

Gergonne warns about the unnecessary difficulties facing the *calculateur* since the new doctrine has replaced the former. He suspects that a double standard is applied: upholding certain principles for the outside world, while not applying them at all in one's own practice because they are so uncomfortable (ibid., 12). Just as Servois was apprehensive about the future development of analysis, Gergonne anticipates threats to the development of algebra. He fears that it will be thrown back to the status of its infancy: "in rejecting [the former conceptions] has not algebra been pushed back to the point where it was in its infancy?" (ibid., 6).

Gergonne's final paragraph confirms Carnot's enormous popularity and his leading role. It presents an apology for having primarily (implicitly) criticized Carnot's *Géométrie de position*. He says that this is only because it presents the new theory in its most developed form, and in no way reduces his respect for Carnot's person and his work: "I ask my readers to believe that I have, nonetheless, the greatest esteem both for the person and the works of the illustrious author of this book" (ibid., 20).

Despite the dominant positions of Carnot and Lacroix and the fact that criticism was expressed only outside Parisian mathematical circles, there was still one critical post within the *École polytechnique*. This was held by a person well known for his reflections on basic principles: Ampère. Indeed, Ampère can be considered to be the true philosophical head of the *École*. He engaged in reflections on basic principles to a far greater depth and with more rigor than his predecessors Prony and Garnier. Moreover, he was much more interested in philosophy and corresponded actively with the philosophers of his time. He had even studied German philosophy, particularly the work of Kant.

Ampère's *Nachlass* contains a large number of manuscript fragments for a *Cours d'analyse mathématique*. Apparently, the cataloguers failed to realize that these many fragments are interrelated. The introductory chapters, though not containing a single finished text, all deal with the relation between quantity and number. In approach after approach, Ampère attempts a rigorous conception of these foundations. This shows a clear analogy to the later attempts by the German mathematician E.H. Dirksen to give a rigorous derivation of the number concept in the first volume of his *Organon*, attempts that he also never completed to his satisfaction (see below, Chapter VII.5.1.). Although a few of Ampère's text segments start off with extensive lists of contents,[76] he does not get beyond basic operations except in a very few of them. There are no finished manuscripts (and therefore no publications on, for example, the number concept), because in his efforts to achieve rigor in basic concepts, he was unable to find any number concept that satisfied him completely. Nonetheless, one basic conception does emerge from the fragments. Ampère was guided by the idea of developing a conception of the number concept that is as abstract as possible, in line with his belief that the most general and most abstract concepts are simultaneously the simplest ones: "in preserving the generality and the simplicity which characterizes the most abstract ideas."[77]

As the first aspect that needs to be clarified, Ampère discusses the relation between *quantité*, *grandeur*, and *nombre* in a number of fragments. He defines *quantité* as that which is composed of parts that correspond to each other; one can increase or decrease a *quantité* only by adding or removing such parts. Under *grandeur*, he understands a variable that can shift from one state to another only if it successively assumes all intermediate states. In another text segment, he therefore distinguishes between a *grandeur* and the values it assumes at each single stage. The only way to measure *quantités* and *grandeurs* is through numbers. Hence, for Ampère, the first signs are those representing equals, is smaller than, and is larger than: =, <, and >.

As a result, the first concepts on which Ampère reflects—in a completely new and comprehensive way— are those of equality and inequality. To measure and determine numbers, one requires a measuring rule. Inequalities determined in this way are needed before one can perform initial operations of decreasing and increasing. It is only after reflecting on equality and inequality and introducing

---

[76]   Académie des sciences Paris; Archives. Nachlass Ampère. The most detailed list of contents extending as far as differential calculus is found in the fragment headed *Principes élémentarires du calcul* (cart. 1, Chapter 1, chem. 4).
[77]   Académie des sciences Paris; Archives. *Nachlass* Ampère. Carton 4, Chap. 4, chemise 75: Cours d'Analyse. Note in chapter 3 (none of these manuscripts are paginated).

their signs that Ampère introduces basic operations as well as the plus and minus signs.

He then immediately explains that the traditional way of labeling positive and negative quantities through their sign is incorrect—a practice continued by both Bézout and Lacroix. This erroneous definition results from the desired meaning of increasing or decreasing:

> It is only because of the name positive quantities that they are given the sign + preceding them, and those that are negative quantities have the sign – preceding them, but this pointless distinction can serve no purpose but to throw one into error.

Ampère is the first to ascertain that such a characterization is not just unnecessary but even leads to error. He shows in great detail that the reason why this definition does not work lies precisely in the nature of algebraic quantities. Specifically, because one is initially dealing with an unknown quantity that is defined as such, one is free to label it as either $+x$ or $-x$, and one is just as free to replace $+x$ by $-y$: $+x = -y$ and $-x = +y$.

Because whether or not a quantity is negative does not depend on its sign, the introduction continues without discriminating between positive and negative quantities:

> We shall not establish for the present any distinction between positive and negative magnitudes; we shall wait to give precise notions to this subject until we have sufficiently developed these first principles.

In a further fragment, Ampère not only emphasizes the equal status of positive and negative quantities, but also goes beyond the level of quantities for the first time and conceives the concept on the level of "abstract numbers." Here—for the first time in French mathematics—he also admits "abstract negative numbers" and goes one step further than Gergonne: "There are no differences between positive and negative quantities, but those that are on the side opposite the unit are marked by abstract negative numbers."

In another text headed *Notes*,[78] Ampère criticizes the recent "objections" to negative quantities and, in particular, to abstract negative quantities. Unfortunately, the various texts are fragments, and these introductory chapters are not continued. Hence, Ampère's conception of negative numbers does not exist in an edited and coherent form. Nonetheless, these fragments do reveal that he rejects Carnot's and Lacroix's *retour* to synthesis on epistemological grounds, and wants, for the first time, to proceed from the number concept rather than the quantity concept.

It is clearly the conception of negative numbers that leads Ampère to keep on trying again. On the one hand, he always finds strong words of criticism for the contemporary empiricist conceptions, referring repeatedly to the *idée du nombre*

---

[78]   Although these notes are numbered, it is not possible to discern which part of the text they belong to. They are written on the back of a brochure on *Polyedres symmetriques*.

négatif. He castigates contemporary authors for the evidentially intentional way they cast a veil of obscurity over the negative numbers:

> It is sufficient for us to have given a precise idea of the distinction between positive and negative numbers, and of the nature of the latter, which modern authors seem intent to cloud with obscurity.

And he also points out how his conception of comparing quantities can be used to obtain not only negative whole numbers but also all other kinds of fractional or irrational ones: "[by use of] the type of comparison which gave us negative numbers, one must in the same way [...] be able to find whole numbers, fractions, or irrationals, and all negative."[79]

According to the *Régistres*, Ampère did not start any of his analysis courses at the *École polytechnique* by introducing the foundations of the number concept. The first-year course generally started with cubic equations. Indeed, these fragments are clearly not lecture manuscripts but first drafts for a future publication. He talks expressly about an *ouvrage*: "[...] the first foundation of mathematical Analysis, that is, the science to which this work is devoted."

As he was planning a book on analysis, it becomes clear that he also wanted to include the foundational aspects not addressed in the courses at the *École*, because they belonged to the knowledge students were assumed to possess at the start of the course. One can assume that Ampère must have discussed his concepts on foundations with his students. This is particularly significant because Ampère served as the *répétiteur* for Lacroix's analysis course after 1805, and this is the year in which Cauchy attended this class (Gilain 1989, 5). Moreover, the fragments can be dated to the period before 1810.[80]

In one footnote to a fragment, Ampère even criticizes Lacroix explicitly: In the section of his textbook on higher algebra, Lacroix swings back to the true theory of negative numbers without giving any explanation for this. Before this, in contrast, he teaches the opposite of a true theory: his rules on changing the sign:

> lacroix n'explique point la vraie theorie des signes litteraux essent.[t][iellement] neg.[s][atives] representant tout ce qu'on veut, sa regle de changer les signes de *c* en est precis[t][ement] le contraire. il en revient à la vraie théorie dans les formules generales mais sans en donner la moindre expl.[on][ication].[81]

---

79   *Nachlass* Ampère, cart. 1, Chapter 1 chem. 4, Livre premier. I. Notions préliminaires.
80   Among the fragments, there is a draft letter to a *citoyen* of Lyon. This title introduced during the Revolution expressing the vision of equality between human beings was largely abandoned after the crowning of Napoleon as emperor in 1804; and Ampère, in particular, had little reason to feel sympathy for the Revolution.
81   Ibid.: *application de l'analyse à la résolution des eqs. à une seule inconnue.*

3.3. A Further Look at England

Up to now, developments in England have been followed up to the end of the eighteenth century, that is, up to the peak in the rejection of negative numbers as admissible objects in mathematics as documented in the writings of William Frend (see Chapter II.2.10.3.). Until about 1830–1840, this dismissive stance continued to be upheld in England and Scotland, although in a less radical form, by mathematicians such as John Walker (1768–1833), John Bonnycastle (1750–1821), John Leslie (1766–1832), and, in his early writings, Augustus de Morgan (1806–1871).

The first criticism of this majority opinion was published at the beginning of the nineteenth century by Robert Woodhouse (1773–1827). Woodhouse remained at Cambridge after graduating as the senior wrangler in the Mathematical Tripos. He became a fellow at two of the university colleges, where he worked as a tutor until called to the chair of Lucasian professor in 1820. His book, *The Principles of Analytic Calculation* (1803), was the first to introduce the "continental" method of differential calculus to England, thus paving the way out of isolation. His students Babbage, Herschel, and Peacock were the founders of the *Analytic Society* at Cambridge, which was responsible for the development of symbolic algebra in England.

In two reviews of texts by Maseres and Frend published in 1801, Woodhouse criticizes the fundamental rejection of negative quantities and defends their further application because of their utility.[82] In memoirs to the Royal Society in 1801 and 1802, and in his book published in 1803, Woodhouse discusses negative quantities and the rule of signs further, but fails to produce a coherent conception (see Pycior 1976, 63 f.). Nonetheless, it seems as if Woodhouse took an isolated stance with his critical opinions.

---

[82]   These reviews have been analyzed by Henley in the work cited in Chapter V.3.1.1.

Chapter VI

Cauchy's Compromise Concept

Cauchy's work on analysis is generally viewed as the decisive new stage that led to the qualitative leap to modern standards of rigor. He is attributed with having worked out the basic concepts of analysis in a way that forms the foundation for modern mathematics. It would therefore seem appropriate to summarize the state of development in basic concepts in France at that time in order to gain a framework in which to categorize and judge Cauchy's contribution properly. However, I shall start with some details on his life.[1]

1. Cauchy: Engineer, Scientist, and Politically Active Catholic

In the first phase of his career, Cauchy pursued a career in engineering rather than science. His first attempts to leave the engineering profession and become a scientist were a failure. He achieved this ambition only after the Restoration in 1815, thanks to his conservative and what could be called fundamentalist Catholicism. Augustin-Louis Cauchy (1789–1857) was born in Paris only one month after the storming of the Bastille. He spent his early years in relatively constrained circumstances in the provinces. As the Secretary of the Parisian Chief of Police, his father, a lawyer, had fled the city. Nonetheless, after the fall of the Jacobins, his father was able to return to Paris and continue his career. He was considerably successful, attaining the post of secretary in the Senate after Napoleon's rise to power. This post came with an official apartment in the Palais du Luxembourg, and as a consequences of Napoleon's policies, many scientists came and went as members of the senate. With the connections this provided, Cauchy's father was finally able to further his son's career decisively.

After receiving private lessons from his father, Cauchy attended the *École Centrale du Panthéon* from 1802 to 1804. Despite doing exceptionally well and

---

[1]   This biographical sketch draws predominantly on Belhoste (1985) as well as his revised and extended English edition (Belhoste 1991). My comments on his impact at the *École polytechnique* draw on Gilain (1989).

winning prizes in the *humanités*, he decided to prepare for the entrance examination to the *École polytechnique*. He received the usual special coaching for this, starting in the fall of 1804. His coach was Dinet, a mathematics teacher at his transformed earlier school, now called the *Lycée Napoleon*. One year later, in the fall of 1805, he took the entrance examination, gaining second place. At only 16 years of age and as one of the youngest students, he commenced studies at what was now a militarized *École polytechnique*. As mentioned in the previous Chapter, his analysis instructor was Lacroix, whose *répétiteur* was Ampère. He completed the two-year course so successfully that he was able to move on to the school of his choice: the best among the French schools of application, the *École des Ponts et Chaussées*, providing access to the highly esteemed engineering profession. During this two-year course, he also learned English, German, and Italian enabling him to read scientific work in these languages. During the practical courses in the summer, he worked on the famous *Canal d'Ourcq* project near Paris.

After completing training, he was accepted into the engineering corps of the *Ponts et Chaussées* at the beginning of 1810. In March of that year, Cauchy was sent to Cherbourg to work on the gigantic project to extend the military harbor. He not only excelled in his engineering works but also carried out mathematical research, though only in fields compatible with an engineering profession. Following Lagrange's advice, he engaged in geometrical research on the theory of polyhedra. In the fall of 1812, after 30 months' service at Cherbourg, he became so ill that he was unable to continue working. He left his post, also giving up his research on geometry, and returned to Paris in the hope of obtaining a scientific post. At the same time, he began to orient his research toward analysis. He first tried to gain Lagrange's post at the *Institut* after his death. This attempt was a failure; the competition was too great. Thanks to repeated grants of leave of absence for health reasons, he was able to remain in the engineering corps. He then spent a long time looking for a post that would enable him to perform research: as an instructor at the École *des Ponts et Chaussées* and then as a librarian. Finally, a post with light duties was found for him in the *Canal d'Ourcq* project, during which he completed his first major work on analysis, his *Sur les intégrales définies* in 1814. The decisive change came about only after the Restoration in 1815. This resulted from his political connections in religious circles. In 1805, Cauchy had been converted to the *Congrégation* by a young *répétiteur* at the *École polytechnique*. This rather secretive, ultra-Catholic, and royalist association organized resistance to the Napoleonic regime, and was consequently trying to gain a foothold in all important public domains. With the Restoration, many of its members gained key positions of power; as an active member, Cauchy could be sure of their support.

In the fall of 1815, when Poinsot was unable to start his analysis course at the assigned time, the governor of the *École polytechnique* forced through a ruling that Cauchy, who still had no teaching experience, should take over the course on the following day. The academic year of 1815/16 ended prematurely in April because of the antiroyalist stance of the students. A reform of the *École* was ordained, and Cauchy was given Poinsot's regular teaching post when it reopened in the fall of 1816. In addition, after three unsuccessful personal applications to the *Institut* up to 1815, Cauchy was offered and immediately accepted the post that had been held by Monge following the latter's political purging in 1816. At the *École polytechnique*, his first step was to radically reform the program for the analysis course. The 1811 decision to base it on the *infiniment petits* was no longer mentioned. However, his influence lasted only about two years due to growing resistance to his theoretical approach. The *École* was increasingly being forced into strict compliance with the requirements of an institute of higher education for engineering, and this even meant a return to explicit instruction in the *infiniment petits*. Cauchy did not just lose all influence over the design of the program; he also had to suffer direct interference in his course. This was further aggravated by major educational and disciplinary problems: in his enthusiasm, he not only continued his lessons beyond the set time allocations but also increased the number needed to complete the course. It seems that his years as a professor at the *École* were a painful experience for everybody involved.

After 1816, and particularly during the 1820s, Cauchy was involved very actively in the projects of the *Congrégation*. He took a very prominent political and religious stance in favor of this papist and reactionary Jesuit–controlled organization that, in some ways, wanted to reverse the Revolution of 1789 through the combined efforts of throne and altar. Cauchy even campaigned for this cause during Academy sessions, which led to his personal isolation. Hence, the revolution in July 1830 was a personal catastrophe for him. He refused to swear allegiance to Louis-Philippe the Citizen King, and followed the papists and ultra-Royalists into self-imposed exile at the beginning of September 1830. His refusal to swear allegiance lost him his post as professor at the *École polytechnique*, the post of *professeur-adjoint* at the *Faculté des Sciences*, and his membership of the engineering corps. He started his life in exile in Switzerland before being called to a chair in mechanics at the University of Turin, the capital of the Kingdom of Sardinia in which the Jesuits were still influential. He resigned this chair in July 1833 when the former King Charles X, now living in exile at Prague, asked him to become the private mathematics and physics tutor for his grandson Henri, *Duc de Bordeaux*. He wanted 13-year-old Henri, who had been declared pretender to the throne by the Legitimist party, to be taught by reliable Catholic teachers. Until the *Duc* was 18 years old in 1838, Cauchy lived with the small court in exile first in Bohemia and later in Gorizia (Friuli). However, his efforts to teach an hour of mathematics and an hour of physics to the young *Duc* each day were evidently a failure. Lacking knowledge of teaching principles, he was unable to overcome his student's disinterest in mathematics. Nonetheless, his loyalty to the Legitimist party led him to persist at this thankless task. It has often been asked whether Cauchy met Bolzano in Prague. This has been confirmed by Rychlík (1962). The ex-king thanked Cauchy for his services by awarding him the title of baron.

In 1838, after completing his private teaching of the *Duc*, his family managed to talk him into returning to Paris. He was able to take up his seat at the Academy again, because this had not required an oath of allegiance. He was now in a position to unleash an enormous barrage of productivity. However, selection and nomination to other offices and functions such as the *Collège de France* were closed to him, first, because of his persistent support for the legitimacy principle, and second, because of his continued support for the Jesuit Catholic party and its campaign for a private Catholic education system in the 1840s under the slogan of *liberté*.

The next revolution, in 1848, liberated Cauchy from his isolation. As the new Republic abolished the oath of office, he was able to become a professor at the *Faculté des Sciences* in 1849, and use his lectures there to commence a synthesis of his theory of functions that had been scattered across so many *mémoires*. Although the Republic soon became a new empire, Napoleon III granted Cauchy a dispensation from the necessary oath of allegiance so that he could continue to teach. He now started to have his own disciples. A final application to the *Collège de France* in 1851 failed again. In 1857, Cauchy died at the age of 68.

French mathematicians belonging to Cauchy's student generation did not attribute an unreserved "rigor" to Cauchy. A typical example is Joseph Bertrand (1822–1900), a leading mathematician in France during the second half of the nineteenth century. He strongly criticized the hagiographic style of the first biography by Valson (1868). Valson claimed that Cauchy had never passed by any object before founding and explaining it completely, and doing this according to the most rigorous standards. Bertrand countered that if there was one famous mathematician to whom this did not apply, this was, without doubt, Cauchy. There was much for which one could praise him, but not specifically for that which everyone considered he lacked completely: "When one can praise such rare and exceptional talents in him, it is a travesty against his memory to cite precisely he who, according to all and evidently in error, had completely let him down" (Bertrand 1869, 206).

Bertrand did not just mean that the sudden freedom to publish in the Paris Academy after 1840 had seduced Cauchy into publishing something every week, even unfinished projects and infertile approaches. Indeed, only 2 of his 20 publications on the theory of substitutions are masterworks; each of the remaining 18, in contrast, contains searches for alternative paths that were finally abandoned. He also meant that Cauchy's work is sometimes distorted through serious inaccuracies and remarkable oversights (ibid., 208). Jacobi makes a very similar comment in a letter to his friend Dirichlet in the middle of 1841, both representatives of the new rigor in Germany:

> Cauchy has become unbearable. Every Monday, broadcasting the known facts he has learned over the week as a discovery. I believe there is no historical precedent for such a talent writing so much awful rubbish. This is why I have relegated him to the rank below us. It is sad when one reaches the point that he announces a great discovery and one hardly even looks in the belief that it may

well be nonsense. And yet, once again, there may be a few pages of great beauty and lasting worth [...].[2]

2. Conflicting Reception of Cauchy in the History of Mathematics

The traditional historiography of mathematics presents Cauchy as the founder of modern mathematical rigor who conjured, as if from thin air, the standards and methods that continue to be valid today. This viewpoint is presented most extensively in Grabiner (1981). She gives a relatively isolated analysis of Cauchy's achievements and limits the study of his context to the "great names" among his predecessors such as Euler and Lagrange. This has contributed to a more recent, one-sided appraisal of Cauchy.

This new fashion of analyzing Cauchy's basic concepts more precisely and reevaluating his achievements has taken two different directions. However, it is based on one common cause: This is an apparently erratic element in Cauchy's conceptual edifice that contradicts the standard of rigor attributed to him. It is his theorem of 1821 stating that the limit function of a convergent series of continuous functions is itself continuous, a theorem that, as a consequence of the further development of the rigor standard, requires *uniform* convergence. The first direction to pick up on this erratic element was a mathematical one. In 1966, one of the two developers of nonstandard analysis (NSA), Abraham Robinson, published an approach in his NSA textbook showing how Cauchy's theorem could be proved by transferring its terminology into NSA and using hyper-real numbers. Robinson (1966, 271) expressly assumes Cauchy's theorem to be false; he even talks about a "famous error." On the other hand, he assumes that even Leibniz had not operated with the standard number domain of real numbers, but with an "extended number system" that included infinitesimals (ibid., 266).

The second direction came from the philosophy of science, from epistemology, and it received considerable attention because of the enthusiasm for revising traditional beliefs in the history of science and reinterpreting the discipline from a theoretical, epistemological perspective generated by Thomas Kuhn's (1962) work on the structure of scientific revolutions. Applying Popper's favorite keyword of fallibilism, the statements of earlier scientists that historiography had declared to be false were particularly attractive objects for such an epistemologically guided revision.

---

2   *Nachlass* Dirichlet, Staatsbibliothek Preußischer Kulturbesitz, Handschriftenabteilung. I am currently preparing an edition of Jacobi's letters to Dirichlet.

The philosopher Imre Lakatos (1922–1972) was responsible for introducing these new approaches into the history of mathematics. One of the examples he analyzed and published in 1966 received a great deal of attention: Cauchy's theorem and the problem of uniform convergence. Lakatos refines Robinson's approach by claiming that Cauchy's theorem had also been correct at the time, because he had been working with infinitesimals. Hence, on the one hand, he disagrees with Robinson, because he no longer claims that the theorem would be correct when transformed into modern NSA, but considers it to have been correct all the time. On the other hand, he assumes that Leibniz and Cauchy had operated with an extended, non-Archimedean number domain (Lakatos 1978, 47). The subsequent discussion, in part provoked by Lakatos, generated a great number of papers leading, in sum, to a much more precise conceptual analysis of the work of Cauchy than before. One notable element in this discussion is the contributions of Detlef Laugwitz, the second developer of NSA, who has been analyzing Cauchy since the 1980s with the goal of ruling out every attribution of error to his work. In a discussion with Pierre Dugac in 1986, he formulated his maxim as follows: Cauchy was such a great mathematician. Everything else he did was correct. Therefore, when faced with a choice between two possible interpretations of his concepts, one has to go for the one that makes his proof correct.[3]

In 1984, Giusti published a decisive article on Cauchy's "errors" that was the first to analyze Cauchy's basic concepts within their own context, examining not only the convergence theorem but also a number of further theorems that are not in line with modern concepts, such as the global continuity and the interchangeability of the order of a function of several variables and the commutability of limit and integral. This spurred Laugwitz to even more detailed attempts to banish the error and confirm that Cauchy had used hyper-real numbers. On this basis, he claims, the errors vanish and the theorems become correct, or, rather, they always were correct (see Laugwitz 1990, 21).

In contrast to Robinson and his followers, Laugwitz (1987) assumes that Cauchy did not use nonstandard numbers in the sense of NSA, but that his *infiniment petits* were infinitesimals representing an extension of the field of real numbers. Laugwitz started his error-banishing program with an active partner: Detlef Spalt. In a series of publications, they tried to dismiss errors in Cauchy by saying that his statements were based on completely special meanings: for example, that by continuity, Cauchy meant what Spalt calls *Feinstetigkeit* (see Laugwitz, 1990, 21). However, Spalt has increasingly distanced himself from this joint program. Remarkably, it is precisely the analysis of the relation between the number, variable, and function concepts in Cauchy that leads him to conclude that he did not use hyper-real numbers at all. In 1996, Spalt published a book on the results of his exegesis of Cauchy's basic concepts and his

---

[3]   In the discussion following Laugwitz's paper on the mathematical infinite in Euler and Cauchy presented on June 2, 1986, at the conference *Konzepte des Mathematisch Unendlichen*, June 2–4, 1986, Bochum, Germany.

conceptual tools. He claims that these interpretations show that Cauchy's statements are correct within the framework of classical mathematics. However, there is one exception in his exegesis: He has failed to attain a coherent interpretation precisely in the case of the convergence theorem (Spalt 1996, 111).[4]

Although I disagree with a number of Spalt's interpretations, his fine-grade analyses have completed part of a task that I had originally set for myself. Hence, in some areas, I shall be able to refer to his book. However, in general, and particularly from a methodological perspective, I cannot agree with Spalt. He pays a high price for his conclusion that Cauchy's theorems exhibit a relative internal coherence. He claims that Cauchy did not establish any innovative concepts that shaped the future development of mathematics; he only brought the analysis of the eighteenth century to completion:

> The world of concepts worked out by Cauchy was not suitable for addressing the new problems facing mathematics. Regarding his concepts, Cauchy was not a teacher and initiator, but a follower (namely, of Lacroix) and one who brought the old to its completion (Spalt 1996, 28).

By retracting Cauchy's importance, Spalt raises just as many methodological problems as Laugwitz with his view that Cauchy developed a singular set of concepts all of his own that other mathematicians did not understand:

> The widespread opinion that it is precisely the fundamental theorems in Cauchy's books that are false is based on a false interpretation of his concepts; they are simply not identical with the ones introduced later (Laugwitz 1990, 20 f.).

The claim that Cauchy produced an isolated mathematics becomes even more problematic when one considers that the controversy basically centers on such fundamental concepts as continuity and convergence. Laugwitz himself talks about the "essentially didactic components of the infinitesimals in Cauchy" (ibid., 18), and it is in textbooks that we find these basic terms and not in isolated research memoirs. Therefore, in the next section, I shall discuss the methodological implications of the hermeneutic problem before proceeding to my own analyses.

3. Methodological Approaches to Analyzing Cauchy's Work

The problems in interpreting Cauchy's texts are essentially methodological. This provides an opportunity to apply the discussion on general methods in Chapter I to a specific concrete case.

---

[4]   One reason for this is Spalt's assumption that Cauchy also succeeded in retaining identical meanings for his concepts over decades.

As mentioned in the previous section, nearly all prior research on Cauchy pays insufficient attention to his historical background. This is why it was rather easy for the "revisionist" positions regarding the attribution of error to introduce alternative interpretations. The outcome of their contributions is that historiography no longer assumes Cauchy's basic concepts to be unequivocal in meaning, or contemporary concepts such as "continuity" to be identical to the concepts bearing the same names today. Grattan-Guinness (1990a, 716 ff.), for example, now avoids talking about Cauchy's "errors," and considers that counterexamples are valid only for quite specific interpretations. He assumes that Cauchy used real and not non-Archimedean values. Bottazzini (1992, LXXXII ff.) also stresses that Cauchy's definitions allow a wide range of interpretations, and states expressly that Abel, for example, assumed a different concept of convergence to that of Cauchy. In the most recent analysis of this much-discussed topic—although without referring to Spalt's book of 1996—Lützen (1999) summarizes that it is "difficult to save Cauchy completely," and it is "hard to avoid the conclusion that Cauchy's concepts were somewhat improper at this point" (ibid., 210).

Although this is a rather remarkable shift compared with the earlier attribution of partial error, it is still a radical rejection of the rigor attributed to Cauchy, particularly in his basic concepts. In this section, I shall try to cast some light on these issues.

In a paper on concepts of the infinitesimal, presented at the Second Workshop on the History of Modern Mathematics (Göttingen July, 1991), I referred to a proposal I made in 1983 on an extended concept of hermeneutics (see Schubring 1986). I did this in order to counter the latest ideas of Spalt and Laugwitz (Laugwitz & Spalt 1988, Laugwitz 1990), who had practically assigned a solipsistic mathematics to Cauchy in order to shore up their claim that he was free of error. Although Laugwitz and Spalt (1988, 7 ff.) do draw on a concept of hermeneutics with their concept of "conceiving" and their rejection of the "classical" methods, this proves to be the abbreviated conception of hermeneutics in the German tradition since Dilthey: Their concept of conceiving remains immanent and descriptive. It does not forge any link between the text and its context.[5] While such an approach may still be justified for literary or philosophical texts, it is inappropriate for mathematical ones. Mathematical activity is always communication as well, at least within the author's own cultural context. In contrast, the "universe of discourse" that Laugwitz and Spalt (1988, 11) claim for Cauchy is more like a hermetic self-discussion, an inappropriate conception for a mathematician who took such pains to disseminate his concepts. This neglect of the context in traditional approaches is not overcome by "revisionist" approaches either. Indeed, one could even say that they reinforce it. They do not apply communication, with, for example, contemporary mathematicians, as a means of clarifying the meaning of concepts.

---

5   Laugwitz (1989, 201) has even categorically rejected statements on the impact of the context of the *École polytechnique* as being "ridiculous."

Since the texts in question come almost exclusively from textbooks for an engineering school, and their specific purpose was to convey the meaning of concepts and not some hermetic closure of a *private* mathematics, Dilthey's "conceiving" approach is particularly inappropriate. The conceptual edifice that Cauchy built up and presented in his lectures was not only exposed to questions from and discussions with students, but was also the topic of final examinations. Since examiners, in line with general practice, were always external, they must also have been particularly familiar with the meaning of the basic concepts. Since every lecture was assigned a *répétiteur*, who went through the content of the lectures with the students, there must have been a permanent intersubjective communication.

As formulated in Chapter I as a program for a material hermeneutics, meaning emerges not only through introspection but also through analyzing the relevant contemporary discussion context and taking into account the contributions of the specific expert community within its cultural and social embedment, without restricting this to isolated "first-class" contributions.

Although, in his concluding analysis, Spalt (1996) declares his maxim to be "embedding Cauchy's conceptual world within the course of historical development," he restricts himself—just like the traditional methods he otherwise rejects—to the great "predecessors (particularly d'Alembert, Carnot)" and to Lacroix, whom he presents as his "teacher" (Spalt 1996, 21). However, even the evaluation of these standard authors is only marginal. This is why Spalt confirms the solipsistic position here as well: Cauchy apparently found himself in a historical situation in which it was possible for "a creative mind to generate its own theory all by itself. With Cauchy, we have such a person before us" (ibid., 26).

If, in contrast, we want to analyze the development of Cauchy's concepts effectively within his context, we can take the following approaches:

1. We can examine his direct context, the *École polytechnique*, and communication with colleagues at this *École*.

2. We can evaluate his communication with other mathematicians.

3. We can place Cauchy's conception in relation to the conceptual developments taking place in the broader expert community—those developments that have already been reconstructed in this book.

I shall take the first approach by starting with Cauchy's famous, repeatedly quoted motto from the introduction to his *Cours d'analyse algébrique* (see Chapter IV.4.). The recourse to geometry proves to be his specific way of dealing with the concept of the compromise between the *limites* and the *infiniment petits* practiced since 1811. In particular, further information will be gained from his close cooperation with Ampère.

I shall tackle the second approach by studying what can be gained from Cauchy's communication with other mathematicians, a field that has been strangely neglected by the history of mathematics up to now. His communication with the Abbé Moigno proves to be particularly informative.

In the third approach, I shall not just analyze Cauchy's conceptual innovations in relation to Lagrange and Euler, as in the work of Grabiner, but in relation to the broader group of mathematicians covered in the present book.

4. Effects of the Context

In the introduction to his *Cours d'analyse algébrique* of 1821, Cauchy characterizes his methodology with what has now become a famous phrase. One can understand it as his programmatic motto; hardly any of the many publications on Cauchy neglects to cite it:

> As for methods, I have attempted to accord them all the rigor that is expected of geometry, in such a way as never to have recourse to arguments drawn from the generality of algebra (Cauchy 1821, ij).

At the same time, most comments on this citation reveal a certain astonishment over why Cauchy not only contrasts the "rigor of geometry" with the "generality of algebra," but also even rates geometry higher as an incarnation of rigor while devaluing algebra and attributing it only "inductions" and an unreliable extension of the applicability of its formulas.[6] After one has taken a close look at the debate on the foundational concepts, particularly in France, the reasons for this stance become completely clear: it is due to the *retour* to *synthèse*, to geometry as the highest value in mathematics, following the peak of the analytic method during the French Revolution, and particularly after the rulings on basic principles at the *École polytechnique* in 1810/1811 that reinstated geometry as the ultimate instance. Moreover, in Cauchy's motto, we recognize not only d'Alembert's campaign against the *inconvéniens* of the generality of algebra but also precisely Carnot's reception and actualization of d'Alembert's concept on the relation between geometry and algebra.

That the concepts on methods in his textbook were shaped by the context of the *École polytechnique* is shown additionally and impressively by the clear change in Cauchy's use of the *infiniment petits*. In the same introduction, Cauchy explains in an apologetic or distanced tone that when introducing continuity, he is unable to avoid presenting the main properties of infinitely small quantities: "In speaking of the continuity of functions, I have not been able to dispense with making the main properties of infinitely small quantities known" (ibid., ij).

The change to the use of infinitely small quantities as foundations becomes very clear when we look at documents from the beginning of Cauchy's teaching activities at the *École polytechnique*. The first such document is the notes taken during Cauchy's first course on analysis in the academic year 1815/16 by

---

6   See, for example, Bottazzini (1992, XVI) or Lützen (1999, 200).

Auguste Comte (1798–1857).[7] These carefully composed notes are preserved—along with further notes on Poisson, Arago, and Poinsot—in Comte's *Nachlass* in the *Maison Auguste Comte* at Paris. Comte belonged to the older of the two student cohorts who were sent down from the *École* in the spring of 1816 and were unable to either complete or continue their training. Cauchy's analysis course was actually aimed at the second-year students and taught integral calculus in line with the existing teaching program. The eighth lesson in this course in the fall of 1815 (and the first documented in Comte's notes) deals with integration based on the theory of series and uses the concepts of Lagrange, Lacroix, and Ampère. Cauchy relies on the concepts of the *limites*; infinitely small quantities, in contrast, are not addressed at all.

He labels the expression

$$\frac{f(x_1 + h) - f(x_1)}{h}$$

as a *rapport* whose *limite* is the *coefficient differentiel* $f'(x_1)$. Here, Cauchy always calls the quantity h a *quantité très petite* or *extrèmement petite*. Because he simultaneously conceives such *rapports* as arithmetic means, Cauchy develops upper and lower estimates for integrals and uses these to test the convergence of series.[8]

The teaching program for the analysis course in 1816/17 is even more typical. Not only did the closure of the *École* from the spring to the end of the academic year provide ample time to revise the program; this was also the only period in which Cauchy was in a position to decisively shape it, his only advisor being Ampère. Ampère's *Nachlass* contains an outline proposal for the analysis program written by Cauchy.[9] Despite some omissions, its wording agrees exactly with that of the finally ratified and printed version (Gilain 1989, 106 f.).[10]

It is not just notable that the *distinction des fonctions continues et discontinues* appears right at the beginning, as the third topic. In addition, infinitely small quantities do not appear at all. Although no method for foundations is mentioned, the structure of the outline shows very clearly that it applies the *limites* and not infinitely small quantities. The outline is divided into four sections corresponding to the classic conception of Euler: *Analyse algébrique*, *Calcul aux différences*, *Calcul différentiel et intégral*, and *Applications du calcul différentiel et intégral à la géométrie*.

---

[7]   Thierry Guitard (1986) was the first to report the existence of these notes. Unfortunately, they start only with Cauchy's eighth lesson (see Gilain 1989, 47).

[8]   Notes of Auguste Comte, Cours d'analyse, 2ᵉ année. Cauchy Professeur, Fol. 1 ff. Maison Auguste Comte, Paris.

[9]   Archives de l'Académie des Sciences, Paris: *Nachlass* Ampère, cart. 4, chap. 4, chem. 74. An English translation of this draft is published in Belhoste (1991, 304 ff.).

It is precisely the section *Calcul aux différences* (even though dropped from the final version) that confirms the classical conception: introduction of operations with finite differences followed by the passage to the limit and operations using the finite as an analogy for the limit. The *Registres d'instruction* for these Cauchy lessons confirm that this, the first analysis course he designed himself, can even be viewed as an apotheosis of the *limites* method (see, for more details, below). At the beginning of the section on differential calculus, he introduces the *coefficient différentiel* as a basic concept in the style of Lagrange and Lacroix (Gilain 1989, 51 ff.). Another new aspect is not only that the integral is already introduced at the beginning of differential calculus, but also that it is not defined as a formal inversion but, independently as a definite integral, as the "sum of the elements" (ibid., 53).

The fact that Cauchy personally introduced his own lesson on infinitely small quantities in the second repetition of the course on analysis in 1820/21—as the third lesson directly before the discrimination between continuous and discontinuous functions—is therefore in no way attributable to a personal decision to change the basic concepts of analysis. Even Cauchy's entry in the *Registres* on the content of this third lesson shows that the compromise concept on linking together the *infiniment petits* and the *limites* was once again in force:

> On infinitely small and infinitely large quantities. The main properties of infinitely small quantities of different orders. Indeterminate expressions. On the ratio between two infinitely small or infinitely large quantities. The limit of the expression $\frac{\sin\alpha}{\alpha}$ (α being an infinitely small quantity) and the expressions $\frac{f(x)}{x}, (f(x))^{\frac{1}{x}}$ (x being an infinitely large quantity) (ibid., 63).[11]

Indeed, since Cauchy's first new analysis course, the *École* had been coming under increasing pressure to teach the method of the *infiniment petits*. As early as the fall of 1817, an amendment was made requiring their use in both the analysis program for the geometric applications and in the mechanics program (previously developed by Cauchy). The explicit return to the resolutions of 1811 was pushed forward not only inside the *École* by physics professors but also outside by the *Conseil de perfectionnement* and, above all, by the examiners. The pressure for this return was not a personal attack on Cauchy; it mirrored the objective development of the *École polytechnique* into a specialized engineering school without any broader scientific tasks. Nonetheless, it has to be admitted that Cauchy also helped to strengthen this pressure.

---

[10]   Rather strangely, Belhoste (1991, 305 ff.) publishes the program of the *previous* year 1815/16 with a structure unchanged since 1811 as the "Official First-Year Program of the Analysis Course (1816)."

[11]   Since Cauchy did not address infinitely small or large quantities in such detail in his textbooks, it is unfortunate that no surviving notes have been found on these later lessons.

What triggered this was Cauchy's first repetition of the first-year analysis course in 1818/19. Instead of the planned 50 lessons, he held 63, extending the end of the course from the beginning of March to the beginning of April 1819. This disrupted the organization of all the other courses. At the *Conseil d'instruction* sessions, Cauchy was criticized strongly, particularly by physics professors complaining about how weak his students were in infinitesimal calculus (ibid., 11). They demanded that he teach the evidently neglected *infiniment petits*:

> He [the professor of physics, Petit] insists further on the consideration of infinitely small quantities, a method with which the pupils are little familiar and which is however so useful for them to know (ibid., 133).

Cauchy defended himself by saying that he certainly wanted to use the *infiniment petits* in the *partie analytique*, but that the students were still not sufficiently familiar with this because he had not yet been able to handle the geometric applications (ibid., 134). In the following months, the pressure on Cauchy increased, not only because of the length of his lessons but also through criticisms from both examiners and internal critics. For example, Arago complained that the analysis course should be a direct preparation for mechanics and should not investigate abstract concepts (ibid., 15). Among the external examiners, it was repeatedly Prony—departing radically from his original stance as a pure algebraist—who criticized Cauchy sharply for insufficient use of the *infiniment petits* and ideas from geometry (ibid., 18 ff.). The pressure finally resulted in the introduction of a factual censorship over the obligatory printed teaching materials. It was agreed that the conception of the courses should not be left to the teachers to decide, but that these had to comply with guidelines laid down by the *Conseils* (ibid., 18).

Under this pressure,[12] Cauchy increasingly incorporated the conception of the *infiniment petits* into his work, in both textbooks and beyond. However, he developed his own version acknowledging Carnot (see below, Section 6.4.).

In his next textbook on differential calculus in 1823, Cauchy points out expressly that he has adopted a compromise concept and that the "simplicity of the infinitely small quantities"—once again, the familiar terminology of the 1811 resolutions—disagrees with the "rigor" that he wished to achieve in his 1821 textbook. This confirms the increasing pressure:

> My main aim has been to reconcile rigor, which I have made a law in my *Cours d'Analyse*, with the simplicity that comes from the direct consideration of infinitely small quantities (Cauchy 1823, v).

Nonetheless, Cauchy's context at the *École polytechnique* did not just consist of hostility and pressure. It also contained a good working relationship with one colleague: Ampère. As the répétiteur in Lacroix's course, Ampère had not just been Cauchy's teacher. The two men also had many religious and political

---

12   Ampère finally succumbed to the continuous pressure and stopped working at the *École polytechnique* in 1828. He then restricted his teaching to the *Collège de France* (Belhoste 1991, 85; Hofmann 1996, 142).

beliefs in common. As they each ran the analysis course in alternate years after 1816, they coordinated their teaching and worked together. The general view in the literature is that this cooperation took the form of Ampère simply acknowledging Cauchy's innovations. This is particularly because the literature refers exclusively to the publication of the beginning of Ampère's analysis course in 1824 (see Bottazzini 1992, LXXXIV) and not to the earlier manuscripts preserved in his *Nachlass*.

We have already got to know Ampère's innovative achievements in earlier Chapters: his technique of nesting intervals, his precise specification of statements by limiting them to intervals, the development of a limit calculus using inequalities, his attempt to prove the existence of the derivation, as well as his development of an independent concept of the definite integral compared with the indefinite integral (Chapter V.2.4.3.). We have also looked at his reflections on the foundations of the number concept and his distinction between quantities and "abstract numbers" (Chapter V.3.2.).

The concept of continuity, in contrast, does not emerge as a basic concept in his lectures before 1815; Ampère continued to follow tradition. Olivier's notes of 1811–1813 do not mention the concept, although the section on functions is written down *sous la dictée de M. Ampère*. Instead, Ampère gave detailed differentiations of the types of functions into algebraic, transcendental, rational, irrational, and so forth.

Manuscript Number 70 in his *Nachlass* contains the same differentiations on the function concept. However, these are joined by the discrimination between *continue* and *discontinue*. This means that it is a later text. However, this note contains no definition of continuity.

Olivier's notes on Ampère's two-year analysis course from 1811 to 1813 confirm, in contrast, that the innovation of basing integral calculus on the definite integral considered so characteristic of Cauchy had actually been developed beforehand by Ampère. The section on integral calculus begins with two new approaches to foundations. The first consists in a previously unattempted proof of the existence of the integral:

> In order better to understand the purpose of this [integral] calculus, and to understand its existence, it is necessary first of all to show the possibility of finding a function by means of its differential.[13]

His proof based on developing into a series leads him to the values of the primitive function at two interval limits, so that *l'integrale définie* already becomes a basic concept, his second new approach to foundations.

---

[13]   AEP, Olivier's notes: Leçon 37, Calcul Intégral, fol. 83.

5. The Context of Cauchy's Scientific Communication: "I'm Far from Believing Myself Infallible"

> Je suis loin de me croire infaillible
>
> —Cauchy to Moigno, 1837 (see Appendix C)

Whereas the last section discussed the context of Cauchy's teaching activities (up to 1830), the present section examines his scientific communication. This was most intensive at the *Académie des sciences* from 1816 onward, interrupted only by his years in exile. Had Cauchy proposed highly anomalous or isolated mathematical conceptions, then sooner or later, this would have to have expressed itself in scientific controversy and explicit statements from one side or the other. Indeed, conflicts with Cauchy can be found, and these did result in public statements. However, they did not concern his mathematics, but his extreme political and religious opinions that were shared by hardly any other members of the *Académie* (see Belhoste 1985, 114 ff.). As far as mathematics is concerned, there seem to have been no differences of opinion regarding basic concepts. Such differences would have had to become apparent, because one of Cauchy's tasks was to compile reports on manuscript submissions, work that he carried out in a large number of commissions with a variety of different colleagues. One conflict over basic principles did emerge at the *Académie*, but this was not with Cauchy but Poisson. Remarkably, it was Poisson's isolated view on the reality of infinitely small quantities that triggered opposition from his colleagues, including Cauchy, whose opinion was published by Moigno (see below).

One of the standard methods in the history of science is to analyze the correspondence of major scientists in order to determine how their achievements were received. The response of contemporaries is an excellent indicator of both the newness of scientific contributions as well as resistance to them. Particularly with an innovator like Cauchy, the use of such methods would even seem essential. One could anticipate that studying the reactions of contemporary mathematicians and controversies would deliver more detailed information on the intended interpretations of concepts. One could expect that modern authors who attribute interpretations of concepts to Cauchy that deviate greatly from the traditional reception would particularly apply these methods to seek confirmation for their hypotheses. However, none of them have done so. Spalt and Laugwitz's approach of seeking meaning exclusively through the internal "conceiving" of a text is typical.

Other historians of mathematics have also not tried to reconstruct Cauchy's communication with contemporary scientists. Despite publication of his complete works—commenced by the *Académie* in 1882 and completed only in 1974—no efforts have been made to compile his *Nachlass* or, in particular, his letters. As Belhoste (1991) determined, Cauchy's descendants submitted parts of his *Nachlass* to the *Académie* in either 1936 or 1937. However, for reasons unknown, the *Académie* simply returned this material, and the family subsequently destroyed it (ibid., 362 f.). In a personal communication, B. Belhoste even told me that this destruction evidently did not take place until the 1950s! All that is left is a few workbooks in the Sorbonne library (ibid., 362). In 1989, B. Belhoste managed to find some remaining family letters. This consists of correspondence with his parents while he was working at Cherbourg from 1811 to 1812, and with his family while he was in exile from 1831 to 1837. However, Belhoste told me that these letters mention his scientific work only in passing. Although C.-A. Valson (1826–1901), himself a professor of mathematics, was granted access to the scientific *Nachlass*, he did not analyze or examine these sources when writing his hagiographic biography (Valson 1868).

Although nearly all letters *to* Cauchy seem to be lost completely, it may still be possible to trace and analyze letters *from* Cauchy to other scientists. I decided to perform an extensive search. Although the outcome is rather modest, it does provide a basis for further research. Appendix C reports the correspondence found in the calendar style used by Roger Hahn to reconstruct the correspondence of Laplace. Here, I shall give a short overview of relevant mathematical personalities.

In France, the main field of correspondence, possible finds are restricted by the lack of a *Nachlass* for more than a small portion of major mathematicians.

1. Ampère. Although his *Nachlass* in the archive of the *Académie des sciences* contains scientific correspondence, none of it is with Cauchy.

2. Jean-Baptiste Biot (1774–1862). The correspondence preserved in his *Nachlass* at the *Bibliothèque de l'Institut* is not very extensive. It does not contain any letters from Cauchy.

3. Carnot. The catalogue for his library at Presles lists several offprints that Cauchy sent to him before 1815. However, any accompanying letters have not been preserved.

4. Gaspard Gustav de Coriolis (1792–1843), *répétiteur* in Cauchy's analysis courses. Although no *Nachlass* has been preserved (Grattan-Guinness 1990a, 1052), one letter he received from Cauchy is printed in Liouville's journal (*Oeuvres*, I, Vol. 4, 38 f.).

5. Jean-Baptiste Fourier (1768–1830). Louis Charbonneau at Montreal, who has performed the most intensive analyses of Fourier's work and assessed his extensive *Nachlass* in the *Bibliothèque Nationale*, has told me that he has been unable to find any letters to or from Cauchy. Nonetheless, Fourier refers to works by Cauchy in a number of manuscripts. In one of these texts—written between

1819 and 1830—Fourier discusses Cauchy's objections to the integrals in his work on heat conduction.

6. Lacroix. Considering the breadth of his activities, the correspondence preserved in his *Nachlass* in the *Bibliothèque de l'Institut* can certainly represent only a small part of that produced. However, it contains no correspondence with Cauchy.

7. Lagrange. His *Nachlass* is preserved at the same location. It also does not contain any correspondence with Cauchy.

8. Laplace. His *Nachlass* has been lost. However, through extensive searches, Roger Hahn has managed to trace a considerable amount of his correspondence (Hahn 1982). Adding to his "Calendar," he told me personally that although there are two letters about Cauchy from before 1815, he has not yet found any direct correspondence between the two.

9. Legendre. No *Nachlass* has been preserved.

10. G. Libri (1803–1869). His *Nachlass* preserved in the *Biblioteca Moreniana* at Florence contains several letters from Cauchy, all from the period before 1830 when Libri fled from Italy to Paris. They are all mathematical in content. Interestingly, when he was in exile, Cauchy also used Libri as a mediator to disseminate his research (see Appendix C).

11. Joseph Liouville (1809–1882), a competitor with Cauchy in several job applications and founder of the journal succeeding that of Gergonne. J. Lützen and E. Neuenschwander have traced various parts of his *Nachlass*. These include the draft of a letter from Liouville to Cauchy in 1856 in which he complains about Cauchy submitting a report on a manuscript for the *Académie* all by himself without consulting the commission established for this purpose (Neuenschwander 1984, 111 f.).

12. Poinsot. Like Lacroix and Lagrange, his *Nachlass* is also preserved in the *Bibliothèque de l'Institut*. However, it contains very few letters and none from Cauchy.

13. Siméon-Denis Poisson (1781–1840), colleague at the *École polytechnique* and competitor on a variety of mathematical research projects as well as propagator of an alternative conception of the *infiniment petits* (see below, Chapter VIII). No *Nachlass* seems to be preserved.

14. Prony. He has an extensive *Nachlass* preserved at the *École Nationale des Ponts et Chausées*. However, this contains only two memoirs from Cauchy containing handwritten dedications.

15. Adhémar-J.-C. Barré de Saint-Venant (1797–1886), an engineer. The archive at the *École polytechnique* contains several letters to him from Cauchy.

Turning to Cauchy's non-French correspondents, the first place to search is among German mathematicians:

16. Gauß. Although an almost complete *Nachlass* has been preserved in the manuscript department at Göttingen University, there is a conspicuous lack of letters from Cauchy.

17. Dirichlet. The *Nachlass* section of the manuscript department of the Berlin *Staatsbibliothek* contains one short letter from Cauchy dated July 25, 1839, in which he invites Dirichlet to join him for dinner on one of his visits to Paris.

18, Friedrich Wilhelm Bessel (1784–1846). The extensive correspondence in his *Nachlass* preserved in the archive of the Berlin Academy contains no letters from Cauchy.

19. C.G.J. Jacobi. His *Nachlass*, also in the archive of the Berlin Academy, contains only a few letters. There are no known letters to him from Cauchy.

The catalogue of autographs at Berlin listing letters from scientists in German libraries contains only the letter to Dirichlet and no other strictly scientific correspondence (see Appendix C).

It should be possible to find further letters from Cauchy in the *Nachlässe* of Italian mathematicians (see also Appendix C).

From the relatively small number of letters traced so far, the only one providing the methodological detail I was seeking for a supplementary discussion or explanation of his work is the letter to Coriolis. It contains, among other things, a further clarification of his continuity concept (see Section 6.5.).

In light of the unsatisfactory situation of having access (at present) to only a minimal part of Cauchy's mathematical communication, we are most fortunate in possessing reports on Cauchy's thinking from a highly communicative popularizer of his concept who was authorized by Cauchy himself. This is François Napoléon-Marie Moigno (1804–1884), known as Abbé Moigno. Trained in the Jesuit seminaries after 1815, he decided to join the Jesuit order in 1822. His superiors ordered him to study mathematics and physics. Therefore, in 1824, he started to study at the *École Normale Supérieure*, attended Cauchy's lectures at the *Faculté des Sciences* and the *Collège de France*, and became his disciple. In 1830, he also fled France. He kept in touch with Cauchy, but did not stay so long in exile. Back in Paris in 1836, he became a mathematics lecturer at the new church-run *École Normale*. He also set up a physics department. This granted him general scientific recognition, and provided good contacts with leading Parisian scientists irrespective of politics and religion. Following Cauchy's return to Paris, the two men worked together closely, both in planning Moigno's mathematics lectures at the *École Normale* and on religious projects (Belhoste 1991, 177 f.). One outcome of this cooperation was Moigno's publication in 1840 of the first volume of a work on differential and integral calculus. This was followed in 1844 by the second volume on integral calculus. The subtitles of the two volumes already show Cauchy's involvement in writing these books: "rédigées d'après les méthodes et les ouvrages publiés ou inédits de M. A.-L. Cauchy" (1840), and "rédigées principalement d'après les méthodes de M. A.-L. Cauchy" (1844).

In these textbooks, Moigno actually brought Cauchy's own earlier textbooks into line with Cauchy's current conceptions by integrating his intervening publications along with some unpublished texts that Cauchy made available to him for that purpose. Moigno could speak as the authorized popularizer of Cauchy, as he himself emphasizes in the introduction to the 1840 volume:

> I know moreover that M. Cauchy, who has allowed me to add to the glory of being his pupil the even greater glory of being his friend, would willingly accept me as an intermediary and as his echo in scientific communications with the public (Moigno 1840, xiij).

It is now extremely interesting to see how Moigno treats the *infiniment petits* in this textbook authorized by Cauchy but produced and published outside the context of the *École polytechnique*. He not only fails to use them as a basic concept, but also even explains explicitly why they are inappropriate as such. This makes Moigno the first writer to pick apart the traditional claim in favor of their purported *simplicité*—evidently with the support of Cauchy. Remarkably, none of the authors who have placed Cauchy in the context of NSA have considered Moigno's assessment of the *infiniment petits* freed from the original *École* context. The only function that Moigno continues to assign to the *infiniment petits* is a minor, instrumental one: in ascertaining the limits of fractions in cases that are described even today as L'Hospital's rule (see, for more detail, Section 6.4.).

In 1840, Moigno became entangled in unsuccessful financial speculations. The Jesuits ordered him to leave Paris and take a teaching post in Canada. However, he refused, and finally left the Jesuit order. This did not impact on his cooperation with Cauchy in any way (Belhoste 1991, 179). After Cauchy's death, Moigno published Cauchy's remaining unpublished lectures from his period of exile in Turin in 1833. Although these texts contain major statements on Cauchy's number concept, they have also not been taken into account by the Cauchy exegetes (see below, Section 6.1.).

Moigno's textbooks are also interesting because they show—particularly explicitly in the prefaces—how intensively he and his master took account of the work of contemporary European mathematicians, evaluated it, and integrated it into their own work. This shows that the image of a Cauchy working in isolation, trapped in his own system, is completely inappropriate.

6. Cauchy's Basic Concepts

Having carefully prepared the ground by reconstructing the context of the mathematical tradition, we are now in a position to present and analyze Cauchy's basic concepts.

6.1. The Number Concept

Cauchy starts his foundational textbook, the *Cours d'analyse algébrique*, by presenting his ideas on the number concept. Whereas most historical literature has simply passed over these sections, Spalt is right to insist that without these basic concepts, it is impossible to correctly understand Cauchy's further concepts such as function and limit. It is particularly interesting to see that Cauchy himself considers it necessary to supplement his short introduction with a 35-page appendix presenting not only his conception of positive and negative numbers but also a detailed account of the resulting operations. The detail of this *Note I* confirms how difficult it was to introduce an alternative basic concept to students and readers trained so one-sidedly in Lacroix's algebra.

Whereas Ampère had been unable to achieve a personally satisfying formulation of the number concept, Cauchy simply bypasses any reflections that he finds too fundamental. He adopts many elements from Ampère's ideas, integrating them into a practicable concept. While differentiating strictly between *nombre* and *quantité*, he also assumes that numbers are an outcome of measurement, and that the basic operations in the quantity system are determined by *accroissements* or *diminutions* (Cauchy 1821, 1 f.). However, his definitions of some central elements differ from those of Ampère: The most fundamental difference is that Cauchy considers the epitome of the number to be exclusively *absolute* numbers. This position, which draws particularly on the tradition of both d'Alembert and Bézout, has two major consequences:

1. The first concerns negative numbers. For Cauchy, these are not numbers, but *quantities*. He views this discrimination between numbers and quantities as being precisely that between absolute numbers and numbers possessing the additional quality of a (preceding) sign. Hence, he considers that positive numbers can appear under two kinds of name: on the one hand, as absolute numbers, as *nombres*; on the other hand, as numbers with a sign, as *quantités*. He defines negative numbers through the concept of *quantités opposées*. He starts this by introducing the *absolute value:* He calls the absolute value of a *quantité* its *valeur numérique*.[14] It should be noted that Cauchy is always satisfied with this verbal descriptor and never introduces any specific sign for the absolute value. He uses this description to declare two quantities as "equal" when they agree not only in their *valeurs numériques* but also in their *signes*; and as *opposées* when the absolute values agree, but the signs are opposite (ibid., 2). With this return to the two *êtres*, however, Cauchy is not following Fontenelle directly, but Buée's adaptation of Fontenelle. Indeed, Cauchy already uses substantive and adjective as an analogy for the two *êtres* in the main text: "the sign + or – placed before a number modifies its sense, just as an adjective modifies the meaning of a noun" (ibid.).

---

[14]   The label *valeur numérique* was in no way new; it had also been used by, for example, Bézout (1781, 10).

The appendix refers directly to Buée's publication (ibid., 403). Cauchy emphasizes that as a result of his conception, positive numbers are always larger than negative ones, and also that negative numbers have to be viewed as becoming smaller as their absolute values increase (ibid., 3). In the main text, he mentions briefly that one could "establish" the sign rule through the concept of opposite quantities, and he refers the reader to the appendix (ibid., 4). This appendix uses the following notation for opposite quantities as "conventions": Uppercase letters such as A to label numbers—thus making $-A$ the opposite quantity to the number A—and lowercase letters such as a for quantities in general.

As equations for numbers, he presents

$$a = +A \text{ and } b = -A.$$

By identifying a with $+a$, he obtains

$$+a = +A \text{ and } +b = -A.$$

Through forming the opposition in a formal manner, he goes on to obtain

$$-a = -A \text{ and } -b = +A.$$

By then entering the (number) values for the quantities a and b, he interprets the composition of the *signes* as their multiplication, thereby obtaining four equations:

$$+(+A) = +A \text{ and } +(-A) = -A,$$

as well as

$$-(+A) = -A \text{ and } -(-A) = +A,$$

which he then generalizes to the "sign rule" as a *théorème* on the product of two signs (ibid., 404 f.). Indeed, Cauchy operates and argues exclusively with algebraic signs here. He neither addresses nor reflects on the discrimination between sign of operation and algebraic sign. Nonetheless, he evidently assumes that he has justified operating with both types of sign. He even assumes that the multiplication of signs impacts exclusively on the signs themselves, but not on the numbers being multiplied. He describes this namely as an "immediate consequence" of his definitions, *que la multiplication des signes n'a aucun rapport avec la multiplication des nombres* (ibid., 406).

Hence, Cauchy's intention is that operating with numbers should not transform a number into a (negative) quantity. This confirms that Ampère's failure to construct a coherent conception of negative numbers was due to the intricate nature of the conceptual problem.

Cauchy goes into remarkable detail when formulating the basic operations of algebra in the further text of the first appendix, right up to raising to the power, extracting the root, and taking the logarithm. He even presents the foundations of trigonometry with a table giving the distribution of signs for the trigonometric functions in the four quadrants (ibid., 431). He probably considered this to be necessary to counteract Carnot's widely favored ideas permitting only positive values in all four quadrants.

2. The second major consequence of the restriction to absolute numbers as the outcomes of measurement processes is that this removes the zero from the

domain of numbers. Although Spalt (1996, 52) notices this special status of the zero, he does not assign any systematic meaning to it. However, when we examine the concept of the limit, we shall already see that the exclusion of the zero has very systematic consequences for analysis. His conception of the zero and the resulting differentiations embeds Cauchy within the tradition of many of the mathematicians already addressed in this book: Fontenelle, Cousin, Martin, L'Huilier, Carnot, and—probably without Cauchy knowing his work—Stockler. Cauchy's concept and its significance for the construction of concepts for analysis will be discussed in more detail below in the section on the limit concept.

This, in contrast, is the appropriate place to present Cauchy's fundamental epistemological ideas on the number concept that, strangely enough, have been neglected in all the hermeneutic exegeses on Cauchy, no matter how extensive. These ideas are to be found in the lectures Cauchy delivered at Turin in 1833. Moigno published them in 1868 under the title of *Sept Leçons de Physique Générale*.

These lectures form an important document: on the one hand, because they show how Cauchy's religiosity influenced his basic concepts in physics and mathematics; on the other hand, because they also confirm that the close relation between physics/mechanics and mathematics in the formation of concepts, which is so important in the present study, applies to the emergence of pure mathematics as well.

The basic theme that Cauchy never tires of varying in these lectures is the study of the divisibility of matter, the number of atoms, and their relation to mathematical points and finally to numbers. Time and time again, Cauchy emphasizes that mathematical concepts exist only as an abstraction of physical reality and have no autonomous existence, that, for example, geometric space is determined by physical bodies:

> Thus the finite space a body occupies is not, as one is often tempted to believe, an object which exists of itself and independently of this body; but it is the existence of the body which creates that space, in the same way that the existence of any object creates its attribute (Cauchy 1868, 442).[15]

The finiteness of matter, just like its only finite divisibility—in contrast to the infinitude of the Almighty—leads simultaneously to the finiteness of geometric concepts:

> To sum up, we should not admit the existence of a length without limits, or of a space which exists on its own, independently of bodies, and which, similar by its nature to that of the bodies which occupy it, would be infinite like God and at the same divisible, that is, even divisible to infinity. Finite spaces, the only ones that can be created, are attributes of bodies and these attributes, just as the bodies themselves can never become infinite. The space occupied by the universe is, and will remain forever, finite, just as God, in virtue of his omnipotence, adds indefinitely to that space, pushing back its limits by new creations. The

---

[15]   It is interesting to see how Cauchy once again uses a grammatical analogy when considering the ontological tie of mathematics: the subject and the attribute.

mathematical point, which has no dimension, is brought into being by the creation of a single atom (ibid., 48).

And the finiteness of geometric objects finally applies to numbers as well: "One can say that mathematical points, distances, areas, and volumes subsist in the way that numbers do" (ibid., 50). He says that one can prove mathematically that the assumption of a number "infinite" would lead to manifest contradictions (ibid., 23). One proof for this, he declares, is that the basic error in Lagrange's theory of functions lies in his assumption that a series does actually extend to infinity:

> If all the genius of Lagrange has not succeeded in founding the theory of analytic functions on a solid foundation, it is because he [Lagrange] regards the principles of that theory as being determined, the sum of the terms which make up any series being prolonged to infinity (ibid., 25).

Moigno gives a further explication of Cauchy's basic belief in an appendix, stating that the concept of numbers and that of the infinite are mutually exclusive; every number is finite:

> These two ideas, *number* and *infinity* contradict each other, each cancels the other out, necessarily and essentially. [...] A number being actually infinite is impossible; every number is essentially finite (ibid., 78).

Cauchy's unshakable religious belief that only God can assume the infinite permitted only finite concepts in physics and mathematics, particularly only finite numbers; any actual, infinitely large and (correspondingly) infinitely small numbers were excluded from his mathematical worldview.

Cauchy's ontological preference for finite, positive numbers simultaneously implies a preference for real numbers as well. He does not assign any number character to imaginary numbers, only discussing these as imaginary *expressions*. His characterization of these quantities reveals a concept analogous to that of Carnot:

> In analysis, we call *symbolic expression* or *symbol* all combinations of algebraic signs which mean nothing of themselves, or to which one attributes a value different from that which it ought naturally to have. [...] The use of these expressions or symbolic equations is often a means of simplifying calculations, and of writing in a shortened form results which are apparently quite complicated (Cauchy 1821, 173).

This is why Cauchy prefers to work with the *modules*, the absolute values of imaginary expressions.

Two further innovations to the number concept need to be mentioned: the first is his introduction of a special sign for many-valuedness. Should the outcome of an operation with a quantity result in several values, double parentheses indicate that Cauchy means each of the possible values. On the other hand, dropping the double parentheses indicates that he means the simplest or the standard value. In a square root, for example, double parentheses indicate that he means either the positive or the negative value; if they are dropped, in contrast, only the positive value:

$$((a))^{\frac{1}{2}} = \pm a^{\frac{1}{2}}, \ a^{\frac{1}{2}} = +a^{\frac{1}{2}} \text{ (Cauchy 1821, 7).}$$

The introduction of this notation shows that Cauchy certainly paid attention to the use of signs in his presentation. Although Ampère also adopted this special notation, it did not gain a broader acceptance.

The second, and far more important innovation is the introduction of an extended calculus with *means* (*moyennes*). Cauchy formulates operations with arithmetic and geometric means in particular, and also proposes a sign for this: $m(a, b)$, for example, for the mean of two quantities a and b (ibid., 14 ff.). Calculations with means form the basis for his central mean value theorems of differential and integral calculus and particularly for his new definition of the definite integral.

6.2. The Variable

Cauchy explicates not just the number concept when considering basic concepts but also that of the variable. He does this by discriminating very precisely between "variable" as an autonomous concept and the values that a variable can assume: "We call a quantity *variable* that which is considered as having to receive successively many different values, one after the other" (Cauchy 1821, 4). There are three notable aspects of this definition:

1. The first concerns the origin of the differentiation between "variable" and "value." Spalt has determined that Cauchy uses this—together with the definition of the limit discussed below—to introduce "*value* as a central concept of a highly developed mathematical theory" (Spalt 1996, 224). He uses it to draw "a concept into the focus of mathematical thinking that had not been a special concept in mathematics before" (ibid.). Because of this purported newness of the value concept, Spalt devotes almost a complete chapter to the question of its origins. His discussions prove to be pure speculation: by calling "value" a "traditional commercial term," he strays into the domain of political economy and searches for social categories for "value" in "commerce," in "capitalist industrialization," and then even in "courtly society," and finally in "feudal absolutism" with a "prestige value" (ibid., 224 ff.).

He would not have needed to engage in such rampant speculation if he were to have studied the contemporary mathematical context. As we have seen, *valeur* was well established in French mathematics—by Lacroix and by Ampère in Cauchy's direct context (Chapter V.3.2.)—and had been developed systematically since Martin's textbook *Élémens des Mathématiques* (1781; see Chapters II.2.10.1. and III.8.4.). Martin was the first to introduce *valeur* into algebra for solving equations. As a member of the first parliamentary chamber, the *Conseil de Cinq-Cents*, Martin had been living in Paris since 1795, where he also published the second edition of his textbook in 1802. Because Cauchy's father cultivated good relations with scientists, artists, and technicians (see

Belhoste 1985, 14 ff.), there may well have been a copy of Martin's book in the private family library.

2. The second aspect concerns Cauchy's emphasis on the term *successivement* in both the definition of variable as well as the subsequent definition of limit: that the single values are assumed one after the other. Giusti (1984, 47) has pointed out that this definition still expresses the traditional kinematic aspect, in the Newtonian sense, and, one can add, also the traditional geometric idea of curves. However, he also points out that it is joined by a new element: the discrete sequence of states, analogous to the strokes of a clock.

The continuous sequence of such clock strokes shows that the variables are defined over discrete sets, for example, over the set of natural numbers. For Cauchy, this means that variables x can be expressed as $x(n)$ with n elements from $\mathbb{N}$ or as x_n, and this shows that variables are already special functions, namely, *sequences*.

3. The third aspect is closely related to the second: As Giusti also noted, the variable is not the only direct premise for the function concept; its definition is supplemented by a complementary one: that of the constant. Cauchy does not follow Euler, for whom constants are a special case of variables (see Chapter II.1.2.), but follows the tradition since L'Hospital of introducing a dichotomy of variables versus constants. Indeed, the above definition of variables is followed by an independent definition of constants: "On the other hand, we call a quantity *constant* [...] every quantity that receives a fixed and determined value" (Cauchy 1821, 4).

Giusti has also pointed out that Cauchy's variables *have to* assume various values, whereas Euler's *can* assume different values. The necessary variability of the values of variables—compared with constants—has, according to Giusti, decisive consequences that Cauchy may well have not intended. For example, if we look at the difference

$$f(x + \alpha) - f(x)$$

in the definition of continuity (see below, Section 6.5), not only is α a variable but also x. This makes this a twofold passage to the limit, and thereby already implies uniformity. Because of the demand for a "successive" assumption of values, the difference can therefore also be conceived as a sequence:

$$f(x_n + \alpha_n) - f(x_n) \text{ (Giusti 1984, 48 ff.).}$$

In the application to the continuity theorem, Giusti has also tried to explain the contradiction between the two proofs of 1821 and 1853 (ibid., 50 ff.).

6.3. The Function Concept

When defining the function concept, Cauchy starts with the concept of variables and also takes Ampère's discrimination between dependent and independent variables as a foundation: The function concept is, in general, completely free from any constraints regarding the corresponding curves:

> When variable quantities are so related to each other that the value of one of them being given, one is able to deduce the values of all the others, one can conceive of the different quantities as being expressed by means of one of them which takes the name of *independent variable*; and the other quantities expressed by means of the independent variable are what are called functions of that variable (Cauchy 1821, 19).

He then goes on to define functions of several variables as well (ibid.).

Cauchy uses a consistent form of sign notation to distinguish between the variable and the value of the variable: Whereas he labels variables with lowercase letters from the end of the alphabet (e.g., x, y, or z), he labels their values with either the corresponding uppercase letters (e.g., X, Y, or Z) or indexed lowercase letters (e.g., x_0, x_1, y_0, y_1).

Likewise, Cauchy also uses signs to distinguish between the function and the value of the function: For example, $f(x)$ labels the function, whereas $f(X)$, $f(x_0)$, and so forth label the value of the function at the position X, x_0, and so forth (see ibid., 41, 44, and passim).

6.4. The Limit and the Infiniment Petit

Immediately after defining variables, Cauchy goes on to define both "limit" and "infinitely small quantity." We shall soon see that this arrangement is not arbitrary but systematic in intent.

The definition of the limit is purely verbal, without signs or inequalities. It now uses the classic term of becoming arbitrarily close, without mentioning, for example, infinitely small quantities:

> When the successive values given to the same variable approach indefinitely a fixed value, so that in the end they differ from each other by as little as one wishes, this last is called the *limit* of all the other values (ibid., 4).

Once again, it is fundamental to the definition here for the variable to adopt single values "successively." Spalt (1996, 47) translates the word *indéfinement* as *unbestimmt*. However, "indefinite approximation" makes little sense; I find "unlimited" more appropriate. Following the French tradition, the only requirement is for the successively assumed values finally to differ from the limit by an arbitrarily small amount; attaining the limit is not considered. The definition of the limit is purely verbal, and is performed without the use or introduction of signs. Cauchy does not introduce the sign *lim*." until the next section; however, only as an abbreviation, as underlined further by his use of the period:

> When a variable quantity converges to a fixed limit, it is often useful to indicate this limit by a special notation, which is what we shall do, by placing the abbreviation

lim.

before the variable quantity being considered (ibid., 13).

In fact, Cauchy uses this sign only as an abbreviation and does not adopt Stockler's and Ampère's approaches toward an operative autonomy. Even Ampère himself did not continue his approaches developed before 1811. This makes it quite clear here that following the resolution of 1811 and the "return" to the *infiniment petits*, it was still possible to carry on teaching the limits approach as part of the practice of the compromise, but that it became "frozen" at the less-developed level found in Lacroix.

Because it was used only as an abbreviation, Cauchy's *lim.*" sign—like that of all its previous users—lacked one completely central element for an operative application: a closer characterization of the limit process, either by indicating the value that the independent variable moves toward, or the variables for which the limit process occurs. Ironically, it is Spalt (1996, 20), so proud of his method of "not adding anything foreign to the sources," who, almost right from the very start, adds the index notation always lacking in Cauchy as well as his French contemporaries. Despite his otherwise copious comments on every change in the text, it is only very late in his work that he follows the reproduction of a statement by Cauchy,

$$0 = \lim_{x=0}\left[\Theta f(x)\right] = \lim_{x=0}\left[\Theta x f'(x)\right],$$

with the remark, "index in 'lim' added as usual" (ibid., 144).

However, it is precisely the neglect of these indices that points to a major conceptual problem field: without using signs to explicate the variables affected by the limit process, it is easy to miss the fact that, for instance, two different limit processes are taking place. Indeed, the lack of any explication with signs proves to be an indication of insufficient reflection on the entire limit process (see below, Section 6.6.).

The definition of the limit is followed by two examples: irrational numbers as a limit to rational ones, and the area of the circle as the limit to which the area of inscribed polygons converges as the number of sides increases. This is followed immediately by the definition of infinitely small quantities:

> When the succesive numerical values of the same variable decrease indefinitely, such as to become less than any given number, this variable becomes what is called *infinitely small* or *an infinitely small quantity.* A variable of this type has zero as its limit (Cauchy 1821, 4).

It can be seen that the definition is constructed completely analogously to that for the limit. The only difference is that here, the condition for the absolute values is made for the values of variables in order to conceive becoming smaller operatively for numbers ("smaller than any given number"). This makes infinitely small quantities special variables, namely, variables with limit zero. Cauchy also expressly confirms this special character. This also implies the opposite conclusion that the "normal" limit definition refers to limits unequal to zero. We have already become acquainted with this widespread tradition of doubling the limit definition because of the special status of zero. Stockler introduced two different limit concepts: for variables with a finite constant differing from zero as a limit and for "variables without a limit of diminution" (see Chapter III.8.5.). Whereas Stockler conceived both concepts as limits, Martin and Carnot, in contrast, restricted the term "limit" to finite nonvanishing values and chose the term "infinitely small quantity" for variables with a limit of zero.

Carnot offers his most explicit explanation of the relation of the definition of a *quantité infiniment petite* to the exclusion of the zero from the *quantités effectives* in his text of 1785, which remained unpublished (see Chapter V.1.6.1. and Appendix B, Part I). However, even in the first published version of 1797, the relation is sufficiently clear (see Appendix B, Part II). He always retains the same basic approach: defining infinitely small quantities as variables with limit zero, even when he replaces the term "variable" with the less specific "quantity" in later editions.

In Martin, whom we have already recognized as a possible source of Cauchy's ideas, the differentiation between the two types of limit is just as clear: Whereas *limite* is reserved for variables that approach an effective value or state, he explains *infiniment petits* as special variables: ones that approach zero (see Chapter III.8.4.).

The analogy between Cauchy's definitions of terms and those of Carnot has been noted before in the literature. However, Cauchy's own dissociation of his work from that of Carnot was considered sufficient to rule out any influences. Indeed, in an article, *Sui metodi analitici*, published in 1831, Cauchy did criticize "the author of the reflections on the metaphysics of infinitesimal calculus" who had viewed infinitely small quantities as variables converging to the limit zero. However, he did not criticize Carnot's definition—because he would then have had to criticize himself—but his use of infinitely small quantities to define the differentials. Cauchy viewed differentials, in contrast, as finite (Cauchy 1974, 163).

One can understand Cauchy's definition of the *infiniment petits* as both a skillful and a characteristic implementation of the compromise conception on the foundations of analysis practiced at the *École polytechnique*: On the one hand, it granted the *infiniment petits* the status of a basic concept; in this way, Cauchy met the demands of the institutional context. On the other hand, they were classified with, or subjugated to, the limit concept, thereby confirming—as already in Carnot—the dominance of the *limite* approach. However, coupling the *limite* with the *infiniment petits* prevented Cauchy from engaging in any further algebraization of the limit concept.

Furthermore, alongside finite limits and the limit zero, Cauchy also introduces a special definition for infinitely large limits. Analogous to the *infiniment petits*, these are *quantités infinies*. Because here as well, the comparison is based only on numbers, that is, absolute or positive numbers, the condition is formulated for absolute values: When the successive *valeurs numériques* of a variable become increasingly larger, so that they are finally larger than any given number, this variable has to be marked by the sign ∓ as the *limite* of the positive infinite (*l'infini positif*), as long as it is a positive variable, and as the negative infinite (*l'infini négatif*), if it is a negative variable (Cauchy 1821, 4 f.).

Cauchy's first application of the *infiniment petits* is the introduction and presentation of the various orders of the infinitesimals. Here as well, Cauchy reveals himself to be a systematizer. With far more precision and detail than earlier authors addressing this topic, he proves at least eight "theorems" on the comparison of single *infiniment petits* of different orders by means of polynomials and—using the possibility of representing infinitely large quantities by means of $1/\mu$ with μ as an *infiniment petit* (ibid., 33)—also for orders of *quantités infinies* (ibid., 26 ff.). Further applications, which will be addressed below, are made particularly in the concepts of continuity, convergence, and the integral.

Because the *infiniment petits* represent only a subconcept of the *limite* concept in Cauchy's textbooks for the *École polytechnique*, we have to ask what systematic significance they held in his concepts. Moigno's textbook of 1840 provides an important source on this. As mentioned before, this book was written under the aegis of Cauchy and was free from the pressure of the context of the *École polytechnique*. What is perhaps no longer so surprising about this analysis textbook is the way it deliberately does *not* present the *infiniment petits* as a basic concept. It is far more the case that Moigno expressly points out that the method of the *infiniment petits* is in no ways more simple or rigorous; in other words, that the justification for the 1811 revision is not correct.

Moigno starts his appraisal with a criticism of S.-D. Poisson (1781–1840), who had made the concept of the *infiniment petits* into his personal program, taking a completely different approach to that of Cauchy. Moigno complains that Poisson revealed an almost missionary zeal to convert all mathematicians to the *infiniment petits* (see, for more detail, Chapter VIII). He wanted to do this with a conception that would justifiably earn him the attribute of being a precursor of NSA: as actual quantities of a new kind with which one may operate mathematically:

> M. Poisson [...] has armed himself with all his powers to bring the scientific world to the infinitesimal method. He claimed that infinitely small quantities are not only a means of investigation imagined by geometers, but that they have a real existence, that is, that there exist magnitudes which are not zero, can even be twice, three-times, four-times other magnitudes, and yet are actually less than any given magnitude (Moigno 1840, xxiv).

Hence, Moigno reveals that there is a consensus on the concept of using the *infiniment petits* heuristically, as a means of research that will subsequently still require a rigorous justification. Moigno not only rejects Poisson on the basis of the concepts shared at that time; he simultaneously discloses the massive contradiction in the French mathematical community. One could also draw an opposite conclusion from this: that an analogous contradiction would have to be discernible if Cauchy were to have practiced the proof and use of hyper-real numbers attributed to him by, among others, the proponents of NSA:

> Although depending on such an imposing authority, these assertions were sharply opposed and repulsed; they not only announced a mystery from which reason took fright without having the right to reject them; many judicious minds saw here an evident impossibility (ibid.).

Moigno's own refutation refers very clearly to Cauchy's proof of the exclusive existence of *finite* numbers given in his physics lectures held at Turin. He uses the question whether quantities that are smaller than any given quantity may also be extended and divisible, or whether they are simple and indivisible. Both alternatives lead to contradictions:

> In effect, either these magnitudes, smaller than any given magnitude, still have substance and are divisible, or they are simple and indivisible: in the first case their existence is a chimera, since, necessarily greater than their half, their quarter, etc., they are not actually less than any given magnitude; in the second hypothesis, they are no longer mathematical magnitudes, but take on this quality, this would renounce the idea of the continuum divisible to infinity, a necessary and fundamental point of departure of all the mathematical sciences (ibid., xxiv f.).

Applying Poisson's concepts would have set the science back to the state of its origins. He said that Poinsot and Cauchy, in contrast, developed more rigorous methods. Moigno cites Cauchy's definition here:

> For these geometers an infinitely small quantity is only a variable or undetermined quantity that has zero as its limit, and which can decrease indefinitely without stopping at any appreciable value (ibid., xxv).

However, he adds that one might also take the traditional approach and study a single value of the variables: "a quantity which, taken on its own, can be conceived as being smaller than any given magnitude" (ibid.). He remarks that if one were to apply this definition and the subsequent rules appropriately, the rules could not lead one astray. However, he has avoided their use because the application of these rules would require difficult antecedent studies through which the method would lose the only advantage that one apparently could not deny it, that of the rapidity with which it leads to the goal:

> This definition sufffices to put beyond doubt the fundamental rules of the infinitesimal method. Easy to apply, these rules cannot lead one astray; if I have avoided making use of them it is only because their use necessarily entails quite tricky preliminary considerations, which causes the infinitesimal method to lose the one advantage that it appears cannot be disputed, the rapidity with which it leads to the result (ibid.).

As an example, Moigno shows that when one wants to obtain the differential of the arc *s* of a curve, it does not suffice to determine that the increment Δs of the arc is practically equal to the chord. It is far more the case that one first needs to show that the increment of the arc differs from the chord by a second-order *infiniment petit*. Although the infinitesimal method could attain rigor in this way, it would then lose its transparency and its rapidity. He considers that the method could lead to results, but it was not exact and "classical" enough as a method of presentation:

> Hindered by this demonstration and drawn back to the proportions of the method we had adopted in our Treatise, the infinitesimal method becomes, it seems, rigorous, but in ceasing to be expedient, one is forced to conclude that, while very advantageous when it is matter of foreseeing or finding results, it does not constitute of itself a sufficiently exact or "classical" method of exposition (ibid., xxvj)

Hence, Moigno's differential calculus is constructed completely on the basis of the limit method. He does not introduce *infiniment petits* and their orders until the sixth chapter, and then for a circumscribed purpose: to determine initially indefinite values such as $\frac{0}{0}, \frac{\infty}{\infty}, 0 \times \infty$, etc. (ibid., 41 ff.).

Further insight into the conceptual horizon of Cauchy's limit concepts can be gained from E.H. Dirksen's comments on the doubling or multiplying of the limit definitions because of the specific number concept and, in particular, the special status of the zero. Dirksen was one of Cauchy's most diligent readers and the most intensive German supporter of Cauchy's innovations in basic concepts (see below, Chapter VII). In his *Organon*, he formulated a systematization of the possible limits of variables that strictly follows Cauchy. It is interesting to note the reference to Cauchy when Dirksen conceives variables as "infinite number series." He uses the possible ways they approach limits to classify them into the following three main groups: (1) infinite number series that remain finite, (2) infinite number series that become infinitely small, and (3) infinite number series that become infinite (Dirksen 1845, 25 ff.).

The basis for the separate identification of series that become infinitely small is that Dirksen also does not consider that the zero belongs to the numbers.

I shall discuss Cauchy's concepts on continuity, convergence, and the integral as characteristic applications of his definition of infinitely small quantities.

6.5. Continuity

At the *École polytechnique*, the continuity of functions had become a universal topic in analysis and mechanics. Garnier, Lacroix, and Prony mentioned it repeatedly in their textbooks and lectures, although predominantly as more of a "metaphysical" basic condition ("law") rather than as an operatively applicable individual concept. In Ampère, we cannot find any such concrete introduction of the continuity concept in his lecture notes *before* 1815. Therefore, Cauchy's definition of continuity as an individual concept represents a great independent innovation within the French context. I shall start by specifying the time frame for this innovation.

Cauchy gave his first lectures at the *École polytechnique* in 1815/16 to the more advanced, second-year students. No use of the continuity concept is mentioned in the *Registres*. After the compulsory break in 1816, he cooperated with Ampère in working out the new teaching program for analysis. This proposed the following topic for the first time: *sur la distinction des fonctions continues et discontinues* (see Gilain 1989, 106).

In contrast to the teaching program for 1816/1817, which scheduled this topic for the third lesson along with the introduction of functions, Cauchy introduced the concept only very late in the course, in the nineteenth lesson (ibid., 52). Since up to now, no notes taken during these lectures have been found, there is no direct way of ascertaining which definition of continuity Cauchy used here. However, some indication can be gained from Cauchy's own entries for the individual lessons in the *Registres*. As already indicated above (Section 4.), this course represents an apotheosis of the *limite* method; one can rightly ascertain that limits are everything here!

Whereas Cauchy generally only summarizes the contents of individual lessons, he occasionally enters the theorems he is addressing in explicit detail. These are all theorems on interchanging limits as a consistent approach to developing an independent limit calculus that would continue the work of Stockler and Ampère:

1. The limit of the sum of several variables is the sum of their limits (2nd lesson, January 20, 1817).

2. The limit of the product of several variables is the product of the limits of the same variables (3rd lesson, January 21, 1817).

3. If two variables x and y have X and Y respectively as their limits, then x^y has the limit X^Y (5th lesson, January 28, 1817).

4. The limit of a continuous function of several variables is a continuous function of the limits of these variables (20th lesson, March 4, 1817) (see Gilain 1989, 51f.; see Figure 17).[16]

Evidently Cauchy optimistically thought that he would be able to establish sound foundations for analysis with the help of just one universal principle: continuity. And this was the Leibnizian continuity principle in its "metaphysical" form, stating that laws retain their validity in the transition from finite to infinite (see Chapter III.5.). This is also the continuity principle that L'Huilier transferred to the limit *lim* of variables in his prize-winning memoir: the principle that the variable possesses the same properties *after* the passage to the limit that it possessed *before* (see Chapter III.8.5.). Here, Cauchy is translating into conceptual operativeness Laplace's epistemological program that all *limites* are interchangeable taken from his lectures at the *École normale*.

Cauchy's famous and controversial theorem that the sum of a convergent series of continuous functions is itself, in turn, a continuous function fits smoothly into this enthusiastic effort to derive theorems from one single

---

[16]   *La Limite d'une fonction continue de plusieurs variables est la même fonction de leurs limites.*

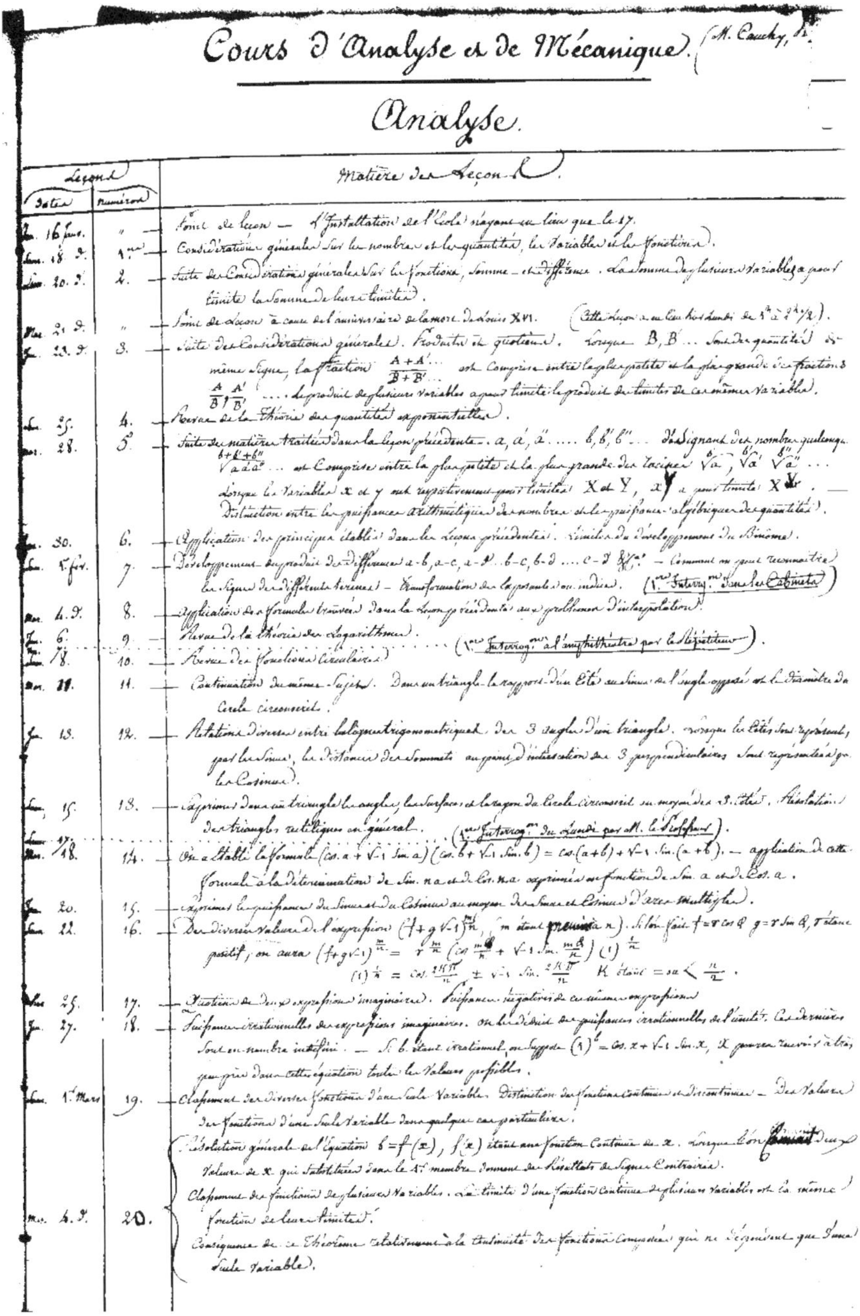

Cours d'Analyse et de Mécanique. (M. Cauchy, ...

Analyse.

Leçons — Dates	Numéros	Matière des Leçons.
J. 16 janv.	"	Point de leçon — L'Installation de l'École n'ayant eu lieu que le 17.
S. 18 d°.	1ère	Considérations générales sur les nombres et les quantités, les variables et les fonctions.
L. 20 d°.	2.	Suite des considérations générales sur les fonctions, somme et différence. La somme de plusieurs variables a pour limite la somme de leurs limites.
Ma. 21 d°.	"	Point de leçon à cause de l'anniversaire de la mort de Louis XVI. (Cette leçon a eu lieu hier lundi de 3h à 4h 1/2).
J. 23 d°.	3.	Suite des considérations générales. Produits et quotients. Lorsque B, B' ... sont des quantités de même signe, la fraction $\frac{A+A'\dots}{B+B'\dots}$ est comprise entre la plus petite et la plus grande des fractions $\frac{A}{B}, \frac{A'}{B'}$ .... Le produit de plusieurs variables a pour limite le produit des limites de ces mêmes variables.
S. 25.	4.	Revue de la théorie des quantités exponentielles.
Ma. 28.	5.	Suite des matières traitées dans la leçon précédente. a, a', a'' .... b, b', b'' ... désignant des nombres quelconques $\sqrt[b+b'+b'']{a a' a''}$ ... est comprise entre la plus petite et la plus grande des racines $\sqrt[b]{a}, \sqrt[b']{a'}, \sqrt[b'']{a''}$ ... Lorsque les variables x et y ont respectivement pour limites X et Y, x^y a pour limite X^Y. — Distinction entre les puissances arithmétiques des nombres et les puissances algébriques des quantités.
J. 30.	6.	Application des principes établis dans les leçons précédentes. Limites du développement du binôme.
S. 1er fév.	7.	Développement du produit des différences a-b, a-c, a-d ... b-c, b-d .... c-d &c. — Comment on peut reconnaître les signes des différents termes — Transformation de l'exposant en indice. (1re Interrog. dans les Cabinets)
Ma. 4 d°.	8.	Application des formules trouvées dans la leçon précédente aux problèmes d'interpolation.
J. 6.	9.	Revue de la théorie des logarithmes. (1re Interrog. à l'amphithéâtre par le Répétiteur).
S. 8.	10.	Revue des fonctions circulaires.
Ma. 11.	11.	Continuation du même sujet. Dans un triangle le rapport d'un côté au sinus de l'angle opposé est le diamètre du cercle circonscrit.
J. 13.	12.	Relations diverses entre les lignes trigonométriques des 3 angles d'un triangle. Lorsque les côtés sont représentés par les sinus, les distances des sommets au point d'intersection des 3 perpendiculaires sont représentées par les cosinus.
S. 15.	13.	Exprimer dans un triangle les angles, la surface et le rayon du cercle circonscrit au moyen des 3 côtés. Résolution des triangles rectilignes en général. (1re Interrog. du Lundi par M. le Professeur).
L. 17/Ma. 18.	14.	On a établi la formule $(\cos. a + \sqrt{-1}\sin a)(\cos. b + \sqrt{-1}\sin. b) = \cos.(a+b) + \sqrt{-1}\sin.(a+b)$. — Application de cette formule à la détermination de sin. na et de cos. na exprimés en fonction de sin. a et de cos. a.
J. 20.	15.	Exprimer les puissances du sinus et du cosinus au moyen des sinus et cosinus d'arcs multiples.
S. 22.	16.	Des diverses valeurs de l'expression $(f+g\sqrt{-1})^{\frac{m}{n}}$, (m étant premier à n). Si l'on fait $f = r\cos\theta$, $g = r\sin\theta$, r étant positif, on aura $(f+g\sqrt{-1})^{\frac{m}{n}} = r^{\frac{m}{n}}\left(\cos\frac{m\theta}{n} + \sqrt{-1}\sin\frac{m\theta}{n}\right)(1)^{\frac{1}{n}}$; $(1)^{\frac{1}{n}} = \cos\frac{2k\pi}{n} \pm \sqrt{-1}\sin\frac{2k\pi}{n}$, k étant = ou $< \frac{n}{2}$.
Ma. 25.	17.	Notation des deux expressions imaginaires. Puissances négatives de ces mêmes expressions.
J. 27.	18.	Puissances irrationnelles des expressions imaginaires. On les déduit des puissances irrationnelles de l'unité. Ces dernières sont en nombre infini. — Si b étant irrationnel on suppose $(1)^b = \cos. x + \sqrt{-1}\sin. x$, x pourra recevoir à très peu près dans cette équation toutes les valeurs possibles.
S. 1er Mars	19.	Classement des diverses fonctions d'une seule variable. Distinction des fonctions continues et discontinues — Des valeurs des fonctions d'une seule variable dans quelques cas particuliers.
Ma. 4 d°.	20.	Résolution générale de l'équation $b = f(x)$, $f(x)$ étant une fonction continue de x. Lorsque l'on connaît deux valeurs de x qui substituées dans le 2e membre donnent des résultats de signes contraires. Classement des fonctions de plusieurs variables. La limite d'une fonction continue de plusieurs variables est la même fonction de leurs limites. Conséquence de ce théorème relativement à la continuité des fonctions composées qui ne dépendent que d'une seule variable.

Figure 17 Cauchy's analysis lecture course (AEP Bibliothèque, $X^{2c}7$, 1817)

epistemological principle that is declared to have general validity. It is quite likely that this enthusiasm impacted negatively on his attention to rigor.

Alongside this recognizably metaphysical continuity principle, there is one further way of gaining access to the formulation of Cauchy's continuity definition of 1817. I managed to gain this from a long-mislaid manuscript in Ampère's *Nachlass*. This manuscript, entitled *Ancien cours d'analyse algébrique*, has recently been rediscovered in the *Nachlass*.

When I first read the title, I thought that this course must belong to the period before 1810/11. However, *Ancien cours* proved to stand for the course in its detailed form as it existed after 1816; in other words, before 1820/21, when drastic reductions in the size and content of algebraic analysis started to be introduced (see Gilain 1989, 12 ff.). When I compared it with the *Registres* for Ampère's first-year course, particularly for the years 1817/18 and 1819/20, I realized that this *Ancien cours* is the text for the 1817/18 lecture course, that is, precisely the course that followed Cauchy's course ending in 1817.[17] Hence, it is highly likely that the definition of continuity given by Ampère in his *Ancien cours* corresponds very precisely with Cauchy's first definition of March 1817. It is conspicuous that Ampère's definition of the function concept here is still tied completely to the geometric curve concept and is completely general:

> There exists yet another distinction of functions. They can be divided into continuous and discontinuous functions. To provide an explanation, not that every curve can be represented by a function. Suppose it is referred to rectangular axes and let the abscissa be increased or decreased in a continuous manner. If the ordinate increases or decreases in the same manner, the function as well as the curve will be continuous, if not, they are discontinuous. Observe that a curve is discontinuous when its description is not subject to one and the same law. When, for example, it is composed of several arcs of circles or parabolas joined together.[18]

The definition is not just completely global and unalgebraic; it explains the continuity of a function through the (not further explained) continuity of the variables. The added explanation of discontinuity confirms the continued strength of the traditional eighteenth-century conception that bound continuity to a standardized algebraic expression. The definition therefore departs quite fundamentally from that of 1821 that was expressed in differences.[19]

---

[17]   Unfortunately, this manuscript is also only a fragment; it breaks off after the nineteenth lesson. Nonetheless, it is a cleanly presented text, written by another hand (except for the title), though, unfortunately, permeated with numerous writing errors.

[18]   Archives de l'Academie des sciences, Paris; *Nachlass* Ampère: carton 4, chap. 4, chem. 76, fol. 2 v – 3 r.

[19]   We cannot completely rule out the possibility that Ampère considered a geometric version to be more appropriate for his students than a (hypothetically) more algebraic first definition by Cauchy. However, since Ampère uses Cauchy's own approach to present several of his other innovations in the same text, this seems rather unlikely.

What is even more informative is that at a later point in the lecture manuscript, Ampère suddenly inserts a second definition of continuity, at the end of a section on polar coordinates. This definition is very close to Cauchy's second definition in his *Analyse Algébrique* of 1821, using the *infiniment petits*: "One says that a function is continuous when to each infinitely small increase of the variable *y* there corresponds also an infinitely small increase of the function" (ibid, fol. 6 v.). This definition, according to which every infinitely small increment in the variables must correspond to an infinitely small increment in the function, diverges particularly from the later ones in that no interval limits are given and the formulation is global.

This was precisely that period when, for the first time since 1815, the *École polytechnique* was demanding a return to the use of the *infiniment petits* and their inclusion in the teaching program for geometric applications of differential and integral calculus (see Gilain, 1989, 9). Hence, one can view the sudden insertion of a definition of continuity using the *infiniment petits* as a coordinated strategy by Cauchy and Ampère in the fall of 1817 to respond to this institutional pressure. Hence, this second definition was new and had not been used by Cauchy in the preceding lecture course.

This makes it clear that the definitions of continuity changed between the beginning of 1817 and 1821, and that the version including the *infiniment petits* concept was a response to the context of the *École*. Indeed, it is conspicuous that the 1821 publication of the *Cours d'analyse algébrique* even presents three different definitions of continuity in succession. They are not identical in meaning, making their interpretation even more difficult.

What is new about all three definitions in 1821 is that they are no longer global statements covering the entire domain of the variable, but restrict themselves to *intervals*: to values "between two given limits." The restriction of statements to intervals was first introduced and practiced by Ampère in his proof of the existence of derivative functions (see Chapter II. 4.3.).

Another innovation is the precondition that all values of the function in the interval in question be unique and finite:

> Let $f(x)$ be a function of the variable x and suppose that for each value of x, intermediate between two given limits, this function assumes always a unique and finite value (Cauchy 1821, 34).

The first definition corresponds in structure to the "mathematical" version of Leibniz's law of continuity (see Chapter III.5.) with the relation between independent and dependent variables and its interpretation by Boscovich (see ibid.). However, it presents the first use of the algebraic notation for differences:

> 1. If, starting from a value of x between these limits, one gives to the variable x an infinitely small increase α the function itself will experience an increase, being the difference

$$f(x + \alpha) - f(x)$$

> which depends both on the new variable α and on the value of x. Given this, the function $f(x)$ will be, between the two limits assigned to the variable x, a

continuous function of this variable, if for each value of x intermediate between these limits, the numerical value of the difference

$$f(x + \alpha) - f(x)$$

decreases indefinitely with α (ibid.).

Inequalities in the sense of Grabiner do not occur in either this or the following definitions. However, Cauchy displays a remarkable conceptual clarity when pointing out that the difference in the function depends on two parameters; on the variable α and on the value of x. As both the difference in the function and the increment variable tend toward zero, Cauchy considers it to be consistent to apply infinitely small quantities in this definition. Because the demand that the absolute value of the difference $f(x + \alpha) - f(x)$ diminish infinitely together with that of α has, according to Cauchy's definition of *infiniment petits*, exactly the same meaning as the demand that $f(x + \alpha) - f(x)$ just like α, be an infinitely small quantity, the following second definition introduced "in other terms" is actually a shorter version of the first:

> 2. In other words, *the function f(x) will remain continuous with respect to x between the given limits if, between these limits, an infinitely small increase of the variable always produces an infinitely small increase in the function itself* (ibid., 34 f.).

The italics in this shorter version come from Cauchy himself, probably to emphasize that it is easier to memorize. The formulation that the variable increment "δ" "produces" an increment "ε" is a clearer depiction of the relation than the less precise "corresponds" in Ampère's adaptation of 1817/18. The long version simultaneously shows what the *toujours* in the shorter version (also used in other definitions and controversial in the literature) stands for: it means "for all values of the interval."

Cauchy adds a third definition of continuity to these first two versions. This is more special, namely, the interval limited to a neighborhood of a specific value of the independent variable. When specifying this here, Cauchy does not even need to mention the *infiniment petits*:

> 3. One says further that the function $f(x)$ is, in the neighborhood of a particular value given to the variable x, a continuous function of that variable, whenever it is continuous between the two limits of x, even when very close to each other, which enclose that particular value (ibid., 35).

Cauchy requires continuity in the neighborhood of the value of a variable in order to define discontinuity. This leads to a new definition of discontinuity as well: by detaching it not only from the traditional tie to an analytic expression but also from the discrimination between *discontigue* and *discontinue*:

> Finally, when a function $f(x)$ ceases to be continuous in the neighbourhood of a particular value of the variable x, one says that it then becomes *discontinuous* and that it has, for this particular value, a *solution of continuity* (ibid.).

This defined discontinuity for a completely general function concept.

In contrast to his predecessors, who, although introducing continuity, had hardly used it, Cauchy introduces extensive operative applications. The definition is followed immediately by an examination of the 11 basic functions he has considered so far (not only simple algebraic ones such as $a + x$ and ax but also transcendental ones such as $L[x]$ and $\sin x$) to determine in the neighborhood of which values they are continuous or discontinuous (ibid., 25 ff.). Cauchy shows that only two of these functions reveal points of discontinuity:

$$\tfrac{a}{x} \text{ and } x^a \quad \left(\text{if } a = -m,\ m \in \mathbf{N}^+\right) \text{ for } x = 0 \text{ (ibid., 37).}$$

He follows this by developing theorems on continuity such as the continuity of a function of several variables that is continuous in the single variables, and the theorem of intermediate values. I shall consider Cauchy's proofs in more detail below.

For the issues in the present book, the priority in this section is on how far the *infiniment petits* are indispensable in Cauchy's continuity concept. Such an indispensability seems unlikely, because Cauchy's definition relegates the *infiniment petits* to the status of a special case of the general limit concept. Support for this assumption comes from two sources.

The first is Cauchy himself. In a letter to Coriolis, his former *répétiteur* in the analysis course, dated January 29, 1837, Cauchy provides introductory comments to some new memoirs that he has sent to Paris from his exile. This letter paraphrases his continuity definitions of 1821. Although he refers expressly to this 1821 text, he now—freed from the context of the *École*—presents the definition *without* using the *infiniment petits*:

> Following the definition given in my *Cours d'Analyse*, a function of a variable is continuous between given limits when, between these limts, each value of the variable produces a unique and finite value of the function, and that this value varies by insensible degrees with the variable itself (Cauchy 1884, 39).

The sweeping verbal formulation "varies by insensible degrees" can be operationalized just as effectively in $\varepsilon-\delta$ inequalities as in *limite* expressions. Cauchy adds a further specification here that functions are generally (except at points of infinity) discontinuous only at points at which multiple values occur: "Given this, a function which does become infinite will not in general cease to be continuous except in becoming multiple [valued]" (ibid.).

This shows that Cauchy still considers the function concept to have a geometric curve concept as its substrate. He is not thinking of completely general functions such as Dirichlet's function taking the values 0 respectively 1 for rational versus irrational arguments.

The second source is the definition of continuity in Moigno's 1840 textbook. This is placed right at the beginning during the introduction of the function concept and even before the introduction of the *limite*. Therefore, although Moigno uses the expression "infinitely small," he assigns it a specific meaning. He starts with the precondition that the values of the function must be finite and unambiguous. However, he drops the limitation to intervals and neighborhoods. In general, Moigno does not use signs to distinguish between the variable and its value:

A function $y = f(x)$ is continuous, when to each value of the variable there corresponds a unique and finite value of the function, and moreover for an infinitely small change $h = \Delta x$ in the value of the independent variable, producing in the value of the function an infinitely small change Δy, the difference $\Delta y = f(x + h) - f(x) = f(x + \Delta x) - f(x)$ is infinitely small. If these two conditions do not hold, the function is discontinuous (Moigno 1840, 2).

Although the formulation takes the same structure as Cauchy's first definition of 1821, there is one important difference: not only is $h = \Delta x$ now a change in the *value* of the variable, but also Δy is a change in the *value* of the function.[20] Therefore, neither Δx nor Δy forms variables, and therefore they also do not form *infiniment petits* in the sense of Cauchy in 1821. Moigno immediately goes on to define the previously unexplained "infinitely small," giving the expression two different interpretations (see above, Section 5.): on the one hand, the meaning of the variable in Carnot and Cauchy; on the other hand, the "epsilon–technique," meaning:

> Here, an infinitely small quantity is a very small quantity which has 0 as its limit, which can decrease indefinitely, without stopping at any appreciable value, or a quantity that can be conceived as being smaller than any given quantity (ibid.).

I shall close this section by discussing the controversies over the meaning of Cauchy's concept of continuity in the literature.

In his important entry on Cauchy in the *Dictionary of Scientific Biography*, Freudenthal (1970) gives a concise formulation of the then dominant conception of the relation between Cauchy's and the modern concepts of continuity: "Cauchy invented our notion of continuity" (ibid., 136). Grabiner (1981, 87 ff.) also assumes that they are identical. If Cauchy were to have meant the modern concept of continuity, then his above-mentioned theorem on the continuity of a function of several variables would be inadequate, because it requires uniform continuity. Giusti (1984) showed that Cauchy's "errors" include not only the theorem on convergent series, as discussed by Robinson and Lakatos, but also a number of further theorems such as the one above.[21] This unleashed an intensive debate on reconstructing Cauchy's intended meaning.

Although the entire debate cannot be reproduced here, I shall summarize some of the main points. In 1981, Bottazzini still saw no problems in the continuity concept (see, for an English translation, Bottazzini 1986, 104 ff.). However, in a detailed analysis of Cauchy's *Cours* carried out in 1992, he no longer considers the two concepts to be identical, but proposes a "*C*-continuity" (a "Cauchy continuity") in order to avoid overhasty identifications (Bottazzini 1992,

---

[20]   A further difference is that Moigno no longer considers the *valeurs numériques* of the differences. Instead, he says that Δx and Δy can be either positive or negative, but that for the sake of simplicity, both cases are summarized under *accroissements*.

[21]   Actually, Burkhardt (1914, 972 f.) had already compiled a list of the problematic theorems 70 years earlier.

LXXXIII). Although calling Cauchy's form of expression "ambiguous," he tends to interpret C-continuity as uniform continuity. His main reason for this comes from comparing Ampère and Cauchy on the basis of the definition of continuity that Ampère gave in his only published teaching script for the analysis course at the *École polytechnique* (Ampère 1824, 11 f.).[22]

Bottazzini's conception is based above all on the opinion that the meaning in Ampère would have to agree essentially with that of Cauchy. This assumption is not generally tenable, however; moreover, his analysis of Ampère's definition shows the problem of historical hermeneutics in mathematics. He asserts: "It is not too difficult to recognize in Ampère's definition what today is called *uniform continuity*" (Bottazzini 1992, LXXXIII). Ampère had defined continuity in the interval, and without infinitely small quantities. Within the limits of the interval given by two values of the variable, he postulated:

> In choosing arbitrarily [...] two other values of the independent variable, whose difference may be as small as one wishes, the difference of the corresponding values of the function also becomes as small as is wished (as cited in Bottazzini's translation, ibid.).[23]

Whether in this purely verbal definition, without any quantifier, it is "not too difficult" to detect a uniformity postulate "for *any couple* of values of *x* in the given interval" (Bottazzini 1992, ibid.) and independent of the choice of a point in the interval does not, in my opinion, seem particularly obvious.

Lützen (1999) has given the best-informed and balanced presentation of the spectrum of meaning in Cauchy's continuity concept. He also considers that "the definition seems ambiguous" (ibid., 207). Nonetheless, on the whole, he likewise tends to favor the interpretation as uniform continuity. This is particularly because this is the only way for the theorem on the continuity of a function of several variables and the theorem over the existence of the integral of a continuous function to be correct (ibid., 207 f.). Lützen hesitates, because he sees approaches toward a pointwise continuity in Cauchy's first definition. He justifies this by noting that it "refers very clearly to the value of the variable *x*" (ibid., 207). However, this is based on a misinterpretation that can be found even more explicitly in the work of Spalt.

Spalt, namely, believes that Cauchy had already defined a pointwise continuity. He achieves this by going beyond Cauchy's text and replacing it with his own version. When translating Cauchy's first continuity postulate, "The numerical value of the difference $f(x + \alpha) - f(x)$ has to decrease indeterminately with that of α ." He adds, "This naturally requires the variable *x* to be assigned a definite value *X* in each case" (Spalt 1996, 86).

---

22   Even this publication remains incomplete. It stops abruptly after 151 pages.

23   "en prenant à volonté [...] deux autres valeurs de la variable indépendante, dont la différence soit aussi petite qu'on peut, la différence des valeurs correspondantes de la fonction devient de même aussi petite qu'on veut" (Ampère 1824, 11 f.).

Evidently, what Spalt means here is that x assumes a definite value, whereas α still runs through values as a variable. Indeed, the prior specification of a value X is clarified by his following equation:

$$0 = \lim_{\alpha=0} f(X + \alpha) - f(X) = \lim_{\alpha=0}\left[f(X + \alpha) - f(X)\right].$$

Though unusual from a modern perspective, the notations in Cauchy's equation nonetheless show unmistakably that x and α both function as variables:

$$f(x + \alpha) - f(x).$$

Giusti (1984) gives the most appropriate interpretation of Cauchy's personal notation style. He poses the precise question, What actually happens to x while α is successively assuming different values? He considers that the answer can be only that this variable changes as well. Otherwise, it would lose its variable character and mutate into a constant for which neither a function nor continuity would be relevant. Because it has the same status as α, one could therefore formalize this in:

> While the variable successively assumes the values α_n decreasing to zero, the other variable x will assume values x_n confined to the interval (a, b) in which f is defined (Giusti 1984, 49).

However, this interpretation lacks any statement on the direction taken by the values of the sequence x_n. To test the continuity of $f(x)$ in the interval (a, b), Giusti states that it is necessary to show that the dependent variable

$$f\left(x_n + \alpha_n\right) - f\left(x_n\right)$$

has zero as limit (ibid., 50). However, this is already equivalent to uniform continuity in the interval (a, b).

Hence, it can be seen that Cauchy's continuity definition is based on a twofold passage to the limit; and it is particularly the problems with such multiple passages to the limit that are also so persistent in the continuity theorem.

6.6. Convergence

> It must be remembered that the word or other notation used to denote a conception is, in itself, of little importance. [...] The name and notation are only outward and visible signs.
> —Jourdain 1914, 668
>
> La formation de la pensée scientifique est inséparable du développement de symbolismes spécifiques pour représenter les objets et leurs relations.
> —Granger 1979 (after Duval 1995, 3)
>
> Il n'y a pas de *noésis* **sans** *sémiosis.*
> —Duval 1995, 5

Cauchy starts his chapter on series and their convergence or divergence by introducing an important notational innovation into French mathematics. This served as the premise for a more general treatment of the properties of series: indexing the terms in a sequence in order to identify them as elements that clearly belong to the whole sequence. Some degree of indexing had already been practiced in France in the context of the *École polytechnique* (see the footnote on indexed sequences in Chapter II.4.3.), but it became a general instrument only through the work of Cauchy.

Cauchy defines a "series" as an unlimited sequence of quantities whose formation from one quantity to the next follows a determined law:

> We call a *series* an indefinite sequence of quantities
>
> $$u_0, u_1, u_2, u_3, \text{ etc.} \ldots$$
>
> which are derived from each other according to a determined law (Cauchy 1821, 123).

He views the quantities themselves as the various terms of the sequence. In his *Ancien cours* of 1817/18, Ampère gives an abbreviated version of the same definition, but without indexing the terms of the sequence: "We call a series a sequence of terms subject to a law."[24]

Cauchy completes this designation with an expression and notation for the general term: "the term which corresponds to the index *n*, namely u_n, is called the *general term*" (ibid.).

Whereas these notations provide the necessary clarity for general operations with "sequences," the signs he introduces for partial sum and sum, in contrast, are suitable only for assessing partial aspects of these concepts, while others are not assessed at all. Cauchy uses

$$s_n = u_0 + u_1 + u_2 + \ldots + u_{n-1}$$

to describe the sum of the first *n* terms of the sequence, with $n \in \mathbb{N}$. With this notation for partial sums, he introduces the concept of *convergence*: as a limit of the sums s_n for ever increasing *n*—as long as such a limit exists:

> If, for ever increasing values of *n*, the sum s_n approaches indefinitely a certain limit *s*, the series will be said to be *convergent*, and the limit will be called the *sum* of the series (ibid.)

If, in contrast, for ever increasing *n*, the sum s_n does not approach any fixed value, he calls the sequence divergent, stating that such sequences possess no sum (ibid.). Regarding these definitions, it should be noted that:

1. Cauchy's basic epistemological stance of permitting only quantities with real and finite values is also evident here. Its consequence is a radical rejection of operations with series that do not possess a definitive finite value.

2. Because he does not need to consider sequences with a sum of zero, he does not need to classify different concepts as a function of whether the sum is zero or a number.[25]

---

[24]   *Nachlass* Ampère, chem. 76, fol. 8 v.

[25]   Dirksen (1845, 420), in contrast, distinguishes between whether convergence leads to an "assignable" limit or to zero, that is, to a quantity that becomes infinitely small.

3. Cauchy introduces this definition only for numerical quantities and provides no definitions or considerations on sequences of functions. This bring us to comments on the notation:

4. Cauchy's definitions are once again predominantly verbal; he uses signs only for the sums. In particular, he does not apply *lim* operatively either in his definitions or, later on, in his proofs for convergence. He uses the *lim* sign only at the end of his extensive chapter on convergence and divergence.

The sum sign s_n or s gives no indications regarding the sequence for which the sum is sought, although a notation such as $s_n(u_n)$ would have seemed appropriate. Although this may still have no negative consequences with simple sequences of numbers, when dealing with sequences of functions, such simple signs, providing no indication regarding the function and the variable or variables involved,[26] make operating and argumentation cumbersome and confused.

Just like the failure to follow the use of the *lim* sign with any notation over the limit process, there is also no notation marking the range of the indices with the signs s_n or s, as available when using the sigma sign:

$$\sum_{n=0}^{m} u_n .$$

There is also no abbreviated notation for partial sums

$$u_n + u_{n+1} + \ldots + u_{m-1} + u_m, \ m > n,$$

as the sigma sign would also have made possible:

$$\sum_{k=n}^{m} u_k .$$

Cauchy introduces his own notation only for the special case of all remaining terms from an index n on:

$$r_n = s - s_n = u_n + u_{n+1} + u_{n+2} + \ldots \text{ (ibid., 130 f.)}.$$

Cauchy's notation is sufficient for introducing and applying his systematically developed convergence criteria such as the ratio test and root test (ibid., 132 ff.).[27] Problems arise only with the controversial continuity theorem in which Cauchy relates the convergence to the continuity of sequences of functions.[28] He argues here with functions of a variable x, with increments of these functions, and with increments of a variable x by an *infiniment petit* α. However, this is all done verbally, because he possesses signs only for the three functions,

$$s_n, \ r_n, \text{ and } s,$$

---

[26]   Here as well, Spalt (1996, 101) simply changes and "enriches" Cauchy's signs without comment; for a value X, he writes $s_n(X)$.

[27]   Ampère had also already presented ratio and root tests in his *Ancien cours.*

[28]   In the literature, to the best of my knowledge, only Pensivy (1988, 12) has pointed out that the notation is ambiguous and therefore the meaning unclear. He mentions the lack of a relation to the variable x in the notation.

but no link from these to x, to α, and to the increments of the functions.

With his theorem on the continuity of the limit function of a convergent series of continuous functions, Cauchy suddenly, without further preparation, shifts from series of numbers and quantities to series of functions. For continuity, he uses his third definition here: continuity in the neighborhood of a value of the variable x. Cauchy's formulation is presented here in its original form to illustrate the paucity of signs:

> Ist Theorem. When the different terms of the series (1) are functions of the same variable x, continuous with respect to this variable in the neighborhood of a particular value s for which the series is convergent, the sum s of the series is also, in the neighborhood of that particular value, a continuous function of x (Cauchy 1821, 131 f.).

The proof is extremely short and also largely lacking in signs. He uses the three sums or functions s_n, r_n, and s for the sequence u_0, u_1, u_2, ..., which he abbreviates to *série (1)* in his theorem without using his sign for the general term. His s_n is clearly continuous in the neighborhood of the x value (according to the premise). He then examines the increments of the three functions when the variable x increases by an *infiniment petit* α. He reports that the increment of s_n is infinitely small for all possible values of n (because of the continuity of u_n, and the increment of r_n—together with r_n itself—becomes "imperceptible" for very large values of n (because of the convergence of the sequence; nonetheless, not only as a function of n but also of x; however, this is not apparent because of his choice of symbols). Hence, he says that the increment of the limit function s is also infinitely small:

> The increase in s_n will be, for the possible values of n, an infinitely small quantity; and that of r_n will become at the same time imperceptible as r_n, if n takes a very considerable value. Consequently, the increase in the function x can only be an infinitely small quantity (ibid., 131).

In a memoir on the general proof of the binomial formula (the main application of this theorem in Cauchy), Abel (1826) remarks in a footnote that "this theorem suffers exceptions," illustrating this with the series

$$\sin \varphi - \frac{1}{2}\sin 2\varphi + \frac{1}{3}\sin 3\varphi - \dots \text{ etc.},$$

whose limit function is discontinuous for each value $(2m + 1)\pi$ of φ, with $m \in \mathbf{N}$. However, he does not name any cause for this error.

Although Fourier had already addressed such series in his *Théorie de la chaleur*—published in 1822, but already familiar to the Academy beforehand—Cauchy continues to use his theorem unchanged for a long time. In 1853, in contrast, he published a short memoir in which he explains that his theorem is valid only for sequences "ordered according to the increasing powers of a variable." For other sequences, it is not valid without restrictions. As proof of a discontinuous limit function, he now names a sequence of the same type as Abel's and Fourier's series. He claims that it is easy to modify the statement in such a way that no further exception occurs (Cauchy 1853, 31 f.). In fact, however, he makes major revisions to the theorem (ibid., 33).

Cauchy himself mentions a "comment" by C.A. Briot and J.C. Bouquet as the motive for his change. As first disclosed by Bottazzini in 1992, this "comment" actually consisted of two memoirs submitted to the Academy for review by, among others, Cauchy. Unfortunately, these memoirs were not published; only shortened versions are available. They contain a broad criticism of Cauchy's work, triggered by controversies over the theory of complex functions (Bottazzini 1992, XCI ff.). One of the changes in Cauchy's theorem is therefore also to relate it explicitly to real variables.

Whereas the theorem was declared to be an "error" in earlier historiography (e.g., Bourbaki 1969, 193), because the premise was only convergence and not uniform convergence, a hefty debate has arisen, as reported above, since Robinson (1966) on whether and under what conditions the theorem may still be "correct."[29] One of the lines of justification was to show that Abel's counterexample could not be a counterexample for Cauchy. This would be because, for example, Cauchy had used another concept of convergence, and therefore, the premise of convergence did not apply to Abel's series (e.g., Spalt 1996, 185 ff.). Giusti (1984, 53) argued that Cauchy's concept formations already express a uniform continuity. Laugwitz (e.g., 1987) long tried to make the proofs appear correct by "discovering hidden lemmas." However, neither Laugwitz or Spalt has managed to confirm that either the 1821 or 1853 theorems are simultaneously correct and have the same meaning, a problem when one views Cauchy as a hero who consistently applies unwavering rigor in his thinking (Laugwitz & Spalt 1988, 15).

Koetsier (1991, 73 ff.) carefully compared the "standard" and "nonstandard" interpretations of Cauchy's continuity theorem(s). However, he applied a purely internalist textual analysis without considering Cauchy's context or the development of the history of concepts in France. He concluded that the nonstandard interpretation raises more problems than it solves, thus making the standard interpretation preferable.

In the latest evaluation, Lützen (1999) has summarized the debate by stating that a definitive decision on which interpretation Cauchy intended is evidently not possible:

> . . . is impossible to refer to Cauchy's definition, since he does not define the sum of a series of functions separately. However, it is difficult to rescue Cauchy entirely, because he himself later acknowledged that his theorem "cannot be accepted without restriction" (Cauchy 1853). It is hard to avoid the conclusion that Cauchy's concepts were somewhat vague at this point. (Lützen 1999, 210; engl.: Lützen 2003, 168).

---

[29]   Jesús Hernández (Madrid) provides a good overview of this debate in a manuscript of 1990 entitled *Cauchy y la convergencia uniforme*. This work is also mentioned by Bottazzini (1992, LXXXV). However, it has been published only in a shortened form (Hernández 1989).

The following will take a closer look at the reasons for this uncertainty, and show that this clearly involves the role of signs. One excellent source of information here is a contemporary criticism of Cauchy's continuity theorem that focuses precisely on these deficits regarding signs. Characteristically, this criticism does not come from France, but from Germany: from a mathematician shaped by the context of mathematics at Göttingen and Berlin, and particularly by the then still dominant practice of combinatorics. The combinatorial school, with its enthusiasm for formal algebraic operations and ever more far-reaching generalizations, had only one way of keeping its rampant combinations under control: by developing appropriate symbols and applying them strictly. The author of this criticism was a mathematician already mentioned repeatedly as the most careful processor and supporter of Cauchy's conceptions in Germany: Enno Heeren Dirksen (1788-1850). Dirksen, a native of East Frisia, first started work in 1807 as a schoolteacher. Between 1817 and 1820, he studied mathematics and astronomy at Göttingen under Mayer, Thibaut, and Gauß. He received his doctorate at Göttingen, and was appointed to associate professor in 1820 at Berlin. Soon afterwards, he completed his postdoctoral habilitation at Berlin, where he worked as a full professor after 1824. Without producing any innovative research achievements of his own, his main field was reflections on foundations. His unfinished work, *Organon*, of which only the middle section was published, is a convincing document for an explication and continuation of Cauchy's program of rigor.

Dirksen's criticism of the theorem is to be found in his review of the first German translation of the *Cours d'Analyse* by C.L.B. Huzler in 1828.[30] Whereas the general tone is strongly enthusiastic over Cauchy's achievements and his new standard of rigor, Dirksen is remarkably reserved about Cauchy's definition of continuity:

> The author preferred to define the concept of the continuity of a function by means of the concept of the infinitely small. It is clear that he could have started from a more obvious perspective, and then derived the proposition presented to explain continuity as a theorem (Dirksen 1829, Column 215).

The sections of his *Organon* dealing with the theory of functions were not published; they are preserved only in manuscript form. Although he does once give an $\varepsilon-\delta$ definition of continuity, most versions contain a definition based on this "more obvious perspective." It proceeds from a conception that deviates from the epistemology dominant in France at that time by maintaining that the limit actually can be reached. In this case, a function f is continuous in x_0 if the value of the function in x_0 agrees with the limit $f(x_0)$:

---

30   In modern literature, the first mention of Dirksen's review is, to the best of my knowledge, in Kurt Richter's (1976) doctoral thesis (University of Halle). Grattan-Guinness (1986, 226) then mentions the review in an article without giving the source. Both publications, nonetheless, fail to note Dirksen's innovative use of signs.

If x designates an independent variable, $f(x)$ an appropriately defined [function], and x_0 a specific value of x, it is said that $f(x)$ is *continuous* for the specific value x_0 and *remains* so insofar as its special value for x_0 of x is completely determined by x and is equal to the value of the limit of $f(x)$ for the limit x_0.[31]

What is particularly notable is how Dirksen's interpretation evidently deviates from Cauchy's first definitions here by taking continuity as local, pointwise continuity. This is in line with Abel, who, at the beginning of his famous memoir on the binomial formula, gives a definition of continuity that deviates from that of Cauchy; or one could say that in contrast to Cauchy's definition, it is unambiguous:

A function $f(x)$ should be called a continuous function of x between the limits $x = 0$, $x = b$ if, for an arbitrary value of x between these limits, the quantity $f(x - \beta)$ approaches the limit $f(x)$ for always decreasing values of β (Abel 1826, 314).

Although plainly modeled on Cauchy's first definition, the difference here is that Abel clearly means pointwise continuity. Some weaknesses in the definition are also unmistakable: β is undefined; no distinction is made between a variable and its value; and the characterization *indéfiniment*, or unlimited, is missing for the approximation. The fact that Abel was living in Berlin when he wrote this memoir is one indication for the existence of a "Berlin context" with its own independent definition of continuity,[32] even though its supporters may well have been scarcely aware that they differed from Cauchy.

Even before this 1829 review, Dirksen had introduced an important innovation in the use of signs. It is already fully developed in a memoir to the Berlin Academy in 1827. As far as I know, Dirksen was the first to provide the limit sign with indications on the limit process and the variables concerned. As German Romanticism was particularly strong at that time, he replaced the abbreviation for the Latin or French *limes* or *limites* with an abbreviation for the German word *Grenze*, resulting in the following notation:

$$\overset{x=x_0}{Gr}.f(x),$$

in which he gives the variable *above* the limit sign. Hence, Dirksen presents his definition of continuity as

$$f(x_0) = \overset{x=x_0}{Gr}.f(x).$$

By specifying the variable in the limit sign, Dirksen could differentiate limit processes for several different variables with signs, and thereby also conceptually.

As far as I know, this made him the first to study multiple limit processes and to discriminate these both analytically and with signs. Figure 18 presents a section from Dirksen's memoir of 1827 as evidence for the process he developed

---

[31]   Dirksen's *Nachlass*, Version "B" of *Functionenlehre*, fourth section, §32, *Erklärung* 21.

[32]   Further, previously unrecognized sources for such a Berlin context are reported in Chapter VII.

for multiple passages to the limit. At the same time, it shows how he discusses the admissibility of *interchanging* passages to the limit.

Interchangeability is also the central point in Dirksen's criticism of Cauchy's continuity theorem. Dirksen's first step is to specify the single processes within the entire passage to the limit. After quoting the theorem, he not only expresses his doubts, but also formulates appropriate symbols to explain it:

Vergleicht man diese Gleichung mit der ihr entsprechenden (55), so sieht man leicht, daſs, insofern beide gleichzeitig statt finden, und also jene eine nothwendige Folge von dieser, sei es unbedingt, sei es auch nur bedingungsweise, bilden soll, sein muſs

$$(68) \ldots \ldots \underset{i=1}{\overset{k=0}{\operatorname{Gr}}} \underset{i=1}{\overset{i=\frac{l-i}{2}}{\Sigma}} \left\{ \operatorname{Gr}.h \underset{a=0}{\overset{h=0}{\Sigma}} \overset{a=l-h}{F(a)} \cos\frac{i x \pi}{l} \cos\frac{i a \pi}{l} \right\}$$

$$= \operatorname{Gr} \underset{i=1}{\overset{h=0}{\left\{ \overset{i=\frac{l-i}{2}}{\Sigma} h \underset{a=0}{\overset{a=l-h}{\Sigma}} F(a) \cos\frac{i x \pi}{l} \cos\frac{i a \pi}{l} \right\}}}:$$

und umgekehrt, insofern diese Gleichung geltend gemacht werden darf, wird auch die Gleichung (67) als eine Folge von (55), und mithin auch von (43), auf einem directen und völlig strengen Wege, zu gewinnen sein.

   Was nun die Gleichung (68) insbesondere anbelangt, so scheint ihre unbedingte Richtigkeit, aus einem allgemeinen Gesichtspunkte betrachtet, mit Grund bezweifelt werden zu können.

Figure 18, operating with double limits (Dirksen 1830, 109 f.)

The reviewer sincerely has to admit that he is not satisfied with the proof suggested for this theorem, and that, on closer inspection, he even doubts whether the theorem itself is correct. Because all the terms of the series are assumed to be functions of x, one obtains, insofar as S_n denotes the sum of the first n terms,

$$S_n = f(x,n) \text{ (Dirksen 1829, column 217)}.$$

This sign now makes it clear that *two* variables are involved in the passage to the limit. By using his own sign for limits, he initially specifies the passage to the limit for n that involves the convergence:

If we now take the special value of x, which is the one considered in the theorem, and denote it by a, and the corresponding sum of the series by S, one obtains in line with the definition for the sum of a series

$$S = \overset{n=\infty}{\operatorname{Gr.}} f(a,n) = f(a,\infty) \text{ (ibid.)}.$$

Continuity, in contrast, involves the other variable, x:

For this quantity to now be continuous in the neighborhood of the value $x = a$, one obtains in line with the concept of continuity

$$S = \overset{\Delta a=0}{Gr.}\ f(a + \Delta a, \infty) \text{ (ibid.).}$$

The explicated theorem reconstructed in this way therefore requires the two passages to the limit to be interchangeable:

As a result, by combining this equation with the previous one,

$$\overset{\Delta a=0}{Gr.}\ f(a + \Delta a, \infty) = \overset{n=\infty}{Gr.}\ f(a,n),$$

that is, if $f(x, n)$ represents the sum of the first n terms of the series, then, if the sum of the series is to be continuous in the neighborhood of the value $x = a$, then

$$\overset{\Delta a=0}{Gr.}\ f(a + \Delta a, \infty) = \overset{n=\infty}{Gr.}\ f(a,n) \text{ (ibid., column 217 f.).}$$

Dirksen is now pointing out that, according to Cauchy's second chapter, the one on continuity, there is no way in which the statement can be generally valid a priori, and he finally formulates a sequence of continuous rational functions as a counterexample:

Now, with $f(x, n)$ and a conceived as given, this equation cannot be maintained with rigorous generality according to the findings of the second chapter, but can, under some circumstances, be subject to exceptions. Therefore, for the theorem to be correct, it is necessary to confirm that such an exception is not possible for $f(x, n)$, insofar as it represents the sums of the first n terms of a converging series that are all continuous as they approach $x = a$. This now seems to be even more difficult, because each function of x and n, $f(x, n)$ for which

$$\overset{n=\infty}{Gr.}\ f(a,n)$$

is finite, and

$$f(x,n+1) - f(x,n) \text{ for } x = a$$

is continuous, can be regarded as the sum of the first n terms of a series corresponding to the assumptions in the theorem (ibid., column 218).

Whereas the first condition implies convergence, the second condition implies the—pointwise—continuity of the $(n+1)$st term in the sequence. Dirksen chooses as a sequence of rational functions with the general term

$$\frac{a - x}{[(n-1)(a-x)+1][n(a-x)+1]},$$

with a positive. These terms form continuous functions for $x = a$. As partial sums, one obtains

$$S_n = f(x,n) = \frac{a - x}{a - x + \frac{1}{n}}.$$

The two limits differ according to whether the passage to the limit for x precedes the passage to the limit for n

$$\overset{n=\infty}{Gr.}\ f(a,n) = 0,$$

and vice versa

$$\overset{\Delta a=0}{Gr}\ f(a + \Delta a, \infty) = 1 \text{ (ibid.).}$$

This leads to "conflicting" results on "the condition necessary for the correctness of the theorem" (ibid., column 219). Dirksen names the differences in the two passages to the limit very precisely:

> The main point overlooked here accordingly seems to be that the concept of the sum of a series for $x = a$, strictly speaking, assumes that x is taken as being determined before n, whereas the judgment on the continuity of the sum demands the opposite, and that both results, when they do not become infinite, can differ from each other (ibid.).

Dirksen accordingly also points out that Abel has already "expressed doubts regarding the correctness of the theorem in question." However, he "satisfied" himself with a counterexample, "without engaging in a closer study of the true reason for the conflict that emerges" (ibid.).

Both Richter and Grattan-Guinness have assumed that Dirksen's review remained unknown because it was published only in a "local" Berlin journal. This assumption is inappropriate, because this journal was in no way just local. It was the *Jahrbücher für wissenschaftliche Kritik*, a project strongly supported by the Prussian Ministry of Culture in order to raise scientific standards and improve communication between all scientific disciplines. Its editors therefore included high-ranking scientists, and one of the editors for mathematics was even Dirichlet. Since Dirksen continued to be active at both the Berlin Academy and the university for decades where he worked intensively on these basic concepts, his arguments were part of the general scientific communication. Indeed, in his meticulous report on "interchanging the sequence of passages to the limit" in the *Enzyklopädie der mathematischen Wissenschaften*, Burkhardt (1914) claims that Dirksen was the first to expressly present "the theorem that the sequence of two passages to the limit cannot simply be interchanged at will" (ibid., 987). He also ascertained that Dirichlet had paid "systematic attention to the sequence of passages to the limit," whereas French mathematics did "not realize for a long time" that the sequence cannot simply be interchanged at will (ibid., 979, 972).[33]

One frequent objection is that Cauchy cannot be accused of an inadmissible interchanging of passages to the limit, because he had already pointed out that results differ when the sequence is changed in his famous memoir on definite integrals in 1814 (see, e.g., Laugwitz 1989, 236 f.). However, this objection is unsound for (at least) two reasons:

1. In this early period of his work, Cauchy still did not view the definite integral as an independent basic concept. In particular, he had not yet defined it as a limit of an infinite number of summands. Hence, the memoir does not study

---

[33]   Indeed, even Carnot's theorem on *équations imparfaites* as discussed by Dhombres reveals a lack of awareness for multiple passages to the limit (see above, Chapter V.1.6.1.).

the interchanging of passages to the limit, but the interchanging application of two variables x and z in double integrals.

2. Cauchy examines such interchanges at points where the first integration leads to an indefinite result such as $0/0$. For both Cauchy and the two Academy reviewers Lacroix and Legendre, it was self-evident that the results are equal for both orders of integration, and that it is precisely this equality that makes it possible to determine the double integral despite the indeterminacy in the one order of integration (Cauchy 1882, 322 ff.). When confronted with the initially different results of two orders of integration, Cauchy's concern was to find a corrective term A in order to reestablish equality. Lacroix and Legendre actually declared it to be one of the main achievements of Cauchy's memoir that he succeeded in determining "exactly the correction necessary to establish the equality between the results obtained by the two ways of carrying out integrations" (ibid., 325). Hence, despite Laugwitz's (1989, 237) claim, Cauchy does not confirm two different results,

$$\int_0^1\int_0^1 \frac{\partial K}{\partial z}dxdz = \frac{\pi}{4}, \text{ and } \int_0^1\int_0^1 \frac{\partial K}{\partial z}dxdz = -\frac{\pi}{4},$$

but determines a corrective term A in order to attain the equality assumed a priori (Cauchy 1882, 322), so that his eventual finding is

$$\int_0^1\int_0^1 \frac{\partial K}{\partial z}dxdz = \frac{\pi}{4} + A = -\frac{\pi}{4} \text{ (ibid., 395)}.$$

This section on convergence will close by commenting on Moigno's lectures on differential calculus of 1840. In his seventeenth chapter on the convergence of sequences, Moigno reports Cauchy's current views on the conditions for developing in particular complex functions into a power series. Bottazzini (1992, CXXV ff.) has described the complicated paths Cauchy took from 1831 onward before finally achieving definitive solutions about 20 years later.[34] The debates with a series of mathematicians on this theme also relate to the development of Cauchy's ideas on the continuity theorem and its eventual reformulation in 1853.

In several memoirs from 1839 and 1840, Cauchy expresses the emphatic belief that he has found such a theorem for the development of explicit, complex functions by which the convergence law is simply reduced to the continuity law (see ibid., CXXX). Moigno adopts this enthusiasm in his book and propagates it directly, saying that Cauchy has managed in recent years to prove a truly remarkable theorem that directly delivers the rules for the convergence of the series obtained from the development of explicit functions. This theorem, he adds in the words of Cauchy, "simply reduces the law of convergence to the law of continuity of functions" (Moigno 1840, 150). This emphasis on being able to reduce convergence to continuity provides an impressive proof of my premise

---

[34]   An essential aspect here was the recognition that the characteristic property for complex functions is analyticity and not continuity.

that Cauchy was guided methodologically by an epistemological continuity principle, as can be seen particularly clearly in his first course in analysis of 1817: his "metaphysical" principle of continuity claiming that laws applying in the finite are also valid after the passage to the limit, and should thus ensure a general interchangeability.

6.7. Introduction of the Definite Integral

The previous sections analyzing Cauchy's development of basic concepts have shown that the *infiniment petits* did not take any indispensable role in their definition and application. In Cauchy's implementation of the compromise concept on the foundations of analysis at the *École polytechnique*, it was the *limite* that formed the higher or central concept to which the *infiniment petits* were subordinated. The first time the *infiniment petits* gain their own independent functional role is in the concept of the integral: The construction of the concept of the definite integral, which Cauchy transformed—on the basis of a concept from Ampère—into the new basic concept of integral calculus, was based essentially on the use of infinitely small quantities. Cauchy used these to obtain the infinite sum of products. The productive influence of the context of the *École polytechnique* is documented by the changes Cauchy made to the concept before its definitive introduction.

The first relevant document is Cauchy's 1814 memoir on definite integrals. However, this still does not contain a distinct definition of the definite integral, although the basic conception can already be ascertained: Cauchy describes definite integrals as the "sum of the elements"—using a terminology recalling the early history of integral calculus when it was based on the indivisibles. However, he does not go into more detail or define this sum aspect. It is clear only that the "elements" are values of functions that are determined by the values of the variables between the limits of integration. Cauchy gives his most detailed statement in this memoir when discussing the double integrals:

> Each of the double integrals which the preceding equation presents is the sum of the elements which correspond to diverse values of x and z comprised between the limits of integration (Cauchy 1882, 379).[35]

The *éléments* here are integrands such as

$$\frac{\partial S}{\partial z}, \frac{\partial U}{\partial x},$$

however, without providing an explicit nesting of intervals and assigning values of functions.

---

[35]   Cauchy went on to revise this 1814 memoir and send it to the publishers in 1825; it came out in 1827. However, these revisions are evidently restricted to supplementary footnotes.

In his first, to all intents and purposes, independently planned analysis course of 1816/17, Cauchy was already using the definite integral to introduce integral calculus. The wording in the *Registres* shows how he used the conception of the "sum of the elements" from his 1814 memoir: "definite integrals considered as the sum of elements" (Gilain 1989, 53). In the next basic course in 1818/19, when the *infiniment petits* were already firmly reentrenched as a foundation of analysis, it is clear that Cauchy already presented the new conception. It was now so well developed that he could introduce not only the "concept" but also the "properties" of the definite integral:

"Notions sur les intégrales définies" (37th lesson);

"Propriétés des intégrales définies" (38th lesson; ibid., 59).

Because there are no known notes taken from this lesson, and the 1821 edition of the *Cours d'analyse algébrique* does not cover differential and integral calculus, the first printed version of Cauchy's definition of the definite integral is in his second textbook, *Résumé des Leçons données à l'École Royale Polytechnique sur le Calcul Infinitésimal* (1823). It is assigned its own chapter: the 21st lesson.

This new definition forms one of the most important applications of the calculus of means that Cauchy developed at the beginning of his *Cours d'Analyse*. For the function to be integrated, he assumes continuity in the interval $[x_0, X]$.[36] He then separates the interval into a (finite) number of subintervals:

$$x_1 - x_0, \; x_2 - x_1, \; \ldots, \; X - x_{n-1}.$$

Deviating from his 1814 and 1817 designation, he now calls these subintervals "elements" and no longer values of functions. He calls the initial point of each subinterval the "origin" of the element in question. Through multiplying each "element" by the value of the function of the "origin," he forms the (finite) sum

$$S = \left(x_1 - x_0\right)f\left(x_0\right) + \left(x_2 - x_1\right)f\left(x_1\right) + \; \ldots \; + \left(X - x_{n-1}\right)f\left(x_{n-1}\right).$$

As Cauchy remarks, the size of S depends on the number n of "elements" in the nest of intervals and on the values of these "elements," and, consequently, on the chosen division of the interval. However, as he subsequently shows in detail, this no longer has any "perceivable influence" on the value of S if the absolute values of the "elements" become very small and the number n very large (Cauchy 1823, 81). By comparing three nests of intervals and applying his calculus of means, Cauchy shows that, independent of the division chosen, S converges toward the definite integral of the function f from x_0 to X. The essential aspect is that the sequences of element values for ever-increasing n form variables with limit zero, in other words, *quantités infiniment petites*:

---

[36]   On the meaning of continuity in Cauchy ranging from simple to uniform continuity, see above, Section 6.5.

Thus, when the elements of the difference $X - x_0$ become infinitely small, the mode of division has only an imperceptible influence on the value of S; and, if the numerical values of these elements are allowed to decrease indefinitely, by increasing their number, the value S will finish by being perceptibly constant, or in other words, it will finish by attaining a certain limit which will depend uniquely on the form of the function $f(x)$ and the values of the extremes x_0, X given to the variable x. This limit is what is called a finite integral (ibid., 83).

From the various proposals for abbreviating the converging sum S available at the time, Cauchy recommends

$$\int_{x_0}^{X} f(x)dy \text{ (ibid., 84)}.$$

The new concept of the definite integral is now proved in its own right, independent from differential calculus. It is clear that in this case, Cauchy has been able to apply his ideas on infinitely small quantities constructively and not just use them to legitimize or represent something else. Cauchy's definition remained the principal foundation until further refinements such as Lebesgue's and Stieltjes's integrals, even though his concept of infinitely small quantities as variables was not adopted widely but used only as a way to express "smaller than every given positive number regardless how small."

One can see that even Riemann formulates his definition of the definite integral parallel to Cauchy, although Riemann does not insert any "origin" values for the function, but means in each subinterval:

Hence, to start with, what should we understand by $\int_{a}^{b} f(x)dx$?

To determine this, let us assume, between a and b, a series of values $x_1, x_2, \ldots, x_{n-1}$ ranked according to increasing quantity, and for the purpose of abbreviation, label $x_1 - a$ as δ_1, $x_2 - x_1$ as δ_2, $\ldots$, $b - x_{n-1}$ by δ_n, while using ε to represent a positive true fraction. Then, the value of the sum

$$S = \delta_1 f\left(a + \varepsilon_1 \delta_1\right) + \delta_2 f\left(x_1 + \varepsilon_2 \delta_2\right) + \delta_3 f\left(x_2 + \varepsilon_3 \delta_3\right) + \ldots + \delta_n f\left(x_{n-1} + \varepsilon_n \delta_n\right)$$

depends on the choice of the intervals δ and the quantities ε. If it now has the property, however δ and ε may be chosen, of indefinitely approaching a fixed limit A, as soon as all δ become infinitely small, this value will be called $\int_{a}^{b} f(x)dx$ (Riemann 1867, 102).

The definite integral founded on the infinitely small quantities not only corresponded to the rational core of the concept of indivisibles but also linked up with Leibniz's original intentions when he introduced his concept of the integral.

6.8. Some Final Comments

Analyzing and understanding the basic concepts in Cauchy's mathematics proves to be a typical reflection of the demands and challenges facing the historiography of mathematics: not only do one-sided applications of modern meanings of concepts prevent an understanding of contemporary work and its contexts, but also one-sided stylizations lead to heroes who can never do wrong. In particular, Spalt's interpretations prove to be a form of "monster-barring" to take the words of Lakatos (1976, 14 ff.). By attributing the strongest possible interpretations to Cauchy's concepts—for example, by interpreting *continuité* as uniform continuity; *convergence* as so-called continuous convergence— "monsters" such as Abel's counterexample are barred as not being affected by the theorem. However, what is specific to the logic of Lakatos's monster barring is that anomalies initially denigrated as mathematically illegitimate eventually become integrated into mathematics as conceptual differentiations that pave the way for further progress.

Indeed, in reply to Giusti, Laugwitz, and Spalt, it has to be pointed out that determining which particular meaning Cauchy intended for continuity or convergence is not all that decisive for understanding the development of concepts. What is more decisive here is that he does *not* differentiate the specific basic concept, that he has *not* given limiting conditions, but that in each case, there was only one meaning of the concept for him, and not yet an unfolding of a new concept field. Ascertaining this should not be taken as a criticism of Cauchy, because given his pioneering achievements such as transforming continuity and convergence into operative concepts, any claim that he should have anticipated all subsequent developments proves to be nothing other than the outcome of such heroizing.

What is more productive, in contrast, is the insight on how directly further conceptual progress is influenced by the explication of the meanings of concepts through appropriate signs.

Finally, we can identify one characteristic of Cauchy's mathematics that has been neglected in earlier literature: Cauchy's research and teaching were guided by a vision: the methodologically and epistemologically leading role of *continuity*. Cauchy himself occasionally expresses this as his belief in the "great law":

> It was seen, in my earlier *Mémoires*, that the great law which limits the existence of formulas is the *law of continuity* of functions (Cauchy 1845, cited in Bottazzini 1992, CXXXVII).

Cauchy's strength lay in his orientation toward this vision; but it is simultaneously also the source of his limitations. It shows how he was actually confined by his adherence to the Leibnizian-Newtonian tradition of the eighteenth century.

Chapter VII

Development of Pure Mathematics in Prussia/Germany

1. Summary and Transition: Change of Paradigm

The previous chapters have shown clearly that the traditional statement in historiography that eighteenth-century mathematics "lacked rigor," was seeking only rapid advances, and left the clarification of basic concepts until the nineteenth century is highly inappropriate. We have already discussed a great number of works on foundations from the eighteenth century, and these also received a wide distribution and reception. The real problem lay elsewhere: the determining "paradigm" of mathematics within which this research was performed had run up against its limits. It was no longer able to solve the backlog of problems. New, progressive solutions became possible only after the transition to a new paradigm.

Moreover, regarding our two concept fields, the previous chapters have shown that the tie to the traditional concept of quantity did not just characterize the dominant paradigm, but simultaneously braked or even blocked conceptual development, leading to the construction of even more convoluted "epicycles."

Hence, advances in concepts on negative quantities in the eighteenth century make it increasingly clear that only absolute numbers were accepted as legitimate number objects. Something that first emerged in a completely implicit way within fields of application finally became formulated increasingly explicitly as a precondition, as a consequence of reflections on foundations. This number concept was the expression of a quantity concept pervading the whole of mathematics that simultaneously aimed to ensure ontological ties to a "reality" (of whatever kind). We have seen how not only the explicitness of absolute numbers as the sole admissible and legitimate foundation but also this ontological tie attains its peak in the work of Cauchy.

Completely analogous to this, we have worked out that the "rational core" making the concept of infinitely small quantities so attractive was to grant the zero an exceptional status: because zero was not admitted as a number, independent conceptual efforts were required when variables approached it.

It is particularly the attempts to justify negative quantities that reveal the impossibility of a general and consistent definition of multiplication as long as mathematics was dominated by the quantity paradigm.

The "solution" to these problems therefore forced its way toward an abstract number concept that would no longer be subordinate to the over-general quantity concept. However, this implied a fundamental change in the complete edifice of conceptual systems in mathematics, in other words, a change of paradigm. Such a change actually did occur: the shift to so-called "pure mathematics." This was noticeable from about 1800 to 1820, though not as a global change, but as a phenomenon initially restricted to one specific cultural domain: northern Germany or Prussia after its internal reforms. Whereas tendencies toward an analogous development can be traced in England shortly after the 1820s with the emergence of symbolic algebra around Babbage, Herschel, and Peacock, the paradigmatic change took a long time to reach completion in France. In line with the engineering context, which initially imposed definite limits on mathematics, the prevailing perspective remained an applied one, and mathematics continued to be determined by the orientation toward *Physico-mathématique* dominant since the *Encyclopédie*. Lacroix formulated this perspective succinctly as a *credo* in a letter to the mathematician F.J. Français in 1810. Français came from Alsace and for this reason had good relations to mathematics in Germany. He had written to Lacroix as a result of his strong criticism of the combinatorial school:

> Pure analysis and geometry are doubtless in themselves very elegant speculations, very suited for exercising the mind, and can provide the opportunity for greatly developing wisdom: but I confess to never having been able to attach a great price to these advantages when they are considered as being the unique purpose for the study of these sciences. [...] After the usual applications, after the "reasoned" expositions of the great methods, which the philosophy of Science reveals and which shows the path which the human mind follows in the research of properties of magnitudes, the science of the calculus appears to me to be no more than a game of chess if it does not provide the key for many phenomena whose laws are inaccessible without its aid. I examine therefore every analytic discovery in the light of the hopes it may offer for the advancement of the physico-mathematical sciences.

Criticizing the page-long combinatorial formula "developments," he added:

> [...] that the difficulties to be overcome seem beyond the development of these polynomials. Such, without doubt, are the motives which have prevented French geometers from taking an interest in the research of German analysts and that of Arbogast (quoted in Schubring 1990, 99 f.).

The present chapter starts off by presenting the emergence of the specific context in which pure mathematics developed. Because the shift to pure mathematics, the transition to theoretical concepts, first started with the negative numbers, the next Section is devoted to their conceptual history within this context. The last part of the chapter then presents selected aspects of "pure" developments in analysis.

2. The Context of Pure Mathematics: The University Model in the Protestant Neohumanist System

Because I have given detailed accounts of the emergence of the institutional context for pure mathematics in northern Germany or Prussia in a number of studies,[1] I shall limit myself to sketching the main findings here. Analyses in the history of science have shown that the characteristic feature of university reforms in Prussia from 1810 onward was the establishment of the "research imperative" (see Turner 1980): the dual role of professors as both teachers and researchers. The core of this reform took place in the philosophical faculties. It started in classical philology before extending to mathematics, the natural sciences, and history. The resulting emergence of the research university in Prussia was initially a unique, isolated phenomenon.

On the one hand, it was due to a precondition common to all Lutheran and protestant northern Germany (see Chapter II.2.6.). The philosophical faculties had managed to improve their status within the universities and even move toward equal rank with the "higher" faculties. Göttingen and Halle were the two universities in which subject specializations had already been introduced in the eighteenth century, and professors were already expected to engage in research. However, this reform was still embedded within the framework of mercantilist policies designed to gain location advantages. When Napoleon closed the University of Halle after his victory over Prussia in 1806, plans were made to found a new university at Berlin.

In contrast, Göttingen, the university of the kingdom of Hanover, followed a completely different path: there, mathematics had transformed itself into what was practically an independent subject area within the faculty with a broad spectrum of predominantly technical, applied courses, starting from a few true university courses and extending to military engineering and civil engineering, to geodesy, and so forth. A gradual reduction in these applied courses and an adaptation of the Berlin model did not occur until the 1830s to 1840s, particularly following the foundation of a polytechnic school at Hanover.

In France, in contrast, the universities—at which mathematics was, in any case, represented only marginally within the *collèges*—had been disbanded during the revolution and replaced by strictly vocational special schools. This vocational character was retained even during the later renaming as *facultés* and integration into the *Université impériale* with its focus on secondary schools. Research, in contrast, was concentrated in the *Institut national* or the *Académie*.

The University of Berlin was founded in 1810 as a deliberate alternative to the

---

[1]   Cf. Schubring 1981a, 1981b, 1983, 1985a, 1985b, 1989, 1990b, 1991, 1992b, 1998.

applied special school model dominating Napoleonic Europe. However, its strong research orientation was the outcome of an initially rather random event: the absorption of the Berlin Academy by the philosophical faculty (see Schubring 1991a, 309 ff.).

One notable effect of the practice of combining research and teaching was that it enabled professors to go beyond their previous elementary and standardized teaching based on textbooks and introduce new research findings into their courses. This was made possible by the new task structure of the philosophical faculties, which were practically organized anew. For the first time, they were assigned their own particular training task: to train teachers for the reformed *Gymnasien*. Because mathematics was a new main subject at these schools, it became an independent study course in Prussian universities as well. By being the first state to make teaching mathematics a profession, Prussia created the basis and the preconditions for the establishment and later expansion of mathematics as a scientific discipline at its universities.

In line with the neohumanist values at these universities, particularly in the philosophical faculties, mathematics became established as a "pure" science, as "pure mathematics." Because teaching at *Gymnasien* was initiated as a scientific profession with a high standard of training in order to ensure the social status of teachers, and because mathematics teachers also represented and practiced the new scientism, their activity encouraged the spread of the ethos of pure mathematics. Indeed, these Prussian teachers in the first half of the nineteenth century made a major contribution to the development of mathematics. A number of them became professors at universities.

The comment of C.G.J. Jacobi (1804–1851), one of the main exponents of pure mathematics, that "it is honorable for science to have no use" is well known. What is less well known is that August Leopold Crelle (1780–1855), engineer and founder of the famous *Journal für reine und angewandte Mathematik* (actually publishing only pure mathematics), although himself a practicing engineer, campaigned for the most pure development of mathematics possible, fully separated from its applications:

> Mathematics by itself, or so-called *pure mathematics*, does not depend on its applications. It is completely idealist; its objects, *number*, *space*, and *force*, are not taken from the external world, but are primitive ideas. They pursue their own independent development from deductions drawn on their basic concepts [...]. Every addition of and ties to applications, on which it does not depend, is therefore disadvantageous to the science itself (quoted in Schubring 1982a, 216).

Crelle presented this conception not only in a memoir of 1828 but also in his 1845 textbook on the theory of numbers: he declared that one should not view the applications either "solely or even *preferably* as the purpose of mathematics" (Crelle 1845, VII), and that pure mathematics takes precedence not only for the development of mathematics but also in the learning process (ibid., X).

One development seems to run counter to this emergence of an autonomous discipline of pure mathematics. This is the attempt over many decades, and under the decisive participation of mathematicians, to found a so-called polytechnic institution in Prussia. The main motivation was evidently the enormous fame and prestige of the Parisian *École polytechnique*. The idea was to set up a similar special institute for mathematics, the natural sciences, and technology, but it failed to pay sufficient attention to the special situation in Prussia.

During the first phase in 1817, the structures in France still had the most decisive impact on planning. Tralles, an expert at the education ministry (see Section 4), conceived an institute in which mathematics should serve as the basic science for all technological disciplines. However, his plans broke down because he was not in a position to start from scratch: Schools under the control of other ministries did not want to be integrated into such a network. The core of the planned institute was to be a reformed version of the existing school of architecture. However, because this had to be handed over to the competing trade ministry in 1820, his plans were left without their foundations.

The next phase from 1822 to 1824 was motivated by a hint from Gauß that he might be interested in accepting a chair at Berlin. Setting up such a special institution would seem to offer an appropriate way of raising the high salary Gauß demanded through getting several ministries to participate. However, it finally emerged that the military, on which the greatest hopes had been placed, had no interest in high-ranking scientific and technological training. During this phase, ideas were already circulating on simultaneously using the institute as a "central seminarium" to train "competent mathematics teachers for the higher teaching institutes of the state" (Schubring 1982a, 213 f.).

The main phase of planning, from 1828 to 1835, was triggered by Alexander von Humboldt's return from Paris in 1827. He once more campaigned for an institute in Berlin in the same mold as the *École polytechnique*. The education ministry adopted his intention insofar as mathematics and chemistry should form the two basic disciplines. However, after such negative experiences in earlier cooperation with other ministries, the ministry was now planning a completely new and autonomous idea to use the institute for *teacher training*, namely, to introduce teacher training as the best preparation for applications. This conception proved to be completely in harmony with Crelle's program of developing mathematics as a pure science in order to provide effective conditions for its applications. It was in his plan for a polytechnic institute in 1828 that the engineer Crelle formulated an apotheosis for pure mathematics:

> It is therefore initially very important *that pure mathematics remain strictly separated from its applications*, particularly in an institute training mathematics teachers while simultaneously helping to attain an important goal of mathematics: to be a means of education. The applications of mathematics are noble and useful fruits of the same; mathematics alone can ripen these fruits and do this fully without disadvantage to itself only if it has previously been allowed to develop purely from within itself without hindrance (quoted in Schubring 1982a, 216).

The main problem in this phase had long been to get the necessary funding, and the plan still finally failed; once again, due to lack of support from the military.

A last phase of planning took place between 1844 and 1850. The driving forces were Jacobi, now working at Berlin, and the influential *Gymnasium* teacher Schellbach. As well as training teachers, they were now also thinking about training "practical scientists." However, the mathematicians and the ministry could no longer agree: because a differentiated school system was now in place, the latter considered that an appropriate foundation for such an institute was already provided by the technical secondary schools (*Realschule*). Schellbach and Jacobi, in contrast, expected that *Gymnasium* education would be needed to produce qualified students.

This neohumanist logic had guided the plans to develop a mathematics teacher training program without its proponents being aware that the necessary institutes were already available: at the universities. The plans for an institute finally became redundant; a high quality of scientific training for mathematics teachers had become established at the six Prussian universities.

Indeed, special institutions fulfilled the intentions of the plans for a polytechnic institute at least since 1828–1835: these were the *Seminare*. Founded initially as institutes for specialized training in philology at the reformed Prussian universities, the model was also adopted for mathematics, first by Jacobi at Königsberg in 1834. These seminars introduced the research character into the study of mathematics while ensuring and developing a high scientific standard in teacher training.

3. Negative Numbers: Advances in Algebra in Germany

> I am not afraid of admitting to the reader that defining the concept of quantity, this basic concept of the science whose first teachings I wish to present here, has indeed cost me more effort than explaining all other concepts in this science; nowhere else have I changed my opinion so often, and, even now, I cannot present what finally seems to be the correct approach with the same degree of confidence that I may well allow myself with some of the other deviations from the norm that one will come across in this book.
>
> —Bolzano 1975, 25

3.1. Algebraization and Initial Reactions to Carnot

H. D. WILCKENS (1800)

The algebraization of basic mathematical concepts progressed rapidly within this pure mathematics context. For the negative numbers, this advance was fueled specifically by the way it contradicted Carnot's work, which had immediately come to the attention of German mathematicians. Directly before this phase, in 1800, a new approach to algebraization was published that went far beyond previous approaches in Germany (the last being that of Hecker; see Chapter II.2.10.4.). This was the work of Heinrich David Wilckens (1763–1832), who personally claimed to have given the theory of opposite quantities "a new look."

The search for biographical data on Wilckens proves to be exceptionally difficult, because he is not to be found in any biographical handbook. Records in the archive of the philosophical faculty at the University of Göttingen give some scattered facts, and Busse (see below) mentions that he was working at Schemnitz, in Hungary, in 1825. However, more information on his biography and career first became available through the discovery of a tribute to Wilckens as a student at the *Bergakademie* Freyberg and later professor in the *Forstakademie* at Schemnitz published by Hiller and Schmidmaier in 1982. Wilckens made his mark as a polymath with a mathematical and scientific orientation. Born at Braunschweig on November 14, 1763, he attended the local *Gymnasium* before going on to study medicine and pharmacy at Helmstedt in 1784. From 1787 onward, he studied mining sciences (*Bergwissenschaften*) at the *Bergakademie Freyberg* in Saxony before finally switching in April 1788 to the study of mathematics at the University of Göttingen.

Wilckens is one of the many young scholars at Göttingen who were encouraged to study the foundations of mathematics by Kästner. However, he went further than most of his colleagues: his ideas on the use of signs led him toward pure mathematics. Wilckens acquired his teaching qualification at Göttingen in 1789 with two memoirs, one reporting physical experiments on electricity, particularly on "positive and negative electricity."[2] However, he had difficulty in obtaining suitable posts, and remained a private scholar at Wolfenbüttel for a long time. He eventually became a mathematics lecturer at the Waltershausen Forestry Institute in Thuringia. When this institute, founded in 1795, closed down in 1799, Wilckens set himself the task of developing mathematical and scientific foundations for teaching forestry. After several unsuccessful attempts at setting up forestry schools in eastern Germany, he was appointed in 1808 the first professor of the new forestry institute at Schemnitz: as a Royal Austrian Engineer and as Professor of "Advanced Forestry." Wilckens died there on May 25, 1832.

At Waltershausen, Wilckens taught algebra to trainee foresters. As he noted, knowledge of algebra was not absolutely necessary for a forester, but could make his job easier for him:

> This will be even truer if he is familiar with the theory of opposite quantities, this superb part of algebra and key to the deepest secrets of mathematics (Wilckens 1800, vii).

Wilckens published his work on foundations in the form of four letters sent in

---

[2]   University archive Göttingen, II. Philosophische Fakultät, Ph 73 (1789/1790).

1797 to Joh. Chr. L. Hellwig, his former mathematics teacher at Braunschweig. The main point for Wilckens was to introduce his own sign for oppositeness and thus to clearly distinguish and separate sign of operation from algebraic sign. The text presents the conception as a transfer of the philosophical–logical concept of oppositeness to a corresponding concept in mathematics.

He presents the logical conception of oppositeness in his first letter: as a set of 10 "explanations and basic principles." The transfer to mathematics is given in the second letter, in which Wilckens introduces his own sign for oppositeness:

For a quantity a, he presents its opposite as $\bar{a}$. The definition of oppositeness could then be presented as $a + \bar{a} = 0$ (ibid., 6). This simultaneously results in $\bar{a} = -a$.

Wilckens uses these differentiations to present the four basic operations in algebra (ibid., 12 ff.). He justifies their legitimacy through the transfer of the philosophical–logical conception to mathematics. He formulates his conception for "quantities," but does not explicate numbers or the relation between numbers and quantities.

The fourth letter addresses operations with "quantities taken as absolute." However, what he means by this is compound expressions otherwise known as "complex" quantities.

F.G. BUSSE (1798, 1801, 1804)

The next German author, chronologically speaking, is a thoroughly odd character: Friedrich Gottlieb (von) Busse (1756–1835). Working in an applied context, he published numerous memoirs on the foundations of mathematics, focusing initially on negative quantities. Although his first memoirs opted for algebraization, they were still completely embedded in the traditional context of geometric quantities: they discussed practical consequences in great detail, but it can be seen that he did not possess his own theoretical concept. However, a later memoir responded directly to Carnot's *Géometrie de Position*, thereby taking further steps in algebraization and acknowledging the work of Wilckens.

Busse, raised to the nobility in 1811, worked for a long time in the small north-German state of Anhalt-Dessau. After attending grammar school in the Prussian city of Magdeburg and studying at the Prussian university of Halle (under, among others, Karsten), he first took a teaching post at the *Philanthropin*, a pietistic education institute at Dessau. After this closed in 1793, he entered the service of the local duke as a technical advisor. He seems to have obtained his doctor's title in 1798 at Leipzig, and, in 1801, he became professor for mathematics, physics, and engineering at the mining academy of the Saxon mining town of Freiberg. In 1825, he published a three-volume work entitled *Darstellung des wahrhaften Infinitesimal-Calculs* (presentation of the true infinitesimal calculus), a work applying his own special hyper-Germanic terminology, in which he operated with quantities not only "becoming infinite" but also "having become infinite."

Busse published numerous textbooks on arithmetic and algebra. Two of his memoirs deal exclusively with negative quantities: the first published in 1801 and the second in 1804. The first, entitled *Neue Erörterungen über Plus und Minus* (new discussions on plus and minus), presents a sharp criticism of Klügel's approach, saying that it actually represents a return to the synthetic methods of single-case treatment and the admittance of exclusively positive or absolute values (see Chapter II.2.10.4.).

In contrast to Klügel, Busse assumes that algebra is superior to geometry. Particularly when applying the former to the latter, he believes that one should take advantage of the greater potentials of algebra instead of "reducing our more comprehensive algebra to a restriction such as that in the geometry of old" (Busse 1801, 8). He finds it unacceptable that in places where one wants to apply algebra to geometry, "one should want to abandon every algebraic $\mp$ [i.e., operating with plus and minus] and restrict oneself merely to the $\mp$ commanding addition and subtraction of common arithmetic" (ibid., 18). He considers that precisely geometry with its multiple relations is clearly an excellent field for applying algebraic operations with plus and minus: "Not only all the absolute quantities of geometry that one has to itemize with numbers in calculations" but, with these, all other relations that increase or decrease each other should "be subjected to a *single* number scale to which end algebra, due to each of these reducing relations, has also been extended to *negative numbers beyond* the zero" (ibid., 19).

Busse presents detailed recomputations of Klügel's 1795 mechanical tasks, proving to Klügel how restrictive it was for him to insist on exclusively absolute negative numbers (ibid., 5ff. and passim, particularly 44). Busse declares Klügel's stance of distancing himself from English mathematics to be purely rhetorical, saying that it was characteristic for English mathematics "to treat all quantities in the task as affirmed" (ibid., 45).

In strict contrast to, for example, d'Alembert, Busse repeatedly emphasizes "the splendid generality and all-encompassing nature of algebra" (ibid., 47), "this algebraic generality, this covering of all the cases that differ only through the $\mp$ of the given quantities" (ibid., 54). For Busse, the existence of negative numbers is a self-evident consequence of his approach. His criticism is that

> instead of having the necessary courage to treat geometry according to the algebraic generality, one was trying piecemeal to reduce this generality itself to that restricted vision to which one was accustomed through the geometry of old and the tangible sort of calculus that followed. Therefore, one did not have the heart to, for example, claim the existence of negative numbers, and that these as such could also be taken as factors (ibid., 59).

The second part of Busse's 1801 memoir addresses calculations with positive and negative quantities in trigonometry. He applies a great deal of verbal effort to work out that the term used to characterize trigonometric lines in Germany up to that time, *Lage* (position), is imprecise, and that one should extend the basic concept to include *direction* as well as length. He develops a complex sign system for this so that he can simultaneously report the direction of the lines in the plane through multiple points on the plus or the minus sign ("high," "deep," etc., ibid., 122 ff.). Indeed, Busse is able to show the benefits of discriminating between position and direction. It enables him to solve a problem in analytical geometry that had been handed down with errors ever since L'Hospital.

This is the formula for the subtangent of a curve. It provides an informative example of how clarifying the concept field of the negative numbers impacted directly on operations in analysis. Controversy over what is the correct formula for the subtangent even delivers the motive for Busse's first memoir of 1801: a disagreement with Klügel.

In 1798, Busse published his own memoir to determine this formula: *Formulae linearum subtangentium*. This memoir shows that in contrast to earlier textbooks, the formula for the traditional line labeled $PT = S$ since L'Hospital has to be assumed to be negative and not positive:[3]

$$S = -y\frac{dx}{dy}, \text{ and not } S = y\frac{dx}{dy}.$$

Klügel protested against Busse's argument within the year: in a letter to C.F. Hindenburg, editor of the short-lived journal *Archiv der reinen und angewandten Mathematik*. Hindenburg published this letter in the *Archiv*, one year later, in 1800, adding a notable commentary.

Klügel's protest shows that with his distinction between position and direction, Busse actually had uncovered a still unresolved issue in analytical geometry. Klügel argues quasi metaphysically: in several possibly "related" cases, the normal or the main form always has to contain *only* positive quantities. Although he personally admits that the subtangent takes the opposite direction to the axis of the abscissa, he still does not want to allow any negative numbers:

> Herr B. expresses doubts regarding the formula for the subtangent $PT = S$ to a curved line $S = y\frac{dx}{dy}$, but only with the intention of determining the position. He posits $S = -y\frac{dx}{dy}$. This is because [...] when constructing a system of lines, he wants to indicate their directions with the signs + and –. However, because in that case for which the formula $S = y\frac{dx}{dy}$ is found, the subtangent decreases from the location of the ordinate in an opposite direction compared with the direction assumed for the abscissa, he inserts the – sign in order to indicate this oppositeness. This procedure alone will elicit confusion. [...]. In all analytical depictions of a combination of quantities, be they geometric or arithmetic, in that case on which the computation is based for all remaining related cases, all quantities have to be viewed as positive; in other words, they must be viewed

---

[3]   This was the usual notation since L'Hospital's textbook of 1696. The original justification given for it was that neither x nor y was designated as an independent variable. However, in the transition to the function concept with $y = f(x)$, the notation would have been $S = -\frac{y}{y'}$.

absolutely in terms of their quantity alone. If one and the same changes its relation in another related case, it is negative, but only in consideration of that bearing the same label in that first case (Klügel 1800, 340 f.).

Klügel declares Busse's theory to be "inconvenient" and his own to be practical. He says that he has "according to my principles, never become embarrassed," and he therefore asks Hindenburg, "Has the same ever happened to you, my esteemed friend?" (ibid., 342).

The answer, which Hindenburg publishes at the end of the letter, does not just confirm Busse and criticize Klügel's stance; it simultaneously contains a rare historical criticism of classical authors. Hindenburg justifies Busse's approach of expressing opposite directions with a minus sign: it is determined exclusively by the specific content and not by a distinction between "normal" case and related case. He also reports that Kästner agreed with Busse when reviewing his memoir in the Göttingen journal *Göttingische Gelehrte Anzeigen* (Hindenburg 1800, 344). Kästner's statement was all the more significant, because of his long-term support for other positions in his textbooks. Although basically following L'Hospital's formula for the subtangent, Kästner had noted its irregularity and made modifications to it. Hindenburg then compares the results on the directions of the subtangent according to the formulas of three mathematicians: L'Hospital, Kästner, and Busse. He does this for

(1) the ellipse with the equation $y^2 = bx - \dfrac{b}{a}x^2$, and

(2) the hyperbola with the equation $y^2 = bx + \dfrac{b}{a}x^2$.

Hindenburg shows that for each of the four possible cases in L'Hospital, results are correct for the ellipse; but for the hyperbola, in contrast, two are false. In Kästner, there are two false results in both the ellipse and the hyperbola, whereas in Busse, all results are correct.

After this unequivocal decision, which simultaneously confirms the importance of discriminating between directions in analytical geometry, Klügel stopped defending his point of view overtly. In the article on opposite quantities in his 1805 dictionary, he expresses himself only relatively vaguely (see Chapter II.2.10.4.).

Remarkable in the light of later developments is Busse's continual emphasis that there exists only *one* basic unity in algebra for the complete number scale (positive and negative numbers), and that "the basic unity has to be assumed affirmatively" (ibid., 139).

Busse's second specialized memoir of 1804 was triggered directly by reading Carnot's *Géometrie de Position* (1803). Once again, it is clear that Carnot is held in high regard; Busse's criticism is very careful and polite. This is the first time that Busse reports algebraic principles for the relation between positive and negative numbers (though in a highly abbreviated manner) before then going on to discuss individual geometric applications.

Busse starts by introducing the distinction between arithmetic and algebra as a basic distinction necessary for the construction of negative numbers. And he presents this as a historical development: In arithmetic and early algebra, one had thought only of absolute numbers. In the more recent, "the already completely accomplished algebra," however, coefficients a, b, c, and so forth may "mean not only any absolute given number, but also any affirmed or negated given number." Likewise, variables x, y, z can take any value sought without previously "knowing whether they will turn out to be affirmed or negated" (Busse 1804, XX f.).

Because of the difference between arithmetic and algebra, the plus and minus signs serve two functions: on the one hand, as "operative" signs for addition and subtraction; on the other hand, as "indicative" signs to characterize the "affirmed and negated" in "pure algebra," and also, in particular, different directions in geometry as "applied" algebra (ibid., XXII f.). However, Busse does not consider the sign functions explicitly.

For the second sign function of minus, as an algebraic sign, Busse now acknowledges the sign $\bar{a}$ proposed by Wilckens. Busse calls this the "opposite quantity" to a, declaring it to be a highly appropriate form of notation (ibid., xviii and passim).

Busse declares his second fundamental discrimination to be that between (pure) algebra and "the applied algebra" or "the application of algebra" (ibid., 2 f.). Difficulties first arise "when one tries to apply the general concepts of affirmed or negated in algebra to [...] geometry" (ibid., 2).

For algebra, he introduces, though unsystematically, such "general concepts of affirmed or negated." He considers this to include that they be defined not only for quantities but also for numbers. However, he neither distinguishes systematically between the two concepts nor pursues the consequences of this distinction:

> A and B are opposite quantities if $A + B = 0$. If they are both already *expressed* through numbers a and b, one also obtains the *opposite numbers* $a + b = 0$ (ibid., 7 f.).

This definition of negative numbers leads to the existence of integer numbers: whereas

> "common arithmetic [...] has only the number scale
> $$1, 2, 3, 4, 5, 6, \text{etc.}$$
> as its measures" (ibid., 8),

in the "more general" or "algebraic" arithmetic, in contrast, measuring is performed "by means of the number scale

$$....-4, -3, -2, -1, 0, +1, +2, +3, +4, ...\text{"} \text{(ibid.).}$$

For this algebraic arithmetic, Busse explains expressly that for measuring "assumed quantities [...] apart from the *affirmed* numbers [...] it also has *negated numbers* in its measuring number scales" (ibid., 10). And Busse declares it to be self-evident that the smaller/larger relations hold for this entire extended number scale: "that every negated quantity is not only smaller than every affirmed one, but is also smaller than 0" (ibid.).

Within his "pure" algebra, Busse does not study the rules of operation any further. He neither considers the extension of arithmetic operations to algebra nor explains operations with operation signs *and* algebraic signs. In particular, he does not discuss multiplication operations for the extended number domain, but only briefly confirms the validity of the rule of signs (ibid., 11).

Busse assumes that concepts of negative numbers based on such principles of algebra would already suffice to reject the theoretical objections of Carnot: "With this, quite a number of the questions that Mr. Carnot lays before the algebraist are already rejected with philosophical laws" (ibid.).

Busse then goes on directly to address a final general "question": the proportions argument. Indeed, he refutes it with the extension concept: Relations are "far more comprehensive in algebra than in common arithmetic (and in the geometry of old)." The proportion

$$+1 : -2 = -2 : +4$$

is clearly valid in algebra. The fact that the last term is algebraically larger than the third, although the second is algebraically smaller than the first, is a problem only for the more limited concepts of common arithmetic (ibid., 12 f.).

In the main part of the text, Busse deals with applications of algebra to geometry, particularly in trigonometry. Busse's approach is to grant that Carnot has proposed "many fine and new principles," while simultaneously showing that "much is evidently also incorrect" (ibid., 43).

On the one hand, he confirms that Carnot's approach is too restricted, particularly through the rejection of negative solutions to equations. Busse concedes that Carnot had "the need [...] to transcend the restriction of the old geometry" (ibid., 47). However, as he points out in a doubtlessly pertinent observation, this leads in practice to a system completely analogous to "Professor Klügel's system of related tasks" (ibid., 48).

Busse takes a number of tasks from Carnot's 1801 and 1803 books and compares Carnot's solutions in detail with his own, in which he derives a general theorem. Indeed, he is able to point out some errors and neglected cases in Carnot (see ibid., 61 f.).

J.F. FRIES (1810)

With Fries, a philosopher for the first time after Kant in Germany took part again in the foundational work on the concept of number. His contribution is the more relevant here, since in 1810 he discussed not only Carnot's book on geometry, but also Busse's criticism at the same time.

Fries is generally of high importance for the establishment of pure mathematics in Germany because he developed a philosophical program of pure mathematics. In addition, Fries was actually able to make conducive contributions to the solving of basic questions of mathematics with this specific philosophy, establishing his own "philosophy of *pure mathematics*" and as its complement a "philosophy of applied mathematics": as "theory of application" whose mediation with mathematics has to be investigated.

Jacob Friedrich Fries (1773–1843) grew up in a pietistic brotherhood at first, but then broke with pietism and followed Kant's philosophy. He studied philosophy and natural sciences, especially chemistry, as well as mathematics, primarily in Jena. After a few years as a private lecturer, he was appointed to Heidelberg in 1805 as a professor of philosophy and elementary mathematics. In 1816, he returned to Jena, as professor of philosophy, but in 1819 he was dismissed because he had participated in the students' Wartburg festival; in 1824 he was re-admitted to lecture on mathematics and the sciences. There is no doubt that Fries was the philosopher with the best-founded knowledge of mathematics and the sciences in Germany of the nineteenth century. The contrast of Fries and Hegel is notorious.

Though strongly influenced by the combinatorial school (cf. Schubring 1996, 370 ff.), he did not understand—as they did—mathematics as a free combinatorial play with signs, but made the relationship of *signs* and *meaning* the basic dimension of his reflection on pure mathematics, since differentiating between the *syntactical* and *arithmetical* meaning of mathematical expressions belongs to the essential elements of his program (Fries 1822, 68 and 157). He thus explicitly criticized mathematicians who did not use summable indefinite series arithmetically (Fries 1822, 286). On the other hand, Fries was not willing to exclude formal series from analysis and to exclusively admit convergent series as mathematical objects, as Cauchy had demanded (Cauchy 1821, iv ff.).

In 1810 Fries published a review of both Carnot's *Géométrie de Position* and Busse's critique on Carnot in the *Heidelbergische Jahrbücher*. He appreciated Carnot's geometric conceptions as "clear and precise," but stated:

> The most important aspect in studies of this kind remains then the establishment of the basic principles. [...] Considering these basic theories, this work cannot be judged in such favorable terms (Fries 1810, 311 f.).

Fries was the first who—generalizing a thought already present in Busse's work—succeeded in clearly identifying the basic conceptual problem in Carnot's refutation of negative numbers to the use of terminology:

> The author mixes two things with each other, namely, the general theory of opposite quantities and the *application* thereof in analytic geometry (ibid., 312; my italics).

In his review of Busse's critical essay, Fries gave this philosophical critique its full mathematical meaning and was the first to explain that the necessary solution to this principal question of admitting negative numbers in mathematics lies in the consequent separation of *numbers* and *quantities*, along the dividing line of pure and applied mathematics.

Fries criticized that Busse's refutation was not complete, especially because he did not make clear conceptual distinctions of numbers and quantities:

> We think that the first flaw, because of which there are so many mistakes in the theory of opposite quantities, lies in the fact that quantity and number are not properly distinguished here, but that rather the designated (*bezeichnete*) numbers ± 7, $\pm a$ are called quantities straightaway. If one regards, however,

positive and negative numbers per se and establishes then the opposite quantities only as cases of application to the calculation with negative numbers, the argumentations of the theory win ease and clarity (ibid., 315).

Fries's critique of Busse's work shows remarkable mathematical competencies; he was able to show in a number of cases—which dealt with the meaning of negative roots—that Busse had not taken up a sufficiently general point of view and had, therefore, rejected useful solutions. Fries's basic concept was that algebra and geometry have to agree with their results:

> The mistake, therefore, never lies in the fact that the assertions of analysis and synthesis do not strictly harmonize, but, where there is one, it is because of the thoughtlessness of the person who is doing the comparison (ibid., 317).

Fries's review and conception did not remain hidden in the review journal, but were received with considerable effect. The astronomer Heinrich Chr. Schumacher (1780–1850), a friend and the main correspondent of Gauß, had translated Carnot's voluminous book and had published the first part of the translation in 1808. In this first part—which formed the basis of Fries's review—Schumacher announced that he was going to add an annotated appendix considering Busse's treatise.[4]

In the second part, published in 1810, Schumacher explained, however, that he could "rightly refrain from" this commentary—he would only have been able to repeat, "what has already been said by the reviewer in the *Heidelbergische Jahrbücher*."[5] Since Schumacher in view of the close circles of communication surely knew the identity of Fries as the reviewer and since he also printed a few additions by Gauß in this second part, it can be assumed that Gauß had learned to value Fries ever since this review of Carnot.

3.2. The Conceptual–Structural Approach of W.A. Förstemann

The solution to the explanation of negative numbers, which in 1810 Fries had only briefly sketched out but not further elaborated, was taken up only shortly afterwards and was turned into a systematically developed foundation: of negative numbers, and also generally of a radically new approach to algebra, which realized not only the independence of algebra from geometry, which had been proclaimed many times before, but also the subordinate status of geometry as application of algebra in concrete terms. It is a work, published in 1817, by the Prussian *Gymnasium* teacher Förstemann (1791–1836), who in a remarkable manner realized the reform conception of the *Gymnasium* teacher as scholar, aimed at in Prussia since 1810: *On the opposition of positive and negative*

---

4   Schumacher: Carnot 1808; Vorrede, IX.
5   Schumacher: Carnot 1810; Vorrede, III f.

quantities.

ON FÖRSTEMANN'S BIOGRAPHY

Since Förstemann's achievements are so outstanding, but his biography is unknown to a great extent, I shall gather here what I was able to find out. He was born in the town of Nordhausen on October 29, 1791; this town, south of the Harz Mountains, was an imperial free city then, until 1803. From 1806 on it belonged to the Napoleonic kingdom of Westphalia; after 1814 it went to Prussia. In 1811, in the middle of the Napoleonic period, Förstemann stated his father's profession: *quatuorvir* in Nordhausen. This means that he was a member of the "four-man committee," an important administrative function.

According to a later letter of Förstemann to Bessel, the school Förstemann went to in Nordhausen was really bad.[6] At the age of nineteen he began studying, which was late for the time then: at the beginning of May 1811, he enrolled at the University of Göttingen, the main university of the kingdom.[7] The leading mathematician in teaching at the time was Bernhard Friedrich Thibaut (1775–1832); Gauß was limited to teaching astronomy. Graduating in mathematics was not customary in Göttingen until the teacher's exam was introduced in 1831. From 1815, Förstemann was a teacher at the *Hundeikersche Schule* in the town of Vechelde, near the town of Braunschweig, i.e., in the same region.

It can be assumed that Förstemann while studying got to know Friedrich Ludwig Wachter, who was a student there from 1809 until 1813 and was in close contact with Gauß. I have described Wachter's great achievements in clarifying the theory of parallels and his tragic disappearance at the beginning of April 1817 (Schubring 1983, 204 f.).[8] From 1816, Wachter was a mathematics teacher at the *Academic Gymnasium* of Danzig. On his recommendation Förstemann was to be hired as mathematics teacher at the *Oberpfarrschule* of Danzig at the beginning of 1817.[9] This means teaching at the middle level, whereas the *Academic Gymnasium* was at the high level (with a crossover to the first years at university).

The year 1817 was also when the Prussian school reform was put into action in Danzig: the *Academic Gymnasium* and the *Oberpfarrschule* were joined and

---

6   AAdW, *Nachlass* Bessel, no. 221, Förstemann's letter of April 18, 1817, fol. 10.

7   Register of the University of Göttingen, no. 23102.

8   In the Bessel *Nachlass* there is a letter by Wachter that is nearest to the date of his disappearing so far: with the letter from March 17, 1817, he sent a treatise on the proof of the parallel postulate; he also explicitly stated that out of the "opposite hypothesis evidently the following theorems develop and the possibility of a geometry opposite to that of Euclid" (Bessel *Nachlass*, no. 392).

9   GStA, Gymnasium Danzig, vol. 1, p. 102: letter of the *Konsistorium* of Westprussia to the interior ministry, May 16, 1817.

transformed to one new *Gymnasium*.[10] At first Förstemann was supposed to take over as a mathematics teacher in the lower and middle grades, according to his job at the *Oberpfarrschule*; for this the school delegation in Danzig had told him to prove himself as qualified through the teachers' exam, the exam *pro facultate docendi*.[11]

The result of the exam in the town of Halle was so excellent that he was not only hired directly, as had been planned, but also substituted for Wachter's position. The school delegation even recommended him for the position of senior master, as Wachter's successor. The respective application—it also reports the opening of the reformed *Gymnasium*—to the ministry not only gives valuable information on Förstemann, it also proves the high esteem for mathematics characteristic of this time of reform: The town council of Danzig

> intends to give the professorship of mathematics and physics, so far held by professor Wachter, to the teacher of mathematics, doctor Foerstemann [...].
>
> According to the report of the Royal scientific exam committee of Halle, and according to the special report by professor Pfaff [...], Foerstemann possesses those qualities in a masterly degree which are necessary for a teacher of mathematics and physics in all grades of an erudite school. Furthermore, he has made himself known as a writer with his treatise on the opposition of positive and negative quantities [...] 1817, and has also given praiseworthy evidence of his knowledge and teaching ability through the teaching in the first grades of the *Gymnasium*, thus we do not have any doubts to ask for his approval as professor of mathematics and physics.[12]

The speedy reply of the ministry, from February 13, shows that it had already formed a high opinion of Förstemann itself:

> In order to speed up the appointing of the teachers' positions and because the suggested men are positively known to it, the same [i.e., the ministry] approves that the professorship vacant because of professor Wachter's leaving, [...] is given to Dr. Foerstemann.

It was added, however, that the new professorships were general titles and no longer "nominal professorships;" because the former "academic" character of the school no longer existed and because teachers—according to the neohumanist concept—were not to be employed in one subject only.[13]

As is clear from the letter, Förstemann already had his doctorate. In fact, the doctorate was awarded to Förstemann with a geometric dissertation, *Theoriae punctorum centralium primae lineae*, at the University of Halle on June 30, 1817. Förstemann turned in his application for the doctorate after the exam for

---

[10]   This process had undergone delays because of evident opposition against deep changes at two schools, especially against the substitution of their two principals with one modernly educated principal of gymnasia.

[11]   On this type of exam and exam committees see Schubring 1983, Chapters 8 and 9.

[12]   GStA, Gymnasium Danzig vol. 1, January 16, 1818, fol. 112. It is the mathematician Johann Friedrich Pfaff (1765–1825).

[13]   Ibid., fol. 114.

the teacher's degree, which had gone so well, on June 28. Pfaff, then dean of the Faculty of Philosophy, suggested to his colleagues to grant the doctorate degree, and they followed him in this point, without an oral exam, so that the date of the diploma is already June 30. In the enclosed *curriculum vitae* of the candidate, he says only briefly about his school days that he went to school in the town of Nordhausen from the age of nine until the age of nineteen. From Easter 1811 until June 1815 he studied in Göttingen, primarily mathematics and also natural sciences (physics and chemistry) and philosophy. As his main teacher he names Thibaut; he also studied mathematics with J.T. Mayer and Gauß.[14]

One must not infer from Förstemann's book on the theory of negative numbers that he was exclusively interested in algebra. Not only does the topic of his dissertation show that he dealt with geometry intensively. His correspondence with Bessel, especially in the years 1819 to 1820, is evidence that working on geometric problems was a favorite activity.[15] In fact, his next book was a textbook on geometry (Förstemann 1827). The two-volume work had an excellent methodical structure: applying the theory of negative numbers to geometric quantities formed a continuous topic.

In the *Crellesches Journal* Förstemann published a number of articles: on questions of both geometry and algebra. Finally, he also presented his arithmetical–algebraic conceptions together in a textbook (Förstemann 1835). In spite of Crelle's reserves as an expert, the ministry expressed to Förstemann its acknowledgment for "the scientific value and the excellent usefulness of this work."[16]

Förstemann was not, however, specialized on mathematics alone. He also published on physics and was an active member of the Danzig Society of Naturalists. He died at the age of 44, on June 27, 1836.

HIS 1817 WORK

Förstemann took up the tendencies toward algebraization dominant in Germany and further developed them independently, also in discussing Carnot's *Géometrie de Position* (Förstemann 1817, VI f.).[17] The starting point for his conception was the separation of the concepts *number* and *quantity*, carried out strictly for the first time. Maintaining the term "quantity" in the title was only a nod to the "customary use of language" (ibid., IX). The innovation of Förstemann's solution was actually not so much this separation, which implied the foundational function of algebra, but the consistent inclusion of the problem

---

[14]   University archive Halle, Philosophische Fakultät: Dekantsakten, Rep. 21 II, no. 10.

[15]   AAdW, *Nachlass* Bessel: no. 221, fol. 1 ff.

[16]   GStA, Gymnasium Danzig, Vol. 2, fol. 155 – 159.

[17]   Förstemann received Carnot's book in the form of Schumacher's translation (cf. Förstemann 1817, 104) and thus got to know Fries's review.

of multiplication.

The basic separation of numbers from quantities is motivated by striving for a general concept of multiplication. Förstemann's point of departure was the criticism of the traditionally "very wide sense" of the term "quantity"; according to it, even numbers were quantities. Förstemann declared such an extension as "harmful"; for him abstract relations were of a different nature than quantities bound to the concrete:

> Numbers are by no means quantities if one does not want to extend the concept of quantity unsuitably over its appropriate limit. Lines, angles, surface areas, solids, weights, intervals, amounts of persons, of books, etc. are q u a n t i t i e s ; n u m b e r s , however, are only expressions of the *ratios* of similar quantities (Förstemann 1817, 1).

With this conceptual separation, he was able to provide the foundation for a general form of multiplication, which so far had been prevented by the mixing with quantities (cf. above especially the "epicycles" in Salimbeni and in Castillon, Chapter IV.1.4.):

> One cannot multiply a quantity by itself, but every number. Thus, there are no powers of quantities, but only of numbers; roots of quantities are, therefore, absurdities. When one usually talks of, for example, powers of straight lines in geometry, one ought to talk only of powers of those numbers which represent these lines in relation to one unity (ibid., 2 f.).[18]

Over and over again Förstemann emphasized the key function of multiplication for the new algebraic approach. In a later work, he pointed to the fact that for addition and subtraction, the "first-level operations," the transition from the concept of quantity to that of number was not necessary: only the operations of the "second level" make this generalization necessary:

> Only in multiplication, the numbers turned out to be necessary; they were not necessary for the derivation of the idea of first-level operations (Förstemann 1825, 5).

The fascinating and innovative aspect in Förstemann's development is his theoretic, conceptual approach. In his first main part, the "arithmetical theory of the opposition of positive and negative numbers," he immediately introduced two new algebraic concepts: "opposition of multiplication" and "opposition of addition," which mean a call for an existence of a neutral and an inverse element concerning two connections and which thus virtually imply an almost axiomatic introduction of the concept of a ring; instead of speaking of "opposite numbers,"

---

[18]   In an appendix Förstemann gave a brief comment on Carnot's concepts. Much of Carnot's critique on the customary theory "falls into pieces if one only has grasped properly the so important distinction of number and quantity" (Förstemann 1817, 161 f.). His critique gains "some appearance most of the time only by confusing the concepts of quantity and number" (ibid., 162). As for the remainder, it is contradictory that he "in the end has the rules for calculating" with negative quantities apply (ibid.). His new "theory of direct and inverse quantities (goes) together with a reasonable theory of positive and negative ones in the end," limited, however, to geometry (ibid., 162 f.).

as before, he thus formed a new object, on an abstract level:

> In addition to the opposition of negative and positive numbers, there is another one in arithmetic whose theory is not subject to difficulties and can contribute very much to clarifying the theory of the opposition of positive and negative numbers. It is the opposition between a number and the quotient of the division of 1 by this number, i.e., generally speaking, between m and $\frac{1}{m}$. This opposition is for multiplication and division what the opposition of positive and negative numbers or quantities is for addition or subtraction. Thus it shall be called o p p o s i t i o n   o f   m u l t i p l i c a t i o n as well as the one between positive and negative numbers o p p o s i t i o n   o f   a d d i t i o n (ibid., 6).

The inverse element was defined by the opposition of multiplication; at the same time Förstemann emphasized again that these definitions are to be introduced only for numbers and not for quantities:

> The multiplication by a number m has the same effect as the division by the multiplicatively opposite $\frac{1}{m}$. One can consider this as the definition of the opposition of multiplication. Only numbers can be multiplicatively opposite to each other, not quantities" (ibid., 6 f.).

Through the existence of inverse elements Förstemann presented the existence of a neutral element, called by him "middle number" or "separating number":

> 1   i s   t h e   m i d d l e   o r   s e p a r a t i n g   n u m b e r   o f   t h e   m u l t i p l i c a t i v e l y   o p p o s i t e   n u m b e r s .   1   i s multiplicatively opposite to itself. Multiplication as well as division by 1 does not change anything. The product of two multiplicatively opposite numbers is = 1; e.g., $4 \cdot \frac{1}{4} = 1$; $\frac{2}{3} \cdot \frac{3}{2} = 1$. A number divided by itself or a quantity measured by itself is 1. Numbers that are greater than 1 are opposite to proper fractions (ibid., 7).

It may be striking that Förstemann did not exclude the case $m = 0$. In fact, he tried to include zero in the multiplicative connection, apparently to meet the traditional claim for the general: by taking zero and infinite to be limits: "The limit of numbers that are greater than 1 is ∞; the one of numbers that are less than 1 is 0; 0 and ∞ are multiplicatively opposite" (ibid.).

After the alienation by introducing the new opposition of multiplication, Förstemann was able to introduce the more familiar additive connection by analogy with the multiplicative one:

> The opposition of addition of numbers is for addition and subtraction what the opposition of multiplication is for multiplication and division.
> T w o   n u m b e r s   a r e   a d d i t i v e l y   o p p o s i t e   i f   t h e s u b t r a c t i o n   o f   o n e   h a s   t h e   s a m e   r e s u l t   a s   t h e a d d i t i o n   o f   t h e   o t h e r (ibid., 7 f.).

Förstemann indicates that he assumes an *extension* of the number domains: in common arithmetic, where the opposition of addition does not occur, there are only "absolute" numbers. In algebra, however, the absolute numbers become positive ones; at first, the same rules of calculation apply to them. The numbers additively opposite are called negative numbers.[19] Förstemann characteristically explains which are the appropriate signs for representing these numbers: The minus sign, customary so far, is inappropriate, firstly, because this sign also means subtraction, and secondly, because "the concept of subtractive numbers (absolute numbers that are to subtracted) is strictly to be separated from the concept of negative numbers" (ibid., 8). Thus, Förstemann not only differentiated clearly between the functions of the two signs, but also—in such an explicit manner seldom seen—demonstrated the difference between subtractive and negative numbers.

As a separate sign for the new kind of number Förstemann took over Wilcken's sign (without stating his name):

> According to the definition the result in signs is
>
> $$a + b = a - \bar{b} \text{ and } a - b = a + \bar{b} \,.$$
>
> e.g., $10 + 3 = 10 - \bar{3} = 13$
>
> $$10 - 3 = 10 + \bar{3} = 7 \text{ (ibid., 9)}.$$

In line with his program of generalization—and in contrast to the French tradition—Förstemann emphasized that letters in algebra are not limited to absolute or positive numbers: "A letter can mean both a positive and a negative number" (ibid.). The bar over a letter indicates "that one is to take the additively opposite number of that one" which is represented by this letter: "If $a = 4$, then $\bar{a} = \bar{4}$; if, however, $a = \bar{4}$, then $\bar{a} = 4$" (ibid.).

For this, Förstemann even developed an operative, involutory calculation:

> $\bar{\bar{a}}$ means the opposite number of $\bar{a}$, and therefore, $\bar{\bar{a}} = a$. Moreover, $\bar{\bar{\bar{a}}}$ means the opposite number of $\bar{\bar{a}}$ or of a; therefore, $\bar{\bar{\bar{a}}} = \bar{a}$; Thus, always two signs of negation, but generally any even number of these signs, cancel each other out; an odd number, however, can be traced back to a single one (ibid.).

From the inverse elements Förstemann also derived the neutral element for addition:

> Between two additively opposite numbers 0 lies exactly in the middle.
>
> 0   i s   t h e   m i d d l e   o r   s e p a r a t i n g   n u m b e r   o f   p o s i t i v e   a n d   n e g a t i v e   n u m b e r s .   0   is additively opposite to itself, $\overset{+}{0} = \bar{0}$; i.e., 0 can be understood both as positive and negative; both have the same result. One has $a + 0 = a - 0 = a$,;i.e., adding or subtracting 0 does not change anything. Moreover, $a + \bar{a} = 0$; i.e., adding two additively opposite numbers is 0; finally, $a - a = 0$; i.e., a number subtracted from itself is 0.
>
> 0 is then for the opposition of addition what 1 is for the opposition of multiplication (ibid., 9 f.).

As Förstemann explains, there is an order relation for the entire domain of

---

[19]   In his 1835 textbook on arithmetic Förstemann introduced negative numbers in the chapter "extension of calculating with numbers" as "a new kind of numbers" (Förstemann 1835, 9).

positive and negative numbers: "One says $2 > \overline{3}$, $3 > \overline{7}$, etc." (ibid., 11).

But he recommended not to read it as "2 is greater than $\overline{3}$"; it is better to say "2 stands above $\overline{3}$ or higher than $\overline{3}$. He therefore did not write the arranged series next to one another, as Euler had done, but on top of or under one another:

One writes the series of positive and negative numbers in the following manner:

$$
\begin{matrix}
\cdot \\
\cdot \\
+3 \\
+2 \\
+1 \\
0 \\
-1 \\
-2 \\
-3 \\
\cdot \\
\cdot
\end{matrix}
$$

." (ibid., 10).

I did not find this manner of writing either before or after Förstemann; it is probably based on an effect of d'Alembert's and Carnot's arguments against quantities that were less than zero.

It is interesting that Förstemann also saw the necessity to introduce an extra sign for the absolute value: in order to evaluate which number—absolutely taken—is the greatest: "I suggest the signs $|\,$ and $|$ whereas the longer line faces the greater number (also the greater quantity). Thus $\overline{5} \,|\,|\,3$ and $\overline{2}\,|\,|\,\overline{3}$" (ibid., 11).

These signs, six years prior to Crelle's suggestion of the absolute value signs used today, were not really suitable and were not taken up: they always referred to an inequality, not to a single term; they presupposed that the direction of the relation was already known, which was not clear in more complicated expressions.

After the presentation of the foundations, Förstemann introduced the basic operations for the new number domain. It is conspicuous that here also he feels it necessary to justify "which meaning" an "[isolated] negative number" yielded as the result of an operation. Indeed, Förstemann had not demanded axiomatically that both operations in the new number domain be closed, so that the traditional question of legitimation remained relevant. His answer was also quite traditional: an isolated negative number stood as a representative of operations that were to be done with it (ibid., 13).

It was novel, however, how Förstemann defined the meaning of the extended operations regarding the original operations. Concerning the most difficult point, the rules of signs, he based the two connections addition and multiplication on the distributive law; he explained that the validity of distributivity for absolute or positive numbers was already known: "It is known that $(a - b) \cdot c = ac - bc$ if *a, b, c* mean absolute or positive numbers" (ibid., 15).[20]

Distributivity hence served as a premise. Förstemann posited the extension of the distributive law for positive *and* negative numbers as a demand or postulate: The extended rules of operation have to correspond with the original ones:

> The applicability of negative numbers that we are looking for demands, however, in such aggregates that for a multiplication like $(a + \overline{b}) \cdot c$ also the rules for the multiplication of aggregates, as are given for absolute numbers, can be applied. Thus one obtains
>
> $$(a + \overline{b}) \cdot c = ac + \overline{b} \cdot c.$$
>
> The result is to be the same as the one above; thus
>
> $$ac + \overline{b} \cdot c = ac - bc.$$
>
> This means that adding $\overline{b} \cdot c$ is the same as subtracting bc, or that $\overline{b} \cdot c = \overline{b \cdot c}$ (ibid., 16).

Förstemann not only realized a remarkably axiomatic structure, but also realized a first practicing of the principle of permanence, which was so named later by Hankel.

Subsequently to his structure of numbers in algebra, Förstemann turned to the problem of application: operating with quantities. In his time this was still the main object of mathematical activity, and not the investigation of algebraic or geometrical structures.

Förstemann developed for such applications to quantities the distinction— known to us since Fontenelle—between a "quantitative nature" of quantities and their "qualitative designation" (ibid., 18 ff.). The traditional problem of operating with negative quantities was the occurrence of inhomogeneous quantities, i.e., quantities of different kinds: quantities characterized by different qualities.

Förstemann made clear that this was a typical problem of application: the impossibility of operating with inhomogeneous quantities was not an argument against the existence of negative quantities; it afforded rather a concrete investigation of single cases in how far "there (are) respective pairs of different qualitative designations whose dissimilarity does not make addition and the other kinds of calculating impossible" (ibid., 21).

Since there are in fact such pairs in many concrete cases, which can be interpreted as additively opposite quantities,[21] Förstemann was able to state as a "result" that the general structures and operations were applicable to quantities of such kind:

> Often it is possible to have two different qualitative designations for two quantities of the same kind, which do not prevent their connection in

---

[20]   Förstemann did not mention the condition $a > b$ here, of which he knew from Carnot.

[21]   Examples for this were "money in possession" versus debts and segments measured in foot, created by movement toward the right versus one created toward the left.

calculations (ibid., 26).

It is surprising to note in what fundamentally structural terms Förstemann thought and argued. It was important to him to show how operating with quantities can be traced back to operating with numbers. Since quantities, after a unity has been choosen, can be expressed in numbers, one of the two qualities can be chosen as measuring unity when one is dealing with opposite quantities: when measuring this quantity one gets positive numbers, when measuring the opposite quality, negative numbers (ibid., 29). Förstemann generalized his result as a structural statement on applications: the solving of problems by the means of reducing them to general structures:

> Now this general statement can be made: a number that expresses the ratio of two quantities of the same kind, but additively opposite, is negative.
>
> By the mediated indication of quantities, one withdraws the expression of the additive oppositeness from the qualitative designations, and transfers it on the numbers that express the ratio of the quantities to be measured to the unity (ibid., 30).

Förstemann had developed these foundational considerations on the operations of addition and subtraction. In the following he presented the remarks on the other basic operations: of multiplication and division for additively opposite quantities. The method was again to "designate the opposition (of quantities) through the opposition of numbers" (ibid., 31). A requirement for division was, of course, that the quantities could be interpreted as of the same kind, for multiplication that only one factor be a quantity, the other a number. Förstemann again explained explicitly "that one cannot multiply a quantity by a quantity"; otherwise, one will "find incomprehensible things in the theory of the opposition of addition [...]" (ibid., 53).

The application of the thus developed foundations of operating with negative numbers forms the main part of Förstemann's book: at first in algebra in order to solve the questions that so far had been discussed in the literature of interpreting negative solutions, but then mainly in geometry, firstly in the traditional Euclidean geometry, then in the so-called application of algebra to geometry, and finally in trigonometry. Subsequently to this, he presented, again as the first, comprehensively how to operate with angular magnitudes as opposite quantities (Förstemann 1817, 70 ff.).

In presenting the applications, Förstemann clarified many of the facts that had been used as arguments against negative quantities in the literature before, e.g., that "a quantity [i.e., here function] that goes through 0 does not necessarily go (over) from + to – or from – to +" (ibid., 78 f.): becoming negative does not have to imply opposite status of the quantities. Thus he was able to criticize Carnot, who

> [wanted] to show the incorrectness of the rule that negative solutions had to be taken in a direction which is opposite that direction in which the positive ones are being taken. That this rule can be applied sometimes, the example I. in § 22 shows. Those who wanted to claim that this rule could always be applied claim incorrectness indeed. This is, however, not the reason to call the usual concept of

negative quantities false or obscure, of which Carnot accuses them (ibid., 104 f.).

With his thorough discussion of the applications of the theory of negative numbers in various fields of mathematics, Förstemann showed in fact that the applications are a problem of their own, neither that they are a simple deduction from theory nor that they can be reduced to theory directly. This independence was not only the result of the fact that many quantities cannot be interpreted as quantities provided with an additively opposite "partner"; moreover, Förstemann explained it in discussing what actually "related" cases in geometrical constructions are and in how far they can serve, therefore, to solve a common formula, and that a theoretical view alone of the mathematical formula does not suffice:

> So, with these constructions also, the greatest variety of those related problems becomes visible, and it was indeed the purpose of this example to show how related geometrical problems, which could be thought to be solved by a single formula, can produce absolutely different formulae whereby only using the opposition of addition does not suffice to integrate these problems and their solutions within one (ibid., 135).

This independence of the applications is indeed the reason that the problem of operating with quantities—after separating numbers from quantities and after the fundamental solution of justifying negative numbers—disappeared from the focus of pure mathematics; in the fields of application, however, in physics for example, the operating was practiced, without further reflection on the foundations.

3.3. Reception Between Refutation and Adaptation

In the conceptions that were developed immediately after the publication of Carnot's book in 1803 and that have been presented so far all his topics were discussed intensively, but as a refutation from the so to speak secure ground of the algebraizing tradition in Germany. Subsequent authors, however, began to vacillate increasingly between refutation and reception, so that now positions following Carnot more closely were also advocated in Germany. The starting point of this change can be defined by geography and politics: it was the region left of the Rhine, which had been a direct part of French territory from 1794/1797 until 1814 and in which the French school and educational system had been established so firmly for a longer time that mathematical basic conceptions had their epistemological effects. Because of the extensive spread of this reception Förstemann's approach remained relatively isolated at first.

M. METTERNICH

The first representative of the left-Rhenish influence set a counterpoint to fend off Carnot's conceptions; his refutation, however, rather harmed his professional career: it is Mathias Metternich (1747—1825), a person with a most extraordinary fate.[22]

Brought up in very simple conditions near the town of Montabaur, in the right-Rhenish area of the electorate of Trier, he was able to go to the *Gymnasium* in the town of Hadamer thanks to promotion by a count and to obtain an education as an elementary school teacher at the *Normal School* in Mainz. Because of autodidactic studies and active participation in groups discussing the ideas of Enlightenment, he was able to work his way up from an elementary school teacher in Mainz to a teacher trainer at the *Normal School*, and in 1784 he finally received the title of *professor matheseos extraordinarius* at the University of Mainz, which had been reformed according to the ideas of Enlightenment. In 1784, too, Metternich took an official sabbatical to study mathematics in Göttingen, rather unusual, since he already was associate professor of mathematics! There Kästner inspired also him to work on the foundations of mathematics. In 1786 he received his doctorate degree at the University of Erfurt, which belonged to the electorate of Mainz and in 1787 even reached the position of full professor of mathematics in Mainz.

The French Revolution completely changed his life. As one of the few German Jacobites and as perhaps the only German mathematician, he dedicated himself entirely to politics and to realizing the revolutionary ideals of liberty when Mainz came under French rule for the first time in 1792. During the capture of Mainz in July 1793 by German troops the Jacobites were maltreated and suffered excruciating prison terms. After varied fates, during which he maintained his political convictions, he finally obtained an administrative job, at first when Mainz definitely became part of the "Republic of Franconia" in 1798, and since 1800 as a mathematics teacher at the *École centrale* of Mainz. When the central school was transformed into a Napoleonic *lycée* in 1803, he applied to this school as teacher and was hired as *professeur de mathématiques speciales* for the two upper forms.[23] During the reorganization of the *lycèes* in 1809 and the reduction of the positions for mathematics teachers, Metternich was not taken over, but was apparently fired, for he did not receive, as he had wished, a pension. Despite his application to have his titles accepted in 1810 for the new formal structure of the *Université impériale*, he did not get a job during the French period. Only since 1814, under the provisional German administration, did he receive a small pension. Projects to get hired at the *Gymnasium* of Mainz failed. Metternich was the author of numerous textbooks; an 1815 work by him on the problem of parallels provoked both Gauss and Wachter to publish the first suggestions on non-Euclidian geometry.

---

[22]   I have presented his biography and an analysis of his writings in more detail in the article: Schubring 1995. The print version, however, is garbled by over a hundred typing errors (no galley proofs had been sent to me for it). The journal published a list of the misleading mistakes in issue 2, 1996, 315–316. Here I also represent results of further biographical research.

[23]   AN, F[17], No. 7441. Affaires générales Mayence.

I think it is likely that in 1809 Metternich had to leave the *lycée* because of his opposition to Lacroix's textbook on algebra and his opinion on negative numbers, which was contrary to that of his own. From exactly 1803, i.e., since the transformation of the *Ècoles centrales* into *lycées,* Lacroix's textbook on algebra was in fact prescribed as the only authorized textbook in all of France, and therefore also in Mainz. The 1803 edition of the textbook also formed, as we have seen (Chapter V.3.1.), an adoption of Carnot's refutation of negative numbers. Since algebra was an essential part of the curriculum in Metternich's classes, and moreover, all teachers were also strictly forbidden to use other teaching materials, a permanent teaching of algebra against his beliefs certainly was a strong nuisance for Metternich.

Soon after 1803 Metternich tried to leave the *lycée* and to acquire a position where he would not have been required to follow Lacroix, such as at the *Ècole normale* being planned in Mainz, and also at the special school of medicine, being planned at Mainz, as professor of physics and hygiene.[24] Neither project materialized, however. Shortly before, in a report to the director of teaching, Fourcroy, in Paris from the 20th Fructidor of Year 13 (September 7, 1805), the responsible school inspectors had already emphasized Metternich's *brillians succès* in teaching on the one hand, while on the other they reported his wish for retirement—after thirty years on the job—and mentioned frequent illnesses, during which he had been replaced by a substitute.[25]

As Metternich wrote in the preface of his translation of Lacroix's algebra—after the seventh edition from 1808—he had taught ten to twelve courses following Lacroix's textbook at the *lycée*, during which most of the comments originated, of which he "informed the students in written form" and which he now published as footnotes or as additions in the German translation. As he stated in the preface right away, he was in fundamental dissent from Lacroix's conception of negative quantities, and had thereby most of the readers, at least the German ones, on his side:

> According to my conviction (and I may believe that most of those who have read Lacroix's textbook share this conviction with me), the concepts of the signs + and − are presented as too vague and too one-sided (i.e., as mere signs), whence the result was that the development of single cases, referring to these signs, could not be of mathematical precision (Metternich 1820, IV).

Thus he felt bound to insert his own paragraphs, in which a general algebraizing theory was developed instead of the single-case method of subsequently correcting the signs: "In §§ 58, a and 58, b, I tried to provide a theory of these signs" (ibid.).

Metternich even went so far as to leave out entire passages of the original. In fact, Metternich's comments begin from the introduction of subtraction; criticizing Lacroix's "nongeneral" presentation, which had limited the

---

[24]   State archive Speyer: G6, Departement Donnersberg, no. 210. Präfekturakten, Abt. I.: Normalschule in Mainz.

[25]   AN, F^{17}, No. 7441.

subtraction of so-called complex terms to absolute quantities, he explained in a long footnote that with negative quantities subtraction has the effect of addition (ibid., 41 f.). But soon the footnotes were not sufficient to present his deviating views: in dealing with multiplication of algebraic quantities, he inserted whole paragraphs in which he explained his own conceptions: beginning with paragraph 30 a (ibid., 51 ff.).

At the point where Lacroix finally—in the context of solving two linear equations in two unknowns (cf. Chapter V.3.1.1.)—had stated his conception of investigating single cases and of changing the signs in the equations he started with in a relatively explicit manner, Metternich inserted two paragraphs, 58 a and 58 b, to develop a general theory:

> Before I proceed with the translation, it is necessary to provide a more complete presentation of the signs + and − , since the author of this whole work did not do it. At too many points in this part, and very clearly in the following, on the application of algebra to geometric objects, one misses a thorough and general theory of these signs; thus the author's explanations become strained and long-winded in single cases, as with the following one, and give only insight into the kind of how one could explain the single result. If one had a general and, as is understood, entirely proven theory of signs, one would only need to use it properly in each single case, and light and order would be brought to difficulties that can be seen in the results. Shouldn't algebra be able to provide this essential theory? It is true, it is missing in Lacroix's textbooks (ibid., 111).

Since this extensive addition, he declared his conception to be superior and Lacroix's continuing discussion of single cases to be "tiringly long" without being able to serve "as general basis in innumerable other cases" (ibid., 119). Finally, a complete translation no longer made sense to him at certain points:

> I did not translate this long § of the author [§ 62, in how far negative solutions can be interpreted] any further, since I can confidently refer to (§ 20 annotation and § 30 a); believing that there is no longer any doubt about the theory of subtraction and that of multiplication [...] (ibid., 121).

And Metternich criticized Lacroix's position on principle from his epistemological view, based on which the permission of negative solutions followed: negative results by no means indicated that "in the conditions of the problem" there are "contradictions." This could be the case if "the results are impossible quantities." This is not the case here, however: "but negative quantities are not impossible, [they] are as understandable as the positive ones," if one applies his theory of signs (ibid., 129).

Probably the most insightful is Metternich's fundamental critique, which so far has been voiced only by him, on manipulating a given mathematical problem and intervening in it, connected with the demand to take the respective problem seriously and not to shy away from difficulties:

> By the way, it is very incorrectly stated if one says that one has to change the conditions of the problem when getting negative results. No, the calculator is not allowed to do that as little as he is allowed to arbitrarily change the signs of the results. A problem is given; and in this description, as given, the algebraist has to turn it into algebraic signs, then he develops the problems (he should do a

proper calculation), the last result, if it is a possible one, contains the answer; [...] one should hold onto the correct reading, and one should not claim at all that one has to change the conditions, because one does not possess any power over it (ibid.).

Calling this book a translation is really contrafactual: it is a refutation and not a translation of Lacroix's approach. As Metternich himself wrote in the preface to the second, practically unchanged, edition, his own contribution to the book was about a half (ibid., VIII).

Metternich presented his own conception of negative quantities coherently, especially in his new paragraphs 58a and 58b. He assumed the distinction of the two different functions of the signs plus and minus: as signs of operation and as algebraic signs. Both functions were connected with the distinction of two characteristics of numbers and quantities: absolute value, called by him "largeness," and quality, called by him "secondary characteristic," i.e., the characteristic of a quantity to possess another one opposite to it, which can "destroy each other alternately if they come together" (ibid., 111 ff.).

Even though Metternich's conception does not contain anything specifically novel compared with the traditional German approaches and no reflection on the relationship between quantities and numbers, still one interesting remark can be found on the frequently seen "compromise" interpretation, i.e., a negative result indicates that operations have to be done with it in order to make sense of it. Against this, Metternich emphasized, according to his epistemological view of necessary internal consistency, that a negative result is not a call for further calculations:

> It is certain that after an accomplished calculation the signs + and − can indicate that no further calculations are to be done; this would be a contradiction, since the calculation has ended; i.e., the value of the unknown number is completely developed and known (ibid., 114).

Metternich was actually not the only user in a Europe influenced by Napoleon who published a dissent from Lacroix's algebra. Recently, D. Beckers presented a similar case from the Netherlands, which were also part of France from 1810 until 1814: I.R. Schmidt (1782–1826) published a translation of Lacroix's algebra to be used as a textbook at the military academy in Delft. He followed the eleventh edition from 1815, which for the most part is identical to the editions since the third from 1803. Schmidt published his translation in 1821, after the end of the Napoleonic period, so that it has to be assumed that Lacroix was still being appreciated at the military academy as a representative of the great success of the French military, which was based on mathematics, since the Revolution.

Schmidt also replaced central statements by Lacroix on negative quantities with his own thoughts, which were based on the reality of negative numbers and the legitimacy of operations with them. The replacements and remarks by the translator show that he also strictly rejected Carnot's and Lacroix's conceptions (Beckers 2000, 116 ff.). The extent of the changes is not, however, comparable to that of Metternich.

J.P.W. STEIN

The counterpart to Metternich is Johann Peter Wilhelm Stein (1796–1831), whose youth and time of education were entirely marked by French influence. He was born in Trier in 1795, in which the old electorate had been restored for a short time, as son of an official and tenant of a count's property. After the definite integration within the French Republic in 1797, the family received this property as their own. Stein grew up in a now French culture; he went to the *école secondaire* in Trier, studied mathematics there with Lacroix's textbooks, and qualified so well in mathematics that he passed the entrance *concours* to the *École polytechnique*. After studying at this famous institution in 1812 and 1813, he was trained as *ingénieur géographe* and began working, practically already assimilated, in French land surveying. After 1815 he returned, however, to the by then Prussian Trier and was hired as mathematics teacher at the Prussian *Gymnasium* there in 1816. An application as principal of the newly founded *Bürgerschule* in Cologne in 1828 was not successful. To compensate for this, the Bonn Faculty of Philosophy awarded him an honorary doctorate degree in 1829. Stein died in 1831 because of "emaciation" (cf. Schubring 1983, 203 f.). Stein published both in Germany and France: on the one hand, many works in Gergonne's *Annales*, among them an important analysis of proofs of the parallel postulate, and on the other in Crelle's *Journal*. For our context here, of his textbooks on arithmetic, algebra, and geometry, the two-volume work on algebra (1828/29) is highly informative, since it is a direct reception of Carnot's and Lacroix's conceptions and an adaptation of Cauchy's modernized approach and had a considerable effect in Germany.

Stein's algebra textbook shows that the concept of negative numbers formed a central break in French and German mathematical culture: for Stein, who was so strikingly in between these two cultures, the clarification of the concept was the main object in his reflection on algebra. Not only was the largest part of the preface dedicated to the theoretical and methodical reflection on negative quantities, but the text of both volumes focused on the presentation of this concept.

Standing between these two mathematical cultures, Stein reflected on the topic by systematically comparing significantly more than his predecessors and contemporaries. Thus, he is the only one in whom we find a categorization of the main types of positions so far advocated (and known to him), subsequently to Karsten's historical analysis of 1786 and before the comprehensive studies of Bolzano (see below):

> The different methods according to which the theory of negative quantities has been treated so far can be divided into three classes, as I believe:
>
> some look at negative and positive numbers as means of designating real quantities, but qualitatively of different kind;
>
> others see negative quantities and basically everything that is not a customary number as a mere, arbitrary sign, which one uses, however, with advantages in calculations;

others, finally, try to do entirely without the use of isolated negative quantities by considering them only in connection with customary quantities, i.e., only as subtrahends (Stein 1828, VII).

At first sight, one might think, because of the expression "arbitrary sign," that the second position corresponds to the dominant French conception from Stein's education. This French position, however, was actually the third. It is striking that Stein did not favor this position at all, although he clearly showed sympathy for it:

> According to this third method, an isolated negative quantity can occur only because of a mistake in the calculation, or, when there is none, it appears because of an inconsistent assumption, which becomes apparent when at the end an impossible operation, that is, the subtraction of a greater number from a smaller one, or even from 0, is still to be done.
>
> It is not improbable that this assumption can be carried out, and I tried it myself with talented students once: however, one develops the results with great difficulty using this method, and one has to constantly admit, especially in the applications, that many conclusions would be overlooked if they did not occur in another way (ibid., VII f.).

As another argument against this conception Stein stated the loss of generality and of using the specific capacities of algebra, which means that in this decisive aspect he followed the tradition in Germany:

> Moreover, the whole method has the essential disadvantage that it occupies the mind with the distinction of a great number of cases that can be recognized only by inner intuition, and thus neutralizes an important part of that which algebra is supposed to accomplish, which is relieving the power of inner intuition. Finally, in such a treatment algebra loses a great part of the generality that it can obtain by the mutual connection of different problems, which becomes evident so easily when one uses isolated negative quantities. [...] Since imaginary quantities have to occur, science would certainly not win that much by avoiding negative quantities than it would lose in terms of clarity and generality (ibid., VIII).

He called the second position his own, which he characterized in greater detail as the "basic opinion that negative quantities and zeros are only signs that present themselves as being necessary in algebraic calculus, but that are not means of representing real quantities" (ibid., IX).

This view of his proves to be an adaptation of Cauchy's conception from the 1821 *Cours d'analyse*, which Stein had immediately received. Stein explained in the preface that he had to deal with a main objection: with the characterization of zero also only as a sign and its exclusion from the field of the "real and intuitive." The discussion of the objection prompted Stein to present the exclusion of zero from the domain of possible numbers much more explicitly than Cauchy (see below).

Stein developed his "basic opinion" in the first part of his second volume as "perfectly consistently" as possible for him. He indeed distinguished clearly, as he had intended, suitable "conventions" from that which is still to prove. This presentation did not come until the second volume, since Stein carefully considered the relationship between the first conception, on oppositeness, and the one he favored also by taking didactical aspects into account. Doing so, he was so honest as to admit that his theory was difficult for students to understand because of the necessary progress from the general to the specific and that one could not "foresee" the results in the applications as easily as in the "theory of oppositeness": Even in his own presentation he would have to

> admit that the theory of oppositeness is the secret guide, and one is reminded of it so frequently that one wants to allow it to have a place in algebra that is more than accidental (ibid., XI).

Thus, Stein presented negative quantities twice in his textbook: once, in the second volume subsequently to his own conception and the other time, in the first, forty-page part of the *first* volume of his textbook following the theory of opposite quantities. Stein's aim was to "apply" this theory "without any metaphysical considerations because of which it is usually disfigured" (ibid., XI). By "metaphysics" Stein probably meant assumptions about the existence of oppositeness. Stein declared himself satisfied to have compiled "something easy to understand" even if this theory was "not as free of all postulates" as his own theory (ibid., XI and XIV).

The starting point for both approaches presented by him was that in subtractions when one calculates with letters, one cannot know from the beginning whether the minuend is greater or the subtrahend. While rejecting fundamentalist approaches, which also in algebra want to apply such distinctions of cases, Stein based his assumptions on a general procedure that includes all of the relations between minuend and subtrahend. Then one "made use of new opinions on the emergence of numbers" (ibid., 2). This innovation causes such essential differences between arithmetical and algebraic operations that theorems from arithmetic can be used only if they have explicitly been proven for the extended domain (ibid., 2 f.). Stein thus took up as foundation the concepts of extending the number domain developed before by Karsten and Hecker among others (cf. Chapter II.2.10.4.).

The first extensive part "on opposite quantities" was dedicated to their "concepts and main properties"; he started from the definition of opposite unities: "Two unities that cancel each other out or destroy each other when taken together are called o p p o s i t e unities" (ibid., 4).

"Opposite quantities" were built on this: as quantities originating from opposite unities. Then Stein defined the "absolute value" of a quantity as its value "regardless of the quality by which it can be distinguished from its opposite" (ibid.). The minus sign serves to characterize the kind of unities. Assuming this he explained the basic properties of opposite quantities:

- Two opposite quantities of the same absolute value neutralize each other.
- If two opposite quantities of different absolute value come together, the one that has the lesser absolute value destroys that part of the other that is absolutely the same (ibid., 5).

He was thus able to deduce the basic operations for the new domain. He was very careful in introducing all extended meanings explicitly. Stein experienced difficulties, however, in keeping up the special status of zero within the framework of the "theory of the opposite": if the positive and negative parts of a whole neutralize each other, one says, "the whole is z e r o , and signifies it with 0"; Stein added that 0 was neither positive nor negative and one could only "call it a quantity because it has parts" (ibid., 7). That zero can have parts resulted from a definition, introduced just for this previously, according to which "parts of a quantity [...] (are called) positive and negative quantities without a distinction, which form the whole when taken together" (ibid., 6).

Stein explained the rules of signs as Arnauld had first argued, by iterated positing of the multiplicand, for numbers: "In the fourth case [negative by negative, e.g., $(-6)\mathbf{x}(-4)$] the opposite of (-6) is to be taken four times, or $(-6)\mathbf{x}(-4) = 6\mathbf{x}4$" (ibid., 13).

With this presentation of the theory of the opposite, Stein followed the traditional German conception of the eighteenth century. He did not receive the new approaches that were based on the distinction of quantity and number. For Stein the concept of number was subordinated to the concept of quantity epistemologically; in this the probably most important influence of his originally French mathematical culture can be seen.

For Stein, numbers did not form proper theoretical objects; they were only a means to operate with quantities signified by them. It was, therefore, a foundation for him that "every number serves as a means of representing a quantity." This becomes possible by an "auxiliary quantity": the unity by which the quantity can be designated with numbers. Thus, Stein came to a very unusual definition of number indeed: "A number is a r u l e or a l a w that describes how one quantity was formed out of another (the unity)" (ibid., XVII). It was Stein's basic conviction that numbers are only subordinate means regarding the actual, primary, quantities:

> To change numbers, increase them, have them develop or form them means to change quantities designated by them, increase them, have them develop, form them (ibid.).

With this basic conviction, there could be no communication between Stein and Förstemann even if Stein had known Förstemann's work. Stein was, however, able to integrate Cauchy's concepts. Stein sharply criticized "the rules of signs" in Cauchy's *Cours d'analyse*, otherwise appreciated as "brilliant":

> because a necessity for his definitions has not been proven indeed and they have not been deduced, therefore [sic: by] the basic idea, but by something secret, which would have needed to come to the light above all else (ibid., X).[26]

But he agreed with him regarding the concept of quantities so fundamentally that he also took over Cauchy's terminology for the most part. Absolute numbers and quantities were clearly distinct because they were understood as representing a relation to reality, to experience. Zero, on the other hand, is "a mere sign" without a number value since it cannot be "specified as something

---

[26]   Cf. our analysis in Chapter VI.6.1.

real and intuitive." The "basic concept of n o t h i n g " cannot result "in a clear knowledge of 0" or at least explain operations with zero (ibid., IX). Only absolute quantities allow number *values*; but zero and negative numbers are— entirely in Cauchy's terms (see Chapter VI.6.1.)—"number quantities."

With this differentiation between number values and number quantities, Stein developed his own theory of negative quantities in the second volume of his textbook on algebra. At each step he made explicit that the theory is to obey a principle of permanence:

> One of the most important difficulties that calculating with numbers has to overcome is that it is to establish rules for those new cases in calculation which are as much as possible in agreement with the customary arithmetical rules and still properly well founded (Stein 1829, 1 f.).

Stein also emphasized over and over again that he rejects the reductionism of avoiding negative quantities:

> If once it is certain that we have to calculate with zeros and negative quantities, then it has to be our wish that these so-called quantities are allowed to be used as much as possible according to the very same rules as numbers, which are the original objects of our calculations (ibid., 26).

The principle of permanence was also to be an economic principle of simplicity:

> Of the rules of arithmetic, "which have indeed been proven for a l l p o s s i b l e number values of letters, but also only for n u m b e r v a l u e s , one has to wish that they were also right if the letters occurring in them, all or partially, meant zero or negative quantities (ibid., 27).

The first new case in calculating with letters is the subtraction "of one number from another equal to it." This type of calculation, which has not been defined so far, could indeed be arbitrarily defined, but it also has to be done conveniently, in greatest agreement possible with customary subtraction. Since the difference $a - a$ is "not to be equated with any number," one could assume "a common short sign for all such differences," that is, 0.

This sign does not have "an actual v a l u e as little as the difference $a - a$ can have an independent value": it is only an abbreviation sign (ibid., 11). For this arbitrary definition, Stein now looked for a convenient determination of the operating rules, which he explained in greatest detail—starting from the operation

$$0 + b,$$

"in which a n u m b e r b is to be added to a s i g n 0, which is not a number" (ibid., 12). For the other case of subtracting a greater number from a smaller one, Stein assumed this "proposition:"

$$a - (a + b) = 0 - b,$$

in which on both sides new, unexplained operations were "indicated." The result is that $0 - b$ was indeed a new sign, to which one must not, however, attribute an actual value. Only the number belonging to it, b, or the absolute value of $0 - b$, possesses a value; the new sign has the name "negative quantity" (ibid., 17). This corresponds exactly to Cauchy's distinction between numbers, those with a number value, and number quantities.

After having posited what the sum of a positive and a negative quantity is to mean, Stein developed in an again very careful manner the operations for negative quantities and for mixed ones, that is, with zero and positive quantities. Especially with multiplication he emphasized that these operations "cannot be deduced from the operations customary so far," that they therefore have to be defined, and he looked for "convenient definitions" (ibid., 23).

He finally summarized the conclusions from these definitions for operations with zero and negative quantities in eight "rules" (ibid., 30 f.). Being satisfied, Stein declared at the end of these comprehensive developments that from these rules all rules can be deduced that are usually given for operating with positive and negative quantities (ibid., 39).

Aside from Stein's epistemology, according to which only positive numbers are numbers for him, so that algebra has "not only *numbers,* but also *arbitrary signs* as "objects," the principle of permanence was used in explaining algebra nowhere else as extensively prior to Hankel, and probably even more explicitly than in Hankel.

Structure and contents of the textbook on algebra are equally interesting and deviating from contemporary schoolbooks. First, it is remarkable that the parallelism of the two introductions of negative quantities in comprehensive parts at the beginning of the first volume and at the beginning of the second is followed by again parallel topics in both volumes: solutions of equations of the first degree. In the first volume he treats this topic according to the theory of opposite quantities, in line with the traditional concept of quantities with detailed explanations of how negative solutions are to be interpreted. In the second volume the presentation is done according to the concept of Bézout/Lacroix, with changes of the original equation when there is a negative solution (cf. Stein 1829, 63 ff.), all in all a remarkable form of "compromise" by parallelizing two approaches.

The other parts contained presentations that were demanding and on the latest scientific level; especially Cauchy's results from his *Cours d'analyse* were integrated. The first volume contained—revolutionary for German textbooks—a part on inequalities and went on with elements of number theory, equations of the second degree, progression, powers, roots, and logarithms. The second volume even made an introduction to analysis, introducing continuity and discontinuity,[27] and also, among others, probability calculus and an introduction to infinite series—with convergence and divergence.

Stein's textbook on algebra did not remain limited to his own *Gymnasium* as many of the schoolbooks back then; in the Rhineland it was paid attention to and praised especially by Diesterweg's students, so that at least a regional

---

27   For the reception of Cauchy's concepts it is relevant that Stein defined continuity globally: a gradual increase of the variable results in a gradual increase of the function (Stein 1829, 86).

impact can be attributed to him. Franz Funck, who first worked at the *Gymnasium* in the town of Recklinghausen and later in the town of Kulm, spoke of Stein's "excellent elements of algebra," and especially quoted his critique on Carnot's third position approvingly (Funck 1841, 3). Anton Niegemann, teacher in the town of Emmerich, also praised Stein's "superb work"[28] and even credited Stein's approach of doing algebra without the theory of opposite quantities in "that he completely achieved his goal while progressing philosophically correctly on his way with circumspection and consequence" (Niegemann 1834, 4 f.). Even in 1852, another of Diesterweg's students, Johann Wilhelm Elsermann, lauded Stein and his concept of negative quantities in a manner quite unusual for his time: "the teacher and scholar Stein, distinguished by scientific rigor, in his excellent work 'the elements of algebra'" (Elsermann 1852, 11).

W. A. DIESTERWEG AND HIS STUDENTS

Another mathematician who worked in the Rhineland and had an effect especially through his students educated by him was Wilhelm August Diesterweg (1782–1835), an older brother of the famous pedagogue Adolph Wilhelm Diesterweg (1790–1866). Diesterweg, from the town of Siegen, studied theology and mathematics in Marburg, Tübingen, and Herborn. After graduation he worked as a private tutor, and from 1808 on he was a mathematics teacher at the *Lyzeum* in Mannheim. When, starting in 1815, the position of mathematics at school was being reduced in the state of Baden, too, he began looking for a job in Prussia. An application at the *Gymnasium* in Cologne failed, since Tralles, as an expert asked by the ministry, could not certify any scientific value of the two elementary writings of 1809 and 1819 submitted by Diesterweg. It is not clear why the ministry appointed him as professor of pure mathematics to the newly founded University of Bonn. Diesterweg now began publishing a greater number of mathematical writings; but they did not contain any new research results. Modern mathematical branches such as analysis and number theory remained unfamiliar to him. He limited himself to the traditional field of elementary geometry. Remarkable are his editions of classical Greek authors, especially of the geometrical writings of Apollonius of Perga. In his education, his teachers' exams, and inspections of mathematics classes in the Rhineland he favored the synthetic method over the analytic one (cf. Schubring 1990c). Because of lectures and textbooks on trigonometry he saw himself challenged to scrutinize and to refute Carnot's exclusion of negative quantities especially in the field of trigonometry. This resulted in his 1831 book *Contributions to the theory of positive and negative quantities.*

---

[28]   Laudatory opinions of mathematics textbooks by other school teachers were extremely rare in Prussia at that time since it was practice that one specific textbook was often used at only one school and that each teacher in general took his own methodological point of view as such an absolute that he strictly rejected textbooks

The book is structured in a most peculiar way: without any preface or motivation and also without a table of contents the book begins immediately. The text neither is a systematic presentation nor a clear discussion; rather, it is a sequence of 74 problems and their solutions. Only in the "supplements" to the solutions is there some annotated text; the "message" of the book results from the sum of these annotations: it was thought to be a refutation of Carnot's rejection of negative quantities and an independent function of algebra, however, without any reflection on the nature of negative quantities or numbers and on the operations with them. This refutation was to be done in a concrete way: by solving the problems in order to show that geometry and algebra do not contradict each other, but actually correspond to each other exactly and that negative solutions do not have to be "thrown out," nor that they do not make sense. The following comments in his supplements are typical of Diesterweg:

> Algebra and geometry agree with each other most exactly. Under the same conditions under which algebra sets up one or two values for the line sought geometry gives one or two solutions (Diesterweg 1831, 26).

In general, Diesterweg stated that he had shown,

> that negative values of the lines sought solve the problem directly and in the same way as the positive ones if the problem is understood in that generality in which algebra understands it, and that the negative values are the second solutions to the problem posed (ibid., 49 f.).

To make his message clear, he solved the problems both with geometry—by showing its construction—and with algebra by establishing its equation and solution, and he discussed the correspondence of the geometrical and algebraic solutions.

Most peculiar was, however, that Diesterweg drew on even older positions in refuting Carnot without taking into consideration the more recent state of discussion, especially of Busse in Germany. In fact, Diesterweg practically based himself exclusively on d'Alembert. Diesterweg took for himself as main position from d'Alembert that algebraically opposite values lie geometrically in opposite directions, an opinion that Carnot had presented as not generally valid.

Diesterweg, however, claimed a very general validity of his view already in the supplement to his fourth problem: "The value x that was found by algebra as negative under all circumstances is put by geometry always in the opposite direction with that what algebra calls positive" (ibid., 15).

One of the few programmatic statements in this book proves not only that Diesterweg meant the specific problem here, but that he wanted to express a general validity:

> It is a main purpose of this work to show [...] that lines indicated by the negative sign never lie in another way than in opposite direction to those designated by the positive sign [...] finally that lines lying in the same direction have the same algebraic sign always and everywhere (ibid., 50).

---

by other authors.

Franz Funck (1803–1886), a former student of Diesterweg in Bonn—from 1821 until 1823—and later a teacher of mathematics in the towns of Recklinghausen and Kulm, sharply criticized Diesterweg's conception: Carnot's attacks "against the concept of the opposite in the application of algebra to geometry" were "as untenable [...] as Diesterweg's defense of him" (Funck 1841, 12). One of Funck's main points of criticism was that one could not deduce a general statement from single examples. Furthermore, Funck criticized a "circular argument" in Diesterweg's proofs: the theorem is presupposed as principle before and then its correctness is deduced from the results, since the theorem

> is included in the treatment of every single problem right from the beginning and is interpreted again from [the problem] afterwards. The generality of the algebraic manner of treatment eludes him [i.e., Diesterweg] in its totality (ibid., 10 and 7).

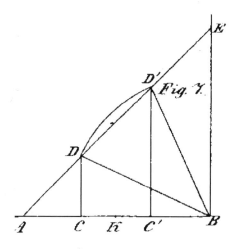

Figure 19: Diesterweg 1831, Table I, Figure 2: interpreting directions in figures

Finally, Funck proved that Diesterweg's proofs were not general regarding themselves either. An example for this may be mentioned here: the problem was to cut a given straight line into two segments so that the sum of its squares is the same as the square of a given straight line P. See Figure 19 for this: AB the straight line given, K its bisecting point, EB a perpendicular on AB and as long as it, the arc of a circle DD' is described out of B with the given line p and DC and $D'C'$ as perpendiculars on AB.

Diesterweg had designated the sought segment AC with y and received two values for it:

$$y = \frac{a}{2} \pm \sqrt{\frac{p^2 - \frac{a^2}{2}}{2}}$$

and commented on this: "The values of y [...] which are both positive, designate the lines AC, AC'; hence the lines with the sign $+$ lie in one direction" (Diesterweg 1831, 9).

This argumentation already shows that the identity of direction to be proven is already presupposed by Diesterweg. Moreover, Funck criticized that the two values are not AC and AC', but the two sought segments AC and CB, and that it was not clear at all whether the two lines are to be considered as unidirectional or opposite: He

> pays no heed to the position of these lines, which one could, by the way, almost with the same right consider as an agreeing one as well as an opposite one without forcing the interpretation especially (Funck 1841, 10).

As for the rest, one also receives negative values for y if the problem is not limited in such a way, but is understood more generally, e.g., for $p > AB$ (ibid.).

Funck did not criticize, however, another of Diesterweg's basic positions, which he also took over from d'Alembert: Diesterweg accepted the correctness of the proportion argument; i.e., he assumed that the proportion

$$1 : -1 = -1 : 1$$

is valid *and* that the traditional conception of proportions applies: according to it— given that the first term is greater than the second—also the third term has to be greater than the fourth. So far, no one had been able to unify both statements as correct. Diesterweg was convinced to have found a solution: by measuring the two ratios of the proportion with different unities!

Diesterweg explained, regarding the proportion $1 : -1 = -1 : 1$, that "the truth of these theorems is not to be doubted, because nothing false can be concluded from true theorems." It was, therefore, the task of a teacher of mathematics, "to try to recognize this truth even if they actually seem to contradict each other" (Diesterweg 1831, 220).

Diesterweg himself had the impression at least that he was stating something paradoxical. His intention, however, of refuting Carnot's argument against the existence of negative quantities was stronger: he had concluded from the validity of this proportion and the greater-smaller argument that $+1$ had to be both greater and smaller than zero. Diesterweg came to the following argumentation in his efforts toward truth: quantities could be measured either with the positive unity $+1$ or with the negative unity -1. If one measures with -1, then -2 is smaller than -3 and $+3$ smaller than $+2$. In order to judge two quantities of the same kind—be they either negative or both positive or mixed—which of them is the greater, it has to be stated before whether the positive or the negative unity is to be taken as measure:

> If in the proportion A, $+1 > -1$ is posited, the first positive unity, that is the first term, is the measure for both terms of the first ratio, it follows that the third term, i.e., the negative unity, has to be the measure for the terms of the last ratio. If one, however, measures -1 and $+1$ with -1, who wants to deny that then $-1 > +1$? (Ibid.).

Diesterweg did not reflect on whether it is consistent to use a different measuring unit in the first ratio from that in the second. Diesterweg further developed this argumentation. When operating with inequalities he explicitly criticized Cauchy for his statement according to which the inequality reverses in multiplication with a negative factor. Rather this holds true: "According to this, if $8 > 7$, [and] $-3 = -3$, also

$$(+8)\,(-3) > (+7)\,(-3)$$

has to be posited" (ibid., 224), since he had operated again with a positive and negative unity at the same time. He accused Cauchy, however, of "letting exceptions to the foundations be valid" with his inequality operations, here the foundation is meant that an inequality remains unchanged in multiplications with the same factor (ibid., 223). Because he had used a positive and a negative unity simultaneously, he even reached a statement that could have been found only in "fundamentalists" until then: $(-3)^2 < (+2)^2$: "One also finds the claim that, if $-3 < +2$ was posited, then this has to be: $(-3)^2 > (+2)^2$."

Against this "the following has to be recalled:" if $+2 > -3$, then also $+1 > -3$, i.e., $+1$ is the measure for both numbers. It follows from this, however, that in the proportion $+1 : -3 = -3 : +9$ the third term had to be greater than the fourth, hence $-3 > +9$. Then, however, -1 was the measure for $+9$, hence also for $+4$, hence $+9 < +4$ and $(-3)^2 < (+2)^2$ (ibid., 224).

Apparently Crelle did not read Diesterweg's book this far, for in his report to the Prussian education ministry he certified him—in contrast to the geometrical editions evaluated by him earlier—a "real scientific value" for the first time (Schubring 1990, 80).

Diesterweg's disciples and other graduates from the study of mathematics in Bonn did criticize a number of mistakes in his concepts. Besides Funck, another Bonn graduate, Franz Heinrich Rump, teacher of mathematics at the *Gymnasium* in the town of Coesfeld, extensively criticized Diesterweg's postulate on opposite signs and opposite directions in 1830: since he published it while Diesterweg was still alive, he wrote it without stating his name and with reference to his Bonn colleague K.D. von Münchow, who as an astronomer was also specialized in trigonometry.

It is strange, however, that to my knowledge, none of them criticized the new form of the proportion argument. Anton Niegemann even explicitly declared that because of Diesterweg's conception "the difficulty is removed," i.e., the interpretation of the proportion $+1 : -1 = -1 : +1$, to which he ascribed "complete correctness," with the traditional greater-smaller argument (Niegemann 1834, 13). Whereas until then only Castillon had proposed the inclusion of the negative unity in 1790, Diesterweg had a certain effect especially with this (see Section 3.4.).

Diesterweg hinted at his motivation for his approach in his book. Discussing the relationship of algebraic and geometrical solutions he stated, "At the same time, however, it is illuminating that geometry, correctly understood, leads to the same generality as algebra" (Diesterweg 1831, 58).

Diesterweg, himself oriented toward elementary geometry, tried to save a place for his geometry against the rapid developments of algebra. To refute Carnot in the field of elementary geometry, he based his arguments only on d'Alembert's older conceptions, although these were not coherently elaborated. Unlike Stein and Metternich, Diesterweg had not directly worked in France; in Napoleonic times he had worked in the state of Baden, which belonged to the *Rheinbund*, and probably felt, therefore, only indirectly the effect of the French mathematical culture.

MARTIN OHM

Finally, one further mathematician has to be treated who became known as author of textbooks especially because he tried to develop an "absolutely consistent system of mathematics," which is the title of his main textbook series: Martin Ohm (1792–1872), younger brother of the physicist Georg Simon Ohm (1787–1854). The central point of the development of such a "consistent" system was also to found a concept of number free of contradictions and to provide an appropriate conception of negative numbers for it. Although Ohm did not name any sources for his conceptions, it is obvious how strongly he received the more recent French works. In fact, during the period of his education Ohm lived in French territory for several years.

Ohm, coming from a family of workmen, spent the first twenty-five years of his life in the town of Erlangen. With the help of scholarships he was able to go to the *Gymnasium*; from 1807 on he studied at the university: mathematics, natural sciences, philosophy, and public finance. In 1811 he became private lecturer of mathematics, with a work on series theory, and gave some lectures. Erlangen was the capital of the duchy of Ansbach and thus belonged to Prussia, but only until 1806. Due to the Peace of Tilsit, Prussia had to cede the duchy to France, which took possession of it immediately. The university also came under French direction. In 1810 France ceded the duchy to Bavaria, which at that time—belonging also to the *Rheinbund*—pursued a policy of modernization strongly oriented toward France.

Since during the restoration in 1816 the teaching of mathematics was being radically reduced in Bavaria and the demand for mathematical lectures greatly declined, Ohm looked for a position in Prussia. In 1817 he became teacher of mathematics there at the *Gymnasium* in the town of Thorn. He considered teaching at school, however, as only temporary, and made a big effort to return to a university. In 1821 he was partially successful; he was transferred to a *Gymnasium* in Berlin and at the same time became "academic lecturer" at the university there after a new postdoctoral qualification. He had proposed as the topic of his *Habilitation* lecture "remarks on opposite quantities."[29]

After being relieved from teaching in 1822, he focused entirely on lectures and working on textbooks. Ohm, who understood academic teaching as a career with promotion according to seniority, achieved through pressure and patronage his

---

29   AAdW, Tralles *Nachlass*, no. 97.

appointment as *professor extraordinarius* in 1824 and in 1839 even as full professor, parallel to Dirichlet (cf. Schubring 1981a, 1983a, 1984a). Like Diesterweg, he at first had many students with his elementary mathematical teaching, but was finally isolated against the modern mathematicians doing research like Dirichlet and Kummer.

Ohm's main textbook series had nine volumes and was published from 1822 on. It was called "Attempt of a completely consistent system of mathematics," which went from arithmetic to integral calculus and saw several editions and revisions. In addition to this, Ohm published a short edition of three volumes as "Pure elementary mathematics" for lectures at engineering and military schools. Contrary to the main goal of a completely consistent system, he modified his conception of negative quantities many times in different editions and in different series of the textbook, even giving contradictory statements.[30]

Already in 1816 Ohm published the core of his conception of the number concept, in a work on the elementary theory of numbers. The foundation was for him the separation of number and quantity: thus number theory forms the basic discipline of all of mathematics; quantities belong to applications and can therefore be taught only after number theory. Ohm especially marked absolute numbers in number theory, assuming an "absolute unity" by whose iterated setting the other numbers are formed. Zero, however, was not a number, but only a sign (Ohm 1816, 3). Differences of two numbers were limited in such a way that the minuend is really greater than the subtrahend, at least "when real numbers are wanted." In the other case the differences do not designate "a number" and do not have, therefore, "any meaning" (ibid., 23). Thus, he based his assumptions on an epistemological priority of absolute numbers and on substantialist positions of existence. The exclusion of the same conceptual status of zero and negative numbers corresponded to the newer French conception following Carnot.

In the first volume of his *Consistent System*, published in 1822, on arithmetic and algebra, he extended the basic structure of 1816. The priority of the "absolute integer numbers" was still confirmed: "there are no other numbers" (Ohm 1822, XI).

Corresponding to this priority the concept of "extension" became the focus of attention. He allowed operating with zero and improper differences as extended operations by defining an appropriate meaning for them. For this he introduced "signs of the form $a - b$" as difference "in the extended meaning of the word." He distinguished the extended meaning of "sum" and "difference'" from their original meaning by especially marking the latter as "real sum" and "real difference" (ibid., 37). Ohm not only extended concepts like summand, minuend, adding, and subtracting, but also the concept of equality. By introducing "subtractive equal" for operating with general differences, he distinguished from it "real" equality. "Subtractive equal'" terms can become

---

[30]   A detailed analysis of these differentiated changes will be published separately.

"really equal ones."

Ohm practically used a principle of permanence according to which all theorems on operations with integer absolute numbers were to hold true also for general sums and differences and the extended equality (ibid., 43). With these extended concepts he at first introduced zero as the subtractive equality $a - a = b - b$, for all a and b, and then the sum $0 + b$ as "additive number" $+b$ and the difference $0 - b$ as "subtractive number" $-b$ (ibid., 44 f.).

In the 1822 work, further extensions of the operations followed: of fractions and hence of rational numbers, and then of positive and negative numbers, which Ohm understood here as integer or fractional either additive or subtractive numbers (ibid., 140). The irrational numbers were the last extension.

Subsequent to number theory, Ohm presented the "general quantity theory." Whereas in number theory the operations by the means of the extensions had been developed in general for the entire domain of real numbers—even given the epistemological priority of absolute numbers—the theory of quantities surprisingly shows a radically reductionist program, which following Carnot and Lacroix accepted only positive quantities and values. Ohm allowed only absolute, integer numbers as measures for quantities, but not negative numbers (ibid., 296). As the only extension he permitted "quotients of absolute integer numbers" (ibid., 299). Corresponding to this rigorous program, zero was also excluded as a measure for quantities (ibid., 348).

Ohm presented the consequences of his reductionist program in the part in which he dealt with the solving of equations. One of the principal rules was the limitation of possible solutions to absolute numbers, be they integer or fractional because of his basic conception: "Such values that are not absolute numbers can, therefore, be disregarded just like that" (ibid., 328).

When solving equations, Ohm therefore had to break the problem into single cases of types of equations in such a way that one required just one value per unknown and that one could subject eventually nonabsolute values to discussing whether it was to be "disregarded" or could be transformed into an absolute value. Ohm adopted the practice of the French mathematical culture of changing the original equation in the case of a negative solution to be his own conception (ibid., 341). He presented the procedure of changing the original equations also for equations with more than one unknown. With trigonometric problems in the plane he distinguished individual cases, separated for each quadrant.

In further editions, however, he partially modified his radical, "French," conceptions. In 1828, he added a paragraph in which he explained operating with "opposite directions"; his inner distance was expressed by the meaningful title: "Numbers called positive and negative are to be treated carefully; hence it is better to not permit them at all" (Ohm 1828, XVI).

The short edition of his textbook for engineering and military education confirms much more strongly his turning toward the conception of opposite quantities. In this textbook for the education of practitioners, *Pure Elementary Mathematics* (1825 – 1826) he practically presented two different conceptions in a parallel manner: his own "view," according to which one "deals only with indicated operations," and a "more material view," according to which operating with zero and negative quantities is allowed (Ohm 1825, 64 f.). He summarized here "positive integer, positive fractional, negative integer, and negative fractional number[s], as well as zero [...] under the name of r e a l n u m b e r s " (ibid., 68). In quantity theory, on the other hand, he kept his view that only a positive number is a valid solution (ibid., 261). In the second edition of this 1834 textbook, he not only introduced the term "opposite" (Ohm 1834, 27), but also added a separate paragraph "Of opposite quantities" in quantity theory. By analogy with the second edition of the "System," he permitted negative measures for quantities $(-m)P$ and quantities $0P$ for "nothing," and, moreover, even granted them "a meaning:" as opposite quantities in the very traditional sense, as debts versus assets etc. (ibid., 36 f.). He also explained here that the basic operations with opposite quantities could be correctly defined (ibid., 97 ff.). Ohm went also so far as to formulate an epistemological *permissibility* for opposite quantities (ibid., 100). Finally, he even stated that such opposite quantities could be justified as of the same kind, with a single unity (ibid.). With this he had left behind both the position of Christian Wolff and the "revisionist" French position.

The main field of proving the efficacy of his conception for dealing with negative quantities was also for Ohm analytical geometry. For this he proposed his own "theory of coordinates," in which he had to distinguish between three different termini to be consistent, at least for himself, regarding his view that quantities can always be only "absolute" and that an existence of opposite quantities is not assumed, although at the same time respective points in the four quadrants could be marked with plus and minus signs:

- coordinates: always absolute quantities,
- coordinate measures: absolute numbers,
- coordinate values: positive or negative numbers or zero; these develop from coordinate measures by putting the sign $(+)$ or $(-)$ in front (Ohm 1826, 49 f.).

Ohm remained with his position that quantities and thus also "coordinate measures" can only be absolute, but the coordinate values could be positive or negative numbers or zero. In the same way, the coordinate axes had "a positive and a negative side." Hence, the four quadrants in the plane can be marked by the signs of the respective axis side (ibid., 59 ff.). Had he excluded negative quantities still in the general quantity theory and demanded the changing of the original equation, now, for example in the equation for the conic section, positive or negative values or zero were equally appropriate for the parameter p to determine the type of the conic section (Ohm 1826, 220). And for the equation of a line, the only explicit criterion was whether the values satisfied the equation, with all real numbers as permitted values (ibid., 84).

3.4. The Continued Dominance of the Quantity Concept

Stein's and Ohm's approaches, which were decisively shaped by a positive reception of the more recent French views, eventually had to adapt more or less to the theory of opposite quantities dominant in Prussia, respectively in Germany. How can this dominance, which is already obvious from what we have said, be proved in more detail?

A detailed proof for Germany cannot be achieved in the frame of this study: up to now, we have analyzed contributions toward clarifying concepts without regard as to whether their authors—as far as they wrote textbooks, and not treatises—intended these to be used in schools, or at universities. To proceed in this way was possible in particular because of the fact that the authors could be essentially counted as belonging to the university level. The neohumanist reforms in education, which began in Prussia, brought great changes to this situation. The encouragement of independent activities and establishment of the scientific profession of mathematics teacher at *Gymnasien* with the status of scholar for all members of the profession, including autonomy with regard

Years	France TOTAL	Prussia	Bavaria	Austria	Other German States	All German States TOTAL
1800–04	6	9	3	–	15	27
1805–09	7	10	2	4	13	29
1810–14	8	6	3	2	18	29
1815–19	1	22	6	5	15	48
1820–24	8	18	1	3	26	48
1825–29	12	20	6	2	15	43
undated	3	2	–	–	–	2
Totals	*71*	*129*	*28*	*27*	*147*	*331*

Figure 20: Comparing the production of algebra and analysis textbooks in France and in the German states(Schubring 1986, 22 ff.)

to method and choice of textbook (even one's own), caused an explosive growth of publications, as can be seen from the numbers of algebra and analysis textbooks edited between 1800 and 1829 in the various German states and territories, and by comparison in France (see Figure 20).

While it is impossible to analyze such a large number of publications here, a detailed analysis is not necessary, since the approaches of the basic alternatives of the time have already been presented. The only additional feature in approaches beyond that is that they try to fathom intermediate positions.

We shall thus just briefly present the textbooks of three authors who were influential in Prussia, and beyond, during the first half of the nineteenth century:

- Johann Andreas Matthias (1761–1831), mathematics teacher and *Gymnasium* director in Magdeburg, provincial school inspector in the Prussian Province of Saxony, and his multi-volume textbook published at the beginning of the reform period;

- Ernst Gottfried Fischer (1754–1831), mathematics teacher in Berlin, member of the Berlin Academy, and *professor extraordinarius* at the University, and his textbook series from the end of the reform period; and

- Adolph Tellkampf (1798–1869), trained in Göttingen, mathematics teacher in Hamm (Westphalia), later director of a higher *Bürgerschule* in Hanover, and his textbook from the later "normal period."

Matthias wrote a *Leitfaden für einen heuristischen Schulunterricht* in 1813, which saw many reprints. In the latter's part on "general quantity theory," he admitted general differences $a - b$, with $b > a$, as "subtractive differences," and operated with them (Matthias 1813, 4 f.). In the respective volume containing "explanations" (*Erläuterungen*), he also introduced, and used, the concept of "opposite" without any restrictions (Matthias 1814, 56 ff.).

Fischer, in his multi-volume *Lehrbuch der Elementar-Mathematik* (first edition: 1820–1824) extensively presented the concept of opposite quantities, in his second volume on *Zahlen- und Buchstabenrechnung*. Fischer started from the concept of quantity, proclaiming quantities to be homogeneous if they can be represented by means of "one and the same unity by numbers" (Fischer 1822, 189). He removed a problem deeply thought about by many others by a stroke of the pen: any limited series of homogeneous quantities, he said, could be counted in two opposite directions—forward and backward. He called two "pieces" counted in such an opposed way opposite quantities (ibid., 189 f.). In his preceding number theory, he had indeed introduced subtraction without any restriction, and declared addition and subtraction to be "opposite modes of calculating," one of which canceled the other (ibid., 60 f.). He did so without adding further reflections, and without explicitly introducing negative numbers, which had been implicitly assumed here as well.

An interesting feature, in particular with regard to Diesterweg's view, is Fischer's strict proposition saying that "opposite quantities are homogeneous any time, can consequently have only one and the same unity," and that "the unity can never be other than positive" (ibid., 192). The concept of extending the number domain does not occur in Fischer.

In the fourth volume of his textbook, in that on algebra and on the theory of conics, virtually an analytic geometry, Fischer had no problem in operating with

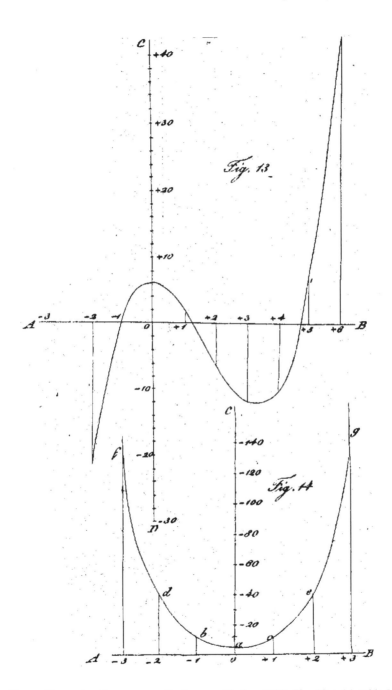

Figure 21, number lines as coordinate axes: Fischer 1829: Fig. 13 + 14, table I
positive and negative coordinates (ibid., 233 ff.). In his figures, he even marked
the coordinate axes with positive and negative numbers, the earliest use of the

so-called numerical axis known to me (cf. Figure 21).

Tellkampf's textbook *Vorschule der Mathematik* (1829) was already significantly oriented toward adapting mathematics instruction at the Prussian *Gymnasien* to rigid boundaries with regard to conception and content as compared to mathematical studies at the university. Tellkampf began with the concept of extending the number domain. In arithmetic, he introduced the basic operations first for integers, then for fractions, and finally for "conflicting (*widerstreitende*) numbers."

He began by explaining the expression "conflicting" he had created himself for quantities, albeit without clearly differentiating between quantities and numbers. With "*widerstreitend*" he referred not only to the fact that a quantity's quality (or type) and quantity (or amount) conflicted with that of another, but also "the ratio, or relation, in which quantities stand to one another." Quantities could very well be homogeneous, but different in their mutual relation, going, for instance in opposite directions (Tellkampf 1829, 22 ff.). From negative quantities, Tellkampf derived negative numbers, albeit also without any conceptual reflection or explication. In contrast to Fischer, Tellkampf made an effort at justifying the rule of signs, but giving no more than one example in verbally explicating how a negative multiplier worked (ibid., 26).

The common feature of all three textbook series is that they distinguished between two functions of the signs of plus and of minus: as signs of operation and as algebraic signs. "Quantity"' and "number," by contrast, had not been conceptually separated yet, and their relation had not been reflected upon.

Because of the manifold conceptual weaknesses of the approaches still adhering to the traditional concept of oppositeness, it is no wonder that there were proponents of sectarian views in Germany as well who condemned every position hitherto developed, and saw only insecure grounds for all the fundamental concepts. Friedrich Schmeißer was one of these proponents. Little is known about his biography; he worked in Saxony, first as private teacher of mathematics in Dresden, then as assistant teacher at the Dresden Knight's Academy, and finally as teacher at the famous *Fürstenschule* (princes' school) Schulpforta. When Schulpforta came to Prussia together with other Saxon territories in 1815, Schmeißer, too, was integrated into Prussian service. In 1820, he changed to the *Gymnasium* of Frankfurt/Oder, where he remained active as a teacher at least until 1855.

In several of his publications, he castigated "commonly defective" conceptions in mathematics, denouncing the "complete uselessness" of some theories, "critically challenging" further theories, or examining their "lack of rhyme and reason" (e.g., Schmeißer 1842, 2). That he accused other authors of lacking rhyme and reason was due, for instance, to his own operations with inequalities, which he multiplied in particular by negative factors. But since he obviously was not familiar with the fact that the signs of an inequality change direction if one does so, he naturally obtained any number of absurd results (ibid., 9 ff.).

The "theory of quadratic equations" did not remain exempt from his castigating "mindless analytic intercourse with signs bare of sense." Anyhow, he made explicit by this what had been implicit in the practice of many earlier mathematicians of admitting at most three types of quadratic equations:

> Of the 4 forms which one finds for the quadratic equations in books on algebra, the $x^2 + ax + b = 0$ quoted in § 22 was declared to be inadmissible, since a sum cannot be nothing; and if a, or b, or both should be *subtractive* for purposes of mutual compensation, but in contradiction to the designation used, this form would be contained in the other forms No. 2, 3, 4, and thus be quite superfluous (Schmeißer 1846, 2).

Two negative solutions were completely inadmissible for Schmeißer. His reflections, however, had at least the effect of informing us that not only the French armies had contributed to disseminating Carnot's ideas, but that Napoleon himself had propagated them in Germany. Schmeißer reported the following incident, which had been confirmed to him by friends and colleagues:

> When the former Emperor Napoleon, traveling back from Tilsit to Paris in July 1807, visited the cadets' house in Dresden, having the teaching staff presented to him, he submitted to these teachers, besides another problem,[31] the question as to what was the sense of $(a - b)$ if $b > a$ was assumed? He did not obtain any answer, perhaps because of a reticence to enter into metaphysical discussions with a ruler so powerful at the time. From this question, however, people think to be able to conclude that even a Napoleon had been irritated by the assumption of absolutely negative quantities upon taking his mathematical studies (Schmeißer 1846, 8).

Compared to the mostly traditional continuations of the notions of oppositeness, and to the more or less intense receptions of the French approaches, there was, after Förstemann, essentially only one other approach of the same orientation in Germany during the first half of the nineteenth century: that of Carl Friedrich Gauß (1777–1855). In his famous paper of 1831 concerning the theory of biquadratic residues, Gauß gave the graphical representation of complex numbers. As basis for that, he developed a conception of negative numbers that assumed the new foundational function of numbers relieving that of quantities. Gauß initially introduced complex numbers by comparing them to real numbers:

> While the arithmetic of real numbers speaks of only two unities, of the positive and the negative, we have in the arithmetic of complex numbers four unities +1, −1, +i, −i (Gauß 1831, 171 f.).

Gauß described the process of generalization of algebra in particular as an extension of the number domains:

> Our general arithmetic, whose extent by far surpasses the geometry of the Ancients, is entirely a creation of the Modern Age. Originally assuming the concept of the absolute integers, it extended its domain step by step; integers

---

31   Note by Schmeißer: Napoleon's other question concerned the proof that the volume of a pyramid is one-third that of a prism with equal base area and height.

were supplemented by fractions, rational numbers by irrational numbers, positive numbers by negative numbers, and real numbers by imaginary numbers. This advance, however, occurred initially with a fearfully hesitant step. The first algebraists preferred to call negative roots of equations false roots, and it is precisely these where the problem to which they refer was always termed in such a way as to ensure that the nature of the quantity sought did not admit any opposite (ibid., 175).

Gauß described the development of algebra as a process of abstraction from number domains to which an object in quantity form can be assigned:

> But as nobody has objections to admitting fractions in general arithmetic, although there are so many countable things where a fraction does not make sense, it was not permissible either to deny negative numbers equal rights with positive numbers for the reason that innumerable things do not admit anything opposite: the reality of negative numbers is sufficiently justified, since they find an adequate substrate in innumerable other cases (ibid.).

By simultaneously separating theory from application, and interrelating them, Gauß legitimized both negative and imaginary numbers: "Imaginary quantities could be supposed to have an underlying object just as well [...] as negative ones could" (ibid.).

Precisely here, in his effort to justify negative numbers, Gauß wrote his well-known formulation to mark the new, theoretical character of mathematics not as a substantialist, but as a relational conception of mathematical objects:

> Positive and negative numbers can find application only where what is counted has an opposite which, imagined combined with it, must be set equal with annihilation. Upon closer look, this condition holds only where what is counted is not substances (objects imaginable for themselves), but relations between two objects (ibid., 175 f.).

Since his main objective was to justify complex numbers, Gauß merely gave one specification for negative numbers:

> The postulate in this is that these objects are ordered into a series in a certain manner, e.g., A, B, C, D, ..., and that the relation of A to B can be considered to be identical with the relation of B to C, etc. Here, the concept of oppositeness involves nothing more than *exchanging* the term of the relation, so that if the relation (or transition) from A to B holds as +1, the relation from B to A must be represented by –1. Inasmuch as such a series is unlimited on both sides, any real integer represents the relation of any term chosen as the beginning to a certain term of the series (ibid., 176).

Surprisingly, this text by Gauß, and his clear exposition of the reality of negative numbers, was not received in the contemporary German literature of the discipline. I did not even find this text mentioned.

I was able to find the sole contemporary reception only at the periphery, in Bohemia, by Bernard Bolzano (1781–1848). Bolzano was very attentive indeed in studying and evaluating precisely the German contemporary positions on negative numbers. He gave his evaluation in two unpublished manuscripts on quantity theory, which remained unpublished and were edited only in 1975. The only non-German mathematicians of the nineteenth century he assessed were

Carnot and Cauchy; all the others are German mathematicians, among them primarily those we have already analyzed here: Fries, Schultz, Busse, Klügel, Tellkampf, Ohm, Stein, Förstemann, and Hermann.[32]

The manuscript *Erste Begriffe der allgemeinen Größenlehre* evaluates the literature published until 1831; I consider it to be the earlier of the two, written soon after 1831. While Bolzano is considered to be one of the pioneers of modern mathematics on the basis of his precise conceptual reflections on analysis, he proves to be, as far as the concept of number is concerned, to have been much under the influence of the then traditional concept of quantity. Bolzano did not conceive of concepts in this field as of relational ones, but admitted only "real" quantities in a substantialist way.

Bolzano demanded that an "object" eventually had to correspond to concepts in the process of abstraction:

> In establishing such a concept, it will not always be necessary to show that some *object* corresponds to it as well, i.e., that a quantity as one imagines it after this concept really exists. Not infrequently, it should be even useful and necessary to introduce notions of quantities with regard to which it is seen later that an object like of the kind they demand is something impossible. To recognize this impossibility alone is often useful. Hence, a proposition which simply states that this kind of object is impossible may be considered to be a truth of its own as pertaining to quantity theory (Bolzano 1975, 227).

As the further text shows, he considered the zero and negative quantities to be such concepts to which no object corresponded.

While Bolzano accepted the concept of oppositeness, introducing additive and subtractive quantities as pairs of "opposite things" (ibid., 248 f.), he declared subtractive quantities to be differences $X - Y$ of quantities X and Y, admitting only absolute quantities for this. The case of $X = Y$, he said, was "a notion of quantity which [...] must be considered to be a notion totally without object" (ibid., 249).

Bolzano declared the case for $Y > X$ to be a nonexecutable operation: "If the quantity designated by $\mathbf{Y}$ [...] is larger than the quantity designated by $\mathbf{X}$; the expression $\mathbf{X} - \mathbf{Y}$ demands an impossible execution" (ibid., 250).

At first glance, a negative result appears to have been so devoid of an object for Bolzano that he demanded to take as representation of quantity in this case the opposite, absolute value:

> Let us thus posit that the expression $X - Y$, which designates, according to the traditional interpretation of its signs, a subtraction of the quantity Y from the quantity X, shall keep this meaning in future only for the case where $X > Y$, but for the case where $Y > X$ shall have the meaning that one shall not subtract Y from X, but X from Y instead (ibid., 251).

Bolzano, however, seems to have created a misunderstanding here, for what he actually meant was that the result was negative, intending nevertheless to confer

---

[32]   I had been puzzled to see that Bolzano had evaluated almost the same authors as those encountered by me.

the character of "real quantity" to such results (ibid.). For this purpose, he inserted his own concept of oppositeness to establish an "extended meaning for the sign of –":

> $X - Y$ shall thus mean in future [...] that if the former is smaller than the latter, one is held to subtract X from Y, and consider the remainder as a quantity opposite to the previous (ibid.).

To explain oppositeness, he referred only to a paragraph (§ 70) in the part *Vorkenntnisse* of his other text *Einleitung zur Größenlehre*. The definition he had given in this text resulted clearly from a context of philosophical reflection of concept, about "representations" **A** and **B**. His abbreviated form for the "concept of opposition" formulated after an extended discussion was also still quite abstract:

> **A** and **B** shall be called opposite to one another if there is any rule representable by mere concepts according to which **B** can be derived from **A**, and if, according to the same rule, **A** can again be derived from **B**[33] (ibid., 141).

There is a total absence of discussion as to how philosophical concepts can be transformed into mathematical ones. Without further ado, and without giving a justification, Bolzano applied (mathematical) opposition as inversion of the signs—not even defining these signs themselves—and likewise as inversion of the signs of operation, also without reflecting on the relation between signs and signs of operation. In particular, there is no justification in Bolzano why a quantity thus placed in opposition to a positive quantity is a "real" quantity for him. An extended concept of quantity, or of number, was not explicated, although he had mentioned the extended meaning of the minus sign.

Bolzano obviously had taken note of Förstemann's sign for opposite sign, replacing, for his own purposes, the latter's dash by the prefix *neg*. Thus, he expressed his suggestion for transforming negative differences into a "real" quantity as **neg (X – Y)**. He also presented the twofold execution as identity:

$$\textbf{neg neg (X – Y) = (X – Y)} \text{ (ibid., 251).}$$

It is typical for how little generalizing Bolzano considered his own opposite quantities that he did not leave the general term **X – Y** to stand for negative differences, but transformed it into **neg (Y – X)** instead, thus actually operating with case distinctions in this manuscript. Besides, he designated, for negative values, the equation **W = neg (Y – X)** as "a merely symbolic equation" (ibid., 251), as distinct from "real" equations. How near he came to conceptions in England, and to Klügel, with this becomes clear from the fact that he believed to have dissociated himself explicitly from a use of the sign for the indeterminate difference, whose meaning he had factually incorporated by segregating symbolic "neg" equations (ibid., 251).

In this manuscript, Bolzano stopped after introducing addition and subtraction, discussing neither multiplication nor division. He tried to compensate for this by criticizing practically almost all the other authors he

---

33   The printed text gives wrongly A instead of B here.

reviewed, voicing the least criticism, however, against M. Ohm. He criticised Carnot for having newly introduced the opposition he had rejected himself under another designation: as direct versus indirect quantities (ibid., 274).

In the other manuscript, *Einleitung zur Grössenlehre*, written about 1840 (ibid., 13), Bolzano, on the one hand, declared opposition to be one of the conditions of the theory of quantities: "(We) permit ourselves to assume even in the pure theory of quantities already" that "quantities of a certain kind also have another kind that is *opposite* to them" (ibid., 32). On the other hand, he took note of Gauß's note of 1831, acknowledging the latter's position while at the same time pointing out the paucity of his formulations:

> From the observations which Herr Hofrat *Gauß* had printed in the April issue of the *Göttinger gelehrten Anzeigen* of July 1831, it is seen that this eminent scholar has arrived at a view quite of his own concerning the nature of opposition and of imaginary quantities as well; but what is being said there does not suffice to become familiar enough with his concept. May it please him not to keep too long from us the exhaustive treatise he has promised. Until then, I believe to be allowed to assume at least that Herr Gauß, just as most of the other mathematicians, assumes that some things *A* and *B* can be called opposite only if *A* relates to *B* like *B* relates to *A*. It is only for this reason that everybody defends the proportion $+1 : -1 = -1 : +1$ so much criticized already by *d'Alembert* (ibid., 144 f.).

In his second manuscript, Bolzano gave a quite philosophical definition of oppositeness, for "representations" (ibid., 141). The possibility of representations, however, continued to be tied to their being "objective." "Objectiveness" (*Gegenständlichkeit*) was a basic expression in this part. Here again, the zero was declared to be "without object" (*gegenstandslos*) (ibid., 164). While introducing subtraction in this manuscript as well (ibid., 162 ff.), Bolzano avoided a discussion and evaluation of negative differences, it remaining obscure how far he received the fact that Gauß had recognized negative numbers to be real. His emphasis on objectiveness, and on the exclusion of the zero, however, suggests that he had not much changed his view expressed in his first manuscript of 1831. Bolzano did not separate the concepts of quantity and of number in these texts. For him, "quantity'" was the more comprehensive and fundamental concept. Furthermore, "quantity" did not contain, for him, all the possible objects (or "things") of mathematics; thus, *directions* were not quantities for him (ibid., 142).

Such a radical position, of course, raises the fascinating question of how Bolzano, in his own theory of numbers, which was based on the theory of quantities for him, coped with negative numbers. His approach proves to be surprising; it has not been taken note of as yet in the literature on Bolzano. In his *Zahlenlehre*, written in the mid-thirties, Bolzano maintained the same substantialist position: numbers are admissible only as "real" ones; the zero is not "real," since it is without object. Even fractions were not real numbers, since no object corresponded to the sign of $\frac{1}{3}$ (Bolzano 1976, 17).

The difference $S - M$ of two numbers was again confined to the case $S > M$ by

Bolzano (ibid., 52). Nevertheless, he created a loophole for himself to be able to recognize negative numbers again as "real" ones, this time by means of simultaneously using the negative unity:

> Under this condition [that the unity be capable of opposition], the expression, or the representation of number, $S - M$ represents a real number even in case $M > S$, although of a unit which is opposite, i.e., negative, to the unit originally assumed (ibid., 53).

Bolzano did not quote Diesterweg; there is possibly an earlier source common to the two for simultaneously operating with two opposite unities.[34] Bolzano probably was not quite convinced of his own position.[35] His text in this section breaks off shortly after the quotation; several subsequent pages have been deleted by Bolzano. In a later section on rational numbers, Bolzano speaks of "a real number to be considered either additively, or subtractively," but refers for justification to these deleted paragraphs. The really telling thing is that Bolzano, in his section on the operations of multiplication and division, avoids talking of multiplying and dividing with negative numbers. He also shirks treating all the cases of the rule of signs (ibid., 55 ff.).

The conceptual battles of the eighteenth century were thus still being fought here. They were by no means "over and done with" yet.

4. The "Berlin Discussion" of Continuity

As has already been shown above (Chapter II.1.5.), Cauchy and Bolzano are traditionally credited with having worked out the concept of continuity. Against that we saw that an understanding of continuity, in the sense of a certain relationship between the value of the independent variable and that of the dependent variable, had already been formulated by Leibniz, and that this way of considering continuity was further developed, for example, by Boscovich. In Berlin, the Leibniz approach became known through extracts from his correspondence prepared by S. König, notwithstanding their doubtful authenticity (Chapter III.5.). We have further seen that continuity was intensively discussed in the eighteenth century, not as a mathematical concept, but as a metaphysical principle in mechanics, especially so in Germany through the contributions of Hausen, Kästner, and Karsten. The nature of intermediate values was always explicitly considered as an element of continuity (Chapter III.9). From the end of the eighteenth century, the concept of continuity was taken up in France, in the context of the *École polytechnique*, initially applied to mechanics with de Prony, then also in analysis, through Garnier and Lacroix,

---

[34]   Applications of this peculiar approach are missing. Bolzano's famous introduction of real numbers is restricted to the positive domain. Zero is the limit of a number that "can diminish indefinitely" (ibid., 95).
[35]   See the epigraph at the beginning of Section 3.

whose work was inspired by Carnot (Chapters V.2.4.3.; V.4.4.).

Finally, we saw that from 1817 Cauchy founded his *Cours d'Analyse* on the concept of continuity, though a formal definition of continuity was not published until 1821, and he himself further amended the definition (Chapter VI.6.5.).

In contrast to the traditional view that the essential results on continuity derive from Cauchy and his little-known contemporary Bolzano, it is surprising to learn that independently of them conceptual work on continuity was achieved in Berlin from the beginning of the nineteenth century. The first exponent of this fundamental work was Johann Georg Tralles (1763–1822). Tralles has for long held a place in the history of mathematics on account of a posthumous slander by Martin Ohm, peddled by Lorey, which described him as a mathematical ignoramus.[36]

Tralles was born in Hamburg, and from 1783 studied mathematics and physics in Göttingen under Kästner and G. Chr. Lichtenberg. After only two years' study he was recommended by both professors for the vacant position of professor of mathematics and physics in Bern at the Academy, a theological training establishment. Two years later he took on the parallel post of professor at the newly founded "Political Institute," the corresponding establishment for the training of civil servants. The focus of his work in Bern was physics; his outstanding research in geodesy led, among other things, to his being sent as delegate for the Swiss republic to the first international scientific conference in Paris in 1798, the purpose of which was to firmly establish the new metric measures of mass and weight. Thus not only did Tralles become the first German mathematician to obtain an accurate view of the French scientific system over the course of his year-long stay, but he also made contact with leading French scientists, with whom he continued to correspond. In 1803 Tralles left Bern to work on geodetics in the Swiss canton of Neufchâtel, at that time a Prussian principality. In 1804 he left Neufchâtel for Berlin, being invited to become a member of the Academy of Science where, in 1810, he was elected permanent secretary of the mathematical class and so became one of the four leaders of the Academy. In the same year, he not only took up the extra duty of becoming the first *ordinarius* for mathematics at the newly founded University of Berlin, but also played a decisive role as member of the Berlin scientific deputation in securing the place of mathematics as a key element in the Humboldt education reforms (cf. Schubring 1983, 44 ff.). In 1822, Tralles died suddenly during a visit to London, where he was engaged in supervising the making of precision scientific instruments.

Apart from his textbook for the Academy in Bern, written in the "neohumanist" style, Tralles left no published account of his fundamental work, so that it remains only in manuscript copies of his lectures and talks. Because of his sudden death, Tralles himself, of course, had no opportunity to review his material. Among his papers, however, there are numerous interesting

---

[36]   The facts of the case will be found in Schubring 1998, 303 ff.

manuscripts containing fundamental work (cf. also Schubring 1998, 317 ff.).

One such manuscript develops the use of continuity as the basis for the theory of functions in analysis. The manuscript bears the title "On the founding of higher analysis," to which is attached the note, "Draft of Lecture (dated 17.3.1808)."

The note "Draft of Lecture" must be in error. Since in 1808 there was not yet a university in Berlin, the numerous scientists then present in Berlin offered a series of public lectures. I have looked through the newspaper *Spenersche Zeitung*, which published notices of such lectures, from January 1808 to April 1809, and could find no evidence of a notice by Tralles.[37] On the other hand Tralles, in the text itself, says that "this treatment" is certainly not "for the beginner in mathematics" (Fol. 2v), so that it was not necessarily the manuscript of a lecture. The date of 1808, on the other hand, certainly refers to an original notice by Tralles himself, so there can be no doubt about the date.

Tralles states his intention as "to set out the fundamentals of transcendental analysis" (Fol. 2). He refused to "base a transcendental analysis on … the theory of infinitely small quantities or other arrangements of these words": "The infinitely large indicates that form and number lead to an incongruence between form and substance" (Fol. 2v). And an infinitely small quantity is not an exact concept, since "quantities or number which are not zero and have no size" deserve no recognition in mathematics (Fol. 3). Tralles's intent was to investigate "whether there might be higher principles for transcendental analysis than algebraic ones" (ibid.). His starting point was the "general form of the function" (ibid.). The concept of a function was for him the building block of the foundations of analysis.

As a particular principle of "transcendental analysis" Tralles introduced the concept of the continuity of a function. As a starting point, he defined the concept of a function. His definition showed not only that it was no longer based on the need for an analytical expression, but also that his concept of a function was constructed on the relation between *values*. We should note further that he gave the name "radical" to the independent variable. His definition runs as follows:

> A function, in the broadest sense, is a quantity that has no determined value, but takes a value when another quantity, on which it is respectively dependent, and which for the sake of brevity I allow myself to call its radical, is itself determined. As soon as the method is given for the value of the function to be determined from the given value of the radical, either exactly or through approximate calculations, it can be considered as given. In other cases, where only the nature of the function itself is given, it is the business of analysis to present it in a form by which its values may be determined (Fol. 7).[38]

---

[37]   Nor is there any reference to such a lecture by Tralles at the Berlin Academy. The full title of the *Spenersche Zeitung* was *Berlinische Nachrichten von Staats- und gelehrten Sachen* [The Berlin news of state and learned matters].
[38]   AAdW, *Nachlass* Tralles, Nr. 24.

In further "consideration of the relation or shape and the value of a function" Tralles restricted himself to single-valued functions, although he considered functions in general as multi-valued (ibid.). Thus he could introduce continuity as a specific principle of higher analysis:

> The condition on the possible values of the function in question consists in that with every change in the value of the radical, it either increases or decreases, as long as the change does not exceed a supposed quantity, which moreover can be as small as one wishes; and that this increase or decrease in the value of the function can be made smaller than any given quantity when the change in the value of the radical is taken to be sufficiently small (Fol. 7v).

Tralles had clearly intended here pointwise continuity, for at the end he explained that the condition always referred to a particular fixed value of the independent variable, or radical. The requirement was that the increase or decrease be valid only locally: "between two values of the radical when one comes sufficiently close to the other, which does not change" (Fol. 8v). Tralles further insisted that continuity required real and finite values:

> that there is here always the basic condition, that when real values of a function correspond to all those of the radical, that lie between two given instances, let it be expressly recalled that we talk about the behavior of the values, but not about impossible values (ibid.).

Tralles's requirement needed his functions to be locally monotonic, since he did not have the use of an algebra of inequalities and absolute values. He began with the proof of theorems about the behavior of continuous functions, which he derived from their definition: that the sum of two continuous functions is continuous, and also their difference, product, and quotient are continuous.[39] He even went so far as to make a general claim: "And in general, every algebraic function of a continuous function is a continuous [function]" (Fol. 11v).

The proof used the local monotonic condition and considered different cases. Tralles was unable to consider oscillating functions, unlike Bolzano, who did so later, though Bolzano's treatment did not appear until his posthumously published theory of functions (cf. Bolzano 2000, 79 ff.).

Tralles described discontinuity as a "jump" in the values of the function (Fol. 10v), but he also gave an algebraic description, in an early use of the letters δ and ε, by which he stated that "for however small a number δ of the radical Δ" (he considers the function to be dependent on an increase Δ) "and for the value α of the function corresponding to δ the inequality $\alpha^e > (1 - \omega)^\varepsilon$ must hold" (Fol. 19v).[40]

In a later manuscript fragment on "functions," catalogued among his papers as "lecture drafts," Tralles gave an algebraic description of discontinuity:

---

39   In his proof, Tralles assumes, without saying so, that the value of the function in the denominator is not zero.

40   Tralles's formulation was very involved, so I shall not explain all of his symbols here.

A discontinuous function is any function which in the expansion of its radical[41] is always capable of possible values of the function, and for however small a change of the radical of any previously fixed value of the function there is a quantity which cannot be made smaller than a given amount.[42]

In this manuscript, Tralles gave a brief statement of continuity: "A function is called continuous when its change can be made smaller than any given quantity when the change in the radical is taken sufficiently small." And he not only explained this for particular values Z, Z', but also symbolized it using the inequality

$$fz + i - fz < g \text{ (ibid. Fol. 28).}$$

Since we have investigated and analyzed, in considerable detail, the different contributions of the mathematical communities of Europe to the subject of continuity, as has so far appeared in the literature, we are in a position to assess the innovative contribution of Tralles. When we compare this with the progressive development in conceptual understanding in Europe around 1800, namely in France, we must understand that the concept of continuity up to 1808—and even up to 1816—had hardly abandoned its position as a set of predominantly "metaphysical" axioms and had not yet been developed into a single scientific operative concept.

This a priori character of the concept was already clear from its being referred to as the "law of continuity." Lacroix, in his textbook on analysis, had already succinctly formulated this character, through comparison with the law of gravitation, whose validity precedes any individual investigation in physics (cf. Chapter V.2.4.2.). Also de Prony had made the law of continuity an epistemological basic requirement; in his *Mechanics* of 1815, however, he had understood continuity as a concrete mathematical requirement and investigated the different cases of continuity and discontinuity (cf. Chapter V.2.4.4.). At the same time, continuity was always identified with differentiability.

The approach taken by Tralles represented for the first time a new conceptual way forward: continuity not only provided him a fundamental concept for the investigation of functions, but was at the same time, corresponding to Tralles's general algebraical method, no longer bound to geometry and to the behavior of curves, and so could also be applied locally.

Certainly there is cause to wonder why this achievement of Tralles remained for so long unknown. However, Tralles published relatively little, and his lecturing style was apparently hardly stimulating; his manuscripts demonstrate a strong inner struggle for clarity and are less concerned with setting forth an engaging presentation of new ideas.[43]

---

[41]   That is, in its domain of definition, an informative first description of "domain of definition"!

[42]   AAdW *Nachlass* Tralles, Nr. 22 Fol. 37.

[43]   Tralles's somewhat clumsy style was also the cause of M. Ohm's denunciation; Ohm had alleged that Tralles claimed $\sqrt{-1} = 0$ (cf. the previous

Actually, we have seen above (Chapter VI.6.5.) two other pieces of evidence pointing to an independent Berlin context for continuity. Dirksen, at the University of Berlin from 1820, had criticized Cauchy's definition of continuity in 1829. In his own work on the theory of functions, Dirksen had, however, used Cauchy's terminology, but had consistently and independently introduced continuity as a pointwise concept. Similarly, we have seen that Abel, in his central treatment of continuity in 1826, if somewhat imprecisely, gave a pointwise definition; at the time of the writing of this account, Abel was living in Berlin and was in close communication with leading Berlin mathematicians (Ore 1957, 89 ff.).[44]

From this I consider it probable that not only was Tralles's concept of continuity current among Berlin mathematical circles, but also Bolzano's contribution to the theory, especially his 1817 treatment of the proof of the intermediate value theorem. It was not until 1826, during his stay in Paris, that Abel made his well known observation, "Bolzano is a clever guy," but it would be reasonable to assume that he had already read his work when in Berlin. Crelle, with whom he mostly communicated in Berlin, possessed the three Bolzano treatises of 1816/1817 in his private library (Schubring 1993, 50). This would also explain why Bolzano's definition of continuity formed a part of the Berlin discussions of the concept. Bolzano's definition of (pointwise) continuity published in the paper of 1817 runs thus:

> [One understands by this] that a function $f(x)$ for all values of x, lying within or outside certain limits, according to the law of continuity varies [to the extent] that whatever the value of x, the difference $f(x + \omega) - f(x)$ can be made smaller than any given quantity, provided ω can always be taken as small as one wishes (here from: Kolman 1963, 179).

Furthermore, Bolzano, in his manuscript from the 1830s on the theory of functions, had also clearly worked out the difference between the concepts of continuity and differentiability. Bolzano proposed that a function having the latter property should be called a "derived function" (*eine abgeleitete Funktion*) (Bolzano 2000, 37 f.). This manuscript was not published, however, until 1930.

Finally, the example of J.P.W. Stein will illustrate that the emerging new

---

footnote).

[44]   Martin Ohm, then active in Berlin, had in 1822, in the analysis section of the second volume of his *Versuch eines vollkommen consequenten Systems der Mathematik* (*Search for a completely consistent system of mathematics*), published a definition of continuity (M. Ohm 1822, 123). Bolzano later, in his manuscript on the theory of functions, criticized this definition and explicitly gave the correct logical dependence of δ (here h or Δx) on ε (here D) (Bolzano 2000, 36). Continuity was not, however, an essential fundamental concept in Ohm's system: he did not define the concept when he introduced functions, or later in the treatment of derivatives of functions, but much later on, in the chapter on entire functions and their integer values; neither was the concept applied. Clearly, in his Berlin context, Ohm felt obliged to make the concept appear somehow.

concepts in the context of the Berlin discussion differed markedly from those concepts that were drawn from French ideas.

We have already shown in Section 3.3. how strongly Stein's concept of negative numbers was affected by French-dominated ideas. In his algebra textbook he had also given an introduction to analysis, in the section "On variable quantities and their limits."

As a necessary prelude to introducing the continuity of functions he gave an explanation of the continuity of an (independent) variable, through the idea of intermediate values:

> One says that a variable quantity changes *gradually* or *continuously* when it cannot pass from one value to another without taking in turn all the values lying between these two values (Stein 1829, 85).

The definition of the continuity of a function that follows uses global, not pointwise, continuity:

> A function of a variable quantity, e.g., $f(x)$, is said to be *continuous* if when the variable quantity x of which it is a function gradually increases, the function also changes gradually (ibid. 86).

A discontinuous function is one for which "the gradual increase of the variable quantity [...] produces a discontinuous change in the function" (ibid.). As an example of a "discontinuous function" he gives $\sqrt{(1-x)x}$, "since for those values of x which are not contained between 1 and 0, it does not at all exist." Obviously in agreement with Cauchy, Stein also gives a definition of global continuity for a particular interval:

> A function can be continuous within certain limits, even when it is not so generally. Thus the function $\sqrt{(1-x)x}$ is continuous between the bounds $x = 1$ and $x = 0$ (ibid.).

Stein comes to define the concept of the limit values only after the definition of continuity. Corresponding to the French view, he excludes the possibility that the variable attains the limit value:

> One says that a variable quantity has a limit when a certain constant quantity is given which the variable can approach as close as is wished, without attaining it (ibid.).

Finally, Stein also followed the approach of Carnot and Cauchy, and defined infinitely small quantities as being variables with limit zero (ibid. 87).

5. The Advance of Pure Mathematics

The working out of the conceptual understanding of continuity serves as an example to illustrate not only that in the emerging new mathematical center of Berlin, effective contributions were made to the foundations of mathematics, but also that these contributions brought a more precise clarification of foundational concepts. In this Section we shall look at the work of two Berlin mathematicians that, in the productive atmosphere of the city, stimulated a rapid improvement in the conceptual modification and improvement of the foundations of analysis. This process of conceptual modification can be seen as a consolidation of the existing pure mathematics.

These two mathematicians were Enno Heeren Dirksen (1788–1845), whom we have already met in Chapter VI.6.6. as a perceptive and critical proponent of Cauchy's foundational concepts, and Peter Gustav Lejeune-Dirichlet (1805–1859), who is the widely recognized representative of the new rigor in pure mathematics.[45]

5.1. Summary of Dirksen's Work

Dirksen had apparently spent at least two decades preparing a monumental work, intended to provide rigorous foundations for analysis, under the title *Organon of the whole of Analysis*. This was conceived as a work with three parts:

* elementary theory of analysis, comprising two sections:
  ➢ elementary algebraic theory of analysis, and
  ➢ elementary transcendental theory of analysis,

and

* analytical theory of functions.

Of these three parts, only the middle part was published in 1845, with the title *Organon of the whole of transcendental analysis*. Of the other parts, a large number of frequently revised drafts were found among Dirksen's papers.[46] This published material alone ran to 940 printed pages, caused by the almost obsessional endeavor of Dirksen to provide a complete systemization of all possible conceptual ramifications. The greater part of the book comprised what was for a long time called "algebraic analysis." In effect, this dealt with the theory of infinite series. Whereas Crelle, in a report to the Ministry of Education, criticized the strangeness of his methods and terminology,[47] Gauß, in a letter of November 5, 1845, in response to being sent a copy of the book, gave it the highest praise, namely saying that he himself had earlier started on a similar approach: Gauß had himself become closer to the approach of Dirksen's work:

> It is one that has always been of great value to me. Even very early, that is, more than 50 years ago, everything I found in books on infinite series was

---

[45]   For some new elements of his biography, especially concerning the beginning of his scientific career, see Schubring 1984.

[46]   The papers are held in private hands in Osterburg, Groothusen, East Friesland. I am indebted to M. Folkerts for allowing me access to the manuscripts.

[47]   Report of 19.9.1845; in: (HA PrKu) Sammlung Darmstädter, H 1818 Crelle; information from M. Folkerts.

unsatisfactory to me and abhorrent to a worthy mathematical mind, and I remember that in 1793 or 1794, I began an attempt to develop the basic concepts in a satisfactory way that, as far as my memory serves me, had considerable similarity with your approach (Folkerts 1983/84, 73 f.).

Of the "numerous folios" Gauß had begun at that time, he said he had not continued with them, on the one hand because "every development, if it should be complete, would require a *great* demand of space," and on the other hand, because at that time it appeared that there was no readership to be found for such treatises, a situation that now seemed to have changed (quotations from ibid.).

It is not possible within the present work to provide a detailed analysis of Dirksen's presentation of the foundations of analysis. This must await another opportunity. Here we can give only a brief overview, emphasizing the key points, relating to how Cauchy's work was received, and the significance of Dirksen's use of symbols for the development of concepts in analysis and in integration.

5.1.1. NUMBER

There are numerous versions of the algebraic elementary theory among Dirksen's papers; some Dirksen himself has marked "from an earlier time," so we have a certain chronological sequence. It is noteworthy that not only was Dirksen not entirely satisfied with any version, but also that the concept of number proves to constitute the kernel of his conceptual problem.

Exactly analogous to Ampère, who engaged himself in what seemed to be a futile attempt at a coherent derivation of the concept of number from the relation between quantity and number (cf. Chapter V.3.2.), Dirksen's problem was to provide a basis for the concept of number. It is indicative that in the first versions, number was a fundamental concept, but that in later versions he moved away from this, and finally number was presented as an intermediately derived concept: in the derivation of the concept of number, appeal was even made to human cognition.

Initially, "number" is a foundational concept, and "quantity" a derived concept. Then this relationship was reversed, and then other basic concepts were inserted *before* "quantity." On the one hand, there is the idea of a "set," and on the other hand, the indeterminate idea of "things," with which Dirksen apparently wished to establish a link with human understanding through imagination.[48] Dirksen had adopted the idea of the set and its incorporation as a fundamental concept of arithmetic and algebra, an idea that had been developed by Kästner and his students Tralles and Metternich, as well as by Schultz;[49] this

---

[48]   "Set'" and "thing" as fundamental concepts are met in two versions, found in "Stack III" in a folder whose cover bears the note, "Theory of elementary analysis. First Section. First Book: On quantities and numbers (first draft and 2 revisions). Beginning of a work of which the printed Organon is a part." The comment is clearly by Fooke Hoissen Müller, Dirksen's friend and chronicler (see Folkerts 1998).

[49]   Johann Schultz had not studied under Kästner, but in Königsberg; it is,

shows that the idea of a set was used as a fundamental concept in Germany well before Cantor (cf. Schubring 1988, 145).[50]

In a fragment on function theory, which appears to be from the middle period of Dirksen's work, there is a page on which is outlined a "Plan of the Work." Under the title "Basic theory of the whole of analysis," the first section of the first part—algebraic elementary theory—was to contain four books:

1. On sets and wholes.
2. On quantities (*quantis*) and their ratios.
3. On numbers.
4. On algebraic quantities.

Later versions of this part were considerably differentiated. In the most polished, and therefore probably latest, manuscript version, the first chapter is called "On sets and their indices," and the first section—"plurality or multiplicity (*Mehrzahl oder Mehrheit*); unity; set, number or moreness (*Vielheit*)"—begins by introducing the "thing."[51]

Dirksen then introduced a "thing" as a generalization of the cardinal aspect and was already able to show elementary operations with sets. In the second chapter he defined quantities as special things: a thing is called a quantity when it contains only a single characteristic (Fol. 15). Finally, in the first chapter of the second book, he succeeded in defining numbers. These were defined as the ratios of quantities, that is, as their measure, Dirksen's formulation, moreover, being even more complicated than this (§ 527, Fol. 70).

If two quantities *A* and *B* were equal, their number was said to be "assignable" (*angebbar*) and if they were unequal, their number "unassignable" (*unangebbar*). It is a feature of the whole of Dirksen's construction of analysis, and especially of his theory of series, that for Dirksen *zero* is not a number, certainly not an assignable number. Dirksen maintained this basic position in all versions of his theory of elementary algebra, and it was also a basic precondition for his concept of series. Also, in the early versions of the first part, which begins with the concept of number, zero held this special status, by virtue of the definition of numbers in terms of the measure of quantities:

§ 1.1. By a *number* we mean here the *geometrical ratio* in which two equal quanta stand in relation to each other as quantities. A number is by this not a

---

however, entirely probable that he had used Kästner's textbook.

[50]   On the concept of a set used by Kästner and Metternich see Schubring 1995, for Tralles see Schubring 1998, and for Johann Schultz see Schubring 1982.

[51]   Where not otherwise stated, the subsequent quotations are taken from the *Organon* manuscript in the Osterburg papers. For the sake of conformity, where Dirksen used underlining for emphasis, we have used italic text, as in the 1845 printed edition of the middle part.
"Stack III," text of Part I: Elementary Theory of Agebraic Analysis, First Book, Chapter I, Section 1.

quantum, but merely the predicate of a quantum.[52]

Dirksen identified zero as the "negation" of any assignable number. A consequence of this approach was that Dirksen explicitly required quotients to be defined only for assignable numbers.

In contrast to the special case of zero, Dirksen had no problem with the definition of negative numbers, nor did he provide much reflection on the subject. He did not introduce them in the succeeding chapter, where numbers were introduced under the arithmetical representation of numbers—here the difference $a - b$ was restricted to the positive case but in the following part, "On algebraic forms of quantities." In this "algebra," Dirksen introduced the differences of two real quantities with no restriction on the outcome. His starting point was the generalization of the concept of number. Completely determined numbers were now called "real algebraic quantities." For these, he introduced the concept of the absolute value—to use Cauchy's term—as its "number value" (*Zahlenwerth*). He also turned to Cauchy in choosing his symbolism. He wrote the absolute value of an algebraic quantity g as

$$\text{v.n. } g,$$

where v.n. stands for *valeur numérique*. Crelle's proposal of 1823 to use bars for the numerical value of a quantity was not, however, adopted. That Dirksen was certainly aware of Crelle's symbolism can be inferred from the fact that he intentionally used Cauchy's symbolism; this can also be seen by his retaining the abbreviation v.n. of the French original, although instead of "number value" he also used the literal translation "numerical value."

Next, Dirksen made a distinction between the numerical value and the sign of a real algebraic number and defined a negative number thus: when a completely determined number a is considered in its numerical value to be as much smaller than any completely determined third number b.[53]

5.1.2. SERIES

The subject of the middle part of Dirksen's foundational work on analysis, the 1845 published *Organon*, was the treatment of infinite series, but exclusively in the sense of number sequences, for which Dirksen used the German word *Reihen* (= series) instead of *Folgen* (= sequence) for the translation of the French "séries." The sum of a series is met only with "simple algebraic expressions for

---

[52]   Folder cover with the note "Erste Entwürfe und Materialien zur Elementarlehre der Analysis (aus früherer Zeit)," in "Stack I": "Entwürfe und Druckfassung des Organon 1845": Introduction, First Chapter.
In defining the concept, Dirksen refers back to Newton and Christian Wolff (Dirksen 1845, X).
[53]   *Elementarlehre der Analysis*, II. Buch. 1. Kapitel; Artikel I.
Distinguishing between completely determined and not completely determined numbers was fundamental to Dirksen's introduction to variables: not completely determined numbers were capable of assuming multiple values, completely determined numbers can have only one value (see Dirksen 1845, XI).

infinite series." Finding sums of infinite series and the existence of the sum of a series were not even indicated as topics in this extensive work.

Dirksen started by explaining the concept of an infinite series as an "act of thinking," in contrast to a finite series, and drew on epistemology and psychology to obtain a "possible" concept (Dirksen 1845, 11). Taking over Cauchy's concept of number (cf. Chapter VI.6.1.), Dirksen distinguished between two types of infinite series: number series, whose terms are positive numbers, and series of quantities, which can have not only negative numbers as terms but also any real numbers, and even imaginary numbers. For both types it was a main task to determine the nature of the behavior of the sequence or series. A starting point was the difference between *remaining-finite* and *not-remaining-finite* series. Dirksen eventually identified eight types, of which the three most important were:

- remaining-finite,
- infinitely small-progressing
- infinite progressing.

In addition, Dirksen also considers for the first time oscillating behavior, i.e., positive–negative–progressing infinite series of quantities, and infinite–infinitely small–progressing infinite number series.

A key element for the understanding of Dirksen's foundation of analysis—as also for the interpretation of Cauchy's *infinitely small quantities*—is the definition of infinitely small. In actual fact, this expression was always used by Dirksen in connection with "progressing" (*werdende*) terms, i.e., never for a single quantity, but only and always for variables. By an *infinitely small-progressing* infinite number series, he understood a completely determined infinite number series, in which

> the progression of terms can be imagined such that they, from the term with some index ϱ [...] all become smaller than any given assignable number β however small that may be imagined (Dirksen 1845, 25).

Since this refers to a number series, it was clear that all the values of the individual terms could only be positive numbers; and since β is given as an assignable number, it was an infinite small-progressing number series of positive terms with the limit zero.

By definition, this is distinctly different from a finite–remaining infinite number series, which is a completely determined infinite number series with finite (positive) limit Q (where the "limit" (*Grenze*) is by definition different from its "limit value" (*Grenzwert*)):

> *Explanation 13a.* By a completely determined assignable number Q is meant that it is the assignable limit of an infinite number series, to the extent that for each assignable number α as small as may be imagined, the series contains a term from which the difference between one of the following terms and Q (the greater of the two numbers always being the minuend) will be respectively smaller than the number α (ibid. 31),

to which Dirksen added,

that a completely determined infinite number series can have an assignable limit Q only to the extent that it is a finite–remaining infinite number series (ibid. 32).

Although Dirksen thus had a finite–remaining and an infinitely small–progressing number series leading to the same type of limit behavior, by virtue of his difference definition, by which $(a_m)—Q$ generates an infinite small–progressing series, the dichotomy of these two series types is a constitutive part of the whole of analysis, in particular since zero *cannot* be an assignable number, and this is also a consequence of his basic epistemological conviction. As we have seen, Dirksen shared this basic conviction with numerous mathematicians of the eighteenth century, especially French mathematicians, and in particular with those mathematicians at the turn of the century and in the early part of the nineteenth century who followed the classical conception of explicitly identifying infinitely small quantities as variables with a limit of zero, that is, with Carnot and Cauchy (cf. Chapters V.1.6.1. and VI.6.4.). It is characteristic of the dichotomy between "assignable" limit values and the limit zero, as with Stockler—who also adopted this dichotomy as a basic assumption—that the same reduction of finite limit values leading to zero limit values occurs (see Chapter III.8.4.).

In contrast to his predecessors, Dirksen explicitly also allowed negative limits (ibid. 39 f.). It underlines his preference for assignable numbers that he defined the concept of convergence for such limits exclusively:

> *Explanation 26.* A finite–remaining infinite series of quantities is an infinite series of quantities with an assignable limit, whenever it has an assignable limit; also for such a case of an infinite series, it is said that it converges to an assignable limit (ibid. 40).

Dirksen stressed that he had developed the foundational part of the *Organon* "explicitly" without the use of abbreviations, so that "the insight into the conceptual interrelations, which construct the essential conditions of scientific knowledge, would not be made more difficult through abbreviations of symbols" (ibid. 275). The introduction and elaboration of such symbols followed in later parts.

The development of this symbolism proves Dirksen's rare skill in identifying the essential nature of a concept by a clear symbolic representation. He was able to draw on the developing technical symbolism of the German combinatorial school, especially as regards the use of indices.

The individual terms of a series of completely determined quantities were represented by letters with subscripts:

$$a_0, \ a_1, \ a_2, \ a_3, \ ...,$$

and a_m was called the *general term* of the infinite series. The quantity m was the *undetermined* value of the series, taking natural numbers and negative numbers as values that were "not completely determined algebraic quantities," and he made it clear that m was the variable of the series and that all the terms of the series could be generated by substituting "particular values of r for m" (ibid. 277).

As an "analytical notation" for the infinite series with the general term a_m, Dirksen introduced the symbol

$$R(a_m), \text{ or simply } (a_m).$$

The ease with which the use of such symbolism was able to represent even complicated operations on sequences and series with clarity was something Dirksen expressly pointed out later in the work. Of particular effect was the introduction of his symbol for a limit, always using "*Gr*" (for *Grenze*) as the German translated abbreviation for *lim*, and his innovation of using the index variable in superscript over the limit sign:

$$\overset{m=\infty}{Gr}\, a_m \text{ or more fully } \overset{m=\infty}{Gr}\, R(a_m) \text{ (ibid. 301).}$$

For general series of quantities, Dirksen, on systematic grounds, did not restrict himself only to recognizing assignable limits as limts. Rather, he attributed to all completely determined infinite series the property "that they *have a limit*" (ibid.).[54] Hence, not only finite–remaining series can have a limit; for infinitely small–progressing series it was thus made explicit that their limit is zero:

> The limit of a completely determined infinite series is zero, whenever the series belongs to the species of infinitely small–progressing series. The analytical symbol for this definition of a limit is 0 (ibid. 302).

By extension, Dirksen not only assigned a limit ∞ to infinitely progressing series, but also recognized a limit for oscillating series, that is, where the limit changes between positive and negative values (ibid. 304). Even in symbolic form, Dirksen held fast to the double nature of his definition of limits. For assignable limits, having a completely determined algebraic quantity g, he used the notation

$$\overset{m=\infty}{Gr}\, a_m = g \text{ (ibid. 309);}$$

for infinitely small–progressing series, he declared its characteristic of having zero as a limit, in the form of a theorem:

> Theorem 4. Let $R(a_m)$ be infinitely small–progressing; then $\overset{m=\infty}{Gr}\, R(a_m) = 0$ or $\overset{m=\infty}{Gr}\, a_m = 0$ and conversely (ibid. 311).

Dirksen is considered to be the one contemporary mathematician to have carefully studied and reconstructed Cauchy's foundation of analysis, and to have made it his own in a critical way—consider his investigations into Cauchy's summation theorem (Chapter VI.6.6.)—and we have here the most telling evidence of this, that Cauchy's *infinitely small quantities*, in the context of the contemporary limited conceptual understanding, are here explained, as Cauchy himself had done, as variables with a limit of zero!

---

54   Dirksen systematically made a distinction between limit (*Grenze*) and limit value (*Grenzwert*). He first introduced limit value (*Grenzwert*) in the theory of functions (see below, Section 5.1.3.).

Further evidence that he considered *infinitely small quantities* to belong to the study of limit values is shown by the fact that Dirksen, in the *Organon*, had thereby concluded the construction of his concept structure, that he explained the "general limit method of analysis" as the "general foundation of the whole transcendental definition of analysis," and in this limit method he saw the characteristic difference between analysis and algebra (ibid. 410). Dirksen further comprehensively established that "infinitely small–becoming" and "possessing an assignable limit" provide mutually complementary views of the same general limit method.

Another example is found in the convergence of infinite series (here, as always, given as sequences). Dirksen also maintained that for the convergence of an infinite series of quantities it is necessary and sufficient[55] for the series "to be either infinitely small–progressing, or an infinite series with an assignable limit" (ibid. 421).

A further conceptual improvement in the *Organon* to be noted is the introduction of double limits, which we owe to Dirksen, such as

$$\overset{m=\infty}{Gr} \left[\overset{\rho=\infty}{Gr} b_{\rho,m} \right],$$

and extensive operations with them, as well as a well established practice of operating with inequalities, for example $\left(a_{\mu+m} - \gamma \right) < \varepsilon$, and the approximation of, for example, two inequalities by use of $\frac{\varepsilon}{2}$ (ibid. 429). Thus Dirksen adopted here the Cauchy criterion for the convergence of series.

5.1.3. THEORY OF FUNCTIONS

The third part of Dirksen's major work, *The Theory of Functions*, is similarly available in many versions. Whereas his theory of functions, in its narrow sense, was set out in five different versions in the foundation section of the third part of the *Organon*, both final parts, on differentiation and integration, were less reworked and some parts only lightly sketched out. In all versions it can be seen that not only are there numerous corrections and even striking out whole passages, but also the author's own criticisms, such as "false" or "not rigorous" written in the margins. In order to provide a view of the structure of the work, here are the chapter headings for version C, a relatively completed text:

Part1: Introduction to the theory of functions

Introductory chapter: discussion of algebra and arithmetic[56]

---

[55]   It is also indicative of the new style of rigor that Dirksen in his propositions, where required, always states explicitly that a condition is "necessary and sufficient."

[56]   Contains a summary of the concepts of quantities and numbers, i.e., Dirksen's "algebra."

Chapter 1   On variables with simple algebraic forms and functions in general

Chapter 2   On infinite function series and their limits

Chapter 3   On analytical expressions, equations and functions together with their classification

From other editions and manuscript fragments, it is clear that this part was also to contain:

On the greatest and least values of a function.

For the second part, manuscripts indicate:

On (finite) differences and integrals of a function[57]

> On the differentials of a function

Final chapter on the differential calculus:

> Discussion of some transcendentals

> Theory of imaginary numbers: on imaginary infinite series

On the integrals of a function:

> definite and indefinite integrals

In what follows, I shall give an overview of key features of the construction of concepts in this third part of the *Organon*, drawing from the most part on Version C.

The idea of a variable, already established as a fundamental concept of Dirksen's work, plays a key role here. While he had managed with the description "not completely determined quantity," namely a quantity capable of taking many values, in the earlier parts of the work, here he explicitly introduces a "variable." His definition shows that he had intentionally taken up Cauchy's definition of a variable "as having to take successively many different values, one after the other" (Cauchy 1821, 4), adapted it to his own use, and pursued the consequences in the greatest detail. One formulation of Dirksen's definition runs thus:

> A completely determined algebraic quantity is called a *variable* whenever it is understood that it can be substituted by another particular unequal value, and that one after the other can be substituted for it.[58]

Dirksen not only understood Cauchy's conceptual differentiation between the variable and its values at a particular position and applied it strictly in his own concepts and practice, but also understood Cauchy's concept of variable as a fundamental concept for *sequences*; and so confirms Giusti's interpretation (cf. Chapter VI.6.5.). Since sequences, or "series' in Dirksen's terminology, are used to construct a foundational concept of the whole of analysis, it is not surprising to find that the theory of series occupies the whole middle part of Dirksen's *Organon*.

Dirksen called a sequence

---

[57]   In one manuscript version, this part is called Book 6 (*6. Buch*); there is no general structuring to explain this numbering.

[58]   Version A; Erstes Kapitel, erster Abschnitt, § 2, Erklärung 1.

$$a_1, a_2, a_3, ..., a_n,$$

"a *system of particular values*" of the variable if the sequence of its terms can be constructed successively from particular values.[59] He defined constants as a special case of variables, in which all the particular values are equal.[60]

After introducing the expressions "original variable" (*ursprüngliche Veränderliche*) and "dependently variable" (*abhängige Veränderliche*), Dirksen explained functions as a special case of variables, namely as a "dependent variable" of some "original variable." The definition of function therefore follows quite generally and purely algebraically, without any geometrical context. In practice this meant that a "function" was understood as a mapping, as a correspondence between argument and value:

> Let (1) $x_1, x_2, ..., x_n$ represent a system of n variables; then a variable y is a *function* of the variables (1), and each of them is an *argument* of y, provided that each in some way is precisely determined.[61]

Typical of Dirksen's careful attention to symbolism is that he also stressed the use of suitable written notation for functions. Symbols such as $\mathfrak{F}$, f, Φ, ... were used as "characteristic" of functions, and could also be used with the arguments of the function as a combined symbol:

$$F(x_1, x_2, ..., x_n).$$

In dealing with functions, the difference between single-valued functions and multi-valued functions was of great importance for Dirksen. He called single-valued functions "uniform" (*einförmig*) and thereby asserted that for these, particular values generated "completely determined algebraic quantities, as soon as the corresponding values of the original variables were respectively regarded as completely determined"; functions that were many-valued he called "multiform" (*mehrförmig*) if their particular values were multi-valued.[62]

As Dirksen had already shown in the published part of the *Organon*, the transcendental form of determination is central to his search for limits of infinite series. His first goal in the theory of functions, after introducing the function, was to introduce the limit values of variables and then the continuity of functions. Since Dirksen had already developed the idea of limits for infinite series in the published material, here he needed only to specify them for functions and from this to proceed to limit values in the proper sense.

The starting point was to consider the limit for a point, a particular value, x_0 of the original variable x. Approaching the limit was, using the Cauchy variable concept, achieved by means of an infinite series

$$R(a_m): \quad a_0, a_1, a_2, ... \text{ with } \overset{m=\infty}{Gr} a_m = x.$$

From the definition of function, a second corresponding series can be

---

59   Ibid., § 5, Erklärung 4.
60   Version C; Erstes Kapitel, erster Abschnitt.
61   Version A; Erstes Kapitel, dritter Abschnitt, § 24, Erklärung 19 a.
62   Version C; erstes Kapitel, Artikel III, § 41, Erklärung 72 a.

constructed:

$$R\big(f(a_m)\big) \quad : f(a_0), f(a_1), f(a_2), \ldots .$$

Consistent with his formal foundational concept, each such series has a limit.[63] Dirksen considers not only the *possibility* of this limit, i.e., whether this is a finite value, but also its *uniqueness*. Here we see another of Dirksen's innovative achievements. He gave examples of different convergence behaviors toward x_0, and so different limits for the function at this point. Based on this clarification, that the simple requirement of the convergence of the sequence of arguments to a particular value of the argument is not sufficient, Dirksen developed his definition of a limiting value of functions in *two* steps. First he stated what was meant by limit, namely a particular value of the infinite series of function values $f(a_0), f(a_1), f(a_2), f(a_3), f(a_n)$, ad infinitum that are associated with the series a_0, a_1, a_2, a_3, a_4, ad infinitum., with $\overset{m=\infty}{Gr}\, a_m = x_0$ (Erkl. 7a). And in a second step, Dirksen requires this limit to be unique:

> *Explanation 8a.* Let x represent an original variable, x_0 any particular given value, and [...] $f(x)$ some single-valued function of x. Then every completely determined algebraic quantity of the *limit value of* [...] $f(x)$, for the limit x_0 of x is notated and represented by $\overset{x=x_0}{Gr}\, f(x)$, provided that it is equal to all the limits of $f(x)$ for the limit x_0 of x.[64]

Consistent with his basic approach of making a distinction between finite–remaining infinite series and infinite small–progressing series (as well as infinitely progressing series), Dirksen formulated his definitions 7 and 8 separately for infinitely small–progressing and for infinitely progressing series, and even separately for their convergence to positive and to negative values; for example, for a "positive–infinitely small–progressing" series of arguments as "*a limit of f(x) for the limit +0 of x*" (7b), and similarly for the uniqueness of the "*limit value of f(x) for the limit +0 of x*," also represented as $\overset{x=+0}{Gr}\, f(x)$ (8b).[65]

Dirksen stressed that "limit" (*Grenze*) and "limit value" (*Grenzwert*) were quite different, and that the "limit value" and its "analytical representation" $\overset{...}{Gr}...$ were exclusively used for uniquely determined quantities.[66] Further, he pointed out that for any function that, for example, has a limit ∞, the "corresponding limit value is impossible."[67]

Along with the concept of limit value, Dirksen also introduced the concepts of the *continuity* and *discontinuity* of (single-valued) functions, initially without the explicit use of sequences or series. Significant here, and indicative of the

---

63   Version D; zweites Kapitel, Artikel IV, § 45.

64   Ibid.

65   Ibid.

66   Ibid., fol. 26.

67   Ibid., fol. 27, Zusatz zu Lehrsatz 48.

character of the Berlin discussion of continuity (cf. Chapter VII.4.), is the conceptual difference between Dirksen's and Cauchy's views of continuity; the difference is the more significant in that in other cases, Dirksen had consistently reconstructed and developed Cauchy's concepts.

A first difference is that Dirksen, in all variations of his definitions of continuity, did not give two forms of the definition, as essentially Cauchy had done, namely for an interval and in the neighborhood of a point, but three forms. In addition to the definition of continuity for an interval and in the neighborhood of a point, Dirksen defined continuity *at a point*. In Chapter VI.6.5. we quoted Dirksen's definition taken from version B. Here is the analogous definition from version D:

> *Explanation 14a.* Let *x* be an original variable, *a* any given particular value, and *y* a single-valued function of them; then of *y* it can be said that it *remains constant for the particular value a of x*, whenever the particular value of *y* corresponds to the particular value *a* of *x*, and the limit value of *y* for the limit *a* of *y* are equal to each other.

In symbols, Dirksen wrote this definition as:

$$f(a) = \overset{x=a}{Gr} f(x).$$

A second difference from Cauchy was that for discontinuity (or "loss of continuity"), he did not just explain this for a special case (with Cauchy, in a neighborhood), but gave definitions for all three cases, even though they were simply negations of the corresponding definitions of continuity.

It is interesting for the comparison between Dirksen and Cauchy to consider the former's definition of continuity over an interval, which is based on *pointwise* continuity:

> Let *x* be an original variable, $a^{(1)}$, $a^{(2)}$ two given particular values, and *y* a single-valued function of them; then it is said of *y* that it *remains continuous from* $a^{(1)}$ *to* $a^{(2)}$, provided that it remains continuous for each particular value that *x* takes between $a^{(1)}$ and $a^{(2)}$.[68]

The definition of continuity in the neighborhood of a value *a* of *x* is formulated exactly the same as by Cauchy: when there is an "assignable quantity *a* , such that *y* remains continuous from $x = a - a$ to $x = + a$."

With the concept of limit values, using the uniqueness of infinite sequences of quantities that Dirksen had previously developed, continuity can now be defined as the continuity of sequences. Such an explicit formulation has not up to now been found among his papers; he did, however, use the continuity of sequences in proofs.[69]

After introducing continuity and discontinuity of functions, Dirksen then formulated the intermediate value theorem:

$$f(\alpha) = A$$

---

[68]   Ibid., Artikel V, § 46.
[69]   E.g., ibid. Lehrsatz 70.

with f continuous between $x = a$ and $x = b$, α lying between a and b, $f(a) \neq f(b)$, and A lying between $f(a)$ and $f(b)$. He was, however, not satisfied with the proof he gave and in the margin commented "not rigorous."[70]

The treatment of continuity and discontinuity in the sections of various versions is relatively brief. In contrast to other sections in which a concept is introduced, where he took care to develop the concept most systematically, exploring all possible types of relations and the particular consequences for functions of however many variables, here he only explored continuity briefly for n original variables, in the three variants, and accordingly added just a few theorems. This brevity is clearly indicative, since he had neither reproduced Cauchy's troublesome theorems nor tried to replace them by his own theorems.

This is particularly the case for Cauchy's theorem that a function of several variables, continuous for each, is continuous overall (Cauchy 1821, 37 ff.), a theorem that was later shown to require uniform convergence. Dirksen, for his part, made no statements about the extent to which the continuity of individual variables may be related to overall continuity. He formulated just one theorem, that the continuity of individual variables follows from the overall continuity of the function, and this for a function of just two variables, with no further generalization:

> Theorem 91. Let [...] $f(x, y)$ be a function of two original variables x and y, continuous in the neighborhood of particular values x_0 and y_0 of x and y: then

$$\underset{x=x_0}{Gr}\ \underset{y=y_0}{Gr}\ f(x,y) = \underset{x=x_0}{Gr}\left\{\underset{y=y_0}{Gr}\ f(x,y)\right\} = \underset{y=y_0}{Gr}\left\{\underset{x=x_0}{Gr}\ f(x,y)\right\}.\text{[71]}$$

This proposition on the separate continuity of x and y is one of the few propositions in Dirksen's extensive manuscripts that uses the commutativity of limits. Since Dirksen had an excellent knowledge of Cauchy's *Cours d'Analyse*, and especially the part on continuity, we may conclude from his silence on the matter that on the one hand, he found Cauchy's theorems problematic, and on the other hand, he saw no way by which he could replace them with correct propositions.

An analogous avoidance strategy is to be seen in Dirksen's text on "infinite function series." Just as in the middle part of his *Organon*, where he set out in complete detail the behavior of infinite series, but did not discuss the concept of the sum of an infinite series (see above, Section 5.1.2.), so in the part on the theory of functions, the nature of the sum of an infinite series of functions is missing. This is particularly surprising, since the sum of an infinite series was of great systematic importance to Cauchy and played a characteristic part in his program for rigor.

As an explanation for this considerable divergence from Cauchy's program it appears that, clearly in the case of infinite series, the weight of the German tradition, especially that of the combinatorial school, was definitely of greater

---

70   Ibid., Lehrsatz 75.
71   Ibid., Bogen 35.

impact than Cauchy's conception. In actual fact, in the middle part of his work on general infinite series (i.e., sequences), Dirksen had considered the sums of such series, but only as part of the technical construction of types of series, sums, differences, products, etc., and even here as part of another technical procedure: the construction of a series of sums as a series of first-order terms, in order to obtain their sum in turn as a series of second-order terms. This process was continued to the nth order series of terms, in order to determine propositions about the relation between the series of terms of different orders (Dirksen 1845, 97 ff.). He only briefly discussed the summability of different orders in the case of finite series (ibid. 13 ff.). However, Dirksen expressly introduced the convergence of infinite series, but generally for operating on such series, and not for investigating the convergence behavior for concrete examples.

It was exactly similar with Dirksen's procedure with the theory of functions. He introduced the concept of a series of functions independently and clearly, again with suitable symbolism, in order to reveal their relevant characteristics.

An infinite series of functions of an original variable (and similarly for more than one variable) was represented by the general term

$$f_m(x),$$

where the index, or number, m gives the position of the term. The value of a term is found from a particular value x_0 of the original variable x.[72]

On the other hand, there is no consideration of the sums of series in the available manuscripts, although Dirksen was naturally familiar with the approximation of functions through the superposition of trigonometric functions. In his treatise of 1827 for the Berlin Academy, he had justly criticized the usability of the mathematical concept of Fourier's *Analytical Theory of Heat* (cf. Dirksen 1830).

However, Dirksen in fact used the expression "sums of infinite series of functions," analogous to the theory of series, but only as part of the technical description of algebraic types of relations of such series. In the short manuscript of the fourth chapter of version D of the theory of functions, which was to deal with infinite series of functions, but contains only the beginning of the chapter, Dirksen had already introduced convergence and continuity (*Erkl.* 9a and 10a), but without drawing out any further connections between these two concepts. Although the title of the concluding part was proposed as "Limits of a function series," the fourth chapter ended with the named definitions. The chapter contained neither theorems nor any reflection on the link between continuity and convergence.

In order to give some explanation for Dirksen's silence on these foundational questions, I see two grounds valid at that time:

* Dirksen was so strongly influenced by the technical concept of series held by the combinatorial school that the idea of a "sum" did not have for him the same meaning as it did for Cauchy. Whereas Cauchy's conceptual view was

---

72   Version D; Viertes Kapitel, Artikel I.

restricted to the idea of finite limits, in Dirksen's understanding, an infinite limit made sense.

- After Dirksen had pointed out, in his 1829 review of Cauchy's *Cours d'Analyse*, that Cauchy's proposition on the continuity of the limit function was not universally valid, and that the cause lay in the non-commutativity of limits, he had clearly accepted the general nonvalidity of the proposition as absolute and afterwards did not look for those mathematical conditions by which such validity could be established.

Thus Dirksen appears to us as one who certainly possessed the necessary "critical" faculty to identify errors and lacunae in the work of others, but as someone who lacked the sufficiently constructive ability to carry out decisive new steps.

To conclude, an overview of the extant parts of Dirksen's work on integration should be given. The outline of the integral calculus is more strongly written, in conceptual terms, than the other sections of the work that exist only in manuscript form. As regards its content, there is a remarkable specific formulation of the integral, compared with the development of this part of mathematics as known to us by historiography. On the positive side, we note that he starts out from the fundamental concept of a definite integral, and he does not use any infinitely small quantities for his definition. Further, the explanation shows Dirksen's awareness of the necessary requirement of continuity of the function in the interval as a condition for the existence of the integral.

On the other hand, as his starting point, Dirksen chose a particular formulation of the definite integral that not only deviated from that of Cauchy, but was also not very fruitful for the development of the concept of the integral, although well received at the time.

He began the "book" on the integral calculus with the definition of the definite integral, without further introduction or explanation, with the following premise:

> *Explanation 1a.* Let x denote an original variable, x_0 and X two particular values, and also let $j(x)$ be a continuous function of x from $x = x_0$ to $x = X$.

Dirksen then set out the series of sums of terms:

$$\frac{X - x_0}{1} \varphi(x_0),$$

$$\frac{X - x_0}{2} \left\{ \varphi(x_0) + \varphi\left(x_0 + \frac{X - x_0}{2} \right) \right\},$$

$$\frac{X - x_0}{3} \left\{ \varphi(x_0) + \varphi\left(x_0 + \frac{X - x_0}{3} \right) + \varphi\left(x_0 + \frac{2(X - x_0)}{3} \right) \right\}, \text{ etc.}$$

and the general term

$$\frac{X - x_0}{n} \sum_{\rho=0}^{\rho=n-1} \varphi\left(x_0 + \frac{\rho(X - x_0)}{n} \right),$$

in order to immediately explain that "the limit of this infinite series"

$$\underset{n=\infty}{Gr} \frac{X - x_0}{n} \sum_{\rho=0}^{\rho=n-1} \varphi\left(x_0 + \frac{\rho(X - x_0)}{n} \right)$$

is "called the definite integral of the differential function $\varphi(x)\, dx$, from $x = x_0$ to $x = X$" and is represented by

$$\int_{x_0}^{X} \varphi(x)dx.[73]$$

Not only was no reason given for introducing this approach, there was also no explanation. Not until what appears to have been the latest version, did he show that for the two trivial cases this definition does indeed give the expected correct result: for $\varphi(x) = 0$ and $\varphi(x) = C$, for all x.[74]

Dirksen's own motivation for this form of definition is found in a much earlier manuscript, in the chapter on finite differences, which was the starting point for the differential calculus. There we find a chapter "On finite integrals of a function." Similarly to the differential calculus, Dirksen gave the finite integral as the sum of the values of the function taken at equidistant intervals,

$$f(x_0) + f(x_0 + h) + f(x_0 + 2h) + f(x_0 + 3h) + \ldots + f(x_0 + (n+1)h))$$

being the definite integral of the first order of the function $f(x)$, from $x = x_0 - h$ to $x = x_0 + nh$, with $\Delta x = h$ and represented by

$$\sum_{x_0 - h}^{x_0 + nh} f(x).[75]$$

Exactly as with the calculus of differences, Dirksen constructed higher-order finite integrals from the corresponding sums. He showed that his definite integral was a generalization of the finite integrals of the first order, for any number of terms with arbitrarily smaller, but still equidistant, steps. Since the step $\frac{X - x_0}{n}$ appears as a factor in all the values of the function, it can be put before the summation. The generalization to arbitrarily many terms is one of the few cases in which Dirksen used sums with infinitely many terms.

Whereas in one version Dirksen merely stated that the definite integral and the limit were "completely the same," in another version he explained that the convergence of the series

$$\frac{X - x_0}{n} \sum_{\rho=0}^{\rho=n-1} \varphi\left(x_0 + \frac{\rho(X - x_0)}{n} \right)$$

---

[73]   Part II, Integral calculus, first chapter. The first chapter is available in many versions, but the introductory part of all of them is almost identical.
[74]   Buch 6, Kapitel 1.
[75]   Teil II, Kapitel II, Definition 1.

was required; for continuous functions, he established the convergence of the series, by means of the Cauchy condition, through considering the difference $\left| a_{\mu(r+1)} - a_{\mu} \right|$.

Not only did Dirksen use this unusual formulation of the definition of definite integrals in the unpublished material we have available to us. He also used this formulation in his lecture programs, e.g., in a lecture to the mathematics class of the Berlin Academy in February 1833 (Dirksen 1835, 130 ff.). It can be seen from this that he would have used this definition of the definite integral in his lecture course on the integral calculus, which was delivered at the same time as Dirichlet was lecturing. A reception of this formulation of the integral can be observed, especially in the context of the Berlin school of mathematics (see the next Section).[76]

In actual fact, there is a conceptual difference between Dirksen's formulation and those of Cauchy and Dirichlet, as well as in the further development by Riemann. This second approach has the purpose of making clear the *existence* of the definite integral, as a unique limit value of an infinite sum. Since the different δ_i and θ_i of the means of the individual terms are not known a priori, this definition is, however, of no use for calculating the value of the terms. Nevertheless the formulation has the advantage of illuminating the approach to the approximation of the area of the integral. Dirksen's formulation on the other hand, as already shown from its genesis in the calculation of finite sums and differences and so also in interpolation, leads to the *construction* of a method for the direct calculation of values of integrals. The formulation, however, completely lacks insight; it is derived solely as a technical procedure for the method of calculating finite sums. The question of existence, however, was not of prime importance, as is basically implied by the way the integral was derived.

On the other hand, the constructive formulation was of little use in practice: the calculation method converged only slowly, and the value of the function at the end of the interval was found only by going beyond the limit.

Finally, Dirksen is himself unfaithful to his constructive method, for he gave a complicated definition of the definite integral in the final version of his writings, with difference series θ_n and θ_n' of integers, apparently in reaction to critics of his too rigid approach. But then, the series were not given directly, but defined implicitly, so that no direct algorithm could be used:

---

[76]   In the differential calculus, other authors also treated the integral as the inverse of differentiation. Lacroix, for example, in both his major manual and in his short textbook, had sections both on "finding the inverse of differences" and on "integration." For a value $x = a + nh$ the integral was given as the finite sum

$\sum f(x) = u + f(a) + f(a+h) + f(a+2h) + \ldots + f(a+(n-1)h)$, where u is a constant (Lacroix 1837, 615). This is used, however, to show the numerical procedure, not as the basis and definition of definite integrals.

$$\overset{n=\infty}{Gr}\, I_n = \overset{n=\infty}{Gr}\,\frac{X - x_0}{\theta_n}\, \overset{\rho=\theta_n'}{\underset{\rho=0}{S}}\; f\!\left(x_0 + \frac{\rho\!\left(X - x_0\right)}{\theta_n}\right).$$

5.2. Dirichlet's Work on Rigor

The final collapse of Dirksen's intended program to provide a complete presentation of the foundations of analysis can also be related to the fact that in Berlin, the very place where he was working, he was caught up and overtaken by a younger, more productive mathematician who, where Dirksen had tried so painfully to systematically lay down foundations, succeeded in producing extensive results through new ideas and energetic activity: Lejeune-Dirichlet.

Dirksen's treatise of 1827 for the Berlin Academy was devoted to the then current problem of the convergence of Fourier series; he criticized not only faults in the published French treatises—in particular he discovered an error in Poisson's proof—but he also announced a particular formulation for it (Dirksen, 1830, 89). His next treatise, of 1829, was intended to provide the long-sought rigorous solution to the problem, but at practically the same moment Dirichlet published his groundbreaking treatise on the convergence of Fourier series (Dirichlet 1829). Dirksen's method was in some ways exactly analogous to that of Dirichlet; but he restricted himself to requiring continuity of the function and made no additional assumptions about the function, so that he only looked at the substitution of different values and did not achieve a real breakthrough (Dirksen 1832). Whereas Dirichlet, through concrete analysis of the subject being investigated, developed new concepts, such as finding the convergence of an apparently not necessarily convergent series by rearranging terms, Dirksen stuck to comparing different variants of the same type and so could not achieve new ideas.

Unlike Dirksen, who is largely unknown, it is not necessary here to give an analysis of Dirichlet's achievements, which are in the main well known.[77] All that is needed is an overview of his contributions to the main matters that interest us here. As an introduction, the description of Dirichlet by C.G.J. Jacobi in 1846 will serve us well, in which he—a Prussian mathematician—is seen as incorporating the new concern for mathematical rigor, and no longer a representative of the long-standing leading French mathematicians:

> Dirichlet alone, not I, nor Cauchy, nor Gauss, knows what a completely rigorous proof is, and we are learning it from him. When Gauss says he has proved something, it seems very probable to me; when Cauchy says it, it is more likely than not, when Dirichlet says it, it is *proved* (quoted from Biermann 1988, 46).

---

[77]   A comprehensive study of his work in mathematics and mechanics, as well as a biography, in preparation for many years   by Jim Cross (Melbourne), is unfortunately not yet finished.

Dirichlet did not make any explicit remarks on the foundation of the number concept, nor is there anything here in his algebraic number theory, which anyway is known to us only through notes of his lectures taken by students, so that it is not possible to exclude the possibility that Dirichlet had made any general propositions. What is clear, in any case, is that it was not at all a problem to accept zero and negative numbers as being numbers themselves, and that the relation between numbers and quantities also no longer presented any fundamental difficulty, and that everything proceeded from integers as a fundamental concept. These he wrote as "0, ±1, ±2, ±3, ±4, ..." (Dirichlet 1893, 434). He also accepted zero as an "integer" (ibid., 5). At the same time it is clear that Dirksen's work had been well received, so we find Dirichlet occasionally using the expression "completely determined" (*vollständig bestimmt*) (cf. Dirichlet, *Werke*, 136), which specific terminology belongs to Dirksen.[78]

For the term *variable*, Dirichlet, just like Cauchy and Dirksen, made a distinction between "fixed values" at particular positions and general variables, capable of taking different values.

It is frequently claimed that Dirichlet was the first to give the modern completely general definition of a *function*. Questions of priority will not be discussed here, but what is relevant is that Dirichlet's definition did not require appeal to any geometrical context. So in 1837, in connection with the definition of a continuous function over an interval, he stated:

> It is not necessary for *y* in the same interval to be dependent on the same law of *x*, and one does not even need to consider that the dependence is expressible by mathematical operations (Dirichlet, Werke, 133).

With his definition of limit values, Dirichlet followed the leading German understanding of the time, by which a variable is able to become equal to its limit:

> In higher analysis one says that a variable has a defined limit, which it continually approaches, when the difference between this fixed value and the variable diminishes without end until zero (Dirichlet 1904, 11).

Dirichlet defined *continuity* in more than one place: the starting point was to use pointwise continuity. The definition in his lectures on the integral calculus is relatively detailed:

> $y = f(x)$ is called a *continuous* and *unique* or *single-valued function* of *x* when to each value of *x* belongs only *one* value of *y*, and when a gradual change in *x* corresponds to a gradual change in *y*, i.e., when for a fixed *x* the difference
>
> $$f(x + h) - f(x)$$
>
> will converge to zero for continually decreasing *h* (Dirichlet 1904, 3).

For *convergence*, Dirichlet's groundbreaking work of 1829 has already been mentioned, where he proved the convergence of Fourier series, not only for

---

[78]   Dirichlet also used the expression "the so-called transition from the finite to the infinite" (der sogenannte Übergang vom Endlichen zum Unendlichen) (Dirichlet, Werke, 139), which at that time was primarily used by Dirksen.

continuous monotonic functions, but also for stepwise continuous and stepwise monotonic functions (Dirichlet, *Werke*, 117 ff.).[79]

For defining an integral, Dirichlet similarly began with the definite integral as the basic idea. It is at first surprising that in 1837 he introduced the definite integral in the same sense as Dirksen had:

> Given A and B as the endpoints of a and b, and ab the curve corresponding to the function $f(x)$, it is clear that this function also defines the area $AabB$, bordered by the coordinates Aa, Bb, the section AB of the abscissa axis and the curve ab, even if it cannot always be indicated exactly. This area is also called the definite integral of the function $f(x)$ from a to b, or between the limits a and b, and is denoted by
>
> $$\int_a^b f(x)dx \ .$$
>
> [...] If the line $AB = b - a$ is divided up into n equal parts, of average width $= \frac{b-a}{n} = \delta$, and through a and the end points of the divisions 1, 2, 3, ... of the corresponding ordinates, parallels with the abscissa axis are drawn, there will be n rectangles whose sum is
>
> (1) $\qquad \delta f(a) + \delta f(a + \delta) + \delta f(a + 2\delta) + ... + \delta f(a + (n-1)\delta),$
>
> which can easily rigorously be proved, and as immediate intuition shows, that by incessant increase of the number n, it finally becomes the area $AabB$ (ibid. 136).

Dirichlet had also explicitly called this a "definition of the definite integral," namely the "limit value of (1)" (ibid. 137). This was, however, a research paper in which he wished to state and use the concept of the definite integral, but without going further into the matter. In his lecture course Dirichlet needed to make this "easily rigorously to be proved" explicit and also had to show the uniqueness of the limit value, independent of the way the interval was divided up. For this, he made appeal to Cauchy's concept of the integral. For a continuous function $f(x)$ he chose a (nonequidistant) division of the interval between a and b,

$$a, x_1, x_2, ..., x_{n-1}, b,$$

and constructed finite sums of rectangles:

$$S = \sum_{\mu=0}^{n-1} f(x_\mu) \cdot (x_{\mu+1} - x_\mu) \ .$$

Dirichlet then used a geometric justification, that with ever greater increase of the intermediate values x_n, the summation S will more closely approach a determined limit:

---

[79] The article of 1837 is an extended version in German (Dirichlet, Werke, 135 ff.).

$$\lim_{n \to \infty} (f(a) \cdot (x_1 - a) + f(x_1) \cdot (x_2 - x_1) + \ldots + f(x_{n-1}) \cdot (b - x_{n-1})) = \int_a^b f(x)dx$$

(Dirichlet 1904, 11).[80]

In addition to the geometric presentation, Dirichlet gave an analytical proof for the convergence of the summation, independent of the choice of interval division. For this he used a modified version of Cauchy's method, substituting the mean value of the function over the corresponding intervals (ibid., 14 ff.). He further pointed out that "the concept of the definite integral is quite independent of the differential calculus" and that one should not, as is "customary," interpret integration as the inverse of differentiation, and that therefore "both concepts, the integral and the differential quotient, should be completely independently [...] justified" (ibid., 20).

The most noteworthy characteristic of Dirichlet's development of the integral is how powerfully he derived the concept, definition, and proof geometrically, in contrast to Dirksen's purely formal presentation, and how thorough and clear is the analytical method of proof. In this way, Dirichlet, at this high point of the development of pure mathematics, impressively showed that intuitiveness and rigor do not have to be in conflict with each other.

5.3. The Reception of Pure Mathematics in Textbook Practice

To conclude this chapter we ought to say something about the way in which innovation and newly established concepts in pure mathematics passed over into the writing of textbooks. The situation differed according to the tradition and practice of different German states at the time, depending on the extent to which analysis was already a subject of study in universities. Following the end of the Napoleonic period, many German states instituted a restoration that not only changed the political system but also impacted on education and culture and, in particular, signaled an end to the "analytical" period with its high respect for mathematics and the natural sciences.

The two northern states of the kingdom of Hanover, whose university in Göttingen offered a wide spectrum of specialized mathematical courses, and the state of Prussia, were places where mathematics and science were relatively better maintained and fostered. For Prussia, however, it is hardly feasible to follow the dissemination of new ideas through a study of textbooks, because of the emergence of a new scientific culture. The rise of the ideas of neohumanism had

---

[80]   As the publisher of the lectures, Gustav Arendt, pointed out in his foreword, he himself had put the variable underneath the limit symbol (Dirichlet 1904, VII). Thus Dirichlet had not adopted this new symbolism from Dirksen.

the effect on leading scholars that the publishing of textbooks became rare: whatever was new became the subject of their teaching, and this had no place in a textbook. Only through student notes of the lectures did some new developments come to be known. This disparity in practice points to the most important difference between mathematics activity in France and Germany at that time. Neither Dirichlet nor Jacobi nor Kummer published any textbooks.

An early example of theoretical reflection was that of Ernst Gottfried Fischer, member of the Berlin Academy and, from 1810, also professor at the university. We have already met him as author of a textbook on elementary mathematics (E 3.4). In 1808 he published reflections on the concept of the infinite in his *Investigations into the Real Meaning of Higher Analysis* (*Untersuchung über den eigentlichen Sinn der höheren Analysis*). Here he showed himself critically "against every attempt to banish the infinite from mathematics" (Fischer 1808, 122). His analysis was admittedly predominantly logico–philosophical in style, especially concerning the relation of extension and intension of concepts, and not so much a development of mathematical concepts. All the same, he was basically in agreement with the limit method (ibid. 180, 163).

The excitement in Germany for the theory of combinatorics was linked to a wide adoption of Lagrangian principles, insofar as this allowed a pure algebraic development of the differential and integral calculus without any "metaphysics." This purely algebraic approach was very strongly received in Germany, so that the mathematics of the combinatorial school was made manifest in analysis, but just as a universal language and without concrete examples of investigations into particular functions and the development of the concept of continuity. Between 1800 and 1825 there was a considerable number of textbooks along these lines.

A productive link between the German combinatorial school and the newer mathematics was achieved by the Viennese academician and professor of mathematics Andreas von Ettingshausen (1796–1878). In a textbook he presented "combinatorial analysis as a preparatory study for theoretical higher mathematics" (1826). In a subsequent two-volume work *Lectures in Higher Mathematics* (*Vorlesungen über die höhere Mathematik*) (1827) he adopted a series of elements from the newer French mathematics. He interpreted "infinitely small proceeding quantities" as variables with a limit zero (Ettingshausen, Bd 1, 1827, 6) and he consistently used the symbol "lim." He discussed the summability and convergence of infinite series and used the Cauchy test for convergence (ibid., 16 ff.).

Continuity did not figure as a topic with Ettingshausen. For integration, he began with the indefinite integral and dealt with the definite integral later (ibid., 336 ff.). It is instructive to note that the definite integral was initially defined in the same way as with Dirksen:

$$\int \varphi(x)dx = \lim.\left[\varphi(x_0) + \varphi(x_0 + \tfrac{1}{n}(x - x_0)) + \varphi(x_0 + \tfrac{2}{n}(x - x_0)) + \right.$$

$$\left. \ldots + \varphi(x_0 + \tfrac{n-1}{n}(x - x_0))\right]\tfrac{x - x_0}{n}$$

(ibid., 337). Subsequently he gave an expression using nonequidistant divisions of the interval for the "general" case:

$$\lim.[\varphi(x_0)\Delta x_0 + \varphi(x_1)\Delta x_1 + \varphi(x_2)\Delta x_2 +...+\varphi(x_{n-1})\Delta x_{n-1}]$$ (ibid., 338).

Another interesting example of combining combinatorial mathematics with the adoption of Cauchy's foundational work was offered by the *Textbook of Algebraic Analysis* (*Lehrbuch der algebraischen Analysis*) (1860) written by the Göttingen professor of mathematics Moritz A. Stern (1807–1894). Stern saw his work as a critical improvement on Cauchy's *Analyse algébrique*. According to Stern, this had not achieved its intended thoroughness because "in its most important part it relied on a theorem whose non validity had been known for a long time," namely the proposition about the continuous limit function. Stern's criticism was that in the newer German works on analysis "instead of investigating whether the above-mentioned Cauchy proposition might not hold under certain restrictions, they rather completely removed it" yet they had "retained the rest of Cauchy's theorems unaltered" (Stern 1860, iv).

The fact that Cauchy's theorem was needed to prove the general binomial theorem, which had become the central plank of the combinatorial school, pointed to the direct importance of combining algebraic analysis with the theory of combinatorics. Stern's approach was that of Dirksen's *Organon*, to develop algebraic analysis without the use of the concept of continuity, and not use continuity until the beginning of the differential calculus. The main subject for him was the theory of series, with detailed treatment of convergence, as well as the general development of binomials and polynomials.

The above-mentioned examples are drawn from Prussia and Austria, as well as an example from the kingdom of Hanover belonging to a somewhat later period. An earlier long-standing textbook used in Göttingen came from the professor of mathematics there, Johann Tobias Mayer, Jr. (1752–1830), *The Comprehensive Theory of Higher Analysis* (*Vollständiger Lehrbegriff der höheren Analysis*, 1818). This two-volume work is firmly in the German tradition of the eighteenth century, especially that of Kästner and Klügel, but had, however, adopted some of the ideas of French authors, principally Lacroix, but also Lagrange and Arbogast. Mayer spoke against the avoidance of the word "infinite" but used it in a realist way. Thus he explained that the proportion

$$\infty \cdot 1 : \infty \cdot 2 = 1 : 2$$

is "completely rigorously true" (Mayer, Bd. 1, 1818, 34). That is, he did not explicitly explore the ratios of functions in the limiting process, but used the limit method implicitly. In this sense he spoke of "limit ratios" (ibid., 59). The subject of the differential calculus was here the study of "approximate or limit ratios" (ibid., 67). He also introduced the ideas of infinitely small quantities and their various orders, but did not use them as part of his foundations of differential and integral calculus. He only mentioned continuity in a declaratory–metaphysical way as "the law of continuity" (ibid., 60) and did not include it in a concrete operative way in order to explore the behavior of functions.

Without doubt the most effective integration of pure mathematics in a German analysis textbook was achieved by Oskar X. Schlömilch (1823–1901). Schlömilch had first studied in Jena under Fries, then in 1840–1841 in Berlin with Dirichlet and afterwards he studied physics in Vienna. Even with his first textbook, *Handbook of Algebraic Analysis* (*Handbuch der algebraischen Analysis*, 1845), he propagated the concepts of pure mathematics, primarily drawing on Cauchy, with the concepts of limits and continuity as basic ideas. Infinitely small quantities did not appear in his work; instead, the concept of a series with zero limit was used:

$$\lim(\tfrac{1}{n}) = 0 \text{ (Schlömilch 1845, 26)}.$$

Also, the zero was not ascribed an exceptional status in any way, and he defined continuity in a pointwise manner (ibid., 41), which he credited to Jacobi and Dirichlet. Fundamental concepts were for him the continuity of functions and the analysis of their behavior through determining limit values. He had thus, in a comprehensive way, established a *limit* calculus.

Again, in the *Handbook of Differential and Integral calculus* (*Handbuch der Differenzial- und Integralrechnung*) of 1847/1848, Schlömilch introduced the integral as a definite integral. It is to be noted that, again, we find that his definition follows that of Dirksen:

$$Lim. \; \delta\left\{f(a) + f(a + \delta) + f(a + 2\delta) + \ldots + f(a + \overline{n-1}\,\delta)\right\},$$

with $\delta = \frac{b-a}{n}$, and the limit being evaluated for unrestricted increase of n (Schlömilch 1848, 2). Since Schlömilch had attended Dirichlet's lectures on the introduction to the definite integral in Berlin in 1840, we may conclude that Dirichlet, who lectured on the definite integral from 1835 (Biermann 1959, 35 ff.), had himself in his early years leaned on Dirksen.

From the selection of the German textbooks given here, we can tell that the grounding of infinitesimal calculus was based on either a purely algebraic approach or the limit method. In contrast to the case of negative numbers, where an, admittedly restricted, adoption of the French concept ensued, no adoption of the French idea of *infinitely small quantities* took place, with the exception of a limited adoption of its classical conception. This holds for other textbooks of the first half of the nineteenth century that I have examined, the details of which would take us beyond the scope of the present work.

Since the polytechnic schools of Germany in the first half of the nineteenth century were not yet institutions of higher education, there was no separate and distinct teaching of the ideas and practice of mathematics, in the sense of "pure" mathematics. There were, however, a number of traditional technical schools, with a long established practice of mathematical instruction, namely the *Bergschulen*, for example the Bergakademie Freyberg in the Duchy of Saxony. Walter Purkert pointed out that in a textbook for engineers used by this school the author had worked "with infinitely small and infinitely large numbers" and saw in it a preliminary stage of non-standard analysis (Purkert 1990, 181). That text was by an eminent engineer, Julius Weisbach (1806–1871), who for his foundation course for engineers and mechanics used his own textbook *The First Foundations of Higher Analysis* (*Die ersten Grundlehren der höheren Analysis*) (1849).

The textbook only ran to 43 pages and offered itself as "mathematics for engineers," a first form of a textbook type in Germany that later became standard; its style was instrumental, with no attempt to present or explore conceptual ideas, simply the claim to be a "study aid for analysis." And in fact, Weisbach did not use infinitely small and infinitely large "numbers," but rather "infinitely small quantities" (Weisbach 1860, 5). He gave no definition of these, and the text in general is strongly characterized by an appeal to intuition. The existence of such a text for technical training, avoiding theoretical reflection and conceptual rigor in favor of appeal to intuition, is naturally hardly surprising. On the other hand, it is interesting that Weisbach at the *Bergakademie* was a successor to Friedrich Gottlieb Busse (see above, Section 3.1.), who published a late work on analysis, the title of which, *Concise and Pure Exposition of the True Infinitesimal Calculus* (*Bündige und reine Darstellung des wahrhaften Infinitesimal-Calculs*) (1825–27), announced that it concerned a hermetic "private" conception. It consisted of an idiosyncratic combining of the calculus with infinitely small and infinitely large quantities of different orders and technical manipulations with limiting procedures. Of course, such an obscure work was unsuited for technical students, but Weisbach, who had studied under Busse, had been won over to the possibility of shortening the limit procedures through the use of infinitesimals.

Chapter VIII

Conflicts Between Confinement to Geometry and Algebraization in France

1. Keeping Up the Confinement of Negative Quantities to Geometry

In the further course of the first half of the nineteenth century, there was no relevant change in the development of the concept of negative numbers in France. The rejection of such theoretical terms of pure mathematics initiated by Carnot, and disseminated by Lacroix in the latter's textbooks, continued to remain formative. An algebraization of this concept field did not take place.

A typical testimony for that is the textbook *Elémens d'Algèbre,* first published by Bourdon in 1817, and then reprinted many times; after 1820 becoming a rival for Lacroix's algebra. It saw sixteen editions in seventy years. Louis-Pierre-Marie Bourdon (1779–1854), one of the first students of the *École polytechnique,* became first a mathematics teacher at a military school, and taught later at several renowned Paris *lycées*. His successful career began in the period of Restoration, as textbook author, as *Inspecteur général*, and as examiner for the *École polytechnique.* His preface already indicates the kind of stumbling block negative quantities still continued to be, and how much space had to be devoted to reflecting on them. Here already, isolated negative quantities were disqualified as "indicators of impossible subtractions," as offering nothing intelligible to the mind (Bourdon 1817, V f.).

With regard to his own conceptions, Bourdon kept to the field of absolute numbers. He introduced his *soustraction algébrique* by recalling arithmetic subtraction, which had the objective of determining the *excès* of a larger number over a smaller number. "Likewise" (*de même*), algebraic subtraction, he said, had the objective of finding the *expression algébrique de l'excès* of one quantity over another quantity. He did not specify this *de même*, however. Bourdon presented his own *théorie des quantités negatives* in connection with the solving of linear equations. He showed no restraint in disqualifying a problem yielding a negative solution as *absurdité* (ibid., 70).

Bourdon did not intend to do completely without these solutions. His approach to integrating them consisted in Bézout's method, or in going back to the original equation, to then change the signs of the unknown, in order to thus obtain a new equation in *langage ordinaire* (ibid., 74). For the legitimacy of these transformations, Bourdon specifically appealed to Carnot (ibid., 73). Negative terms appearing in operations were assigned neither the character of numbers nor that of quantities. Considered to be situated beyond the area of legitimate mathematical concepts, they were deprecatingly called *expressions negatives* (ibid., 75).

Nevertheless, two new elements can be recognized in Bourdon's presentation. Accepting Carnot's doubts published in 1813 regarding whether the rule of signs could be proved (see Chapter V.1.6.3.), he gave up entirely on proving it. He criticized it as a weak point of presentation in many textbooks that the proof of the rule of signs for "polynomials"—which required only subtractive quantities, but not negative numbers—had been transferred to monomials, and proclaimed, without stooping to details, that all these proofs by transference were defective (ibid., 74). Bourdon's conclusion, however, is surprising. If one did not accept extending the rule of signs to polynomials, he said, one would deprive oneself of essential advantages provided by the *langue algébrique*, namely of expressing the solutions of several questions of the same kind by a single formula (ibid., 74). In his foreword, he got even more specific: if one refused to extend the rule of signs to polynomials, negative solutions had to remain without explication. This, however, would be a greater evil than admitting a simple rule that so naturally flowed from other rules already most rigorously proven (ibid., vi), an argumentation trying to find a balance between lesser and greater evils in applying mathematics.

The second new element, however, stands in clashing contrast not only to Carnot, but also to a majority view that had been most common since d'Alembert: Bourdon dared to declare negative expressions to be smaller than zero! The necessity of using negative expressions in algebraic calculus, and of operating with them as with absolute quantities, had led the algebraists, he said, toward two propositions that were going to be frequently applied in what was to follow: "Every negative quantity $-a$ is less than 0; and of two negative quantities, the least is the one whose numerical or absolute value is the larger" (ibid, 75 f.).

Hence, $-a < 0$, and $-a < -b$, if a was absolutely larger than b. While Bourdon nevertheless continued to insist, in agreement with tradition, that in case of negative solutions one had to *rectifier l'énoncé du problème* (ibid., 90), he gave no new advice with regard to conceptions. In an obvious effort to remove negative solutions from the limelight as unique representatives of *absurdité*, he was the first to call attention to the fact that *valeurs positives* could themselves be *valeurs absurdes*": in cases in which the problem called for integral solutions, whereas the values obtained were fractions (ibid., 89).

In his introduction to equations of second degree, Bourdon admitted for "monomials"—that is, not generally for numbers—that square roots might have positive and negative values. Thus

$$\sqrt{9a^4} = \pm 3a^2 \text{ (ibid., 160)}.$$

Besides, he was able to give, as general solution of the quadratic equation,

$$x = -\frac{p}{2} \pm \sqrt{\frac{p2}{4} - q} \text{ (ibid., 171)}.$$

This did not prevent Bourdon, however, from conducting an extensive discussion for every concrete equation of second degree as to whether negative values could be prevented by transforming x into $-x$ in the original equation (ibid., 176 ff.). In his *discussion générale*, he went so far as to proclaim such clarifications to be generally always necessary (ibid., 183).

Just as continuing the traditional is permeated with some novel elements in this author, it is also remarkable that he did not only include a separate section on inequalities, but also profited from this occasion to explain—even before Cauchy—that inequalities are inverted upon multiplying them by a negative quantity (ibid., 199).

Lacroix already had been unable to stick to his own foundational position. In the higher parts of algebra, he had no problem with operating on negative numbers. Bourdon also did not hold consistently to his own epistemological position. He later spoke of *nombres négatifs* (ibid., 232), using, in subsequent parts of algebra, *nombres reels* as a concept encompassing negative numbers.

The epistemological rejection of negative numbers as a dominant trait in the French mathematical culture, in particular in that of the first half of the nineteenth century, is confirmed by many instances. Typical evidence is a booklet of 1828 that belongs among the first forms of elaboration of vector theory: *La vraie théorie des quantités négatives et des quantités prétendues imaginaires*, by C. V. Mourey, an author about whom no biographical data at all are known, and no further mathematical publications either (cf. Schubring 1997, 12).

Mourey's publication confirms that the epistemological orientation was to reject any generalizability of algebra, and in claiming geometry to be the authoritative instance of mathematics that guaranteed sense and meaning. Mourey admitted subtraction only in arithmetic, provided that the subtrahend was smaller than the minuend. Consequently, expressions like $4 - 6$ were *absurde*. Mourey was so radical in this that he excluded subtraction from the operations of algebra altogether. Since both unknown and undetermined terms occurred in algebra, he argued, one could not know beforehand whether in the expression $x - a$, for instance, x might be smaller than a: "It follows from this that the sign $-$, considered as expressing subtraction, cannot be admitted in algebra. Algebra [...] does not admit subtraction" (Mourey 1861, 1 f.).

The point thus was to find a means to supplant (*suppléer*) subtraction in algebra (ibid., 2). The conception Mourey developed for this purpose was that of calculating with directed line segments, which he called *chemins orientés*, hence a tentative form of vector calculus. Mourey was a bit equivocal in calling this *science* that he had proposed both an *algèbre directive* and a *géométrie directive* (ibid., 39).

In his concepts and in his epistemological orientation, Mourey had been clearly influenced by Carnot. Explicitly appealing to Carnot, however, was the author of an influential textbook on pure mathematics: Louis Benjamin Francœur (1773–1849), professor at the Paris *Faculté des Sciences*, and at the *École normale*. His multiply reprinted textbook was addressed to the students of the *École polytechnique* and the *École normale*, as well as to candidates preparing for the *concours* of admission. Here again, we find—at first in the discussion of equations of first degree—a host of propositions of the kind that say that subtractions like $d - b$ for $b > d$ were *impossibles* and *absurdes*, and that algebraic operations were admissible only if their *exécution est possible* (Francœur 1819, 149 f.). The new thing with Francœur is that he extended the argumentation to rational terms:

$$x = \frac{d-b}{a-c},$$

showing in detailed case distinctions for which kind of larger and smaller relations among the four coefficients the solution is permissible, or absurd, or re-interpretable as permissible by switching signs in the original equation. In agreement with Carnot, he declared that isolated negative quantities were only *êtres de convention*, nothing but symbols *qui n'ont aucune existence par eux-mêmes* (ibid., 150 f.). To make things abundantly clear, he proclaimed that one was entitled to switch all the signs of an equation: *On a le droit de changer tous les signes d'une equation*, thus distinguishing rectifiable solutions from those definitively absurd (ibid., 151).

At the same time, Francœur's book supplies an example for the attitude criticized by Bourdon for its absolute command to reject negative solutions as unmathematical, while at the same time not incriminating fractions, for instance, where integral solutions were sought. Francœur had posed the question after how many years a son would have reached a fourth of his father's age, their present ages being 12 and 40 years. For the solution $x = -\frac{8}{3}$, he declared the problem to be *absurde*. Upon replacing x by $-x$, he was satisfied with obtaining the fractional result $x = 2\frac{2}{3}$, although he had asked for years of age (ibid.).

In the part of his textbooks on analytic geometry, Francœur also explicitly reflected on signs used in applying algebra to geometry. For doing so, he adopted Carnot's conception of correlating direct and indirect figures. He showed in detail that one had to replace the direct figure by the correlating indirect figure in case of a negative solution (ibid., 337 ff.). Nevertheless, he had no difficulty in admitting coordinate systems containing four quadrants in which the x- and y-coordinates might also assume negative values (ibid., 346).

A former student of the *École polytechnique*, and later well-known historian of mathematics, Maximilien Marie (1819–1891), confirmed with his treatise of 1843 on the nature of negative quantities that the epistemological rejection of negative quantities continued to remain a determinant for French theory formation. While he distinguished between numbers and quantities, admitting *nombres négatifs* for exponents, and for negative logarithms (Marie 1843, 11), *grandeur* remained his basic notion, and negative *grandeurs* were excluded (ibid., 6). As far as they appeared as solutions of equations, he said, they required reinterpretation, in the tradition already familiar (ibid., 16). Marie remained just as conservative in the application to geometry. All his figures were situated purely in the positive area, within the first quadrant.

The epistemological rejection of negative numbers did not remain confined to a more or less intimate mathematical *community*, rather, it was continuously disseminated as the basic attitude within the French culture, and thus was able to maintain its decisive influence for a long time to come.

A testimony of its broad cultural anchoring is supplied by the book of a non-expert dabbling in mathematics instruction. F. C. Busset (deceased in 1847), a topographer and civil service surveyor, published a book on mathematics instruction in secondary education in 1843, the first of its kind after Lacroix's reflections of 1805 on the method of how to teach mathematics. In his book, Busset expressed his profound concern with the contemporary quality of mathematics instruction. He saw the principal danger for the quality of instruction and the level of education in the effect of false scientific concepts and theories. His most prominent example for such false theories was negative numbers, and Euler, for Busset, was the principal culprit.

Busset declared it to be *le comble de l'aberration de la raison humaine*—the upshot of aberration of human thought—to inquire whether *quantités plus petites que rien* existed at all. Actually, however, there were indeed scholars and teachers, he complained, who even answered this question in the affirmative; the foremost being Euler, in his textbook on algebra. In spite of his own respect for Euler's genius, he was compelled to declare that such mental aberrations could prevent gifted minds from studying mathematics (Busset 1843, 47).

Beyond such tracts, there is a famous literary testimony as to how the rule of signs for negative quantities indeed plunged a young Frenchman enthused with mathematics into utter confusion of his soul, the autobiography of Henri Beyle (1783–1842), a writer who later became famous under the pen name Stendhal. Stendhal, born and raised in Grenoble, was desperately intent on fleeing the circle of his family, which he felt to be oppressive. His aspiration was to depart for Paris in order to take up studies at the recently founded *École polytechnique*.

To be able to do so, he first had to engage in acquiring solid mathematical knowledge in Grenoble itself. He was very desirous of this subject not only because of his goal of life in Paris, but also because of the character of this science as a stronghold of truth and certitude, and thus as a protection against the hypocrisy of which he reproached and accused his own family.

He was overwhelmed by a deep crisis, however, when nobody could tell him satisfactorily why minus times minus gave plus: *une des bases fondamentales de la science qu'on appelle algèbre* (Stendhal 1961, 311). His two mathematics teachers—his official one at the *École centrale*, and his private instructor— referred him to mathematical authorities like Euler and Lagrange, or simply smiled down condescendingly at the fourteen-year-old. They were in no position to explain (ibid., 311 ff.). A drawing by Stendhal himself—a rectangle on one side of a straight line, and two rectangles on its other side – shows that he considered the positive and the negative to be two basically separate domains (ibid., 313).

Stendhal was close to ascribing to mathematics, too, the same lack of principle as to the Jesuits, whom he held in contempt, until he decided to override his doubts in favor of the opportunity of escaping from Grenoble by the subterfuge of mathematics: *vraies ou fausses les mathématiques me sortiront de Grenoble* (ibid., 317).

A further instructive testimony for the fact that the problem of negative numbers was being generally debated in the French culture is supplied by a publication of 1820 containing an nonexpert's claim to having discovered the true foundations of algebra, and in particular of negative quantities: *Doctrine de l'Algèbre, basée sur ses vrais principes encore inconnus.* Its author Bonnefin had been treasurer for French navy invalids in St. Malô, and had devoted four years after his retirement to elaborating the ultimately true theory of algebra, prompted by his own children's complaints upon learning algebra that they were being compelled to believe word for word things inexplicable, and realizing that their teacher with whom he was well acquainted was not to be blamed for this, he concluded that the cause for such inexplicabilities must lie in the science as such.

Bonnefin's remarkable approach lies in that he intended to clarify and ensure the foundations of algebra by means of its applications, more precisely with its application in his own former field of work, in bookkeeping. In particular the concept of negative quantities, which he conceived of as opposite quantities, became clear, he argued, as soon as one grasped them on the basis of the notion of double bookkeeping. To the affirmative and to the negative of algebra, assets and debts corresponded in everyday commerce (Bonnefin 1820, 2, 33).

Certainly, no (mathematical) culture is perfectly uniform, and thus there was indeed (at least) one exception in France to the generally dominant view. It is expressed in the algebra textbook by Louis Lefébure de Fourcy (1785 until ca.

1864), the first edition of which appeared in 1833, and which saw seven printings until 1862. Lefébure had been student at the *École polytechnique* after 1803, and taken over multiple functions of *répétiteur* there. Later, he became mathematics teacher at a Paris *collège*, admission examiner for the *École polytechnique*, and finally Lacroix's successor at the *Faculté des Sciences*. His major works were concerned with descriptive geometry.

While Lefébure did not reflect on the relationship between quantities and numbers in his textbook, introducing negative quantities without establishing any prior theoretical structure of algebra, he was the first in France to recognize that operations in algebra had to be extended as compared to arithmetic, and that this had to be effected by means of appropriate *conventions*.

He introduced negative quantities by admitting the unlimited execution of subtraction. He thus explicitly admitted isolated negative quantities—excluded elsewhere as *impossibles* or *absurdes*—but defined them in the traditionally equivocal mode as quantities *précédées du signe* – (Lefébure de Fourcy 1838, 9 f.). He had no problem with stating, *on convient de les regarder comme plus petites que zero* and with operating *d'après ces conventions*, with inequalities like $-2 < 0$, and $-5 < -2$ (ibid., 10). Lefébure's main achievement consists in his having worked out that algebraic operations involving negative quantities require a definition of their own, precisely by *conventions nouvelles* as compared to arithmetic operations, either by extending the traditional meaning, or by new conventions (ibid., 11).

Among the four operations, his discussion of multiplication is of course the most fascinating. The innovation here was that Lefébure—in particular for the case of minus by minus—did not aim at any "proof," but rather at a new definition of multiplication that he designated as a convention to complement the definition of old:

> One is led naturally to add to this as a complement of the definition, this new convention that, in the case where the multiplier is negative, the multiplication must be carried out as if it were positive, and the sign of the product should be changed afterwards (ibid., 15).

Upon presenting the equations of first degree, Lefébure discussed the status of negative solutions in detail. Here again, he was able to realize innovations for the French community. He made clear that negative solutions of equations were not from the very outset *impossibles*, but only *si l'énoncé d'un problème exige que l'inconnue soit positive* (ibid., 88). Hence, negative solutions could very well exist without the formulation of the question being *impossible* (ibid., 89). Lefébure also was the first to explicitly and generally challenge the culturally compulsory abhorrence of negative solutions. There might be a host of other limiting conditions in a problem which could make a solution impossible, as in case of fractions, if integral solutions were called for, or if values larger than permissible were obtained (ibid., 98).

Obviously, however, Lefébure remained isolated among his contemporaries in France with his "normalizing" innovation. This can be seen in any case from the 23rd printing of Lacroix's algebra, which was published in 1871 by Eugène Prouhet, a mathematics teacher and *répétiteur* at the *École polytechnique*. Prouhet had left Lacroix's text unchanged, and had praised the textbook's high quality in his preface to the new edition:

> Few elementary books have been conceived with so much maturity, produced with so much care, and revised and improved with so much perseverance: this is what explains its success (Prouhet, in: Lacroix 1871, iii).

In a voluminous "Note I" of eleven pages in his appendix, Prouhet cautiously undertook to show—while at the same time expressing his veneration for the late master and his doctrines—that it was nevertheless possible to sensibly operate with isolated negative quantities under appropriate conventions. Before setting out on this impromptu crash course, Prouhet again pledged his full adherence to Carnot's and Lacroix's epistemology. Negative quantities, he avowed, were *véritables symboles d'impossibilité*. Because of this *impossibilité*, they could be only of very limited usefulness in algebra, he regretted (ibid., 357). After having covered himself by this lip service to hallowed principles, Prouhet, however, went on to ascribe yet some use to negative quantities, provided this use was based on a convention stated beforehand. Prouhet also established here that the "meaning" of the originally arithmetic operations had to be "extended" for algebraic applications (ibid., 361). Immediately after listing the advantages of using positive and negative quantities, in particular for assembling all the algebraic expressions of the same type, Prouhet closed with the pious reminder that negative quantities were in principle a *symbole d'impossibilité*.

Prouhet's appendix thus had the function of giving a free hand, on the one side, to operating with negative quantities, and on the other side of keeping alive the bad conscience concerning their epistemological inadmissibility, thus complicating the transition to a new level of generality in algebra, and making it more difficult to achieve.

2. Last Culmination Points of the Infiniment Petits

2.1. Poisson's Universalization of the Infiniment Petits

As shown in Chapters IV and V, the *École polytechnique*—originally founded to realize the *méthode d'analyse*, and as the elite institution of higher education in France the model of algebraization in mathematics—had been forced in 1810 to return to the *méthode de la synthèse* practiced by the military schools of the *Ancien Régime*, and to give priority to the geometrical and intuitive methods. For analysis, this had meant to change from the *méthode des limites* to the method of the *infiniment petits*. The original, purely formal concession of demonstrating, in simple cases of application, that the new basic method agreed, say, with the method of *limites*, had been elaborated, in particular by Cauchy, into the dualism of a method of compromise in which the *infiniment petits* virtually became a case in which the limit method was to be applied.

This section shall study how the *infiniment petits* fared in the ongoing French development of foundations. Practically no such studies have as yet been undertaken in the historiography of mathematics. Instead, there is an overwhelming consensus that after Cauchy and Weierstraß, the ε-δ approach generally prevailed in the nineteenth century.[1]

Typical for that view is, for instance, how M. Kline assumes a direct transition from Cauchy to Weierstraß in developing analysis:

> To remove the vagueness in the phrase "becomes and remains less than any given quantity," which Bolzano and Cauchy used in their definition of continuity and limit of a function, Weierstrass gave the now accepted definition that $f(x)$ is continuous at $x=x_0$ if given any positive number ε, there exists a δ such that for all x in the interval $|x - x_0| < \delta$, $|f(x) - f(x_0)| < \varepsilon$ (Kline 1972, 952).

In view of the strong cultural and social pressure brought to bear on the *École polytechnique* to induce it to change its fundamental conceptions, and in view of not only the continued existence of this institutional context but also the increasing confirmation of its character as an engineering college, it was absolutely improbable from the very outset that the *infiniment petits* should suddenly disappear from French mathematics. It may indeed be said that they did not really start to "bloom" until 1820. Remarkably, this was combined with propagating different and competing conceptualizations of the *infiniment petits*, which may permit us to obtain further insights into the reception of Cauchy's conception of the fundamental concepts of analysis.

It has gone quite unobserved in the historiography of mathematics that there was a propagation of the *infiniment petits* quite deviant from Cauchy's conceptualization in France, an attempt to popularize them as *actually* infinitely small quantities. This version would hence have had a claim on anticipating non-standard analysis if its propagators had been able to know that the latter would some day exist. The conceptions were those of Siméon-Denis Poisson (1781–1840), who was able to promote them to far-reaching dissemination and impact because of his central position within the French educational system.

---

[1]   This consensus not only assigns Cauchy to one of the two sides without further ado, but at the same time makes no allowance for any differentiation in the views on the foundations of analysis in the nations concerned, (Western) Europe being assumed as a region mathematically uniform.

Poisson, both mathematician and physicist, represented almost like no other the postrevolutionary practice of *cumul,* of simultaneously occupying posts attached to different institutions. As he additionally occupied a post at the switchboard of educational policy for more than two decades, he was able to execute most efficient *patronage.* Poisson began his studies at the *École polytechnique* in 1798, and became immediately after his graduation in 1800 *adjoint* of a *répétiteur* in analysis. After 1802, he stood in for Fourier in the latter's course of analysis, and became a regular lecturer at this school already in 1806. Until 1815, he taught mechanics, for the most time alternating with de Prony. In 1816, he changed to the position of an *examinateur permanent.* From 1812, he was at the same time a member of the *Institut* in its physics class, and his third simultaneous post was from 1809 that of professor at the new *Faculté des Sciences* in Paris, for the subject of mechanics. The real basis of his power in educational politics, however, was that he remained a member of the *Conseil royal de l'instruction publique* for decades, from 1820 to 1840, across several changes of *régime* (Bru 1981, 81 ff.).

Originally, the focus of Poisson's research was more on analysis. Almost as if competing with Cauchy, he made numerous contributions during the first half of the 1810s, in particular on integral theory. Gradually, however, his attention shifted to probability and to mechanics. He turned into a propagator of the *infiniment petits* in his lectures on mechanics at the *Faculté*, and in his textbook on mechanics. In the first edition of his mechanics textbook of 1811, Poisson still had been able to do without such declarations of principle, but in the second edition of 1833, he inserted a part on the foundations of analysis in which he explicitly rejected the program of "dualism," declaring himself in favor of an "exclusive use" of the *infiniment petits* as foundational method. This claim at exclusiveness, together with his attempt to use his central position of influence to make his own view prevail as the generally accepted one, for the first time provoked a public debate and critique of this conception of *infiniment petits.*

In his first edition of 1811, Poisson still had discussed the relation between analysis and synthesis that was being heatedly debated at the time. He had made a point of not opting for a definite position, but declared instead that he himself was using the analytic or the synthetic method according to which was most appropriate to the object concerned (Poisson 1811, vij ff.). In an introductory part on *Notions Préliminaires*, he introduced basic concepts of mechanics such as solids, mass, force, and direction.

In his second, considerably restructured edition of 1833, however, he used a comprehensive *Introduction* not only to define basic concepts of mechanics, but also to present his own view of the fundamental concepts of analysis. The program was contained in the heading of the section introducing this part: "n° 12. In this work, we shall exclusively employ the method of infinitely small quantities; the fundamental principles of infinitesimal analysis" (Poisson 1833, iij).

In the text itself, Poisson repeated this rejection of the dualism compromise, thus explaining why he gave a brief presentation of his own of the foundations of analysis in this introductory part:

> In this work I shall employ exclusively the method of infinitely small quantities; this is why it is necessary to recall, in this introduction, the principles of infinitesimal analysis (ibid., 14).

As core of his conception, he presented a definition of infinitely small quantities, a definition that made Poisson the first to explicitly introduce the latter as actually infinitely small quantities: "An infinitely small quantity is a magnitude smaller than any given magnitude of the same nature" (ibid.).

To justify the existence of such infinitely small quantities that were definitively smaller than any quantity of the same kind, Poisson referred to the continuous change of quantities, and in particular to the intermediate value property:

> One is led necessarily to the idea of infinitely small quantities when one considers the successive variations of a magnitude subject to the law of continuity. Thus time increases by degress less than any interval than can be assigned to it, no matter how small. Space passes through by different points of a body increase also by infinitely small degrees; for each point cannot move from one position to another without passing through all the intermediate points (ibid.).

With these grounds, Poisson believed himself able to justify the "real existence" of the *infiniment petits*, far beyond a purely heuristic function: "Infinitely small quantities thus have a real existence and are not just a means of investigation dreamed up by geometers" (ibid.).

The remarkable new element in Poisson's conception is that it quickly provoked sharp public criticism, and that the critics, for the purpose of refuting Poisson's doctrine, appealed to the theories of Carnot and Cauchy, which were understood to be in complete agreement: *infiniment petits* as special variables! The most acute and most comprehensive critique was published in 1836 by J.-J. V. Guilloud. I have been unable to find any data on his biography. At least, he is easier to grasp as a real person than, say, Mourey. He published several other writings, the frontispieces of which show him to be a mathematics teacher at various institutions. His first book, of 1827, is a *Traité de physique appliquée aux arts et métiers.* Its German translation appeared already in 1828. In his book of 1849 on the theory of approximation, he is called an *ancien chef d'institution*, that is, a former headmaster of a private school. One of his publications, of 1850 on cosmography, is intended for candidates for *l'école militaire*, suggesting that he probably had become a teacher at a military school. He must have died before 1853, since this was the year an editor named Terrien published a revised version of Guilloud's handbook on physics. Guilloud addressed his critique both against the second edition of Poisson's textbook on mechanics, and against the latter's lectures on mechanics at the Paris *Faculté des Sciences*. With his critique, he "demolished" Poisson's entire mathematical introduction in a way that may serve as a guideline here to present it. On questions of principle,

Guilloud challenged Poisson's basic definitions. All evidence spoke against their admissibility. However small a quantity might be, there existed always yet smaller ones, and there were always additional ones between it and zero. Only the zero, he said, was smaller than any given quantity: *il n'y a vraiment que zéro qui soit au dessous de toute grandeur donnée* (Guilloud 1836, 4).

As *grandeur*, and *quantité*, according to their definition, were capable of being augmented and diminished, there were always, beside a quantity capable of diminution, still smaller ones. Because of this definition of quantity, a quantity that was smaller than any given one must coincide with zero: "il ne peut pas exister de quantité au dessus de zéro plus petite que toute grandeur donnée" (ibid., 5).

Guilloud did not exclude the use of the term of *infiniment petit*, but inquired instead what meaning it should be given. He had found the metaphysics of the *infiniment petits*, as he had elaborated it for himself, clearly demonstrated by Carnot, and the same had been applied by Cauchy in the latter's teaching: "I found it clearly stated by Carnot. It was used in his teaching at the École Polytechnique by M. Cauchy" (ibid., 3).

Guilloud confirms our analysis of the reception of how Cauchy took up Carnot's concepts (Chapter VI.6.4.), as Guilloud explicitly defines *infiniment petits* in Carnot's/Cauchy's sense as special variables: "An infinitely small quantity is a variable which, being already very small, is susceptible of approaching zero as close as one wishes, or which has zero as its limit" (ibid., 5).

As the pertinent element of this definition, Guilloud explicitly emphasized the property of variable, for only by its having the limit of null, an *infiniment petit* could become smaller than any given quantity:[2]

> An infinitely small quantity is essentially a variable, and it is only on account of its variablility that it can become smaller than any given amount; the moment it becomes a constant, one would be able to find even smaller constants (ibid.)

At the same time, Guilloud stressed, in agreement with Carnot, that infinitely small quantities need not be from the outset smaller than any given quantity, but that one could attribute correspondingly smaller values to them (ibid.).

In what followed, Guilloud challenged Poisson's legitimization for creating his concept. Of course, it was correct that the successive variations of a continuous quantity contained the ideas of the infinitely small, but in a meaning quite different from that intended by Poisson: If one said that time grew in infinitely small intervals, this meant that the intervals could be divided into ever smaller parts. If smallest units of time existed, analogously to those of space, this meant that there were ultimate, no longer divisible, particles of time as well. Poisson's *infiniment petits* were thus not entitled to possessing an *existence réelle* (ibid., 6 f.).

---

[2]   For given, that is, constant, quantities, Guilloud used, like Poisson, the expression "grandeur."

Like Guilloud, Lacroix rejected the existence of actually infinitely small quantities. Since the third edition of his *Traité élementaire* on analysis, which appeared in 1820, he inserted a completely reedited observation on the metaphysics of analysis with which he explicitly responded to the new boom of the *infiniment petits*. He kept to this text in the fourth (1828) and fifth editions (1837), stressing that the foundations of analysis became rigorous only with the method of limit, and that only this method made the claims of the infinitely small quantities make sense. An actual existence could not be ascribed to these quantities, all the less because they did not offer any clear purchase to the mind:

> It is always in respect of this consideration [of limits] that one has a need to explain the difficulties which may arise from mentioning infinite or infinitely small quantities, which mathematicians do not need to suppose to have an actual existence (Lacroix 1820, 629).

If one carefully analyzed the foundations of differential calculus, he said, one would thus always hit on the concept of limit. In the last, fourth, edition of his methodological handbook *Essais sur l'Enseignement ...* of 1838, Lacroix criticized Poisson even more incisively and directly: He began by referring once again to Lagrange's rejection of the *infiniment petits*, whose "definition was little satisfying." Quantities, however, that were to be smaller than any given quantity, he said, could only be absolute nullities, as was also evident in one of the most basic proofs of the method of exhaustion:

> In effect, when one says that there are magnitudes less than ALL given magnitudes, one is stating in truth, only absolute zero (0): it is stated just like this in the proof of one of the fundamental propositions of the method of exhaustion (Lacroix 1838, 345).

Guilloud thus also showed in particular that Poisson's avoidance of the concept of limit made his foundation of analysis unfounded. Guilloud proved this first for Poisson's concept of the orders of infinitely small quantities. For Poisson had not only abstained from defining what an *infiniment petit* of first order was. Since because of his abhorrence of the concept of limit he had also omitted to define what a ratio between infinitely small quantities really was, his entire concept of orders of infinitely small quantities lacked foundation. Since Poisson had required only, for two *infiniment petits a and b*, that their ratio be infinitely small as well, in order that b might be an *infiniment petit* of second order (Poisson 1833, 14 f.), Guilloud proved, for the example $b = a\sqrt{a}$, that $a\sqrt{a}$ already was an *infiniment petit* of second order according to Poisson's own definition (Guilloud 1836, 9).

Poisson's principal avoidance of the concept of limit indeed proved to be the major defect of the foundations of analysis he propagated. It is really remarkable that Poisson believed himself able to rely, for his concept of *infiniment petits*, on only the two original principles established by L'Hospital for deriving the entire analysis required in mechanics—against the entire foundational progress meanwhile achieved—i.e., to rely on the principle of extended equality, and on the principle of identifying curves with polygons. Basically, he even conceived of the first of L'Hospital's principles as of a consequence drawn from the first:

> The fundamental principle of infinitesimal analysis consists in that two finite quantities, differing from each other only by an infinitely small amount, can be regarded as equal, since one would not be able to assign any inequality between them, no matter how small (Poisson 1833, 15 f.)

With regard to this, Guilloud observed that this definition was not exact, as long as the concept of limit was not used. As an example for two finite quantities, he gave $x = 1 + a$, $y = 1 + 4a$, a being an *infiniment petit* in this. While the difference $x - y$ then was an *infiniment petit,* but without $x = y$ being valid, only the limit of x was equal to that of y (Guilloud 1836, 14 f.).

Consequently, one had to differentiate the proposition:

- For two constant magnitudes (*grandeurs*), the difference could not be infinitely small. If the difference was smaller than any assignable quantity, then it was zero.
- If, conversely, two variables differed by an *infiniment petit,* their limits were equal, or the limit of their ratio was 1 (ibid., 15).[3]

For Poisson, the possibility of identifying curves with polygons was a conclusion from the postulate of extended equality: in the theory of *infiniment petits*, one considers curves "comme des polygones d'un nombre infini de côtés infiniment petits" (Poisson 1833, 23 f.).

This meant that the chord of an infinitely small arc was identical with the arc (ibid., 24). Guilloud did not enter on this principle in detail, since it again assumed actually infinitely small quantities. Something making sense resulted only as a proposition about the limits of these quantities (Guilloud 1836, 56).

It confirms the key role of the definite integral for this stage of the development of the foundations of analysis that Poisson's concept of the *infiniment petits* without any concept of limit fails precisely in establishing the concept of the definite integral.

At first, Poisson posited differentials also as infinitely small quantities in his own sense: dx as *l'accroissement infiniment petit* of an independent variable x, and dy as analogous increment of a function y of x. For a function fx of x, he then posited the latter's definite integral from a to b as

$$Fb - Fa = \int_a^b fx\,dx.$$

If one continued now by assigning to x an infinite number of values one after the other, increasing from a to b by infinitely small differences, and if one took these differences—be they equal or unequal—as value of dx,[4] the sum of all these

---

[3]   Guilloud thus also implies the principle of substitution (see below).

[4]   Here, Poisson suddenly got mixed up with Carnot's-Cauchy's meaning of *infiniment petits* as variables that assume values.

values of the differential *fxdx* was equal to the definite integral $Fb - Fa$ (Poisson 1833, 17).

Poisson considered this proposition to be a *théorème* that he had proved, in his own view, by means of infinite sums of infinitely small quantities, bare of any reflection on admissibility and existence of such sums. For his proof, Poisson assumed that the finite length of interval $b - a$ resulted as sum of *un nombre infini de quantités infiniment petites*, that is to say, as

$$\delta_1 + \delta_2 + \delta_3 + \ldots + \delta_n = b - a \text{ (ibid.)}.$$

By neglecting the *infiniment petits* of higher order, one has $F(x + dx) - Fx = fxdx$, and one could form in an analogous way, for the respectively adjoining values a and δ_1, $a + \delta_1$, and δ_2 $a + \delta_1 + \delta_2$ and δ_3, up to $b - \delta_n$ and δ_n, the equations

$$F(a + \delta_1) - Fa = fa\delta_1,$$
$$F(a + \delta_1 + \delta_2) - F(a + \delta_1) = f(a + \delta_1)\delta_2,$$
$$F(a + \delta_1 + \delta_2 + \delta_3) - F(a + \delta_1 + \delta_2) = f(a + \delta_1 + \delta_2)\delta_3,$$

$$Fb - F(b - \delta_n) = f(b - \delta_n)\delta_n.$$

la somme of these equations was

$$Fb - Fa = fa\delta_1 + f(a + \delta_1)\delta_2 + f(a + \delta_1 + \delta_2)\delta_3 + \ldots + f(b - \delta_n)\delta_n;$$

and hence the theorem was proven (ibid., 17 f.).

This formal mode of operation already contained the foundation of the definite integral, without losing a word about the existence of the sum, its uniqueness, etc. Not only does this crude return to notions of indivisibles contrast sharply with the elaborate conceptual instruments of analysis, as they had at this time been developed in particular by Cauchy, the virtual postulate of a factually universal validity of continuity in mechanics stands in just as striking contrast to Fourier's mathematization of the processes of heat distribution.

While Poisson had indeed abstractedly admitted for the case of the differential that if the differential coefficient X became infinite, the differential Xdx became indefinite, he had simultaneously proclaimed that such a circumstance could not occur in mechanics (ibid., 16 f.). And while he declared his "theorem" to be invalid in a wholesale way for definite integrals if a function became infinite at one point, he excluded such an exceptional case from occurring in his mechanics (ibid., 18).

In addition to that, Guilloud was to show to Poisson that upon methodical use of the concept of limit, functions containing individual singularities in many cases possess a finite definite integral, as Guilloud had shown in great detail.

How can Poisson's conceptual "fundamentalism" be understood? He was extremely productive: a bibliography of 1981 lists 180 of his publications. In his mathematical education, Lagrange was formative for him. Later, he placed himself under Laplace's patronage, working, in the frame of Laplace's program, on a broad spectrum of topics from mathematical physics. His publications beginning in 1802 were at first primarily devoted to questions of analysis, until about 1815. In the years after 1810, he published as it were in a race with Cauchy, in particular about definite integrals (1811, 1813, and 1815). After 1815, his focus shifted to themes of mathematical physics, but he also continued to publish on analysis, primarily on integral theory, and delivered in 1825 a rather rare contribution on foundational issues: toward clarifying the multiplicity of the values of de Moivre's formula in the case of fractional exponents.[5]

The striking thing in Poisson's mathematical works is not only that they do not contain any discussion or conceptual introduction of basic concepts. Beyond that, I realized that he did not apply the limit concept in these research works either. Nor does any sign of "lim" appear. Only once was I able to observe, in a treatise of 1823 concerning definite integrals, the appearance of the term *limite*. This is where Poisson designates a definite series as *limite* of another, convergent series (Poisson 1823, 429).[6]

It may be assumed that Poisson's avoidance of the *limite* method was determined by his original position as an adherent of Lagrange. The hardening of this position toward internalizing the decision on methods taken at the *École polytechnique* in 1811 to favoring *infiniment petits* to the extreme of propagating its exclusiveness, however, has hitherto remained inexplicable. In any case, this "fundamentalization" did not happen before 1815. As has been said, the first edition of his textbook on mechanics of 1811 did not contain any such positions on mathematical foundations, and his lectures on statics, as first part of mechanics of March and April 1815, still adhered to his textbook text of 1811, as Auguste Comte's written record shows.[7]

Poisson was harshly criticized for his partisanship in favor of actually infinitely small quantities not only in France, but also abroad. Moritz A. Stern, whom we have got to know in Chapter VII.5.3. already as author of a textbook on algebraic analysis, and for his critical reception of Cauchy, translated the second edition of Poisson's mechanics into German in 1835, but he dissociated himself in his notes from almost all of the author's positions on foundations, thus also from Poisson's principle of considering material points as solids of

---

[5]   On the history of this question, which remained obscure for a long time, see Jahnke 1987. The total of Poisson's production has been comprehensively analyzed in Grattan-Guinness 1990a.

[6]   "Limite" otherwise occurred with him only to designate interval boundaries.

[7]   The handwritten records on Poission's course of mechanics, and on the courses held by other professors, are found in the archives of the Maison Auguste Comte in Paris.

infinitely small dimensions, and finite solids hence as a collection of an infinitely large number of material points (Stern 1835, 555 ff.). He was strongly opposed, in particular, to postulating the existence of infinitely small quantities in real life, giving a characteristic philosophical justification for that which was based on Kantian philosophical categories. In this, Stern referred to Poisson's justification of the reality of such quantities, according to which time and space accrued by infinitely small steps:

> If one asks, however, whether space and time are quantities which accrue? The author answers (§ 112), time and space are not explained [*on ne définit ni le temps ni l'espace*]. We are accustomed to considering the concepts of time and space as forms of thought which have no reality beyond us, which causes the proof of the reality of the infinitely small to collapse by itself (ibid., 558).

Of considerably greater systematic significance than this appeal to German philosophical tradition is Paul Mansion's (1844–1919) assessment in his analysis textbook of 1887. With this textbook, Mansion, mathematics professor at the Belgian University of Gand, realized the first steps toward an evaluation of positions concerning the fundamental concepts of analysis that was no longer nationally confined, but oriented internationally and independently, thus providing approaches to axiomatization. His analyses will for this reason be treated separately below, and in particular in our closing part. At this point, they will be drawn upon only insofar as they are relevant for his assessment of Poisson.

As one of the major positions taken with regard to the *infiniment petits*, Mansion presented the conception of Carnot-Cauchy, according to which these variables have the limit of zero, and according to which the method of infinitesimals consists in looking for the limits of sums, and of ratios, of such quantities. Mansion also took up Carnot's notion of preferring to call the *infiniment petits* instead *indéfiniment petits*, and thus was able to incorporate Leibniz into this current, which he called dominant in his time, relying on Cauchy (Mansion 1887, 212 f.).

Mansion confronted this view with that of *certains géomètres*, according to which infinitely small quantities really existed: "According to these, there exist quantities different from zero and which are however less than any assignable quantity" (ibid.).

Mansion called quantities thus conceived of as *pseudo-infiniment petits*, with the intention of recalling the obvious contradiction this definition contained (ibid., 214). According to these geometers, he said, infinitesimal calculus was based on the principle that one could replace a quantity by another, provided that the two differed by a *pseudo-infiniment petit*. Fortunately, practice mostly operated, instead of with these pseudo quantities, with the *indéfiniment petits*. And, not quite surprisingly, Mansion presented Poisson as the main propagator of the *pseudo-infiniment petits* in the nineteenth century, with some of his disciples after him (ibid., 223 f.). Interestingly, Mansion ascribed to him, as historical predecessors, Johann Bernoulli, and the Marquis de l'Hospital (ibid., 222).

Mansion's independent evaluation thus confirms anew that Cauchy's concept of compromise of the *infiniment petits* was understood, by his contemporaries, and generally by mathematicians of the nineteenth century, as a variant of the *limites* method, and that conceptions of actually infinitely small quantities were deemed to be beyond the mathematically admissible.

Further insights into the contemporary understanding of the concept can be obtained from the fact that Cauchy's "partisans," in particular, voiced most incisive criticisms against Poisson. Moigno, in many respects Cauchy's *porte-parole* (cf. Chapter VI.6.4.), not only demonstrated the mathematical unreliability of Poisson's concepts in the preface to his analysis textbook of 1840, but also criticized Poisson's claim that his approach was universally valid. In presenting his own chapter on infinitely small quantities, he took a position of principle against Poisson: During the last years of his life, the latter had declared war on the theory of functions as developed by Lagrange, harnessing his entire institutional power for an attempt at making his own method of infinitesimals compulsory for the entire world of scholars (Moigno 1840, xxiv). Moigno confirmed that the claim at an *existence réelle* of infinitely small quantities had met with a broad and strong rejection:

> Although based on so imposing an authority, these assertions were fiercely attacked and repulsed; they announced but a mystery of which reason took fright without having the right to reject it; many judicious minds saw here an evident impossibility (ibid.)

Moigno made use of his critique of Poisson to highlight Cauchy's concept of variables with their limit of zero as the appropriate view. For to assume quantities that were actually smaller than any given quantity would imply to give up the idea of the *continu divisible à l'infini*, an idea that for him was the necessary basis of all the mathematical sciences. Moigno noted with satisfaction that "some rather unfortunate applications" of Poisson's principles, whose greatest drawback lay in their reducing the science too openly to its primitive stages, had made manifest that the more rigorous methods of Poinsot and Cauchy were called for (ibid., xxiv f.).

It was actually an understatement, however, to accuse Poisson of having attempted to swear the entire "world of scholars" to his own principles. On the basis of his position of power as the unique expert on mathematics within the *Conseil royal de l'instruction publique* since 1820, he also included the schools into his campaign for universalization.

The *Conseil royal* prepared all the essential decisions of the ministry of education in their very frequent sessions. Poisson profited from the ministry's request of September 24, 1837, addressed to the *Conseil*, asking the latter to present a proposal for geometry instruction at the *collèges* (Belhoste 1995, 147), to deviate from the traditional custom, which consisted in listing teaching subjects only quite summarily in the curricula,[8] to nail down methodological prescriptions as well. While the *Conseil*'s decision of September 26, 1837, immediately decreed by the ministry, upheld the structural frame of two geometry lessons for the *troisième* form, Poisson prevailed in having his favorite method of using the *infiniment petits* centrally prescribed as compulsory of all the *collèges*: "The two geometry lessons will remain appended to the *troisième* class; but this teaching will be based on the method of infinitely small quantities."

Since this decree making Poisson's method the basis of elementary geometry instruction could be counted on to raise eyebrows, it was complemented by a justification saying that it had the combined advantage of simplifying proofs in geometry, and of preparing for the superior course in mathematics: "[...] which combines with the advantage of simplifying proofs the advantage of preparation for the advanced course in the mathematical sciences."

Of course, the mathematics instruction of the time did not extend to any infinitesimal calculus. At the same time, the decree made Poisson's conception compulsory for all French schools of secondary education: *La méthode d'enseignement de la Géométrie sera applicable à tous les Collèges du royaume.*[9]

As at the *École polytechnique* 1811, this provoked resounding objections from some *collèges*. The response to these was again a formula of compromise: while future geometry instruction was to be based on the method of *infiniment petits*, it was permissible to tolerate the old methods for some time (Belhoste 1995, 148).

In the year that followed, Poisson's method was explicitly confirmed: for plane geometry in the same form, the *troisième,* a decree dated September 29, 1838, presented a justification taken from the field of the concept of number. The method of the *infiniment petits* allegedly dispensed from making allowance for the case of incommensurable quantities in the proofs:

> Plane geometry [will be taught], based on the method of infinitely small quantities already prescribed by the council; a method which will require, in proofs, consideration of the case of incommensurables."[10]

A subsequent decree dated October 9, 1838, specified, by presenting a first detailed mathematics curriculum, how geometry was to be taught on the basis of the *infiniment petits*:

* In plane geometry for the *troisième,* curves were to be conceived of as polygons having an infinite number of sides; in particular, circles were to be conceived of as regular polygons.

---

8   The curricula for mathematics instruction valid since the French Restoration dated of 1814, and of 1826 (Belhoste 1995, 88 f. and 113 f.).

9   AN, F[17], no. 12863. See also: Belhoste 1995, 147 f.

10   Bru 1981, 86, and supplementary information by letter from Bernhard Bru.

- In space geometry for the *seconde,* the sphere's surface is to be correspondingly conceived of as formed by rotating a regular polygon having an infinite number of sides, and the sphere's volume as composed of an infinite number of pyramids with summits lying in the sphere's center (Belhoste 1995, 148 ff.).

With his universalization of the *infiniment petits*, and in particular making the latter compulsory for mathematics instruction in secondary schools, Poisson virtually indeed returned, as Moigno says, to the "cradle" (*berceau*) of the science: from the intended developments in analysis and mechanics back to the stage of departure, that is, back to questions of geometry concerning areas and volumes as they had been treated in the theory of indivisibles.

The impact of a mathematician as influential as Poisson did of course not remain confined to his lifetime. In particular, two of his disciples continued to bear the torch for his basic conceptions: Navier and Cournot. Mansion counted both among the propagators of the *pseudo-infiniment petits* (Mansion 1887, 224).

Claude Louis M. H. Navier (1785–1836) is known primarily for his research into elasticity, but he also gave lectures on analysis at the *École polytechnique*, in particular after Ampère and Cauchy had left. The elaboration of Navier's lectures on analysis published in 1840 attempts another compromise between an explicit use of the *limites* method, and of Poisson's *pseudo-infiniment petits*: "[…] la quantité dx supposée infiniment petite, c'est-à-dire plus petite que toute grandeur donnée" (Navier 1840, 10).

Antoine-Augustin Cournot (1801–1877), known for his works on probability, enjoyed Poisson's resourceful protection, also assuming leading positions in the educational administration. After having been mathematics professor at the *Faculté des Sciences* in Grenoble for some time, he published his lectures on analysis in 1841. He was not able to do without using the *limites* method either, but foundered in trying to prove them inferior to Poisson's conception of the *infiniment petits.*

Cournot's basic tenor was a praise of Poisson's conception: however clever one applied the *méthode des limites*, one would always reach questions for which one would have to dispense with this method, replacing it by using *infiniment petits* of various orders. For this method was no ingenious artefact, he said, but rather the natural expression of the mode of generation of physical quantities that grew around elements that were smaller than any given finite quantity (Cournot 1841, 86). One was thus correct in stating that the *infiniment petits* existed in nature (ibid., 87). This was true not only for physical quantities, however, but just as well for quantities of pure geometry. The method of infinitesimals corresponded better to the nature of mathematical concepts, it was the direct method.

On the other hand, Cournot was forced to admit that the concept of the infinitely small lent itself logically only to an indirect definition, one mediated by the *limites*.[11] The *rigeur demonstrative* hence belonged immediately to the *limites* method, and only indirectly to the method of infinitesimals, the latter thus becoming a simple translation of the former. The consequence from this was that one had to make evident, in the simplest cases, the identity of the results of both methods. Cournot, however, continued by taking up the turn of decisions of 1810/11 at the *École polytechnique*: as soon as one had grasped the translation between the two methods, he said, one had to abandon the method of *limites*, as, according to Cournot's postulate, only the infinitesimal method permitted to solve the difficult problems, without unnecessary technical frills (ibid., 88).

This superficial rejection of technical, conceptual differentiation shows how right Servois had been in warning in 1814 against the possibility of (French) mathematics placing itself at a disadvantage by preferring intuitively accessible concepts (cf. Chapter V.2.5.2.). In our last section, this will be made additionally evident from the so-called principle of substitution of infinitely small quantities, which became effective only in France.

2.2. The Apotheosis of the Dualist Compromise: Duhamel

We had already said, in Chapter I.3., how systematically important Duhamel's reflections concerning the methodology of science for the basic concepts of analysis became in France since about 1850. This will also serve as a suitable conclusion of this study. Duhamel raised Cauchy's program of a dualism encompassing both the methods of *limites* and the methods of the *infiniment petits* to the highest level of analysis, which acquired a pervasive structure by his effort. In this effort, he rejected Poisson's one-sided program, curtailing its impact. At the same time, he was convinced that he would be able to concur with the ongoing general effort at conceptual differentiation by elaborating a principle of substitution, which had already been tentatively implied in earlier approaches, into a general principle of substitution. In doing this, Duhamel unwittingly drew to light implicit limitations and weaknesses in Cauchy's program.

Jean-Marie-Constant Duhamel (1797–1872), born in St. Malô, attended the *lycée* in Rennes, studied since 1814 at the *École polytechnique*, and was forced, like all students of this year, to return to his hometown in 1816. After studying law in Rennes, he became a teacher of mathematics in Paris, primarily at private schools. After 1830, he taught at the *École polytechnique,* first filling in for Coriolis, and after 1836 independently, lecturing on analysis and mechanics. Later, he also assumed teaching obligations at the *École Normale (Supérieure),*

---

[11]   Besides Poisson's definition proper, Cournot obviously also had Cauchy's definition in mind as well.

and at the *Faculté des Sciences*. He was much esteemed for his lucid lectures. His research pertained primarily to topics from mechanics.

On analysis, he published several textbooks whose sequence documents his changes of view with regard to foundations. He moved from Cauchy's position, which in itself already was a "compromise," toward a further "compromise": to adopting Poisson's program. In his *Cours d'analyse*, first printed in 1840/41, he assumed priority for the *méthode des limites*; while infinitely small quantities were introduced in Cauchy's sense, they were not assigned any prominent position.[12] The transition to a renewed dualist program, and to a compromise with Poisson, was marked in 1856 by the likewise two-volume textbook *Éléments de calcul infinitésimal*, which saw a second edition in 1860–61.[13]

Duhamel claimed to have given a presentation with this textbook that deviated strongly from traditional texts (Duhamel 1856, XIII). His structure is indeed subject to a quite particular logic, as systematic development both of the *méthode des limites* and of the *notion des infiniment petits*. According to his approach, his book was intended as *développement de ces deux conceptions*, which were closely linked. Duhamel strictly refused to assign priority, or dominance, to one of the two conceptions. Instead, he conferred on both the honor of having been developed by the ancients, in the classical age. He professed that both *idées générales* were "the most fruitful concepts of mathematics" (ibid., XIII f.). Only in the definition of the *infiniment petits*, according to Carnot and Cauchy, they happened to present themselves as two special cases of *limes* formation.

One of the consequences of Duhamel's approach was that he had to precede the differential and integral calculi proper by an exhaustive chapter on foundations that consumed 40 percent of the first book's volume,[14] but not in the sense of an "algebraic analysis," as in other authors, but rather to develop the issue as to how quantities could considered to be *limites* of variables. Consequently, this first chapter bore the heading *Des quantités considérées comme limites*.

Duhamel saw three different types of how quantities could appear as limits of variables:

- as limits of sequences,

---

[12]   The statement in S. Dostrovsky's DSB entry on Duhamel claiming that the latter had changed his foundational position concerning infinitesimals in the second edition (DSB, vol. IV, 222 f.) is not correct. Actually, Duhamel had explained that he had moved from using the differential to using the differential quotient (Duhamel 1847, *avertissement*). Conversely, while trying to assess, in his first edition, whether the differential, or the differential quotient, was better suited as the basic concept, he had decided on the differential as the *plus convenable* (Duhamel 1841, 21 f.).

[13]   A third and fourth edition followed after Duhamel's demise in 1874 and in 1886.

[14]   The second volume, on the integration of differential equations, cannot be treated here.

- as limits of sums of infinitely small quantities, and
- as limits of ratios (*rapports*) of infinitely small quantities.

Duhamel did not intend to treat the theory of sequences and series in this textbook, and thus only listed some of the principal results in an appendix (*Note 1*). The central objects, in contrast, were the two other kinds of quantities, whose last was that discipline newly generated by the modern mathematics (namely determining the limits of the ratios of two infinitely small differences: the values of functions to the values of variables).

For both of these two latter types of quantity, Duhamel established a very simple common general principle on whose fundamental importance and general validity he was fond of congratulating himself, the so-called principle of substitution, according to which it was permissible to substitute infinitely small quantities by certain others. This worked to simplify operations, he said, and at the same time permitted one to check whether exactness had been maintained in the simplification. We shall treat this principle in detail below. On the basis of this principle, differential calculus, for Duhamel, proved to be identical with the third type of quantity, and integral calculus identical with the second type of quantity. Duhamel's headings of chapters and of sections hence already show a surfeit of the expressions of *limites* and of *infiniment petits* that is without precedent in any other textbook. He declared that there existed a general theory of the second and third types of *limites*, which made it feasible to develop that general theory before proceeding to derive the rules for differential and integral calculus, in order to ensure more simplicity and generality for that calculus (ibid., XIX). His chapter on foundations was indeed exclusively devoted to presenting this approach.

After an introduction of the concept of number, which focused on legitimizing incommensurable numbers as limits of larger or smaller rational ratios, thus corresponding to the introductory sections on real numbers in textbooks published in later periods, there followed, as announced, foundations of a limit theory, introduced by the definitions of *limite* and *quantité infiniment petite*. The two definitions kept closely to Cauchy's concepts. For the arbitrary approximation of the variables to the limit, Duhamel likewise used the expression *indéfiniment*. For "infinitely small," he strictly adhered to Carnot-Cauchy: "On appelle *quantité infiniment petite, ou simplement infiniment petit, toute grandeur variable dont la limite est zero* (ibid., 9).

There followed a first *principe fondamental* for limits—if two variables identical everywhere tend toward their respective limits, the two limits are identical—as well as first propositions concerning a limit calculus: limits of sums, products, quotients, and powers. Remarkably, all these propositions were presented only in verbal form. Their proofs did not use any sign for limit, a fact that prevented the question of an algebraizing limit calculus from arising (ibid., 9 ff.).

These elements of the limit method were immediately followed by a section on applications: "application to simple cases of elementary geometry." In a direct sense, this section was a rejection of Poisson's program of universalization. The very application of *infiniment petits* in elementary geometry Poisson had postulated, and been able to have decreed for all French schools, Duhamel now presented, in this section of his own book, as an application of the *limites* method, for the problem of determining the circle's circumference as limit of the length on inscribed and circumscribed polygons, and the corresponding calculations of volumes (ibid., 15 ff.). The circle's circumference thus proves to be the limit of a sequence, and accordingly belongs among the first of the above three types of quantities.

Duhamel's challenge to Poisson, however, remained on the whole rather more superficial. In his own conception of mathematics, Duhamel actually tried to improve on fundamentals of Poisson's program: As had been suggested already by his refraining from algebraizing the *limes* calculus, Duhamel did anything but continue Cauchy's program of creating a conceptual, theoretical edifice of analysis. Nothing from Cauchy's tentative algebra of inequalities was taken up, or even continued, by Duhamel. His orientation remained rather more confined to the frame of *physico-mathématique:* the principal part of his chapter on foundations was devoted to the application of the *limites* method and of the *infiniment petits* to problems posed by geometry, and by mechanics. Doubtlessly in marked contrast to Poisson, Duhamel studied for instance there, in applications to statics, points of gravity by means of determining limits (ibid., 75 ff.). He remained, however, as it were locked in Poisson's embrace; he did not accede to conceptually promising analytic problem formulations.

This imprisonment is particularly evident from the so-called principle of substitution, which Duhamel was so proud of, and which shall now be presented in more detail.

2.2.1. THE PRINCIPLE OF SUBSTITUTION FOR INFINIMENT PETITS

Zerner, who systematically examined the French analysis textbooks of the nineteenth century, categorized them according to their methodological-conceptional approach into three "generations."

As middle, or second, generation, he identified a group of textbooks that without exception used the principle of substitution as their *principe fondamental* for differential and integral calculus. The crucial model for this generation was Duhamel's book (Zerner 1986, 10 f.).

What is this principle, which fell into utter oblivion by the beginning of the twentieth century, actually about? We have already become acquainted with it in one of its two versions. Indeed, the first of the two conditions, or principles, posited by L'Hospital for his new *calcul*, actually is a principle of substitution. This first condition, which I have also called extended relation of equality (see Chapter III.7.), precisely requires substitutability. According to this principle, it should be admissible to exchange two quantities for one another, provided they differ by an infinitely small quantity.

In his *Principia*, Newton used in his central Lemma I a proposition about quantities becoming identical at the limit which can be interpreted as a tentative formulation according to which the ratio of such quantities is assigned the value 1 (Chapter V.1.6.1.). Carnot systematically expanded this conception: he called two quantities or two systems "infinitely close" if the limit of their ratio was 1 (ibid.). In his treatise for the contest called by the Berlin Academy, he established several theorems concerning such a substitution of quantities.

In France, Francœur was the first to take up Carnot's concept of substitution, in his textbook first published in 1809 (cf. Section 1. above), declaring the substitution of quantities, provided their ratio became equal to 1, to be the basis of the infinitesimal method: The method consisted in substituting the original quantities by other quantities in their proximity, without thereby changing the result. For this, Francœur explicitly added a reference to Carnot's concept of error compensation:

> The infinitesimal method consists, as can be seen, in substituting in the calculation, for true magnitudes of which the object is made, other quantities which are close to them but such that the error made does not alter the result; instead of the proposed magnitudes, which may be difficult to treat and would complicate the operations, one takes in their place other quantities which are simpler and which lend themselves better to the research one has in view and to the calculations one has to carry out. But to be entitled to use false values, one must above all be assured that it will not result in any error and that if one adds to them that which they lack, these parts will eliminate each other (Francœur 1819, vol. 2, 350).

The means to ensure that no error occurred in substituting the original quantities by other quantities close to them was to ensure that the ratios between the former and the latter must become the identity:

> Thus, in order that the method can be used with complete safety, an indispensable condition must be fulfilled, that of the equality of last ratios, which consists in comparing the true magnitudes with those that have been substituted, having them vary together and see whether, in their progressive diminishing, their ratios tend without interruption to unity, if unity is the limit, or the last ratio (ibid.).

While this *calcul* presented itself as a means of approximation, it attained even the rigor of algebra, he said, since its point was to determine the ultimate ratios (ibid., 352).

Duhamel expanded the concept of substitution, proclaiming it as the general basis of differential and integral calculus. His chapter on foundations is hence almost exclusively devoted to presenting and evaluating it. As has been mentioned, the principle of substitution consisted of two parts for Duhamel. One of them amounted practically to adopting the already known principle for *raisons* and *rapports* of variables whose difference had the limit of zero:

> Second Theorem.—The limit of the ratio of two infinitely small quantities is not changed when one replaces these quantities by others, not equal to them, but whose ratios to them have respectively unity as their limit (Duhamel 1856, 37).

Concerning this principle of substitution for ratios, Zerner, an active French mathematician of our time, observes that it does not contain any problems and is being taught in France even today, albeit not as a principle of foundational character: "It poses no particular problem; we still teach it today but without according it a fundamental character" (Zerner 1986, 10). Zerner herewith confirms how long-lived this concept's tradition has been in France, whose restricted validity has been discussed in Chapter V.1.6.1.

More important for Duhamel, however, was the second, new part of the principle of substitutions for sums. The objective was to simplify "measuring" quantities of the second type, by substituting, in an infinite sum of infinitely small quantities, the aforementioned quantities by other infinitely small quantities that were "close" to them:

> One of the most important conceptions in the measure of magnitudes is that by which they are considered as limits of sums of quantities of the simplest type which tend indefinitely toward zero at the same time as the number of them increases indefinitely (Duhamel 1856, 35).

This introduction already raises suspicion, for Duhamel himself speaks here of two different limit processes running simultaneously: increasing the number of summands, and reducing their amount. According to Duhamel, the *indéfiniment petits* were to a large degree indeterminate, since they were subject only to the condition that the difference between their sum and the quantity proper have the limit of zero. There was thus much latitude, he said, in selecting these elements, and one could profit from that to simplify operations. This having been said, he formulated the *principe fondamental* that justified these simplifications:

> Theorem – The limit of the sum of infinitely small positive quantities is not changed when one replaces these quantities with others whose ratios to them have respectively unity for their limit (ibid.).

What was meant was this: Let there be given a sequence $\alpha_1, \alpha_2, \alpha_3, ..., \alpha_m$ of *infiniment petits* whose sum tends toward a limit, while the number of their terms *augmente indéfiniment*. Let there be given, at the same time, another such sequence of *infiniment petits*:

$$\beta_1, \beta_2, \beta_3, ..., \beta_m.$$

If the respective ratios

$$\frac{\alpha_1}{\beta_1}, \frac{\alpha_2}{\beta_2}, \frac{\alpha_3}{\beta_3}, ..., \frac{\alpha_m}{\beta_m}$$

themselves always have unity as limit, the principle permits to substitute the α_i by the β_i: the second sum, that with the β_i, converges to the same limit as the first sum with the α_i. Zerner added the sardonic comment that any question inquiring into the nature of this convergence would be "indiscreet" (Zerner 1986, 19).

One cannot but wonder at the high esteem in which this principle of substitution was held. According to Duhamel himself, it served to found integral calculus or, more precisely, the definite integral. Actually, however, it had a large role already in his introductory chapter on foundations. This is where it appeared for the first time, directly after the presentation of the fundamental principle just mentioned, in a section on determining areas: "How the areas of plane curves can be understood as limits of sums of *infiniment petits*."

Duhamel began with the arc of "arbitrary curves," subdividing the unspecified area formed by such a curve and an axis into parallelograms with ever diminishing distances. The sum of these subareas, he said, was identical with the total area sought. If one now proceeded to substitute the delimiting curve arcs of these elementary areas by polygons, adding, or subtracting, infinitely small parts, the sum of these new, simpler subareas was identical to the total area sought because of the fundamental principle. His conclusion is circular inasmuch as he had already assumed the existence of the area or of the sum of infinitely many terms (Duhamel 1856, 41 ff.).

It is seen as a gain here that while Duhamel seeks to replace Poisson's pseudo-*infiniment petits* by Cauchy's *infiniment petits*, he at the same time shares some essential conceptual premises of Poisson, i.e., that of assuming the existence of sums of infinitely many parts without close inspection.

This is once more evident in his formal introduction of the integral as a definite integral. For this purpose, Duhamel generalized his view of curve areas taken from elementary geometry, using the sign expression

$$\lim. \sum F(x)\Delta x,$$

while only verbally speaking of an interval nesting. He mentioned a second possible interval nesting without, however, making a comparison of limits. This probably was to have been performed by the principle of substitution beforehand. Duhamel then explicitly proclaimed the necessity of demonstrating that *une limite déterminée* of the above sum existed. Although he did not stipulate any limiting condition with regard to the function to be integrated, only requiring *une fonction quelconque* instead, he declared after only a few phrases, *il est donc certain* that such a limit existed. Again, the circular assumption had been made that the area under the curve existed (ibid., 418 f.). Not even Cauchy's requirement of continuity had been adopted.

As a conceptual problem behind the principle of substitution not recognized by the "second generation" authors, we meet again the issue, manifest since Carnot, of omitting to differentiate between various limit processes, i.e., of failing to discover the substitutability of the limit transitions and the uniformity that are necessary for convergence.

A typical example for the potential awareness of the necessity of differentiating between the various limit processes, and of more exactly defining the properties of uniformity is Mansion's book. Its internal structure is illustratively contradictory. In a first part, it briefly presents the basic theories of infinitesimal calculus, essentially in line with the French "second generation's" view. In a second part elaborated later, in contrast, the same basic theories are reviewed again historically and critically. This is where the book attains an independent function of examining and evaluating these theories on an international scale. Mansion showed that contradictory mode of presentation in the case of the principle of substitution, too. In the first part of his book, he presents this principle as *principe fondamental* of infinitesimal calculus (Mansion 1887, 11 ff.).

In the second, critical, part, however, Mansion presents "examples of a false application of the principle," by means of three series whose terms, while tending in their ratio toward unity as limit, do so only in a determinate sequence of limit processes. Mansion indeed explicitly inserted two different index variables, p and n:

$$A_n = \alpha_1 + \alpha_2 + \alpha_3 + \ldots + \alpha_n, \ \alpha_p = \frac{1}{n}$$

$$B_n = \beta_1 + \beta_2 + \beta_3 + \ldots + \beta_n, \ \beta_p = \frac{1}{n}(1 + x), \ x = \frac{p}{n}$$

$$C_n = \gamma_1 + \gamma_2 + \gamma_3 + \ldots + \gamma_n, \ \gamma_p = \frac{1}{n}\left(1 + \frac{n^2 x}{c^{n^2 x^2}}\right), \ x = \frac{p}{n}$$

(ibid., 241).

In case of a fixed p and of an arbitrarily increasing n, $\lim \frac{\beta_p}{\alpha_p} = 1$ held, and in case of a fixed x and an arbitrarily increasing n, $\lim \frac{\gamma_p}{\alpha_p} = 1$, but $\lim \frac{\beta_p}{\alpha_p} = 1 + x$. For the former assumption, it thus followed that one could apply the principle of substitution for A_n and B_n, hence that $A = B$. For the latter assumption, however, the principle of substitution was applicable for A_n and C_n, with the consequence $A = C$. However, since $A_n = 1$, there is also $A = \lim A_n = 1$, and conversely, $B = \lim B_n = 1\frac{1}{2}$, hence $B_n = 1\frac{1}{2} + \frac{1}{2n}$, as well as

$$C_n = 1 + \frac{1}{e} + \frac{2}{e^4} + \frac{3}{e^9} + \ldots + \frac{n}{e^{n^2}}, \text{ and hence } C = \frac{e}{e-1}.$$

All sums were thus unequal, and the principle of substitution was disproved (ibid., 242 f.).

Mansion thus concluded that the principle's condition was only a *sufficient*, but not a *necessary* one. The principle held only if all the limit processes were conducted *simultanément*. Mansion did not find the cause for the principle of substitution's lack of validity. He thus did not reflect, or operationalize, the concept of uniformity in his text of 1887. There was also no reception of Dirksen's or Weierstraß's conceptual achievements.

In Duhamel's textbook, the basic definitions of analysis for variables and functions, and for continuity in the chapter (*livre* II) on the *calcul des derivées*, are treated only after the comprehensive foundational chapter on *limites* and *infiniment petits*. It is remarkable that all these definitions are made in a purely verbal mode, without any use of signs, and continuity, too, is presented without any inequalities, etc. Duhamel introduced functions as dependent variables, leaving the question as to the type of dependence quite open (Duhamel 1856, 221). His definition of continuity was still completely traditional, having taken up neither Cauchy's explications of the concept, nor the developments in Germany, in particular not those made by Dirichlet. To provide a basis, Duhamel first defined traditionally and globally when a variable is continuous: if it satisfies the condition of having the property of intermediate value (ibid., 222). According to that, a function is continuous if its independent variable is continuous, if it always has real values, and if it satisfies the intermediate value property itself:

> A function is said to be continuous when, in making the quantities on which it depends vary in a certain manner, it is constantly real, and varies itself in a continuous manner, that is, that it cannot pass from one value to another without passing through all the intermediary values (ibid., 223).

Only an afterthought implied continuity in the interval, when he mentioned that a function could become discontinuous beyond definite limits.

The basic concept of differential calculus, for Duhamel, was the differential quotient, which he called *derivée*:

$$\lim . \frac{\Delta y}{\Delta x} = F'(x),$$

thus making the concept of function the basis of analysis for him (ibid., 227). He derived, however, the rules of differentiation for differentials. Duhamel obtained the derivation of a product of functions via an unusual path. Immediately after having introduced derivation, he went on to treat functions of several variables and composite functions, and introduced partial derivations. He then obtained the differentiation of products as a special case of the differentiation of homogeneous functions of several variables (ibid., 237 f.).

He was obviously intent on avoiding the usual proofs concerning the evanescence of higher terms. Thus, he went on to give proofs of his own for the differentiation of x^m, offering even two: one about developments into series (including case distinctions for the exponent; for positive integers and rational numbers, for negative, and for incommensurable numbers), and a second, tricky proof about the logarithm of the function, including case distinctions for positively and negatively valued functions (ibid., 242 ff.).

In his analysis textbook, neither the zero nor negative numbers assumed the character of exceptions. Duhamel operated with them without reserve. In our introduction in Chapter I.3., in contrast, we had mentioned that his goal in his five-volume handbook on methods of the theoretical sciences had been to clarify the *obscurités* he had taken note of in his youth during mathematics instruction at school, and during his studies at the *École polytechnique*.

Indeed, he devoted almost the entire second volume of this work to the development of the concept of number. For this purpose, Duhamel attached the concept of number epistemologically closely to the concept of quantity (*grandeur*) in order to express relations of *pluralité* between quantities. He already made a concession here in including *unité* among *pluralité*, and thus had an unlimited sequence of numbers with *le nombre un* as smallest number (Duhamel 1878, 3).

The basic operations with these natural numbers were followed by a first *extension de l'idée de nombre*, by the introduction of fractions, actually via *rapport* as a basic concept (ibid., 41 ff.). After extending this to decimal fractions, there followed further *extensions* for positive numbers: roots and incommensurable numbers. Only after introducing signs was it possible for him to form equations, and to introduce "letter quantities." With this, he acceded to equations of first degree, and this is where he now began to reflect on negative numbers. Curiously, he had not yet defined the zero—which actually does not fit into his derivation from the concept of quantity—up to this point. It simply appears now without any foundation proper.

Duhamel had introduced the zero before merely implicitly as a placeholder, at the occasion of explaining the place value system, but without giving it a name and a sign, and in particular without endowing it with the character of number, i.e., merely as a digit (Duhamel 1878, 27).

From his transition to discussing equations, on the other hand, he used the name *nul* and the sign 0, without epistemological reserve, and obviously assigning to it the character of a number (ibid., 115 ff.). His eventual admission of negative quantities implied an even more surprising turn. Remarkably, the starting point for this admission was Duhamel's approach to obtaining *generality* of results, as he expressed it himself in his chapter heading: "Des moyens d'obtenir la généralité dans les résultats" (ibid., 97).

The objective of determining the values of unkowns must always be, he said, to obtain "the general solution," and this general solution consisted in indicating the operations that were required in each single case for determining the solution. This indication was called "formula" (ibid., 97 f.). According to Duhamel, it is thus the operative moment that ensures generality.

On the basis of this understanding of "formula," Duhamel discussed multiplication as a problem, since the rule of signs occurs in it. He insisted on claiming that the rule of signs was proved, and applicable, only for "polynomials"—i.e., terms like $(a - b)$ (ibid., 105). Since he denied mathematical legitimacy as "quantities" to negative quantities on principle,[15] he

---

[15]   This is why he criticized d'Alembert and Carnot for having become involved in the discussion of the alleged proportion $1 : -1 = -1 : 1$. Since -1 was not

was obliged to study cases with negative solutions individually, and to transforming them according to the approaches known to us since Clairaut and Bézout until the result appeared as a positive one (ibid., 121 ff.).

At least, Duhamel declared it to be *inconvénient* that so many different forms had to be examined to obtain a complete solution of one and the same question. Since these various forms differed essentially only on presenting different combinations each of the signs of plus and minus, one was led to attempting to assemble all of them within a *type unique* from which the single formulas with their respective combinations of signs could be derived according to definite rules. One would thus obtain a *formule générale*. Duhamel then inquired "at what price" such an advantage could be attained (ibid., 126). The price for this *généralisation des formules* turned out to be the introduction of the *quantités négatives*.

This price at first seemed to be very high, as Duhamel never tired of explaining anew, the formulas requiring the multiplication of isolated negative terms. His solution repeated just as often was to have this multiplication occur according to the rules proved for "polynomials," while at the same time insisting that this did not mean at all to accord mathematical legitimacy to negative quantities: "Nous ne saurions trop de rappeler que dans ce qui précède nous n'avons attaché aucun sens aux opérations sur les quantités negatives" (ibid., 132).

These were only "modes of speaking" borrowed from operating with "polynomials." They were by no means *véritables operations* with isolated negative quantities. Duhamel thus also criticized earlier mathematicians for having believed themselves able to prove such rules of operation, or for having acknowledged an existence to these *êtres fantastiques* (ibid.).

With regard to negative quantities, Duhamel thus proved to be a just as eager a disciple of Carnot as in the case of infinitely small quantities. While he permitted operating with negative quantities, he did so under a fundamentalist epistemological general reserve, without extending the domain of numbers. The "generality" he strove for was actually quite limited, quite in the sense last propagated by Klügel in Germany one hundred years before. The generality of the *formule générale* was illusionary, mathematically legitimate was only its specification for single cases having different combinations of signs (ibid., 179). It was not based on any unified concept of number.

While Duhamel formulated the traditional reserve against the status of zero only weakly, and hardly applied it in practice, he was probably the one to formulate most radically and systematically the epistemological rejection of negative quantities that was still strong in the France of his generation. He did not cease to warn against assigning any meaning to negative quantities. There

---

a quantity, there was no such proportion (Duhamel 1878, 164 ff.), he said, taking up Christian Wolff's argument of 1710.

was no provable rule, he said, for carrying out mathematical operations with them (e.g.: ibid., 135).

At the same time, however, Duhamel systematically examined the fields of application for negative quantities like no other before him in France: thus for exponents, in polynomials, in solutions of equations, and in the application of algebra to geometry. It is probably because of this systematic study that Duhamel ended up removing, in a surprising about face, what had been an important epistemological hurdle before his time. In contrast to d'Alembert and Carnot, he said, it could very well be shown that negative quantities could be considered to be smaller than zero (ibid., 166). This followed from the basic concepts of calculating with inequalities. From $a < c$ alone, $a - c < 0$ resulted. This inequality expressed that a negative quantity was smaller than zero. Duhamel showed how inequalities with positive terms resulted immediately from inequalities with negative terms: From $-5 < -3$ results $3 < 5$, and from $-2 < 0$ results $0 < 2$ (ibid., 169).

Multiplication by a negative quantity inverted the inequality's direction. Duhamel's own explicit *conclusion* from this question was that if one wanted to have general methods in the frame with inequalities, one was compelled to state that negative quantities were treated as being smaller than zero, and smaller by just as much as the absolute value was larger (ibid., 169 f.). At the same time, the unified operating with positive and with negative quantities practiced by this working with inequalities factually implied recognition of negative quantities. In that respect, Duhamel established even another dualism compromise: besides the compromise between *limites* and *infiniment petits*, a second between rejecting negative quantities on epistemological principles, and a new form of justifying their practical use in mathematics.

While this split of consciousness shows up in several additional places in his handbook on foundations,[16] Duhamel operated in his analysis handbook, as has been said, without reserve with zero and with negative numbers.

2.3. The End of the Classical Infiniment Petits

Duhamel's work not only represents a highly elaborated version of the epistemological rejection of negative quantities on the basis of the traditional concept of quantity, and not only forms the climax in the intertwining of the

---

[16]   E.g., in discussing equations of second degree, where he says for two solutions $x = 5$ and $x = -1$ that there was no "number" satisfying the second, only the number 5 was a solution. The equation was satisfied, however, by an "expression négative -1" as well (ibid., 174 f.), and in his application to geometry in vol. 3, he presented generalization by means of negative quantities in an analogous way as a summary of single cases (Duhamel 1868, 55 ff.).

conceptions of *limites* and *infiniment petits*, but is at the same time the last work of a French mathematician to exhaustively reflect on and present the principles for how to conceive of limit processes. In several aspects, Duhamel was the last of an era, even if elements of his concepts survived for some time in various textbooks, in particular in those for training engineers.

His work forms a closing point for the very reason that in the highest rise of its concepts it already contained the conditions for their fall. We have already seen from the reprints of Lacroix's algebra in the 1870s that the prevailing rejection of negative numbers in France had been on the decline, and that operating with them nevertheless had become ever more natural in practice, even if the epistemological reserves persisted. Also, justifying the principle still dominant in France in the eighteenth century of banning the zero from the field of quantities—and hence from the concept of number as well—had become ever less present and convincing among the mathematicians of the nineteenth century. Cauchy's sophisticated differentiations invented to maintain a systematic distinction between the zero and "genuine" quantities were difficult to grasp for his own contemporaries, and their reception by the next generation was even less enthusiastic.

This leveling of the boundaries of principle between positive numbers as representatives of "genuine" quantities on the one hand, and of those between the nonquantity of zero and negative numbers on the other, however, did away with the necessity of conceptually distinguishing between between a "normal" *limite* and an infinitely small quantity. The zero became a limit just as legitimate as any positive number, and even as a negative number. The *classical* notion of *quantités infiniment petites*, as developed in the eighteenth century since Varignon, and as brought into its classical form by Carnot and Cauchy in succession to Martin and Stockler, had lost its legitimacy, and become obsolete.

The main field of applying *infiniment petits*, that of comparing *infiniment petits* of different order, had actually served primarily to study how rational expressions behaved when approximating zero, or the infinite. Of course, it was possible to transform this application into limit considerations without drawing on any infinitely small quantities. One such model were the so-called Landau symbols O and o.

Due to Cauchy's efforts at founding the definite integral, the classical *infiniment petits* had received a new, fruitful field of application. The principle of substitution had absorbed this concept field, extending the effective life span of the *infiniment petits* for some decades. Finally, however, the principle of substitution became obsolete as well, thus pulling the carpet from under the classical *infiniment petits*. As Zerner has shown, the French textbooks of the "third generation," which commenced with Jules Tannery's work of 1886, received the new concepts of rigor from Germany and Italy as well, for the first time introducing the distinction between ordinary and uniform convergence (Zerner 1986, 10, and passim).

The end of the classical infinitely small quantities, as variables, came almost simultaneously with the first publications on the "modern" infinitely small quantities, as numbers (Veronese 1891).

This transition has been acutely remarked on by Otto Stolz, the brilliant analyst of the foundations of mathematics (cf. Chapter IX.2.). While he made the first attempts at justifying infinitely small quantities as a new type of quantity, he at the same time delivered a eulogy on their classical version:

> As has been remarked at the outset, one is in no need at all of the infinitely small in differential and integral calculus. For Cauchy already, this term serves only to abbreviate [...] and could be suppressed completely without leaving any gap. While Cauchy's presentation is not unchallengeable, the improvements added to it are only a more exact formulation of his ideas partly acknowledged by himself. In this respect, nothing more is thus to be expected from any kind of infinitely small quantities (Stolz 1883b, 36).

Duhamel's work probably also was one of the last great efforts at establishing mathematics and mechanics on the same principles. This program not only corresponded to the venerated French tradition of *physico-mathématique*. It also agreed with the basic concept of rationalism in France (cf. Otte 1998 II). Duhamel's approach, to combine, among other things, mathematics and mechanics into the common *sciences du raisonnement*, is also an apotheosis of this rationalism.

Chapter IX

Summary and Outlook

One of the strongest impressions upon surveying the conceptual developments reported here is what staying power is exerted by traditions in the shape of time-honored concepts, attitudes toward concepts, and epistemological values. It is obviously the fundamental concepts that provide the strongest support for traditions. This is probably an expression of the fact that the basic concepts were acquired at an early stage of the participants' biographies, contributing as it were to the structure of their personalities. This is evidence of the formative character of the basic attitudes conveyed by the respective educational systems. Because of the linkage between the educational system concerned and its specific culture, this confirms our initial thesis that mathematical concepts have a cultural history. Quite different courses taken can be established for nationally and culturally specific moldings and developments from the sixteenth to at least the nineteenth century.

This closing view on Germany and France extending until the end of the nineteenth and the beginning of the twentieth century shall study how far the contrast between maintaining traditional concepts and innovative concepts became preserved in cultural attitudes.

1. Principle of Permanence and Theory of Forms

Let us first turn to the negative numbers, whose development after the 1840s we have not described as yet. Here, we can establish a radical and almost pervasive generalization of the concept of number. Typical evidence for that is contained in a text by Hermann Hankel, who initiated this new stage: *Theorie der complexen Zahlensysteme* (1867).

While Hankel adopted Förstemann's conception of additive and multiplicative "oppositeness" (Hankel 1867, 9), without, however, naming Förstemann, he generalized the approaches toward a principle of permanence developed at his time for extending operations. This generalization had two foundations:

- Firstly, the concept of number was his basic concept. As he explained himself, the concept of number can be conceived of "without considering the concept of quantity." Recalling Gauß's famous remark of 1831, he accepted the concept of quantity "only as an intuitive substrate of the number forms" (ibid., VIII).

- Secondly, he based his theory on Hermann Graßmann's conception of introducing the operations in a quite general or formal way by means of a "pure theory of forms" serving as a foundational discipline (ibid., 16). Because of Hankel's reception of Graßmann, the formal study and introduction of synthetic (*thetische*) and analytic (*lytische*) operations (ibid., 18 ff.) was used by many of the subsequent modern works treating the foundations of analysis.

Hankel explained his *Princip der Permanenz der formalen Gesetze* (principle of the permanence of the formal laws) as follows:

> Lest we fall into the abstruse, we shall subject the operations involving mental objects to such formal rules that they can contain, within themselves, the actual operations involving intuitive objects, and the numbers expressing their ratios as subordinate ones (ibid., 11)

By means of the principle of permanence, he derived, for addition and subtraction as special synthetic and analytic operations, the rules for adding and subtracting negative "objects" from the associativity and commutativity of these operations. In an analogous mode, he derived the rules of signs for multiplication and division as further synthetic or analytic operations by means of distributivity (ibid., 29ff. and 40 ff.). In familiar terms, Hankel stressed, as a matter of principle, the simultaneously arbitrary and necessary character of these posits:

> It cannot be too sharply stressed against a view widely spread that these equations [the rules of signs] can nevermore be proved in formal mathematics; they are a r b i t r a r y   c o n v e n t i o n s in favor of preserving formalism in the calculus. [...] Once these conventions have been established, however, all the other laws of multiplication will follow from them b y   n e c e s s i t y (ibid., 41).

As the relation of number to quantity had still to be considered, too, in this stage of developing pure mathematics, Hankel's chapter on the theory of forms was succeeded by another on quantities, in which operating with integers and with negative numbers was derived within quantity theory by means of an operating referring to a unity e, that is, for instance, by the transition from

$$A+B \text{ to } (A+B)e \quad \text{(ibid., 60 ff.)}.$$

A further generalization was implemented by Carl Weierstraß. The first broad international reception and recognition of the oppositeness approach, which had at first been concentrated in Germany, occurred precisely in this Weierstrassian form toward the end of the nineth century. For his purpose, Weierstraß had applied the mode of designation using pairs of numbers (a, b) for their real and imaginary parts, a mode that had been developed for complex numbers, to opposite numbers as well, positing

$$(A, B) = (A', B'), \quad \text{if} \quad A + B' = A' + B.$$

In this way, arbitrarily many pairs of numbers define the same number. By designating $(0, A)$ with $-A$, one obtains the usual notation for negative numbers.

These elaborated concepts were disseminated in the textbook *Theoretische Arithmetik* authored by Weierstraß's Austrian disciple Otto Stolz (1842–1905). It is well known that Weierstraß himself aired his own conceptions only in his lectures, but not in textbooks. It was the very textbook by Stolz of 1885, and its new edition by Stolz and J.A. Gmeiner of 1900, that essentially took care of disseminating Weierstraß's ideas on an international scale. Since Weierstraß was also very critical of other authors' publishing his own orally presented ideas, it is quite an irony that the above volume by Stolz/Gmeiner drew on two French works, of all sources, for describing Weierstraß's conception!

> The method of pairing numbers to introduce new kinds of numbers, however, was thought of first by W.R. Hamilton. Later, C. Weierstraß used it for deducing relative numbers from absolute numbers (cf. H. Padé, *Premières leçons d'algèbre élém.*, 1892, S. XII, and C. Bourlet, *Leçons d'algèbre*, 1894 [sic! 1896], 59) (Stolz/Gmeiner 1911, 78).

The new, formal view of mathematics encountered less resistance to its reception in other mathematical cultures than within its own, among German *Gymnasium* teachers, for instance, who remained faithful to the traditional view where mathematics was a science in which everything could be rigorously proved. An illustrative example is the extended furious debate over the years of 1883, 1884, and 1885 in the *Zeitschrift für mathematischen und naturwissenschaftlichen Unterricht* (ZfmnU), a journal catering to the professional needs of German *Gymnasium* teachers, this debate having been prompted by a letter to the editors sent in by a junior teacher fresh from the university and trained in the modern foundations of arithmetic, who provoked his betters by stating that the rule of signs was not provable, but represented a convention. When the junior teacher tried to defend himself against strident criticism by appealing to his training at the University of Bonn, and in particular to Rudolf Lipschitz and the latter's *Lehrbuch der Analysis* (1877), the journal's editor J.C.V. Hoffmann addressed himself to Lipschitz to inquire whether he really meant this, and how he hoped to justify it.

When Lipschitz did not stoop to a debate with mere classroom hacks, but politely referred to the presentation in his textbook instead, Hoffmann denounced the "high-faluting position of present-day university teaching" (Hoffmann 1884, 113), declaring it to be irrefutable that there could never be any tampering with "propositions like $(-a)\cdot(+b) = -ab$ and $(-a)\cdot(-b) = +ab$ [...] because their underlying truths are n e c e s s a r y and not a r b i t r a r y . " And he added self-assuredly, flaunting his lack of historical knowledge, the following:

> Where, by the way, is the mathematician who has taught, after calculating with letters was invented, that $(-a)\cdot(-b)$ is not $= +ab$, but rather $= -ab$? Let him please be pointed out to me! (Ibid., 111).

Many *Gymnasium* teachers vilified the upstart junior teacher. One of them titled his reply that it was "child's play" to prove the rule of signs thus: "An Admonition to Young Mathematics Teachers and to Those Hoping to Become Such" (ZfmnU 1884, 106).

The traditionalists were so embittered in this debate as to qualify the innovators as their "adversaries" (e.g., ibid., 107). As the debate continued, Hoffmann himself had to admit gradual insights of the kind that the time-honored definition of multiplication might be in need of modification (ibid., 344 f.). But nevertheless he, and with him the overwhelming majority of *Gymnasium* teachers, went to extremes to resist recognition of any "arbitrary" element in mathematics (e.g., ZfmnU 1884, 582 and 591). They considered it a crass violation of the epistemology indispensable, in their own view, for teaching mathematics: of the inherent necessity of concepts, and of modes of inference.

How developments went specifically in many countries has for the most part not been investigated as yet. Studies for the present volume had to be mostly restricted to the cultures existing in three particularly advanced countries: France, Germany, and England. Some glimpses of Italy have been added (Chapters II.2.3., and IV.1.4.). Studies concerning other European countries have to my knowledge been undertaken only for the Netherlands, by D. Beckers for the eighteenth century, and for the beginning of the nineteenth (cf. Chapter VII.3.3.).

It was also in the Netherlands where the probably most voluminous (243-page) treatise on the nature and application of positive and negative quantities was published in 1815, by the then leading mathematician Jacob de Gelder (1765–1848), professor at the Delft school of engineering. Its text would merit a detailed analysis, but I have to confine myself to some brief remarks. It is marked by an anti-French resentment, since it was written just after the Netherlands had been "liberated" from having been incorporated into Napoleon's Empire. It discusses French conceptions critically and in detail, once more here primarily the dominant views of d'Alembert and Carnot. Although de Gelder's intention had been two refute both, and although he assumed an independent role for algebra, using the theory of opposite quantities and stressing the twofold function of the plus and minus signs as algebraic signs and as signs of operation, thus doing without demanding limitations of operations, he nevertheless developed a view that did not differ in principle from d'Alembert's and Carnot's. Without differentiating between quantities and numbers, he admitted only absolute quantities as mathematical objects. What were called positive and negative numbers, he said, were not different kinds of numbers, but only two opposite "states" of an absolute quantity (de Gelder 1815, 48 ff.). A new element in the text, however, is found in the beginning of an axiomatic approach in algebra, for the fundamental concepts of algebra are established by way of *axioms* ("Axioma") (ibid., 60 ff.).

I was finally able to find some evidence of how things developed in Portugal. During the last third of the 18th century, two different receptions confronted one another, both represented by professors at the University of Coimbra:

- José Monteiro da Rocha (1734–1819) accepted and disseminated the French positions of Clairaut and d'Alembert, as well as Bézout's ambivalent view;
- José Anastàcio da Cunha (1744–1787), in contrast, received and propagated the English rejection of negative quantities which was new at the time. In particular, he used the sign ~ for indeterminate subtraction, appealing to Thomas Simpson (cf. Chapter II.10.3.) for doing so (Schubring 2001, 107).

Even one hundred years later, da Cunha's position was aggressively held in Portugal: António José Teixeira (1830–1900), mathematics professor at the University of Coimbra, published documentary material concerning a controversy between the two mathematicians. In his own commentary on d'Alembert's argumentation with regard to the proportion $1 : -1 : : -1 : 1$ which was mentioned there, Teixeira declared:

> Everybody knows that the so-called isolated negative quantities are not quantities, and that they thus cannot be compared to the proper quantities: the latter are neither larger nor smaller than the former, since the negative quantities do not possess any arithmetical existence (Teixeira 1890, 190).

It was thus inadmissible, he said, to compare 1 to –1 with the intention of determining which of the two was larger (ibid., 191).

I was able to find further indications of the wide dissemination and long-livedness of realism-oriented views of mathematical concepts, which might be called everyday life epistemologies, and which led to rejecting negative quantities or numbers, in Brazil. I shall confine myself to presenting two texts dating from the year 1868. The first is an algebra textbook, *Apostilla de Algebra* by the private teacher Luis Pedro Drago. His textbook is essentially based on Bézout's and Bourdon's views: negative quantities are introduced only as "expressions" (*expressões*), not as quantities having the same status as positive ones, and they are merely "considered" to be smaller than zero (Drago 1868, 5). The presentation becomes more extensive in the context of treating linear equations. In this context the author declared that while a solution like $x = -3$ satisfied the equation, it did not fit the problem, since the "expression" (-3) indicated that something was to be subtracted instead of being added, as the problem required (ibid., 124). For Drago, negative solutions were indeed "impossible" (*impossiveis*); they were interpretable only as indication of the need to reformulate the problem (ibid., 126 ff.).

Even more radical was Benjamin Constant's publication of the same year, 1868: *Theoria das Quantidades Negativas.* While Constant (1836–1891), engineer, teacher educator, and later one of the most influential proponents of positivism in Brazil, admitted opposite states for quantities in algebra as *modos de existencia* (Constant 1868, 6), he declared the notion of negative quantities smaller than zero to be "absurd" (ibid., 12). The zero already was no longer a quantity, he said, and something that had diminished until its vanishing point could not be disminished further (ibid., 13). On this basis, he sharply criticized contemporary French and Belgian textbooks such as those by Bourdon, Fourcy,

Cirodde, and Paque, who had at least still conceded that one could compare negative quantities to positive ones (ibid., 14 ff.). Since he did not use, however, any sign of his own for the absolute value, he experimented with equations like

$$a = -a,$$

a in the second term having another meaning than in the first term (namely the absolute one) (ibid., 58 ff.).

2. On the New Rigor in Analysis

As we have seen in Chapter VI, Cauchy's concept structure of analysis remained relatively isolated in France. The *École polytechnique* as the principal institution for teaching mathematics had been increasingly confined to its task of training engineers; work going beyond that was no longer encouraged. The lectures on analysis preserved insofar a "metaphysical" character as they continued, in their presentation of foundations, to be marked by their endeavor to grapple with the infinite, in particular concretely within their quite different efforts at meeting the contemporary cultural and social pressure toward integrating the concept of the *infiniment petits* into their conceptual edifice.

Even after the end of the "second" generation of analysis textbooks, numerous textbooks were still being published that were marked in this sense "metaphysically" by the confrontation with the infinite, and which propagated the concept of the classical *infiniment petits*.

A typical example is presented by J. N. Haton de la Goupillière's (1833–1927) *Éléments du Calcul Infinitésimal* of 1860, which properly speaking still belongs to the second generation. The author, an *ingénieur des mines*, and professor for analysis and mechanics at the *École des Mines,* had the intention of orienting his work to the applications for mining engineers. He thus limited the subject matter to a certain selection. As appropriate method, he chose neither that of Leibniz, nor the contemporary *méthode mixte*, that is, the compromise between *limites* and *infiniment petits*, but declared it to be evident that he had not hesitated to select the method of the *infiniment petits* as perfected by Carnot, giving the well known grounds. Not only was the method the sole method suited for applications, he said, but it also facilitated research. In his conception, the author adhered so closely to Carnot that he believed himself able to "rigorously" justify the method with the principle of error compensation (Haton de la Goupillière 1860, V f.). For this purpose, he faithfully kept to Carnot's argumentation in the latter's *Métaphysique,* according to which the seeming approximation provides exact values (ibid., 5 ff.). *Infiniment petits* were also introduced in the classical way as variables having the limit of zero (ibid., 9).

The tendency toward conserving traditions is primarily evident in fields of application. A quite analogous approach is offered by Paul Haag (1843–?), an engineer of the other important civil branch of the *Ponts et Chausées,* and likewise professor at its school, the *École des Ponts et Chaussées,* and, like Haton de la Goupillière, connected to the *École polytechnique* as *répétiteur.* His textbook of 1893, too, relied exclusively on the *infiniment petits,* but primarily in Duhamel's sense: based on the principle of substitution, they served, in their application to *rapports* and to *sommes,* to found differential and integral calculus (Haag 1893, 1 ff.).

The textbook of a member of the French Academy, too, is marked by the context of applications despite the fact that it was written for the *Faculté des Sciences,* which was attaining a function more independent of the *École polytechnique.* J. Boussinesq (1842–1923), professor for *Mécanique physique* at the Paris Faculty, confined the object of analysis to *continuous* functions (Boussinesq 1887, 1 f.). He likewise assumed *infiniment petits* as the basic concept in Duhamel's sense in order to be able to conceive of quantities as of limits of variables. Just as analogously, he assumed the principle of substitution (ibid., 62 ff.).

The confinement to continuous functions is typical for the tradition and practice of *physico–mathématique* in France. Charles Méray (1835–1911), a mathematician known for his work on differential equations and on function theory, formulated the dominant view decidedly in his textbook on analysis of 1894 in a polemic against the concepts formed by the new rigor. Thus, he declared functions without derivatives, nonintegrable functions, and even discontinuous functions to be only of speculative interest:

> Functions discontinuous, without derivatives, not integrable, etc. *are encountered only in metaphysical dissertations*; it is therefore pointless to be concerned with them (Méray 1894, XIII).

He not only affirmed that it was pointless to be concerned with them, but also challenged the connected requirements of rigor. He qualified them as *bizarrerie,* as "complicated detours"; they provided in no way the necessary instruments to study even the most minor function. He thus considered it unnecessary to examine them in depth. On closer inspection, one would certainly find new weaknesses with them, but without ever reaching firm ground (ibid., XIV). To a considerable part, Méray's insistence on benign, not malignant objects of mathematics was already a reaction to the dramatic change that began to prevail in French mathematics from the 1880s. The dominance of *physico–mathématique* was waning, and an increasing number of scientists were concerned with questions of theoretical mathematics. This change was closely tied to a further change of paradigm. After the 1880s, the hitherto rather modestly functioning *facultés des sciences* attained an ever increasing independent function for scientific training in the French Republic newly established after 1871, their core being ENS, the *École Normale Supérieure,* which offered ever more highly qualified teaching. This process led to newly founded universities in 1896 that integrated the hitherto isolated faculties. This not only resulted in the *École polytechnique* losing its up to that point dominant role in French mathematics, but also shifted the foci of interest and research away from topics of geometry and mechanics toward analysis as a new center of gravity (Gispert 1991, 206 ff.).

This change in concept and institutions became manifest in several textbooks on analysis that received, and elaborated on, the new concepts developed in particular in Germany. Two typical works of this kind are the *Introduction à la Théorie des Fonctions d'une Variable* of 1886 by Jules Tannery (1848–1910), lecturer at the ENS in Paris, and in particular the second edition of the *Cours d'analyse de l'École Polytechnique* of 1893 by Camille Jordan, (1838–1922), professor at this school, since the latter's first edition of 1882 had still been immersed in the *mainstream* tradition in France (Gispert 1982). The completely revised second edition, like Tannery's textbook, shows a profound reception of the new works on conceptual rigor coming in particular from Germany.

This reception is particularly explicit in Tannery. He was the first to rely on an international mathematical production—on Cauchy, Abel, Dirichlet, Riemann, Heine, Weierstraß, Dini, and Lipschitz—most heavily, however, on Cantor, Weierstraß, and Dedekind. Typical is his cautious reference to Weierstraß's lectures: "as far as my information is exact" (Tannery 1886, IX). In the context relevant here of the linkage between fundamental contexts, the following points in this revised edition merit being highlighted:

- Negative and irrational numbers are introduced as part of the preceding establishment of real numbers.
- Limit and convergence considerations are based on an unfolded calculus enveloping inequalities involving the signs for absolute values (Crelle's dashes); by means of these signs, operating and estimating with e and d now becomes practice.
- The *limes* sign is now generally used, including giving the index variable, in Jordan only, if confusion might arise:

$$\lim_{n=0} \frac{f(x+n) - f(x)}{n} = f'(x).$$

In Tannery, in contrast, it is generally

$$\lim_{n=\infty} u_n = U,$$

being identical with $\left| U - u_p \right| < \varepsilon$ (Tannery 1886, 25).

- Continuity of functions was now clearly differentiated as to its various properties. Tannery first introduced continuity in the interval, and then pointwise continuity (ibid., 103 ff.). Subsequent to that, he formulated the continuity of sequences (ibid., 111), while not explicating uniform continuity. Jordan, for his part, introduced continuity as pointwise, and then generalized to continuity in the interval, more precisely on a set E, the domain of definition. After that, he introduced uniform continuity on this set

(Jordan 1893, 46 ff.).

- In a quite analogous way, the two also differentiated between (ordinary) convergence and uniform convergence of series of functions (Tannery 1886, 132 f.; Jordan 1892, 310 f.). Both works, however, maintained Cauchy's unspecific mode of notation for series of functions:
$$u_1, u_2, ..., u_n, ... \text{ and } u_1 + u_2 + ... + u_n + ...$$

- Integral calculus was again based on the definite integral. For its definition, functions were no longer assumed to be continuous, but the new class of integrable functions was introduced instead. The definition of the definite integral followed that of Riemann's integral (Tannery 1886, 269f.; Jordan 1893, 37 f.).

If we consider, in addition to this pervasive modernization in France, that, say, Ulisse Dini's (1845–1918, mathematics professor at the University of Pisa) Italian textbook of function theory of 1878 is structured in an analogously rigorous and conceptual way,[1] we will be able to conclude that an international community at least of top mathematicians formed for the first time during the last third of the nineteenth century, a community that intensely advanced research on the basis of shared conceptual notions of analysis.

When and where the actual convergence to an *international* community took place, however, remains an open question.

3. On the Rise of Modern Infiniment Petits

After the end of the classical infinitely small quantities—or anyhow of their productive use—tentative approaches to the modern infinitely small developed, mainly in Germany. There exists, to my knowledge, no study of the history of their emergence other than isolated remarks in diverse publications. As in the other concluding parts of this chapter, we intend only a sketch here, as a view from the classical perspective.

One of the reasons for the lack of interest in this stage of emergence obviously is that historians have traditionally taken no note of the difference between infinitely small quantities and infinitely small numbers. As we have seen, the still imprecisely conceived preclassical *infiniment petits* had arisen from the

---

[1]   It may well be said that Dini's work showed the most rigorous structure of analysis of the time. For instance, he differentiated continuity into pointwise continuity, continuity in the neighborhood of a point, continuity in the interval, piecewise continuity, distinction between uniform and nonuniform continuity. Exemplary is also his discussion of necessary and sufficient conditions for the integrability of functions, and his explication of the Riemann conditions (Dini 1892, 318 ff.).

indivisibles. They became the clearly defined classical *infiniment petits*, i.e., variables as special quantities. The step in modern times, on the other hand, was to extend the concept of number, and to establish the *infiniment petits* as a new type of number. This step, however, was being prepared only at the end of the nineteenth century, and at the beginning of the twentieth, and did not yet become completely explicit. The protagonists favoring this direction remained ambivalent as well in their approaches, and marked by the tradition of the concept of quantity. The starting point of these new approaches, after the classical concept had virtually lost its functions by the modern extension of the limit methods, had been to look for a new solid foundation of infinitely small quantities in the preclassical, or in Poisson's, sense. The primary interest in this research was not operative, as in the preclassical and classical concepts, but in agreement with the new leitmotiv of rigor in analysis, in particular in Germany, was an explorative one that aimed at extending foundations.

Very revealing is thus a statement by Schlömilch of 1845 on the concept of infinitely small quantities. Schlömilch, one of the most active propagators of Cauchy's innovations in Germany, on the one hand, makes it abundantly clear that these quantities continued to be understood in the frame of the traditional concept of quantities, and that something like a "nonstandard" interpretation was completely beyond the conceptual horizon even of mathematicians fascinated with foundations:

> Such experiments [of applying calculational operations to the actually infinite] give indeed rise to the strongest contradictions against the foremost principles of quantity theory; thus, for instance, the proposition $\infty + a = \infty$ violates the principle that every quantity is larger than one of its parts. If one wishes to resolve this contradiction, there is nothing left *) than to declare for ∞ that it is no quantity, but doing this makes one escape from Scylla to fall prey to Charybdis, for now one has added two things of different kind, one of them belonging to the domain of quantities, while the other does not.

As a footnote, he had added at *):

> One could try another means of resolution by declaring the ordinary definition of "quantity" to be too special, giving another. Well, then give it! But this has not yet been done by any advocate of this view, and for very simple reasons (Schlömilch 1845, XI).

On the other hand, this is the first instance indicating a germinating awareness of the fact that such operating cannot be legitimized with the familiar concepts, but rather requires extending the theoretical frame by concepts hitherto unknown.

It is typical that Paul du Bois-Reymond, one of the major partisans of the new approaches, discusses the latter as part of a "metaphysics of the analytic foundational concepts" (Du Bois-Reymond 1882, 285). Du Bois-Reymond (1831–1889) first studied medicine, and then mathematical physics in Königsberg, but stayed aloof from the mathematical schools under formation in Germany. As professor in Freiburg, Tübingen, and at the Technische Hochschule Berlin, he maintained an intense communication with Weierstraß, without, however, identifying with the latter's views.[2] Indeed, his work *Die allgemeine Functionentheorie* is structured along a dialogue on foundations between an "idealist" and an "empirist." The author takes a stand against "a separation of number [...] from quantity," since mathematics would then be no more than a "formalist–literal skeleton," a "mere game of signs." Accordingly, he also obscured, in discussing infinitely small quantities, the difference between quantities and numbers: He only spoke in a neutral tone of the "infinitely small" (cf. ibid., 71, and passim). Du Bois-Reymond's intention was to "prove the existence" of the infinitely small. For him, this proof followed from geometry, from the unlimited character of the "number of points on the unit line segment." This, he said, called forth "with logical necessity the belief in the infinitely small" (ibid., 71): "hence the infinitely small really exists" (ibid., 72), at least in the idealist system. While calling the actually "infinitely small" a "new type of quantity," he nevertheless claimed that it "had in common with the finite all the latter's properties" (ibid., 75). Since du Bois-Reymond's primary interest in this discussion was metaphysics, he did not apply the new type of quantity operatively. He only explained that it met L'Hospital's requirements for infinitely small quantities (ibid., 74).

Du Bois-Reymond had declared his interest in the *infiniment petits* in 1873 already in a letter to Weierstraß, but clearly with "metaphysical" intention, noting, after having quoted Weierstraß's famous example of a continuous, nowhere differentiable function:

> What moves me so much in your examples is that they seem to me to be a proof for *the existence of the infinitely small.* I am not yet able to formulate this clearly, but its outline drifts ever more distinctly before my soul. The infinitely small is incomprehensible, but does anything speak against its existence? The infinitely large is not graspable, either, and who dares to doubt its existence![3]

Du Bois-Reymond did not so much achieve something conceptually new as to have prompted mathematicians to reflect on new types of quantities. This motivated Otto Stolz (1842–1905), who received—already a *Privatdozent* in Innsbruck—a scholarship from the Austrian government for studying three terms in Berlin (1869/70 to 1870/71), and afterwards one term in Göttingen. In these studies, Stolz was profoundly influenced by Weierstraß, and became the first to present the model of Weierstrassean rigor of analysis in a textbook in the German language (see Section 2. above).

Stolz was deeply interested in the foundations of analysis, studying the contributions to it by eminent and less well known authors on the basis of his own comprehensive knowledge of the literature, exerting an essential influence

---

[2]   Gerhard Hessenberg, one of the prominent mathematicians of the Nelson-Fries school, called him one of the "adherents" of the "mathematical mysticism" still existent among the mathematicians of the time, but as one who "kept his distance from all mysticism in the field of productive activity" (Hessenberg 1904, 137).

[3]   GStA, *Nachlass* Weierstraß, Mappe 4, letter dated December 13, 1873 (cf. Schubring 1998a); original emphasis is by underlining.

by this on Felix Klein as well (Chr. Binder 1989). His research into foundations has found far too little attention as yet. His point of departure, however, probably due to his classical education, was to assign priority to the concept formations of the Greek, and in particular attributing to them that they had achieved the model of rigor. One of the consequences of this was that he also deemed the concept of quantity to be more basic than the concept of number (cf. Stolz 1891).

Stolz's special achievement in this connection is that he showed, on the basis of one of du Bois-Reymond's contributions of 1877, the general starting point from where new types of quantities and numbers, in particular infinitely small numbers, can be constructed. At first only in a note to his historically well-informed article of 1881 on Bolzano, but subsequently in an extensive separate article of 1883, Stolz elaborated the fundamental importance of the Archimedean axiom. To my knowledge, he was the first to show what kinds of operations are possible with systems of quantities without the Archimedean axiom having to be satisfied, and for which operations this axiom must be assumed. Thus, the continuity of a system of absolute quantities requires that it satisfy the axiom. He called such systems, in accordance with du Bois-Reymond, *linear* systems, and non-Archimedean systems *nonlinear* ones (Stolz 1883a, 506 f.). In contrast to du Bois-Reymond, Stolz was aware of the fact that not all the properties of finite quantities hold for the new types of quantities. This is why he noted for the constructed actually infinitely small and infinitely large quantities, "Several of the properties of the ordinary absolute quantities, however, had to be given up for this" (Stolz 1891, 16).

Stolz thus seems to have been the first to apply the method of axiomatization, which had hitherto been invented and applied for non-Euclidean geometries—i.e., the hypothetical positing of geometrical systems that satisfy certain axioms, or not—productively to "theoretical arithmetic."[4] It was the subsequent unfolding of this method of axiomatization by Hilbert and his disciples that made it the new form of generalization in mathematics.

Stolz used examples from du Bois-Reymond to construct two types of infinitely small quantities. These were a class of functions, which Stolz called *Momente der Funktionen*, showing specific limit behavior. Stolz defined calculating operations for these "things" by first introducing equality and smaller/larger relations for them, and then proceeding to addition, subtraction, and multiplication. Dividing them, in contrast, was not possible unrestrictedly. By "extending" this system of moments, it was possible to set the "things" in relation to absolute numbers (Stolz 1885, 205 ff.). Stolz did not intend to develop an operational system of new quantities. He, too, only wanted to demonstrate—in a more "metaphysical way"—the possibility in principle of the

---

[4]   Mircea Radu has shown in his doctoral thesis (2000) *Nineteenth Century Contributions to the Axiomatization of Arithmetic*, that the development of axiomatics for arithmetic was not a consequence of the axiomatization of geometry toward the end of the nineteenth century, but had set in earlier and independently.

actually infinitely small quantities, and this meant for him, of non-Archimedean quantities. His even more modest aspiration was to show "that they do not always obey the axiom of Archimedes" (ibid., 213). Stolz had no problem admitting that he considered these systems by no means "indispensable in analysis," and that "their practical exploitability was inferior" to that of traditional concepts of quantity (ibid., 206 f.), for instance because negative numbers are not attainable in the extension, and because no general multiplication was definable for his second type of quantity (ibid., 213 f.).

Stolz's examples indeed did not initiate any further intramathematical development, remaining isolated reflections on mathematics. They were at first discussed only by Georg Cantor, as a mathematical admissibility of infinitely small quantities contradicting the basic concept of his own set theory, according to which the infinitely large, while being mathematizable and subordinable to a calculus, is not reciprocal to infinitely small quantities of any kind. In particular, the works of Stolz just quoted provoked Cantor to make explicit his long-held conviction according to which infinitely small quantities were paradoxical, and hence impossible. Cantor did so in a letter to Weierstraß of 1887, which he subsequently included in his treatise *Mitteilungen zur Lehre vom Transfiniten* in a periodical of the same year. Since it had been taken from a letter, he only communicated and explicated a "proposition," but without proof: According to that, an assumed infinitely small linear quantity ζ could not be made into a finite one, not even by multiplication by a transfinite order ν, however large. This contradicted the concept of linear quantities; the Archimedean axiom thus was not an independent one, but followed by necessity from the concept of linear quantities (Cantor 1966, 407 ff.).

This reasoning of Cantor, however, completely ignored that Stolz's examples explicitly referred not to linear, but rather to nonlinear, quantities. In a brief answer, Stolz quite modestly pointed out this fallacy (Stolz 1888, 601 ff.).

Cantor's brief hints prompted the Italian mathematician Giulio Vivanti (1859–1949) to publish a complete version of the proof against the existence of actually infinitely small quantities in 1881. In doing so, however, he even bowdlerized Cantor's argumentation by explicitly using, as linear quantities, geometrical entities, namely segments. Without much problem, he was thus able to conclude that infinitesimal calculus required only finite quantities (Vivanti 1891, 146).

Even Giuseppe Veronese (1854–1917), mathematics professor in Padua, to whom the literature attributes the introduction of infinitely small numbers, confined himself, in his concept formations, to geometry. While Veronese spoke of "numbers" in connection with actually infinitely large and small quantities (Veronese 1894, XXV), all his arguments really refer to straight-line segments. By using such segments, he attempted to prove the existence "of the infinitely small" (ibid., 96 ff.). He also founded negative numbers on geometrical concepts, again via segments (Veronese 1894, 198 ff.). At the same time, he made the qualification that it was "not necessary *to believe in the concrete reality* of the really [actually] infinitely large and infinitely small" (ibid., XII). For Veronese's approach, too, it was essential to examine systems of quantities as to which axioms they minimally satisfy, and in particular to investigate quantities for which the Archimedean axiom did not hold.

Veronese's approaches did not have any substantial effect, just as those of Stolz did not. This was also due to the fact that Giuseppe Peano, in an incisive review of 1892, demonstrated that the author had fallen prey to fundamental inconsistencies in his definitions and hypotheses (Peano 1892a). In a second article, Peano showed—on the basis of Cantor and Vivanti—that constant and actually infinitely small segments are contradictory in themselves (Peano 1892b).

It is remarkable that after Stolz's first attempts at conceiving of abstract objects as infinitely small quantities—which did not show any further operational capacity, however—attention again concentrated on geometrical objects. Weierstraß, however, had characterized this already in 1873, in his answer to du Bois-Reymond's views, as a path not suitable for generalizaton:

> Regarding the infinitely small, I should like to note only that the views on them will be essentially different according to whether one enters the domain of analysis from geometrical and physical notions, that is, with the concept of extensive quantity, or from algebra, that is, from the concept of number and the necessarily given arithmetical basic operations attached to it. I consider the latter path to be that on which alone analysis can be founded with scientific rigor, and where all difficulties can be removed (Mittag-Leffler 1923, 203 f.).

Nevertheless, there subsisted for a long time a developmental stage of analysis in which infinitely small numbers did not belong to the conceptual horizon, and where geometry remained the field where non-Archimedean objects could be made concrete. An impressive example of that is Hilbert, who unfolded the method of axiomatization in its most advanced form in his famous *Grundlagen der Geometrie* (first edition 1899). This is indeed where he discusses the independence of the axioms of continuity of other axioms, constructing non-Archimedean geometries for this purpose (cf. Hilbert 1999, 47 ff.).

Felix Klein, in his *Elementarmathematik*, also used geometrical objects to give his teacher students a model of non-Archimedean quantities, the hornlike angles—hence for the first time nonlinear quantities (see Chapter II.1.1.; Klein 1925, 221 ff.). Since the mathematically interesting in them had for a long time been developed further in the concept of curvature in analysis, this modeling did not give rise to new impulses.

In contrast, Gerhard Hessenberg (1874–1925), an eminent representative of the Nelson-Fries school, which discussed foundational issues in close cooperation with Hilbert,[5] explicitly listed the possibility of infinitely small numbers within arithmetic in his treatise *Das Infinite in der Mathematik*:

---

[5]   Since my long-prepared study on the philosophy of mathematics in Fries and in Nelson and the latter's school is not yet available in print, I should like, for

> As a matter of principle, it must be observed that an arithmetic is imaginable without any logical contradiction in which a number *a*, together with all its multiples, is smaller than another number *b* and the totality of the latter's aliquot divisors (Hessenberg 1904, 189).

He did not see any promising program in that, however:

> In such an arithmetic, however, ordinary numbers are contained as elements, and everything added to that is ballast which does not solve any problem, does not simplify any proof, but only raises anew all the difficulties which fortunately have been removed today in ordinary arithmetic (ibid.).

Hilbert himself, however, later showed, in a lecture of 1919–20 on mathematical cognition, the existence of such numbers as constructible in a process of progressive generalization of the number concept. He presented the system of rational numbers as a "continuum of first order," and the system of real numbers as a "continuum of second order," which he established in such a way "that the ordinary rules of calculus are all preserved" (Hilbert 1992, 8). He then inquired whether the process of generalization could be continued by inserting ever new numbers, in particular by infinitely small numbers. Hilbert gave a model of such new "numbers," showing that all the elementary operations can be executed in the extended number domain. He also showed, however, that this "continuum of third order" did not satisfy the Archimedean axiom (ibid., 8 ff.). Since giving up this axiom resulted in the axiom of completeness being no longer satisfied, the continuum of second order, for Hilbert, had both against the narrower continuum of first order, and against the wider continuum of third order, the "intrinsic advantage" of completeness. With it, "the formation of numbers hence found its natural conclusion." The priority on completeness led Hilbert to believing

> that from the logical–systematic point of view as well there are grounds for proceeding, in forming the concept of the continuum, to go precisely up to the continuum of second order, and to stop there (ibid. 13).

Contrasting these approaches oriented toward foundations, Ludwig Neder, in a contribution of 1941, was the first to present a concrete approach to modeling differential calculus with actually infinitely small quantities, which probably did not receive any attention at the time. Carl Schmieden and Detlef Laugwitz, as well as Abraham Robinson, then definitively founded the modern infinitely small numbers in the 1950s.

Leonard Nelson's (1882–1927) activity in Göttingen, to refer to Peckhaus's book (1990).

4. Some Closing Remarks

At the close of this study and of presenting a long-term history of the conceptual developments, structural parameters become salient—across all cultural and national differences—that took effect again and again, and are still virulent today. Charles Méray's categorical rejection of an entire current of research because of its bizarre objects of functions was not simply reactionary polemics. On the one hand, his view evidences that he was profoundly marked by a vision of continuity as a principle determining mathematics in its entirety, the powerful and lasting effect of which we have exemplified by Cauchy's work, and which precisely ensured that mathematics and mechanics could be "made to fit." On the other hand, Méray's plea in favor of intuitiveness expressed the concern that mathematical activity should remain culturally anchored. In earlier stages, too, rejecting conceptual generalizations had been most closely coupled with the intention of keeping mathematical research legitimized within the understanding of mathematics that was culturally and socially shared. There are only isolated cases in the context of such cultural traditions in which some elements of mathematical concepts were held to be absolute or unidirectional, tendencies that have been called "fundamentalism" here, where no fruitful confrontation in debate was possible any longer. On the other hand, there were only rare stages in which the cultural and social context itself urged in favor of generalization in science; just recall the self-confident belief in the superiority of the moderns over the *anciens* before 1700, the enthusiasm for the analytic method at the time of the French Revolution, and the scienticism both in the German Empire and in the French Third Republic after 1871—together with the aspiration toward new rigor in analysis.

Struggling against the weight of tradition, generalization generally found it rather difficult to prevail. Even in its modern, most developed, form, in the axiomatic method masterfully elaborated as a means of cognition by Hilbert as its best known representative, it saw itself again and again confronted with the reproach of formalism. It has often been shown, also by Hilbert himself, that his program of formalization did not mean just any formalism, but rather a consistent elaboration of the theoretical character of mathematical concepts, which at the same time expands their meaning and content. Hilbert, in particular, stressed that "mathematical concept formation is prompted by intuition and guided by experience" (Hilbert [1919–1920] 1992, 11).

The historical agents again and again understood, and formulated, the polarity between intuitiveness and generalization as the opposition between *synthesis* and *analysis*. The immense variance of the meanings assumed for this contrast in the respective constellations confirms the broad scope that has been elaborated in the volume *Analysis and Synthesis in Mathematics* (Otte, Panza 1997). What is surprising in view of this variance, however, is how consistently the historical agents respectively concerned agreed in interpreting the opposition between synthesis and analysis as one of geometry and mechanics versus arithmetic and algebra. Just as surprisingly, the concept of quantity finally turned out to be a pervasive structural parameter—both for the concept field of negative numbers and for the concept field of the *quantités infiniment petites*, as the epistemological support of a holistic character in mathematics against a specialization that appears to place culture in jeopardy.

Rigor as a central value, in contrast, was claimed by both sides for their respective positions, in an exemplary fashion by the partisans of synthesis in the 1811 decision of the *École polytechnique* in favor of introducing the *infiniment petits*, and by Cauchy's claiming for himself the rigor of geometry as a model in 1821. Rigor, obviously, is the strongest subjective moment in mathematics.

Appendix

A. The Berlin Contest of 1784 Reassessed

As shown in Chapter III.8.4., Roger Martin had emphasized three results of his 1786 *mémoire*, quite in line with the requirement of the famous 1784 Berlin Academy contest. He had given, he said, a clear and precise theory of the mathematically infinite, he had shown how true theorems resulted from contradictory assumptions, and he had also shown that the concept of ultimate ratios was a safe principle suitable to replace the concept of the infinite (Martin 1788, 71 f.).

That Martin referred so directly to the Berlin Academy's contest question of 1784 suggests that he himself submitted a treatise. My careful scrutiny of the treatises still present in the academy archives turned up a negative result. Martin never sent in anything.[1]

At the same time, this negative result prompted me to examine this contest task's significance. None of the publications on the history of analysis (including earlier contributions of my own) omits highlighting this contest as a milestone in clarifying foundations. Its question was indubitably provoking, and productively addressed foundations. Hence, a host of qualified submissions should have been expected, emanating in particular from eminent mathematicians.

Gillispie, at the occasion of his edition of the paper submitted by Carnot, already reported that the other submissions did not contain any answer proper to the contest's question (Gillispie 1971, 157). He did not provide any systematical evaluation, however.

From the initially 23 submissions, 21 are still present in the archives. L'Huilier's prize-winning work was returned to him at his own request for use in print, and one contribution was withdrawn. The overwhelming impression is that a large majority (15) among the 22 texts was submitted by laymen, lay authors in the sense that they were unable to argue according to mathematical standards, and to move forward the contemporary level of knowledge in mathematics. Many of the texts comprise only a few pages, four papers are glaringly unmathematical.

Among the seven contributions in German, five are from laymen. The only original and interesting German text was submitted by a professional mathemati-

---

[1]   AAdW, Preisschriften der Akademie; I-M 838-858 [submissions for the contest of the mathematical class 1784].

cian. Besides one text in Latin, there were fourteen texts in French. Except for four interesting texts, the majority here also are without significance. Since the text that was awarded the "accessit," that is, second prize, contains neither a reflection on the problem, nor an explicit answer,[2] it may be assumed that awarding it the "accessit" was due to an error and that this distinction was probably assigned to Carnot's treatise. Carnot's text indeed represents a top achievement among the answers submitted that at the same time really worked on the question in a conceptionally interesting mode. Incidentally, other submitters also argued with error compensation (cf. Chapter V.1x.).

What is surprising about this contest's effect is that professional mathematicians seem not to have felt at all concerned. While Karsten, for instance, took note of the contest, he preferred to publish his work on his own (Karsten 1786) instead of submitting it. Since Martin's second *mémoire* is also a text that might well have been submitted, I had to ask myself whether members of other academies had been excluded from participation. Karsten was a member of the Munich Academy, Martin of the Academy of Toulouse. But the director of the Berlin Academy's archives, Wolfgang Knobloch, kindly communicatied to me in detail that there was no such exclusion. The academy's statutes of 1744 prescribed an open contest for the first time; they even permitted the Academy's own members to participate. This rule was changed in 1746: members of the Berlin Academy were excluded henceforth, but otherwise, there were no limitations.

It must thus be concluded that the contemporary mathematicians by no means considered the question to be as significant and foundational as it has hitherto been seen by historiography.

B. Carnot Defining Quantités Infiniment Petites

Part I

[1.] En effet, qu'est-ce qu'une quantité infiniment petite? Ce n'est autre chose que la différence de deux grandeurs qui ont pour limites une même troisième grandeur (par ce terme de grandeur, j'entends ici, une quantité effective. C'est-à-dire qui ne soit ni 0, ni $\frac{1}{0}$) voilà en admettant la notion des limites une réponse claire, exacte et précise (Carnot 1785, note 1, 174).

---

2   Ibid., Nr. 856.

[2.] Toute quantité dont la dernière valeur est une quantité quelconque effective, c'est-à-dire dont la limite n'est ni 0 ni $\frac{1}{0}$, se nomme *quantité finie*; toutes celles au contraire, qui n'ont point de limites effectives, c'est-à-dire, dont les dernières valeurs sont 0 ou $\frac{1}{0}$ se nomment *quantités infinitésimales*.

Parmi les quantités infinitésimales, celles dont la limite ou dernière valeur est 0 se nomment *infiniment petites.** *

  * Nous avons dit (note 1) qu'une quantité infiniment petite n'est autre chose que la différence de deux autres quantités qui ont pour limite commune une troisième grandeur quelconque effective, ce qui revient au même que la définition présente. Car soit x une quantité qui ait 0 pour dernière valeur, et X une autre quantité qui ait pour limite une grandeur quelconque effective Y, il est clair qu'on a $x = (X + x) - X$, c'est-à-dire que x est la différence des quantités $X + x$ et X, lesquelles ont évidement pour limites la même grandeur Y. On trouvera [...] la preuve que les quantités dont les géomètres font usage sous la dénomination de quantités infiniment petites sont en effet telles que nous venons de les définir.

  Ce nom porteroit cependant à croire, que les grandeurs auxquelles on le donne, doivent non seulement avoir 0 pour limite, mais que de plus elles ne peuvent avoir elles-mêmes que des valeurs très petites, c'est-à-dire qu'on ne seroit pas maître de leur attribuer des valeurs égales à telle ou telle grandeur donnée; mais cela n'est pas, ce sont des quantités variables auxquelles de même qu'à toute autre variable on peut attribuer diverses valeurs déterminées: ainsi l'expression d'infiniment petite est assez impropre, et ne contribue pas peu à faire prendre une idée fausse de ces quantités (ibid., Nr. 11, 12 and note, 181).

Part II

[1.] On nomme en général quantité *infiniment petite* la différence d'une quantité quelconque auxiliaire à sa limite; ainsi, par exemple, RZ, qui est la différence de RS à MP, est ce qu'on appelle une quantité infiniment petite (Carnot 1797, 25).

[2.] Ainsi on peut dire *en général qu'une grandeur infiniment petite n'est autre chose qu'une quantité dont la limite est 0, et qu'au contraire, une quantité infiniment grande n'est autre chose qu'une quantité dont la limite est* $\frac{1}{0}$ (ibid., 26).

[3.] Il suit encore de là qu'on peut regarder toute quantité infiniment petite comme la différence de deux quantités auxiliaires qui ont pour limite une même troisième quantité désignée (ibid., 27).

[4.] Donc on peut dire en général *qu'une quantité infiniment petite n'est autre chose que la différence de deux quantités auxiliaires qui ont la même limite* (ibid., 27 f.).

[5.] Donc, en général, on peut dire *qu'une quantité infiniment petite est le rapport de la différence de deux grandeurs qui ont pour dernière raison une raison d'égalité à chacune de ces grandeurs* (ibid., 28).

[6.] Enfin, il est évident qu'on peut dire encore *qu'une grandeur infiniment petite n'est autre chose qu'une quantité non désignée, à laquelle on attribue d'abord une valeur quelconque arbitraire, et qu'on suppose ensuite décroître insensiblement jusqu'à zéro* (ibid., 28 f.).

Part III

[1.] On nomme en général ~~quantités~~ infiniment petites ~~toutes celles qui~~ dans un système quelconque de quantités toutes celles qui peuvent être supposées aussi petites qu'on le veut et rendues simultanément égales à zéro sans rien changer à la valeur des autres quantités du système soit données soit variables.[3]

[2.] Dans un système quelconque de quantités, on nomme infiniment petites celles qu'on est maître de supposer en même tems aussi petites qu'on le veut et qui peuvent etre rendues simulanément egales à zéro sans rien changer aux données ou conditions proposées.

[3.] On appelle, en général, infiniment petites toutes les quantités qu'on est maître de supposer aussi petites qu'on le veut et de rendre simultanément égales à zéro sans rien changer aux données ou conditions proposées.

[4.] On nomme en général infiniment petites dans un systéme quelconque de quantités; celles qu'on est maître de rendre simultanément aussi petites qu'on veut et même égales à zéro, sans rien changer à la valeur des autres quantités soit constantes soit variables.

[5.] [On nomme en général infiniment petites dans un systeme quelconque de quantités, toutes celles qu'on a la faculté de faire decroître insensiblement et simultanément jusqu'à zéro sans rien changer à la valeur des autres quantités du systeme proposé].[4]

---

3   Deleted by Carnot himself.
4   Brackets by Carnot himself.

[6.] Dans un systeme quelconque de quantités, on nomme infiniment petites celles qu'on est maître de supposer toutes en meme tems aussi petites qu'on le veut sans altérer la valeur des autres ou changer rien aux données de la question.

[7.] On nomme infiniment petites dans un systeme quelconque des quantités celles qu'on est maitre de rendre aussi petites qu'on le veut ou aussi approchantes qu'on le veut de zéro, toutes en meme tems mais sans rien changer à la valeur des autres quantités ou conditions de la question proposée.

[8.] Dans un systeme quelconque de quantités, on nomme <u>infiniment petites</u> toutes celles qui peuvent être supposées simultanément et par changement insensible aussi petites qu'on veut ou aussi approchantes qu'on veut de zéro sans rien changer aux quantités proposées ou données de la question.

[9.] Dans un système quelconque de quantités qui sont des ~~constantes ou~~ variables on nomme infiniment petites toutes celles de ces dernières qui peuvent etre rendues simultanément et par changement insensible aussi petites qu'on veut ou aussi approchantes qu'on le veut de zero sans rien changer aux autres quantités proposées ou données de la question.

Source: *Nachlass* Carnot, Nolay, carton 28, no. 4.

Part IV

[1.] J'appelle *quantité infiniment petite*, toute quantité qui est considérée comme continuellement décroissante, tellement qu'elle puisse être rendue aussi petite qu'on le veut, sans qu'on soit obligé pour cela de faire varier celles dont on cherche la relation (Carnot 1813, 16).

[2.] Les différences des quantités qui se correspondent entre tous ces systèmes peuvent donc être supposées aussi petites qu'on le veut, sans rien changer aux quantités qui composent le premier, et qui sont celles dont on cherche la relation. Ces différences sont donc de la nature des quantités que nous appelons *infiniment petites*: puisqu'elles sont considérées comme continuellement décroissantes, et comme pouvant devenir aussi petites qu'on le veut, sans que, pour cela, on soit obligé de rien changer à la valeur de celles dont on cherche la relation (ibid., 17).

[3.] Lorsque, parmi ces quantités auxiliaires, il s'en trouve d'une nature telle qu'on soit maître de les rendre toutes à la fois aussi petites qu'on le veut, sans faire varier en même temps les quantités proposées, cette circonstance donne lieu à des simplifications accidentelles très importantes, et ce sont précisément ces simplifications qui ont fait naître cette branche de calcul qu'on nomme *analyse infinitésimale*, laquelle n'est autre chose que l'art de faire choix de semblables auxiliaires, les plus convenables suivant les différents cas, de s'en servir de la manière la plus avantageuse, pour exprimer les conditions des diverses questions, et pour opérer ensuite l'élimination de ces mêmes quantités, afin qu'il ne reste plus dans les formules, que les seules quantités dont on voulait connaître les rapports (ibid., 23).

[4.] Car ce que dans celle-ci [la méthode infinitésimale] on nomme quantités infiniment petites n'est évidemment autre chose, d'après la définition que nous en avons donnée (14),[5] que la différence d'une quantité quelconque à sa limite, ou, si l'on veut, une quantité dont la limite est 0 (ibid., Vol. 2, 35).

[5] Or, la différence d'une quantité quelconque à sa limite est précisément ce qu'on nomme ou ce qu'on doit nommer une quantité infiniment petite (ibid., Vol. 2, 69).

C. Calendar of Cauchy's Traceable Correspondence

SEPTEMBER 3, 1815

Letter from Cauchy to the Abbé Auger in Le Havre: report on the search for the politicalreligious writings that Abbé Auger had requested. At the same time, report on measures for the restoration of the monarchy and on Cauchy's own activities for the planned elections.

Published in: Durry: Mariemont 1959, 216217.

MAY 30, 1816

Letter from Cauchy to Paolo Ruffini.

Ruffini Opera, Vol. 3.

MARCH 1820

Letter from Cauchy to Charles Babbage: delighted reaction to Babbage's sending of three treatises on the *calcul des fonctions* and on the summation of series.

Published, with disfiguring reading errors, in: J.M. Dubbey 1978, 90.

FEBRUARY 3, 1821

Letter from Cauchy to Libri: remarks about a manuscript by Libri on number

---

5   That is, the definition reproduced here as no. 1 in part IV.

theory.

Location: Biblioteca Moreniana, Florence. Engl. Translation published in Belhoste 1991, 321–323.

MAY 18, 1821

Letter from Cauchy to Libri: announcing receipt of manuscript about prime numbers.

Location: Biblioteca Moreniana, Florence. Engl. translation published in Belhoste 1991, 323.

JULY 24,1821

Letter from Cauchy to Sophie Germain: thanks for a work sent by her. Sending himself a copy of his *Analyse Algébrique*.

Location: Bibliothèque Nationale, Paris: Ms f. Fr. 9118, fol. 4.

SEPTEMBER 20,1821

Letter from Cauchy to Ruffini.

Ruffini Opera, Vol. 3.

JULY 23, 1826

Letter from Cauchy to Sophie Germain. Thanks for letter and accompanying *mémoire* .

Location: Bibliothèque Nationale, Paris: M f. 9118 Frs, fol. 6.

MARCH 3, 1828

Letter from Cauchy to Libri: news that he has succeeded in nominating Libri as a *membre correspondant* of the *Académie*.

Location: Biblioteca Moreniana, Florenz. Engl. translation published in Belhoste 1991, 324.

JULY 21, 1829

Letter from Cauchy to Libri: sending him more specimens of the *Exercises de Mathématiques*.

Location: Biblioteca Moreniana, Florenz. Engl. translation published in Belhoste 1991, 324–325.

AUGUST 6, 1829

Letter from Cauchy to Ricardi: Thanks for communicating the admission into the *Académie Royale des Sciences, Lettres et Arts* of Modena .

Location: published in: Terracini 1957.

OCTOBER 7, 1830

Letter from Cauchy to the Emperor of Austria: request for support for the project of a Jesuit academy in Fribourg.

Location: published in: A. Terracini 1957, 195.

JUNE 27, 1831

Letter from Cauchy to the president of the *Académie des Sciences*, Paris: sending a text published in the *Biblioteca Italiana*. Explaining his extended absence.

Location: Personnel file of Cauchy in the Academy archives. Published in: A. Terracini 1957, 183. Engl. (inexact) translation published in Belhoste 1991, 325.

JULY 30, 1832

Letter from Cauchy to Charles Konig: Thanks for his appointment as *membre étranger* of the Royal Society in London. Announcing the sending of *mémoires*.
Location: D.E. Smith Collection, Butler Library, New York.

MARCH 25, 1833

Letter from Cauchy to the president of the *Académie des Sciences*, Paris: Sending a lithographed *mémoire* on the *calcul des limites* in celestial mechanics.
Location: Bibliothèque de l'Institut, Paris. Engl. translation published in Belhoste 1991, 325–326.

JULY 18, 1833

Letter from Cauchy to the Count de l'Escarène: information on his travel to Prague to begin his task of educating the prince.
Location: Musée National de l'Éducation, Rouen. Engl. translation published in Belhoste 1991, 326-327.

SEPTEMBER 24, 1834

Letter from Cauchy to the Count de l'Escarène: postponing the decision whether he will continue the teaching assignment in Turin or remain in Prague.
Location: Musée National de l'Éducation, Rouen. Engl. translation published in Belhoste 1991, 327–328.

OCTOBER 9, 1835

Letter from Cauchy to Giorgio Bidone: Sending him his text *Mémoire sur l'interpolation*.
Location: Library of the *Accademia delle Scienze di Torino*. Excerpts and discussion in P. Dupont 1982.

NOVEMBER 4, 1835

Letter from Giorgio Bidone to Cauchy: Thanks for sending a treatise and reminding of the scientific conversations in Turin important for him.
Location: Library of the *Accademia delle Scienze di Torino*. Excerpts published in A. Terracini 1957, 202.

NOVEMBER 13, 1835

Cauchy's answer to Bidone's letter of November 4: Arguing for the superiority of his interpolation method as compared with the method of least squares.
Location: Library of the *Accademia delle Scienze di Torino*. Excerpts published in P. Dupont 1982, 467.

PUBLISHED FEBRUARY 22, 1836 (IN COMPTES RENDUS DE L'ACADÉMIE DES SCIENCES [C.R.])
Letter from Cauchy to the president of the *Académie des Sciences*, Paris: comments on the new issues of the *Nouveaux Exercises*.
Location: Cauchy Œuvres, Série 1, tome 4, 5–9.

PUBLISHED FEBRUARY 9, 1836 (IN C. R.)
Letter from Cauchy to Ampère: supplement to the letter of last Friday, on *théorie de la lumière*.
Location: Cauchy Œuvres, Série 1, tome 4, 9–11.

PUBLISHED APRIL, 11, 1836 (IN C. R.)
Letter from Cauchy to Ampère: explaining different phenomena of light in a wave system.
Location: Cauchy Œuvres, Série 1, tome 4, 21–30.

PUBLISHED MAY 2, 1836 (IN C. R.)
Letter from Cauchy to Libri: on the *théorie de la lumière*.
Location: Cauchy Œuvres, Série 1, tome 4, 30–32.

PUBLISHED MAY 9, 1836 (IN C. R.)
Letter from Cauchy to Libri: continuation of the same subject.
Location: Cauchy Œuvres, Série 1, tome 4, 32–36.

PUBLISHED OCTOBER 3, 1836 (IN C. R.)
Letter from Cauchy to Libri: continuation of the same subject.
Location: Cauchy Œuvres, Série 1, tome 4, 36–38.

JANUARY 29, 1837 (PUBL. IN C. R. FEBR. 13TH, 1837)
Extract from letter of Cauchy to Coriolis: on his *calcul des limites*.
Location: Cauchy Œuvres, Série 1, tome 4, 38–42.

PUBLISHED MARCH 6, 1837 (IN C. R.)
Letter from Cauchy to Libri: on solutions of equations.
Location: Cauchy Œuvres, Série 1, tome 4, 42–45.

PUBLISHED APRIL 22, 1837 (IN C. R.)
Letter from Cauchy to an addressee not named: accounts of his researches on integral equations.
Location: Cauchy Œuvres, Série 1, tome 4, 45–48.

PUBLISHED MAY 22, 1837 (IN C. R.)
(open) letter, without addressees: "Première Lettre," about the solution of equations of an arbitrary degree
Location: Cauchy Œuvres, Série 1, tome 4, 48–60.

PUBLISHED MAY 29, 1837 (IN C. R.)
(open) letter, without addressees: "Deuxième Lettre," about the solution of

equations of an arbitrary degree.

Location: Cauchy Œuvres, Série 1, tome 4, 61–80.

JUNE 12, 1837

Letter from Cauchy to Moigno: answer to a letter by Moigno of May 17. Will examine whether the critique by Sturm and Liouville of his theorem of 1833 is well founded. On priority problems. Urges quick publication of three of his letters (6, 13, and 18 May).

Location: Bibliothèque de la Sorbonne. Engl. translation published in Belhoste 1991, 328–330.

DECEMBER 15, 1837

Letter from Cauchy to Moigno: Sending manuscript on the theory of light.

Location: notebook *Mélanges*, in the possession of Cauchy descendants. Engl. translation published in Belhoste 1991, 330.

JULY 25, 1839

Letter from Cauchy to Liouville: invitation for dinner and request to transmit the same invitation to Dirichlet.

Location: *Nachlaß* Catalan, Bibliothèque Générale of the University of Liège. Engl. translation published in Belhoste 1991, 331. (According to Belhoste, this letter was addressed to Catalan. As the entire context shows, however, this does not apply.)

JULY 25 (29?), 1839

Letter from Cauchy to Dirichlet: Invitation for dinner at his home, together with Liouville.

Location: *Nachlass* Dirichlet, Handschriftenabteilung der Staatsbibliothek Preußischer Kulturbesitz Berlin.

MARCH 19, 1841

Letter from Cauchy to the president of the Göttingen Academy of the Sciences: Thanks for the appointment as a foreign member.

Location: Handschriftenabteilung der Staats- und Universitätsbibliothek Göttingen, 4° Cod. Ms. hist. lit. 116:3.

OCTOBER 16, 1842

Letter from Cauchy to Moigno: supplement to a verbal conversation about the reflection of rays.

Location: Personnel file of Moigno in the library of the Centre des Fontaines of the French Jesuit province. Engl. translation published in: Belhoste 1991, 331–332.

FEBRUARY 6, 1843

Letter from Cauchy to Saint-Venant: hint at own researches on mechanics, at the occasion of a treatise submitted by Saint-Venant to the *Académie*.

Location: Archive of the *École polytechnique* (AEP), collection Saint-Venant. Engl. translation published in: Belhoste 1991, 332–333.

JUNE 30, 1843
Letter from Cauchy to Isadore Geoffroy-St. Hilaire: explanation on the affair around the choice of a successor for Libri.
Location: D.E. Smith Collection, Butler Library, New York

SEPTEMBER 29, 1843
Letter from Cauchy to Pingard (*Institut de France*).
Location: Sammlung Darmstädter, F1c 1836 (1).

JULY 31, 1845
Letter from Cauchy to Saint-Venant: request to communicate him his results on the behavior of curve systems on surfaces, for comparing them with own results.
Location: AEP, collection Saint-Venant. Engl. translation published in: Belhoste 1991, 333–334.

APRIL 18TH, 1847
Letter from Hermann G. Graßmann to Cauchy (draft): priority claim because of Cauchy's *clefs algébriques*.
Published in: Hermann Grassmann's *Gesammelte Werke*, Vol., III, 2nd part (Leipzig: Teubner, 1911), 120–121.

UNDATED, 1848
Letter from Cauchy to Herschel: recommendations for Padre Vico. Transmission of own *mémoires*.
Location: archives of the Royal Society, London, ms. 328. Engl. translation published in: Belhoste 1991, 335–336.

OCTOBER 5, 1850
Letter from Cauchy to Saint-Venant: answer to a biographical inquiry (about a third person).
Location: AEP, collection Saint-Venant. Engl. translation published in: Belhoste 1991, 336.

DECEMBER 12, 1850
Letter from Cauchy to Eugène Burnouf.
Location. Bibliothèque Nationale, Paris: Ms Cote 10593, fol. 283–284.

MAY 14, 1852
Letter from Cauchy to the dean of the *Faculté des Sciences*, Paris: explaining the new oath affair.
Location: Personnel file Cauchy, archive of the *Académie des Sciences*. Engl. translation published in: Belhoste 1991, 336–337.

JUNE 3, 1852
Note from Cauchy to J. A. Adolphe Régnier: returning two theses examined by him, as a professor of the *Faculté of the Sciences*.
Location: D.E. Smith Collection, Butler Library, New York.

MARCH 16, 1856

Letter from Cauchy to Leverrier: request to be accompanied, for the arranged visit of the observatory, by his grandson during the latter's school holidays.

Location: Bibliothèque de l'Institut, ms. 3710, fol. 196.

JULY 9, 1856

Letter from Liouville to Cauchy (draft): protest against the fact that Cauchy had delivered a report on a manuscript submitted to the *Académie* without consulting the appointed committee.

Location: Bibliothèque de l'Institut. Publ. in: Neuenschwander, "Joseph Liouville (18091882): Correspondance Inédite," *Bollettino di Storia delle Scienze matematiche*, 4 (1984), 111–112.

NOVEMBER 23, (S.A.), PROBABLY 1856

Letter from Cauchy to Leverrier: request for an immediate encounter still the same morning, before the meeting of the *Académie*, for transmitting his calculations and results to the *Académie* by Leverrier.

Location: Bibliothèque de l'Institut, ms. 3710, fol. 198. Engl. translation published in: Belhoste 1991, 334–335.

DECEMBER 7, 1856

Letter from Cauchy to Leverrier: communication that he accepts the latter's private invitation.

Location: Bibliothèque de l'Institut, ms. 3710, Fol. 197.

S.D., ca. JANUARY 1857

Draft letter (in German) from Kummer to Cauchy. Having heard by Hermite of *objections* Cauchy had made in an Academy session against the rigor of the main theorem in Kummer's paper on the theory of complex numbers, Kummer tells that he found already the error himself and inserted a note in Crelle's Journal which he would send off immediately.

Location: Archiv der Berlin-Brandenburgischen Akademie der Wissenschaften Berlin, *Nachlass* Dirichlet, No. 63.

1 FEBRUARY 1857

Cauchy's answer to Kummer, assures the latter of his own high esteem mentioning his role in having Kummer awarded the *Académie*'s "médaille" for his work on number theory. While stating that he had not been able to find in Kummer's paper on complex numbers "toute la rigueur que j'aurais desiré", he admits to a possible too fast or superficial reading, assures him that he would convey Kummer's then hopefully rigorous theorem to the *Académie*.

Location: ibid, No. 58.

NOVEMBER 27, (S.A.)

Letter from Cauchy to Comguet (*agent de change*).

Location: Handschriftenabteilung der Staatsbibliothek Preußischer Kulturbesitz Berlin: Sammlung Darmstädter, F1c 1836 (1).

References

Sources

ÉCOLE POLYTECHNIQUE, PALAISEAU [AEP]
Archives
Personnel files: VI.1.b2:
 J. G. Garnier
 S. F. Lacroix
Bibliothèque
A IIa, 56: A.-M. Ampère, Cours de calcul différentiel et intégral, donnés à l'École
   Polytechnique, années 1811–12–13. Notes prises par L.-Th. Olivier
A IIIa 51: Jean Guillaume Garnier, Leçons d'Analyse Algébrique, Différentielle et
   Intégrale, données en l'an 9 à la première division de l'École Polytechnique
   (Paris: Baudouin, Floréal an 9)
III 3b, J. G. Garnier, Programme de Calcul Intégral et de Calcul differentiel, [..], an 7
X^{2b} 10, Programmes d'Enseignement
X^{2c} 7, Registres d'Instruction

ARCHIVES DE L'ACADÉMIE DES SCIENCES, PARIS
Papiers de A.-M. Ampère
 chem. 74, Cours de calcul intégral et différentiel
 chem. 75, Cours d'analyse
 chem. 76, Ancien cours d'analyse algébrique

ARCHIVES NATIONALES DE FRANCE, PARIS [AN]
F ^{17}, 7441: Affaires générales Mayence
F^{17}, 12818 Commission livres classiques
F^{17}12863: Conseil royal de l'instruction publique
108 AP, Archives Carnot

BIBLIOTHÈQUE DE L'INSTITUT, PARIS
Papiers de S.-F. Lacroix, mss. 2396, 2397

MAISON AUGUSTE COMTE, PARIS
Cours d'analyse, Seconde année. Cauchy Professeur (lecture notes taken by A.
   Comte)
Cours de mécanique: Poisson Professeur (lecture notes taken by A. Comte)

ARCHIVES OF THE FAMILY CARNOT
 Presles
 Nolay/Chateau de La Rochepot

ARCHIV DER BERLIN-BRANDENBURGISCHEN AKADEMIE DER
   WISSENSCHAFTEN, BERLIN (former: Archiv der Akademie der
   Wissenschaften der DDR), [AAdW]
Preisschriften der Akademie: I-M 838-858, Preisaufgabe der mathematischen Klasse
   1784

Nachlass F. W. Bessel. Nr. 221: Letters from W. A. Förstemann
   Nr. 392: Letter from F. L. Wachter
Nachlass J. G. Tralles: Nr. 22; Nr. 24; Nr. 97

STAATSBIBLIOTHEK PREUßISCHER KULTURBESITZ BERLIN, HANDSCHRIFTENABTEILUNG
Nachlass Dirichlet

GEHEIMES STAATSARCHIV PREUßISCHER KULTURBESITZ, BERLIN-DAHLEM [GSTA]
(former: Deutsches Zentralarchiv, Abt. Merseburg)
Akten des preußischen Kultusminsteriums:
Rep. 76 VI, Sekt. IV, z, Nr. 3, Das Gymnasium in Danzig, vol. 1 and vol. 2

UNIVERSITÄTSARCHIV GÖTTINGEN
II. Philosophische Fakultät, Ph 73

UNIVERSITÄTSARCHIV HALLE
Rep. 21 II Nr. 10: Bestand Philosophische Fakultät, Dekanatsakte

OSTERBURG, GROOTHUSEN
Nachlass Enno H. Dirksen

LANDESARCHIV SPEYER
G 6, Departement Donnersberg. Präfekturakten, Abt. I: Nr. 210. Normalschule in Mainz

BIBLIOTECA MORENIANA FLORENZ
L. F. A. Arbogast, Essai sur des nouveaux Principes de calcul différentiel et de calcul intégral, indépendans de la théorie des infiniment-petits et de celle des limites. Manuscript 1787

Publications

Mahdi Abdeljaouad, "Le manuscrit mathématique de Jerba: Une pratique des symboles algébriques maghrébins en pleine maturité," Paper presented at the Septième Colloque Maghrébin sur l'histoire des mathématiques arabes. Marrakech 30 mai–1er juin 2002.

Niels Henrik Abel, "Untersuchungen über die Reihe [...]," *Journal für reine und angewandte Mathematik*, 1826, *1*: 311–339.

Jean le Rond d'Alembert, *Traité de Dynamique* (Paris: David l'aîné, 1743).

Jean le Rond d'Alembert, *Opuscules Mathématiques*, Tome Premier, Sixième Mémoire: "Sur les Logarithmes des quantités négatives," Paris 1761, 180–230.

Jean le Rond d'Alembert, *Opuscules Mathématiques*, Tome VIII, Cinquante–Huitiéme Mémoire: "Sur les quantités négatives," Paris 1780, 270–279.

Jean le Rond d'Alembert, Article "Courbe," *Encyclopédie, ou Dictionnaire Raisonné des Sciences, Arts et des Métiers,* tome IV, 1754, 377–389.

Jean le Rond d'Alembert, "Differentiel," tome IV, 1751, 985–989.

Jean le Rond d'Alembert, "Élémens des sciences," ibid., tome V, 1755, 491–497.

Jean le Rond d'Alembert, "Équation," ibid., tome V, 1755, 842–855.

Jean le Rond d'Alembert, "Infini," ibid., tome VIII, 1765, 702–704.

Jean le Rond d'Alembert, "Limite," ibid., tome IX, 1765, 542.

Jean le Rond d'Alembert, "Négatif," ibid.,tome XI, 1765, 72–74.

Jean le Rond d'Alembert, "Quantité," ibid., tome XIII, 1765, 653–655.

Jean le Rond d'Alembert, "Variable," ibid., tome XVI, 1765, 840.

Jean le Rond d'Alembert, *Essai sur les élémens de philosophie,* XI.: Eclaircissement sur les élémens d'algèbre, Paris 1805. Edition Olms, Hildesheim 1965.

André-Marie Ampère, "Mémoire: Recherches sur quelques points de la théorie des fonctions dérivées qui conduisent à une nouvelle démonstration de la série de Taylor, et à l'expression finie des termes qu'en néglige lorsqu'on arrête cette série à un terme quelconque," *Journal de l'École Polytechnique,* tome 6, cahier 13, 1806, 148–181 [1806a].

André-Marie Ampère, "Démonstration générale du principe des Vitesses virtuelles, dégagée de la considération des infiniment petits," *Journal de l'École Polytechnique,* tome 6, cahier 13, 1806, 247–269. [1806b]

André-Marie Ampère, Précis des leçons sur le Calcul Différentiel et le Calcul Intégral (s.l., s.a. [Paris 1824]).

Daniel Amson, *Carnot* (Paris: Perrin, 1992).

Kirsti Andersen, "Cavalieri's Method of Indivisibles," *Archive for History of Exact Sciences,* 1985, *31*: 291–367.

L. F. A. Arbogast, *Essai sur de nouveaux principes du calcul différentiel et de calcul intégral, indépendans de la théorie des infiniment-ptetis et de celle des limites* (1789; manuscript; Biblioteca Moreniana Florence).

L. F. A. Arbogast, Mémoire sur la nature des Fonctions arbitraires qui entrent dans les Intégrales des équations aux différentielles partielles. St. Petersburg 1791.

A. Arcavi, M. Bruckheimer, *The negative numbers. An historical source-work collection for in-service and pre-service mathematics teachers courses.* The Weizmann Institute of Science. Department of Science Teaching. Rehovot (Israel), 1983.

Aristoteles, *Kategorien: Lehre vom Satz* (Peri hermeneias) Übers., mit e. Einl. u. erklärenden Anm. vers. von Eugen Rolfes (Hamburg: Meiner, 1968)

Aristotle, Aristotle's *Metaphysics*: a revised text with introduction and commentary . By W. D. Ross. Vol. 2 (Oxford : Clarendon Press, 1970).

Aristotle, *Aristotle. The Physics.* Engl. Transl. by Ph.H. Wicksteed, F.M. Cornford. Vol. 2 (Cambridge: Harvard Univ. Pr., 1980)).

[Antoine Arnauld], *Nouveaux Elémens de Géométrie* (Paris: Ch. Savreux, 1667).

[Antoine Arnauld], *Nouveaux Elémens de Géométrie.* Seconde édition [1683] (La Haye: H. van Bulderen, 1690).

Antoine Arnauld, Pierre Nicole, *La Logique ou l'Art de Penser* (Original 1662), Introduction de Louis Morin, Paris: Flammarion, 1970.

Sylvain Auroux, "La philosophie mathématique de Condillac," *Bulletin de la Société Française de Philosophie,* 1981, *75*, no. 1, 7–17.

Sylvain Auroux, "Idéologie et Langue des Calculs," *Histoire, Épistémologie, Langage,* 1982, *4*, 53–57.

Sylvain Auroux, Anne Marie Chouillet, "Condillac ou la vertu des signes. Notice," eds.: S. Auroux & A.M. Chouillet: E. B. de Condillac, *La Langue des Calculs* (Lille 1981), I–XXXVIII.

Gaston Bachelard, *La formation de l'esprit scientifique* [1938] (Paris: Vrin, 1975).

Patrice Bailhache, "Lazare Carnot entre d'Alembert et Lagrange: Originalité de sa Mécanique," in: Charnay et al. 1990, 543–555.

Evelyne Barbin, "Méthode cartésienne et figure géométrique dans les Eléments de Géométrie de Lamy," Questions de Méthodes au XVIIème Siècle, éd. I.R.E.M. des Pays de la Loire, Centre du Mans (Le Mans, 1994), 141–153.

H. Barreau, "Lazare Carnot Défenseur des Infinitésimaux," éd. Jean Dhombres, Faire de l'histoire des mathématiques: documents de travail. Cahiers d'Histoire et de Philosophie des Sciences No. 20 (Paris: Belin, 1987), 1–10.

Giovanna Baroncelli, "Bonaventura Cavalieri tra matematica e fisica," eds. Massimo Bucciantini, Maurizio Torrini, Geometria e Atomismo nella Scuola Galileiana (Firenze: Leo S. Olschki, 1992), 67–102.

Danny Beckers, "Positive Thinking. Lacroix's theory of negative numbers in the Netherlands," Revue d'Histoire des Mathématiques, 2000, 6: 95–126.

Bruno Belhoste, Cauchy. Un mathématicien légitimiste au XIXe siècle (Paris: Berlin, 1985).

Bruno Belhoste, Augustin-Louis Cauchy. a Biography (New York: Springer, 1991).

Bruno Belhoste, "L'enseignement des mathémtiques dans les collèges oratoriens au XVIIIe siècle," Le Collège de Riom et l'enseignement oratorien au XVIIIe siècle, Textes réunis et présentés par Jean Ehrard (Paris: CNRS éditions 1993), 141–160.

Bruno Belhoste, Les Sciences dans l'Enseignement secondaire français. textes officiels. Tome I: 1789–1914 (Paris: Institut national de recherche pédagogique, 1995).

Bernard Forest de Bélidor, Nouveau Cours de Mathématique, à l'usage de l'artillerie et du génie. Nouvelle édition (Paris: Jombert, 1757).

Johann Bernoulli, "Discours sur les Loix de la Communication du Mouvement" (1727), id., Opera Omnia, tomus III, 7–107.

Johann Bernoulli, Die Differentialrechnung aus dem Jahre 1691/92. Nach der in der Basler Universitätsbibliothek befindlichen Handschrift übersetzt, mit einem Vorwort und Anmerkungen versehen, von Paul Schafheitlin (Leipzig: Akademische Verlagsgesellschaft, 1924).

Johann Bernoulli, Opera Omnia, tam antea sparsim edita, quam hactenus inedita. Tomus secundus; tomus tertius (Lausanne, Genève: Bousquet 1742). [Reprint, ed. J. E. Hofmann. Hildesheim: Olms 1968].

Johann Bernoulli, Der Briefwechsel von Johann I Bernoulli. Band 2: Der Briefwechsel mit Pierre Varignon. Erster Teil: 1692–1702. Edited and annotated by Pierre Costabel und Jeanne Peiffer (Basel: Birkhäuser, 1988).

Johann Bernoulli, Der Briefwechsel von Johann I Bernoulli. Band 3: Der Briefwechsel mit Pierre Varignon. Zweiter Teil: 1702–1714. Bearbeitet und kommentiert von Pierre Costabel und Jeanne Peiffer (Basel: Birkhäuser, 1992).

Joseph Bertrand, "La vie et les travaux du baron Cauchy, par C. A. Valson," Journal des Savants, 1869, 205–215.

Étienne Bézout, Cours de Mathématiques à l'usage des Gardes du Pavillon et de la Marine. Quatrieme Partie. Contenant les Principes généraux de la Méchanique, précédés des Principes de Calcul qui servent d'introduction aux Sciences Physico-Mathématiques (Paris: Musier, 1770) [1770a].

Étienne Bézout, Cours de Mathématiques à l'usage du Corps Royal de l'Artillerie. Tome second. Contenant l'Algebre et l'application de l'Algèbre à la Géométrie (Paris: Imprimerie Royale, 1770) [1770b].

Étienne Bézout, Cours de Mathématiques à l'usage du Corps Royal de l'Artillerie. Tome Troisieme. Contenant les Principes généraux de la Méchanique, et *l'Hydrostatique; précédés des Principes de Calcul qui servent d'introduction aux Sciences Physico-Mathématiques* .(Paris: Imprimérie Royale, 1772).

Étienne Bézout, *Cours de Mathématiques à l'usage des Gardes du Pavillon et de la Marine. Troisieme Partie. Contenant l'Algebre et l'application de cette science à l'Arithmétique et la Géométrie* (Paris: Pierres, 1781) [2nd edition].

Étienne Bézout, *Cours de Mathématiques à l'usage du Corps de l'Artillerie. Tome Troisieme. Contenant les Principes généraux de la Méchanique, et l'Hydrostatique; précédés des Principes de Calcul qui servent d'introduction aux Sciences Physico-Mathématiques.* Nouvelle Édition, revue et corrigée (Paris: Richard, Caille et Ravier, an VII [1799]).

Étienne Bézout, *Cours de Mathématiques à l'usage des Gardes du Pavillon, de la Marine, et des Élèves de l'École Polytechnique. Seconde Partie, contenant les Elémens de la Géométrie, la Trigonometrie rectiligne, et la Trigonometrie rectiligne.* Trosième édition (Paris: Courcier, an XII) [1803–04].

Kurt-R. Biermann, *Johann Peter Gustav Lejeune Dirichlet. Dokumente für sein Leben und Wirken.* Abhandlungen der Deutschen Akademie der Wissenschaften zu Berlin, Klasse für Math., Physik und Technik; Jahrgang 1959, Nr. 2 (Berlin: Akademie, 1959).

Kurt-R. Biermann, *Die Mathematik und ihre Dozenten an der Berliner Universität 1810-1933* (Berlin: Akademie, 1988).

Christa Binder, "Über den Briefwechsel Felix Klein–Otto Stolz," *Wissenschaftliche Zeitschrift der Ernst-Moritz-Arndt-Universität Greifswald. Mathem.-Naturwiss. Reihe*, 1989, 38/4: 3–7.

Paul René Binet, "Mémoire sur la fonction dérivée, ou coefficient différenciel du premier ordre, lu par M. Binet" [report by Ampère], *Nouveau Bulletin des Sciences, par la Société Philomatique.* Tome Premier, Paris 1807: No. 16, Seconde Année. Paris Janvier 1809, 275–278.

Jean-Baptiste Biot, *Essai sur l'histoire générale des sciences pendant la Révolution Française* (Paris, an 11 - 1803).

Errett Bishop, "Review of Elementary calculus, by H. Jerome Keisler. Boston 1976," *Bulletin of the American Mathematical Society*, 1977, *83*: 205–208.

Michel Blay, "Deux moments de la critique du calcul infinitésimal," *Revue d'histoire des sciences*, 1986, *39*: 223–253.

Michel Blay, *Les raisons de l'infini. Du monde clos à l'univers mathématique* (Paris: Gallimard, 1993).

Michel Blay et Alain Niderst, "Introduction," Fontenelle, *Éléments de la Géométrie de l'Infini* (Paris: Klincksiek, 1995), 7–33.

Théodore Le Pouillon de Boblaye, *Notice sur les écoles du génie de Mézières et de Metz* (Metz 1867).

Georg Bohlmann, "Übersicht über die wichtigsten Lehrbücher der Infinitesimal-Rechnung, von Euler bis auf die heutige Zeit. Bericht, erstattet der Deutschen Mathematiker-Vereinigung," *Jahresberichte der Deutschen Mathematiker-Vereinigung*, 1897, *6* : 91–110.

Jean Bollack, "Un futur dans le passé. L'herméneutique matérielle de Peter Szondi," *Introduction à l'herméneutique littéraire*, Peter Szondi (Paris: Éd. du Cerf, 1989), I–XVII.

Bernard Bolzano, *Bernard-Bolzano-Gesamtausgabe*, eds. Eduard Winter et al. (Stuttgart-Bad Cannstatt: Frommann):

Reihe 2, Nachlass A, Nachgelassene Schriften Bd. 7 *Grössenlehre*. [1], "Einleitung zur Grössenlehre" and "Erste Begriffe der allgemeinen Grössenlehre" 1975.

Reihe 2, Nachlass A, Nachgelassene Schriften Bd. 8 *Grössenlehre*. 2, "Reine Zahlenlehre" 1976.

Reihe 2, Nachlass A, Nachgelassene Schriften Bd. 10 *Functionenlehre*. *Verbesserungen und Zusätze zur Functionenlehre*, 2000.

Baldassare Boncompagni, "Intorno ad un'opera dell'Abate Nicolò Luigi de La-Caille 'Leçons élémentaires de Mathématiques,' ecc.," *Bullettino di Bibliografia e di Storia delle Scienze Matematiche e Fisiche*, 1872, *5*, 278–293.

Bonnefin, *Doctrine de l'Algèbre, basée sur ses vraies principes encore inconnus* (Paris/St. Malo: Courcier, 1820).

Karl Bopp, *Der große Arnauld als Mathematiker* (Leipzig, 1902).

Maria Teresa Borgato, Luigi Pepe, "Lagrange a Torino (1750–1759) e le sue lezioni inedite nelle R. Scuole di Artiglieria," *Bollettino di Storia delle Scienze Matematiche*, 1987, 7: 3–200.

Henk J. M. Bos, "Differentials, Higher-Order Differentials and the Derivative in the Leibnizian Calculus," *Archive for the History of Exact Sciences*, 1974, *14*: 1–90.

Henk J. M. Bos, "Newton, Leibniz and the Leibnizian Tradition," ed. Ivor Grattan-Guinness, *From the Calculus to Set Theory, 1630–1910* (London: Duckworth, 1980), 49–93.

Henk J. M. Bos, *Lectures in the History of Mathematics*. American Mathematical Society, 1993.

Roger Joseph Boscovich, *De Natura, et usu Infinitorum, et Infinite parvorum Dissertatio* (Rom: Collegium Romanum, 1741).

Roger Joseph Boscovich, *De Continuitatis Lege et ejus Consectariis pertinentibus ad prima Materiae Elementa eorumque Vires Dissertatio* (Rom: Collegium Romanum, 1754) [1754a].

Roger Joseph Boscovich, *Elementorum Universae Matheseos*. Tomus III., continens Sectionum Conicarum elementa, Nova quadam methodo concinnata et Dissertationem de Transformatione Locorum Geometricorum ubi de Continuitatis lege, ac de quibusdam Infiniti Mysteriis (Rom: G. Salomon, 1754) [1754b].

Charles Bossut, *Traité élémentaire d'algèbre* (Paris: Jombert, 1773).

Charles Bossut, *Cours de Mathématiques à l'usage des élèves du corps Royal du Génie*. Tome premier: *Arithmétique et Algèbre* (Paris: Jombert, 1781).

Charles Bossut, *Cours de Mathématiques à l'usage des écoles royales militaires*. [Tome II] *Mechanique et Hydrodynamique* (Paris: Jombert, 1782).

Charles Bossut, *Traités de Calcul Différentiel et de Calcul Intégral*. 2 tomes (Paris: Imprimérie de la République, an VI [1798]).

Umberto Bottazzini, *The Higher Calculus: A History of Real and Complex Analysis from Euler to Weierstrass* (New York u.a.: Springer, 1986).

Umberto Bottazzini, "Editor's Introduction," *A.-L. Cauchy. Cours d'analyse de l'École Royale polytechnique. Première partie: analyse algébrique*," ed. Umberto Bottazzini (Bologna: CLUEB, 1990), IX–CLXVII [the book gives 1992 as its date, too].

Jean-Louis Boucharlat, *Anfangsgründe der Differenzial- und Integral-Rechnung*. Translated from the French by F. J. Goebel (Frankfurt a. M.: Andreaeische Buchh., 1823).

Jean-Louis Boucharlat, *Eléments de calcul différentiel et de calcul intégral*. Septième édition (Paris: Mallet-Bachelier, 1858).

Louis Antoine de Bougainville, *Traité du Calcul Intégral*, pour servir de suite à l'Analyse des Infiniment-Petits de M. le Marquis de L'Hôpital. 2 vol. (Paris: Desaint & Saillant, 1754).

Nicolas Bourbaki, *Éléments d'histoire des mathématiques*. Seconde édition: revue, corrigée et augmentée (Paris: Hermann, 1969). Nouvelle édition: Paris 1974.

Louis-Pierre-Marie Bourdon, *Élémens d'Algèbre* (Paris: Courcier, 1817).

J.-B.-E. du Bourguet, *Traités élémentaires de Calcul différentiel et de Calcul intégral, indépendans de toutes notions de quantités infinitésimales et de limites*. 2 tomes (Paris: Courcier, 1810).

Carlo Bourlet, *Leçons d'Algèbre élémentaire*. Cours complet pour la Classe de Mathématiques A, B, publié sous la direction de M. Darboux (Paris: Colin, 1896).

J. Boussinesq, *Cours d'analyse infinitésimale*. Tome 1. Calcul différentiel (Paris: Gauthier-Villars, 1887).

Margaret Bradley, *Gaspard-Clair-François-Marie Riche de Prony: his career as educator and scientist*. Ph.D. thesis, Coventry Polytechnic 1984.

Margaret Bradley, "Civil engineering and social change: The early history of the Paris *Ecole des ponts et chaussées*," *History of Education*, 1985, *14*: 171–183.

Margret Bradley, "Prony the Bridge-builder. The Life and Times of Gaspard de Prony, Educator and Scientist," *Centaurus*, 1994, *37*: 230–268.

Margaret Bradley, *A career biography of Gaspard Clair François Marie Riche de Prony, bridge-builder, educator, and scientist* (Lewiston, N.Y: E. Mellen Press, 1998).

Herbert Breger, "Über den von Samuel König veröffentlichten Brief zum Prinzip der kleinsten Wirkung," ed. Harmut Hecht, *Pierre Louis Moreau de Maupertuis. Eine Bilanz nach 300 Jahren* (Berlin: Nomos, 1999), 363–381.

Herbert Ernst Brekle, Hans Jürgen Höller, Brigitte Asbach-Schnitker, "Der Jansenismus und das Kloster Port-Royal," *Die Philosophie des 17. Jahrhunderts. Band 2: Frankreich und die Niederlande*, ed. Jean-Pierre Schobinger (Basel: Schwabe, 1993), 475–559.

Gordon G. Brittan, "The Role of the Law of Continuity in Boscovich's Theory of Matter," ed. P. Bursill-Hall, *R. J. Boscovich. Vita e Attività Scientifica - His Life and Scientific Work*. Atti del Convegno Roma, 23–27 maggio 1988 (Roma: Istituto della Enciclopedia Italiana, 1993), 215-227.

L. W. B. Brockliss, *French Higher Education in the Seventeenth and Eighteenth Centuries. A Cultural History* (Oxford: Clarendon Press, 1987).

Bernard Bru, "Poisson, le calcul des probabilités et l'instruction publique," eds. M. Métivier, P. Costabel, P. Dugac, *Siméon-Denis Poisson et la science de son temps* (Palaiseau, 1981), 51–94.

Adrien-Quentin Buée, "Mémoire sur les Quantités imaginaires," *Philosophical Transactions of the Royal Society of London*, 1806, Part I, 23–88.

Heinrich Burkhardt, "Trigonometrische Reihen und Integrale, bis etwa 1850," *Enzyklopädie der mathematischen Wissenschaften*. Band II, 1: Analysis (Leipzig: Teubner, 1914), 819–1354.

Piers Bursill-Hall, ed., *R. J. Boscovich. Vita e Attività Scientifica - His Life and Scientific Work*. Atti del Convegno Roma, 23–27 maggio 1988 (Roma: Istituto della Enciclopedia Italiana, 1993).

Friedrich Gottlieb Busse, *Neue Erörterungen über Plus und Minus, Tadel einiges bisherigen und Darstellung eines genaueren Gebrauches desselben für die Trigonometrie ...* (Cöthen: Aue, 1801).

Friedrich Gottlieb Busse, *Vergleichung zwischen Carnots und meiner Ansicht der Algebra und unserer beyderseitig vorgeschlagenen Abhelfung ihrer Unrichtigkeit* (Freyberg: Craz und Gerlach, 1804).

Friedrich Gottlob von Busse, *Bündige und reine Darstellung des wahrhaften Infinitesimal-Calculs*. 3 vols. (Dresden: Arnoldi, 1825–1827).

F. C. Busset, *De l'enseignement des mathématiques dans les collèges, considéré sous le double point de vue des préscriptions réglementaires de l'Université, et des principes fondamentaux ce la science* (Paris: Chamerot, 1843).

Pierre Jean Georges Cabanis, *Oeuvres philosophiques*. Tome 2 (Paris: Pr. Univ. de France, 1956).

Cach, "Essai sur la théorie des quantités négatives," *Annales de Mathématiques pures et appliquées*, 1813, 4: 1–6.

Florian Cajori, "Algebra," ed. Moritz Cantor, *Vorlesungen über Geschichte der Mathematik*, Band IV (Leipzig: Teubner, 1908), 72–153.

Florian Cajori, *William Oughtred. A Great Seventeenth-Century Teacher of Mathematics* (Cicago/London: Open Court, 1916).

Florian Cajori, *A history of the conceptions of limits and fluxions in Great Britain from Newton to Woodhouse* (Chicago: Open Court, 1919).

Florian Cajori, "Controversies on Mathematics between Wallis, Hobbes, and Barrow," *Mathematics Teacher*, 1929, 22: 146–151.

Étienne Camus, *Cours de Mathématique. Premiere Partie. Élémens d'Arithmétique* (Paris: Ballard, 1749).

Étienne Camus, *Cours de Mathématique*. Troisieme Partie. Tome Premier: *Élémens de Mechanique statique*. Troisieme édition (Paris, Prault, 1766).

Georg Cantor, "Mitteilungen zur Lehre vom Transfiniten" (1887), Nachdruck in: *Gesammelte Abhandlungen mathematischen und philosophischen Inhalts*, ed. Ernst Zermelo (Hildesheim: Olms, 1966), 378–439.

Moritz Cantor, "Petrus Ramus, Michael Stifel, Hieronymus Cardano, drei mathematische Charakterbilder aus dem 16. Jahrhundert," *Zeitschrift für Mathematik und Physik*, 1857, 2: 353–376.

Moritz Cantor, *Vorlesungen über Geschichte der Mathematik*. Zweiter Band: Vom Jahre 1200 bis zum Jahre 1668 (Leipzig: Teubner 1900); Dritter Band: Vom Jahre 1668 bis zum Jahre 1758, 1901, Vierter Band: Vom Jahre 1759 bis zum Jahre 1799, 1908.

Girolamo Cardano, "Ars Magna Arithmeticae," in: *Opera Omnia*, vol. 4, 1663, 303–376.

Girolamo Cardano, "De Regula Aliza," in: *Opera Omnia*, vol. 4, 1663, 377–434.

Girolamo Cardano, *Opera Omnia. The 1662 Lugduni edition*. With an introduction by August Buck. Vol. 4, 1663 (New York and London: Johnson Reprint, 1967).

Lazare Carnot, "Essai sur les Machines en général" (1783), in: Carnot 1797b, 1–124.

Lazare Carnot, "Dissertation sur la théorie de l'infini mathématique," 1785, in: Gillispie 1971, 169–267.

Lazare Carnot, *Réflexions sur la Métaphysique du Calcul Infinitésimal* (Paris: Duprat, an V [1797]).

Lazare Carnot, *Œuvres Mathématiques* du Citoyen Carnot (Basle: J. Decker, 1797) [1797b].

Lazare Carnot, *Betrachtungen über die Theorie der Infinitesimalrechnung*, von dem Bürger Carnot. Aus dem Französischen übersetzt und mit Anmerkungen und Zusätzen begleitet von Johann Karl Friedrich Hauff (Frankfurt am Main: Jäger, 1800) [1800a].

Lazare Carnot, "Lettre du citoyen Carnot au citoyen Bossut, contenant quelques vues nouvelles sur la trigonometrie," in: Charles Bossut, *Cours de Mathématiques. Tome second: Géométrie, et Application de l'Algèbre à la Géométrie*. Nouvelle édition revue et augmentée (Paris: Firmin Didot, an IX – 1800), 401–421 [1800b].

Lazare N. M. Carnot, *De la Corrélation des Figures de Géométrie* (Paris: Crapelet, Duprat, an X = 1801).

Lazare N. M. Carnot, *Géométrie de Position* (Paris: Crapelet, Duprat, an XI – 1803) [1803a].

Lazare N. M. Carnot, *Principes Fondamentaux de l'Équilibre et du Mouvement* (Paris: Crapelet, Deterville, an XI – 1803). [1803b]

Lazare N. M. Carnot, *Mémoire sur la Relation qui existe entre les distances respectives de cinq points quelconques pris dans l'espace; suivi d'un essai sur la Théorie des Transversales* (Paris: Courcier, 1806). [also contains: Digression sur la nature des quantités dites négatives].

Lazare N. M. Carnot, *Geometrie der Stellung*. Translated by H. C. Schumacher. Erster Theil (Altona: Hammerich, 1808). Zweyter Theil (Altona: Hammerich, 1810).

Lazare Carnot, *Réflexions sur la Métaphysique du Calcul Infinitésimal* [Reprint der Ausgabe von 1813]. 2 tomes (Paris: Gauthier-Villars, 1921).

Lazare Carnot, *Reflexions on the Metaphysical Principles of the Infinitesimal Analysis*. Transl. by W. R. Browell (Oxford: Parker, 1832).

Jean de Castillon, "Examen philosophique de quelques principes de l'Algèbre," *Nouveaux Mémoires de l'Académie des Sciences et Belles-Lettres de Berlin pour 1790 et 1791*. Premier mémoire: 331–341. Second mémoire: 342–363.

Augustin-Louis Cauchy, *Cours d'Analyse de L'École Royale Polytechnique. Première Partie. Analyse algébrique* (Paris: Imprimérie Royale, 1821).

Augustin-Louis Cauchy, *Résumé des Leçons données à l'École Royale Polytechnique sur le Calcul Infinitésimal* (Paris: Imprimérie Royale, 1823).

Augustin-Louis Cauchy, "Mémoire sur les intégrales définies," [1814/25] *Œuvres Complètes*, Ire Série, tome I (Paris: Gauthier Villars, 1882), 319–506.

Augustin-Louis Cauchy, "Note sur les séries convergentes dont les divers termes sont des fonctions continues d'une variable réelle ou imaginaire, entre des limites données," [1853] *Œuvres Complètes*, Ire Série, tome 12 (Paris: Gauthier Villars, 1900), 30–36.

Augustin-Louis Cauchy, *Sept Leçons de Physique Générale* (Paris: Gauthier Villars, 1868), ed. Moigno.

Augustin-Louis Cauchy, "Sui metodi analitici," [1830/31] *Œuvres Complètes*, II^{de} Série, tome 14 (Paris: Gauthier Villars, 1974), 149–181.

Augustin-Louis Cauchy, "Extrait d'une lettre à M. Coriolis," [1837] *Œuvres Complètes*, Ire Série, tome 4 (Paris: Gauthier Villars, 1884), 38–42.

de la Chapelle, *Traité des Sections Coniques* (Paris: Quillau, 1750).

de la Chapelle, *Institutions de Géométrie, enrichies de notes critiques sur la nature et les développements de l'Esprit humain*. Quatrieme edition. 2 volumes (Paris: Debure, 1765).

de la Chapelle, "Limite (Mathémat.)," *Encyclopédie*, tome 9, 1765, 542.

de la Chapelle, "*Quantité, en termes d'Algèbre*," *Encyclopédie*, tome 13, 1765, 655.

Jean-Paul Charnay (ed.), *Lazare Carnot. Révolution et Mathématique*. Vol.s I et II (Paris: L'Herne, 1984–1985).

Jean-Paul Charnay et al. (éd), *Lazare-Carnot ou le Savant-Citoyen*. Actes du Colloque tenu en Sorbonne les 25-29 Janvier 1988. Université de Paris-Sorbonne, Série Historique - Vol. II (Paris: Presses de l'Université de Paris-Sorbonne, 1990).

Fritz Chemnitius, *Die Mathematiker, Astronomen und Physiker an der Universität Jena* [1558-1914]. Edition of a manuscript, edited by Gert Schubring (München: Institut für Gessch. d. Naturwissenschaften, 1992).

Alexis Clairaut, *Élémens d'Algèbre* [Original 1746], see: fifth edition by Lacroix 1797.

Alexis Clairaut, *Élémens de Géométrie* [Original 1741] (Paris: Gauthier-Villars, 1920).

Henry Thomas Colebrooke (ed.), *Algebra, with arithmetic and mensuration: from the Sanscrit of Brahmegupta and Bhascara*. Transl. by Henry Thomas Colebrooke. Unaltered reprint of the edition of 1817. (Walluf: Sändig, 1973).

Le Collège de France, 1530–1930. Livre Jubilaire, par Lefranc et al. (Paris: Presses univ. de France, 1930).

Randall Collins, *The sociology of philosophies: a global theory of intellectual change* (Cambridge, Mass.: Belknap Press of Harvard Univ. Press, 1998).

Étienne Bonnot de Condillac, *La Logique ou Les Premiers Développemens de l'Art de Penser*. Texte établi et présenté par Georges Le Roy, Œuvres Philosophiques de Condillac. Volume 2 (Paris: Presses Universitaires de France, 1948).

Étienne Bonnot de Condillac, *Die Logik oder die Anfänge der Kunst des Denkens. Die Sprache des Rechnens*, ed. Georg Klaus (Berlin: Akademie, 1959).

Étienne Bonnot de Condillac, *Essai ueber den Ursprung der menschlichen Erkenntnisse*. Edited and translated by Ulrich Ricken .(Leipzig Reclam 1977).

Étienne Bonnot de Condillac, *La Langue des Calculs*. Texte établi et présenté par Anne-Marie Chouillet. Introduction et Notes par Sylvain Auroux (Lille: Presses Universitaires de Lille, 1981).

Benjamin Constant Botelho de Magalhães, *Theoria das Quantidades Negativas* (Petropolis: B. P. Sudré, 1868).

Leo Corry, *Modern algebra and the rise of mathematical structures* (Basel: Birkhäuser, 1996).

Pierre Costabel, "Leibniz et la notion de 'fiction bien fondée'," *Leibniz - Tradition und Aktualität. V. Internationaler Leibniz-Kongreß Hannover 1988. Vorträge* (Hannover: G. W. Leibniz-Gesellschaft, 1988), 174–180.

Pierre Costabel, "Introduction. Analyse et mécanique. Quelques questions abordées dans la correspondance," in: Bernoulli 1992, 3–24.

Antoine-Augustin Cournot, *Traité élémentaire de la Théorie des fonctions et du calcul infinitésimal* (Paris: Hachette, 1841).

Jacques-Antoine-Joseph Cousin, *Leçons de Calcul Différentiel et de Calcul Intégral* (Paris: Cl. A. Jombert, 1777).

Jacques-Antoine Joseph Cousin, *Traité de Calcul Différentiel et de Calcul Intégral* (Paris: Regent & Bernard, an 4 – 1796).

August Leopold Crelle, *Enzyklopädische Darstellung der Theorie der Zahlen und einiger anderer damit in Verbindung stehender analytischer Gegenstände* (Berlin: Reimer, 1845).

Jean Pierre de Crousaz, *Commentaire sur l'analyse des infiniments petits* (Paris: Montalant, 1721).

Christian August Crusius, *Anleitung über natürliche Begebenheiten ordentlich und vorsichtig nachzudenken*. Anderer Theil. Die andere und vermehrte Auflage (Leipzig: J. F. Gleditsch, 1774).

José Anastácio da Cunha, *Principios Mathematicos* (Lisboa 1790). Reprint Departamento de Matemática, Universidade de Coimbra: Coimbra, 1987.

José Anastácio da Cunha, *Principes Mathématiques* (Bordeaux 1811). Reprint Departamento de Matemática, Universidade de Coimbra: Coimbra, 1987.

Arthur Czwalina (ed.), *Arithmetik des Diophantos aus Alexandria*. Translated from the Greek and explained by Arthur Czwalina (Göttingen: Vandenhoeck & Ruprecht, 1952).

François de Dainville, *La Naissance de l'Humanisme moderne* (Paris 1940; quoted from reprint Slatkine: Genève 1969).

L'Abbé Deidier, *Suite de L'Arithmétique des Géométres, contenant une introduction à l'Algèbre et à l'Analyse; avec la résolution des Equations du second et du troisieme degré;* [...]; *l'Arithmétique des Infinis; les Logarithmes; les Fractions décimales, etc.* [Vol. 2 of: L'Arithmétique des Géométres] (Paris/Amsterdam: Mortier 1739).

L'Abbé Deidier, *La Mesure des Surfaces et des Solides, par l'Arithmetique des Infinis et les Centres de Gravité* (Paris: Ch.-A. Jombert, 1740).

L'Abbé Deidier, *Suite de la Mesure des Surfaces et des Solides: Le Calcul Differentiel et le Calcul Integral, Expliqués et Appliquées a la Geometrie. Avec un Traité Preliminaire, contenant la maniere de resoudre les Equations de quelque degré qu'elles soient, les proprietés des Series, les Equations qui expriment la nature des Courbes* [...] (Paris: Ch.-A. Jombert, 1740).

Jean-Baptiste Delambre, *Rapport historique sur les progrès des sciences mathématiques depuis 1789* (Paris: L'Imprimérie Impériale, 1810).

Giuseppina Dell'Aquila, Mario Ferrari, "La lunga storia dei numeri interi relativi," *L'insegnamento della matematica*: Parte seconda, 1994, 17A, no. 3, 248–265. Parte terza, 1994, 17A, no. 4, 335–361. Parte quarta: 1995, 18A, no. 4, 340–363. Parte quinta, 1996, 19A: 311–339.

De l'Usage des Expressions Négatives dans la Langue Française. Par C** [François Collin d'Ambly,] (Paris: Bailly, an X [1802]).

Sergei S. Demidov, "Zakon nepreryvnosti G.-V. Lejbnica i ponjatie nepreryvnosti funkcii u Ejlera [Russian; G. W. Leibniz's 'continuity law' and Euler's concept of function continuity]," *Istoriko-matematiceskie issledovanija*, 1990, No. 32-33, 34-39.

de Prony: see Prony.

René Descartes, *Discours de la Méthode* (Original 1637), edition used: Introduction et Notes par Ét. Gilson Paris: Vrin 1970.

René Descartes, *Œuvres*, eds. Charles Adam, Paul Tannery. Nouvelle Présentation. Tome VI (Paris, Vrin, 1973). Tome X (Paris: Vrin, 1974).

Desponts, "Mémoire analytique sur les illimitables et les évanoissantes, ou Le Guide des Commençans dans l'étude du calcul infinitésimal," *Mémoires du Lycée du Département de L'Aube*, No. Ier, Troyes, an X [1802], 81–120.

Desponts, "Nouvelle Exposition des Principes du calcul qu'on appelle infinitésimal, ou les vraies principes de ce calcul puisés dans la plus simple analyse, et dégagés de toute hypothèse métaphysique quelconque," *Mémoires du Lycée du Département de L'Aube*, No. III, Troyes, an XII [1804] [Annexe].

Antoine-Louis-Claude Destutt de Tracy, "Mémoire sur la faculté de penser," *Mémoires de morale et politique de l'Institut* an 6 [1798], 283–450.

Antoine-Louis-Claude Destutt de Tracy, *Élémens d'Idéologie. Première Partie: Idéologie proprement dite*. Seconde édition (Paris: Courcier, an XIII = 1804).

Jean Dhombres, *Nombre, mesure et continu : épistémologie et histoire* (Paris: CEDIC/Nathan, 1978).

Jean Dhombres, "Quelques aspects de l'histoire des équations fonctionnelles liés à l'évolution du concept de fonction," *Archive for History of Exact Sciences*, 1986, *36*: 91–181.

Jean Dhombres, Réédition de: Ambroise Fourcy, *Histoire de l'École Polytechnique* [1828]. Avec une introduction et des annexes (Paris: Belin, 1987).

Jean Dhombres (ed.), *Leçons de mathématiques*. Édition annotée des cours de Laplace, Lagrange et Monge avec introductions et annexes Avec introd. et annexes par Bruno Belhoste ... (Paris: Dunod, 1992).

Jean et Nicole Dhombres, *Lazare Carnot* (Paris: Fayard, 1997).

Wilhelm Adolf Diesterweg, *Beitraege zu der Lehre von den positiven und negativen Groessen* (Bonn: Habicht, 1831).

Jean Dieudonné, *Geschichte der Mathematik 1700–1900*. With collaboration of Pierre Dugac et al. (Braunschweig: Vieweg, 1985).

Ulisse Dini, *Grundlagen für eine Theorie der Functionen einer veränderlichen reellen Grösse*. Mit Genehmigung des Verfassers deutsch bearbeitet von Jacob Lüroth und Adolf Schepp (Leipzig: Teubner, 1892).

Johann Peter Gustav Lejeune-Dirichlet, "Sur la convergence des séries trigonometriques quiservent à représenter une fonction arbitraire entre des limites données," *Journal für reine und angewandte Mathematik*, 1829, *4*: 157–169.

Johann Peter Gustav Lejeune-Dirichlet, *Vorlesungen über Zahlentheorie*. Herausgegeben und mit Zusätzen versehen von Richard Dedekind. Vierte Auflage (Braunschweig: Vieweg, 1893).

Johann Peter Gustav Lejeune-Dirichlet, *Vorlesungen über die Lehre von den einfachen und mehrfachen bestimmten Integralen*. Ed. Gustav Arendt (Braunschweig: Vieweg, 1904).

Johann Peter Gustav Lejeune-Dirichlet, *Werke*. Zwei Bände. Ed. L. Kronecker, continued by L. Fuchs (Berlin 1889 – 1897)

Enno H. Dirksen, "Über die Darstellung beliebiger Funktionen mittelst Reihen, die nach den Sinussen und Cosinussen der Vielfachen eines Winkels fortschreiten," *Abhandlungen der Königlichen Akademie der Wissenschaften zu Berlin*. Aus dem Jahre 1827. Mathematische Klasse (Berlin, 1830), 85–113.

Enno H. Dirksen, Rezension: "A. L. Cauchy's Lehrbuch der algebraischen Analysis. Aus dem Französischen übersetzt von C.L.B. Huzler, Königsberg 1828," *Jahrbücher für wissenschaftliche Kritik*, 1829, Band 2: col. 211–222.

Enno H. Dirksen, "Über die Summe einer, nach den Sinussen und Cosinussen der Vielfachen eines Winkels fortschreitenden, Reihe," *Abhandlungen der Königlichen Akademie der Wissenschaften zu Berlin*. Aus dem Jahre 1829. Mathematische Klasse (Berlin, 1832), 169–187.

Enno H. Dirksen, "Über die Anwendung der Analysis auf die Rectification der Curven, die Quadratur der Flächen und die Cubatur der Körper," *Abhandlungen der Königlichen Akademie der Wissenschaften zu Berlin*. Aus dem Jahre 1833. Mathematische Klasse (Berlin, 1835), 123–168.

Enno H. Dirksen, *Organon der gesammten transcendenten Analysis. Erster Theil. Transcendente Elementarlehre* (Berlin: Reimer, 1845).

Willis Doney, "Malebranche, Nicolas," *The Encyclopedia of Philosophy*. Volume Five, ed. Paul Edwards (New York, London: Macmillan, 1972), 140–144.

Luiz Pedro Drago, *Apostillas de Algebra* (Rio de Janeiro, 1868).

Paul du Bois-Reymond, *Die allgemeine Functionentheorie. Erster Teil: Metaphysik und theorie der mathematischen Grundbegriffe: Größe, Grenze, Argument und Function*. (1882) Neudruck von Detelf Laugwitz (Darmstadt: Wiss. Buchgesellschaft, 1968).

DuBourguet: see du Bourguet.

J.M. Dubbey, *The Mathematical Work of Charles Babbage* (Cambridge: Cambr. Univ. Press, 1978,).

Pierre Dugac, "Euler, d'Alembert et les fondements de l'analyse," ed. J. J. Burckhardt, E. A. Fellman, W. Habicht, *Leonhard Euler 1707–1783. Beiträge zu Leben und Werk. Gedenkband des Kantons Basel-Stadt* (Basel: Birkhäuser, 1983), 171-184.

Jean-Marie-Constant Duhamel, *Cours d'Analyse*. 2 vol. (Paris: Bacheliers, 1841/40). Séconde édition: Paris, Bachelier 1847.

Jean-Marie-Constant Duhamel, *Éléments de Calcul Infinitésimal*. 2 vols. (Paris: Mallet-Bachelier, 1856).

Jean-Marie-Constant Duhamel, *Des Méthodes dans les sciences de raisonnement.*, 5 vol. (Paris: Gauthier-Villars, 1865-1873) (2nd ed.: 1878).

Pierre Duhem, *Medieval Cosmology. Theories of Infinity, Place, Time, Void, and the Plurality of Worlds*. Edited and Translated by Roger Ariew (Chicago and London: University of Chicago Press, 1985).

G. D. Duncan, "Public Lectures and Professorial Chairs," in: McConica 1986a, 335–361.

P. Dupont, "Osservazioni di A. Cauchy su un metodo d'interpolazione, a confronto col metodo dei minimi quadrati, in uno scambio di lettere con G. Bidone dell'Accademia delle Scienze di Torino", L. Grugnetti, O. Montaldo (eds.), *La Storia delle Matematiche in Italia* (Cagliari 1982), 461–468.

Alain Duroux, "La valeur absolue: difficultés majeures pour une notion mineure," *petit x*, no. 3, 43-67.

Marie-Jeanne Durry, *Autographes de Mariemont*. Deuxième Partie. Tome I: de Boufflers à Lamennais (Paris: Nizet, 1959).

Raymond Duval, *Sémiosis et Pensée Humaine* (Bern etc.: Lang, 1995).

Rudolf Eckart (ed.), *Abraham Gotthelf Kästner's Selbstbiographie und Verzeichnis seiner Schriften, nebst Heyne's Lobrede auf Kästner* (Hannover: Ernst Geibel, 1909).

C. H. Edwards, Jr., *The historical Development of the Calculus* (New York, Heidelberg: Springer, 1979).

Johann Wilhelm Elsermann, *Ueber die fortschreitende Verallgemeinerung der arithmetischen Operationen und die Natur der damit zusammenhangenden verschiedenen Zahlformen*. Schulprogramm Gymnasium Wetzlar 1852.

Moritz Epple, "Genies, Ideen, Institutionen, mathematische Werkstätten: Formen der Mathematikgeschichte," *Mathematische Semesterberichte*, 2000, *47*: 131–163.

Andreas von Ettingshausen, *Vorlesungen über die höhere Mathematik* (Wien: Gerold, 1827).

Euclid, *The thirteen books of Euclid's "Elements"*. Transl. from the text of Heiberg with introd. and commentary by Th. L. Heath .(New York: Dover Publ., 1926).

Leonhard Euler, *Introductio in analysin infinitorum* (Lausanne 1748). Deutsche Übersetzung von H. Maser: *Einleitung in die Analysis des Unendlichen* . Erster Teil (Berlin 1885). Reprint, mit einer Einführung von Wolfgang Walter (Berlin/Heidelberg: Springer, 1983).

Leonhard Euler, *Vollständige Anleitung zur Algebra* (St. Petersburg 1770) Edition used: Neue Ausgabe mit den Korrekturen der Opera Omnia Leonhard Eulers, I. Serie, Band I, herausgegeben von Andreas Speiser (Leipzig: Reclam, 1940).

Leonhard Euler's *Einleitung in die Analysis des Unendlichen.* Aus dem Lateinischen übersetzt und mit Anmerkungen und Zusätzen begleitet von Johann Andreas Christian Michelsen. Erstes Buch (Berlin: S. F. Hesse, 1788).

Leonhard Euler's *Vollständige Anleitung zur Differenzial-Rechnung.* Aus dem Lateinischen übersetzt und mit Anmerkungen und Zusätzen begleitet von Johann Andreas Christian Michelsen. 2 Bände (Berlin und Libau: Lagarde und Friedrich, 1790/93).

Leonhard Euler, *Vollständige Anleitung zur Integralrechnung* [3 Bde., 1768–1770] Aus dem Lateinischen ins Deutsche überetzt von Joseph Salomon, 4 Bde. (Wien: Gerold, 1828-1830).

Leonhard Euler, *Opera Omnia,* series I. Opera Mathematica. Volumen XVII. *Commentationes Analyticae ad Theoriam integralium pertinentes.* Volumen primum. Ed. August Gutzmer (Leipzig, Berlin: Teubner, 1915).

Leonhard Euler, *Opera Omnia.* series I. Opera Mathematica. Volumen XVII. *Commentationes Analyticae ad Theoriam integralium pertinentes.* Volumen Tertium. Ed. A. Liapunoff, A. Krazer, G. Faber (Leipzig, Berlin: Teubner, 1932).

Leonhard Euler, *Opera Omnia,* series I. Opera mathematica Volumen 23 *Commentationes analyticae ad theoriam aequationum differentialium pertinentes.* vol. 2 (Leipzig, Berlin: Teubner, 1938).

Leonhard Euler, *Opera Omnia,* series 1. Opera mathematica Volumen IX *Introductio in analysin infinitorum .* Tomus 2 (Leipzig, Berlin: Teubner, 1945).

Leonhard Euler, *Opera Omnia.* series Quarta A: Commercium Epistolicum. Volumen Quintum: *Commercium Epistolicum cum A. C. Clairaut, J. D'Alembert et J. L. Lagrange.* Ed. Adolf P. Youschkevitch et René Taton (Basel: Birkhäuser, 1980).

Leonhard Euler, *Elements of algebra.* Transl. by John Hewlett. (New York: Springer, 1984).

Leonhard Euler, *Introduction to Analysis of the Infinite.* Transl. by J.D. Blanton. 2 vol.s (New York: Springer, 1988)

Leonhard Euler, *Foundations of Differential Calculus.* Transl. by J.D. Blanton (New York: Springer, 2000).

Johann Albert Eytelwein, *Grundlehren der höhern Analysis* (Berlin: Reimer, 1824).

Lenore Feigenbaum, "Brook Taylor and the Method of Increments," *Archive for History of Exact Sciences,* 1985, *34*: 1–140.

J. V. Field, "The infinitely great and the infinitely small in the work of Girard Desargues," *Desargues et son temps,* eds. Jean Dhombres, J. Sakarovitch (Paris: Blanchard, 1994), 219–230.

Ernst-Gottfried Fischer, *Untersuchung über den eigentlichen Sinn der höheren Analysis: nebst einer idealischen Übersicht der Mathematik und Naturkunde nach ihrem ganzen Umfang* (Berlin: Weiss, 1808).

Ernst Gottfried Fischer, *Lehrbuch der Elementar-Mathematik.* Zweiter Theil: *Die Elemente der Zahlen- und Buchstabenrechnung* (Berlin: Nauck, 1822).

Ernst Gottfried Fischer, *Anmerkungen zu seinem Lehrbuch der Mathematik. Zweites Heft, welches allgemeine Untersuchungen und Anmerkungen zu der Arithmetik enthält* (Berlin und Leipzig: Nauck, 1822) [1822a].

Ernst Gottfried Fischer, *Lehrbuch der Elementar-Mathematik. Vierter Theil: Anfangsgründe der Algebra und der Lehre von den Kegelschnitten* (Berlin/Leipzig: Nauck, 1829).

J. M. Fletcher, "The Faculty of Arts," McConica 1986a, 157–199.

Wilhelm August Förstemann, *Ueber den Gegensatz positiver und negativer Größen* (Nordhausen: Happach, 1817). Reprint: LTR-Verlag Wiesbaden 1971.

Wilhelm August Förstemann, *Bemerkungen über verschiedene Begriffe und Theorien aus der allgemeinen Grössen- und Zahlenlehre.* Schulprogramm Gymnasium Danzig 1825.

Wilhelm August Förstemann, *Lehrbuch der Geometrie, besonders als Hilfsmittel zum Unterrichte an höheren Bildungsanstalten.* Zwei Theile (Danzig: Anhuth, 1827-1829).

Wilhelm August Förstemann, Arithmetisches Uebungsbuch: ein Hülfsmittel zu einem zweckmäßigen Unterrichte in der Zahlenrechnung, Buchstabenrechnung und Algebra (Königsberg: Bornträger, 1835).

Menso Folkerts, "Der Mathematiker E. H. Dirksen und C.F. Gauss," *Mitteilungen der Gauss-Gesellschaft Göttingen*, 1983/84, Nr. 20/21, 66–76.

Bernard le Bouyer de Fontenelle, *Éléments de la géométrie de l'Infini* (Paris: Imprimerie Royale, 1727).

Fourcy 1828: see Dhombres 1987.

Louis Benjamin Francœur, *Cours Complet des Mathématiques Pures.* Seconde édition, révue et considérablement augmentée. 2 tomes (Paris: Courcier, 1819).

Craig G. Fraser, "The Calculus as Algebraic Analysis: Some Observations on Mathematical Analysis in the 18th Century," *Archive for History of the Exact Sciences*, 1989, 39, 317–335.

William Frend, *Principles of Algebra.* 2 vols. (London 1796).

Hans Freudenthal, "Cauchy, Augustin-Louis," ed. C.C. Gillispie, *Dictionary of Scientific Biography*, vol. III (New York: Scribner's, 1971), 131–148.

Jean-Pierre Friedelmeyer, *Le calcul des dérivations d'Arbogast dans le projet d'algébrisation de l'analyse à la fin du XVIIIe siècle.* Cahiers d'Histoire et de Philosophie des Sciences. No. 43. Société Française d'Histoire des Sciences et des Techniques (Paris: Blanchard, 1994).

Jacob Friedrich Fries, [Rezension von:] Geometrie der Stellung von L. N. M. Carnot, [...] übersetzt von H. C. Schumacher, [...] Erster Theil, Altona 1808; Vergleichung zwischen Carnots und meiner Ansicht der Algebra von F. G. Busse, Freyberg 1808, *Heidelbergische Jahrbücher der Literatur*, Jahrgang 3, Abt. 4, Heft 1, 1810, 28–36. Reprinted in: L. Geldsetzer, G. König (ed.): Jacob Friedrich Fries, Sämtliche Schriften, Bd. 25 (Aalen: Scientia, 1996), 308–317.

Jacob Friedrich Fries, Die mathematische Naturphilosophie, nach philosophischer Methode bearbeitet (Heidelberg: Winter, 1822).

Franz Funck, *Beleuchtung der Carnot'schen Einwürfe gegen die bisherige Theorie der entgegengesetzten Größen und der Diesterweg'schen Widerlegung derselben.* Schulprogramm Gymnasium Culm 1841.

Athanasios Gagatsis, "Eine Studie zur historischen Entwicklung und didaktischen Transposition des Begriffs 'absoluter Betrag'," *Journal für Mathematik-Didaktik*, 16, 1995: 3-46.

François de Gandt, "L'évolution de la théorie des indivisibles et l'apport de Torricelli," eds. Massimo Bucciantini, Maurizio Torrini, *Geometria e Atomismo nella Scuola Galileiana* (Firenze: Leo S. Olschki, 1992), 103–118.

Dominique-J. Garat, "Analyse de l'entendement. Programme du cours," Séances des écoles normales... , vol. 1 1795 (Paris, 1800), 138-151.

Jean-Guillaume Garnier, *Leçons de Calcul Différentiel.* Troisième édition (Paris: Vve. Courcier, 1811).

John Gascoigne, "Mathematics and Meritocracy: The Emergence of the Cambridge Mathematical Tripos," *Social Studies of Science*, 1984, *14*: 547–584.

John Gascoigne, *Cambridge in the age of the Enlightenment. Science, religion and politics from the Restoration to the French Revolution* (Cambridge University Press 1989).

Carl Frierich Gauß, "Theoria residuorum biquadraticum. Commentatio secunda. [Selbstanzeige]," *Göttingische gelehrte Anzeigen*, 23. 4. 1831 (Nachdruck in: Gauß, *Werke*, Bd. II, 1863, 169–178).

Peter Gay, *The Enlightenment: an Interpretation; the Rise of Modern Paganism* (London: Weidenfeld and Nicolson, 1966).

Peter Gay, *Freud for historians* (Oxford: Oxford Univ. Pr., 1985).

Gehler, *Physikalisches Wörterbuch*, Band 4. Neue [= 2.] Auflage (Leipzig: Schwickert, 1798) [1. Auflage 1791].

Jacob de Gelder, *Proeve over den waren Aard van den positieven en negatieven Toestand der Grootheden in de Stelkunst en in de Toepassing van dezelve op de Meetkunde* ('s Gravenhage/Amsterdam: van Cleef, 1815).

Joseph-Diez Gergonne, "Réflexions sur le même sujet ["Essai sur la théorie des quantités négatives"], *Annales de Mathématiques pures et appliquées*, 1813, 4: 6–20.

Helmuth Gericke, *Geschichte des Zahlbegriffs* (Mannheim: Bibliograph. Inst., 1970).

Christian Gilain, *Cauchy et le Cours d'analyse de l'Ecole Polytechnique*. Bulletin de la Société des Amis de la Bibliothèque de l'Ecole Polytechnique. no. 5, Juillet 1989.

Christian Gilain, "Sur l'histoire du théorème fondamental de l'algèbre: théorie des équations et calcul intégral," *Archive for History of the Exact Sciences*, 1991, *42*, 91–136.

Donald Gillies, ed., *Revolutions in mathematics* (Oxford: Clarendon Pr., 1992).

Charles C. Gillispie, *Lazare Carnot Savant*. A monograph treating Carnot's scientific work, with facsimile reproduction of his unpublished writings on mechanics and on the calculus, and an essay concerning the latter by A. P. Youschkevitch (Princeton: Princeton University Press, 1971).

Charles C. Gillispie, "Carnot, Lazare," *Dictionary of Scientific Biography*, vol. 3, 1971, 70–79 [1971a].

Charles C. Gillispie, "Laplace, Pierre-Simon, Marquis de," *Dictionary of Scientific Biography*, Vol. XV, Supplement I (New York: Scribner's, 1978), 273–403.

Charles C. Gillispie, Adolf P. Youschkevitch, *Lazare Carnot Savant et sa contribution à la Théorie de l'Infini Mathématique*, avec trois mémoires inédits de Carnot (Paris: Vrin, 1979).

Girault de K[er]oudou, *Leçons analytiques du calcul des fluxions et des fluentes ou calcul différentiel et intégral* (Paris: 1777).

Hélène Gispert, *Camille Jordan et les fondements d'analyse*. Thèse de doctorat 3e cycle. Université Paris-Sud Département de Mathématique. Publications Mathématiques d'Orsay 82–05. Orsay 1982.

Hélène Gispert, "Features from the French mathematics development and the higher education institutions (1860–1900)," in Schubring (ed.), 1991, 198–213.

Enrico Giusti, "Dopo Cavalieri. La discussione sugli indivisibili." eds. O. Montaldo, L. Grugnetti, *Atti del Congresso "La Storia delle Matematiche in*

Italia." *Cagliari, 29 settembre–1 ottobre 1982* (Università di Cagliari, s. d. [1982]), 85–114.

Enrico Giusti, "Gli 'errori' di Cauchy e i fondamenti dell'analisi," *Bollettino di Storia delle Scienze Matematiche*, 1984, *4*: 24–54.

Enrico Giusti, "Image du Continu," *The Leibniz Renaissance*. International Workshop Firenze 1986 (Firenze: Leo S. Olschki, 1989), 83–97).

Enrico Giusti, "Quelques Réflexions sur les <Principios> de da Cunha," ed. Luis Manuel Ribeiro Saraiva, *Proceedings of the International Colloquium: José Anastácio da Cunha (1744–1787), the mathematician, and the poet* (Lisboa: Imprensa Nacional/Casa da Moda: 1990), 87–111.

Enrico Giusti, "*Ratio* e *Proportio* tra Geometria e Filosofia naturale," eds. M. Fattori, M.L. Bianchi, *Ratio*. *VII Colloquio Internazionale*, Roma 1992 (Firenze: L. S. Olschki, 1994), 415–438.

Georges Glaeser, "Épistémologie des nombres relatifs," *Recherches en didactique des mathématiques*, 1981, *2*: 303–346.

J. A. Gmeiner, "Nachruf: Otto Stolz," *Jahresbericht der Deutschen Mathematikervereinigung*, 1906, *15*: 309–322.

Catherine Goldstein, "On a Seventeenth Century Version of the 'Fundamental Theorem of Arithmetic'," *Historia Mathematica*, 1992, *19*: 177–187.

Judith V. Grabiner, "The Origins of Cauchy's Theory of the Derivative," *Historia Mathematica*, 1978, *5*: 379–409.

Judith V. Grabiner, *The Origins of Cauchy's Rigorous Calculus* (Cambridge, Mass.: The MIT Press, 1981).

Judith V. Grabiner, *The Calculus as Algebra. J.-L. Lagrange, 1736–1813* (New York and London: Garland Publ., 1990).

Judith V. Grabiner, "Was Newton's Calculus a Dead End? The Continental Influence of MacLaurin's Treatise of Fluxions," *American Mathematical Monthly*, 1997, *104*: 393–410.

Anthony Grafton, "Polyhistor into Philolog: Notes on the transformation of German Classical Scholarship, 1780–1850," *History of Universities*, 1983, *3*: 159–192.

Gilles-G. Granger, *Langages et Épistémologie* (Paris: Klincksiek, 1979).

Ivor Grattan-Guinness, "The emergence of mathematical analysis and its foundational progress, 1780–1880," Ivor Grattan-Guinness (ed.), *From the Calculus to Set Theory, 1630–1910. An Introductory History* (London: Duckworth, 1980), 94–147.

Ivor Grattan-Guinness, "The Cauchy–Stokes–Seidel Story on Uniform Convergence again: was there a fourth man?," *Bulletin de la Société Mathématique de Belgique*, 1986, *38*: 225–235.

Ivor Grattan-Guinness, "The *Società Italiana*, 1782-1815: A Surcey of its Mathematics and Mechanics," *Symposia Mathematica*, 1986, *27*: 147-168 [1986a].

Ivor Grattan-Guinness, *Convolutions in French Mathematics* (Basel, etc.: Birkhäuser, 1990) [1990a].

Ivor Grattan-Guinness, "Work for the Hairdressers: The Production of de Prony's Logarithmic and Trigonometric Tables," *Annals for the History of Computing*, 1990, *12*: 177–185 [1990b].

Emmanuel Grison, "Lazare Carnot, Fondateur de Polytechnique," in: J.-P. Charnay 1990, 168–183.

Gros, "Un Professeur D'Autrefois. Roger Martin," *Mémoires de l'Académie des Sciences, Inscriptions et Belles Lettres de Toulouse*, 1919, 7: 107–135.

Angelo Gueraggio, Marco Panza, "Le *Reflexions* di Carnot e le *Contre-Réflexions* di Wronski sul Calcolo Infinitesimale," *Epistemologia*, 1985, *8*: 3–32.

Niccolò Guicciardini, *The development of Newtonian Calculus in Britain 1700–1800* (Cambridge: Cambridge University Press, 1989).

J. J. V. Guilloud, *Examen critique de la doctrine des infiniments petits, exposée par M. Poisson à la Faculté des Sciences, suivi de quelques remarques sur le cours de mécanique du même professeur* (Paris: Beaujouan, 1836).

Thierry Guitard, "La querelle des infiniment petits à l'École Polytechnique au XIXe siècle," *Historia Scientiarum*, 1986, *30*: 1–61.

Paul Haag, *Cours de Calcul Différentiel et Intégral* (Paris: Dunod, 1893).

S. Haegel, *Les nombres negatifs ont une histoire* (IREM Strasbourg, 1992).

Roger Hahn, *Calendar of the correspondence of Pierre Simon Laplace* (Berkeley: Office for History of Science and Technology, 1982).

Roger Hahn, "L'enseignement scientifique aux écoles militaires et d'artillerie," in: Taton 1986, 513-546.

Hermann Hankel, *Theorie der complexen Zahlensysteme, insbesondere der gemeinen imaginären Zahlen und der Hamiltonschen Quaternionen nebst ihrer geometrischen Darstellung* (Leipzig: L. Voss, 1867).

Hermann Hankel, "Grenze," eds. Johann Samuel Ersch, Johann Gottfried Gruber, *Allgemeine Encyclopädie der Wissenschaften und Künste*. Bd. 90 (Leipzig: Gleditsch, 1871), 185–211

Thomas L. Hankins, "The Concept of Hard Bodies in the History of Physics" [Review of W. L. Scott 1970], *History of Science*, 1970, *9*: 119–128.

J.-N. Haton de la Goupillière, *Éléments du Calcul Infinitésimal* (Paris: Mallet-Bachelier, 1860).

Johann Karl Friedrich Hauff 1800: see L. Carnot 1800.

Christian August Hausen, *Elementa Matheseos*. Pars Prima (Leipzig: Marcheana, 1734).

Thomas Hawkins, *Lebesgue's Theory of Integration: its Origins and Development* (Madison: Univ. of Wisconsin Press, 1970).

Johann Peter Hecker, *Über den gewöhnlichen Vortrag der Anfangsgründe der Lehre von den entgegengesetzten Zahlen*. Erste Abhandlung. Programmrede des Rektors der Universität Rostock zur Feier des Weihnachtsfestes. Rostock, den 25sten December 1799 (Rostock: Adler, 1799).

Johann Peter Hecker, *Über den gewöhnlichen Vortrag der Anfangsgründe der Lehre von den entgegengesetzten Zahlen*. Zweyte Abhandlung. Programmrede des Rektors der Universität Rostock zur Feier des Osterfestes. Rostock, den 13ten April 1800 (Rostock: Adler, 1800) [1800a].

Johann Peter Hecker, *Über den gewöhnlichen Vortrag der Anfangsgründe der Lehre von den entgegengesetzten Zahlen*. Dritte Abhandlung. Programmrede des Rektors der Universität Rostock zur Feier des Pfingst-Festes. Rostock, den 1ten Jun. 1800 (Rostock: Adler, 1800) [1800b].

Karl Hengst, *Jesuiten an Universitäten und Jesuitenuniversitäten* (Paderborn: Schöningh, 1980).

Jesús Hernandez, "Cauchy y la convergencia uniforme: algunas notas," *Actas de la Reunión Matemática en Honor de A. Dou* (Madrid: Editorial Univ. Complutense, 1989), 327–339.

Anton Herrmann, *Abhandlung über die wahre Natur des Positiven und Negativen nebst einer leicht faßlichen Berichtigung der Begriffe von den sogenannten unmöglichen Größen* (Wien: Gerold, 1818).

Heinz-Jürgen Hess, Fritz Nagel (eds.), *Der Ausbau des Calculus durch Leibniz und die Brüder Bernoulli.* Studia Leibnitiana, Sonderheft 17 (Wiesbaden: Franz Steiner, 1989).

Gerhard Hessenberg, "Das Unendliche in der Mathematik," *Abhandlungen der Fries'schen Schule. Neue Folge*, 1904, Erstes Heft, 135–190.

David Hilbert, *Grundlagen der Geometrie.* Mit Supplementen von Paul Bernays. 14. Auflage. Hrsg. und mit Anhängen versehen von Michael Toepell (Stuttgart/Leipzig: Teubner, 1999).

David Hilbert, *Natur und mathematisches Erkennen. Vorlesungen, gehalten 1919– 1920 in Göttingen*, Ed. David E. Rowe (Basel: Birkhäuser, 1992).

István Hiller, Dieter Schmidmaier, *Heinrich David Wilckens. Student an der Bergakademie Freiberg und erster Professor des Forstinstituts in Schemnitz* (Freiberg und Sopron: Bergakademie Freiberg, 1982).

Bernard R. Hodgson, "Le Calcul Infinitésimal," eds. D. F. Robitaille, D. H. Wheeler, C. Kieran, *Selected Lectures from the 7th International Congress on Mathematical Education* (Sainte-Foy: Les Presses de l'Université Laval, 1994), 157–170.

Jean Chr. Ferd. Hoefer, *Nouvelle Biographie Générale depuis les temps les plus reculés jusqu'à nos jours.* 46 tomes (Paris: Firmin Didot Frères, 1852–1866).

James R. Hofmann, *André-Marie Ampère* (Cambridge: Cambridge Univ. Press, 1996).

J.C.V. Hoffmann, "Des Herausgebers Bemerkungen und Mitteilungen, sowie Stimmen andrer Fachkollegen über Härters Artikel," *Zeitschrift für mathematischen und naturwissenschaftlichen Unterricht*, 1884, *15*: 108–113.

Frederick A. Homann, "On Boscovich's *De natura et usu infinitorum* and Other Mathematical Works: Translation and Commentary," ed. P. Bursill-Hall, *R. J. Boscovich. Vita e Attività Scientifica - His Life and Scientific Work.* Atti del Convegno Roma, 23–27 maggio 1988 (Roma: Istituto della Enciclopedia Italiana, 1993), 407–436.

R. Hooykaas, *Humanisme, Science et Réforme. Pierre de la Ramée (1515–1572)*, (Leyden: Brill, 1958).

Jens Høyrup, *Lengths, widths, surfaces: a portrait of old Babylonian algebra and its kin* (New York: Springer, 2002).

M. Hube, *Versuch einer Analytischen Abhandlung von den Kegelschnitten* (Göttingen: V. Boßigel, 1759).

Erika Hültenschmidt, "Enzyklopädien, Wissensdifferenzierung und Sprachwissenschaft um 1800 (Frankreich)," in: Schubring, *Einsamkeit u. Freiheit*, 1991, 57–90.

Institut de France, *Académie des Sciences, Procès-Verbaux des Séances de l'Académie tenues depuis la fondation de l'Institut jusqu'au mois d'août 1835.* Tome I, an IV–VII (1795–1799) (Hendaye, Imprimérie de l'Observatoire d'Abbadia, 1910); tome IV, an 1808–1811 (ibid., 1913).

Hans Niels Jahnke, "Motive und Probleme der Arithmetisierung der Mathematik in der ersten Hälfte des 19. Jahrhunderts," *Archive for History of Exact Sciences*, 1987, *37*: 101–182.

Camille Jordan, *Cours d'analyse de l'École Polytechnique*. Deuxième édition, entièrement refondue. Tome 1 (Paris: Gauthier-Villars, 1893).

Philip Jourdain, "The origin of Cauchy's conceptions of a definite integral and of the continuity of a function," *Bibliotheca mathematica*, 1914, 3e série, *6*: 661–703.

Dominique Julia, "Universités et Collèges à l'époque moderne: Les institutions et les hommes," *Histoire des Universités en France*, éd. Jacques Verger (Paris: Privat, 1986), 141–197.

Immanuel Kant, "Versuch, den Begriff der negativen Größen in die Weltweisheit einzuführen," ed. Wilhelm Weischedel, *Werkausgabe Immanuel Kant*. Band II: Vorkritische Schriften bis 1768, 2 (Frankfurt am Main: Suhrkamp, 1968).

Immanuel Kant, *Attempt to introduce the conception of Negative Quantities into Philosophy*. Transl. David Irvine (London: Watts, 1911).

Wenceslaus Johann Gustav Karsten, *Lehrbegriff der gesamten Mathematik. Der zweyte Theil. Weitere Ausführung der Rechenkunst. Die Buchstabenrechnung. Die ebene und sphärische Trigonometrie, ...* (Greifswald: A. F. Röse, 1768).

Wenceslaus Johann Gustav Karsten, *Lehrbegriff der gesamten Mathematik. Der vierte Theil. Die Mechanik fester Körper* (Greifswald: A. F. Röse, 1769).

Wenceslaus Johann Gustav Karsten, *Lehrbegriff der gesamten Mathematik. Des zweyten Theils zweyte Abtheilung. Anfangsgründe der Mathematischen Analysis und höheren Geometrie, mit Rücksicht auf eine Preisfrage vom Mathematisch-Unendlichen* (Greifswald: A. F. Röse, 1786) [1786a].

Wenceslaus Johann Gustav Karsten, *Mathematische Abhandlungen, theils durch eine Preisfrage ..., theils durch andere neuere Untersuchungen veranlasst* (Halle: Renger, 1786).

Abraham Gotthelf Kästner, *De lege Continui in Natura Dissertatio* (Leipzig. Langenheim, 1750).

Abraham Gotthelf Kästner, "Betrachtungen über die Beschaffenheit und den Gebrauch des analytischen Vortrages," Vorrede, in: M. Hube, *Versuch einer Analytischen Abhandlung von den Kegelschnitten* (Göttingen 1759), [xv–xxxviii].

Abraham Gotthelf Kästner, "Vorlesung über den wahren Begriff des mathematischen Unendlichen," *Göttingische Anzeigen von Gelehrten Sachen*, 1759, 497–500 [1759a].

Abraham Gotthelf Kästner, *Anfangsgründe der Arithmetik, Geometrie, ebenen und sphärischen Trigonometrie, und Perspectiv*. Der mathematischen Anfangsgründe 1ten Theils erste Abtheilung. Fünfte vermehrte Auflage (Göttingen: Vandenhoek und Ruprecht, 1792) [1. Auflage 1758].

Abraham Gotthelf Kästner, *Anfangsgründe der Analysis endlicher Grössen*. Der mathematischen Anfangsgründe IIIter Theil; erste Abtheilung. Dritte, stark vermehrte Auflage (Göttingen: Vandenhoek und Ruprecht, 1794) [1. Auflage 1760].

Abraham Gotthelf Kästner, *Anfangsgründe der Analysis des Unendlichen*. Der mathematischen Anfangsgründe IIIter Theil; zweyte Abtheilung. Zweyte Auflage (Göttingen: Vandenhoek und Ruprecht, 1770) [1. Auflage 1761].

Abraham Gotthelf Kästner, *Anfangsgründe der höhern Mechanik*. Der mathematischen Anfangsgründe IVter Theil; erste Abtheilung (Göttingen: Wwe. Vandenhoek, 1766).

H. Jerome Keisler, *Elementary calculus* (Boston: Prindle, Weber and Schmidt, 1976).

Johannes Kepler, *Nova stereometria doliorum* (Original 1615): *Neue Stereometrie der Fässer* [translated from the Latin and edited by R. Klug] (Leipzig, 1908). Reprint: Leipzig: Akad. Verl.-Ges. Geest u. Portig, 1987.

Philip Kitcher, "Fluxions, Limits, and Infinite Littleness," *ISIS*, 1973, *64*: 33–49.

Felix Klein, *Elementarmathematik vom höheren Standpunkt*, Bd. 2 (Berlin: Springer, 1925). Engl. Transl. (E.R. Hedrick, C.A. Noble): *Elementary Mathematics from an Advanced Standpoint. Geometry* (New York: Dover, 1939).

Jacob Klein, *Greek Mathematical Thought and the Origin of Algebra* (Cambridge, Mass.: The MIT Press, 1968).

Morris Kline, *Mathematical Thought from Ancient to Modern Times* (New York: Oxford University Press, 1972).

Morris Kline, *Mathematics. The Loss of Certainty* (New York: Oxford University Press 1980).

Morris Kline, "Euler and Infinite Series," *Mathematics Magazine*, 1983, *56*: 307–325.

J. Klostermann, *Le quarré d'une quantité négative est négatif et non positif* (St. Petersbourg, 1804).

J. Klostermann, *Démonstration que la règle: moins multiplié par moins donne plus, induit en erreur et qu'elle ne s'accorde pas avec les opérations de l'esprit humain*. Avec permission de la Censure (St. Petersbourg: Iversen, 1805).

Georg Simon Klügel, *Anfangsgründe der Arithmetik, Geometrie und Trigonometrie, nebst ihrer Anwendung auf praktische Rechnungen, das Feldmessen und die Markscheidekunst*. Zweyte verbesserte Auflage (Berlin, Stettin: F. Nicolai, 1792).

Georg Simon Klügel, "Über die Lehre von den entgegengesetzten Größen," *Archiv der reinen Mathematik und Physik*, Erster Band, Leipzig 1795, 309–319, 470–481.

Georg Simon Klügel, "Aus einem Schreiben des Herrn Professor Klügel an den Herausgeber," *Archiv der reinen Mathematik und Physik*, Eilftes Heft [Vierter Band]. Stück V, 1800, 340–342 [Mit einem "Zusatz des Herausgebers" (= C. F. Hindenburg), 343–348].

Georg Simon Klügel, *Mathematisches Wörterbuch oder Erklärung der Begriffe, Lehrsätze, Aufgaben und Methoden in der Mathematik*. Erste Abtheilung: Die reine Mathematik, vol. 1, 1803; vol. 2 1805; vol. 4, 1823; vol. 5, 1831 (Leipzig: Schwickert).

Teun Koetsier, *Lakatos' Philosophy of Mathematics. A Historical Approach* (Amsterdam. North-Holland, 1991).

Arnost Kolman, *Bernard Bolzano* (Beerlin: Akademie, 1963).

Sybille Krämer, *Sprache, Sprechakt, Kommunikation: sprachtheoretische Positionen des 20. Jahrhunderts* (Frankfurt am Main: Suhrkamp, 2001).

Wolfgang Krohn/Günter Küppers, *Die Selbstorganisation der Wissenschaft* (Frankfurt am Main: Suhrkamp, 1989).

Heinrich Kühn, "Meditationes de quantitatibus imaginariis construendis et radicibus imaginariis exhibendis," *Novi Commentarii Academiae Scientiarum Imperialis Petropolitanae*. Tomus III, ad Annum 1750 et 1751, 1753, 170–223.

Nicolas-Louis de LaCaille, *Lectiones Elementares Mathematicae, seu Elementa Algebrae, et Geometriae; ex Editione Parisina Anni MDCCLVI in Latinum traductae a C. S. e S. J. cum correctionibus ab ipso Authore communicatis* (Wien, Prag: Trattner, 1758).

M. Lacoarret, Ter-Menassian, "Les Universités," in Taton 1986, 125-168.

Sylvestre-François Lacroix, *Traité du Calcul différentiel et du Calcul intégral.* Tome premier (Paris: Duprat, an V = 1797) [1797a]. Tome second (Paris: Duprat, an VI = 1798). Tome troisième (Paris: Duprat, an VIII = 1800). Seconde édition: Tome premier (Paris: Courcier, 1810). Tome second 1814. Tome troisième 1819.

Sylvestre-François Lacroix, *Lehrbegriff des Differential- und Integralcalculs.* Aus dem Französischen übersetzt und mit einigen Anmerkungen und Zusätzen begleitet von Johann Philipp Grüson. Zwei Teile (Berlin: Lagarde, 1799-1800). [this is the translation of Lacroix 1797: only of volume 1].

Sylvestre-François Lacroix, *Traité élémentaire de Calcul différentiel et de Calcul intégral* (Paris: Duprat, 1802). Deuxième édition: Paris: Courcier, 1806; Troisième édition: Paris 1820; Quatrième édition: Paris 1828; Cinquième édition: Paris: Bachelier 1837.

L. C. [= Sylvestre-François Lacroix], Élémens d'Arithmétique et d'Algèbre. *Élémens d'Algèbre, par Clairaut,* cinquième édition. Avec des Notes et des Additions tirées en partie des Leçons données à l'École normale par Lagrange et Laplace, et précédée d'un Traité élémentaire d'Arithmétique. Tome premier, tome second (Paris: Duprat, 1797) [1797b].

L. C. [= Sylvestre-François Lacroix], *Élémens d'Algèbre,* à l'usage de L'École Centrale des Quatre-Nations (Paris: Duprat, an VIII) [1799].

L. C. [= Sylvestre-François Lacroix], *Élémens d'Algèbre,* à l'usage de L'École Centrale des Quatre-Nations. Seconde édition, revue et corrigée (Paris: Duprat, an IX = 1800).

Sylvestre-François Lacroix, *Élémens d'Algèbre*, à l'usage de L'École Centrale des Quatre-Nations. Troisième édition, revue et corrigée (Paris: Duprat, an XI = 1803) [1803a]. Cinquième édition, revue et corrigée (Paris: Courcier, an XIII = 1804). Septième édition, revue et corrigée (Paris: Courcier, 1808). Vingt-troisième édition, revue, corrigée et annotée par E. Prouhet (Paris: Gauthier-Villars, 1871).

Sylvestre-François Lacroix, *Traité élémentaire de Trigonométrie rectiligne et sphérique, et d'application de l'Algèbre à la Géométrie.* Seconde édition, revue et augmentée (Paris: Duprat, an VIII). [1799/1800]. Troisième édition, revue et augmentée (Paris: Duprat, an XII = 1803) [1803b].

Sylvestre-François Lacroix, *Essais sur l'Enseignement en général, et sur celui des Mathématiques en particulier.* Quatrième édition, revue et corrigée (Paris: Bachelier, 1838).

Joseph Louis Lagrange, "Note sur la métaphysique du calcul infinitésimal," *Miscellanea Taurinensis,* tome II, 1760–1761, 17–18 = *Œuvres,* tome I (Paris: Gauthier-Villars, 1867), 597-599.

Joseph Louis Lagrange, "Sur une nouvelle espèce de calcul relatif à la différentiation et à l'intégration des quantités variables," *Nouveaux Mémoires de l'Académie royale des Sciences et Belles-Lettres de Berlin* pour 1772 (Berlin 1774, 185–221) = *Œuvres,* tome III, 1869, 441–476.

Joseph Louis Lagrange, "Discours sur l'objet de la Théorie des Fonctions analytiques," *Journal de l'École Polytechnique,* an VII (1799), tome 2, cahier 6, 232–235.

Joseph-Louis Lagrange, *Leçons sur le calcul des fonctions.* Nouvelle édition (Paris: Courcier, 1806) = *Œuvres,* tome X (Paris: Gauthier-Villars, 1884).

Joseph-Louis Lagrange, *Méchanique analitique* (Paris: Desaint, 1788) Deuxième édition: *Mécanique analytique.* Tome I, Paris 1811; tome II, Paris 1815 = *Œuvres,* tome XI–XII (Paris: Gauthier-Villars, 1869).

Joseph-Louis Lagrange, *Théorie des Fonctions analytiques, contenant Les principes du Calcul différentiel, dégagés de toute considération d'infiniment petits, d'évanouissans, de limites et de fluxions, et réduits à l'analyse algébrique des quantités finies.* Nouvelle édition (Paris: Courcier, 1813) = *Œuvres*, tome IX (Paris: Gauthier-Villars, 1881).

Joseph Louis Lagrange, *Mathematische Werke.* Deutsch herausgegeben von August Leopold Crelle. *Erster Band, die Theorie der analytischen Functionen enthaltend* (Berlin: Reimer, 1823).

Joseph Louis Lagrange, *Mathematische Werke.* German edition by August Leopold Crelle. *Zweiter Band, die Vorlesungen über die Functionen-Rechnung enthaltend* (Berlin: Reimer, 1823).

Imre Lakatos, *Proofs and Refutations. The Logic of Mathematical Discovery* (Cambridge: Cambridge University Press, 1976).

Imre Lakatos, "Cauchy and the continuum: the significance of non-standard analysis for the history and philosophy of mathematics," *Imre Lakatos: Mathematics, science and epistemology. Philosophical papers Volume 2,* eds. John Worrall and Gregory Currie (Cambridge. Cambridge University Press, 1978), 43–60.

Paul Lallemand, *Histoire de l'Éducation dans l'ancien Oratoire de France* (Paris: Ernest Thorin, 1888).

Pierre Lamandé, "Les manuels de Bézout," *Rivista di Storia della Scienza*, 1987, *4*, 339–375.

Pierre Lamandé, "Deux manuels mathématiques rivaux: le Bézout et le Lacroix - ancien contre nouveaux régime en calcul infinitésimal," *Wissenschaftliche Zeitschrift der Universität Rostock, Gesellschaftswissenschaftliche Reihe*, 1988, *37,9*: 16–25.

Pierre Lamandé, "Des différents rôles de l'écriture dans un manuel mathématique: L'exemple de Bézout," *Sciences et Techniques en Perspectives*, no. 19, 1990–91, 31–40.

Bernard Lamy, *Traité de la Grandeur en général qui comprend l'arithmétique, l'algèbre, l'analyse et les principes de toutes les sciences qui ont la grandeur pour objet* (Paris 1680). Quoted here the second edition, in the Amsterdam edition,, given incorrectly as the 3rd edition (Amsterdam: Desbordes, 1692).

Bernard Lamy, *Nouveaux Elemens de Geometrie, ou de la Mesure des Corps, qui comprennent ..* (Amsterdam: George Gallet, 1692).

Bernard Lamy, *Les Elemens de Geometrie, ou de la Mesure des Corps, qui comprennent.* Sixieme édition, revue et augmentee (Amsterdam: Pierre Mortier, 1734)

Janis Langins, *La République avait besoin de savants* (Paris: Belin, 1987).

Detlef Laugwitz, "Die Messung von Kontingenzwinkeln," *Journal für reine und angewandte Mathematik*, 1970, *245*: 133–142.

Detlef Laugwitz, "Infinitely small quantities in Cauchy's textbooks," *Historia Mathematica*, 1987, *14*: 258–274.

Detlef Laugwitz, "Definite Values of Infinite Sums: Aspects of the Foundations of Infinitesimal Analysis around 1820," *Archive for the History of Exact Sciences*, 1989, *39*: 195–245.

Detlef Laugwitz, "Das mathematisch Unendliche bei Cauchy und bei Euler," ed. Gert König, *Konzepte des mathematisch Unendlichen im 19. Jahrhundert* (Göttingen, Vandenhoeck u. Ruprecht, 1990), 9–33.

Detlef Laugwitz, Detlef Spalt, *Another view on Cauchy's Theorem on Convergent series of Functions - an essay in the methodology of historiography in mathematics*. Technische Hochschule Darmstadt, Fachbereich Mathematik, preprint Nr. 1133. Darmstadt, April 1988.

Lam Lay-Yong, Ang Tian-Se, "The earliest negative numbers: how they emerged from a solution of simultaneous linear equations," *Archives internationales d'histoire des sciences*, 1987, *37*: 222–262.

Hans Lausch, "Moses Mendelssohn: 'Ein Algebraist würde das Gute in seinem Leben mit positiven Grössen vergleichen.' Zur Unwirklichkeit des Negativen im 18. Jahrhundert," *Mendelssohn Studien*, Band 8 (Berlin: Duncker und Humblot, 1993), 23–36.

Damian Riehl Leader, *A History of the University of Cambridge*. Volume 1, The University to 1546 (Cambridge: Cambridge University Press 1988).

Louis Lefébure de Fourcy, *Leçons d'Algèbre*. Troisième édition (Paris: Bachelier, 1838).

Gottfried Wilhelm Leibniz, *Mathematische Schriften*. Herausgegeben von C. I. Gerhardt. Band I: Briefwechsel zwischen Leibniz und Oldenburg, Collins, Newton Reprograf. Nachdr. der Ausg. Berlin 1849: Hildesheim 1962. Band III/2: Briefwechsel zwischen Leibniz, Jacob Bernoulli, Johann Bernoulli und Nicolaus Bernoulli (Halle 1856), Reprint Hildesheim 1962. Band IV: Briefwechsel zwischen Leibniz, Wallis, Varignon, Guido Grandi, Zendrini, Hermann und Freiherrn von Tschirnhaus (Halle 1859), Reprint Hildesheim 1962; Band V: Die mathematischen Abhandlungen (Halle, 1858), Reprint Hildesheim 1962. Band VI: Die mathematischen Abhandlungen (Halle 1860), Reprint Hildesheim 1962.

Gottfried Wilhelm Leibniz, *Die philosophischen Schriften*. Herausgegeben von C. I. Gerhardt. Zweiter Band (Halle 1879), Reprint Hildesheim: Olms, 1960; Dritter Band (Halle 1887), Reprint 1960; Siebenter Band (Halle 1890), Reprint 1961.

Gottfried Wilhelm Leibniz, *Hauptschriften zur Grundlegung der Philosophie*. Herausgegeben von Ernst Cassirer. Band I (Leipzig: Meiner, 1904).

Gottfried Wilhelm Leibniz, *La naissance du calcul différentiel. 26 articles des* Acta Eruditorum. Introduction, traduction et notes par Marc Parmentier (Paris: Vrin, 1989).

Gottfried Wilhelm Leibniz, *Philosophical papers and letters*. Translated and ed. by Leroy E. Loemker. 2nd edition (Dordrecht. Reidel, 1969).

Lexikon bedeutender Mathematiker, eds. Siegfried Gottwald et al. (Leipzig: Bibliograph. Institut, 1990).

[Marquis de L'Hospital] *Analyse des infiniment petits, pour l'intelligence des lignes courbes* (Paris: Imprimerie Royale, 1696).

Simon L'Huilier, *Exposition Élémentaire des Principes des Calculs Supérieurs, qui a remporté le prix proposé par L'Académie Royale des Sciences et Belles-Lettres pour l'année 1786* (G. J. Decker: Berlin, 1786).

Simon L'Huilier, *Principiorum Calculi Differentialis et Integralis Expositio Elementaris*, ad normam dissertationis ab Academia Scient. Reg. Prussica anno 1786 praemii decoratae elaborata (J. G. Cotta: Tübingen, 1795).

Wilhelm Lorey, *Das Studium der Mathematik an den Deutschen Universitaeten seit Anfang des 19. Jahrhunderts* (Leipzig: Teubner, 1916).

Jesper Lützen, "Grundlagen der Analysis im 19. Jahrhundert," *Geschichte der Analysis*, ed. H. N. Jahnke (Heidelberg: Spektrum, 1999), 191–244. Engl. Transl.: "The Foundation of Analysis in the 19th Century", *A History of Analysis* (Providence/RI: AMS, 2003), 155–195.

J. G. E. Maaß, *Grundriß der reinen Mathematik zum Gebrauche bei Vorlesungen und beim eigenen Studium* (Halle: Renger, 1796).

Colin MacLaurin, *A Treatise of Fluxions. In Two Books* (Edinburgh: Ruddimans, 1742).

Colin MacLaurin, *A Treatise of Algebra* (London: Millar and Nourse, 1748).

Michael S. Mahoney, "Infinitesimals and transcendent relations: The mathematics of motion in the late seventeenth century," *Reappraisals of the Scientific Revolution*, eds. David C. Lindberg, Robert S. Westman (Cambridge: Cambridge University Press, 1990), 461–491.

Luigi Maierú, "John Wallis: Lettura della Polemica fra Peletier e Clavio circa l'angolo di contatto," ed. Massimo Galuzzi, *Giornate di Storia della Matematica* Cetraro (Cosenza)1988, (Commenda di Rende: Editoria Elettronica, 1991), 315–364.

Luigi Maierù, "Considerazioni attorno alla polemica fra Peletier e Clavio circa l'angolo di contatto (1579–1589)," *Archive for History of Exact Sciences*, 1990, *41*: 115–137.

Luigi Maierù, *Fra Descartes e Newton: Isaac Barrow e John Wallis* (Soveria Manelli, Messina: Rubbettino, 1994).

Jacques Maillard, éd., *L'Oratoire à Angers aux XVIIème et XVIIIème siècles* (Paris: Klincksiek, 1975).

Marie-François-Pierre Maine de Biran, "Mémoire sur les rapports de l'idéologie et des mathématiques," ed. Pierre Tisserand, *Œuvres.* Tome III, Première Section (Paris: Alcan, 1924), 1–26.

Malebranche, *Œuvres Complètes.* Tome XVII-2: *Mathematica*, éd. Pierre Costabel (Paris: Vrin, 1968)

Malebranche, *Œuvres Complètes.* Tome XX: Documents biographiques et bibliographiques, ed. André Robinet (Paris: Vrin, 1967).

Antoni Malet, *Studies on James Gregorie (1638–1675).* Ph.D. dissertation Princeton University, 1989.

Antoni Malet, *From Indivisibles to Infinitesimals. Studies on Seventeenth-Century Mathematizations of Infinitely Small Quantities* (Bellaterra: Universidad de Barcelona, 1996).

Carlo Felice Manara, "Il problema del continuo geometrico nel pensiero di Ruggero Boscovich," eds. M. Bossi, P. Tucci, *Bicentennial commemoration of R. G. Boscovich*, Milano, September 15-18, 1987. Proceedings (Milano: Unicopli, 1988), 171–188.

Paolo Mancosu, *Philosophy of Mathematics and Mathematical Practice in the Seventeenth Century* (New York, Oxford: Oxford University Press, 1996).

Jonathan Mandelbaum, *La Société Philomathique de Paris de 1788 à 1835: essai d'histoire institutionnelle et de biographie collective d'une société scientifique parisienne*. Paris, École des Hautes Études en Sciences Sociales, Dissert., 1980 (UMI, Ann Arbor, Michigan).

Paul Mansion, *Résumé du Cours d'analyse infinitésimale de l'Université de Gand. Calcul différentiel et principes de calcul intégral* (Paris: Gauthier-Villars, 1887).

Joseph-François Marie, *Leçons Élémentaires de Mathématiques*. Par M. l'Abbé de La Caille. Nouvelle édition, avec de nouveaux Élémens d'Algèbre, de Géométrie, de Trigonometrie rectiligne et sphérique, de Sections coniques, de plusieurs autres Courbes, de Calcul Différentiel et de Calcul Intégral (Paris: Desaint, 1778).

Maximilien Marie, *Discours sur la nature des grandeurs négatives et imaginaires et interprétation des solutions imaginaires en géométrie* (Paris: Carilian-Gœury, 1843).

Maximilien Marie, *Histoire des sciences mathématiques et physiques* (Paris: Gauthier-Villars, 1883).

Zeljko Markovic, "Boscovich's Theoria," ed. Lancelot Law Whyte, *Roger Joseph Boscovich, S. J., F.R.S., 1711–1787. Studies of his Life and Work on the 250th Anniversary of his Birth* (New York: Fordham University Press, 1961), 127–152.

Roger Martin, *Élémens de mathématiques*, à l'usage des écoles de philosophie du Collège Royal de Toulouse (Toulouse: Robert, 1781).

Roger Martin, "Mémoire sur la maniere de démontrer, par les méthodes des Anciens, les hypotheses de Leibnitz dans le Calcul Différentiel," *Mémoires de l'Académie Royale des Sciences, Inscriptions et Belles-Lettres de Toulouse*. Tome I (Toulouse 1782), 43-64.

Roger Martin, "Mémoire contenant l'application des principes tirés de la méthode des limites aux diverses parties du calcul de l'Infini," *Mémoires de l'Académie Royale des Sciences, Inscriptions et Belles-Lettres de Toulouse*. Tome III (Toulouse 1788), 29–72.

Roger Martin, *Éléments de Mathématiques, à l'usage des écoles nationales* (Paris: Didot, an X [1802]).

Francis Maseres, *Dissertation on the Use of the Negative Sign in Algebra* (London: Richardson/Payne, 1758).

Johann Andreas Matthias, *Leitfaden für einen heuristischen Schulunterricht über die allgemeine Größenlehre, Elementargeometrie, ebene Trigonometrie, gemeine Algebra und die Apollonischen Kegelschnitte* (Magdeburg: Heinrichshofen, 1813).

Johann Andreas Matthias, *Erläuterungen zu dem Leitfaden für einen heuristischen Schulunterricht über die allgemeine Größenlehre, etc.* Erste Abtheilung: Die Elemente der Allgemeinen Größenlehre (Magdeburg: Heinrichshofen, 1814).

Jürgen Mau, *Zum Problem des Infinitesimalen bei den antiken Atomisten* (Berlin: Akademie Verlag, 1954).

Johann Tobias Mayer, *Vollständiger Lehrbegriff der höheren Analysis* (Göttingen: Vandenhoek u. Ruprecht, 1818).

Jean-Mathurin Mazéas, *Éléments d'arithmétique, d'algèbre et de géométrie, avec une introduction aux sections coniques*. Seconde édition (Paris, 1761).

James McConica (ed.), *The History of the University of Oxford*. Volume III, The Collegiate University (Oxford: Clarendon Press 1986) [1986a].

James McConica, "The Rise of the Undergraduate College," McConica 1986a, 1–68 [1986b].

James McConica, "Elizabethan Oxford: The Collegiate Society," McConica 1986a, 645–732 [1986c].

F. A. Medvedev, *Razvitie Ponyattiya Integrala* [The development of the concept of the integral] (Moskva 1974).

Fyodor A. Medvedev, *Scenes from the History of Real Functions* (Basel: Birkhäuser, 1991).

Herbert Mehrtens, *Moderne-Sprache-Mathematik* (Frankfurt/M.: suhrkamp, 1990).

Charles Méray, *Leçons nouvelles sur l'analyse infinitésimale et ses applications géométriques*. Première partie: Principes généraux (Paris: Gauthier-Villars, 1894).

Robert K. Merton, *Science and Technology in Seventeenth Century England* (New York: H. Fertig, 1970).

Mathias Metternich, *Anfangsgründe der Algebra von Sylvestre-François Lacroix*. Nach der siebten Auflage übersetzt und mit erläuternden Anmerkungen und Zusätzen vermehrt (Mainz: Kupferberg, 1811). (2nd edition: Mainz: Kupferberg, 1820).

Andreas Metz, *Handbuch der Elementar-Arithmetik in Verbindung mit der Elementar-Algebra* (Bamberg, Würzburg: J. A. Göbhardt, 1804).

Johann A. Chr. Michelsen, *Anfangsgründe der Buchstabenrechnung und Algebra* (Berlin: S. F. Hesse, 1788) [1788a].

Johann A. Chr. Michelsen, [Anmerkungen und Zusätze zur deutschen Übersetzung von L. Eulers *Introductio in Analysin Infinitorum*] (Berlin: S. F. Hesse, 1788) [1788b] - see: Euler 1788.

Johann A. Chr. Michelsen, *Gedanken über den gegenwärtigen Zustand der Mathematik und die Art, die Vollkommenheit und Brauchbarkeit derselben zu erhöhen* (Berlin: S. F. Hesse, 1789).

Gösta Mittag-Leffler, "K. Weierstraß: Briefe an Paul du Bois-Reymond," *Acta Mathematica*, 1923, *39*: 199–225.

Jürgen Mittelstraß, "Malebranche, Nicole," *Enzyklopädie Philosophie und Wissenschaftstheorie*. Band 2: H-O, ed. Jürgen Mittelstraß (Mannheim; Bibliograph. Institut, 1984), 751–752.

Herbert Möller, *Vereinfachte Analysis*. Skriptum (Münster, 1981).

F. Moigno, *Leçons de Calcul Différentiel et de Calcul Intégral, rédigées d'aprés les méthodes et les ouvrages publiés ou inédits de M. A.-L. Cauchy*. 2 vols. (Paris: Bachelier, 1840–1844).

F. Moigno, "Préface" und "Impossibilité du nombre actuellement infini," A.-L. Cauchy, *Sept Leçons de Physique Générale* (Paris: Gauthier Villars, 1868), I–XII, 77–104.

Gaspard Monge, *Application de l'Analyse à la Géométrie*. Quatrième édition [1809]. Reprinted in: J. Liouville (ed.), Application de l'Analyse à la Géométrie. Cinquième édition (Paris: Bachelier, 1850).

Sergio Moravia, *Il tramonto dell'illuminismo* (Bari: Laterza, 1968).

Dimitri Mordukhai-Boltovskoi, "Geschichte und Genese der Limestheorie," *Archeion* (Santa Fé), 1933, *15*: 45–72.

Thomas Mormann, *Argumentieren, Begründen, Verallgemeinern: zum Beweisen im Mathematikunterricht* (Königstein/Ts.: Scriptor, 1981).

C. V. Mourey, *La vraie théorie des quantités négatives et des quantités prétendues imaginaires*. Deuxième édition (Paris: Mallet-Bachelier, 1861).

Conrad H. Müller, *Studien zur Geschichte der Mathematik: insbesondere des mathematischen Unterrichts an der Universität Göttingen im 18. Jh.* (Leipzig: Teubner, 1904).

Claude L. M. H. Navier, *Résumé des leçons d'analyse données à l'École polytechnique*, suivi de notes par M. J. Liouville (Paris: Carillan-Gœury, 1840).

Ludwig Neder, "Modell einer Leibnizischen Differentialrechnung mit aktual unendlich kleinen Größen sämtlicher Ordnungen," *Mathematische Annalen*, 1941–1943, *118*: 718–732.

G. H. F. Nesselmann, *Versuch einer kritischen Geschichte der Algebra. Erster Theil. Die Algebra der Griechen* (Berlin, 1842). Reprint Minerva: Frankfurt, 1969.

Otto Neugebauer, *Vorlesungen über Geschichte der antiken mathematischen Wissenschaften. Band 1: Vorgriechische Mathematik* (Berlin: Springer, 1934).

Erwin Neuenschwander, "Joseph Liouville (1809-1882): Correspondance inédite et documents biographiques provenant de différentes archives parisiennes," *Bollettino di Storia delle Scienze Matematiche*, 1984, *4*: 55-132.

Erwin Neuenschwander, *Riemanns Einführung in die Funktionentheorie. Eine quellenkritische Edition seiner Vorlesungen mit einer Bibliographie zur Wirkungsgeschichte der Riemannschen Funkttionentheorie* (Göttingen: Vandenhoeck und Ruprecht, 1996).

Isaac Newton, *Mathematische Prinzipien der Naturlehre*. Mit Bemerkungen und Erläuterungen herausgegeben von J. Ph. Wolfers (Berlin: 1873) [reprint: Darmstadt: Wissenschaftliche Buchgesellschaft, 1963].

Isaac Newton, *Opera quae exstant omnia*, ed. Samuel Horsley. Band 1. Reprint (Stuttgart-Bad Cannstatt: F. Frommann, 1964).

Isaac Newton, *Philosophiae Naturalis Principia*. The Third Edition (1726), with variant readings. Assembled and edited by Alexandre Koyré and I. Bernard Cohen. Volume I (Cambridge: Cambridge at the University Press, 1972).

Sir Isaac Newton's *Mathematical principles of natural philosophy and his system of the world*. Translated into English by Andrew Motte in 1729. The transl. revised and supplied with an historical and explanatary apparatus, by Florian Cajori (New York: Greenwood, 1969) [1969a].

Isaac Newton, *The Mathematical Papers*. Ed. D. T. Whiteside (Cambridge: Cambridge at the University Press). Volume II: 1667–1670 (1968); Volume III: 1670–1673 (1969); Volume IV: 1674–1684 (1971); Volume VII: 1691–1695 (1976); Volume VIII: 1697–1722 (1981).

Anton Niegemann, *Durchführung der Theorie der entgegengesetzten Größen durch die Grundoperationen der allgemeinen Arithmetik - und Beseitigung der von Carnot dagegen erhobenen Einwendungen*. Schulprogramm Gymnasium Emmerich 1834.

Martin Ohm, *Elementar-Zahlenlehre zum Gebrauch für Schulen und Selbstlernende* (Erlangen: Palm u. Enke, 1816).

Martin Ohm, *Lehrbuch der Arithmetik, Algebra und Analysis*. Nach eigenen Prinzipien (Berlin: Reimer, 1822) (Versuch eines vollkommen consequenten Systems der Mathematik, Erster Theil).

Martin Ohm, *Lehrbuch der niedern Analysis*. - 2. umgearb., durch viele neue erl. Beisp. verdeutlichte Ausgabe (Berlin: Riemann, 1828/29). (Versuch eines vollkommen consequenten Systems der Mathematik, 1/2) Erster Theil: Arithmetik und Algebra. 1828. - Zweyter Theil: Algebra und Analysis des Endlichen. 1829.

Martin Ohm, *Die reine Elementar-Mathematik*: weniger abstrakt, sondern mehr anschaulich u. leichtfaßlich ... zunächst für seine Vorles. an d. Kgl. Bau-Akademie zu Berlin, dann auch zum Gebrauche an anderen ähnlichen Lehranstalten, bes. aber an Gymnasien u. zum Selbst-Unterrichte (Berlin: Riemann):
Erster Band: *Die Arithmetik bis zu den höhern Gleichungen* 1825. Zweiter Band: *Die allgemeine Grössenlehre und die ebene Raumgrössenlehre* 1826.
Zweite Auflage. Erster Band: *Die Arithmetik bis zu den höhern Gleichungen* (Berlin: Jonas, 1834).

Martin Ohm, *Die analytische und hoehere Geometrie in ihren Elementen: mit vorzueglicher Beruecks. d. Theorie d. Kegelschnitte. Erste Forts. seiner reinen Elementar-Mathematik* (Berlin: Riemann, 1826).

Gratien Olléac, *Mémoire sur des théories nouvelles des nombres opposés, des imaginaires et des équations du troisieme degré* (Toulouse: Desclassan, an II [1794]).

Oystein Ore, *Niels Henrik Abel. Mathematician Extraordinary* (Minneapolis: Univ. of Minnesota Press, 1957).

Michael Otte, *Mathematik in der Philosophie*. Institut für Didaktik der Mathematik, Universität Bielefeld: I. *Naturalisierte Erkenntnistheorie*. Occasional paper 170 (Bielefeld, November 1998); II. *Repräsentation und Möglichkeitsgedanke.Occ.* paper 173 (Bielefeld, Dezember 1998).

Michael Otte, *Mathematik als Prozeß*. Institut für Didaktik der Mathematik, Universität Bielefeld: occasional paper 178 (Bielefeld, Januar 2000).

Michael Otte, Marco Panza, *Analysis and Synthesis in Mathematics. History and Philosophy* (Dordrecht: Kluwer, 1997).

Jacques Ozanam, *Cours de Mathématiques*. Tome premier. Nouvelle edition (Paris: Jombert, 1697).

Jacques Ozanam, *Nouveaux Elemens d'Algèbre* (Amsterdam: Gallet, 1702).

Marco Panza, "Il manoscritto del 1789 di Arbogast sui principi del calcolo differenziale e integrale," *Rivista di storia della scienza*, 1985, *2*: 123–157.

Pappus of Alexandria, *Book 7 of the Collection*. Edited with translation and commentary by Alexander Jones (New York: Springer, 1986).

Enrico Pasini, *Il reale e l'immaginario. La fondazione del calcolo infinitesimale nel pensiero di Leibniz* (Torino: Sonda, 1993).

Giuseppe Peano, "Dimostrazione dell'impossibilità di segmenti infinitesimi costanti," *Rivista di Matematica*, 1892, *2*: 58–62 [1892b].

Giuseppe Peano, "Recensione: G. Veronese - Fondamenti di geometria a più dimensioni e a più specie di unità rettilinee, ecc. 1891," *Rivista di Matematica*, 1892, *2*: 143–144 [1892a].

Volker Peckhaus, *Hilbertprogramm und kritische Philosophie: das Göttinger Modell interdisziplinärer Zusammenarbeit zwischen Mathematik und Philosophie* (Göttingen: Vandenhoeck & Ruprecht, 1990).

Jeanne Peiffer, Amy Dahan-Dalmédico, *Wege und Irrwege - eine Geschichte der Mathematik* (Basel: Birkhäuser, 1994) [orig.: *Routes et Dédales*].

Jeanne Peiffer, "La conception de l'infiniment petit chez Pierre Varignon, lecteur de Leibniz et de Newton," *Leibniz - Tradition und Aktualität. V. Internationaler Leibniz-Kongreß Hannover 1988. Vorträge* (Hannover: G. W. Leibniz-Gesellschaft, 1988), 710–717. [1988a].

Jeanne Peiffer 1988b, see Johann Bernoulli 1988.

Henry Pemberton, "Some Considerations on a late Treatise intituled, *A new Set of Logarithmic Solar Tables*, etc. intended for a more commodious method of finding the Latitude at Sea, by Two Observations of the Sun," *Philosophical Transactions*, vol. LI, part II for the Year 1760, London 1761, 910–929.

Michel Pensivy, "Jalons historiques pour une épistémologie de la série infinie du binôme," *Sciences et Techniques en Perspective*, 1987–88, *14*.

Michel Piclin, "Antoine Arnauld, 1612–1694," ed. Denis Huisman, *Dictionnaire des Philosophes, A–J* (Paris: Presses Universitaires de France, 1993), 148–152.

Antoine Picon, *L'Invention de l'Ingénieur moderne. L'École des Ponts et Chaussées 1747–1851* (Paris: Presses de l'École Nationale des Ponts et Chaussées, 1992).

Gaston Pinet, *Histoire de l'école polytechnique* (Paris: Baudry, 1887).

John Playfair, "On the Arithmetic of Impossible Quantities," *Philosophical Transactions of the Royal Society of London*, 1778, 68:318–343.

Louis Poinsot, "Des principes fondamentaux et des règles générales du calcul différentiel," *Correspondance sur l'École Polytechnique*, tome III [1814–1816], No. 2, Mai 1815, 111–123.

Siméon-Denis Poisson, *Traité de Mécanique*. Tome premier, tome second (Paris: Courcier, 1811).

Siméon-Denis Poisson, *Traité de Mécanique*. Seconde édition, considérablement augmentée. Tome premier, tome second (Paris: Bachelier, 1833).

Siméon-Denis Poisson, *Lehrbuch der Mechanik*. Nach der zweiten sehr vermehrten Ausgabe übersetzt von Moritz A. Stern. Erster Theil (Berlin: Reimer, 1835). Zweiter Theil (ibid., 1836).

Siméon-Denis Poisson, "Suite du Mémoire sur les intégrales définies et sur la sommation des séries," *Journal de l'École royale polytechnique*, Tome XII, dix-neuvième cahier, 1823, 404–507.

[François Daniel Porro], *Exposition du Calcul des Quantités négatives; dans laquelle on prouve qu'en Algèbre, il n'y a ni Multiplicateur, ni Quotient, ni Exposant, ni Logarithme négatif, ni Racine imaginaire,* [...] (Avignon [Besançon], 1784).

Bruce Pourciau, "Newton and the notion of limit, *Historia Mathematica*, 2001, *28*: 18-30.

Jean Prestet, *Elémens des Mathématiques ou Principes Generaux de toutes les sciences qui ont les grandeurs pour objet* (Paris: Pralard, 1675).

Jean Prestet, *Nouveaux Elémens des Mathématiques ou Principes Generaux de toutes les sciences qui ont les grandeurs pour objet*. Seconde édition, plus ample et mieux digerée. 2 vols. (Paris: Pralard, 1689).

Proclus (Diadochus), *Kommentar zum ersten Buch von Euklids "Elementen."* Eingel., mit Kommentaren u. bibliogr. Nachweisen vers. u. in d. Gesamtedition besorgt von Max Steck (Halle/S.: Dt. Akad. d. Naturforscher, 1945).

Gaspard Riche de Prony, "Cours d'Analyse appliquée à la mécanique," *Journal de l'École Polytechnique*, an III (1795), tome 1, cahier 1, 92–119 [1795a].

Gaspard Riche de Prony, "Suite des leçons d'analyse," *Journal de l'École Polytechnique*, an III (1795), tome 1, cahier 2, 1–23. [1795b]

Gaspard Riche de Prony, "Suite des leçons d'analyse. Considérations sur les principes de la méthode inverse des Différences," *Journal de l'École Polytechnique*, an III (1795), tome 1, cahier 3, 211–273 [1795c].

Gaspard Riche de Prony, "Suite des leçons d'analyse. Des Suites Recurrentes," *Journal de l'École Polytechnique*, an IV (1796), tome 1, cahier 4, 459–569.

Gaspard Riche de Prony, "Cours de Mécanique de l'an V," *Journal de l'École Polytechnique*, an VI (1798), tome 2, cahier 5, 1–19.

Gaspard Riche de Prony, "Introduction aux Cours d'Analyse pure et de d'Analyse appliquée à la Mécanique," *Journal de l'École Polytechnique*, an VII (1799), tome 2, cahier 6, 213–218.

Gaspard Riche de Prony, *Mécanique philosophique, ou analyse raisonnée des diverses parties de la science de l'équilibre et du mouvement.* Journal de l'École Polytechnique. Tome III, septième et huitième cahiers (Paris, Imprimérie de la République, an VIII [=1800]).

Gaspard Riche de Prony, *Leçons de Mécanique Analytique,* données à l'École Polytechnique. Première partie, Seconde partie (Paris: Imprimérie de l'École Royale des Ponts et Chaussées, 1815).

Helmut Pulte, *Das Prinzip der kleinsten Wirkung und die Kraftkonzeptionen der rationalen Mechanik.* Studia Leibnitiana, Sonderheft 19 (Stuttgart: Franz Steiner, 1989).

Walter Purkert, "Infinitesimalrechnung für Ingenieure—Kontroversen im 19. Jahrhundert," ed. D. Spalt, *Rechnen mit dem Unendlichen* (Basel: Birkhäuser, 1990), 179-192.

Helena M. Pycior, *The role of Sir William Rowan Hamilton in the development of British modern algebra.* Ph.D. thesis, Cornell University (Michigan: University Microfilms, 1976).

Helena M. Pycior, "Mathematics and Philosophy: Wallis, Hobbes, Barrow, and Berkeley," *Journal of the History of Ideas,* 1987, *48*: 265–286.

Helena M. Pycior, *Symbols, Impossible Numbers, and Geometric Entanglements. British Algebra Through the Commentaries on Newton's Universal Arithmetick* (Cambridge: Cambridge University Press, 1997).

João Filipe Queiró, "José Anastácio da Cunha: a forgotten forerunner," *Mathematical Intelligencer,* 1988, *10,* no. 1, 38–43.

Petrus Ramus, *Algebra* (Paris: apud Andream Wechelum, 1560).

Petrus Ramus, *Arithmetices Libri Duo, et Algebrae totidem:* à Lazaro Schonero emendati et explicati. Ejusdem Schoneri libri duo: alter, De Numeris figuratis; alter, De Logistica sexagenaria (Francofurdi: apud heredes Andreae Wecheli, 1586).

Petrus Ramus, *Scholarum Mathematicorum Libri unus et triginta* (Paris 1569). Here quoted from the edition: A Lazaro Schonero recogniti et emendati (Frankfurt: Andreas Wechelus, 1599).

Raymond-Roux, *Leçons élémentaires de Calcul Infinitésimal. Pour servir de suite au Livre de M. Mazéas, et d'Introduction aux Sciences Physico-Mathématiques* (Paris: Brocas, 1784).

Otto Julius Rebel, *Der Briefwechsel zwischen Johann (I.) Bernoulli und dem Marquis de L'Hospital in erläuternder Darstellung* (Bottrop: Postberg, 1934).

Marcel Reinhard, *Le Grand Carnot.* 2 vol. (Paris: Hachette, 1950–1952).

Oskar Reuleaux, *Die geschichtliche Entwickelung des Befestigungswesens, vom Aufkommen der Pulvergeschütze bis zur Neuzeit* (Leipzig: Göschen, 1912).

Charles-René Reyneau, *Analyse demontrée ou la Méthode de résoudre les problêmes des mathématiques, et d'apprendre facilement ces sciences.* Seconde édition, augmentée des remarques de M. de Varignon. Tome I:(Paris: Quilliau, 1736). Tome II: *Usage de l'Analyse, ou de la maniere de l'appliquer à découvrir les propriétés des figures de la Geometrie, ..., en employant le calcul ordinaire de l'Algebre, le calcul differentiel et le calcul integral* (Paris: Quillau, 1738).

[Charles-René Reyneau], *La Science du Calcul des Grandeurs en général, ou Les Élémens des Mathématiques,* par l'Auteur de l'Analyse Démontrée (Paris: J. Quillau, 1714).

Charles-René Reyneau, *La Science du Calcul des Grandeurs en général, ou Les Éléments des mathématiques*. Seconde édition. 2 vol. (Paris: G.-G. Quillau, 1739).

Kurt Richter, *Zur Herausbildung, Entstehung und Entwicklung des Begriffs der gleichmäßigen Konvergenz* (Dissertation Pädagogische Hochschule Halle, 1976).

Robin E. Rider, *A Bibliography of Early Modern Algebra*. Berkeley papers in History of Science, no. VII (Berkeley: University of California, 1982).

Bernhard Riemann, "Ueber die Darstellbarkeit einer Function durch eine trigonometrische Reihe," *Abhandlungen der Königlichen Gesellschaft der Wissenschaften zu Göttingen*. Dreizehnter Band, von den Jahren 1866 und 1867 (Göttingen 1867), 87–132.

Bernhard Riemann, *Einführung in die Funktionentheorie* 1861: s e e Neuenschwander 1996.

Dominique-François Rivard, *Élémens de Mathématique*. Quatrième édition, revue et augmentée de nouveau par l'Auteur (Paris: Lottin et al., 1744).

Dominique-François Rivard, *Abrégé des Élémens de Mathématique*. Cinquième édition, revue et augmentée par l'Auteur (Paris: J. Dessaint/Ch. Saillant 1761).

André Robinet, "Jean Prestet, ou la bonne foi cartésienne," *Revue d'histoire des sciences et de leurs applications*, 1960, *13*: 95–104.

André Robinet, "Le groupe malebranchiste, introducteur du Calcul infinitésimal en France," *Revue d'histoire des sciences et de leurs applications*, 1960, *13*: 287–308. [1960a].

André Robinet, "La philosophie malebranchiste des mathématiques," *Revue d'histoire des sciences et de leurs applications,* 1961, *14*: 205–264.

André Robinet, "Introduction," in: Malebranche *Œuvres*, tome 6/7 (Paris: Vrin, 1966), I–XXVII.

André Robinet, "Le Groupe malebranchiste de l'Académie des Sciences," Nicholas Malebranche, *Œuvres*, tome XX: Malebranche Vivant. Documents Biographiques et Bibliographiques (Paris: Vrin, 1967), Chap. IV.

Abraham Robinso*n, Non-Standard Analysis (Amsterdam: North-Holland, 1966).*

Daniel Roche, Le siècle des lumières en province: académies et académiciens provinciaux, 1680–1789 (Paris: Mouton, 1978).

Nicolas Rouche, *Le sens de la mesure: des grandeurs aux nombres rationnels* (Bruxelles: Didier Hatier, 1992).

Walter W. Rouse Ball, *Origin and history of the mathematical Tripos* (Cambridge, 1880).

Walter W. Rouse Ball, *A History of the study of Mathematics at Cambridge* (Cambridge: at the University Press 1889).

Walter W. Rouse Ball, *A short account of the history of mathematics* (1908; reprint New York: Dover Publ., 1960).

Franz Heinrich Rump, *Über den Gebrauch der entgegengesetzten Aggregationszeichen bei den goniometrischen Funktionen*. Schulprogramm Gymnasium Coesfeld 1830.

Karel Rychlík, "Sur les contacts personnels de Cauchy et de Bolzano," *Revue d'histoire des sciences et de leurs applications*, 1962, *15*: 163–164.

La Sabretache (ed.), *Centenaire de Lazare Carnot 1753–1823. Notes et Documents inédits* (Paris 1923).

Erik Lars Sageng, *Colin MacLaurin and the foundations of the method of fluxions.* Ph.D. Dissertation Princeton University 1989.

Leonardo Salimbeni, "Intorno alla Moltiplicazione ed alla Divisione Algebraiche," *Memorie di Matematica e Fisica della Società Italiana*, Tomo VII (Verona: D. Ramanzini, 1794), 482–507).

Luis Saraiva, "On the first History of Portuguese Mathematics," *Historia Mathematica*, 1993, *20*: 415–427.

Luis Saraiva, "Garção Stockler e o 'Projecto sobre o estabelecimento e organisação da instrucção publica no Brazil'," Sergio Nobre (ed.), *Actas - Anais II Seminario Nacional de História da Matemática*, Aguas de São Pedro, Março 1997 (Rio Claro, SP: 1997), 25–43.

Nicholas Saunderson, *The elements of algebra: in ten books; to which is prefixed, an account of the author's life and character, collected from his oldest and most intimate acquaintance* (Cambridge : University Pr., 1740).

Jean Sauri, *Institutions Mathématiques*, servant d'introduction à un cours de philosophie, à l'usage des Universités de France. Seconde édition (Paris: Valade, 1772). Trosieme édition (Paris: Frouillé, 1777).

Schafheitlin 1924: see Johann Bernoulli 1924.

Oskar Xaver Schlömilch, *Handbuch der algebraischen Analysis* (Jena: Frommann, 1845).

Oskar Xaver Schlömilch, *Handbuch der Differenzial- und Integralrechnung* (Greifswald: Otte, 1847–1848).

Friedrich Schmeißer, *Kritische Beleuchtung einiger Lehren der reinen Analysis, welchen der Vorwurf der Ungereimtheit gemacht worden ist.* Schulprogramm Gymnasium Frankfurt/Oder 1842.

Friedrich Schmeißer, *Kritische Beleuchtung einiger Lehren der reinen Analysis, welchen der Vorwurf der Ungereimtheit gemacht wird.* Schulprogramm Gymnasium Frankfurt/Oder 1846.

Christoph Schöner, *Mathematik und Astronomie an der Universität Ingolstadt im 15. und 16. Jahrhundert* (Berlin: Duncker und Humblot, 1994).

Paul Schrecker, "Arnauld, Malebranche, Prestet et la théorie des nombres négatifs," *Thales*, 1935, *2*: 82–90.

Gert Schubring, *Das genetische Prinzip in der Mathematik-Didaktik* (Stuttgart: Klett-Cotta, 1978).

Gert Schubring, "The Conception of Pure Mathematics as an Instrument in the Professionalization of Mathematics," *Social History of Nineteenth Century Mathematics*, eds. H. Mehrtens, H. Bos, I. Schneider (Basel: Birkhäuser 1981), 111-134 [1981a].

Gert Schubring, "Mathematics and Teacher Training: Plans for a Polytechnic in Berlin," *Historical Studies in the Physical Sciences*, 1981, *12/1*, 161-194 [1981b].

Gert Schubring, "Ansätze zur Begründung theoretischer Terme in der Mathematik - Die Theorie des Unendlichen bei Johann Schultz (1739-1805)," *Historia Mathematica*, 1982, *9*, 441-484.

Gert Schubring, "Pläne für ein Polytechnisches Institut in Berlin," *Philosophie und Wissenschaft in Preußen*, eds. F. Rapp, H.W. Schütt (Technische Universität Berlin 1982), 201-224. [1982a]

Gert Schubring, "Die Mathematik an der Ecole Normale des Jahres III - Wissenschaft und Methode," *Wissen und Bewußtsein. Studien zur Wissenschaftsdidaktik der*

Disziplinen, ed. F. Schmithals. (Hamburg: Arbeitsgemeinschaft für Hochschuldidaktik, 1982), 103-133 [1982b].

Gert Schubring, *Die Entstehung des Mathematiklehrerberufs im 19. Jahrhundert: Studien und Materialien zum Prozeß der Professionalisierung in Preußen (1810-1870)* (Weinheim: Beltz, 1983).

Gert Schubring, "Das mathematische Leben in Berlin. Zu einer entstehenden Profession an Hand von Briefen des aus Erlangen stammenden Martin Ohm an seinen Bruder Georg Simon," *Erlanger Bausteine zur Fränkischen Heimatforschung*, Jahrbuch 30 (1983), 221-249 [1983a].

Gert Schubring, "Essais sur l'histoire de l'enseignement des mathématiques, particulièrement en France et en Prusse," *Recherches en Didactique des Mathématiques*, 1984, *5*, 343-385.

Gert Schubring, "Martin Ohm und Friedrich August Pfeiffer. Eine Ergänzung zu "Das Mathematische Leben in Berlin", *Erlanger Bausteine zur Fränkischen Heimatforschung*, Jahrbuch 31 (1984), 203-205 [1984a].

Gert Schubring, "Das mathematische Seminar der Universität Münster, 1831/1875 bis 1951," *Sudhoffs Archiv*, 1985, *69*, 154-191 [1985a].

Gert Schubring, "Die Entwicklung des mathematischen Seminars der Universität Bonn, 1864-1929," *Jahresberichte der Deutschen Mathematiker-Vereinigung*, 1985, *87*, 139-163 [1985b].

Gert Schubring, "Ruptures dans le statut mathématique des nombres négatifs," *petit x*, no. 12, 1986, 5-32.

Gert Schubring, "Wilhelm Lorey (1873-1955) und die Methoden mathematik-geschichtlicher Forschung," *mathematica didactica*, 1986, *9*, 75-87 [1986a].

Gert Schubring, "On the methodology of analysing historical textbooks: Lacroix as textbook author," *for the learning of mathematics*, 1987, *7*, 41-51.

Gert Schubring, "Epistemologische Debatten über den Status negativer Zahlen und die Darstellung negativer Zahlen in deutschen und französischen Lehrbüchern 1795-1845," *Mathematische Semesterberichte*, 1988, *35*, 183-196.

Gert Schubring, "Historische Begriffsentwicklung und Lernprozeß aus der Sicht neuerer mathematikdidaktischer Konzeptionen (Fehler, "Obstacles", Transposition)," *Zentralblatt für Didaktik der Mathematik*, 1988, *20*, 138-148 [1988a].

Gert Schubring, "Pure and Applied Mathematics in Divergent Institutional Settings in Germany: the Role and Impact of Felix Klein," *The History of Modern Mathematics. Volume II: Institutions and Applications* eds. David Rowe, John McCleary (Boston: Academic Press 1989), 171-220.

Gert Schubring, "Les échanges entre les mathématiciens Français et Allemands sur la rigueur dans les concepts d'Arithmétique et d'Analyse," *Echanges d'Influences Scientifiques et Techniques entre Pays Européens de 1780 à 1830*, Colloques du C. T. H. S., No. 5: Actes du 114e Congrès National des Sociétés Savantes. Paris 1989 (Paris: Editions du CTHS 1990), 89-104.

Gert Schubring, "Das mathematisch-Unendliche bei J. F. Fries," *Konzepte des mathematisch Unendlichen im 19. Jahrhundert*, ed. G. König (Göttingen: Vandenhoeck u. Ruprecht 1990), 152-164 [1990a].

Gert Schubring, "Zur strukturellen Entwicklung der Mathematik an den deutschen Hochschulen 1800-1945," *Mathematische Institute in Deutschland 1800-1945*, ed. Winfried Scharlau (Braunschweig: Vieweg 1990), 264-278. [1990b]

Gert Schubring (ed.), *"Einsamkeit und Freiheit" neu besichtigt: Universitätsreformen und Disziplinenbildung in Preußen als Modell für Wissenschaftspolitik im Europa des 19. Jahrhunderts*; Proceedings of the Symposium of the XVIIIth International Congress of History of Science at Hamburg-Munich, 1-9 August 1989 (Stuttgart: Steiner, 1991).

Gert Schubring, "Spezialschulmodell versus Universitätsmodell: Die Institutionalisierung von Forschung," in: *Einsamkeit u. Freiheit*. 1991, 276-326 [1991a].

Gert Schubring, *Konzepte des Infinitesimalen in Frankreich und Deutschland zu Beginn des 19. Jahrhunderts*. Vortrag beim Second Workshop on the History of Modern Mathematics, Göttingen, 15.–17.7.1991. Duplicated manuscript [1991b].

Gert Schubring, "Zur Modernisierung des Studiums der Mathematik in Berlin, 1820–1840," *AMPHORA. Festschrift für Hans Wußing zu seinem 65. Geburtstag*. Eds. S.S. Demidov et al. (Basel: Birkhäuser 1992), 649–675 [1992b].

Gert Schubring, "Bernard Bolzano—Not as Unknown to His Contemporaries as is Commonly Believed?" *Historia Mathematica*, 1993, *20*, 45–53.

Gert Schubring, "Evolution du concept d'infiniment petit aux 18ème et 19ème siècles," *Histoire d'Infini*, Actes du 9ᵉ Colloque Inter-IREM 'Epistémologie et Histoire des mathématiques', ed. IREM de Brest, (Brest: IREM de Brest 1994), 317-326.

Gert Schubring, "Differences in the Involvement of Mathematicians in the Political Life in France and in Germany," *Bollettino di Storia delle Scienze Matematiche*, 1995, *15*: 61–83.

Gert Schubring, "Changing cultural and epistemological views on mathematics and different institutional contexts in 19th century Europe," *L'Europe mathématique - Mythes, histoires, identités. Mathematical Europe - Myths, History, Identity*, eds. Catherine Goldstein, Jeremy Gray, Jim Ritter (Paris: Éditions de la Maison des Sciences de l'Homme, 1996), 361–388.

Gert Schubring, "L'interaction entre les débats sur le statut des nombres négatifs et imaginaires et l'émergence de la notion de segment orienté," *Le nombre, une hydre à n visages: entre nombres complexes et vecteurs*, Hrsg. Dominique Flament (Paris: Éditions de la Maison des Sciences de l'Homme, 1997), 1–14.

Gert Schubring, "Johann Georg Tralles: Der erste Ordinarius für Mathematik an der Universität Berlin - Eine Edition seiner Antrittsvorlesung 1810," *Mathematik in Berlin. Geschichte und Dokumentation*. Zweiter Halbband, ed. Heinrich Begehr (Aachen: Shaker Verlag 1998), 297–343.

Gert Schubring, "An unknown part of Weierstraß's *Nachlaß*," *Historia Mathematica*, 1998, *25*: 423–430 [1998a].

Gert Schubring, *Analysis of historical textbooks in mathematics. Lecture notes. 2., rev. ed.* (Rio de Janeiro: PUC do Rio de Janeiro, Dept. de Matemática, 1999).

Gert Schubring, "Novas Fontes e Abordagens na História dos Números Negativos: Uma Análise de um Manuscrito do Monteiro da Rocha," *IV Seminário Nacional de História de Matemática. 8 a 11 de Avril de 2001. Anais*, John A. Fossa (ed.), (Editora da SBHMat: Rio Claro/SP, 2001), 95–108.

Gert Schubring, "Argand and the early work on graphical representation: New sources and interpretations," *Around Caspar Wessel and the Geometric Representation of Complex Numbers*. Proceedings of the Wessel Symposium at The Royal Danish Academy of Sciences and Letters, Copenhagen, August 11–15 1998, Jesper Lützen (ed.), (C. A. Reitzel: Copenhagen, 2001), 125–146.

Gert Schubring, "Aspetti istituzionali della matematica," *Storia della scienza*, Vol. VI: *L'Etá dei Lumi* (Istituto dell'Enciclopedia Italiana: Roma, 2002), 366–380.

Gert Schubring, *Le retour du refoulé – Der Wiederaufstieg der synthetischen Methode an der École polytechnique* (Augsburg: ERV Rauner, 2004).

Gert Schubring, "A Case Study in Generalisation: The Notion of Multiplication," *Activity and Sign - Grounding Mathematics Education. Festschrift for Michael Otte*, eds. M. Hoffmann, J. Lenhard, F. Seeger (Dordrecht: Kluwer, 2004), 283–293 [2004a].

Ernst Schulin, *Traditionskritik und Rekonstruktionsversuch: Studien zur Entwicklung von Geschichtswissenschaft und historischem Denken* (Göttingen: Vandenhoeck & Ruprecht, 1979).

Johann Schultz, *Anfangsgründe der reinen Mathesis* (Königsberg: Hartung, 1790).

J. F. Scott, *The Mathematical Work of John Wallis* (New York: Chelsea, 1981) [Original: London 1938].

Wilson L. Scott, *The Conflict between Atomism and Conservation Theory 1644–1860* (London: Macdonald, 1970).

Johann Andreas von Segner, *Deutliche und vollständige Vorlesungen über die Rechenkunst und Geometrie.* Zweite, verbesserte Auflage (Lemgo: J. H. Meyer, 1767).

Gaston Serbos, "L'École Royale de Ponts et chaussées", in: Taton, *Enseignement* 1986, 345-363.

François-Joseph Servois, "Reflexions sur les divers systèmes d'exposition des principes du calcul différentiel, et, en particulier, sur la doctrine des infiniments petits," *Annales de Mathématiques pures et appliquées*, 1814, *5*: 141–170.

Jacques Sesiano, "Une Arithmétique médiévale en langue provençale," *Centaurus* 1984, *27*: 26–75.

Jacques Sesiano, "The Appearance of negative solutions in medieval Mathematics," *Archive for History of Exact Sciences*, 1985, *32*: 105–150.

Gerard and Wiebe Sierksma, "The Great leap to the Infinitely Small. Johann Bernoulli: Mathematician and Philosopher," *Annals of Science*, 1996, *56*: 433–449.

Thomas Simpson, *A Treatise of Algebra; Wherein the Fundamental Principles Are fully and clearly demonstrated, and applied to the Solution of a great Variety of Problems* (London Nourse, 1745).

Robert Simson, "De Limitibus Quantitatum et Rationum. Fragmentum," Roberti Simson, M. D., Matheseos Nuper in Academia Glasguensi Professoris *Opera Quaedam Reliqua* (Stanhope: Glasgow, 1776), IV, 1–33.
Reprint in: Francis Maseres (ed.), *Scriptores Logarithmici*; or a Collection of several curious tracts on the nature and construction of Logarithms, mentioned in Dr. Hutton's Historical Introduction to his new edition of Sherwin's mathematical tables: together with some tracts on the binomial theorem and other subjects connected with the doctrine of logarithms. Volume VI (R. Wilcks: London, 1807), 87–110.

Detlef D. Spalt, "Die mathematischen und philosophischen Grundlagen des Weierstraßschen Zahlbegriffs zwischen Bolzano und Cantor," *Archive for History of Exact Sciences*, 1991, *41*: 311–362.

Detlef D. Spalt, *Die Vernunft im Cauchy-Mythos* (Thun/Frankfurt a. M.: H. Deutsch, 1996).

Johann Peter Wilhelm Stein, *Die Elemente der Algebra.* Erster Cursus (Trier, Lintz, 1828); Zweiter Cursus (Trier: Lintz, 1829).

Stendhal [Henry Beyle], *Vie de Henry Brulard.* Texte établi avec introduction, bibliographie et notes par Henri Martineau (Paris: Garnier Frères, 1961).

Moritz Abraham Stern, "Zusätze des Übersetzers," in: Poisson 1835, 552–567.

Moritz Abraham Stern, *Lehrbuch der algebraischen Analysis* (Leipzig/Heidelberg: winter, 1860).

Simon Stevin, *The principal works of Simon Stevin*, ed. by Ernst Crone. 5 vols. (Amsterdam: Swets & Zeitlinger, 1955–1966).

Rudolf Stichweh, *Zur Entstehung des modernen Systems wissenschaftlicher Disziplinen. Physik in Deutschland, 1740–1890* (Frankfurt/M.: suhrkamp, 1984).

Michael Stifel, *Arithmetica Integra* (Norimbergae: Petreius, 1544).

Ernest Stipanic, "Sur le continu linéaire de Boscovich," ed. P. Bursill-Hall, *R. J. Boscovich. Vita e Attività Scientifica - His Life and Scientific Work*. Atti del Convegno Roma, 23–27 maggio 1988 (Roma: Istituto della Enciclopedia Italiana, 1993), 477–489.

Francisco de Borja Garçaõ Stockler, *Compendio da Theorica dos Limites, ou Introducçaõ ao Methodo das Fluxões* (Offic. da Academia das Sciencias: Lisboa, 1794).

Otto Stolz, "B. Bolzanos Bedeutung in der Geschichte der Infinitesimalrechnung," *Mathematische Annalen*, 1881, *18*: 255–279.

Otto Stolz, "Zur Geometrie der Alten, insbesondere über ein Axiom des Archimedes," *Mathematische Annalen*, 1883, *22*: 504–519 [1883a].

Otto Stolz, "Die unendlich kleinen Größen," *Berichte des Naturwissenschaftlich-Medizinischen Vereins in Innsbruck*, 1883, *XIV*: 21–43 [1883b].

Otto Stolz, *Vorlesungen über allgemeine Arithmetik*. Band I (Leipzig: Teubner, 1885).

Otto Stolz, "Ueber zwei Arten von unendlich kleinen und von unendlich grossen Grössen," *Mathematische Annalen*, 1888, *31*: 601–604.

Otto Stolz, *Grössen und Zahlen*. Rede zur Kundmachung der gelösten Preisaufgaben (Leipzig: Teubner, 1891).

Otto Stolz/J. A. Gmeiner, *Theoretische Arithmetik*. Band I, zweite Auflage (Leipzig: Teubner 1911) [first edition 1900].

Antoine Suremain-Missery [Suremain de Missery], *Théorie purement algébrique des Quantités imaginaires et des Fonctions qui en résultent, où l'on traite de nouveau la question des Logarithmes des Quantités négatives* (Paris: Firmin Didot, an IX – 1801).

R. C. H. Tanner, "The Ordered Regiment of the Minus Sign: Off-beat Mathematics in Harriot's Manuscripts," *Annals of Science*, 1980, *27*: 127–158 [1980a].

R. C. H. Tanner, "The Alien Realm of the Minus: Deviatory Mathematics in Cardano's Writings," *Annals of Science*, 1980, *27*: 159–178 [1980b].

Jules Tannery, *Introduction à la Théorie des Fonctions d'une Variable* (Paris: Hermann, 1886).

René Taton, "Sylvestre-François Lacroix (1765-1843), Mathématicien, Professeur et Historien des Sciences," *Actes du VIIe Congrès international d'Histoire des Sciences*, Jerusalem 1953, 588–593 [1953a].

René Taton, "Laplace et Sylvestre-François Lacroix," *Revue d'Histoire des Sciences et de leurs Applications*, 1953, *6*: 350–360 [1953b].

René Taton, "Condorcet et Sylvestre-François Lacroix," *Revue d'Histoire des Sciences et de leurs Applications*, 1959, *12*: 127–158, 243–262.

René Taton (ed.), *Enseignement et diffusion des sciences en France au XVIIIe siècle* (Paris: Hermann, 1986).

René Taton, "Diversité et Originalité de l'Œuvre Scientifique de Lazare Carnot," in: J.-P. Charnay 1990, 455–470.

Brook Taylor, *Methodus Incrementorum Directa et Inversa* (London: Pearson, 1715).

Antonio José Teixeira, "Questão entre José Anastasio da Cunha e José Monteiro da Rocha," *O Instituto* [de Coimbra], 1890, 20–27, 119–131, 187–202, and passim.

Adolph Tellkampf, *Vorschule der Mathematik* (Berlin: Rücker, 1829).

George Friedrich Tempelhoff, *Anfangsgründe der Analysis des Unendlichen. Erster Theil, welcher die Differential-Rechnung enthält; zum Gebrauch der Königlichen Preußischen Artillerie* (Berlin, Stralsund: G. A. Lange, 1770).

Alessandro Terracini, "Cauchy à Torino," *Rendiconti del Seminario Matematico dell'Università e del Politecnico di Torino*, 1957, 16: 159203.

J. N. Tetens, "Eine allgemeine Formel für die Potenzen mehrtheiliger Größen," *Der polynomische Lehrsatz, das wichtigste Theorem der ganzen Analysis*, ed. Carl Friedrich Hindenburg (Leipzig: Fleischer d.J., 1796), 1–47.

Text in Context. Contributions to Ethnomethodology, eds. Graham Watson/Robert M. Seiler (Newbury Park: Sage, 1992).

[Bernhard Friedrich Thibaut], "St. Petersburg" [Rezension von Klostermann 1804], *Göttingische gelehrte Anzeigen*, 1805, I. Band, 167–168.

Roger Tisserand, *Au Temps de l'Encyclopédie: L'Académie de Dijon de 1740 à 1793* (Paris: Boivin, 1936).

Johannes Tropfke, "Zur Geschichte der quadratischen Gleichungen über dreieinhalb Jahrtausend," *Jahresbericht der Deutschen Mathematikervereinigung*, 1934, 43: 98–107; 1934, 44: 26–47 and 95–119.

Johannes Tropfke, *Geschichte der Elementarmathematik*. Band 1: Algebra, 4. Aufl., vollständig neu bearb. von Kurt Vogel (Berlin: de Gruyter, 1980).

Clifford A. Truesdell, *The rational mechanics of flexible or elastic bodies, 1638-1788. Introduction to Leonhardi Euleri Opera Omnia, vol. X et XI seriei secundae* (Zürich: Füssli, 1960).

R. Steven Turner, "The Prussian Universities and the Concept of Research," *Internationales Archiv für Sozialgeschichte der deutschen Literatur*, 1980, 5: 68–86.

C.-A. Valson, *La Vie et les Travaux du Baron Cauchy* [1868]. Réimpression augmentée d'une Introduction par René Taton (Paris: Blanchard, 1970).

Pierre Varignon, *Eclaircissemens sur l'Analyse des Infinimens Petits* (Paris: Rollin, 1725).

Pierre Varignon, *Elemens de Mathematique* (Amsterdam: F. Changouin, 1734).

Jacques Verger (éd.), *Histoire des Universités* (Paris: privat, 1986).

J. J. Verdonk, *Petrus Ramus en de Wiskunde* (Assen: Gorcum/Prakke, 1966).

Jean-Luc Verley, "La controverse des logarithmes des nombres négatifs et imaginaires," *Fragments d'histoire des mathématiques*. Brochure A.P.M.E.P., no. 41 (Paris, 1981), 121–140.

Jean-Luc Verley, "Le statut des nombres complexes chez d'Alembert," Monique Eemery, Pierre Monzani (eds.), *Jean d'Alembert, savant et philosophe: Portrait à plusieurs voix*. Actes du Colloque 1983 (Paris: èditions des archives contemporaines, 1989), 361–375.

Giuseppe Veronese, *Grundzüge der Geometrie von mehreren Dimensionen und mehreren Arten gradliniger Einheiten in elementarer Form entwickelt*. Transl. by Adolf Schepp (Leipzig: Teubner, 1894). [Italian Orig.: 1891]

Giuseppe Vivanti, "Sull'infinitesimo attuale," *Rivista di Matematica* (Torino), 1891, 1: 135–153.

Vincenzo Vita, "Le Definizioni del Continuo in Aristotele," *Cultura e Scuola*, no. 111, 1989, 218–227.

Klaus Volkert, "Zur Differenzierbarkeit stetiger Funktionen - Ampère's Beweis und seine Folgen," *Archive for History of Exact Sciences*, 1989, *40*: 37–112.

P. G. J. Vredenduin, *De geschiedenis van positief en negatief* (Groningen: Wolters-Noordhoff, 1991).

Christian Friedrich Wagner, *De angulis, quos dicunt, mixtilineis. Ad Aristotel. analyt. pr. I.I. cap. XXIIII. § 1.* Schulprogramm Gymnasium Gumbinnen (Königsberg: Hartung, 1815).

John Wallis, *Opera Mathematica* I, with a foreword by Christoph-J. Scriba, (Hildesheim: Olms, 1972).

John Wallis, *De Algebra Tractatus; Historicus et Practicus* (Oxoniae 1693). Reprint in: Wallis, *Opera Mathematica*, Vol. II (Hildesheim: Olms, 1972).

Julius Weisbach, *Die ersten Grundlehren der höheren Analysis oder der Differenzial- und Integralrechnung. Für das Studium der praktischen Mechanik und Naturlehre möglichst populär bearbeitet* (Braunschweig: Vieweg, 1860).

Ed. Lancelot Law Whyte, *Roger Joseph Boscovich, S. J., F.R.S., 1711–1787. Studies of his Life and Work on the 250th Anniversary of his Birth* (New York: Fordham University Press, 1961).

Heinrich David Wilckens, *Die Lehre von den entgegengesetzten Größen in einem neuen Gewande* (Braunschweig: Reichard, 1800).

Friedrich August Wolf, *Vorlesung über die Encyclopaedie der Alterthumswissenschaft*; edited by J.D. Gürtler (Leipzig, 1839).

Rudolf Wolf, "Simon Lhuilier von Genf," in: Wolf, *Biographien zur Kulturgeschichte der Schweiz* (Zürich: Oxell u. Füßli, 1858), 401–422.

Christian Wolff, *Der Anfangs-Gründe Aller Mathematischen Wissenschaften Letzter* [= 4.] *Theil, Welcher so wohl die gemeine Algebra, als die Differential- und Integral-Rechnung, und einen Anhang von den vornehmsten Mathematischen Schriften in sich begreift.* Die andere [= 2.] Auflage, hin und wieder verbessert und vermehrt (Halle: Renger, 1717). Neue [= 4.], verbesserte und vermehrte Auflage (Halle: Renger, 1750) = Gesammelte Werke, eds. J. École et al., Band 15, 1 (Hildesheim: Olms, 1973).

Christian Wolff, *Elementa Matheseos Universae.* Tomus I. qui commentationem de Methodo Mathematica, Arithmeticam, Geometriam, Trigonometriam planam, et Analysin tam finitorum, quam infinitorum complectitur. Editio novissima [= x.], multo auctior et correctior (Halle, Renger, 1742). = Gesammelte Werke, Hrsg. J. École et al., Band 29 (Hildesheim: Olms, 1968)

Christian Wolff, *Mathematisches Lexikon* (Leipzig: Gleditsch, 1716).

Michael Wolff, *Der Begriff des Widerspruchs. Eine Studie zur Dialektik Kants und Hegels* (Königstein/Ts.: Hain, 1981).

Hoëné de Wronski, *Réfutation de la théorie des fonctions analytiques de Lagrange* (Paris 1812).

Adolf P. Youschkevitch, "The concept of function up to the Middle of the 19th Century," *Archive for History of Exact Sciences*, 1976, *16*: 37–85.

Martin Zerner, *Sur l'analyse des traités de l'analyse: Les fondements du calcul différentiel dans les traités français, 1870–1914.* Cahier de didactique des mathématiques. IREM de Paris Sud. Université Paris VII. numéro 30 (Paris, 1986).

Martin Zerner, *La Transformation des traités français d'analyse (1870–1914)*. Laboratoire J.-A. Dieudonné, Université de Nice-Sophia-Antipolis. Prépublication no. 389 (Juin 1994).

Index of Names

Sources and Studies in the
History of Mathematics and Physical Sciences

Continued from page ii

Continued from previous page

J. Sesiano
Books IV to VII of Diophantus' *Arithemetica*: **In the Arabic Translation Attributed to Qustā ibn Lūqā**

L. Sigler
Fibonacci's Liber Abaci: A Translation into Modern English of Leonardo Pisano's Book of Calculation

B. Stephenson
Kepler's Physical Astronomy

N.M. Swerdlow/O. Neugebauer
Mathematical Astronomy In Copernicus' De Revolutionibus

G.J. Toomer (Ed.)
Apolonius Conics Books V to VII: The Arabic Translation of the Lost Greek Original in the Version of Banū Mūsā ,Edited, with English Translation and Commentary by G.J. Toomer

G.J. Toomer (Ed.)
Diocles on Burning Mirrors: The Arabic Translation of the Lost Greek Original, Edited, with English Translation and Commentary by G.J. Toomer

C. Truesdell
The Tragicomical History of Thermodynamics, 1822–1854

I. Tweddle
James Stirling's Methodus Differentialis: An Annotated Translation of Stirling's Text

K. von Meyenn/A. Hermann/V.F. Weiskopf (Eds.)
Wolfgang Pauli: Scientific Correspondence II: 1930–1939

K. von Meyenn (Ed.)
Wolfgang Pauli: Scientific Correspondence III: 1940–1949

K. von Meyenn (Ed.)
Wolfgang Pauli: Scientific Correspondence IV, Part I: 1950–1952

K. von Meyenn (Ed.)
Wolfgang Pauli: Scientific Correspondence IV, Part 2: 1953–1954

J. Stedall
The Arithmetic of Infinitesimals: John Wallis 1656